Mastering Autodesk®
Architectural Desktop 2006
for Architecture

Mastering AutoCAD®
Architectural Desktop 2006
for Architecture

PAUL F. AUBIN

Autodesk®

THOMSON
—TM—
DELMAR LEARNING

Australia • Canada • Mexico • Singapore • Spain • United Kingdom • United States

Mastering AutoCAD®Architectural Desktop 2006

Paul F. Aubin

Autodesk®

Autodesk Press Staff:

Vice President, Technology and Trades SBU:
Alar Elken

Editorial Director:
Sandy Clark

Senior Acquisitions Editor:
James DeVoe

Senior Development Editor:
John Fisher

Marketing Director:
Dave Garza

Channel Manager:
Dennis Williams

Marketing Coordinator:
Stacey Wiktorek

Production Director:
Mary Ellen Black

Production Manager:
Andrew Crouth

Production Editor:
Stacy Masucci

Technology Project Manager:
Kevin Smith

Technology Project Specialist:
Linda Verde

Editorial Assistant:
Tom Best

Library of Congress Cataloging-in-Publication Data:
Card Number:

ISBN: 1-4180-2052-4

NOTICE TO THE READER

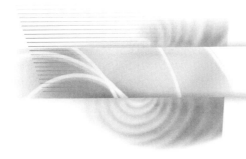

CONTENTS

CHAPTER 6 COLUMN GRIDS AND STRUCTURAL LAYOUT...... 322

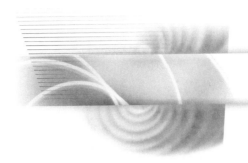

PREFACE

WELCOME

"The purpose of this primer is to acquaint the beginning student with the range of graphic tools which are available for conveying architectural ideas. The basic premise behind its formulation is that graphics is an inseparable part of the design process, an important tool which provides the designer with the means not only of presenting a design proposal but also of communicating with himself and others in the design studio."

Frank Ching – Architectural Graphics, Copyright © 1975 Van Nostrand Reinhold Company, Inc.

This passage prefaces the book "Architectural Graphics" by Frank Ching. The goal set out by Mr. Ching in this venerable resource is to convey the intimate relationship between the tools of architectural drafting and the process of architectural design. Over the course of the last two decades, the "graphic tools" available for the conveyance of "architectural ideas" have undergone dramatic change. For most architects, AutoCAD has been at the center of this change. AutoCAD has historically been a "horizontal" product, meaning that it targets a broad base of users—*anyone* needing to produce technical drafting with accuracy. However, as the simple existence of Mr. Ching's reference can attest, generating architectural graphics requires more than simply having a T-square. A "vertical" approach in CAD software is needed, one that specifically addresses the uniqueness of architectural graphics and design. In Autodesk Architectural Desktop we have such a tool; a computerized tool that can truly live up to the goal set out by Mr. Ching 30 years ago. The purpose of *Mastering Autodesk Architectural Desktop 2006* is to acquaint the user at all levels with this new breed of graphic tool for the conveyance of architectural ideas.

There are two basic goals to this book: First, to shorten the Architectural Desktop learning curve and to help the reader in developing a good sound method. All one needs for success is the proper understanding of how the program functions and a clear understanding of what it can and cannot do. This coupled with good procedure is perhaps the magic key to success in *mastering* Architectural Desktop.

AUTOCAD AND ARCHITECTURAL DESKTOP: WHAT IS THE DIFFERENCE?

Autodesk Architectural Desktop 2006 (ADT) offers a variety of tools not available in the base AutoCAD drafting package. ADT includes a collection of objects representing the most common architectural components such as Walls, Doors, Windows, Stairs, Roofs, Columns, Beams and much more. All of these objects are able to take advantage of *Display Control* (purpose-built display based upon object function and architectural drawing conventions), *Anchors* (physical rule-based linkage between one object and another) and *Styles* (collections of parameters applied to objects as a group) to drive design. AutoCAD does not offer such objects or functionality and relies instead on generic geometric components such as lines, arcs and circles, which need to be assembled by the operator to represent the architectural (or non-architectural) items being designed and organized manually through often complex layer and file schemes.

In Architecture (or otherwise), to accomplish most tasks requires a combination of understanding of one's goals, ample time and planning, and access to the right tools. Although knowledge and planning are critically important, having the proper tool for the job can often levy a dramatic impact on the overall success or failure of a given undertaking. A handsaw and a power saw are both capable of cutting wood. However, the power saw is generally capable of creating a better cut in less time, provided the operator knows how to use it properly. The situation is the same with creating architectural documents. While both AutoCAD and Architectural Desktop are able to accomplish the job, ADT is designed specifically for architectural design/drafting (AutoCAD, while capable of producing architectural documents, is not designed specifically for this task) and will generally do a better job in less time, provided of course that the user knows how to use it properly. Having purchased this book, you most likely already own or have access at work to ADT. Read on, and upon completion of this book, you will have the knowledge needed to use it properly

WHAT IS AN INTELLIGENT OBJECT?

An *intelligent* object is an entity within ADT that is designed to behave as the specific "real world" object after which it is named. The creation of a floor plan in generic AutoCAD involves a process of drafting a series of lines and curves parallel to one another to represent walls, doors and other elements in the architectural plan. This process is often time-consuming and labor intensive. When design changes occur, the lines must be edited individually to accommodate the change. Furthermore, a plan created this way is two-dimensional only. When elevations and sections are needed, they must be created from scratch from additional lines and circles, which maintain no relationship to the lines and circles that comprise the original plan.

In contrast, Architectural Desktop includes *true* architectural objects. Rather than draft lines as in the example above, ADT includes a true Wall object. This object has all of the parameters of an actual wall built directly into it. Therefore, one need only assign the values to these parameters to add or modify the wall within the drawing. In addition, the Wall object can be represented two-dimensionally or three-dimensionally, in plan or in section, using a single drawing element. This means that unlike traditional drafting, which requires the wall to be drawn several times: once for plan, once for section, and again for elevation; an ADT Wall need only be drawn once, and then "represented" differently to achieve each type of drawing (plan, section and elevation). An even greater advantage of the ADT object is that if it is edited, it changes in all views. This is the advantage of its being a single object, and it provides a tremendous productivity boon. With lines, each view remains a separate drawing; therefore, edits need to be repeated for each drawing type, a definite productivity drain.

Objects also adhere to rules built into them that control their behavior under various circumstances. Doors know they should cut holes in Walls. Spaces (rooms) know to grow and shrink when their controlling edges are reshaped. Columns know to move when the column grid line they are attached to moves. Stairs remain constrained to restrictions placed on them by building codes. Tags remain attached and continue to report their associated data even across XREFs. These and many other relationships are programmed into the software. The intelligence of the object extends even further. ADT objects may have graphical and non-graphical data attached to them, which can be linked directly to schedules and reports. All of these features allow us to elevate our ordinary model to a "Building Information Model" (BIM).

Intelligent objects make the process of creating architectural drawings much more efficient and streamlined. Mastery of objects begins with understanding their properties, their styles, their rules. Mastery with ADT begins with mastery of individual objects, but more importantly, requires mastery of the interrelationship of objects and the procedures and best practices required to take fullest advantage of them. Through the process of learning ADT, you will learn to construct a *Building Information Model*—an interconnected series of objects and rules used to generate all of the required architectural documentation and communication, which is greater than the sum of its parts. In general, while both are critical, best practice generally dictates greater emphasis on the "Information" rather than the "Model."

WHO SHOULD READ THIS BOOK?

The primary audience of this book is users new to ADT with some AutoCAD experience. However, it is equally suited to the existing ADT user as well, with reference throughout to those items that are new to 2006 from the previous release. This specifically includes anyone who currently uses AutoCAD to produce Architectural Construction Documentation or Design Drawings, Facilities Layouts, or Interior Design studies and documentation. Architects, Interior Designers, Design

Build Professionals, Facilities Planners and Building Industry CAD Professionals stand to benefit from the information contained within.

YOU SHOULD HAVE SOME AUTOCAD BACKGROUND

Although no prior knowledge of ADT is required to read and use this book, this book assumes a basic level of AutoCAD experience. You should be familiar at least with the basics of drafting, layers, blocks, object snaps and plotting.

FEATURES IN THIS EDITION

Mastering Autodesk Architectural Desktop 2006 is a concise manual focused squarely on the rationale and practicality of the ADT process. The book emphasizes the *process* of creating projects in ADT as an interconnected series of objects, rather than a series of independent commands and routines. The goal of each lesson is to help the reader complete their building design projects successfully. Tools are introduced together in a focused process with a strong emphasis on "why" as well as "how." The text and exercises seek to give the reader a clear sense of the value and potential of each tool and procedure. *Mastering Autodesk Architectural Desktop 2006* is a resource designed to shorten your learning curve, raise your comfort level, and, most importantly, give you real-life tested practical advice on the usage of the software to create architectural designs.

WHAT YOU WILL FIND INSIDE

Section I of this book is focused on the necessary prerequisite skills and underlying theory behind ADT. The section is intended to get you acquainted with the software and put you in the proper mindset. Section II relies heavily on tutorial-based exercises to present the process of creating a building model in Architectural Desktop, relying on the software's built-in Project Management functionality. Two projects are developed concurrently throughout the tutorial section: one residential and one commercial. Detailed explanations are included throughout the tutorials to clearly identify why each step is employed. Annotation and other features specific to construction documentation are covered in Section III. Section IV is devoted to output both in print and in the exciting VIZ Render application.

WHAT YOU WON'T FIND INSIDE

This book is not a command reference. This book approaches the subject of learning ADT by both exposing conceptual aspects of the software and extensive tutorial coverage. No attempt is made to give a comprehensive explanation of every command or every method available to execute commands. Instead, explanations cover broad topics of how to perform various tasks in Architectural Desktop, with specific examples coming from architectural practice. There are dozens of Auto-CAD command references on the market, and any one of them is a good complement to this book. In addition, references are made within the text wherever appropriate to ADT's extensive on-line help and reference materials available on the Web. The focus of this book is the Design Development and Construction Documentation phases of architectural design. The Mass Modeling and other Conceptual Design tools of Architectural Desktop are not extensively covered in this edition.

STYLE CONVENTIONS

Style Conventions used in this text are as follows:

Text	Autodesk Architectural Desktop.
Step-by-Step Tutorials	1. Perform these steps.
Exploration Tutorial	1. Explore these steps.
Menu picks	**View > 3DViews > Top**
Dialog box and palette input	For the length type **10'-0" [3000]**.
Keyboard input	Type **COMMAND** and press ENTER. Type **599** and press ENTER.
Tools on tool palettes	Click the *Concrete 12 Wall tool*
ADT Object Style Names	Assing the *Brick 4 Brick 4 Wall style*
File and Directory Names	*C:\MasterADT2\Chapter01\ Sample File.dwg*

CAD Manager Note: Especially for CAD Managers—there are many issues of ADT usage that are important for CAD Managers and adherence overall to office standards. Throughout the text are notes to the CAD Manager titled "CAD Manager's Note." If you are the CAD Manager, pay particular attention to these items because they are designed to assist you in performing your CAD Management duties better. If you are not the CAD Manager, these notes can help give you insight into some of the salient CAD Management issues your firm may be facing. If your firm does not have a dedicated CAD Manager, pay close attention to these points because these issues will still be present, only there will not be a single individual dedicated to managing these issues and solving relevant related problems as they arise.

FEATURES NEW TO RELEASE 2006

If you are currently using a previous release of Autodesk Architectural Desktop, there are a host of new features in this release. It is important to note that there is no file format change between ADT 2004, 2005 and 2006. ADT 2006 uses the ADT 2004 file format. Therefore, you can seamlessly share files with users of the previous release. However, you will not be able to share files in this way with releases earlier than 2004. Please see Appendix D for more information. Features new to ADT 2006 from 2005 will be indicated by a small "New in ADT 2006" icon. If you are a proficient ADT 2005 user, you can skim the book for these sections to quickly get up to speed on what's new. If you have a little more time, go ahead and read through the complete chapters. You may find some items that escaped your notice in 2005.

UNITS

This book is written in both Imperial and Metric units. Symbol names, scales, references and measurements are given first in Imperial units, followed by the Metric equivalent in square brackets [ ]. For example, when there are two versions of the same symbol or file, they will appear like this within the text:

Aec4_Room_Tag_P [**M_Aec4_Room_Tag_P**], or this "Open the file named *First Floor Imperial.dwg* [*First Floor Metric.dwg*]."

When the scale varies, a note like this will appear: **1/8"=1'-0"** [**1:100**].

If a measurement must be input, the values will appear like this: 10'-0" [3000]. Please note that in many cases, the closest logical corresponding metric value has been chosen, rather than a "direct" mathematical translation. For instance, 10'-0" in Imperial drawings translates to 3048 millimeters; however, a value of 3000 will be used in most cases as a more logical value

 NOTE Every attempt has been made to make these decisions in an informed manner. However, it is hoped that readers in countries where Metric units are the standard will forgive the American author for any poor choices or translations made in this regard.

All project files are included in both Imperial and Metric units on the included CD ROM. See the "Files Included on the CD ROM" topic below for information on how to install the dataset in your preferred choice of units.

HOW TO USE THIS BOOK

The order of chapters has been carefully thought out with the intention of following a logical flow and architectural process. If you are relatively new to ADT, it is recommended that you complete the entire book in order. However, if there are certain chapters that do not pertain to the type of work performed by you or your firm, feel free to skip those topics. However, bear in mind that not every procedure will be repeated in every chapter. For the best experience, it is recommended that you read the entire book, cover to cover. As mentioned above, if you currently use ADT 2005, you can look for the "New in 2006" icons throughout the text. CAD Managers can skip to the "CAD Manager's Note" to find sections relevant to them. Most importantly, even after you have completed your initial pass of the tutorials in this book, keep *Mastering Autodesk Architectural Desktop 2006* handy, as it will remain a valuable desk resource in the weeks and months to come.

FILES INCLUDED ON THE CD ROM

Files used in the tutorials throughout this book are located on the included CD ROM at various stages of completion. Therefore, you will be able to load the file for a given chapter and begin working. When you install the files from the CD, the files for all chapters will be installed automatically. The files will install into a folder on your C drive named *MasterADT 2006*. Files MUST be installed on the C Drive in the "MasterADT 2006" folder. The default installation automatically uses this folder. Inside this folder will be a folder for each chapter. Please note that in some cases, a particular chapter or subfolder will not have any drawing files. This is usually indicated by a text file (TXT) within this folder. For example, the *Chapter03* folder contains no drawing files and instead contains a text document in a sub-folder named: *There are no files for Chapter 3.txt.*

INSTALLING CD FILES

Locate the *Mastering Autodesk Architectural Desktop 2006* CD ROM in the back cover of your book. To install the dataset files, do the following:

1. Place the CD in your CD drive.

   An installer window should appear on screen after a moment or two.

2. To install the dataset files in Imperial units, click the Imperial Dataset button. To install the dataset files in Metric units, click the Metric Dataset button.

Installation will commence automatically and all files will be installed to a folder named *C:\MasterADT 2006* on your hard drive.

 CAUTION Please do not move the files from this location, or the Tool Catalogs and Project Palettes will not function properly. Moving any of the other files can also cause issues with project files. See the "Re-Pathing Projects" topic below.

If you do not intend to perform the tutorials in certain chapters, it is OK to delete the files for those chapters. Simply delete the entire folder for the chapter(s) that you wish to skip. If you wish to install both the Imperial and Metric datasets, return to the installer and repeat the steps above for the other units. Installation requires approximately 240 MB of disk space per unit type (480 MB if you install both).

PROJECTS

The ADT Drawing Management (Projects) tools are used exclusively throughout this text. Please do not open and save files directly outside the Project Navigator. Although there is no physical difference between a drawing file created inside a project and one created outside, procedurally there are large differences. Please follow the instructions at the start of each chapter regarding the installation of files and loading of the correct current project files. Furthermore, even though it is theoretically possible to continue in the files created in Chapter 5 throughout the entire book, it is recommended that you re-load the files and projects for the chapter that you are working in from the installed datasets at the start of each chapter. Sometimes edits have been made to the files after the completion of the previous chapter that you will not have if you do not start fresh with the CD version.

For an excellent example of an ADT Project, load and explore the included sample project. From the File menu, choose **Project Browser** and load the **Autodesk Architectural Desktop Sample Project** found in the *C:\Program Files\Autodesk Architectural Desktop 2006\Sample* folder on your hard drive.

PROJECT BULLETIN BOARDS AND DATASET UPDATES

When you load a Project in ADT, a bulletin board for that Project will load in the ADT Project Browser window (on the right). Review this page as it loads. If you have a live Internet connection, you will be informed on this page if updates to the Dataset (or the book text as a PDF) are available for download from www.paulaubin.com. If an update is available, you will be able to click the link directly in this page.

In addition, a DWF (Design Web Format) of each chapter is provided at the root of the project folders. You can click the icon in the bulletin board page to view it directly in Project Browser. If you wish to see it in the full Autodesk DWF Viewer application (included with ADT and available as a free download from Autodesk.com), click the icon to load the DWF in the Project Browser and then right-click and choose Full View.

RE-PATHING PROJECTS

In some cases when you load a project, you will be prompted to re-path the project. This occurs when the project has been moved from its original location. If you move the CD files to a location other than *C:\MasterADT5*, a message like the one in Figure P–1 will appear. If you receive this message, click Yes. This is very important, because the project files will not function properly if you choose No.

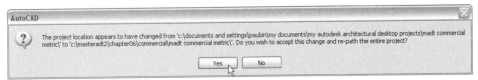

Figure P–1 *If a project has been moved, you will be prompted to re-path project files. Always answer Yes.*

WINDOWS 2000 AND XP COMPLIANT PATHS

Autodesk Architectural Desktop 2006 is Windows XP logo compliant. Part of achieving this distinction means that the default paths to many resource files are buried deep in the *Program Files* or *Documents and Settings* folder structures. Your CAD Manager may have opted to move these resources out of these locations and to a central location on the server. This book assumes that all files are in the default installed locations. Check with your IT or CAD support personnel for more information on this issue.

If you need to browse the *Documents and Settings* folder to locate ADT resources, the default location (for English language versions) is as follows:

C:\Documents and Settings\All Users\Application Data\Autodesk\ADT 2006\enu

It is important to note that the *Application Data* folder is a hidden folder in Windows. Therefore, by default you will be unable to browse this location. To turn on the display of hidden files, choose **Tools > Folder Options** in Windows Explorer. On the View tab, choose Show Hidden Files and Folders and then click OK. Again, check with your IT or CAD support person before making this change.

ONLINE COMPANION

Additional resources related to the content in this book are available online. Log on to our Web site for complete information at:

http://www.autodeskpress.com

SERVICE PACKS

It is important to keep your software current. Be sure to check online at **www.autodesk.com** on a regular basis for the latest updates and service packs to the Architectural Desktop software. Having the latest service packs installed will help ensure that your software will continue to run trouble-free. Autodesk Architectural Desktop 2006 also has the Communication Center icon in the bottom right corner of the Application status bar. This icon will light up when updates and information are available.

WE WANT TO HEAR FROM YOU

We welcome your comments and suggestions regarding *Mastering Autodesk Architectural Desktop 2006*. Please forward your comments and questions to:

The CADD Team
Delmar Learning
Executive Woods
5 Maxwell Drive
Clifton Park, NY 12065-8007
Website: www.autodeskpress.com

ABOUT THE AUTHOR

Paul F. Aubin is the author of *Mastering Autodesk Architectural Desktop* and *Autodesk Architectural Desktop: An Advanced Implementation Guide*, and co-author of *Mastering VIZ Render – A Resource for Autodesk Architectural Desktop Users*, published by Autodesk Press. Paul has a background in the architectural profession spanning more than 16 years. In addition to writing, Paul is an independent consultant and travels the country speaking and providing implementation, training and support services. He currently serves as the Moderator for *Cadalyst* magazine's online CAD questions forum, and has been published in their print magazine. He is a regular speaker at Autodesk University and was the recipient of the Autodesk Central Region 2001 Architectural Award of Excellence. Paul has been a guest speaker at Chicago area events for The American Institute of Architects Northeast Illinois, Chicago Area Users' Groups and The Association of Licensed Architects. Prior to becoming an independent consultant, Paul was an Associate with a leading Autodesk Systems Center in Chicago where he successfully trained over 1100 professionals in Architectural Desktop, AutoCAD and 3D Studio VIZ. While in architectural practice, Paul served as CAD Manager with an Interior Design and Architecture firm in downtown Chicago and amassed many years of hands-on project architect level experience. The combination of his experiences in architectural practice as a CAD Manager and an Instructor give his writing, classroom instruction and consultation a fresh and credible focus. Paul is an associate member of the American Institute of Architects. He received his Bachelor of Science in Architecture and his Bachelor of Architecture from The Catholic University of America. Paul lives outside Chicago with his wife Martha, their sons Marcus and Justin, and daughter Sarah Gemma.

DEDICATION

This book is dedicated to my daughter Sarah Gemma. I love you princess.

ACKNOWLEDGEMENTS

The author would like to thank several people for their assistance and support throughout the writing of this book.

Thanks to Jim Devoe, John Fisher, and all of the Delmar team. It continues to be a pleasure to work with so dedicated a group of professionals.

Thanks to Eric Stenstrom of Wiss, Janney, Elstner Associates, Inc. for technical editing, Sue Gaines for copy editing, both Joanne Sprott and Sue Gaines for indexing and Mike Boyd, Melissa Colbert, and the folks at ATLIS, for composition.

A special acknowledgement is due the following instructors who reviewed the chapters in detail:

Debra Dorr–Phoenix College, Phoenix, AZ

Joseph Liston–University of Arkansas at Ft. Smith, Fort Smith, AR

Jeff Porter–Porter and Chester Institute, Watertown, CT

Margaret Robertson–Lane Community College, Eugene, OR

Edward Rother–Cerritos Community College, Norwalk, CA

Charles West–Dakota County Technical College, Rosemount, MN

Jon McFarland–CAD Instructor, Virginia Marti College of Art and Design, Lakewood, OH

Susan M. Sherod–Engineering Department Chair, Santa Ana College, Santa Ana, CA

There are far too many folks in Autodesk Building Solutions Division to mention. Thanks to all of them, but in particular, Jim Awe, Jim Paquette, Kelcy Lemon, Julian Gonzalez, Brook Potter, Chris Yanchar, Paul McArdle, Mark Webb, James Smell, Rob Finch, Bill Glennie, Dennis McNeal, William (Fitz) Fitzpatrick, Anna Oscarson, Scott Reinemann, Scott Arvin, Bryan Otey, Matt Dillon, John Janzen, Tatjana Dzambazova, Simon Jones, Michael Nachtsheim and all of the folks at Autodesk Tech Support.

I am ever grateful for blessings I have received in lifelong friends: Mark Zifcak and Ron Bailey and for my family: My parents Maryann and Del, my brothers Marc and Tom. My wonderful children: Marcus, Justin and Sarah Gemma. You three are a constant reminder of what is most important in life. Finally, I am most grateful for the constant love and support of my wife, Martha.

Quick Start

General Architectural Desktop Overview

INTRODUCTION

This Quick Start provides a simple tutorial designed to give you a quick tour of some of the most common objects and features of Autodesk Architectural Desktop 2006 (ADT). You should be able to complete the entire exercise in 45 minutes or fewer. At the completion of this tutorial, you will have experienced a first-hand look at what ADT has to offer.

OBJECTIVES

- Experience an overview of the software.
- Create your first Architectural Desktop model
- Receive a first-hand glimpse at many ADT tools and methods.

 ### CREATE A SMALL BUILDING

Let's get started using ADT right away. For the next several minutes, we will take the "whirlwind" tour of the ADT tool set. All of the tools covered in the following steps use default Architectural Desktop settings. The chapters that follow cover each of these items and settings in detail. This book was authored using Microsoft Windows XP Professional but will work equally well in Microsoft Windows 2000 Professional. Basic AutoCAD knowledge is assumed but not required. Please refer to the Preface for complete details on these and other assumptions.

LAY OUT A SIMPLE BUILDING

1. Launch Autodesk Architectural Desktop 2006 from the icon on your desktop.

   You can also launch ADT by choosing **Start > All Programs > Autodesk > Autodesk Architectural Desktop 2006 > Autodesk Architectural**

Desktop 2006. (If you are using Windows 2000, "All Programs" will be just "Programs.")

2. From the File menu, choose **New**.

▶ Choose the *Aec Model (Imperial Stb).dwt* [*Aec Model (Metric Stb).dwt*] template (see Figure Q–1).

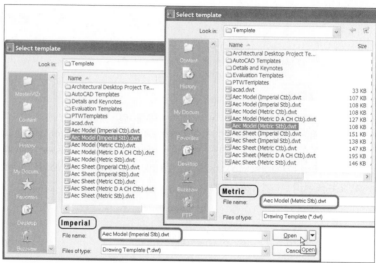

Figure Q–1 *Choose an AEC Model template file*

 NOTE Throughout this book, Imperial units and files will be listed first, followed by metric in brackets, for example, **Imperial [Metric]**. See the Preface for complete details on Style Conventions used throughout this book.

For this exercise, be sure to use the "**New**" command from the File menu. The "New" Toolbar icon calls the AutoCAD QNEW command, which does not prompt for a choice of template file. For more information on QNEW and ADT Templates, refer to Chapter 3.

New in ADT
2006

3. On the Application Status Bar, (at the base of the ADT screen), right click the "DYN" toggle and choose **Settings**.

The Drafting Settings dialog should open with the Dynamic Input tab active.

▶ Place a checkmark in the all three options in this dialog to turn them all on and then click OK (see Figure Q–2).

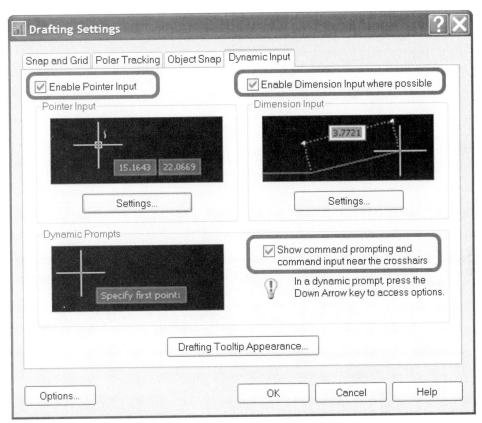

Figure Q–2 *Enable all Dynamic Input options*

These settings, which are new to AutoCAD and Architectural Desktop 2006, display prompting and input instructions directly onscreen at the cursor location. Refer to Chapter 1 for more information on these settings.

4. On the Shapes toolbar, click the Rectangle icon (see Figure Q–3)

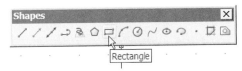

Figure Q–3 *Click the Rectangle icon on the Shapes toolbar*

▶ At the "Specify first corner point" prompt, click a point on screen.

▶ At the "Specify other corner point" prompt, press the down arrow key on your keyboard to access command options of rectangles.

▶ Arrow down till "Dimensions" is selected and then press ENTER (see Figure Q–4).

Figure Q–4 *Access rectangle command options with the down arrow key*

▶ At the "Specify length for rectangles" prompt, type **30′ [9000]** and then press ENTER.

▶ At the "Specify width for rectangles" prompt, type **20′ [6000]** and then press ENTER.

Move your mouse side to side and up and down. Notice that the rectangle is sized based on the dimensions you indicated and your mouse movements will indicate where it is placed relative to the first corner point.

▶ Click your mouse to complete placement (see Figure Q–5).

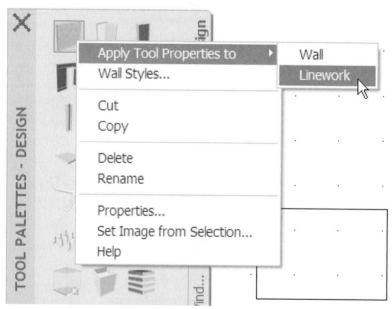

Figure Q–5 *Click the mouse to place the rectangle*

5. Zoom in on the rectangle.

If you have a wheel mouse, you can roll the wheel to zoom, and drag with the wheel pressed in to pan. You can also use any standard AutoCAD method to zoom. Zoom commands are located on the Zoom flyout of the Navigation toolbar. To access flyout icons, click and hold down the mouse on the Zoom icon. Other options will appear. Select the one you wish to use such as Zoom Window. For more information on Zoom and Pan, refer to the online help.

Most ADT object creation commands are located on tool palettes. The tool palettes can be floating or docked to the sides of the screen. To open them if they are not already open, press CTRL+3 or use the Tool Palettes command on the Window menu.

6. On the Design tool palette, right-click the **Wall** tool and choose **Apply Tool Properties to > Linework** (see Figure Q–6).

 NOTE If you don't see a Walls tab on the Tool Palettes, right-click the Tool Palettes title bar and choose **Design** (to load the Design Tool Palette Group) and then click the Design tab.

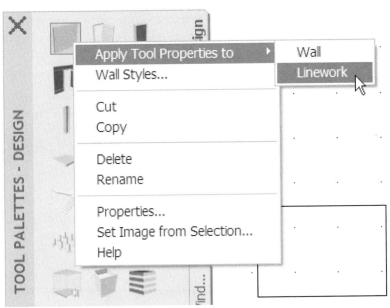

Figure Q–6 *Convert Linework to Walls by applying Tool Properties*

▶ At the "Select lines, arcs, circles, or polylines to convert into walls" prompt, select the rectangle on screen and then press ENTER.

▶ At the "Erase layout geometry" prompt, choose **Yes** (see Figure Q–7).

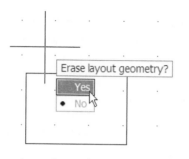

Figure Q–7 *Create Walls and erase the layout rectangle*

 TIP If you prefer, you can also right-click in the drawing window and choose Yes, or type Y at the command prompt and then press ENTER.

If you have the command line window open (usually it is docked at the bottom of the screen) it will indicate: "4 new wall(s) created." You can also press F2 to make a text window appear which will reveal this message and show a history of past commands. In any case, the rectangle on screen will be replaced with four Walls that will remain selected. More information on Walls can be found in Chapters 4, 9 and 10.

7. With the Walls still selected right-click and choose **Insert > Door** (see Figure Q–8).

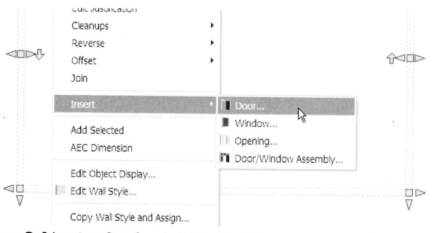

Figure Q–8 *Inserting a Door from the Wall's right-click menu*

A Door with some associated "Dynamic Dimensions" will appear, and the Properties palette will appear if it was not already open on screen. (If only the title bar of the Properties palette appears, move your mouse pointer over it, and it will *pop* open.)

▶ On the Properties palette, set the Width to **3'-0"** **[910]** and the Height to **7'-0"** **[2110]**.

▶ In the Location grouping, choose **Offset/Center** from the Position along wall list (see Figure Q–9).

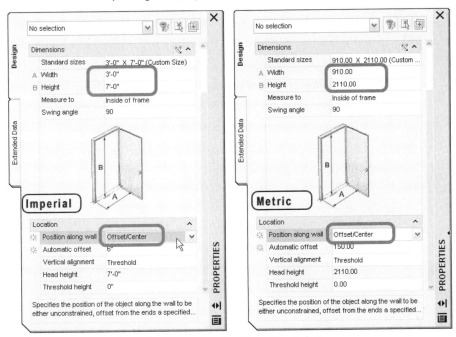

Figure Q–9 *Designate Door Parameters on the Properties palette*

▶ Move the mouse near the center of the left vertical Wall in the drawing.

NOTE Notice that the Dynamic Dimension on either side of the Door reads the same value.

▶ Move the mouse slightly side to side. When the Door swings to the outside of the building, click the left button to place the Door.

▶ Right-click and choose **Enter** (or press ENTER at the keyboard) to end the Add Door routine.

8. Select the bottom horizontal Wall, right-click and choose **Insert >** **Window.**

▶ On the Properties palette, set the Width to **4'-0"** **[1200]** and the Height to **5'-0"** **[1500]**. Leave the Position along wall set to **Unconstrained** this time.

▶ Click anywhere along the bottom Wall to place a Window and then place two more in the top horizontal Wall.

The dynamic dimensions surrounding the Windows and Doors as they are being added are "live" and can be used to input precise values. The size of the Window itself is currently the active dimension and cannot be edited directly, but rather must be changed in the Properties palette. However, if you press the TAB key the "active" dimension will cycle and you can type in a value into either of the other dynamic dimensions.

9. With the Window command still active, move to the right vertical Wall and then press the TAB key on the keyboard.

Note that one of the two offset dimensions is now highlighted.

▶ Type in a value for this highlighted dimension such as **4′-0″ [1200]** and press ENTER (see Figure Q–10).

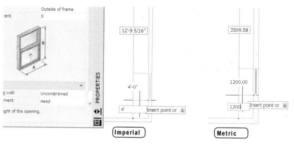

Figure Q–10 *Add some Windows using Dynamic Dimension input*

 NOTE Imperial dimensions throughout this text use the 'Feet and Inch' format for clarity. However, when typing these values into ADT, neither the inch symbol (′) nor the hyphen (-) separating the feet from inches is required. 4′-0″ can therefore be typed in ADT as either **4′** or **48** to achieve the same result. As you can see, zeros can also be omitted. Hyphens are required only when separating inches from fractions. Therefore 6 Ω is typed as **6-1/2**. You can also use decimal equivalents instead of fractions like: **6.5**. Metric values in this text are in millimeters and can be typed in directly with no unit designation required. More information on Style Conventions used in this book can be found in the Preface.

The Window appears in the Wall offset from the corner by the exact amount typed in.

▶ When you have finished adding Windows, right-click and choose **Enter** to end the command (or simply press the ENTER key).

For more information on adding Walls, Doors, and Windows, refer to Chapter 4.

EDIT THE MODEL

Objects built with Architectural Desktop can be manipulated quickly and easily via the Properties palette or via direct manipulation. "Direct manipulation" refers to editing that takes place directly on the object within the drawing window.

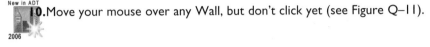

New in ADT 2006

10. Move your mouse over any Wall, but don't click yet (see Figure Q–11).

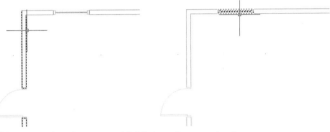

Figure Q–11 *Objects under the cursor highlight prior to selection*

Notice that the objects under the cursor will highlight temporarily as the cursor hovers over them. This behavior can be customized and turned on and of in the Options dialog. See Chapter 1 for more information.

11. Click on the top horizontal Wall to select it.

The selection preview will disappear and a series of grips will appear in various shapes. (Grips are the small geometric shapes placed at strategic editing points when an object is selected).

12. Hover over one of the grips—do not click it (see Figure Q–12).

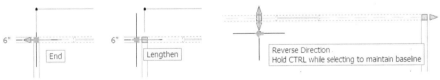

Figure Q–12 *Grips provide on screen tips indicating their function*

A small tip will appear on screen indicating the function of the highlighted grip and any CTRL key toggle options it may have. If you click a grip, it activates the grip and the tip disappears. At first, don't click the grips yet; simply move the mouse over each one to see what their functions are.

13. First hover over and then click on one of the triangular shaped grips at the midpoint of the Wall.

These grips affect the width of the Wall.

▶ Make the Wall thicker by dragging away from its center and then clicking (see Figure Q–13).

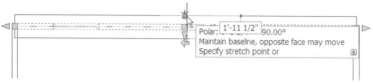

Figure Q–13 *Using Grips to edit the thickness of the Wall*

14. With the Wall still selected, right-click and choose **Select Similar**.

This will select all four Walls, but not the Doors or Windows.

▶ On the tool palettes, click on the Walls tab to show the Walls tool palette.

This reveals several different types of Walls ready to use in your drawings.

▶ Right-click on the ***Brick-4 Brick-4 [Brick-090 Brick-090] tool and choose* Apply Tool Properties to > Wall** (see Figure Q–14).

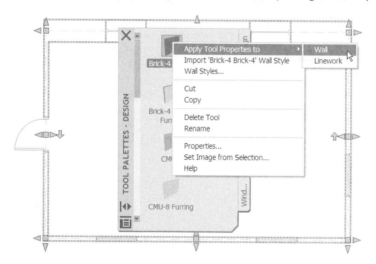

Figure Q–14 *Apply a new Wall Style with all four selected Walls*

Hatching should appear within the Walls, and our manually edited Wall should now be the same width as the other Walls. This particular Wall type (called a Wall Style) had a built-in fixed width. For more information on working with and swapping Wall styles, refer to Chapter 10.

▶ Right-click and choose **Deselect All**.

15. Select the Door. (Click on it to select).

▶ Click one of the small arrow-shaped grip points to flip the Door swing. Repeat with the other arrow grip (see Figure Q–15).

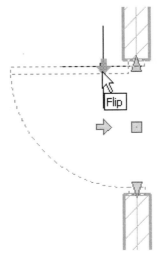

Figure Q–15 *Flip the Swing of the Door with Grips*

▶ Right-click and choose **Deselect All** to complete the change.

16. Right-click in the drawing window and choose **Basic Modify Tools > Move**.

▶ At the "Select objects" prompt, click on any Window and then press ENTER.

▶ At the "Specify base point" prompt, click anywhere in the drawing.

▶ At the "Specify second point" prompt, click a random point inside the building (see Figure Q–16).

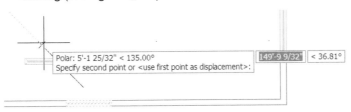

Figure Q–16 *Attempt to move the Window out of the Wall*

Notice that the Window did move, but did not move to the point we picked. Instead, it remains constrained to the Wall, and therefore it simply moved along the length of the Wall closest to the point that we clicked. This behavior is determined by an "anchor." Anchors control the relationship of objects to one another, such as keeping the Window attached to a Wall. For more information on anchors, refer to Chapter 2.

17. Choose **Undo** from the Edit menu to return the Window to its original location.

ADD A FLOOR SLAB AND ROOF

We have given our simple building some walls and openings, let's complete the enclosure with a roof and a floor slab.

18. Click the Design tab of the Tool Palettes; find the *Slab* tool (scroll down if necessary) and then right-click and choose **Apply Tool Properties to > Linework and Walls.**

TIP If your tool palettes have a scroll bar, you can use it to scroll a palette, or if you move your mouse over an unused portion of the palette, a small 'hand' icon will appear. Click and drag with this hand icon to scroll.

▶ At the "Select walls or polylines" prompt, select all of the Walls and then press ENTER.

TIP You can click a point outside the model and then surround the entire drawing with a box and click again. Only the Walls will highlight as they are the only object within the selection box eligible to convert to Slabs.

▶ At the "Erase layout geometry" prompt, choose **No.**

NOTE If dynamic input is OFF (use the DYN toggle at the base of the screen), right-click and choose **No** or type **N** at the command line prompt and then press ENTER.

▶ At the "Specify slab justification" prompt, choose **Top.**

▶ At the "Specify wall justification for edge alignment" prompt, choose **Right**.

▶ At the "Specify slope Direction" prompt, right-click and choose **Left**. (see Figure Q–17).

Figure Q–17 *Answering the command line prompts of the Apply Slab Properties to Walls command*

19. With the Slab still selected, right-click and choose **Properties**. (If the Slab is no longer selected, click it near the door opening to select it.)

   ▶ On the Properties palette, within the Dimensions grouping, change the Thickness to **1′-0″** [**250**].

   ▶ Scroll down to the Location grouping, and type **-1′-0″** [**-250**] for the Elevation parameter (see Figure Q–18).

   This will make the Slab thicker and move it to the correct location below the bottom edge of the Walls.

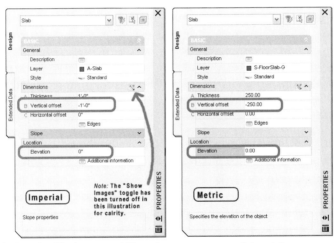

Figure Q–18 *Edit the Thickness and Elevation parameters of the slab.*

20. On the Design palette, right-click the Roof tool and choose **Apply Tool Properties to > Linework and Walls**.

21. At the "Select Objects" prompt, click to select all of the Walls, and then press ENTER.

 NOTE The window selection method mentioned above will not work well here.

22. At the "Erase layout geometry" prompt, right-click and choose **No**.

See Figure Q–19 for the result. For more information on Slabs and Roofs, refer to Chapter 12.

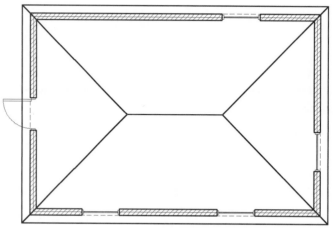

Figure Q–19 *Adding a simple Roof based on the existing Walls*

New in ADT
2006

EDIT THE MODEL IN ELEVATION

Our top wall contains only one Window (if you added more, delete the extras for this exercise) let's explore how that elevation would look with a few additional windows.

23. Click a point outside and above the model to the right, move the mouse to the outside left of the model just below the upper Wall and click again (see Figure Q–20).

Figure Q–20 *Select the upper portion of the model with a crossing window*

This is called a "crossing selection window." When you click from right to left, all objects touched by the selection window are selected. When you click form left to right instead, only those surrounded completely by the window are selected. Search for 'select objects' in the online help for more information on object selection.

▶ With the upper Wall, its Window the Roof and Slab selected, right-click and choose **Isolate Objects > Edit in Elevation** (see Figure Q–21).

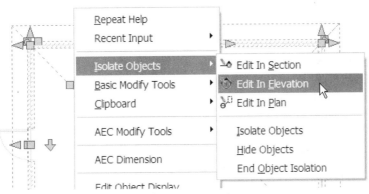

Figure Q–21 *Choose Edit in Elevation from the right-click menu*

▶ At the "Select linework or face under the cursor" prompt, move the mouse around and note the various edges and surfaces that highlight—do not click yet.

▶ Highlight the upper edge of the Roof object (a blue line will appear to indicate that the edge is selected) and then click (see Figure Q–22).

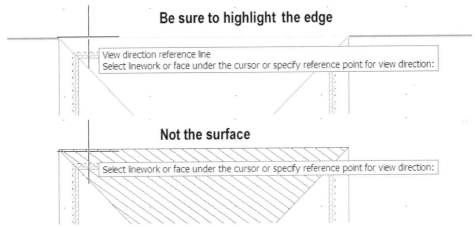

Figure Q–22 *Highlight geometry to determine the elevation vantage point*

▶ At the "Specify elevation extents" prompt, drag down slightly (enough to include the upper Wall) and then click (see Figure Q–23).

The model will change views and zoom to the selected area. All non-selected objects will be hidden temporarily to make editing easier. A small toolbar with a single icon will also appear. Use this icon to restore the previous view when editing is complete.

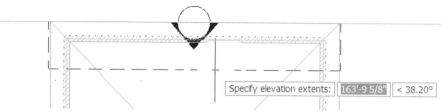

Figure Q–23 *Drag to indicate extent of elevation view*

24. Select the Window (now shown in elevation).

    Take note of the grips again.

    ▶ Click the square (Location) grip at the bottom edge and drag the Window to a new location on the Wall.

All grip editing functions equally well in elevation view as it did in plan view.

    ▶ Repeat this process and drag it to the left placing it a short distance from the left side of the Wall.

25. With the Window still selected, right-click and choose **AEC Modify Tools > Array**.

    ▶ At the "Select an edge to array from" prompt, move the mouse over each edge of the Window and notice the same type of edge highlighting that we saw above.

    ▶ Highlight the right vertical edge of the Window and then click (see Figure Q–24).

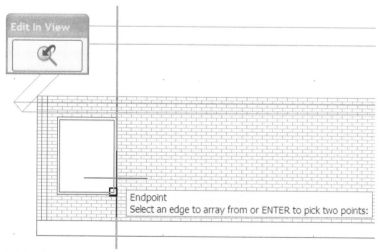

Figure Q–24 *Highlight geometry to determine the elevation vantage point*

    ▶ Move the mouse to the right and note the dimension that appears.

▶ At the "Drag out array elements" prompt, type **8'-0"** **[2400]** and then press ENTER.

▶ Drag the mouse to the right until two new Windows appear and then click.

There will now be three windows in this elevation. Notice that they cut holes in the Wall just as they did in plan. It might be nice to space them equally on this wall.

26. Click to select each of the three Windows in this elevation.

27. Right-click and choose **AEC Modify Tools > Space Evenly**.

▶ At the "Select an axis to space evenly on" prompt, hover over the bottom edge of the Slab to highlight it and then click.

▶ At the "Select the first point along the axis" prompt, click the outside bottom corner of the Slab.

▶ At the "Select the second point along the axis" prompt, click the opposite bottom corner of the Slab (see Figure Q–25).

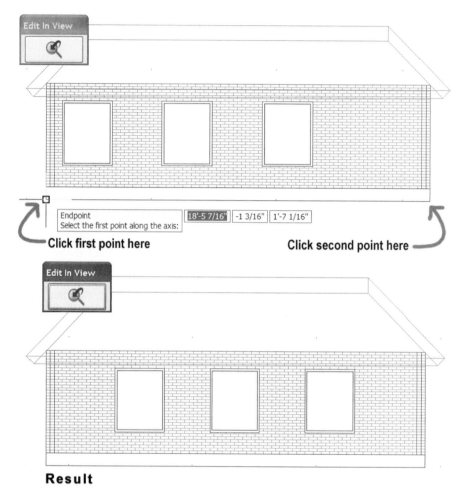

Result

Figure Q–25 *Snap from one endpoint to the other to indicate spacing*

28. Click the Exit Edit in View icon to return to the original plan view.

SET UP A PROJECT WITH FLOOR LEVELS

We can quickly take the geometry that we have started here and turn it into a multi-story building. Let's take a quick look at the Drawing Management System.

1. From the File menu, choose **Project Browser**.

   ▶ Using the drop-down list on the left, choose *My Documents*.

   ▶ Double click the *Autodesk* folder and then the *My Projects* folder.

▶ At the bottom of the window, click the New Project icon (see Figure Q–26).

Figure Q–26 *Add a new project and give it a name*

▶ In the Add Project dialog box, type **Simple Building** in the Project Name field.

▶ If you wish, also type a Project number and Description (these are optional) and clear the checkmark from the "Create from template project" checkbox (see Figure Q–26).

2. Click OK to create the project and then click the Close button to dismiss the Project Browser.

The Project Navigator palette should appear on screen. If it does not, choose Project Navigator Palette from the Window menu or press CTRL + 5.

3. Click the Project tab (if it is not already active). In the Levels grouping, click the small Edit Levels icon (see Figure Q–27).

Figure Q–27 *Click the Edit Levels icon on the Project tab to change and add floor levels to the project*

▶ In the Levels dialog box, change the Floor to Floor Height for Level 1 to **9′0″ [3000]**.

▶ Click the Add Level icon and then change the Floor to Floor Height for Level 2 to **8′-0″ [2750]**.

▶ Click the Add Level icon again to add a Level 3 and then click OK.

If an alert dialog box appears asking if you wish to "regenerate all views," simply click Yes to accept this and dismiss the dialog box.

4. On the Project Navigator palette, click the Constructs tab.

For complete information on Views, regenerating Levels, Constructs, and the Project Navigator, see Chapter 5.

5. Right-click on the *Constructs* folder and choose **Save Current DWG as Construct**.

▶ In the Name field, type **First Floor** , place a check mark in the Level 1 box, and then click OK (see Figure Q–28).

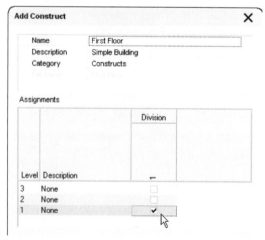

Figure Q–28 *Save the current drawing as the First Floor–Level 1 Construct*

6. Click on the Roof object in the drawing, right-click and choose **Properties**.

▶ In the Dimensions grouping, choose **Plumb** from the Edge cut list.

▶ In the Dimensions: Lower Slope grouping, change the Plate Height to **0** (zero) and the Rise to **6″ [50]**.

7. Click again on the Roof (be sure to click a highlighted edge and not a grip), hold down the mouse, and drag it on top of the Constructs folder of the Project Navigator palette. When the mouse changes shape to an arrow with a small box, release the mouse (see Figure Q–29).

This action will move the Roof from the First Floor file and create a new Construct file from it. If your Roof is changing shape instead of moving, then you selected a Grip (the small colored squares). Press the ESC key, undo any changes to the Roof, and try again. Drag from the highlighted edge, *not* the grip.

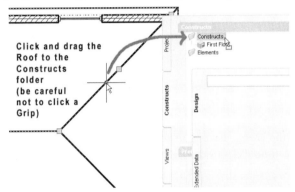

Click and drag the Roof to the Constructs folder (be careful not to click a Grip)

Figure Q–29 *Drag the Roof to the Project Navigator palette to create a new Construct*

▶ In the Add Construct dialog box, name the new Construct **Roof**, check Level 3 for its Assignment, and then click OK.

8. The First Floor Construct file will still be open on screen. Choose **File > Save** to save the file (or press CTRL + S).

9. On the Project Navigator palette, right-click on the *First Floor* Construct and choose **Copy Construct to Levels**.

▶ In the Copy Construct to Levels dialog box, place a check mark in Level 2 and then click OK (see Figure Q–30).

We now have an exact copy of the *First Floor* to use for the *Second Floor*.

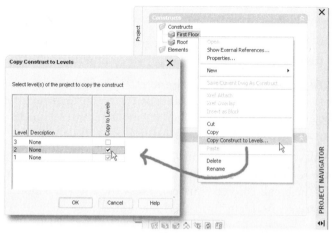

Figure Q–30 *Copy the First Floor Construct to the Second Floor Level*

▶ Right-click on the new Construct—currently named *First Floor(2)*—and choose **Rename**.

▶ Rename it to **Second Floor** .

10. Right-click on *Second Floor* and choose **Open** (or simply double-click it).

11. Select the Door, place your mouse over the highlighted edge (not a grip) press and hold down the right mouse button and then drag it to the other side of the plan (near the vertical Wall on the right) and release.

▶ A small menu will appear at the cursor. Choose **Copy here**.

Notice that the new Door is automatically attached to and cuts a hole in the right side Wall.

12. On the tool palettes, click the Windows tab.

▶ Right-click the *Picture – Arched* tool and choose **Apply Tool Properties to > Door, Door/Window Assembly, Opening**.

▶ At the "Select doors, door/window assemblies, and/or openings to convert" prompt, click the Door in the left Wall (the original one, not the new copy) and then press ENTER.

▶ With the Window still selected, change the Height to **5′-0″ [1500]** and the Sill height to **2′-0″ [600]**.

13. Close the *Second Floor* file, when prompted, click yes to save the file.

CREATE A MODEL VIEW

Having created all the separate parts of our building (Constructs) let's see how it all looks together.

14. On the Project Navigator palette, click the Views tab.

15. Right-click on the *Views* folder and choose **New View Dwg > General**.

   ▶ For the Name type **Model** and then click Next.

   ▶ On the Context screen, check all three Levels and then click Next.

   ▶ On the final screen, click Finish.

16. On Project Navigator, double-click *Model* to open it.

   ▶ From the View menu, choose **3D Views > NW Isometric** (see Figure Q–31).

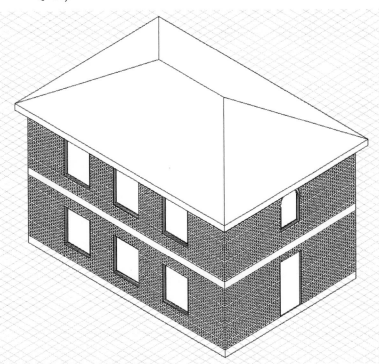

Figure Q–31 *View the Model from an Isometric viewpoint*

Feel free to try other isometric views and shading modes from the 3D Views and Shade menus.

EDIT THE MODEL

As you view the model in 3D, suppose we wanted to make a modification to something; it is simple to make changes and then quickly view the results.

17. Click on the second floor in the *Model* file to select it.

Notice how the entire floor plate highlights; the *Model* is comprised of External References (XREFs). An XREF is a link to the original file that updates when changes are made. Project Navigator automatically creates all XREFs for use when we setup a View. See Chapter 5 for more information.

▶ Right-click and choose **Open XREF**.

18. Select the vertical Wall on the right.

If you highlight the Slab instead, press the ESC key and try again.

▶ Click the square Location grip at the center of the Wall. Begin moving it to the left.

▶ Type 5′-0″ [**1500**] and then press ENTER (see Figure Q–32).

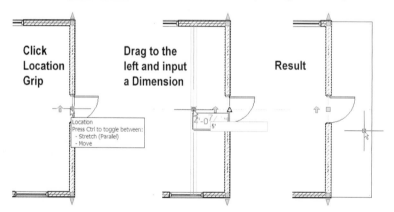

Figure Q–32 *Use the Location Grip to move a Wall and simultaneously Stretch its neighbors*

Notice how the two horizontal Walls at the top and bottom stretched to stay connected to the one we moved. If any of your Windows are now in undesirable locations, you can move them the same way. Select them, then click the square Location grip in the center and move to a new location. You can also use the AEC Modify tools discussed above to re-center and space Windows if you like.

19. On the Design Tool palette, click the **Structural Column** tool.

▶ On the Properties palette, change the Logical length to **8′-0″** [**2750**].

> ▶ At the "Insert point" prompt, click a point near the corner of the Slab where the Wall previously was, and then press ENTER to accept the default rotation.

> ▶ Repeat at the other corner, and then press ENTER to complete the command.

ADD ROOF RAFTERS

Let's make one additional edit (this time to the Roof file) before we return to the model.

20. On the Project Navigator palette, click the Constructs tab and then double click the *Roof* file to open it.

21. From the View menu, choose **3D Views > SE Isometric**.

22. Select the Roof object, right-click and choose **Convert to Roof Slabs**.

23. In the Convert to Roof Slabs dialog, select the Erase Layout Geometry checkbox and then click OK.

24. On the Design tool palette, click the **Structural Beam** tool.

> ▶ On the properties palette, set "Trim automatically" to **Yes**. From the "Layout type" list, choose **Fill**.

> ▶ From the Justify list, choose **Top Center**.

> ▶ In the Layout grouping, choose **Yes** for "Array." Set the "Layout method" to **Repeat** and the "Bay size" to **2'-0"** [**600**] (see Figure Q–33).

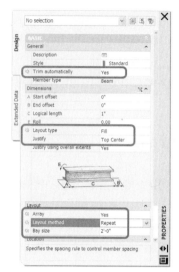

Figure Q–33 *Set the parameters or a Structural Beam to repeat in an Array pattern*

> 25. Move the mouse over the bottom edge (at the eave) of the Roof Slab.

Notice that a series of beams will appear ghosted in oriented to the edge highlighted by the mouse. You can move the mouse around on screen and highlight other edges to see other options. Return to highlighting the bottom edge again before you click.

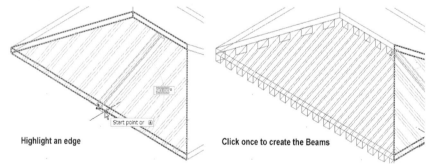

Highlight an edge   Click once to create the Beams

Figure Q–34 *Highlight an edge of the Roof Slab and then click to add Beams*

> ▶ Click to create a series of equally spaced Beams automatically trimmed to the shape of the Roof Slab (see Figure Q–34).

> ▶ Repeat on the other three Roof Slabs and then press ENTER to complete the routine.

> 26. Save and Close the *Roof* file.

The *Second Floor* file should still be open from the edits above. Let's close it and save it now.

27. Save and Close the *Second Floor* file.

Back in the *Model* file (which should still be open on screen) a small balloon will appear in the lower right corner of the screen (see Figure Q–35).

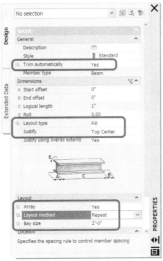

Figure Q–35 *After editing the Roof and the Second Floor, an alert will appear in the Model to reload them*

28. Click the Reload Modified Xrefs link in the External Reference alert balloon.

29. From the View menu, choose **3D Views > SE Isometric**.

Notice how the edits made to the *Second Floor* and the *Roof* are now visible in the *Model*.

30. From the View menu, choose **Shade > Hidden**.

31. From the View menu, choose **Shade > Gouraud Shaded.**

32. Click the 3D Orbit icon on the Navigation toolbar.

▶ Click anywhere in the 3D viewport and drag the mouse.

Notice the free-form rotation of the model in 3D (see Figure Q–36).

TIP It is often easier to view the model without the grid displayed. To turn this grid off, click the GRID toggle button in the task bar at the bottom of the screen.

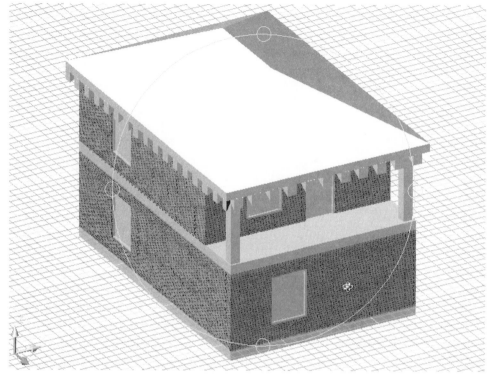

Figure Q–36 *View the Model and its edits in 3D with shading turned on*

▶ Press the ESC key when finished orbiting.

33. From the View menu, choose **Shade > 2D Wireframe**.

34. From the View menu, choose **3D Views > Top**.

▶ Using the wheel on your mouse, roll down two clicks.

 NOTE If you don't have a wheel mouse, use the Zoom Real-time icon on the Navigation toolbar to zoom out a bit (drag down)

Notice the different ways the objects, particularly the Doors, are displayed in each view. This is the result of ADT display control (refer to Chapter 2 for more information). Consult an AutoCAD Command reference or the online help for more information on zooming, panning and shading modes.

CREATING CALLOUTS

Let's finish our quick tour of Architectural Desktop by generating some elevations, placing them on a sheet and creating some output.

Continue in the *Model* file.

1. Right-click the title bar of the tool palettes, choose the Document group and then click the Callouts tab.

2. Click the **Exterior Elevation Mark A3** tool (see Figure Q–37).

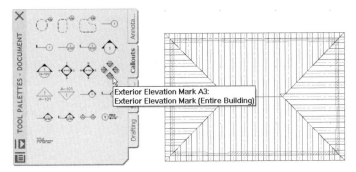

Figure Q–37 *Load the Callouts tool palette and click the* **Exterior Elevation Mark A3** Tool

▸ At the "Specify first corner of elevation region" prompt, click a point below and to the left of the lower left corner of the building.

▸ At the "Specify opposite corner of elevation region" prompt, click a point above and to the right of the upper right corner of the building.

The Place Callout Worksheet will appear.

▸ At the bottom of the worksheet, choose **1/4"=1'-0"** [**1:50**] from the Scale list.

▸ In the "Create in" area, click the Current Drawing icon (see Figure Q–38).

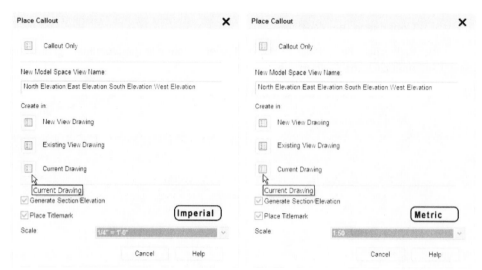

Figure Q–38 *Use the Place Callout Worksheet to place Elevations within the Current Drawing*

> ▶ At the "Specify insertion point for the 2D elevation result" prompt, click a point to the right of the model.

> ▶ At the "Pick a point to specify the spacing and direction of elevations" prompt, move the mouse up slightly and click again.

3. When the operation is complete, Zoom out to see the results. (Double-click the wheel of your mouse or use Zoom Extents on the Zoom toolbar).

There are now four overall building elevations of the model, complete with callout references. The numbers within these references currently display question marks (?). When we add these elevations to a plotting sheet, these question marks will automatically be replaced with actual numeric references. We will do this below. First, we need to designate the part of the drawing that we wish to use as a plan.

4. On the Project Navigator palette, right-click the *Model* file and choose **New Model Space View**.

An Add Model Space View worksheet will appear.

> ▶ Type Floor Plan for the Name and then click the Define View Window icon on the right (see Figure Q–39).

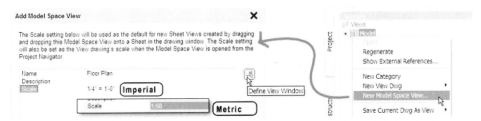

Figure Q–39 *Create a Model Space View to designate a portion of the Model to display as a Floor Plan*

▶ Following the command prompts, click to define a rectangular region that surrounds the building plan and is large enough to include all of the callouts.

The Add Model Space View worksheet will reappear.

▶ Click OK to complete the operation and dismiss the worksheet.

5. Save and Close the *Model* file.

CREATING SHEETS

6. On the Project Navigator palette, click the Sheets tab.

7. Right-click Simple Building at the top of the list, and choose **New > Sheet**.

▶ In the New Sheet dialog, type **A101** in the Number field and **Simple Building** in the Sheet Title field, and then click OK (see Figure Q–40).

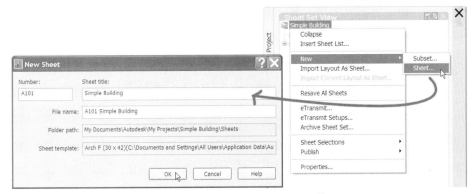

Figure Q–40 *On Project Navigator, Sheets tab, create a new Sheet*

8. In the Project Navigator, double-click *A101 Simple Building* to open it.

 NOTE If *A101 Simple Building* did not appear in the list, click the Refresh icon at the bottom of Project Navigator.

9. On the Project Navigator palette, click the Views tab.

 ▶ From Project Navigator, drag and drop *Model* directly onto the Sheet in the drawing window.

A ghosted image of the first elevation will appear ready to be placed at a specific location on the Sheet.

 ▶ At the "Specify insertion point" prompt, click a point on the sheet to place the first Elevation. Repeat as prompted for each drawing.

Notice how all of the callouts now display drawing numbers and proper Sheet references. (Zoom in as necessary to see this.) See Figure Q–41 for an example of the final Sheet.

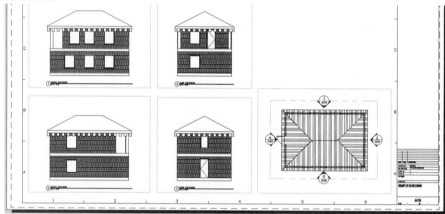

Figure Q–41 *Your first complete ADT model presented on a Sheet!*

10. On the Project Navigator, click the Views tab and then double-click the *Model* file to open it.

11. From the View menu, choose **3D Views > SE Isometric**.

12. At the bottom of the drawing, click the small Open Drawing Menu icon and choose **Publish to 3D DWF** (see Figure Q–42).

Figure Q–42 *Publish the Model to 3D DWF*

13. In the "AEC 3D DWF Publishing Options" dialog, accept all defaults and simply click OK.

14. At the "Select objects to publish to 3D DWF" select the entire 3D model (do not include the elevations) and then press ENTER.

You will be alerted when the publish operation is complete.

15. When prompted, click Yes to open the 3D DWF file.

The Autodesk DWF Viewer application will open and load the 3D DWF file. You can use the icons on the toolbar to zoom, pan and orbit the model in much the same way you can in ADT. The wheel of your mouse will zoom and pan as it does in ADT. You can switch to perspective with the tool on the toolbar and orbit around the model. On the left is a tree listing all of the objects in the *Model* file. Expand the various items on the tree and click on them to see them highlight in the model. You can right click for menu options such as isolate and hide objects. The Autodesk DWF Viewer application is included free with Architectural Desktop and is available as a free download from Autodesk.com. If you choose to publish models and drawings to this format, you can send them to your clients and consultants who can then view them using the viewer. You can also create DWF files of your 2D construction document files. To do this, go back to ADT click the Sheets tab of Project Navigator right-click the sheet and choose **Publish > Publish to DWF**. Open the DWF in Autodesk DWF Viewer when finished.

Congratulations! You've just completed your first Architectural Desktop model and project (albeit a very simple one), complete with a ready-to-plot Sheet and DWF files to view and send to clients and consultants. You now have a very brief idea of some of the objects and tools included with ADT 2006 and their respective capabilities. You are now ready to explore the ADT interface in more detail and begin following the many tutorials in this book. If you are new to ADT, read the Chapters 1-3 now. This will indoctrinate you in the ground rules to successful ADT usage. Read on

16. Save and Close all project files.

 NOTE Later when you quit ADT, if a prompt appears requesting that you repath your project, simply accept this and click the Re-path button. Re-Pathing and Projects will be discussed in detail in Chapter 5.

SUMMARY

Getting started with ADT is as simple as clicking a tool on the palettes and locating points in the drawing editor.

Add Walls, Doors, Windows and Roofs from the Design tool palette. Edit the parameters in the Properties palette or using grips.

Objects can be created directly or converted from other objects using a tool's right-click options.

Projects help you to formally establish the overall parameters of the entire building project, such as level management and file organization.

You can easily make an existing drawing part of the current project.

Once you have set up a project, it is easy to generate a series of elevations and compose a sheet layout.

Projects can be published to 2D and 3D DWF formats that can be shared with outside consultants and clients using a free viewer from Autodesk.

Introduction and Methodology

This section introduces the methodology of Autodesk Architectural Desktop 2006. Many concepts will be familiar to the seasoned AutoCAD user; many will be new. If you are a current AutoCAD user, skim through this section looking for concepts unique to ADT, particularly in Chapter 2.

If you do not have AutoCAD experience, then please read this entire section. It may also benefit you to complete some basic AutoCAD tutorials prior to reading it.

Section I is organized as follows:

Chapter 1	User Interface
Chapter 2	Conceptual Underpinnings of Architectural Desktop
Chapter 3	Workspace Setup

User Interface

INTRODUCTION

This chapter is designed to get you acquainted with the user interface and work environment of Autodesk Architectural Desktop 2006 (ADT). Collectively, all aspects of the user interface and work environment are referred to as the 'Workspace.' In addition to the Workspace, this chapter will also explore any necessary AutoCAD skills required for successful usage of ADT. If you did the Quick Start tutorial prior to this chapter, then you are already familiar with some of the objects and features of ADT. Read on to begin understanding the logic of the Workspace and what user interface skills are required to be successful with ADT.

OBJECTIVES

- Understand the ADT Workspace.
- Gain comfort with the user interface.
- Explore the prevalence of the right-click.
- Assess your existing AutoCAD skills.

UNDERSTANDING THE ADT WORKSPACE

The Workspace of Autodesk Architectural Desktop 2006 offers a clean and streamlined environment designed to put the tools and features that you need to use most often within easy reach, while allowing for endless customization for those whose needs vary. ADT's Workspace shares many similarities with AutoCAD 2006. However, there are some distinct differences as well. For instance, ADT has its own collection of highly specialized tool palettes and a very different set of pull-down menus and icons. We'll explore the ADT Workspace here, and later cover some of the traditional AutoCAD elements. The focus of the AutoCAD items is specifically on those things that are critical to ADT usage and success. For more detailed information of AutoCAD Workspace, commands and features, consult the online help or book on AutoCAD. Follow along in Architectural Desktop as you read following topics.

THE DRAWING EDITOR

The Autodesk Architectural Desktop drawing editor includes many features and controls. Presented here is a simple overview of the most important features (see Figure 1–1). For a complete discussion of all interface features, refer to the online help topic "The Workspace" in the Architectural Desktop User's Guide.

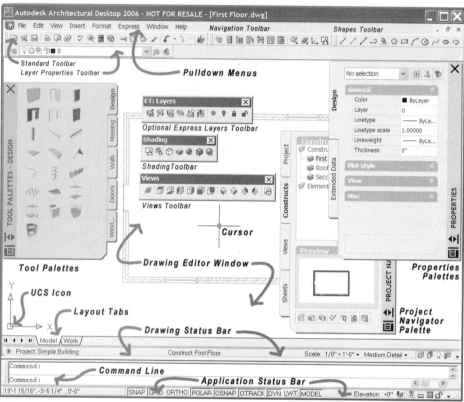

Figure 1–1 *Major components of the Architectural Desktop drawing editor*

Consistent with most Windows software, the ADT screen is framed with pull-down menus along the top edge, the Windows minimize, maximize and close icons in the top right corner, an application status bar along the bottom edge, and a variety of toolbars along various edges of the screen. In addition to these Windows standards, the ADT screen also includes the command line, typically docked along the bottom edge of the screen just above the application status bar, and tool palettes. Above the command line sits the drawing status bar, which is similar in appearance to the application status bar, but differs in function (see Figure 1–2). The command line and tool palettes are critically important interface elements in ADT and will be elaborated on in topics below. If you are a seasoned AutoCAD user, you are already very familiar with the command line. However, as we will see in the topic below, ADT 2006 offers a very viable alternative to the command line called dynamic input. If you have used ADT 2004 or 2005, (or AutoCAD 2004 or 2005), tool palettes will also be familiar. Other key elements of the ADT screen include the layout tabs, the UCS icon and the main drawing editor window (see Figure 1–1). Several of these key interface items warrant further discussion and are elaborated on in the topics that follow.

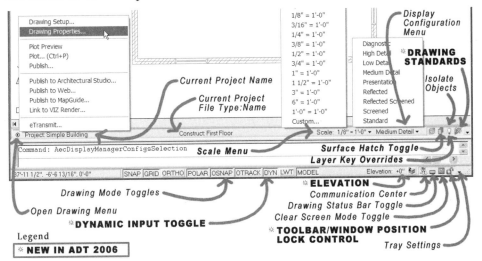

Figure 1–2 *The drawing and application status bar areas*

The application status bar runs across the length of the bottom edge of the drawing editor and includes a series of Drawing Mode toggles such as GRID, SNAP, POLAR and the new DYN. Each of these modes helps you to control cursor movements and make drawings more accurate. Many of these are covered below and elsewhere in this manual; you can also look them up in the online help. In addition, the application status bar also includes the Communication Center. This little icon will pop up a message periodically alerting you to news and information about ADT direct from Autodesk (this feature requires a live Internet connection.) Click this icon for live updates to the software, articles of interest and other information.

 NOTE After installation, a message will pop up here until you configure your personal preferences. Simply click here and choose what if any messages you wish to receive and how frequently.

New in ADT
2006

Grouped next to the Communication Center are some new items: Elevation control and the Toolbar/Window position lock controls. Items that are new to the 2006 Workspace are indicated in Figure 1-2 with a small "new" icon as well as different font.

With Elevation control, you can quickly set the current Z Elevation in the drawing and then toggle the automatic substitution of this Z value for all clicked points. This can be very handy when working in 2D to keep things "flat." It can also be helpful in 3D to avoid inaccurate Z snapping based on view direction. Examples of the use of this new tool can be found in Chapter 6.

The Toolbar/Window position controls enable you to lock certain elements of your workspace and prevent them from accidentally being moved or turned off. If you want to maintain the look of your custom user interface, this tool can be a big help.

The drawing status bar stays attached to the bottom edge of the drawing window and reveals information about the current drawing. As you can see in Figure 1–2, this includes quick access to the Open Drawing menu, Surface Hatch toggle, Scale menu and the Current Display Configuration menu. In addition, if the current drawing belongs to a project in ADT, its drawing type and name will appear. (In Figure 1-2, on the left, the current project is "Simple Building" and in the middle, the drawing is a Construct named "First Floor.") The Drawing Management tools in ADT were touched upon briefly in the Quick Start tutorial, and are covered in detail in Chapter 5. The Open Drawing menu gives quick access to Drawing Setup and publish commands. Through it, you can quickly set the Units and Scale of a drawing using the Drawing Setup command, assign Title, Subject and Keywords to a drawing using the Drawing Properties command, Publish, Plot, eTransmit, Link and Export the drawing. (Each of these last few options outputs the drawing data in some way.) On the right side you will find the Current Display Configuration menu. This menu allows you to change the currently active Display Configuration within the current drawing window (in model space, or viewport in a paper space layout.) Display Configurations are covered in greater detail in Chapter 2. Next to this menu, a small quick pick icon will appear if the drawing has external references. External references (XREFs) are links to other drawing files. Details and techniques on their usage will be covered throughout this book. The final icon on the right, new to ADT 2006, is the Drawing Standards quick pick.. Use it to configure standards in a project and synchronize the current drawing to the standards.

TOOL PALETTES

Tool palettes provide instant access to a complete collection of ADT tools organized in logical groupings. Tool palettes combine the user-friendly visual icon-based interface of toolbars with the flexibility, power and customization potential of pull-down menus. Simply click on a tool to execute its function (you do not need to drag it). Tools are interactive, and many parameters can be manipulated on the Properties palette while the tool is active. Furthermore, properties can be preassigned to the tools so that default settings are automatically assigned on tool use. Using the Content Browser, you can add tools and complete palettes to your personal Workspace at any time—more on the Content Browser below).

The default installation of ADT loads several basic tool palettes populated with a variety of the most commonly used tools. The palettes are organized into tool palette groups (see below). The Design tool palette group contains the most basic architectural object tools. The Design palette (part of the Design tool palette group) contains a basic tool for each of ADT's architectural object types. The remaining palettes contain tools with more specific parameters. All tool palettes are contained on a single palette in the Workspace. Groups are loaded by right-clicking the title bar.Individual palettes are loaded by clicking their tab on the tool palettes. ADT includes other palettes as well, such as the Properties palette, DesignCenter and the Project Navigator. All of these palettes share the same basic interface and behavior.

UNDERSTANDING TOOL PALETTES

1. Launch Architectural Desktop 2006 if it is not already running.

2. On the Standard toolbar, click the QNew icon (see Figure 1–3).

 TIP If you are not sure which icon is Qnew, hover your mouse over each tool until a tool tip appears. The Qnew icon looks like a small piece of paper with a dog-eared corner and the default location is the upper left hand corner.

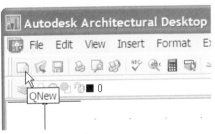

Figure 1–3 *Create a New drawing using QNew*

QNew will automatically create a new drawing file using your default template. If the Select Template dialog box appears when you click QNew, choose the template: *Aec Model (Imperial Stb).dwt* [*Aec Model (Metric Stb).dwt*].

CAD **Manager Note:** If QNew fails to load a template automatically, then choose **Format > Options** and click the Files tab. There, expand the Drawing Template Settings item and then the Default Template File Name for QNEW item. Finally, select the entry listed there, click the Browse button, and choose your preferred default template. These steps need only be done once, and will remain in place in the current profile on your machine. For more information on profiles, see Chapter 3.

3. If the tool palettes are not loaded, choose **tool palettes** from the Window menu, or press CTRL + 3. (You can also click the tool palettes icon on the Navigation toolbar.)

Tool palettes can be left floating on screen or can be docked to the left or right side of the drawing editor. Simply drag the palettes by the title bar to the left or right side of the screen. The title bar will dynamically shift from left to right as you move the tool palette close to either edge of the screen or it will dock to the edge of the screen. (see Figures 1–4 and 1–5).

4. Right-click the title bar of the tool palettes and check the setting of "Allow Docking."

A checkmark next to Allow Docking indicates that the palette will dock (attach) when close to the edge of the screen. No checkmark means that it will stay floating (See Figure 1–4).

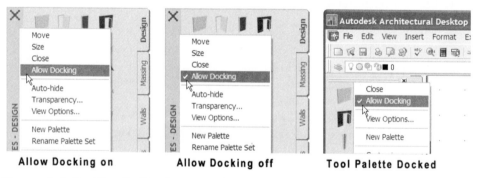

Allow Docking on   **Allow Docking off**   **Tool Palette Docked**

Figure 1–4 *The Allow Docking feature toggles docking of the tool palettes*

5. Test the behavior with Allow Docking on and then with it off.

If you dock a tool palette and wish to return it to floating, you can right-click on the double gray bars and remove the "Allow Docking" checkmark or click the double horizontal gray bars and drag the palette into the drawing window.

6. After experimenting, turn off Allow Docking.

7. Drag the tool palettes first to the left edge of the screen, and then to the right to see the title bar flip.

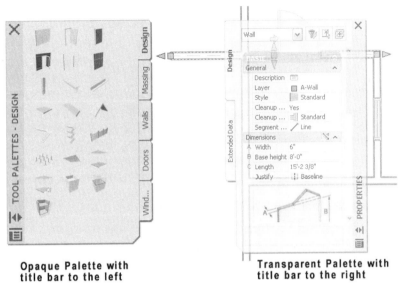

**Opaque Palette with
title bar to the left**

**Transparent Palette with
title bar to the right**

Figure 1–5 *Palettes dynamically justify their title bar to the appropriate edge of the screen*

 NOTE Figure 1–5 shows transparency turned on for the palette on the right. To do this, right-click the title bar (or click the small Properties icon in the bottom corner of the palette's title bar—shown in Figure 1–8) and then choose **Transparency**. However, this feature can cause a slowdown in performance on some systems, so be sure to test on your system to gauge performance before using regularly.

Many of the palettes (tool palettes, properties, etc) have tabs along the edge (or along the top for DesignCenter). Click these tabs to see other tools and options. For the tool palettes, you can customize these tabs and configure their properties; to do so, right-click on a tab (make it current first by clicking on it). When all tabs are not visible, there will be several tabs "bunched up" at the bottom of the tool palette; click there to reveal hidden tabs (see Figure 1–6).

Tool palette tabs can be grouped. A tool palette group includes a small subset of the total available tool palettes. The default installation includes three groups: Design, Document and Detailing.

8. Click on one or more tabs to switch between different palettes.

9. On the tool palettes, right-click on a tab.

   Note the menu options.

10. If all tabs are not shown, click on the bunched up group of tabs at the bottom to see menu revealing the hidden tabs (see Figure 1–6).

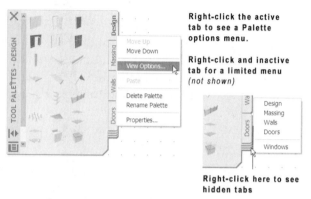

Figure 1–6 *Accessing palette options and hidden tabs*

▶ **Move Up & Move Down**—Shift the location of the selected tab relative to its neighbors.

▶ **View Options**—Opens a dialog with options for changing the icon size and configuration displayed on the palette(s) (see Figure 1–7).

▶ **Paste**—Only available after a tool (from this or another palette) has been copied or cut.

▶ **Delete & Rename Palette**—Allows you to delete or rename the selected palette.

▶ **Properties**—Allows you to change the Name and Description of the current palette.

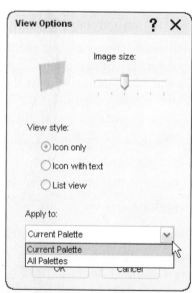

Figure 1–7 *View options change icon style and size for this or all palettes*

Another group of options is available for the entire palette group. In the bottom corner of the title bar in every palette are two small icons. The first toggles on and off the "Auto-hide" feature of palettes. When this feature is enabled, the palette will automatically collapse to just its title bar whenever the mouse pointer is moved away from the palette. The palette will "pop" back open when the pointer pauses over the title bar again. This same feature can be controlled with the Auto-hide option in the palette Properties menu available by clicking the second icon (in the bottom corner) or right-clicking on the title bar.

11. On the tool palettes, click the small Auto-hide icon (see Figure 1–8).

12. Move your mouse away from the palette.

    Notice that the palette collapses to just the title bar (see Figure 1–8).

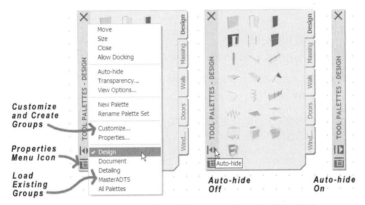

Figure 1–8 *Access the Properties menu, load Groups and toggle Auto-hide*

13. Move your mouse back over the collapsed title bar.

    Notice that the palette expands again.

14. Click the Auto-hide icon again to turn it off.

    NOTE For the remainder of this chapter, please turn off the Auto-hide. At ˊ pletion of the exercise, you may set it whichever way you prefer.

15. Click the Properties icon (or right-click the title bar) to ˊ menu.

    Note the various options.

CAD **Manager Note:** Later in Chapters 4 and 6 we will explore sˋ
New Palette. Palettes can be made that include any combination of stock
Complete palettes of project-specific tools can be created and subsequentlɣ

of the project team. Furthermore, these palettes can be linked to a remote catalog location and set to refresh each time ADT is loaded. This will guarantee that project team members always have the latest tools and settings. The customization potential of tool palettes is nearly limitless.

TOOL PALETTE GROUPS

As mentioned above, tool palettes can also be organized into groups. Right-click the tool palettes title bar to access other groups. By default, ADT installs three tool palette groups: Design, Document and Detailing. In addition, new to ADT 2006, when a project is loaded, a tool palette group uniquely named for the project will be added (and potentially made current).

16. Right-click on the tool palettes title bar and choose **Document** (to load the Document tool palette group) from the menu (see the left panel of Figure 1–36).

    Notice that all of the tool palette tabs change to Documentation functions (see the middle of Figure 1–9).

17. Right-click on the tool palettes title bar again and choose **Detailing** (to load the Detailing tool palette group) from the menu (see the right of Figure 1–9).

18. Right-click on the tool palettes title bar again and choose **All Palettes** (to load palettes from all tool palette groups at once).

    Notice that now all of the tool palette tabs from all groups appear (not shown in the figure).

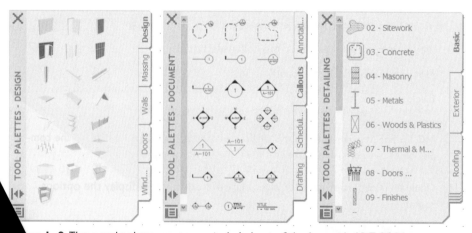

re 1–9 Three tool palette groups are included out of the box with ADT 2006

n create your own groups if you wish. To do this, right-click the tool pal-
e bar and choose **Customize**. In the Customize dialog, you can create new

groups by right-clicking on the right. Add and remove items from each group by drag and drop.

USER INTERFACE

Now that you have explored some of the common elements of the ADT Workspace, it is important to have a look at the most common interface elements. Chief among them is the prevalence of right-click context menus. Nearly every function in ADT makes use of at least some options on a right-click menu. In addition to the right-click, you will also frequently interface with objects directly on screen using dynamic dimensions and grip editing. As you interact with your drawings and models, it will be necessary to move fluidly around your screen and be comfortable viewing the model from all views, zoomed in and out. All of these items will be addressed in this topic.

RIGHT-CLICKING

In Autodesk Architectural Desktop, you can right-click on almost anything and receive a context-sensitive menu. In fact, we have just seen several examples in the previous topic on tool palettes. These menus are loaded with functionality and will be used extensively. Please note that alternative methods of command execution, though they often do exist, will rarely be elaborated on in these pages.

 TIP As a general rule of thumb, "When in doubt, right-click."

The next several figures highlight some of the more common right-click menus you will encounter in ADT. Do take a moment to experiment with right-clicking in each section of the user interface. You will also discover the typical Windows 2000/XP right-click menus appear in all text fields and other similar contexts (this is used for Cut, Copy, Paste and Select All). Let's explore the right-click.

RIGHT-CLICK IN DRAWING EDITOR (DEFAULT MENU)

1. If Architectural Desktop is not running, launch it now.

2. Press the ESC key to clear any commands or object selections.

3. Move the mouse to the center of the screen and right-click. Notice the menu that appears (see Figure 1–10).

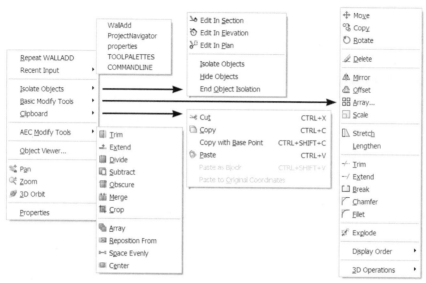

Figure 1–10 *Default right-click menu*

The default right-click menu appears when you right-click in the drawing editor with no commands active and no objects selected. It is divided into sections of function. The first item will always show the last command executed and beneath that a flyout list of recent commands. Repeat COMMAND, (where COMMAND is the last command run), will give a shortcut to executing the last command. (Figure 1–10 shows the WALLADD command.) The next section includes Isolate Objects (used to control visibility of selected objects and access the new Edit in View functionality), all of the common AutoCAD Modify commands (such as Move, Copy and Rotate) as well as Clipboard functions (Cut, Copy and Paste). AEC Modify Tools includes a collection of special ADT profile editing tools (this list includes many new commands for 2006). Object Viewer is a separate viewing window for quick study of selected objects. This will be explored in more detail in Chapter 4. Pan, Zoom and 3D Orbit are the standard AutoCAD navigation commands, and finally, Properties will open the Properties palette if it is not open and make it active it is open already.

CAD Manager Note: Many Veteran AutoCAD users will no doubt lament the loss of the single right-click to ENTER and repeat previous commands. Although the behavior of the right-click can be changed back to this style, it is recommended that you *not* do this. In doing so, a great deal of necessary ADT functionality will be lost. Please try the default setting throughout the duration of this book. If after completing the lessons in this manual you are still convinced you will be more productive with the right-click set to ENTER, then at least consider 'Time-sensitive right-click' (available on the User Preferences tab of the Options dialog) as an alternative. The Time-sensitive right-click option makes the right-click behave like an ENTER with a Quick click of the right button. A 'longer click' will display a shortcut menu. This feature will offer a good compromise to many seasoned AutoCAD users.

Please remember that both the ENTER key and the SPACE bar on the keyboard function as ENTER within the AutoCAD environment. For veteran AutoCAD users, the old 'rule of thumb' still applies. Keep your left thumb on the SPACE bar for a quick ENTER. If you do decide to change the behavior of the right-click, choose **Options** from the Format menu; click the User Preferences tab, and then the Right-click Customization button. You can configure 'Time-sensitive right-click' here.

RIGHT-CLICK ON TOOLBARS

1. Move your mouse over any visible toolbar icon, and then right-click. Notice the menu that appears.

   Toolbars that are *currently* displayed on screen will have a check mark next to them. Toolbars that are *not* displayed will have no check mark. (See Figure 1–11.)

2. Choose an item without a check mark. Notice the appearance of a new toolbar on screen.

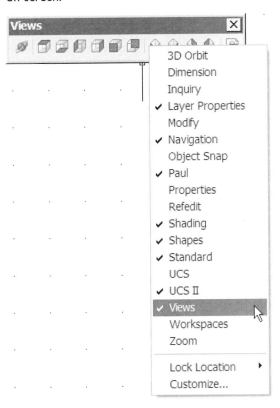

Figure 1–11 *Loading toolbars*

3. Right-click any icon on the newly loaded toolbar.

4. Choose the same name from the list to remove the check mark. Notice the disappearance of the toolbar from the screen. (You can also close it by clicking the close box in the top right corner.)

Right-clicking on toolbars gives access to a variety of toolbar functions: verification of which toolbars are active, ability to load and unload toolbars, the ability to customize toolbars and the ability to lock the location of toolbars on screen. Customization of toolbars is beyond the scope of this book; for more information on this topic, refer to the online help.

 Locking toolbars is a new function in ADT and AutoCAD 2006. You can access the lock options from the same right-click menu that we just explored for toolbars. You can also access it from the small lock icon at the bottom right corner of the screen (shown in Figure 1–2). When you choose to lock an interface item, you will not be able to move or close the item until it is unlocked. Test out this functionality to see how it might work for you.

RIGHT-CLICK TO SEE TOOLBAR GROUPS

Toolbars are saved in different groupings to make them more manageable. Right-clicking an icon as in the procedure above will show only the toolbars available in the same group. To see all of the available groups and their corresponding toolbars, locate any blank unused portion of the docked toolbars and right-click there. In the default ADT installation, there will be only one toolbar group: ADT. However, if you have installed other applications, such as the optional Express Tools, the optional menus on the **Window > Pulldowns** menu or built and loaded custom toolbars, those will show here as well.

1. Locate a blank section of any docked toolbar. (It will appear as an unused gray colored portion of the toolbar row or column).

2. Right-click in this blank area to reveal a menu with all toolbar groups available in your version of ADT.

   The ADT menu group will always appear; other choices vary from machine to machine depending on what additional menu groups are loaded or additional third-party software is installed.

3. Choose a toolbar group to reveal all of the toolbars available in that group (see Figure 1–12).

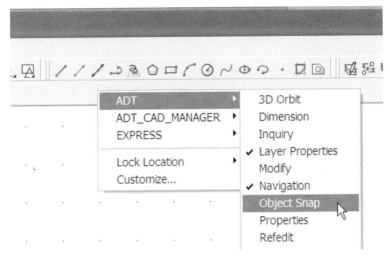

Figure 1–12 *Loading toolbars from other menu groups*

4. Choose a toolbar to load, as detailed above. Close it when finished.

If you decide to load additional toolbars in your workspace, you can leave then floating on screen or you can drag them and dock them to any side of the screen. You can also use the Lock Location options to lock the positions of all toolbars (and optionally other interface elements) if you choose.

RIGHT-CLICK IN THE COMMAND LINE

When you right-click in the command line, a small context menu appears (see Figure 1–13). Choosing Recent Commands shows a menu of the last several commands executed. Use this menu as a shortcut to rerun any of these commands. The Copy History command puts a complete list of all command line activity on the clipboard that can then be pasted into any text editing application. You can also access the Options command from this menu. (The Options command is also available on the Format menu.)

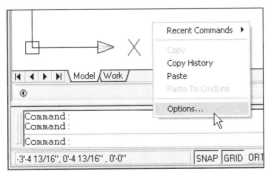

Figure 1–13 *Right-click in the command line*

New to ADT 2006, you can now close the command line window. To do this, make the command line a floating window. You can float it by dragging the small double gray bar on its edge, and then releasing when the command line has "un-docked" from the edge of the screen. Once the command line is floating, you will see the standard Windows close box (looks like an "X"). Click this box to close the command line. When you do this, a warning dialog will appear.

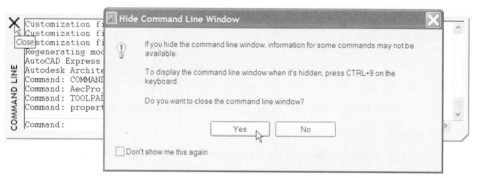

Figure 1–14 *You can close (Hide) the Command Line Window—use* **CTRL + 9** *to re-display it*

It is highly recommended that you use either the command line window or the dynamic input dynamic prompts (see below) option. If you disable both of these, it will be very difficult to use the software effectively (see Figure 1–14).

RIGHT-CLICK WITH OBJECTS OF THE SAME TYPE SELECTED

When one or more objects of the same type are selected in the drawing editor, right-clicking will reveal a menu of editing commands specific to the object type in question.

1. On the Design tool palette, click the **Wall** tool.

   If the Design tool palette is not visible, refer to the "tool palette groups" heading above for information on how to make it appear.

2. Click a point anywhere on the left side of the screen within the drawing editor.

3. Move the mouse position to the right side of the screen and click again.

4. Right-click and choose **Enter**.

Notice that "Enter" is the default option at the top of the menu, but that several other options appear as well. Most of the options shown are also available on the Properties palette and the command line.

5. Click directly on the newly created Wall object. It will be highlighted, with several grips along its length.

6. Right-click (see Figure 1–15).

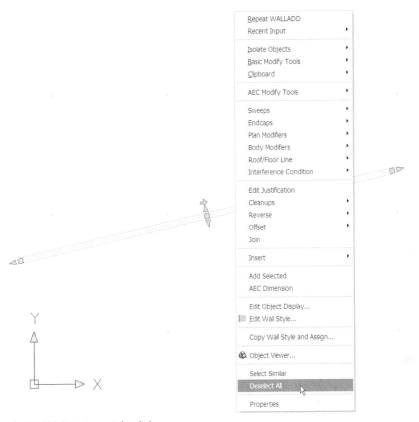

Figure 1–15 *Wall options right-click menu*

Notice that several different Wall-specific commands appear on the right-click menu.

7. Choose **Deselect All** from the menu.

RIGHT-CLICK WITH OBJECTS OF DIFFERENT TYPES SELECTED

When more than one object of different types are selected in the drawing editor, right-click will reveal a menu with basic editing commands not specific to any particular object type. This menu will be similar to the default menu seen above when no items were selected.

1. On the Shapes toolbar, click the Polyline icon.

2. Click a point anywhere on the left side of the screen within the drawing editor.

3. Move the mouse position to the right side of the screen and click again.

4. Right-click and choose **Enter**, (or press ENTER).

5. Click somewhere in the upper right corner of the screen (being careful not to click directly on top of any object).

6. Move the pointer to the lower left corner of the screen and click again. (Both objects should be highlighted. Look up "Crossing Window Selection" in the online help for more information).

7. Right-click and study the menu that appears (see Figure 1–16).

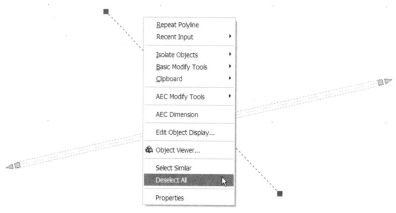

Figure 1–16 *The generic right-click edit menu*

The menu shows only those commands that are common to both types of objects selected, mostly simple AutoCAD editing commands in this case.

RIGHT-CLICK WHILE A COMMAND IS ACTIVE

Most ADT commands have one or more options. These options can be accessed by typing directly in the command line, using dynamic input onscreen prompts or from the right-click menu.

1. On the Shapes toolbar, click the Polyline icon.

2. Click a point anywhere on the lower left side of the screen within the drawing editor.

3. Move the mouse position to the bottom right side of the screen and click again.

4. Move the mouse to the upper right corner of the screen and click a third time.

5. With the command still active, right-click (see Figure 1–17).

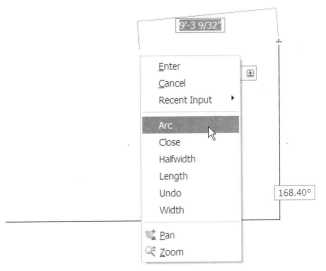

Figure 1–17 *Right-click within a command (polyline in this case) to access its options*

Compare the menu that appears with the options shown in the command line. You will see many of the same options are available in both places. (The same options are also listed in the onscreen prompting if you have dynamic input enabled—see below.)

6. From the right-click menu, choose **Arc**.

7. Move the mouse to the left of the screen and click again.

8. Right-click and choose **Close**.

RIGHT-CLICK IN THE APPLICATION STATUS BAR

The application status bar gives quick access to many of the drafting settings available in Autodesk Architectural Desktop. If you wish to customize the default settings of any of these drafting modes, simply right-click the button and choose **Settings** (see Figure 1–18).

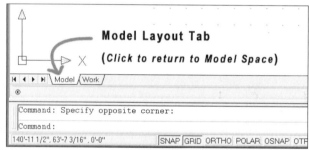

Figure 1–18 *Right-clicking the controls on the application status bar to access options*

Note that ORTHO only has the choices of On and Off. MODEL/PAPER actually toggles the screen from model space to Layout (paper) space. There is no right-click menu associated with this button. We will work in model space for most exercises in this book. For the time being, do not click this control. If you have clicked it, the only way to return to the correct mode is to click the layout tab labeled "Model" (see Figure 1–19).

Figure 1–19 *Click the Model tab to return to model space if necessary*

DYNAMIC INPUT

As noted above, the command line is only one way that we can interact with and access command options. Dynamic Input—new in Architectural Desktop 2006—gives us many onscreen cues and prompts to make the interactive process of creating and manipulating objects more fluid and user-friendly. Dynamic Input has a simple toggle button in the application status bar alongside the other drafting modes like SNAP, GRID and POLAR. If you right click this toggle, you will find many options to customize the Dynamic Input behavior. Let's explore some of those now.

POINTER INPUT

1. At the bottom of the screen on the Application Status Bar, right click the DYN toggle and choose **Settings** (see Figure 1–20).

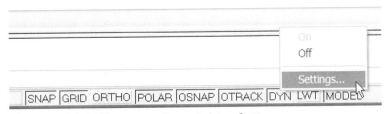

Figure 1–20 *Right-click DYN to access Dynamic Input Settings*

The Drafting Settings dialog will appear with the Dynamic Input tab active.

2. Deselect all checkboxes in this dialog, place a checkmark only in the "Enable Pointer Input" checkbox and then click OK (see Figure 1–21).

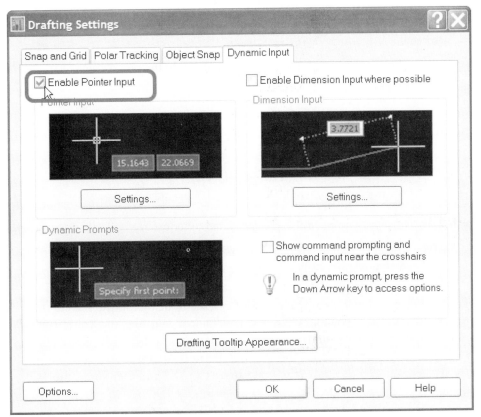

Figure 1–21 *Enable only the Pointer Input option*

This option provides text input fields at the cursor where you can type in coordinates as you draw. All objects in ADT exist in a coordinate grid (referred to as the "World Coordinate System" or "WCS"). There are two systems of coordinate input used by the software to allow you to indicate precise locations in the drawing relative to this coordinate system—Cartesian and Polar. In the Cartesian system, you input locations using "X" (horizontal) and "Y" (vertical) coordinates. In the Polar system, input is based on a distance (measured in units) and a direction (measured in degrees around the compass). Either system is valid for input in ADT and you can switch on the fly by varying your input syntax. The syntax for Cartesian input is: X,Y—where X and Y are input as positive or negative numbers in the current unit system (inches, feet, meters, etc). The syntax for Polar is D<A—where D equals the distance (nearly always a positive number in the units of the drawing) and A is the angle at which this distance is measured in degrees. Both systems can optionally add a third coordinate for the Z direction when working in 3D. View the topic: "Use Coordinates and Coordinate Systems" in the online help for more information on coordinate input.

3. On the Design palette, click the **Wall** tool and then click a point on the screen.

4. Move the mouse around slowly on screen and note the two dynamic prompts that appear (see Figure 1–22).

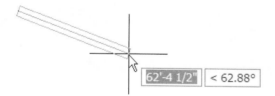

Figure 1–22 *Pointer Input gives coordinate prompts at the cursor onscreen*

5. Type a number such as **10′ [3000]** on your keyboard and note the number will automatically appear in the first coordinate field—do not press ENTER yet.

By default, ADT uses Polar coordinates as you can see indicated in the second onscreen prompt. However, you can change this default if you like and you can always input values in either system at any time. After you indicate the first value, type a '<' (less than sign) to input the Polar angle next, or for Cartesian coordinates, type a comma (,) to interpret the first value as an "X" and then input the "Y" value.

6. On your keyboard, type a comma (,)

Notice that the first value "locks" and the second prompt activates.

7. Move the mouse around a bit.

Notice that the first value of 10′ [3000] X is locked in so that the Wall is constrained in width and your mouse movements only affect the vertical position. Do not click yet.

8. Type in a second value such as **10′ [3000]** again and then press ENTER (see Figure 1–23).

Figure 1–23 *Pointer movement is currently constrained to the first locked field use the mouse or type to set the other value*

9. Repeat these steps, but instead of locking the first value with a comma, type the less than symbol (**<**) this time.

10. Move the mouse around on screen.

Notice that this time the length of the Wall has been locked and you can rotate freely around the first point. Your second value will be interpreted as the rotation angle of the Wall this time. Try it

11. If you have drawn any Walls, erase them. (Select them, and then press the DELETE key).

DIMENSION INPUT

Let's continue to explore the dynamic input settings by enabling the Dimension input. We will disable Pointer input and then enable Dimension input to understand it better.

1. Right click the DYN toggle and choose **Settings**.

2. Clear the Enable Pointer Input checkbox and place a checkmark in the "Enable Dimension Input where possible" checkbox and then click OK (see Figure 1–24).

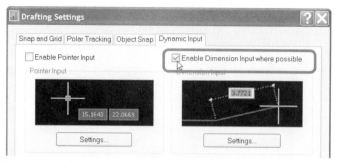

Figure 1–24 *Enable only the Dimension Input option*

3. On the Design palette, click the **Wall** tool and then click a point on the screen.

4. Move the mouse away from the first point.

Notice the dynamic dimension that appears attached to the length of the Wall and another indicates the angle of rotation. These dimensions will appear at key points, the content will vary depending on the type of object that you are drawing.

5. Try typing in a value such as 10′ [3000] again.

Notice that the value will input into the active dimension. To apply the typed value in dynamic dimensions, you must press ENTER.

6. With the Wall command still active, press the TAB key once.

You might need to move the mouse slightly after the TAB key. Notice that the angle is now highlighted and ready to receive input. Each time you TAB, the focus cycles to the next dynamic dimension.

7. Press ENTER or ESC to complete the Wall command and then erase any Walls drawn.

DYNAMIC COMMAND PROMPTING

The third option in the Dynamic Input dialog enables command prompting at the crosshairs. You can use this in place of, or in addition to, the command line window. Enable this setting with either or both of the other two settings.

1. Right-click the DYN toggle once more and choose **Settings**.

2. Place a checkmark in all three boxes this time including the "Show command prompting and command input near the crosshairs" box and then click OK.

3. On the Design palette, click the *Wall* tool.

Notice the command prompt (matching the one shown in the command line window) directly at the cursor (see Figure 1–25).

Figure 1–25 With *Command Prompting enabled, prompts show directly at the cursor*

4. Click a point to start the Wall.

   Notice the End Point prompt.

5. Click another point.

6. Press the down arrow on your keyboard.

   A list of command options will appear at the cursor. (This is the same list you would get if you right-clicked at this point).

7. Press the down arrow again to begin moving down the list of options. Press ENTER to choose an option.

Continue experimenting as much as you wish. You can return to the Dynamic Input settings at any time and choose different options. For the remainder of this text, it will be assumed that all three dynamic input options are active. If you want to learn more about Dynamic Input, return to the settings dialog and then click the Help button at the bottom of the dialog.

DIRECT MANIPULATION

Direct manipulation refers to the ability to manipulate object geometry directly in the drawing editor without the need to visit a dialog box or even a palette. To do this, you interact with the various grips of the objects. In ADT 2004, we saw many purpose-specific grip shapes appear in ADT. That trend has continued and additional Grip shapes and functions have been introduced in ADT 2005 and 2006.. Let's take a look at some of the basic direct manipulation techniques.

1. Using the **Wall** tool on the Design palette, create a single horizontal Wall on screen.

2. Click on the Wall just created to reveal its grips.

3. Hover your cursor over each grip.

   Do not click the grips yet, simply pass the cursor over each one and wait until the tool tip appears (see Figure 1–26).

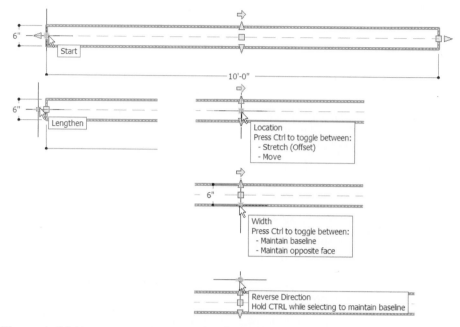

Figure 1–26 *Hover over a grip to reveal its function and options*

Notice that each grip shape serves a different function. Hovering over the grip reveals this function in a tool tip. If there are options to this grip (usually invoked by the CTRL key,) they will be revealed in the tool tip as well.

4. Click one of the Width grips.

    It is shaped like an isosceles triangle and points away from the width of the Wall.

5. Drag the grip and click to set a new width for the Wall (see Figure 1–27).

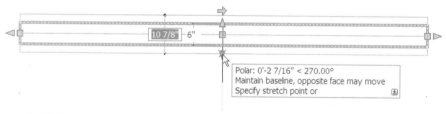

Figure 1–27 *Dragging to a new width with the "Width" grip*

6. Try it again, only this time, type a new width into the value field (highlighted onscreen) and then press ENTER to set the width to that value.

7. Try one more time, but press the CTRL key once after you click the triangular grip.

Notice how this method changes the width to distance away from the opposite face of the Wall instead of equally about the center.

8. Try the End and Lengthen grips successively.

Note the difference in behavior between these two grips. With the End grip, you have full range of motion and can change the endpoint location as well as the angle of the Wall segment. With the Lengthen grip, you can change only the length of the Wall, without affecting its orientation (see Figure 1–28). These are both very powerful tools.

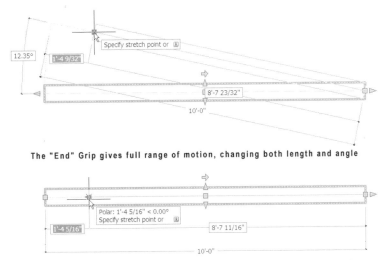

The "End" Grip gives full range of motion, changing both length and angle

The "Lengthen" Grip changes *only* the length, not the angle

Figure 1–28 *Noting the difference between the End and Lengthen grips*

This quick overview of grips on Wall objects gives you just some idea of the potential of this very powerful interface item. Look for unique grip shapes on every ADT object. Hover your mouse over them to reveal the tool tip of their function. We will see several more examples throughout this book.

EDGE GRIPS

Profile-based ADT objects (such as Spaces, AEC polygons and In-Place Edit Profiles) have round grips at their corners and long thin grips at the edges.

1. If you did not reload the Design Palette group above, right-click the tool palettes title bar and load it now.

2. On the Design tool palette, click the Space tool. Accept all of the defaults and add a Space anywhere in the drawing at a 0° rotation.

3. Click on the Space to reveal its grips.

4. Hover your cursor over each grip.

Do not click the grips yet; simply pass the cursor over each one and wait until the tool tip appears (see Figure 1–29).

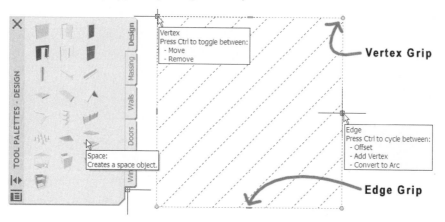

Figure 1–29 *Explore the Grip Shape Functions of a Profile-based Object*

5. Click one of the circular shaped grips and move it. Click again to complete the move.

6. Click on the same circular shaped grip and then press the CTRL key once.

    Notice that the shape of the Space changes to reflect the removal of this grip. The operation is not complete, however, until you click the mouse again.

7. Click anywhere to complete the removal of the vertex.

8. Click one of the thin rectangular shaped grips and move it. Click again to complete the move.

    This operation moves the selected edge while stretching the attached edges.

9. Click one of the thin rectangular shaped grips again.

10. Press the CTRL key once.

    Notice that a new vertex is formed.

11. Press the CTRL key again.

    Notice that arc segment is formed.

12. Continue experimenting with any of these functions before continuing.

All of these grip functions work on any ADT object type that uses closed profile shapes. Feel free to draw other objects and experiment.

THE COMMAND LINE, MENUS AND TOOLBARS

We have already discussed the Command Line Window above. This text-based interface, typically docked at the bottom of the ADT screen (see Figure 1–30), can be set to display one or more text lines at a time; the default configuration displays three lines. The word "Command" will be displayed when the there is no active operation (see Figure 1–30). Most AutoCAD and Architectural Desktop commands can be typed into this command line area and executed by means of the ENTER key. As we have seen in the previous topic, we can enable onscreen dynamic prompting as an alternative or in addition to the command line window prompts. Most common commands can also be executed by choosing them from palettes, menus or icons. The exact method of command execution is largely a matter of personal preference. As a general rule of thumb, palettes, menus, icons and onscreen prompting tend to be more user friendly, while the command line tends to provide the most options and allow the fullest automation potential as well as a certain familiarity and comfort for experienced AutoCAD users.

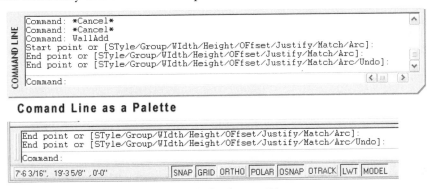

Comand Line as a Palette

Comand Line docked to the default position

Figure 1–30 *The ADT command line*

TIP As any veteran AutoCAD user can attest, there really is only one golden rule to AutoCAD/ADT usage: 'Always read your command line.'

Despite the much heavier reliance in ADT 2006 on modern interface items such as palettes and direct manipulation, it is still important to understand and pay attention to the command line or the onscreen prompts. Remember this rule, and you will be on your way to success with this software package. Disregard this rule and you will surely struggle and be frustrated. If you need to see more than three command lines or want to read back through previously executed commands, press F2 to toggle the AutoCAD text window. The AutoCAD text window is a history window allowing you to scroll back through prior commands. By default, this window is limited to 400 lines and is for viewing only.

Pull-down menus are also in ADT, as in all Windows applications. Menus in ADT have been reserved for the utilitarian functions, such as file management operations, and formatting the drawing. Many of the items on the Format menu will be explored throughout this book. If you are familiar with a release of ADT prior to 2004, you may be wondering what has become of the Concept, Design, Documentation and Desktop menus. Rest assured that none of the commands or functionality contained within those pull-down menus has been lost. The interface has undergone a significant overhaul starting with ADT 2004 and 2005, and refinements continue in this release, making it much more consistent and simple to use. Commands and tools are now organized in more logical locations to provide easier interaction. For instance, much of the former Design menu is now located on the Design tool palette (see the tool palettes section above).

For those who prefer pull-down menu interaction, go to the Window menu and choose **Pulldowns > Design Pulldown** and **Pulldowns > Document Pulldown** to instantly load two menus containing most of the commands that were previously contained on the Concept, Design, Documentation and Desktop menus. Also, you can still freely customize all pull-down menus and toolbars as you always have in AutoCAD/ADT. Please note that tool palettes are also *fully* customizable. See Chapters 4 and 6 for more examples.

New to ADT and AutoCAD 2006 is an entirely new user customization interface. Custom user workspace files that contain menus, toolbars, palette locations and settings and much more can be saved and loaded with the click of an icon. All such customizations are saved in a new file format with a CUI extension. CUI stands for "Custom User Interface." For a good overview of the customize Workspace features and working with CUI files, view the "How Customization Has Changed" topic in the online help.

INTELLIMOUSE

If you have a Windows Intellimouse, (or any third-party mouse with a middle button wheel and the proper driver,) ADT provides instant zooming and panning using the wheel! If you don't have a wheel mouse, this might be a good time to get one. This modest investment in hardware will pay for itself in time saved and increased productivity by the end of the first day of usage. Using the wheel you have the following benefits:

> ▶ To **Zoom**—Roll the wheel.

> ▶ To **Pan**—Drag with the wheel held down.

> ▶ To **Zoom Extents**—Double-click the wheel. (Do the same type of double-click that you would do with the left button, only on the wheel instead. It takes a little practice at first.)

The ZOOMFACTOR command controls the rate of zooming with the wheel. Type ZOOMFACTOR at the command line and then press ENTER. Type the percentage you wish to magnify with each rotation of the wheel on your mouse, and then press ENTER again. This setting applies globally, so you only need to do this once. The default setting is 60.

If you have tried to use your wheel to zoom and pan and instead you get a menu with Object Snap settings, you will need to adjust the setting for MBUTTONPAN. This command is a toggle setting that turns on and off the wheel zooming and panning feature. At the command line type MBUTTONPAN and then press ENTER. Be sure the value is set to **1** and press ENTER again. A setting of "1" turns this feature on, while "0" turns it off. If MBUTTONPAN is set properly and the wheel still does not function properly, you may need to adjust the settings in your Mouse applet in the Windows Control Panel. Usually the wheel works best when the wheel button is set to "Autoscroll." You may also need to update or re-install your mouse driver. Check your mouse manufacturer's website for complete details on mouse driver installation.

PREREQUISITE AUTOCAD SKILLS

The following list of "rules" serves two purposes. For the beginner, it will give you a focused list of topics to research and explore to prepare yourself for the chapters ahead. For the experienced AutoCAD user, treat this list as a self-assessment checklist of prerequisite AutoCAD skills. Use the list to identify areas where you might need a little brushing up.

AUTOCAD SKILLS – "THE RULES"

▶ **Always read the command line**—This was covered in the previous section, but repetition always aids in retention. Remember: read the command line or onscreen prompts when using dynamic input and you will be on your way to success with ADT. Disregard this rule and you will surely struggle and be frustrated.

▶ **When in doubt, right-click**—This was also covered earlier, but it is no less important than the others are; therefore it is also repeated here. This is especially critical for ADT commands and functions.

▶ **Draw accurately and cleanly**—The benefits of a cleanly drafted model, where all corners meet, shapes close, and dimensions make sense, and where double lines and broken lines have been avoided, cannot be overstated. In ADT you will be drafting with Walls instead of lines, but the rule remains the same. Always use tools like Trim, Extend, Offset, Fillet, Chamfer, (right-click and select Basic Modify Tools) and Object Snaps, and the payoff throughout the process will be tremendous.

▶ **Draw once; use many**—This rule embodies a major purpose of any CAD software package. This is certainly true for AutoCAD/ADT. Always look for ways to reuse what you have already created. In ADT this goes beyond the concept of copy and array. In Chapter 2, when we explore the Display System, and Chapter 5 when we look at projects, you will see that "draw once; use many" has numerous applications and interpretations throughout your use of ADT.

▶ **Work ByLayer**—Working ByLayer has been the norm and the rule since the earliest releases of AutoCAD.

NOTE This rule is included here because in many firms it is still the norm and a critical rule. However, in ADT, this is not as critical as it once was for all tasks. Many things can now be managed equally well (and often better) using the ADT display system instead of layers, making the ByLayer rule moot. See Chapter 2 for more information on the Display System.

▶ **Never explode objects**—Never explode an ADT object, hatch or dimension. *Never!* If you explode an ADT object, all its "smarts" are gone. This is almost never a good idea.

▶ **Save often**—You only need to lose everything you have done in the last three hours once to realize that Auto-save (in the Options dialog box) isn't good enough. Here is a simple test you can do any time while you work to see if it is time to save. Any time you can say to yourself: "boy, I'd hate to draw that again..." it is time to QSAVE! Creating regular backups is always a good idea.

▶ **Work smarter, not harder**—*Just because you can, doesn't mean you should.* Sure you can make a Wall style that includes 20 components, but is this always a *good* idea? ADT is a wonderful tool, but it can't do *everything* well yet. Learn what it can and can't do well, and learn to discern the often "fine line" between the two. This takes practice, but it is worth the effort. Many examples of this principle will be covered throughout this book.

▶ **Be consistent**—You will make many small decisions as you work through a project. You will name objects, configure parameters, and establish procedures. There is always more than one way to do things; whatever method you choose, implementing it consistently in similar situations will make it much easier to revisit the issue later. It will also make it much easier to work with other team members on the same project.

▶ **There are always exceptions to every rule**—Saved the best for last. Rules were made to be broken, were they not? These rules can be ignored or broken, but often at a tremendous loss of productivity. There are times when breaking a rule will make sense; with practice and experience you will learn when it is appropriate. Like Mother always used to say about falling in love: "When the time comes, you'll just know." Till then, however, best to stick to the rules.

TOOLS AND ENTITIES

Your experience with Architectural Desktop will be much more fruitful if you are already familiar with each of the following concepts. While it is beyond the scope of this book to cover each of these topics in detail, there are dozens of good Auto-CAD books and manuals available. The online help system is also a good resource.

▶ **Object Snaps**—All ADT (or AEC) objects take advantage of the standard AutoCAD Object Snaps. The Node, Center and Insertion Point Object Snaps are used extensively by ADT to mark special reference points within AEC objects. Because the Node and Center Object Snaps are not as frequently used in standard AutoCAD applications, veteran AutoCAD users should train themselves to make use of this powerful ADT feature.

▶ **Layers**—Layers are like *categories* that help organize all of the data within a drawing file. AutoCAD has long used layering to keep drawings organized and manageable. ADT objects benefit from "automatic" layering. The specific layer used for an object will depend on the layer Standard used in ADT. For instance, in the United States, the default Layer Standard in ADT is the "AIA (256 Color)" published by American Institute of Architects. (Actually the Layering Standard installed by default is compliant with recommendations of the most recent version of the AIA Layering as published in the U.S. National CAD Standard.) All layering defaults can be customized to meet your office's layer standard needs. If you use ADT in a country other than the United States, several other Layer Standards are provided.

▶ **External References**—An external reference establishes a file link between two or more drawings in a project set. Architectural Desktop is designed to take full advantage of the XREF functionality of AutoCAD. (And with the built-in Drawing Management system, it takes it much further.) This allows you to leverage your existing strategies to their fullest advantage (XREFs are used extensively in this book—see Chapter 5 for more information).

▶ **Layouts (Paper Space)**— Paper space received a major overhaul back in AutoCAD 2000 and even got a new name: Layouts. This powerful tool is most often associated with setting up sheets for plotting. ADT takes full advantage of this functionality as well. In addition, the advanced display functionality of Architectural Desktop can be combined with the functionality of paper space layouts. This allows even more control over output than layouts and viewports alone would allow. With the AutoCAD Sheet Set functionality built right into the ADT Drawing Management system, Layouts take on an even more important role. If you are upgrading from an AutoCAD release prior to 2000, take the time to explore the many resources available that cover the topics of Layouts, Page Setup, Sheet Sets and Plotting. Also, refer to Chapter 19.

▶ **Template Files**—Template files are a critical piece of the ADT puzzle. This subject is covered in detail in Chapter 3.

▶ **Polylines**—Polylines are used extensively in ADT to help create custom objects. A long list of Architectural Desktop objects can have their shapes customized to match the shape of a Polyline. You can even right-click on a Polyline to convert it to certain ADT objects such as Profiles and Mass Elements. When customizing ADT objects, you must use *lightweight* polylines; line and arc segments only, no splines or fit curves.

▶ **Blocks**—Blocks have been a huge part of AutoCAD productivity for years. A block is a collection of objects that have been grouped together and given a name. They then behave as a single object. Blocks continue to play an important role in customizing ADT objects. Another important issue associated with Blocks is that existing libraries of Blocks can be incorporated into an ADT work environment fairly seamlessly.

CAD Manager Note: Your existing block libraries can be converted to ADT content (making them 'draggable') using the AEC Create Content Wizard on the Format menu. Blocks, multi-view blocks, mask blocks, entire drawings and custom commands can be converted with this wizard. Refer to the online help or a copy of 'Autodesk Architectural Desktop: An Advanced Implementation Guide' for exact steps.

IMPORTANT AUTOCAD TOOLS

This topic includes some important AutoCAD functionality that seasoned AutoCAD users may have overlooked. If you are not familiar with the following tools, you should consult an AutoCAD resource or the online help and become conversant with them.

▶ **Auto Snap Options**—Parallel and Extension are two Auto Snap features that often go unused by veteran AutoCAD users. Parallel draws a line parallel to an existing line. This can be set as a running snap, but it works best when used as an override. (To use as an override, hold the SHIFT key down and right-click—this calls the OSNAP cursor menu—and then choose **Parallel**.) Extension allows the selection of a point along the extension (trajectory) of an existing line or arc. This provides functionality akin to a *virtual* Extend command.

▶ **Polar Tracking**—Polar tracking (right-click the POLAR button on the status bar and choose **Settings**) tracks the cursor movement along increments of a set angle, similar to the way ORTHO forces lines to move at multiples of 90°; however, the angle is a user-defined increment such as 45° or 15° (see Figure

1–31). Custom user angles can also be added. Additionally, unlike ORTHO, POLAR does not limit movement to the constraint angles. Instead, it snaps to those angles when the cursor gets close; otherwise it moves freeform. PO-LAR and ORTHO are not available at the same time. Toggling one on will toggle the other off and vice versa. Access Drafting Settings from the Format menu. The default setting for ADT 2006, is an Increment Angle of 30°, with Additional Angles of 45°, 135°, 225° and 315° added. This gives your cursor all of the positions of the traditional 45 and 30/60/90 triangles used in hand drafting.

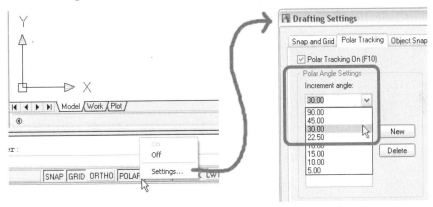

Figure 1–31 *Angle choices available in Polar Tracking*

▶ **Object Snap Tracking**—Uses Object Snap points to set up temporary tracking vectors to align geometry to precise points. On the Object Snap tab of the Drafting Settings dialog box (right-click the OSNAP button on the status bar and choose **Settings**), put a check mark in the "Object Snap Tracking On" check box to turn it on. A temporary alignment path will project from the various Object Snap points in the drawing. To use this feature, both the OTRACK and OSNAP toggles must be on (buttons pushed in on the status bar.) To track, first activate a command, and then hover the mouse over a snap point for a moment. Do NOT click the point. As the cursor moves away from the point, a small plus (+) sign will remain indicating that the point has been "acquired" (see Figure 1–32). Moving horizontally or vertically from this acquired point will enable the tracking feature and keep the cursor lined up with the point. Several points can be acquired, and multiple tracking vectors can be used simultaneously to achieve very precise alignments. Check the online help for more information.

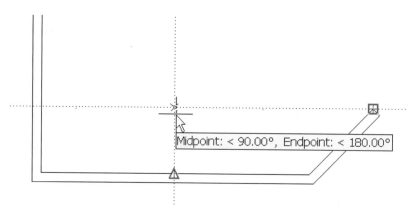

Figure 1–32 *Acquiring and tracking from temporary track points*

SUMMARY

The ADT 2006 interface is designed to be streamlined, logical and easy to use.

ADT relies heavily on the functionality of the right-click menu for editing commands.

When in doubt, right-click.

The command line can be hidden and dynamic onscreen prompting enabled instead.

A good foundation in basic AutoCAD skills will enhance your learning of ADT and help you start off on the right foot.

CHAPTER 2

Conceptual Underpinnings of Architectural Desktop

INTRODUCTION

Architectural Desktop (ADT) is an object-based Computer Aided Design (CAD) software package. It seeks to facilitate the creation of "Building Information Models" (BIM) from which plans, sections, elevations as well as quantities and other data can be readily extracted. The extracted drawings serve as two-dimensional "reports" of the "live" 3D model data. The advantages of this approach are many. From a 2D production point of view, this means less time drafting and coordinating building data, because plans and sections are both being generated from the same source data. If the data changes, both plan and section receive the change. Schedules and data reports of quantities, component sizes, materials used, and scores of other property data are also within the realm of possibility and fully accessible. To achieve this level of functionality, it is important to understand a bit about what makes ADT (and the Building Information Model that it creates) tick. That is the goal of this chapter. In particular, the focus will be on three major ADT concepts that are not available in the underlying AutoCAD drafting package. These are: Display Control, Anchors, and Object Styles. Drawing Management and Property Sets also offer power and flexibility not available within standard AutoCAD; these topics will be covered extensively in Chapters 5 and 15, respectively.

OBJECTIVES

In this chapter, we will explore the meanings of parametric design, Building Information Modeling and object-oriented CAD. Following the steps of a tutorial on the display system, you will learn how to display a single drawing model in many different ways which serve a variety of architectural drawing and documentation needs. By exploring the Style Manager and Anchors, we will begin to gain comfort with some of the critical conceptual underpinnings of the Architectural Desktop software package. Upon completion of this chapter you will be able to:

■ Understand objects and their properties.

- Work with the Display System.
- Understand object Styles.
- Understand Anchors.

PARAMETRIC DESIGN

Creating models in Architectural Desktop is an exercise in the practice of parametric design. *Parametric design* refers to the direct manipulation of various object characteristics (or "parameters") available within each of the objects as design decisions are being formulated. Objects in ADT are programmed to represent the real-life objects for which they are named. All real-life objects have a series of defining characteristics that determine their shape, size, and behavior. Parametric Design allows us to design while manipulating those real-world parameters directly on the objects.

To get a better sense of what these parameters might be, think of the characteristic to which you would refer if describing the object verbally to a colleague without the benefit of a drawing. Consider, for instance, a door. If discussing a particular door needed in a project with the contractor over the telephone, we would rely on descriptive adjectives and verbal dimensions such as "the door is a particular width, and a particular height, it is solid core and has a hollow metal frame." Once we had settled on the door required and hung up the phone, we would then need to convey graphically in our drawing documents the decisions we had just made regarding that door (and any others like it). In traditional CAD (AutoCAD), this would mean translating dimensions and materials into corresponding lines, arcs, circles and/or blocks that represent the required dimensions and materials in the drawing. In BIM, (ADT), dimensions and other specifications are simply input into a series of fields and stored with the data for the object. This data remains accessible throughout the life of the object and the drawing via the object's properties. Therefore, the next time we phone the contractor and realize that circumstances on the site have forced us to spec a different door and size, rather than redraw the door and manually adjust the wall in which it sits, as we would in traditional CAD, we now simply re-access the properties for that door and input the new values. Not only does the door itself update because of this change, but the wall in which it is inserted updates as well. Other linked views such as Elevations and Schedules will also receive the change. This is just one example of parametric design. Object parameters are always available for editing; data never needs to be recreated, only manipulated. Some principles of parametric design are as follows:

▶ **Draw Once**—In traditional CAD, each object needs to be drawn for each required view; therefore the same door or wall may need to be drawn two, three or more times. With ADT, objects need only be drawn once. They are then "represented" in each of the required views of Plan, Section and Elevation.

▶ **Progressive Refinement**—Complete or final design information is rarely known at the early stages of a design project. Changes occur frequently and often several times. In traditional CAD, it is easy to add new information. However, when major design changes occur, drawings must often undergo time-consuming redrafting. With ADT, designs can be progressively refined over the life of the project. As new data is learned or design changes occur, ADT object parameters may be adjusted appropriately without the need to erase and re-create the drawing. The objects are drawn once, and then modified and refined as required.

▶ **Style-Based versus Object-Based**—Most ADT objects make use of styles. A style is a collection of object parameters saved in a named group. When styles are assigned to objects, all properties of the style are transferred to the object in one step. If the style parameters change later, all of the objects using the style will change as well. This is similar to the behavior exhibited by Text and Dimension styles in traditional AutoCAD. ADT simply utilizes many more styles, and manages them and their relationships to objects much more completely than the corresponding AutoCAD counterparts. In some cases however, object parameters are assigned directly to the individual objects and *not* controlled by the style. Consider again the Door object as an example. The Door style would be used to designate the *type* of door, such as a hinged double door. However, double-hinged doors can come in a variety of sizes; the door type double-hinged is a style-based parameter, while the size is an object-based parameter.

▶ **Live versus Linked**—Some drawing types in ADT are edited directly on the "live" model data. This is the case with floor plans or live sections. The display system (see below) controls what displays on the screen as we are working in ADT. Plans views are live. If changes are made to the objects within the plan, those changes will be seen simultaneously on the live model in all other live views. However, some drawing types, namely 2D sections, 2D elevations, and schedules, are "linked." Rather than being a live view of the model, these separate drawings function as "reports" of the model that maintain a link to the live model data. These views must be periodically updated.

THE DISPLAY SYSTEM

Early in our architectural careers, we are taught the traditional rules of architectural drafting. These rules govern such things as what a plan or elevation drawing represents, how to create one, and most importantly, what to include and what not to include in making the drawing "read." Although there are accepted universal rules in place, a large part of the process involves personal style. Therefore, the rules need to be consistent enough to allow them to convey information reliably, and flexible enough to allow for stylistic variation. Amazingly enough, although CAD software such as AutoCAD has revolutionized the way design drawings are created, prior to Architectural Desktop, the software offered no specialized tools to assist the architect in achieving the unique graphical look required by architectural documents. Rather, lines were still painstakingly laid out one at a time as they had been in hand drafting, following the internalized prescriptions learned in architecture school. If a plan, section, and elevation were required to convey design intent, three completely separate drawings needed to be created and, more importantly, coordinated. The display system in ADT addresses this situation by incorporating the *rules* of architectural drafting directly into the software. Plans, sections, and elevations can now be generated *directly* from a single building model. This reduces rework and redundancy by requiring one set of objects, with three different modes of display (see Figure 2–1). The tools are flexible and fully customizable, so we can fully benefit from this powerful tool and still introduce the nuances of our own personal style into the process.

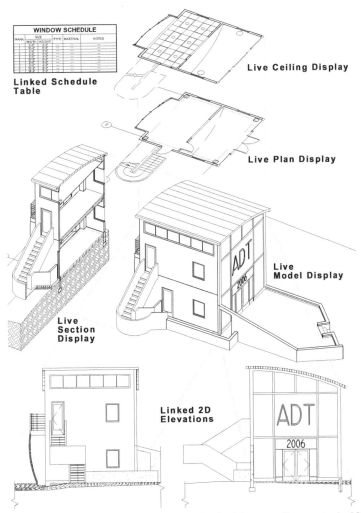

WINDOW SCHEDULE					
MARK	SIZE		TYPE	MATERIAL	NOTES
	WIDTH	HEIGHT			

Linked Schedule Table

Live Ceiling Display

Live Plan Display

Live Model Display

Live Section Display

Linked 2D Elevations

Figure 2–1 *Generating plan, section, elevation and schedule views from a single ADT model*

CAD Manager Note: The best way to quickly understand the display system is to start with the drawing template files (DWT) provided with the software. Much of the display system has already been configured for typical design situations in these templates. For most firms, these templates can be used 'as-is' or with minor customization. If you are new to ADT, begin with the out-of-the-box templates and display system configurations and then slowly begin customizing them to suit your firm's particular needs as required. This approach guarantees a complete understanding of the tools as you learn by example using suggested settings. The display system is complex, but it is also extremely powerful. In order to *fully* master the use of Architectural Desktop, it is important to become very comfortable with the display system.

THE DISPLAY SYSTEM'S RELATIONSHIP TO LAYERS

Display control determines how ADT objects are displayed under different view-ing conditions and circumstances. (Display control has no effect on AutoCAD en-tities such as lines, arcs and circles.) Layers are used as a global organizational tool for the management of drawing data (both AutoCAD and ADT), much like a drawing-wide categorization system. Display control supplements layers in helping you control what is seen and how it is displayed on the screen and in print. Each ADT object contains a series of components. The display control tools determine the display characteristics of each of these components. In some cases, the display properties of individual ADT object subcomponents are in fact handled by layers, although this is certainly not required (and in many cases not desired either). To summarize, Layers know nothing of Display, however, Display settings can op-tionally include Layers. Layers work on all entities; AutoCAD and ADT alike, but Display works only with ADT objects. Both can be used to control what is seen on screen and in print. Layers do this globally in an absolute way—they cannot re-spond to the condition of the drawing. Display settings can change if the condition of the drawing meets certain criteria. The Display system has been designed spe-cifically with Architectural drawing needs in mind, Layers have not.

OVERVIEW AND KEY DISPLAY SYSTEM FEATURES

The display system offers many features and benefits:

- ▶ **ADT objects display differently under various viewing conditions**—Display control settings can dynamically change the display of a building model from plan to elevation to section or 3D model, with a simple change in the viewing direction on screen.

- ▶ **Fully customizable**—Configuration of the display system components and their individual object properties can be customized to suit specialized needs. Customization can be as simple as modifying a setting or two in the configura-tions provided in the default templates, or as complex as a completely cus-tom-built solution tailored to a project-specific or office-wide need.

- ▶ **Understands the nuances of architectural drafting**—Object compo-nents such as "Cut Plane," (Walls) "Defining Line" (Sections) and "Muntins" (Windows) allow an architect to configure in a very specific way the precise visual expression a component ought to have in a particular display circum-stance. Display modes such as Plan, Reflected, Plan High Detail, and Plan Low Detail allow object components to appear differently and in greater or lesser detail under different scale and presentation conditions.

The Display System Tool Set

The display system tool set consists of a collection of interconnected components. It is important to understand some concepts and terminology related to each of these components before you begin to work with the display system.

▶ **Object/Subcomponents**—All ADT objects contain one or more subcomponents. These are simply the individual pieces of the object. A door, for example, contains (among others) the following *Plan* subcomponents: Door Panel, Swing, Stop and Frame. Just like traditional drafted AutoCAD entities (lines, arcs and circles) the entire object (the Door in this case) will be assigned AutoCAD properties such as Layer and Lineweight, while the individual subcomponents may also receive their own individual property settings through ADT Object Display.

▶ **Display Properties**—Display properties are the collection of display settings for a particular object. These include Visibility mode (On or Off), Layer, Color, Linetype, Lineweight, LTScale, and Plot Style. They also include many object-specific settings like Cut Plane height and Swing Angle. These are applied as "Drawing Default," "Style Override" or "Object Override" level.

  ▶ **Drawing Default**—In the hierarchy of display settings, the Drawing Default settings come first and establish the baseline for a particular type of object: all walls, all stairs, etc. within a particular drawing.

  ▶ **Style Override**—Style Override affects a subset of drawing objects that belong to the particular style in question only. Establishing the display settings at the object's style level offers many benefits: logical grouping of similar objects, global point of control to make changes, and consistency throughout the drawing and the project. Style-level settings override the Drawing Default settings.

  ▶ **Object Override**—Object Override affects only a single object selected in the drawing. Object-level settings override those of both the Drawing Default and, if present, the Style Override. In general, frequent application of object-level overrides should typically be avoided.

▶ **Display Representation (Display Rep)**—As dictated by the conventional rules of architectural drafting, each type of object has one or more ways in which it must be displayed. Display representations control the behavior of objects under various drawing situations such as Plan, Elevation and Reflected. Representations also control the specific display characteristics of an object's individual subcomponents within a particular view (see Figure 2–2).

▶ **Set**—Set describes which objects ought to display in a particular drawing situation, and in which mode they ought to be displayed. A set closely approximates a particular type of drawing, such as "Floor Plan" or "Building Section" (see Figure 2–2).

▶ **Configuration**—Configuration controls which set will appear on the screen, as determined by a particular viewing direction. Configurations can be "Fixed View" (same set appears regardless of current view direction) and "View Direction Dependent" (loads a different set based upon view direction). See Figure 2–2.

Putting It All Together

Summarizing how all of these components fit together, objects have one or more representations (display modes). These representations control the Layer, Color, Linetype, Lineweight, and Plot Style of each of the object's internal subcomponents assigned in one of three ways: Drawing Default, Style Override, or by Object Override. Individual objects and their appropriate representations are grouped together in a set. Sets are loaded into the viewport based on the conditions outlined in the Display Configuration. Changing the viewport view direction will trigger the display of individual sets based on these conditions. Figure 2–2 shows these relationships. The far bottom left shows the Views toolbar. Next to it is the Display Configuration menu on the drawing status bar (first shown in Figure 1–2). The remaining items along the bottom of Figure 2–2 are from the Display Manager, available on the Format menu.

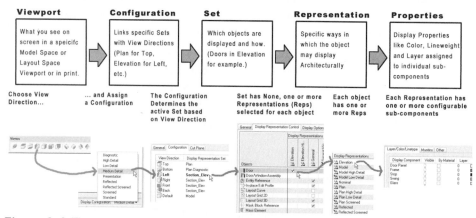

Figure 2–2 *The relationship of components in the display system*

The display system is dynamic. Changes made to its settings are immediately evident in the drawing. Many aspects of its configuration can be set up once and used from one drawing to the next, but certain settings will be in a constant state of flux as project needs dictate. The goal of the display system is to provide a single interface and tool set for controlling the myriad of display needs in architectural drawings. The goal here is two-fold: reduce some of the tedium associated with architectural drafting, and create a single building model that is capable of representing itself in all the ways necessary to display all the required drawings in a document set. With the display system, it is no longer necessary to draw the same data twice or three times to accommodate plans, sections, elevations and schedules. Features such as those offered by the display system are central to the concept of Building Information Modeling.

WORKING WITH THE DISPLAY SYSTEM

We begin our exploration of the display system by exploring the settings contained in the default template files. The display system is configured in much the same way in each of the out-of-the-box templates. The Imperial unit sample file used here was created from the *Aec Model (Imperial Stb).dwt* template file. Normally, the metric sample files use the *Aec Model (Metric Stb).dwt* as the starting point. However, in this chapter, since there is no explicit mention of units either Imperial or Metric, there is only one dataset.

INSTALL THE CD FILES AND LOAD SAMPLE FILE

Since the skills covered in this tutorial do not rely on measurement, the dataset used in this chapter is provided in Imperial units only. Both Imperial and Metric datasets have been provided for all subsequent chapters. Please see the "Units" heading in the Preface for additional information.

1. Install the dataset files located on the Mastering Autodesk Architectural Desktop 2006 CD ROM.

   Refer to "Files Included on the CD ROM" in the Preface for information on installing the sample files included on the CD.

2. Launch Autodesk Architectural Desktop 2006.

3. On the File menu, choose **Project Browser**.

4. From the drop-down folder list, choose your *C:* Drive.

5. Browse to the *C:\MasterADT5\Chapter02* folder.

6. Double-click *MADT6 Chapter02* to make this project current (you can also right-click on it and choose **Set Project Current**). Then click Close in the Project Browser.

 NOTE The Project Navigator Palette should appear on screen. If it does not appear, simply choose **Project Navigator** from the Window menu or press CTRL + 5.

7. On the Project Navigator Palette, click the Views tab and then double-click to open the *Floor Plan* View to open it.

LOADING A DISPLAY CONFIGURATION

We are going to load several different Display Configurations. Be sure to Zoom and Pan around the drawing a bit after each change to see the complete effect.

8. On the drawing status bar, on the right side, open the Display Configuration pop-up menu (it currently reads "Medium Detail").

A menu list of all Display Configurations appears (see Figure 2–3).

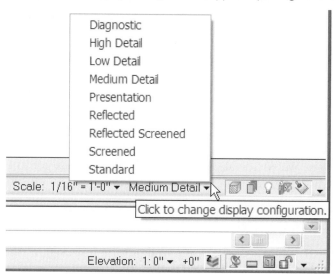

Figure 2–3 *The Display Configuration pop-up menu on the Drawing Status Bar*

9. Choose **High Detail** (see Figure 2–4).

Zoom and pan around the drawing to see the change. Notice that the hatch patterns in Walls and Spaces got closer together. The column bubbles got smaller, as did the text and arrows of the dimension strings. The Stair and Railing on the left side of the plan also got more detailed.

 NOTE The images in Figures 2–4 through 2–7 are composites including cropped portions of the entire file.

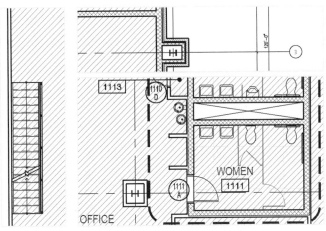

Figure 2–4 *Load High Detail from the Display Configurations list*

10. From the Display Configuration pop-up menu on the drawing status bar, choose **Low Detail** (see Figure 2–5).

    Zoom and pan around the drawing to see the change. Notice that all the hatching changed again, Wall hatching disappeared, Stairs got simpler and the column bubbles got bigger.

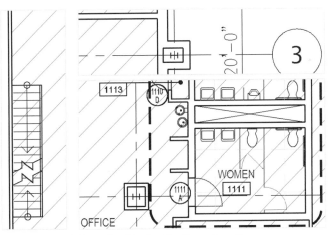

Figure 2–5 *Load Low Detail from the Display Configurations list*

11. From the Display Configuration pop-up menu on the drawing status bar, choose **Reflected** (see Figure 2–6).

 NOTE In each case, you may notice the slight delay while the Display Configuration loads and makes the change to the drawing.

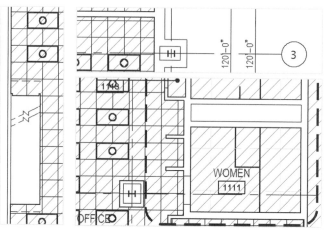

Figure 2–6 -*Load Reflected from the Display Configurations list*

Notice the change to the drawing: the toilet room fixtures and doors have disappeared and the ceiling grids and lighting have appeared. This configuration is intended for working on reflected ceiling plans, however the level of detail is equivalent to Medium Detail.

12. From the Display Configuration pop-up menu on the drawing status bar, choose **Presentation** (see Figure 2–7).

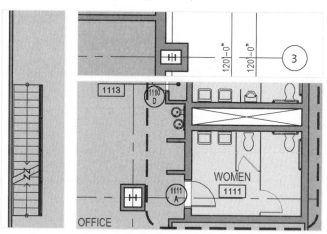

Figure 2–7 *Load Presentation from the Display Configurations list*

In this configuration, all of the Walls and Spaces are now filled with a solid color poché. None of the internal components within the Walls are displayed.

As you can see, loading Display Configurations has a broad effect upon the entire drawing. The goal of a Display Configuration is to adjust the display of all ADT components within the drawing for a particular situation. Those situations range from Floor Plans to Reflected Ceiling Plans to Presentation Plans, at a range of detail levels. Each of the individual configurations provided with ADT by default is designed to serve a particular drawing, modeling or printing task.

CAD Manager Note: If the solid fill patterns are displaying on top of the other geometry, you may need to adjust the value of Object Sorting Methods. You may do this at the command line (or when Dynamic input is turned on, directly on screen). To do so, type **SORTENTS** , press ENTER , and then type **127** and press ENTER again.

13. Reload the **Medium Detail** Display Configuration.

    Notice the drawing's change back to the default Display Configuration.

By loading these Configurations you can begin to see some of the power of the Display System; particularly when it comes to displaying the same information in more or less detail.

14. On the Project Navigator Palette, click the Sheets tab and then double-click to open the *A101 Chapter 2* Sheet.

Sheets and Sheet Sets are used for plotting ADT projects. Projects will be discussed in more detail in Chapter 5, and Plotting is covered in Chapter 19. Most Sheets have a single Layout tab using a name similar to the drawing. This Sheet is no exception and loads with the "Chapter 2 – Sheet Layout" tab active. These Layout tabs are used to compose the organization of items within the title block border for printing. This particular Sheet also has an additional layout tab. It was added specifically to showcase the Scale-Dependent qualities of the out-of-the-box Display Configurations and is not intended for plotting.

 NOTE Adding a new layout tab, like the "Scale Dependent" one seen here, is as easy as right-clicking on an existing layout tab and choosing **New layout**. Layouts are typically used in Sheets for plotting, but they can also be used to show and work in several simultaneous views of the model in any drawing.

15. At the bottom of the drawing window, click the layout tab named Scale Dependent (see Figure 2–8).

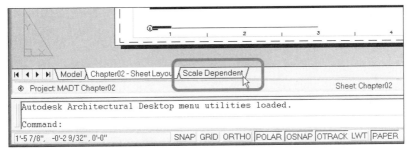

Figure 2–8 *Switch to Scale Dependent layout in the Sheet file*

16. Zoom and Pan around the drawing to compare the three viewports to one another (see Figure 2–9).

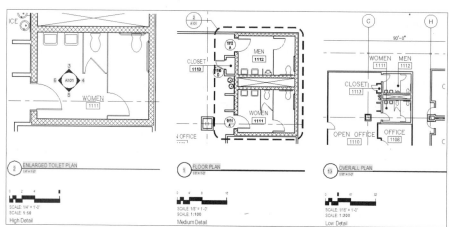

Figure 2–9 *Scale Dependent Display Configurations configured at proper scales in a Paper Space Layout*

Changing Display Configurations as we did in Model Space directly in the Walls file is useful and gives a good sense of what the differences are, but it is not until you view them in a properly configured Paper Space Layout as we have here, with viewports assigned to the appropriate plotting scales and the Lineweight display turned on, that you really see how striking the differences are. On the Sheet Layout tab, Elevations and a Schedule are also displayed. Feel free to explore further in this drawing before continuing.

VIEW DIRECTION DEPENDENT CONFIGURATIONS

So far we have looked only at the changes that occur in a Plan (Top) view. Display control affects the display of objects in all views. A "View Direction Dependent" configuration is tied into the viewing direction in the drawing or the active viewport. This means that as the view direction is changed, the Display Configuration will automatically adjust what is displayed. View Direction Dependent Configurations contain a default setting and at least one other condition tied to any of the six orthographic views: Top, Bottom, Left, Right, Back, and Front. A maximum of six special conditions can be specified, one for each orthographic view. There is always a setting configured for "Default" which will be displayed either when an orthographic view does not have its own override setting, or if a viewing angle other than the six orthographic views is chosen.

1. On the Project Navigator Palette, click the Views tab and then double-click to open the *Floor Plan* View.

NOTE If you left the *Floor Plan* View open above, then this action will simply make that file active.

2. At the bottom of the drawing window, click the layout tab named View Direction Dependent. There will be four viewpoints with labels, all of which are currently empty.

3. Double-click in the top right viewport (labeled SE Isometric) to activate it.

   When activated, its border will become bold.

To determine which Display Configuration is currently active on screen or within a particular viewport, simply glance at the Display Configuration pop-up menu that we used above.

Note that Medium Detail is currently active.

4. From the View menu, choose **3D Views > SE Isometric** or click the SE Isometric icon on the Views toolbar (see Figure 2–10).

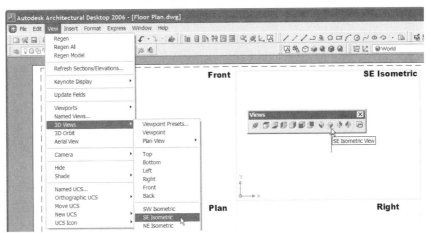

Figure 2–10 *Changing viewport view direction using either the View menu or the Views toolbar*

   A three-dimensional model of the drawing will appear.

5. Click in the top left viewport (labeled Front) to activate it.

6. From the View menu, choose **3D Views > Front** (or on the Views toolbar, click the Front icon).

7. Zoom in on the toilet room area (particularly the fixtures and the door swings).

   The 2D Plan display graphics have been replaced with graphics appropriate for an elevation.

8. Click in each of the two remaining viewports and set the view direction to match the labels in the corner. (For the Plan viewport, set the view direction to Top.) Zoom in to get a better look at the changes.

The details may be difficult to make out in the elevation and model viewports. To make them easier to see, we can toggle the surface hatching off.

9. Click inside one of the Elevation viewports (Right or Front). At the bottom right corner of the screen on the Drawing Status bar, click the small Toggle Surface Hatch icon next to the Display Configuration pop-up (see Figure 2–11).

Figure 2–11 *Medium Detail Display Configuration from several viewpoints*

Notice that this affects both the Front and the Right viewports. The Display Set loaded by the Medium Detail Configuration is the same for all elevation views, which is why both Front and Right changed; but it is different for Model, which is why SE Isometric did not change. If you wish, you can activate the SE Isometric view and repeat the same process to toggle hatching off there as well.

Even with the hatching disabled, it may still be difficult to see clearly since in wireframe display we are seeing through several walls, which makes discerning certain details difficult. In this file are two Building Section Line objects. They are labeled with text leaders in the drawing. These objects can be used to generate 2D Section/Elevation objects; these are linked 2D drawings derived from the live 3D model or Live Sections that actually cut away part of the model's display to reveal more clearly the internal objects. Both of these topics are explored in detail in Chapter 16. For now, let's just enable the Live Section display and see the effect on our model.

10. Click inside the SE Isometric view to make it active. Select the Section Line object labeled as Section Line 1, right-click and choose **Enable Live Section**.

Do not select the text or leader, but rather the rectangular object to which the leader points.

TIP If you have trouble selecting the Section Line Object, try Regenerating the Model. Click the Regenerate Model icon on the Standard toolbar (next to Undo and Redo), press **ENTER** and then try selecting again.

11. Click in the Right viewport. From the View menu, choose **Shade > Hidden** (or click the Hidden icon on the Shade toolbar). (See Figure 2–12.)

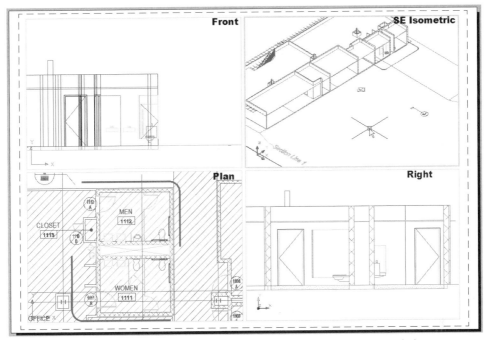

Figure 2–12 *Medium Detail Display Configuration from several viewpoints, including an enabled Live Section*

Note the change in the SE Isometric and Left viewports. Although each of these viewports is actually displaying a different Display Set, the Live Section operates at the Display Configuration level. Therefore, all Medium Detail views are affected, including the Front viewport. We do not, however, see any changes to the Top viewport. This is because Live Sections do not display in Plan views regardless of the Display settings. If you wish to test the behavior witnessed here, try loading High Detail (or any other Display Configuration) in the SE Isometric viewport. Notice that the model will no longer be sectioned. There is also a second Section Line object. Right-click Section Line 1 again, choose **Disable Live Section** to restore the complete model, and then repeat the Enable Live Section steps on the other section line (Section Line 2). If you wish, you can also enable both at the same time. If you wish to see the sectioned portion of the model displayed transparent, right click the section line after Live Section has been enabled, and choose Toggle Sectioned Body Display.

12. Double-click outside the viewports to toggle back to paper space (or click the MODEL toggle on the status bar at the bottom of the screen).

    None of the viewport edges should be bold anymore.

As you can see, with a View Direction Dependent configuration, simply changing the viewing direction on screen will load a different graphical display of the model.

FIXED VIEW DIRECTION DISPLAY CONFIGURATIONS

A Display Configuration need not be view direction dependent. It can be assigned a single fixed configuration. In this case, changing viewing direction has no effect on the Display Configuration. There is only one Fixed View Display Configuration provided with the default template files. The "Diagnostic" Display Configuration shows all objects the same regardless of view direction. The Diagnostic Display Configuration uses special display modes for several important objects that help you analyze object relationships and troubleshoot problems in the drawing. For instance, Walls are displayed with both the Graph and Sketch Display Representations active, but not the normal Plan or Model Representations. Both of these modes display useful information about the Walls that is not likely to be printed; Graph displays Wall cleanup information (see Chapter 9 for more information), and Sketch shows the Wall's baseline, centerline and direction. All of these terms are explored later in Chapters 4 and 10. When Diagnostic is active, you will also notice that each of the Door and Room Tags has a small curved line attached to another object. These are called "Tag Anchors" and are usually invisible, since we typically don't want them to print. However, they are very useful when you are trying to verify data within Schedule Tags or troubleshoot problems with tag attachments and Schedules. Schedules and Tags are covered in Chapter 15; anchors are discussed below.

13. On the drawing status bar, click the Display Configuration pop-up menu.

14. At the "Select viewports or RETURN for Paper Space viewport" prompt, select all four viewports (Click the light blue edges, or use a crossing window selection), and then press ENTER.

15. From the Display Configuration pop-up that appears, choose **Diagnostic**.

    Notice that regardless of the active view direction, the model will use the same configuration. In addition, notice that the Live Section display is no longer active.

Zoom and Pan within any viewport if you wish to explore the changes made by the Diagnostic Display Configuration. It is not important to completely understand the use of the Diagnostic Display Configuration at this time; the point of this exercise is to show the differences between a View Direction Dependent Configuration and a Fixed View Configuration. In the next sequence, we will learn how to set up our own Display Configurations.

16. Click the Model tab to return to model space for the next sequence.

USE THE DISPLAY MANAGER

The Display Manager is the primary interface for managing the display system. The Display Manager can be a little overwhelming at first. Stick with it; once you get the hang of it, you will see that it is quite logically organized and very powerful. In this lesson, you will learn the basics of the ADT Display Manager. Our aim in the steps that follow is to give a glimpse at the potential this powerful tool offers.

1. From the Format menu, choose **Display Manager**.

2. Click the plus sign (+) next to Configurations.

Study the icons next to the various configurations in the list (see Figure 2–13).

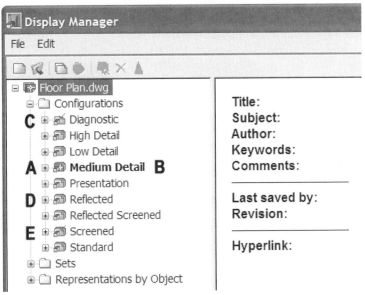

Figure 2–13 *Understanding the icons used for configurations*

A) **A dark blue icon**—Indicates the default display configuration for the drawing (tiny AutoCAD cursor on it). The default is used automatically when creating new layouts. To set the default for the drawing, choose **Drawing Setup** from the Open Drawing pop-up menu on the drawing status bar and then click the Display tab. You can also right-click a configuration in the Display Manager and choose **Set as Drawing Default**. In Figure 2–13, Medium Detail is the default.

B) **Bold text**—Indicates that this configuration is active in the current viewport. In Figure 2–13, Medium Detail is current.

C) **A single sheet of paper icon**—Indicates that it is a *Fixed View* Display Configuration. In Figure 2–13, only Diagnostic is Fixed. (See the "Fixed View Direction Display Configurations" topic above.)

D) **A stack of sheets icon**—Indicates that it is a *View Direction Dependent* Display Configuration. In Figure 2–13, all Configurations except Diagnostic are View Direction Dependent. (See the "View Direction Dependent Configurations" topic above.)

E) **A red check mark in the icon**—Indicates that the configuration is in use in one or more viewports within the drawing, but not necessarily visible on screen. In Figure 2–13, Diagnostic and Screened are indicated as "in use." On your screen, Screened will not appear as "in use."

 NOTE If it is not obvious to you where a particular Display Configuration is "in use," check your various layout tabs. It is likely that this is where you will find most of them in use; for instance, this is why the Diagnostic has a checkmark, since we left it assigned to the viewports on the View Direction Dependent Layout tab.

THE CONFIGURATION TAB

3. Click on the High Detail Display Configuration in the tree view at left to select it.

4. On the right side of the Display Manager, click the Configuration tab (see Figure 2–14).

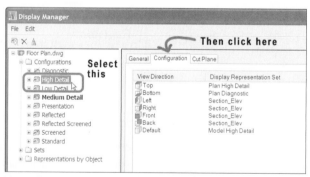

Figure 2–14 *View the Configuration tab for the selected Display Configuration*

The Configuration tab lists each of the six orthographic views and also contains an entry for Default under the heading View Direction. Next to each view direction, a Display Representation Set can be loaded. Display Representation Sets are described in detail below. If there is an entry next to the Default *only*, the Display

Configuration is fixed. If more than one view direction contains a Set entry, then the configuration is View Direction Dependent (see above). In this case, we can see that High Detail is a View Direction Dependent Display Configuration, because there is more than one entry.

> 5. Click on Diagnostic in the tree view at the left to select it.

Notice the difference on the Configuration tab. Here, there is only an entry next to the Default view direction. This makes the Diagnostic Display Configuration a Fixed Display Configuration (see Figure 2–15). You will also notice a check mark in the Override View Direction checkbox. This allows you to control the behavior of multi-view blocks displayed in this configuration. Multi-view blocks are ADT objects with user-defined graphics. See the "Create a Custom Multi-View Block" topic in Chapter 11 for more information on multi-view blocks.

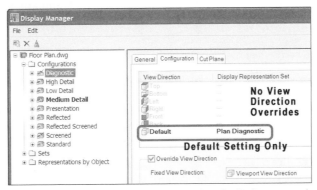

Figure 2–15 *A Fixed Display Configuration contains only a Default entry and no others.*

There are two other tabs for Display Configurations: General and Cut Plane. The General tab contains simply the names and description of the configuration. You can use this tab to edit those values if you wish. The Cut Plane tab establishes a single cut plane height that is used by all objects within the drawing. Many objects (Walls, Windows, Curtain Walls, etc.) use a cut plane when determining what to draw in Plan displays. This "Global Cut Plane" is used to help synchronize all of these objects relative to a baseline height in the drawing. If you click the Cut Plane tab, you will see that most of the Configurations use a Cut Plane of 3'-6" [1400] except the two Reflected ones, which use 7'-6" [2300]. We will discuss this feature more in Chapter 8. For now, let's take a closer look at the active Display Configuration.

> 6. Click on the Medium Detail Configuration in the tree view at left to select it.

In this particular configuration, the Top view direction is configured to use the Plan Display Set. (Display Sets determine which ADT objects display and how; see the "Understanding Sets" heading below.) This indicates that whenever the drawing is viewed from the Top, the *Plan* Display Representation Set will be loaded in the viewport. Next to each of the Left, Right, Front and Back view directions in the Medium Detail Configuration, the *Section_Elev* set is loaded. Diagnostic is used if the drawing is viewed from Bottom. Finally, if none of these conditions is met (the view direction is set to something other than Top, Left, Right, Front, Back or Bottom) the default set will be loaded; in this case, *Model* (see Figure 2–16).

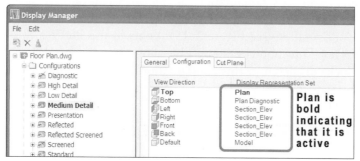

Figure 2–16 *Examine each of the assignments in the Medium Detail Configuration*

As you can see from the settings contained in these two (and any other) Display Configurations, they simply reference Display Sets based upon a given viewing direction(s). What then, determines why the objects change their graphics when we switch? Let's have a look at Sets to begin to learn the answer

UNDERSTANDING SETS

A set simply indicates which objects are displayed on screen and what Display Representation(s) each should use (see the "The Display System Tool Set" heading above).

NOTE Not all Representations will be available for all object types.

1. Continuing in the Display Manager, click the plus sign (+) next to the Sets folder (see Figure 2–17).

   This will reveal all of the Display Representation Sets (or just "Sets") available in this drawing.

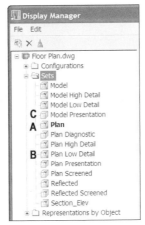

Figure 2–17 *Expand the Sets folder*

As noted above, the Plan set is the active display set in the drawing. This is evident both on the Configuration tab of the Medium Detail Display Configuration and by the fact that Plan in the Sets list at left is bold. The Model tab (model space) is currently set to Top view in the drawing.

Notice the icons next to each Set name:

A) **Bold Text**—Indicates that this is the current Set. In Figure 2–17, Plan is current.

B) **A red check mark in the icon**—Indicates that the Set is "in use" by one or more Configurations. In Figure 2–17, all Sets are in use except Model Presentation.

C) **A small box (without a check mark)**—Means that the Set is not currently being used by a Configuration. In Figure 2–17, Model Presentation is the only Set not in use.

CAD Manager Note: Like most named objects in ADT, Configurations and Sets can be purged from a drawing only if they are unused. Highlight the item you wish to delete, and then click the Purge icon at the top of the Display Manager dialog box (it looks like a little broom). Be cautious, as no dialog will appears to confirm deletion. Purged items can be restored with Undo. Please note: the Sets with the small bit of green shading in the corner cannot be purged even if they are unused. These are default Sets that are auto-created by the software.

2. Click on the Plan Set in the tree view at left to select it.

Just like the Configuration folder above, each Set has three tabs. The General tab serves the same purpose as it did for Configurations. You can use it to change Name and Description. (Default Sets, the ones with the small bit of green shading, cannot be renamed.) There is also a Display Representation Control tab and a Display Options tab. Let's take a look at the Display Representation Control tab.

3. On the right side of the dialog box, click the Display Representation Control tab (see Figure 2–18).

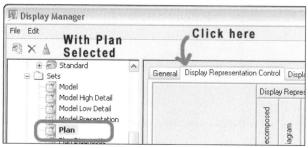

Figure 2–18 *Access the display control settings of Sets*

4. Scroll both horizontally and vertically on the right side of the dialog box.

On the left of the Display Representation Control tab is an alphabetical list of all ADT objects. At the top are all of the Display Representations within the current drawing. Notice that a Set is comprised of a series of check boxes. Objects that have one or more of their Display Representations checked in this Set are visible on screen or in the current viewport. Objects with no boxes checked are invisible.

 TIP It is usually a good idea to stretch out the size of this dialog as large as your monitor will allow.

For example, perhaps you have decided that Spaces, while very valuable for attaching Room Tags, holding Room Finish data and their contribution to sections, are not very useful in plans. Using the power of Sets, we can simply turn off Space objects in the current Display Set (Plan) while leaving them displayed in the other Sets such as Model.

5. Locate the Space entry, and click on it.

This will highlight the complete row in blue.

6. Clear the check box in the Plan column (see Figure 2–19).

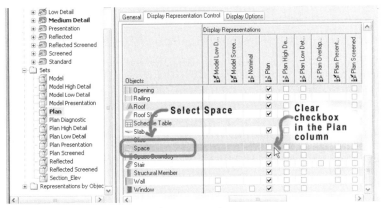

Figure 2–19 *Turning off the spaces in the current view*

7. Click OK to accept changes and return to the drawing.

   There will be a slight delay as the drawing regenerates the current display. Notice that the Spaces (the hatching inside of each room) have disappeared (see Figure 2–20).

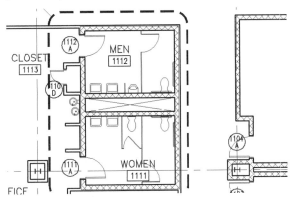

Figure 2–20 *The floor plan with Spaces turned off*

It is important to note that this technique is very different from simply freezing the Space object's layer. Had we frozen the layer, Spaces would be invisible in *all* Display Configurations. This might be all right for Display Configurations like Medium and High Detail, but it would not be acceptable for Presentation. See for yourself; try loading other Display Configurations now.

8. Load the **Presentation** Display Configuration using the technique covered above in the "Loading a Display Configuration" heading.

   Notice that the Spaces appear here, shaded with solid color fills. Had you frozen the Walls A-Area-Space Layer, they would have been invisible here as well. Try it if you like to see for yourself.

9. Load the **High Detail** Display Configuration.

Notice that the Spaces are still visible in this (or any other) Display Configuration as well. If you returned to the Display Manager, you would see that, although we cleared the active Display Representation (Plan) for Spaces in Medium Detail, there is still a check mark in Plan High Detail for the now active High Detail Display Set.

10. Return to the Display Manager (**Format > Display Manager**).

11. Expand Sets again and select Plan High Detail (the active set, indicated by bold).

12. Highlight Space on the right side again.

    Scroll over and take note of the check mark that appears in the Plan High Detail column.

13. With Space still selected, clear Plan High Detail and instead put a check mark in the Plan Screened column (see Figure 2–21).

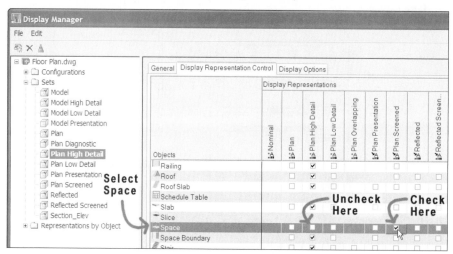

Figure 2–21 *Making Spaces display using the Plan Screened Display Representation*

14. Click OK to accept changes and return to the drawing.

Spaces now display using the Plan Screened Display Representation. In model space they will appear solid black. To see the halftone effect, click the Plot Preview icon on the Standard toolbar or choose **Plot Preview** from the File menu. If you wish to preview a zoomed-in window region, choose **Page Setup Manager** from the File menu and then click New. In the New Page Setup dialog, choose *Model* and then click OK. From the What to Plot list, choose **Window** and then designate a region of the plan to preview. Back in the Page Setup dialog, click Preview (see Figure 2–22). When you are finished previewing, press ESC three times to return to the drawing.

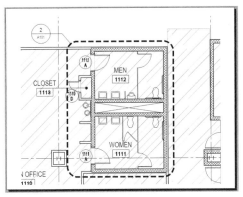

Figure 2–22 *Spaces displayed screened back in 50% halftone*

This simple exercise shows the versatility of ADT object display. With the simple selection of a different display mode in the current Set, the drawing can take on a dramatically different look.

UNDERSTANDING DISPLAY REPRESENTATIONS

Objects in ADT have many different display modes. These are called "Display Representations" (Reps). Each Representation corresponds to a particular drawing type such as Plan or Elevation.

1. Return to the Display Manager.

2. Click the plus sign (+) next to Representations by Object (see Figure 2–23).

   This will reveal the complete list of ADT objects.

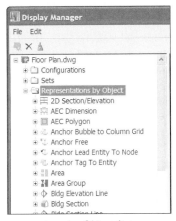

Figure 2–23 *Expand the Representations by Object list*

3. From the list of objects in the tree view at left, select Space.

This will reveal a list of all the Display Representations available for this object type in a column on the right side with all of the available Sets listed across the top right. Check boxes again appear indicating specifically which Space object Reps are loaded in each Set.

On the right side of the dialog box in the Display Representations column, study the icons next to each of the entries in the list. There are several built-in Display Representations. Additional user-defined representations can be created to meet needs not addressed by the default ones. You can configure the settings of the default representations, but they may not be renamed or deleted. Create user-defined representations by duplicating any of the default ones. These can be renamed and, provided they are not being used, deleted. A couple of the Reps available for Spaces (and several other object types in this drawing) are in fact custom-created Reps (see Figure 2–24).

▶ **A rainbow-colored "properties" icon**—Indicates that it is one of the default (permanent) representations (see Figure 2–24).

▶ **A black and white "properties" icon with a small black arrow**—Indicates that it is a custom-created representation (see Figure 2–24).

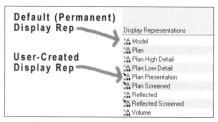

Figure 2–24 *Icons differ for default permanent Display Representations and user-created ones*

CAD Manager Note: To create a user-defined Display Representation, you must duplicate an existing one that is similar to the one you wish to create. Select the Rep that you wish to duplicate, right-click it and choose **Duplicate.** Give the new Rep a unique name. As with all named objects, choose your name carefully. Once you have created the custom Display Rep, you can configure its parameters and assign it to a Set. Usually, you will also create a custom Set and Configuration along with your new Rep, but this is not required. See the steps that follow for information on building custom Sets and Configurations.

4. Click OK to return to the drawing without making changes.

EDIT THE PROPERTIES OF A DISPLAY REPRESENTATION

The parameters of a Display Representation control all of the individual subcomponents of each ADT object. For instance, Walls might have brick, concrete block and drywall components. Doors have the door panel, the frame and the swing. The specific subcomponents that an object type contains may vary for each of its Display Reps. For instance, the Space objects that we have been working with here have a Floor component and a Ceiling component. However, these components are accessible only within the Model Rep. In the Plan Rep for Spaces, you will instead see Net Boundary, Gross Boundary and Hatch. (New to ADT 2006 is the ability to also enable Net, Gross and Usable boundaries for Space Objects. More information can be found on this feature in Chapter 18). This means that, although the Configurations and Sets determine which objects are visible and invisible on screen, it is the Display Reps that determine exactly *how* those objects will be drawn graphically. One other aspect of Display Representations is important to note before we begin to explore them: the display settings of Externally Referenced objects come first from the XREF. In other words, if we wish to edit the way a particular object or objects in the drawing behaves, we must open the file in which that object actually lives. In this case, we will edit the Space objects. They are contained in an XREF named Walls.

 NOTE It is also possible to assign Display Overrides to XREFs. Examples of this will be explored in chapters that follow.

5. Click anywhere on the plan.

   The entire plan will highlight. This is an Externally Referenced (XREF) drawing. See the online help and Chapter 5 for more information on XREFs.

6. Right-click and choose **Open XREF**.

   On the Drawing Title Bar, the name of this file reads: Walls.

7. From the View menu, choose **3D Views > SE Isometric** (or click the SE Isometric icon on the Views toolbar).

8. Return to the Display Manager; expand Representations by Object and then Select Space.

9. On the right side, right-click on Model (row, not column) and then choose **Edit Drawing Default**. (You can also double-click the Model Rep to edit it.)

10. Click the light bulb icon next to Ceiling component to turn it off (see Figure 2–25).

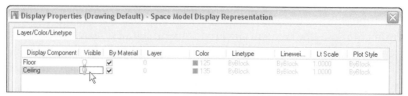

Figure 2–25 *Turn off the Ceiling subcomponent in the Model rep for Spaces*

11. Click OK twice to accept the change and return to the drawing.

Notice that the Spaces now display their floors only. You will still see Ceiling Grids and light fixtures. These are actually separate object types. The Ceiling Grid is not part of the Space. It is a 2D grid only, intended for use in the layout of Reflected Ceiling Plans. Typically you would use the Ceiling component of Space (which we just turned off), for portraying the ceiling plane in 3D. If you like, you can turn off Ceiling Grids and Masking Blocks in this Display Set as well. (The lights are actually Masking Blocks; see Chapter 13 for more information on ceiling grids and lights.) You would do this by repeating the steps in the "Understanding Sets" topic. Give it a try now if you like. You will be able to see everything better if you shade the model.

12. From the View menu, choose **Shade > Gouraud Shaded** (or click the Gouraud Shaded icon on the Shading toolbar). See Figure 2–26.

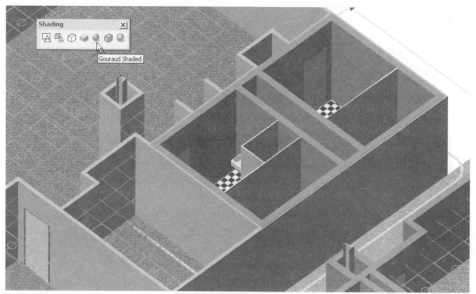

Figure 2–26 *Viewing the Model with Space's ceiling subcomponent turned off*

TIP If you want to turn off the ceiling plane in just *some* of the Spaces in a file, select the Spaces and on the Properties palette, change the Ceiling thickness to 0.

ADD A SET

The default template provides a good mix of Display Configurations that achieve different common views of the model. However, from time to time, you may find that you wish to see a different type of display that is not included with the default template. Once you have gained some comfort with the Display Manager, your display possibilities are virtually endless.

1. Return to the Display Manager.

2. Right-click on the Sets folder and choose **New**.

   A new set will appear ready to be named. The default name is New Display Set.

3. Type **Space Volumes** and then press ENTER.

   With Space Volumes selected, scroll around on the Display Representation Control tab and note that all check boxes are empty.

4. On the right, on the Display Representation Control tab, select Space, and then place a check mark in the Volume column (see Figure 2–27).

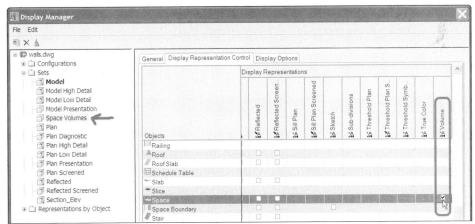

Figure 2–27 *In the Space Volumes set, place a check mark in Volume for Space objects*

New sets can be added by right-clicking either the Sets folder or an existing Set. Normally you will prefer to right-click on an existing Set; this will create a duplicate of the selected Set that can then be edited. Creating a new Set while right-clicking on the Sets folder creates an empty Set with nothing checked (as we have done here) that must be configured from scratch. For demonstration purposes, we have created an empty Set, but it is usually easier to copy from an existing one.

ADD A CONFIGURATION

Creating a Set by itself is not enough to see it in the drawing. We must also create a Configuration that references the new Set. (Refer to the "View Direction Dependent Configurations" and "Fixed View Direction Display Configurations" topics above for more information on Configurations.) As with Sets, you can create a new Configuration by right-clicking the Configurations folder or an existing Configuration. The difference between the two methods is also the same: right-clicking an existing Configuration copies it, while right-clicking the Configurations folder creates one based on the default settings. This time, let's copy an existing one.

Be sure that the Display Manager is still open. If it is not, choose **Display Manager** from the Format menu.

5. Expand Configurations, and then right-click on Medium Detail and choose **New**. Type **Space Volume Model** for the name and press ENTER.

6. On the right side, click the Configuration tab.

Notice that all of the assignments are the same as Medium Detail. This is because we created the new set by duplicating Medium Detail.

7. Next to Default (in the View Direction column) click on the existing entry in the Display Representation Set list (currently Model).

A pop-up menu will appear.

8. From the list that appears, choose Space Volumes (see Figure 2–28).

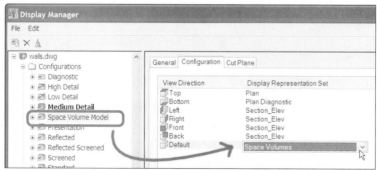

Figure 2–28 *Modify the default view condition for the Space Volume Model Configuration*

9. On the left side in the tree, right-click on Space Volume Model and choose **Set to Current Viewport**.

NOTE This is the same as choosing the Display Configuration from the pop-up menu on the Drawing Status Bar.

10. Click OK to view the results (see Figure 2–29).

    It may take a few moments for the drawing to be updated. Notice how all ADT objects have disappeared except Spaces. Furthermore, the Spaces are now displayed as solid volumes.

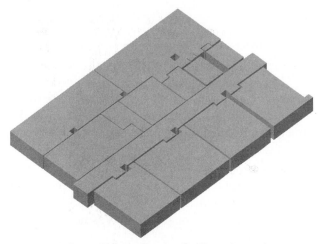

Figure 2–29 *Drawing with Space Volume Display Configuration active*

When customizing the display system, it is worth the extra effort to perform the additional steps outlined in the "Add a Set" and "Add a Configuration" topics. The reason for this is that we can now switch easily from Medium Detail to Space Volume Model in a single and simple operation on the Drawing Status Bar, as covered at the start of this tutorial. Space Volume Model is now just another choice on the drawing status bar list. Most everyday users of the software will not need to interface with the Display Manager to the extent covered here, if at all. The purpose of the exercise is to illustrate the power and functionality available in the ADT display system. There is much more to cover in the specifics of manipulating the individual Display Representations of the various ADT objects. There is plenty more that we could do with our custom Display Configuration and the Display Manager in general. However, we will end our tour here for now. We will explore more Display topics in upcoming chapters. As we move forward you will appreciate the exposure that this display control primer has given as you begin see the pervasiveness of the ADT display system.

CAD Manager Note: Much of the configuration of the display system is best left in the realm of office standards. Poll your team and try to determine what Configurations and associated Sets they will require. Try to use the offerings provided out of the box as much as possible. Set up any custom displays required and save them in the office standard templates. (Again, using the default templates as a starting point is an excellent way to begin.) This will greatly enhance overall office productivity. You should seek to create the majority of custom user-defined Display Representations that are required by your team as well. However, individual configuration of the properties of each Display Representation will most likely be tweaked on a regular basis by the users as project needs dictate. Procedures and reasons for doing so are covered throughout the remainder of this book as circumstances dictate.

OBJECT STYLES

Virtually all ADT AEC objects use styles to define global object parameters. (These can include both physical and display parameters.) Much like text and dimension styles in core AutoCAD, object styles control all of the formatting and configuration of the object. Using styles is a powerful way to control the behavior of objects and quickly make global changes when the design changes. For instance, at any time during project design, a user could simply go back to a style describing a wall and make a modification, such as changing the size of the concrete block from 6″ [152] to 8″ [203]. The change would be reflected throughout the drawing on all wall objects that were associated with that wall style. In most cases, it is best to think of styles as "types" in the same way we commonly distinguish wall types and door types in a construction document set. Each type needed in the set would have a corresponding ADT style.

WORKING WITH STYLES AND CONTENT BROWSER

Following is an overview of key Object Style features:

> Editing parameters in the style globally updates all objects within the drawing referencing that style.

> Styles can control physical parameters, display parameters and data property set information used for schedules.

> Styles can be shared between drawings using the tools available in the Style Manager, Tool Catalogs and Tool Palettes.

Object styles are saved in individual drawing files. These drawings can be part of a particular project or a central library accessible to all people in the office. If you create a style in one drawing and wish to use it in another, you can do so easily by saving the style to a tool catalog. This catalog can be accessed by other users and its tools used in any drawing file, making your style easy to use across multiple project files or throughout the entire office. You can also access a large collection of out-of-the-box tool catalogs provided with ADT 2006. All tool catalogs are accessed from the Content Browser. Content Browser is a web browser–like tool that is designed specifically to browse, store and access ADT Styles and Content items. Before going to Content Browser, it is useful to determine what Styles if any are contained within the current drawing file.

STYLES IN THE CURRENT DRAWING

It is easy to determine the name of a particular style that is applied to an existing object within the drawing. To do so, we simply click to select the object in the drawing editor and then view the listing for Style on the Properties palette. You can continue on with the same project loaded from the previous tutorial. If you did not complete the previous tutorial, follow the steps in the "Install the CD Files and Load Sample File" heading above to install and load the dataset.

1. On the Project Navigator Palette, click the Constructs tab and then double-click to open the *Walls* Construct.

 NOTE If you left the *Walls* Construct open from the previous tutorial, then this action will simply make that file active.

If the Space Volume Model Display Configuration is still active from the previous exercise, load Medium Detail now. From the View menu, choose **3D Views > Top**, and from the View menu choose **Shade > 2D Wireframe**.

2. Zoom in on the Toilet Rooms.

3. Select any Wall within the drawing.

4. Right-click and choose **Properties**.

5. On the Properties palette, toward the top, take note of the Style name.

You can use the same technique to see what other styles are available within the drawing.

6. With the Wall still selected, click on the Style list on the Properties palette.

This will activate a pop-up menu for styles.

7. Open the Menu and view the list of styles (see Figure 2–30).

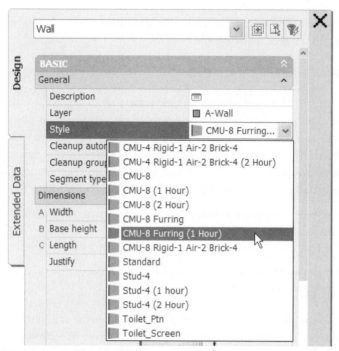

Figure 2–30 *View the list of Wall styles contained in the current drawing*

All of the styles on this list are currently available within the current drawing. If you need a style that is not on this list, you will need to create or import it. See the "Working with the Content Browser" heading next for more information on importing styles, and refer to Chapter 10 for more information on creating and editing styles. You can choose a different style from the list to change the style of the selected Wall to a different style that already exists in this drawing.

8. With the Wall still selected, click on the Style list on the Properties palette.

9. From the Style list, choose a different style.

Note the change to the Wall within the drawing.

10. Undo the change by repeating the same steps and then choosing the original name, or press CTRL + Z to undo.

The same technique could be used for any architectural object in the drawing. Try it on the Doors, or Windows if you like. Be sure to undo any changes when you are done experimenting.

WORK WITH THE CONTENT BROWSER

1. On the Navigation toolbar, click the Content Browser icon (or press CTRL + 4).

The Content Browser window will open. Content Browser is very similar to a Web browser. It is organized in two panes. Navigation is on the left and the content is displayed on the right. Standard Web browser navigation buttons (Back, Forward, Refresh, etc.) are arrayed across the top of the left pane. Action buttons for creating new catalogs and such are placed at the bottom of the left pane. In the Library home (the main page) there are two such buttons: one creates a new catalog and the other modifies the Content Browser view options (see Figure 2–31).

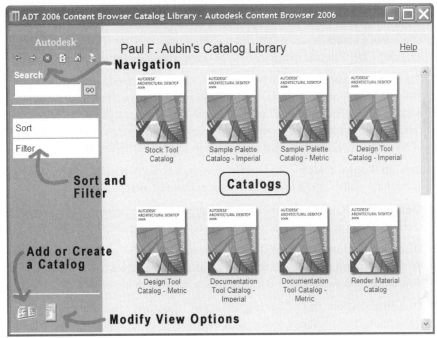

Figure 2–31 *The Content Browser main home page (varies depending on installed options).*

ADT ships with a vast library of pre-made content. Content items include Object Styles, Symbols and Annotation Routines. All of these items can be accessed from the Content Browser. The following is a list of each of the catalogs provided with a brief description.

 NOTE The exact list of Catalogs available in your Content Browser varies depending on the options chosen during installation of Architectural Desktop 2006. If you performed a Full Install of the US English version, the following Tool Catalogs should be available:

▶ **Stock Tool Catalog**—Contains all the standard tools that come with Architectural Desktop. There is a tool for each architectural object type and several other core commands. These tools do not reference any particular unit system.

▶ **Sample Palette Catalog - Imperial**—Contains three categories corresponding to the three basic tool palette groups installed in the standard out-of-the-box installation. Each of these categories contains a backup copy of the installed palettes belonging to the Imperial units Installation. The palettes contain a mix of stock tools, styles and documentation content.

▶ **Sample Palette Catalog - Metric**—Contains three categories corresponding to the three basic tool palette groups installed in the standard out-of-the-box installation. Each of these categories contains a backup copy of the installed palettes belonging to the Metric units Installation. The palettes contain a mix of stock tools, styles and documentation content This catalog also includes a UK Palettes category with palettes especially for work in the U.K.

▶ **Sample Palette Catalog - Metric D A CH**—Contains several palettes in Metric Units dedicated to the needs of Germany (D), Austria (A) and Switzerland (CH). The palettes contain a mix of stock tools, styles and documentation content specialized to needs of the building industry in those nations. (This option is available only if chosen as an option during installation).

▶ **Design Tool Catalog - Imperial**—Contains tools that refer to all the architectural object styles and AEC design content in Imperial units.

▶ **Design Tool Catalog - Metric**—Contains tools that refer to all the architectural object styles and AEC design content in metric units.

▶ **Documentation Tool Catalog - Imperial**—Contains tools that refer to all the documentation object styles, such as schedule tables and area calculation objects, as well as AEC documentation content in Imperial units.

▶ **Documentation Tool Catalog - Metric**—Contains tools that refer to all the documentation object styles, such as schedule tables and area calculation objects, as well as AEC documentation content in metric units.

▶ **Render Material Catalog**—Contains a large collection of material definitions for use in both ADT and VIZ Render.

▶ **My Tool Catalog**—This catalog is empty and ready to customize for your own use. Use it to store custom tools or your favorites from the standard tools.

▶ **Architectural Desktop & VIZ Render Plug-ins**—Contains links to Web sites for VIZ Render and other utilities and plug-ins.

To open a catalog, simply click on it. You can then use the links at the left (or right) to navigate through the categories and palettes to the tools.

ACCESS A TOOL CATALOG

For most exercises in this book, both Imperial and Metric units and catalogs will be referenced. However, the dataset for Chapter 2 is provided in only Imperial units; therefore this sequence will reference only the Imperial Catalog.

> 2. With Content Browser open, click on the Design Tool Catalog - Imperial (see Figure 2–32).
>
>   An introduction page will appear on the right. Read through it and then proceed to the next step.

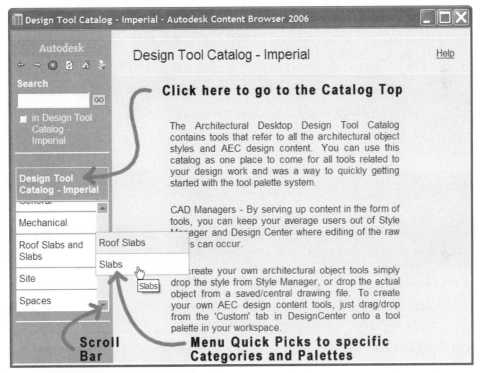

Figure 2–32 *Navigating through a Catalog*

> 3. On the left, scroll the list down to locate the Walls category.
>
> 4. Hover your mouse over Walls.
>
>   A submenu will appear.
>
> 5. Click on the CMU item in the submenu.

Here you will find several pages of CMU Wall styles ready to drag and drop into your drawings. You can scroll through each page if you wish. We are going to drag and drop one of these CMU Wall style tools into the drawing; it will execute the Wall command and draw a Wall in the selected style.

6. Locate the Wall style named CMU-8 Furring. (It is on the fourth page—use Next and Prev to navigate through the pages.)

Each of the tools located in the Content Browser uses an Autodesk technology called "iDrop." iDrop is a web-based technology that allows content to be dragged into drawings from web pages, complete with all required data, associated parameters and files. In the case of ADT Styles, all style, material and schedule data properties will be included.

7. Place your mouse over the small eyedropper icon, click and hold down the mouse button and then drag the CMU-8 Furring style tool into the drawing window (see Figure 2–33).

Figure 2–33 *Using iDrop to drag and drop styles into the drawing*

The Add Wall command will begin, and all of the parameters on the Properties palette will be set to match the tool just dropped.

8. Following the command prompts, draw a segment or two of Wall with this tool and then press ENTER.

Notice that this style matches the one used for the Toilet Rooms.

ADD A PALETTE

As easy as it is to navigate and drag from the Content Browser, it will be more convenient (and require far fewer clicks) to have a palette with the entire collection of styles required for a project set up as tools and readily at hand.

1. Right-click the title bar of the tool palettes and choose **New Palette** (see Figure 2–34).

A new tool palette will appear ready to be named.

Figure 2–34 *Create a new tool palette*

2. Type **Chapter02** for the name and then press ENTER.

3. Bring the Content Browser to the front (press CTRL + 4, or click its button on the Windows task bar).

 TIP If you wish, you can right click the title bar (or the button on the task bar) of the Content Browser and choose "Always on Top" to keep it perpetually in front of other windows.

4. Click and hold down on the eyedropper icon next to the CMU-8 Furring.

    As before, wait for the eyedropper to appear to "fill up."

5. Drag and drop the tool onto the Chapter02 palette in ADT. When a plus (+) sign appears on the cursor release it to add the tool to the palette (see Figure 2–35).

 NOTE You can also add and remove tools from any of the out-of-the-box palettes, not just those that you create.

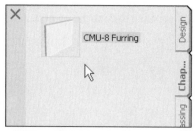

Figure 2–35 *Drop a tool on your custom tool palette*

6. On the same page of the Content Browser, locate the CMU-8 Air-2 Brick-4 Wall style, and drag it to the Chapter02 palette as well.

Your custom Chapter02 palette now contains two tools.

CAD **Manager Note:** Custom tool palettes offer a great deal of customization potential. Custom palettes are built in the ADT work space as we have done here, and then can optionally be dragged and dropped to a catalog in the Content Browser. In this way, a library of Office Standard tool palettes and tools can be built up and saved to the server. These palettes can then be dragged and dropped by users to their own work spaces. Most importantly, the palettes can maintain a back link to the source palette in the Content Browser if so configured. When you make changes to the tools contained on these linked palettes, individual users need only click a small 'refresh' icon to update their own version of the palettes to the latest version on the server. You can even set the refresh to occur automatically each time ADT is launched. This technique can also be used with great success for project-specific palettes as well. New to ADT 2006 is the ability to automatically swap out groups of project-based palettes when the current project is switched. We will explore these topics in further detail in later chapters.

APPLY TOOL PROPERTIES TO DRAWING OBJECTS

Tools can be clicked to execute their built-in function directly in the drawing. You can also use the tool as a means to import or re-import its associated style. A tool property can also be applied directly to objects in the drawing. (You may recall this use of tools from the opening tutorial in Quick Start.) The two tools that we have here are "Object" tools. An Object tool creates a particular class of ADT object and requires a certain amount of user interaction, such as changing parameters on the Properties palette or clicking points in the drawing (as with a Wall tool). An important parameter of an Object tool is its style reference. Both of the tools that we built here are Wall tools with the Style setting being the parameter that distinguishes them from one another.

1. Right-click on the CMU-8 Air-2 Brick-4 Tool.

Take note of the menu items that appear. In particular, note the option to: Import "CMU-8 Air-2 Brick-4" Wall Style. Recall the exercise above where we viewed the list of Wall styles that were currently resident in this drawing. When you right-click a tool and the option reads "Import" rather than "Re-import" (as shown below) this indicates that the style is *not* currently available within this drawing (see Figure 2–36). If you were to choose this option, it would import this style into the drawing. It would not however, *apply* the style on any particular Wall. To do that, we would choose the first menu option: Apply Tool Properties to. In either case, the style would be imported automatically from the remote library file where it resides to the current drawing.

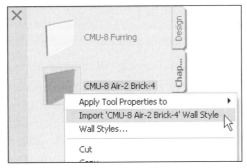

Figure 2–36 *Right-click a tool to see options, in particular the Import Style option*

2. Choose **Apply Tool Properties to > Wall**.

3. At the "Select Wall(s)" prompt, click on one or more of the exterior Walls to the right of the plan and then press ENTER.

    Notice the change to the Wall component makeup: the insulation component has been removed.

4. Repeat steps 3 through 7 above in the "Styles in the Current Drawing" exercise.

    Notice that the CMU-8 Air-2 Brick-4 style is now part of the list.

    Repeat the "Apply to" sequence on other Walls if you like.

5. Zoom in on the toilet rooms in the plan.

6. On the Chapter02 palette, right-click on the CMU-8 Furring Tool (see Figure 2–37).

Take note of the menu items that appear. In particular, note the option to Re-import 'CMU-8 Furring' Wall Style. As noted above, if the style does not currently exist within the drawing, this item will read "Import" and will import that style definition to the current drawing. However, if the style is already present in the drawing, as is the case with the CMU-8 Furring Style, then this option will "re-import" the style from the library file and replace the version of it that exists in the current file. This will "update" the style to the latest design. This tool is very valuable and extremely powerful, but it should be used with caution. Remember, that styles are "types." When you re-import the style, all instances of that style in the current drawing will be affected; they will *all* update

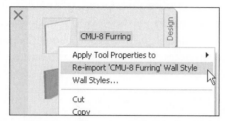

Figure 2–37 *Right-click a tool to see options, in particular the Re-Import Style option*

7. Choose **Re-import 'CMU-8 Furring' Wall Style** from the context menu.

Notice the change to *all* of the Toilet Room Walls. The version of the Style that was in this drawing originally had only a single component to represent the furring. The one re-imported from the library has two components: the furring and the drywall (see Figure 2–38). This gives you an example of the potential of the Re-import feature and some of the variety possible with object styles; again, use with caution.

Figure 2–38 *Comparing the original style on the left with the re-imported version on the right*

8. Repeat the step above in the "Access a Tool Catalog" heading to draw a segment or two of Wall with the CMU-8 Furring Style from the Content Browser.

Notice that, even though we dragged the Wall tool from the catalog as before, the look of the Style was different in each instance. A tool will not automatically import its Style from the remote drawing if it already resides in the current drawing. Rather, it will simply use the version resident in the drawing. In this case, the version used in the current drawing originally had less detail, but because the name was the same, importing a more detailed version from the library was as simple as a single right-click.

If this version of the style does not suit you, undo the change. For instance, with the new style you may experience plotting difficulties or Wall Cleanup problems. Wall Cleanup is the automatic interaction between Wall segments that forms "clean" corners and intersections (see Chapter 9 for complete details on troubleshooting Wall Cleanup).

CAD Manager Note: Another important interface element for working with styles is the Style Manager. You can access the Style Manager on the Format menu. Like the Content Browser, it can also be used to retrieve styles saved in remote library files. However, the Style Manager is more than a tool to browse and import styles. You can use it to create new styles and edit existing styles as well. It also provides a convenient way to quickly assess all of the styles contained in the current file. If you have used Architectural Desktop before, Style Manager will be very familiar to you. We will use the Style Manager to help build and edit styles in later chapters. Feel free to use it if you wish as an alternative to the methods showcased here.

New in ADT 2006   Style Manager also provides access to Project Standards library files if being used in a Project environment. Project Standards provides a way to manage all common Styles (and Display settings) for a project team in one or more library files. These files are typically stored on a network server and made read-only to all team members except the project's team leader or data coordinator. When Project Standards are enabled, library files will appear in a node at the top of the tree listing in Style and Display Manager. Project Standards 'synchronize' and 'update' functionality is also integrated throughout the interface of both managers. Project Standards will be explored in greater detail in the chapters that follow.

The potential of object Styles in ADT is nearly limitless and we have only begun to scratch the surface here. Again, the goal of this chapter is to familiarize you with the concepts that are important to working in ADT. As with display control shown above, throughout the remaining chapters of this book, styles will play an important part. Here we focused on retrieving and applying styles to objects within our drawing. In Chapters 10 and 11 we will look at editing existing styles and building custom styles. In that discussion, we will also explore the Style Manager mentioned here. The Style Manager is a tool that is used to create, edit and manage styles.

Remember: think of styles as "types" akin to those types we typically indicate on a construction document set. This will help you in deciding when and how to build and apply styles.

ANCHORS

Anchors are used to build intelligent links between two objects. Anchors enable two objects to be linked together in a logical manner. Usually, this is a physical relationship where one object controls the location and orientation of another. As one object moves or is transformed, the anchored object will move and be transformed in relation to it. The way in which the anchored object moves and is transformed is governed by the rules established within the anchoring parameters. For instance, doors automatically anchor to walls. If the wall moves, the door moves with it. The relationship between the two objects in an anchored relationship is hierarchical. The door is anchored to the wall, not the other way around. Therefore, moving the wall *will* force the door to move, but moving the door *will not* move the wall. However, the door's position as it moves must remain within the confines of the wall (or a neighboring wall). This behavior is controlled by the rules and constraints built into the anchor.

Understanding and fully exploiting the usage of anchors while using ADT is one of the key ingredients to success with the software. Anchors have the following key features:

▶ Anchors ensure that logical relationships between objects are maintained automatically as edits are made to project files.

▶ There are many anchor types available, each exhibiting a set of unique parameters. (These include: Wall, Column, Railing and Tag Anchors).

▶ Anchoring is often built into other routines in the software; therefore, the anchor is often attached by means of performing a related function (for instance, when a door is inserted in a wall, the anchor is automatically attached).

TYPES OF ANCHORS

As stated above, most anchoring occurs automatically as various objects are added to project files. This is true of Doors anchoring to Walls, Railing anchoring to Stairs, Columns anchoring to Grids and Lights anchoring to Ceiling Grids. However, there are several more generic anchors (which are listed here) which are accessed from the Content Browser in the Stock Tool Catalog in the Parametric Layout & Anchoring category. These Anchor commands can be used to establish logical and intelligent relationships between your ADT objects. Brief descriptions of each type are listed here. However, the best way to understand anchors is to use them in real situations. Use the following list as a guide to types of anchors. The tutorial that follows will help to further illustrate the concept of anchors.

▶ **Curve**—Used to anchor an object along the *length* of another object. A wall anchor used by doors is a type of Curve anchor. (Refer to Chapters 4 and 10 for more information on Walls and Doors.)

▶ **Leader**—Used to anchor a specific point on an object to a specific point on another object and connect these two points with a *leader* line. A Column Bubble anchor used by column bubbles tools when anchoring to column grids is a type of Leader anchor. (Refer to Chapter 6 for more information on Column Grids and Bubbles.)

▶ **Node**—Used to anchor an object from its insertion point to a *specific point* (or node) on another object. A column object uses a Node anchor to attach itself to a Column Grid object, as do lighting fixtures to Ceiling Grids. (Refer to Chapter 6 for more information on Column Grids and Chapter 13 for more information on Ceiling Grids.)

▶ **Cell**—Must be used in conjunction with a 2D or 3D Layout Grid (Layout Grids are defined next). The anchored object will be anchored to the center of a *2D cell* formed by the grid. The anchored object can optionally *reshape* to conform to the shape of the grid cell. Nested objects within Curtain Wall objects use an anchor similar to a Cell anchor. (Refer to Chapter 8 for more information on curtain walls.)

▶ **Volume**—Must be used in conjunction with a 3D Layout Grid (Layout Grids are defined next). The anchored object will be anchored to the center of a *3D cell* formed by the grid. The anchored object can optionally *reshape* to conform to the shape of the grid cell volume. Nested objects within Curtain Wall objects use an anchor similar to a Volume anchor. (Refer to Chapter 8 for more information on curtain walls.)

▶ **Object**—Used to anchor one ADT directly to another ADT object. The original position, rotation and offsets are maintained when using this anchor, thereby making anchoring a simple one-step process.

LAYOUT TOOLS

When you work with anchors, there are always three objects involved: the object to which you are anchoring, or the "parent," the object being anchored, or the "child," and the anchor itself, which contains the "rules of the relationship." Sometimes the parent object is not an obvious building component (like a Wall, Stair or Curtain Wall), but more of a design relationship itself. Provided with ADT are tools called Layouts whose purpose is to assist in establishing repetitive design relationships such as equal spacing or dimensional repetition. There are three basic types of Layout tools.

▶ **Curve**— A Layout Curve is a *one-dimensional* Layout tool following the path of a line, polyline, spline, Wall or other linear shape. Even closed shapes like a circle or an ellipse may be used. Anchor points called "nodes" occur along the curve at parametrically defined intervals. Any ADT object can be anchored to these nodes using a Node anchor. Layout Nodes can be spaced equally, repeat a certain distance, or be placed manually.

▶ **Layout Grid 2D**—The Layout Grid 2D is a *two-dimensional* Layout tool. The X and Y dimensions of the layout form a grid. Anchor points called "nodes" occur at each grid line intersection. Any ADT object can be anchored to these nodes using a Node anchor. ADT objects can also be anchored to the center point of each cell using a Cell anchor, and optionally resize to the cell dimensions. Grid lines can be spaced equally, repeat a certain distance, or be placed manually in both directions.

▶ **Layout Grid 3D**—The Layout Grid 3D is a *three-dimensional* Layout tool. The X, Y and Z dimensions of the layout form a volume. Anchor points called "nodes" occur at each 3D grid line intersection. Any ADT object can be anchored to these nodes using a Node anchor. ADT objects can also be anchored to the center point of each 2D cell formed by grid lines running in any two of the three dimensions using a Cell anchor. Finally, the geometric center point of each volume can be anchored to using a Volume anchor, and optionally resize to the cell dimensions. Grid lines can be spaced equally, repeat a certain distance, or be placed manually in all three directions.

EXPLORE OBJECTS WITH ANCHORS

Continue working in the same file from the previous exercise. If you closed the file, click the Constructs tab of project Navigator and then double click the Walls file to re-open it now.

1. Zoom in to the upper right corner of the classroom plan.

2. Click on the Door to the classroom.

3. Right-click and choose **Basic Modify Tools > Move** (see number 1 in Figure 2–39).

4. At the "Specify base point or displacement" prompt, click a point near the center of the Door (see number 2 in Figure 2–39).

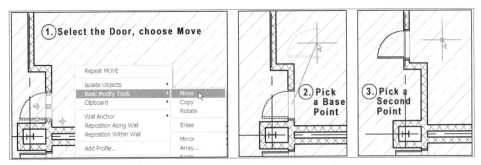

Figure 2–39 *Testing a Wall anchor*

    5. At the "Specify second point of displacement" prompt, click a random point anywhere within the classroom space (see number 3 in Figure 2–39).

Notice that the door moves, but stays constrained to a Wall; albeit a different Wall in this case. The most important point is that, even though we picked a point within the classroom space, the Door still "jumped" to the closest neighboring Wall.

NOTE Your screen may not look exactly like Figure 2–39. If you followed the procedures above and edited the current Display Configuration, your Spaces may not be displayed. Also, since we did not use Object Snaps or dimensions to make this move, your Door may have jumped to a different Wall or location.

    6. From the Edit menu, choose **Undo** (or press CTRL + Z) to return the Door to the original location.

    7. In the same Classroom, select the Window.

Suppose that you want to shift the position of the Window within the thickness of the Wall. As you saw with the Door, a regular AutoCAD Move command would move the Window, but would not allow you to move it within the Wall's thickness. (Go ahead and try it if you like. Just be sure to undo before continuing). This is because the Window's anchor controls its position relative to the Wall (including its position within the thickness of the Wall). You can however, move it within the Wall by editing the anchor parameters. The easiest way to do this is by using grips.

    8. Hover over the square grip in the middle of the Window and wait for the tool tip to appear (see Figure 2–40).

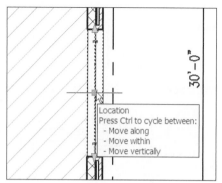

Figure 2–40 *Hover over a grip to reveal the various* **CTRL** *key functions*

Notice that there are three options to cycle through on this grip.

9. Click the Location grip.

10. Press the CTRL key once.

Notice that the Window is now moving perpendicular to the Wall. Notice also the Dynamic Dimensions that track its position as you move your mouse.

11. Move your mouse to the left, type in **3″** and then press ENTER (see Figure 2–41).

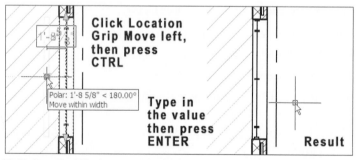

Figure 2–41 *Shifting the Window position Anchor within the Wall with grips*

12. Right-click and choose **Deselect All**.

13. Zoom in to the lower left corner of the same classroom, at the column near the Door.

14. Select the column at the lower left corner of the room.

15. Right-click and choose **Basic Modify Tools > Move**.

16. Use any base point.

17. At the "Specify second point of displacement" prompt, click a random point in the hallway nearby (see Figure 2–42).

The MOVE command will appear to have failed. This is because the Column is an-chored to the grid and cannot move off the grid line intersection.

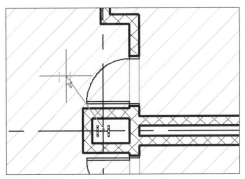

Figure 2–42 *Try to move an anchored column*

18. Repeat the Move command on the same Column, but this time click the second point much farther away from the original location.

The column will seem to disappear. In fact, what has happened is similar to the door's behavior before. In this case, the column is anchored to the Column Grid object (a type of layout grid, as described earlier). The column has moved to a dif-ferent node. Most likely, that node was already occupied by another column, and there are now two columns on a single node. This accounts for its apparent disap-pearance. Be careful of doubling up anchored objects like this.

19. Undo the last MOVE command.

    Anchors control all aspects of an anchored object's position and orientation.

20. Select the column again, right-click, and choose **Basic Modify Commands > Rotate**.

21. Click a base point at the center of the column.

22. At the "Specify rotation angle" prompt, type **90** and then press ENTER .

    Notice that the column has not rotated. To rotate an anchored column, we must manipulate the anchor properties. Fortunately, like the window above this is easy to do with grips.

23. Select the Column and take note of the two grips.

    The square grip is the location grip for the Column; the diamond-shaped one is the "Roll" (or rotation) grip.

24. Click on the Roll grip.

    Note that the existing Roll for this column is 90°.

25. Begin moving the mouse and note the dynamic dimensions that appear.

26. Type **0** and then press ENTER (see Figure 2–43).

The Column is now oriented the opposite way.

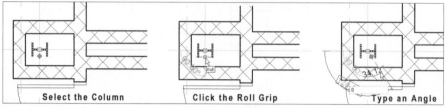

| Select the Column | Click the Roll Grip | Type an Angle |

Figure 2–43 *Change the orientation of a column with the Roll grip*

OBJECT ANCHORS

We have explored several anchors that are already present in this file. Now let's use the Object Anchor to establish a new anchor relationship in our file.

In the "Add a Palette" heading above, we created a Chapter02 Tool Palette. We will now follow a similar process to add tools to this palette. If you did not complete that exercise, please do so now. You only need the palette for this exercise; it is not necessary to add the Walls tools to it, if you have not done so already.

Make sure that the Chapter02 Tool Palette (created above) is active (click on its tab if it is not).

1. Open the Content Browser (press CTRL + 4, or choose it from the Window menu).

2. Click on the *Stock Tool Catalog*, and then from the left navigation bar, choose *Parametric Layout & Anchoring Tools*.

A collection of Layout and Anchor tools will appear on the right (see Figure 2–44).

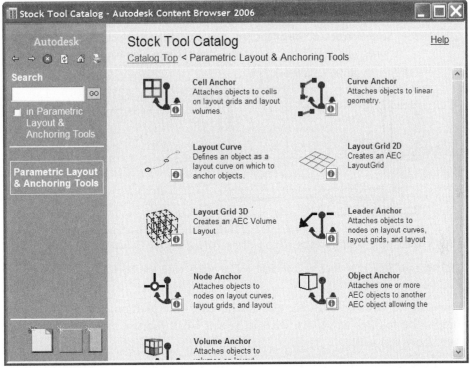

Figure 2–44 *Navigate to the Parametric Layout & Anchoring page of the Content Browser*

 NOTE All of the tools in the Stock Tool Catalog are unit independent, meaning that they work in both Imperial and Metric units.

3. Click and hold down the eyedropper icon next to the Object Anchor tool (see Figure 2–44).

Wait for the eyedropper to appear to fill up.

4. Drag and drop the tool onto the Chapter02 palette in ADT. When a plus (+) sign appears on the cursor, release it to add the tool to the palette.

 NOTE Later we will need two more tools, so let's add them to the palette now, while we are here.

5. Repeat this process to drag and drop the Layout Curve tool onto the Chapter02 palette. Be sure to drag from the eyedropper icon.

6. Repeat again for the Node Anchor tool.

7. Zoom in on the toilet rooms.

8. On the Chapter02 Tool Palette, click the newly added Object Anchor tool.

9. At the "Select objects to be anchored" prompt, click to select all of the toilet room fixtures adjacent to the horizontal Men's Room plumbing Wall and then press ENTER (see Figure 2–45).

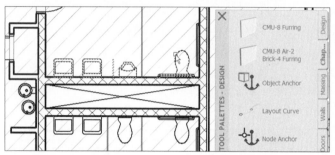

Figure 2–45 *Select all of the fixtures on the plumbing Wall of the Men's Room*

10. At the "Select an object to anchor to" prompt, click the horizontal Men's Room plumbing Wall.

The fixtures in the Men's Room are now anchored to the plumbing Wall. If this Wall moves, the fixtures will move with it. If this Wall rotates, the fixtures will also rotate. Let's test this.

11. Click to select the horizontal Men's Room plumbing Wall.

12. Click the square Location grip at the center of the Wall. Move the Wall up a bit and click.

Notice that all of the fixtures remained attached to the Wall and moved with it.

13. Undo the move and try rotating the Wall. (Right-click and choose **Basic Modify Tools > Rotate**).

Notice that all of the fixtures rotated with the Wall. Undo this transformation as well. If you wish, you can also anchor the fixtures in the Women's Room to its plumbing Wall. Notice that in both transformations, the Space object and the "X" within the plumbing Wall cavity were unaffected. Moving Walls does not, unfortunately, affect the shape of these objects, so a second edit would be necessary. If later you decide to remove the anchor relationship, click the anchored objects; a small arc will appear connecting the anchored object to the parent object. A small minus (-) sign grip will appear along this arc. Click it to remove the relationship.

EXPLORE LAYOUTS

In our final tutorial in this chapter, we will explore the use of a Layout Curve. A Layout Curve adds nodes to any linear object such as a line, a polyline or a Wall.

Other ADT objects can then be anchored to those nodes. This provides a valuable architectural design tool.

1. Pan over to the Open Office space in the upper left corner of the plan adjacent to the toilet rooms. (There is a single chair in the upper left corner of the room.)

2. In the upper left corner of the room, draw an arc (use Arc 3 Points on the Shapes toolbar) as shown in Figure 2–46.

   The exact dimensions are not important; simply match the shape shown in the figure. This will be easiest to accomplish if you temporarily disable OSnaps. To do this click the OSnap icon on the drawing status bar or press F3.

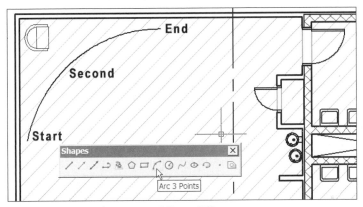

Figure 2–46 *Draw an arc in the corner of Open Office*

Above at the start of the "Object Anchors" heading, we added several tools to the Chapter02 Tool Palette. We will now use the remainder of those tools to copy and anchor the chair provided here along the arc that we just drew.

3. On the Chapter02 Tool Palette, click the Layout Curve tool.

4. When prompted to "Select a curve," pick the arc just drawn.

5. At the "Select node layout mode" prompt, choose **Space evenly**. (If you have Dynamic Input disabled, right click to access this option).

6. Press ENTER at each of the "Start Offset" and "End Offset" prompts to accept the defaults.

7. At the "Number of nodes" prompt, type **6** and then press ENTER to complete the sequence (see Figure 2–47).

   Six small circles will appear along the length of the arc.

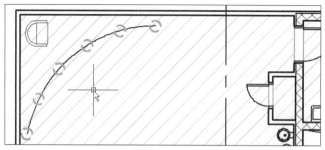

Figure 2–47 *Create a Layout Curve along the arc*

We now have a Layout Curve that references the shape of the arc. The Layout Curve and the arc are separate entities, but are linked together. If you change the shape of the arc, the Layout Curve will follow. To change the parameters of the Layout Curve itself, such as the spacing of the nodes, click one of the nodes, right-click and choose **Properties**. Change whatever parameters you wish in the Dimensions grouping of the Properties palette. Let's now anchor something to these nodes. We could anchor equipment symbols, plumbing fixtures, lighting fixtures, Walls, Columns, planting symbols or just about anything we wished to have spaced evenly along this arc. In this example, we will anchor chair symbols along the Layout Curve. A chair has already been provided, in this file, in the corner of this room. We will explore browsing, inserting, and creating AEC Content (like this chair) in Chapters 4, 10 and 11.

8. On the Chapter02 Tool Palette, click the Node Anchor tool.

9. At the "Node anchor" prompt, choose **Copy to each node**. (Right click if Dynamic input is off).

10. At the "Select object to be copied and anchored" prompt, pick the chair.

11. At the "Select layout tool" prompt, pick one of the six magenta circles (do not pick the arc).

    Notice that, although there are now six chairs anchored to the Curve, they are pointing the wrong way.

12. Press ENTER to complete the Node Anchor command.

13. Select all six chairs, right-click and choose **Node Anchor > Set Rotation**.

    Be careful to select only the chairs and not the Layout Curve, the arc or the Space. If you select other objects in addition to chairs, the Node Anchor option will not appear on the right-click menu.

14. At the "Rotation angle about X axis" and "Rotation angle about Y axis" prompts, accept the default of 0 by pressing ENTER .

15. At the "Rotation angle about Z axis" prompt, type **90** and then press ENTER (see Figure 2–48).

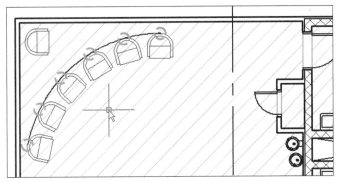

Figure 2–48 *Rotating all of the chairs to point into the room*

16. Click on the arc to select it. (Be careful not to select the chairs or the Layout Curve nodes.)

17. Change the shape of the arc using the grips.

    Notice the effect on the Layout Curve and the anchored chairs. Continue to experiment if you wish.

 NOTE Although the Arc must remain in the file since the Layout Curve is attached to it and it governs the shape and location of the Layout tool, you can place the Arc on a non-plotting layer such as A-Anno-Nplt, which is included in the file. In this way, the Arc remains readily available on screen, but will not plot on final printed drawings.

Anchors ensure that critical design relationships are maintained even as the typical edits and design changes are occurring in the drawing. There are other examples of anchors within this drawing file that you can explore. For instance, try moving the Stair object and note that the Railing goes with it. Switch to the Reflected Display Configuration (see the "Loading a Display Configuration" heading above) and explore the Anchor parameters used to anchor the lighting fixtures to the Ceiling Grid. As you explore, you will get a better sense of the many anchoring possibilities within ADT.

DISPLAY THEMES

New in Architectural Desktop 2006 are Display Themes. A Display Theme is actually an ADT object that can change the way other ADT objects display. This occurs independently of the current Display Configuration. A Display Theme queries the drawing for certain properties, when the values of these properties meet the conditions outlined within the Display Theme Style, the display of the affected objects is modified. The modified display remains in effect as long as the Display Theme is active. While you can insert as many Display Theme objects into the drawing as you wish, only one can be active at any given time. Previously active Display Themes are automatically disabled when a new one is inserted. You can disable a Display Theme at any time. The topic of Display Themes is introduced here because it related to both the display system and styles: A Display Theme is a Style-based object that affects object Display.

USING DISPLAY THEMES

The Scheduling Tool Palette (in the Documentation Tool Palette group) contains a few sample Display Theme Style tools. There are additional examples in the Content Browser in the Documentation catalogs. Use them like any other tool.

1. On the Project Navigator palette, click the Views tab and then double click the *Floor Plan* View file to open it.

 NOTE If you left the *Floor Plan* View open from the previous tutorials, then this action will simply make that file active.

2. Right click the Tool Palettes title bar and choose Document to load the Document tool palette group.

3. Click the Scheduling tab.

4. Click the Theme by Space Size tool.

5. At the "Upper left corner of display theme" prompt, click a point next to the plan and then press ENTER at the next prompt (see Figure 2–49).

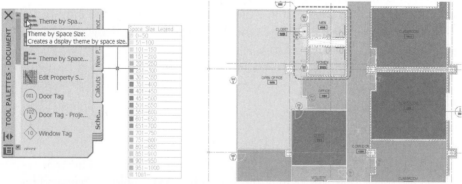

Figure 2–49 *Add a Space Size Display Theme*

All of the Space objects in the drawing will be color-coded based on their area. The Display Theme object displays a legend defining the range of square footage attributed to each color. This Display Theme will control the color and shading of the Space objects as long as it is active in the file. You can disable or delete it at any time. To disable a Display Theme, right click it and choose Disable Display Theme, delete the Display Theme object (the legend) or add another Display Theme (which will automatically disable the existing one upon insertion).

6. On the Scheduling tool palette, click the Theme by Fire Rating tool.

7. Click a point to place it in the drawing and then press ENTER to accept the default size.

Notice the changes to the drawing. The spaces return to their previous hatched display and the colors return to normal. The existing Display Theme now has a line drawn through it to indicate that it is disabled. The Walls and Doors in the file are now color coded to indicate their respective fire ratings. There are other Display Theme styles included in the Content Browser, however they would not yield the desired results in this file as they rely on the names of the Space Style in order to color code the Spaces. If you use the Space Styles found in the Content Browser in the Design Catalog, these additional Display Themes can color code them based on Style name. The potential of Display Themes is nearly limitless. In future chapters we will explore them further and build some custom Display Theme Styles.

8. Erase both Display Theme objects (the legends) to disable and remove them from the file.

9. Close and Save all files to complete the tutorial.

CONTENT LIBRARY

Once you have gained an understanding of the concepts covered in this chapter, you will no doubt begin to amass a collection of Styles and Display Configurations that will be useful in future projects. In addition, there are thousands of pre-built styles, blocks, and multi-view blocks that have been included with Architectural Desktop to get you started. These resources are accessible via the Content Browser, as we have seen. We have discussed styles, the display system and anchors here. You should already be familiar with AutoCAD blocks. A multi-view block is an AEC object that uses one ore more AutoCAD blocks in different viewing conditions. This allows custom objects to be created that need to look different from various viewing angles. For instance, many architectural elements are drawn differently in plan view than in elevation, such as a plumbing fixture, furniture, medical equipment, or millwork. Use custom multi-view blocks to represent these elements intelligently from these different viewing angles. In the display system described earlier, multi-view blocks respond automatically to these different viewing needs.

The plumbing fixtures and the chair in this file are examples of Multi-View Blocks. You may want to examine them further.

By default all Styles and Content items are stored in a single root folder on the hard drive. This folder is referred to as the Content Library. Your CAD Manager will set its exact location either on your local machine or more commonly on your office network. As new styles and content are accumulated, it is wise to develop a process for adding them to this library. The topics of multi-view blocks and other design content will be covered in more detail in the chapters ahead.

SUMMARY

Usage of Architectural Desktop is predicated on a process of parametric design and progressive refinement.

Create objects quickly with whatever information is available, and then continuously modify the objects as the design progresses.

The display system allows building models to automatically represent a variety of display modes with a click of the mouse.

Object styles deliver on the promise of progressive refinement by allowing design data to be updated globally as the design evolves.

Anchors provide complex and robust rule-based relationships between various individual components within the design.

Object styles, display configurations and custom content can be stored in a central library that grows over time.

The Content Browser allows you to quickly retrieve tools, styles and content.

CHAPTER 3

Workspace Setup

INTRODUCTION

Autodesk Architectural Desktop is a robust program with many user-customizable options and settings. The available settings cover a wide variety of functions. Items as disparate as the color of the screen's background to the specific template file that ought to be used when creating a new drawing are among the vast collection of configurable settings.

In many cases, these settings are referred to as "system variables." System variables typically change some aspect of the working environment and include a list of two or more specific choices. The simplest system variables are basic *on/off* settings. More complex variables can include a long list of possible settings from which to choose. Some system variables are "global" and need only be set once, and they remain in effect permanently on the computer until changed, regardless of which drawing may be open on screen. Other system variables are "drawing specific," meaning that their value can change from one drawing to the next. Due to the large quantity of system variables available to ADT, it is useful to have some convenient way to manage them. ADT offers two ways to manage system variables and establish user preferences, *profiles* and *template files*. AutoCAD users will already be familiar with these two time-tested tools. Their function in ADT differs from generic AutoCAD only in the inclusion of additional settings that are specific to Architectural Desktop.

Depending on the specific installation options chosen, Autodesk Architectural Desktop creates one or more of the following profiles: Architectural Desktop – Imperial, Architectural Desktop – Metric and Architectural Desktop – Metric D A CH. Naturally the difference between these three is primarily in the active units settings. ADT 2006 also will automatically load a template named *Aec Model (Imperial Stb).dwt* [*Aec Model (Metric Stb).dwt*]. The default profile loads the menus, toolbars, palettes, units and system variables for general ADT use. The default template includes the most common drawing settings and display control parameters, and in some cases can also include selected AEC content, such as wall and door types, layers and so forth. However, despite all of these carefully configured defaults, a few settings should be modified. These settings and the reasons for editing them will be covered in this chapter. In addition, we will explore the Drawing Setup command, Layer Standards, and default template files.

 NOTE Your company may already have standards for many of the following topics. Please check with your CAD Manager or CAD Support Professional regarding the existence of CAD standards and proper procedures to follow when using them. If you are the CAD Manager for your office, work with office team members to establish the best methods to manage these policies and procedures.

 NOTE Each localized (non-US) version of ADT installs slightly differently. The exact version of ADT that you have installed may vary slightly from that which is indicated here.

OBJECTIVES

The goal of this chapter is to expose you to many of the issues involved in setting up your work space. The following topics will be explored in this chapter:

- Create a profile.
- Assign a profile to a custom ADT desktop icon.
- Explore Drawing Setup.
- Understand layer standards.
- Work with template files.

CREATING A NEW PROFILE

Profiles are used to manage all of the permanent settings. These settings, once in place, usually do not need to change often or at all. The profile also stores the set of pull-down menus available at the top of the screen, the currently loaded toolbars and their positions on screen, and your tool palette set and its onscreen position. If necessary, more than one profile can be created to facilitate easy swapping of different menu sets, palettes, toolbars, and system variables for different design needs or for several users sharing the same computer. The office CAD Manager or other CAD power user usually handles the creation and management of profiles. If your office has personnel in these roles, check with them before modifying or creating profiles. Profiles are accessed from the Options dialog box. Options can be found on the Format menu or by right-clicking in the Command line.

CREATE A NEW PROFILE

 NOTE The following sequence on creating a Profile is included should you desire to learn to build or customize ADT profile(s). However, it is not required that you build your own Profile to use ADT 2006. If you wish, you may skip this topic and simply use the default Architectural Desktop Profile in lieu of the "MasterADT" Profile built here with no detriment to the remaining tutorials in this book.

1. Launch Autodesk Architectural Desktop 2006.

2. From the Format menu, choose **Options**.

 TIP You can also right-click in the Command line area and access the Options command from the context menu.

3. In the Options dialog box, click the Profiles tab.

   You will see a list of existing profiles; there might be several. At the top of the dialog box, the currently active profile name is listed.

4. If the active Profile at the top of the dialog is not currently Architectural Desktop – Imperial or Architectural Desktop – Metric, select one of those and then click the Set Current button. (You can also double-click it.)

   Choose *Imperial* if you are in the United States and use Imperial units, otherwise choose *Metric*.

5. Click the Add to List button.

 NOTE The Add to List function actually copies the current profile.

6. In the Add Profile dialog box, type **MasterADT** for the Profile name.

7. For the description type: **Mastering Autodesk Architectural Desktop 2006 Profile**.

8. Click Apply and Close.

9. The new entry will appear in the list. Double-click it. (You can also select it and choose **Set Current**.)

   At the top of the dialog box, next to Current Profile, MasterADT should now appear (see Figure 3–1).

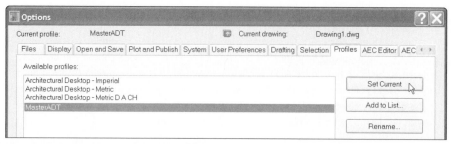

Figure 3–1 *Set the new profile MasterADT as the current profile*

CONFIGURE THE NEW PROFILE

10. Click the Files tab.

11. Expand the Template Settings and then the Default Template File Name for QNEW item.

12. Verify that *Aec Model (Imperial Stb).dwt [Aec Model (Metric Stb).dwt]* is selected, otherwise, select the listed entry and then click the Browse button to choose it.

The "QNEW" command, accessed by clicking the Qnew icon on the Standard Toolbar (or typing **qnew**), will begin a new drawing by automatically loading the template chosen here without a confirmation dialog box. To choose an alternate template or start from scratch, choose **File > New**.

13. Click the Display tab.

Included on this tab are the options for User Interface (UI) Colors and Crosshair size. Many users have their own preferences for UI colors and many prefer the use of a full screen cursor.

14. Make your preferred choices on the Display tab.

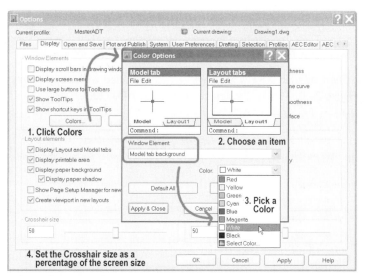

Figure 3–2 *Make Color and Crosshair size selections (the author prefers a white background)*

15. Click the Open and Save tab.

16. In the File Safety Precautions area, Under Automatic save, change the Minutes between saves value to **20**.

 Caution: Automatic Save actually functions as an automatic backup. If the file has not been saved by the time entered in the Minutes between saves box, a backup file will be created in the temporary folder on your system, (which is listed in the Files tab). It is better to think of the Automatic Save feature as a "Save Reminder." Should the Auto-Save message appear at the command line, perform an actual save (**File > Save**) command immediately to avoid losing work. Also, note that ADT creates an additional backup file with a BAK extension each time you save the file. This file will be saved to the same directory as the DWG file. If a DWG file should become corrupt, rename the BAK to a DWG extension (in Windows Explorer) and open this file. (You will need to turn off the Hide extensions for known types in the Windows Folder Options control panel settings to do this. See Windows help for more information.) You should be able to recover anything up until your last save. No matter what your settings on the Open and Save tab, just remember to save often!

New in ADT
2006

 NOTE new to Architectural Desktop 2006 is the AutoCAD Drawing Recovery Manager. This tool makes it much easier to recover files after a system failure. If ADT crashes, the Drawing Recovery Manager will appear the next time you launch the program. The most recent DWG and BAK file will be listed for each drawing that you had open at the time of the crash. You can recover and save these drawings from this interface. The Drawing Recovery Manager will allow you to open BAK files directly without requiring you to rename them in Windows Explorer first. If you do crash, be sure to send the report request that appears. These reports help Autodesk troubleshoot problems in the software and devise fixes for them.

17. In the Demand load XREFs area, verify that "Enabled with copy" is chosen (see Figure 3–3). This is the default setting for ADT 2006.

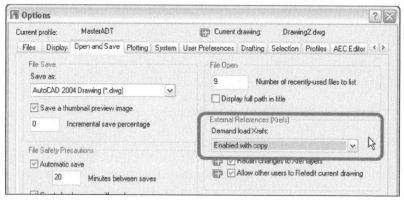

Figure 3–3 *Make certain that XREFs use Enabled with copy or Disabled*

CAD Manager Note: This setting is very important to team environments. If 'Disabled' is chosen, the Demand Load feature for XREFs will be turned off and performance may suffer. If the 'Enabled' setting is chosen, the user may enjoy better performance, but they will be locking *every* XREF file they load, thereby preventing any one else on the project team from opening those files! 'Enabled with Copy' is a good compromise. However, depending on your specific network environment, either 'Disabled' or 'Enabled with Copy' may be suitable. The only critical thing to remember is that 'Enabled' *never* be chosen when working in a networked environment.

18. Take note of the Blue drawing icon next to some of the items in the Options dialog box (see Figure 3–4).

This icon indicates a setting saved only in the drawing file. These settings are not saved globally by the profile. Use template files to manage these settings.

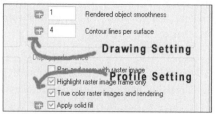

Figure 3–4 *Drawing level settings indicated by the colored icon*

Many veteran AutoCAD users have long been accustomed to being able to use the right mouse button as a quick way to press ENTER. However, since ADT makes such heavy use of the right mouse button for context menus, many seasoned Auto-CAD users have lamented the loss of the simple yet cherished right-click equals ENTER functionality. Fortunately, ADT offers the perfect compromise solution to this dilemma: "Time-sensitive right-click." With this feature enabled, the right-click will behave as it used to in previous versions of AutoCAD by issuing an ENTER. To display the shortcut menu, hold the right-click button down a bit longer. To enable this option, click the Right-click Customization button on the User Preferences tab. If you wish to use this setting, perform the following steps:

20. Click the User Preferences tab.

21. Click the Right-click Customization button.

22. Place a check mark in the Turn on time-sensitive right-click option and then click Apply and Close.

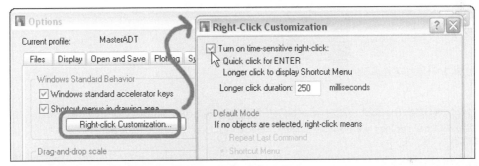

Figure 3–5 *Turning on the time-sensitive right-click option*

Remember that using this feature is a matter of personal preference. You are strongly urged to not disable the right click menu settings. Doing so will make it difficult or impossible to execute many of ADT's core functions. As an alternative to the time sensitive right click, remember that you can also SPACE BAR on the keyboard for ENTER. In this way, you could leave the default right click behavior as is and rather than right click to ENTER, simply press the SPACE BAR instead.

That completes our profile settings.

23. Click the Apply button, and then click OK.

VERIFY AND LOAD PALETTES AND TOOLBARS

As we saw in the previous chapter, most ADT commands may be accessed from tool palettes or toolbars. In this topic, we will be certain that the most common tool palettes and toolbars are loaded.

24. Verify that the ADT tool palettes are loaded (see Figure 3–6).

If they are not loaded, from the Window menu choose **Tool Palettes**, or press CTRL + 3.

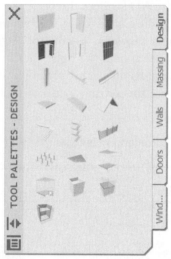

Figure 3–6 *The ADT tool palettes*

 NOTE The tool palettes can be left floating, or docked on the edges of the screen. Other settings are available as well. Refer to Chapter 1 for more details.

CAD Manager Note: ADT installs with a collection of sample Tool Palettes available. The Content Browser is a tool that provides access to copies of all of these standard tool palettes. It can also be used to set up new, or customize existing tool palettes for use by the project team. To open it, click the Content Browser icon on the Navigation toolbar (or press CTRL + 4). For more information on working with the Content Browser, see the 'Working with the Content Browser' heading in Chapter 2.

25. Load and position on screen any toolbars you use frequently. Load toolbars easily by using the techniques covered in the "Right-Click on Toolbars" heading in Chapter 1. These will be "remembered" by the MasterADT Profile.

The Standard, Navigation, Layer Properties, and Shapes toolbars are loaded by default. It is suggested that you consider loading the Views, Shading, and Zoom toolbars in addition to these. UCS II can also be handy, as can the Express Layer Toolbar if you have the Express tools installed. UCS II provides tools to work with User Coordinate Systems, which are reference planes that can be helpful for creating geometry in elevations and 3D.

26. From the File menu, choose **Exit**. (You can also press CTRL + Q. If asked to save a drawing, click No.)

27. On the Windows desktop, locate the Autodesk Architectural Desktop 2006 icon.

28. Right-click it and choose **Copy**.

29. Right-click next to it and click **Paste**.

30. Right-click the new copy and choose **Rename**. Type **MasterADT** and then press ENTER (see Figure 3–7).

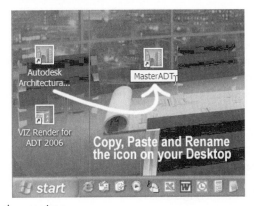

Figure 3–7 *Rename the new icon*

31. Right-click the newly named MasterADT icon and choose **Properties**.

32. Click the Shortcut tab.

Figure 3–8 *Replace the default profile call with MasterADT*

In the Target field locate the following string:

```
/p ''Architectural Desktop - Imperial'' [/p ''Architectural Desktop -
    Metric'']
```

33. Replace "Architectural Desktop – Imperial" ["Architectural Desktop – Metric"] with **"MasterADT"**. Be sure to leave the "/p". The quotes around MasterADT are optional. See the note below.

The complete text in the Target field will read as follows:

```
'C:\Program Files\Autodesk Architectural Desktop 2006\acad.exe'/ld
    'C:\Program Files\Common Files\Autodesk Shared\AecBase47.dbx'
    'MasterADT'
```

The first part of the Target field tells the Windows shortcut which application to load, in this case "*acad.exe*" (or AutoCAD). Following that, "*/ld*" instructs the shortcut to load the ADT functions immediately upon launch. Without this switch, we would be loading just standard AutoCAD. The "/p" switch in the Target box tells ADT to launch into the profile listed after */p*, in our case "MasterADT."

NOTE The use of quotes in the text strings of the Target box is required if spaces occur in any of the paths or names. In the default strings, both "Program Files" and "Autodesk Architectural Desktop" include spaces, and therefore require the use of quotes around the entire string. In our case, "MasterADT" is one continuous word. You can use quotes if you wish, but they are not required.

34. Click OK to close the MasterADT Properties dialog box.

TEST THE NEW ICON

35. Double-click the new icon to test it.

36. Verify that the MasterADT profile is active by returning to the Options dialog box.

    Visually inspect the screen to see if all of your toolbar setup remained from the previous exit.

TIP If any error messages appeared, go back through the steps above to troubleshoot the problem or contact your CAD or IT support person.

DRAWING SETUP

Once you have a profile in place to manage all of the global settings, you can turn your attention to the drawing-specific settings. These include items such as units, layers, and scale. The Drawing Setup command is used to configure these settings within the current drawing. All settings in the Drawing Setup dialog box are saved within the drawing. (There is also an option to save the settings as the default for future drawings.)

SET THE BASE UNIT OF MEASURE TO INCHES

In this sequence, we will explore the interaction of the Units and Scale settings by switching between Imperial and Metric units. The steps in this sequence apply to users of *both* Imperial and Metric units.

1. On the drawing status bar, click the Open Drawing Menu and choose **Drawing Setup**. (The Open Drawing Menu is depicted in Figure 1–2 in Chapter 1.)

2. Click the Units tab (see Figure 3–9).

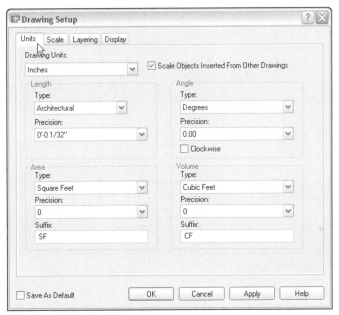

Figure 3–9 *The Units tab of Drawing Setup*

3. In the top left corner under Drawing Units, choose **Inches**.

   The settings available in the other four sections, Linear, Angular, Area, and Volume, will adjust as required to be consistent with the base Drawing Unit (in this case Inches).

4. In the Area grouping, open the Type list.

   Notice that there are three choices, all listed in Imperial units of measure.

5. Choose **Square Inches**.

   Notice the change in the Suffix field to: Sq.In.

6. Experiment with the settings in the Angular and Volume groupings.

SET THE DRAWING SCALE FACTOR

7. At the top of the Drawing Setup window, click the Scale tab.

   Drawing Scale lists all of the most common architectural scales. ADT annotation objects, such as targets, tags, symbols, and schedule tables, automatically scale to the value assigned here. In order to choose the right drawing scale, it is necessary to know at which scale the final drawing will plot.

8. Choose **1/4″ = 1′-0″** (see Figure 3–10).

   The field below Custom Scales changes to **48.000** to reflect the new choice.

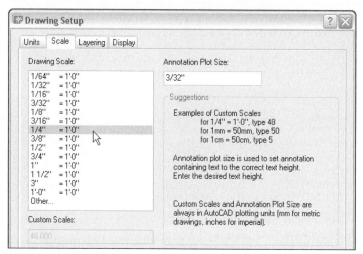

Figure 3–10 *Assign a drawing scale*

If the scale you wish to use is not listed, enter a custom scale. Determine it by dividing the drawing unit by the scaled unit. For example, to work at 1″=10′, divide 10′ (120″) by 1″ to arrive at the scale factor of 120.

9. Click on "Other" in the Drawing Scale list.

10. In the Custom Scales box, type **120** (see Figure 3–11).

Figure 3–11 *Entering a custom scale*

NOTE If the scale that you want is not already on the list, you can click the Edit Scale List button to add to or edit the list.

11. Type **1/8″** in the Annotation Plot Size field.

The Annotation Plot Size value controls the plotted height of text contained within symbols inserted in the drawing. The value represents the height of the text portion of the symbol in the *final* printed output. The most typical text height sizes are 1/8″ [3] and 3/32″ [2.5]. The graphic portion of the symbol will scale proportionally to whatever factor is necessary to achieve the correct Annotation Plot Size for text.

SET THE BASE UNIT OF MEASURE TO MILLIMETERS

12. Return to the Units tab.

13. In the Drawing Units section, choose **Millimeters** (see Figure 3–12).

Notice the change in the Length section. The unit Type has changed to Decimal.

Figure 3–12 *Changing the unit changes the available options of the other settings*

14. In the Area grouping, open the Type list again.

Notice the change in available choices. They are now all metric settings.

15. Choose **Square Millimeters** (see Figure 3–13).

Notice the change in the Suffix field to mm2.

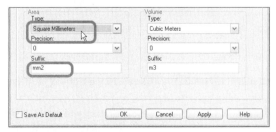

Figure 3–13 *The suffix reflects the unit chosen*

16. Experiment again with the settings in the Angle and Volume sections.

17. Click the Scale tab again.

    The choices under Drawing Scale now represent the most common metric architectural scales.

18. In the Drawing Scale list, choose **1:100**.

    As before, notice that the field below Custom Scales changes to 100 to reflect the new choice. As before, if the scale you wish to use is not listed, click Other, and type in a scale factor in the Custom Scale field or click the Edit Scale List to add to the preset list.

19. Change the Annotation Plot Size to a value appropriate for plotted symbols in Metric such as **2.5** or **3**.

SET THE LINEAR UNIT TYPE

20. On the Units tab, change the Drawing Units to **Feet**.

    Notice that the Length unit Type is Decimal, *not* Architectural.

 NOTE FOR IMPERIAL UNITS ONLY Architectural and Engineering unit formats are available only when the base unit is inches. Both feet and inches can be entered when you work in Architectural or Engineering as long as you distinguish feet from inches at the command line and in text input boxes with an apostrophe (') following the number. To enter inches, typing only the number is required. Hyphens are not required to separate feet from inches. Hyphens are used to separate fractions from whole numbers. For more complete information on valid input of Imperial units in the ADT, refer to Table 3–1 and the online help.

Table 3–1   *Acceptable Imperial Unit Input Formats*

Value Required	Type This:	Or This
Four feet	4′	48
Four inches	4	4″
Four feet four inches	4′4	52
Four feet four and one half inches	4′4-1/2	4′4.5 or 52.5

 NOTE Typing the inch (″) mark is acceptable as well; however, it is not required. Dimensions throughout this text use the Feet and Inch format for clarity. However, feel free to enter dimension values in whatever formats you prefer. Eliminating the inch mark reduces keystrokes and is recommended despite the inclusion of it in this text.

LAYER STANDARDS

Layers provide an important organization and management tool to AutoCAD and Autodesk Architectural Desktop alike. Part of the core functionality of AutoCAD, layers provide a means to centrally control object color, visibility, linetype, lineweight, and plotting properties. Because a typical architectural drawing can contain several dozen layers, several tools have been devised over the years to make layer management easier and more useful. Layers and layering strategy have kept many a CAD Manager busy and have been the topic of many books, including the popular *CAD Layer Guidelines, Second Edition* published by The American Institute of Architects and its successor, the US National CAD Standard 2nd Edition. For reasons such as these, CAD Managers and users alike will be pleased to learn that layering of ADT objects is handled automatically by the software. (Please note that drafted entities such as lines, arcs and polylines are *not* auto-layered.)

Layer Usage in ADT

Layers are an important part of any ADT drawing. There are many popular notions regarding the definition and usage of layers. The most common description of layers is the "sheets of acetate" analogy. This description of layers and layering uses the metaphor of several sheets of acetate, each with a different piece of drawing information, stacked on top of one another. When taken together as a composite, they represent a complete drawing. This metaphor, though illustrative, is not complete. While effectively conveying a good mental picture of a collection layers, it fails to convey any information as to why we might use layers in the first place. It also implies properties that are not characteristic of layers such as "stacking" or the implication that one layer could "cover" the information on another. This is not possible with layers. A more evocative way to comprehend the potential of layers is to think of them as categories. Layering is simply a drawing-wide categorization system. When we think of layers this way, it is easy to understand their full potential. It is also easy to decide when a new layer is appropriate and what it ought to be named.

Thinking of layers this way is appropriate in generic AutoCAD, but it is especially helpful when we attempt to reconcile their role together with display control in ADT. As detailed in the previous chapter, display control offers a completely new way to control visibility, color, and lineweight of objects. Determining when to use the display system and when to use layering can sometimes be tricky. However, if we treat layering as a global drawing-wide categorization system, the distinction becomes much clearer. Display control addresses very specific object-level display needs, while layers are much more general and pervasive. As such, it is appropriate to approach layering in ADT in "broad brush" fashion. In the default configuration, all walls are placed on a Wall layer regardless of type or style, likewise with doors, and all other objects. Each major ADT object class has a unique default layer assignment. Internal object subcomponents of ADT objects typically rely less on layers and more on Display Control parameters. For example, the control of individual object components like the swings of doors, the row lines of a schedule table, or the color of 3D components is often better handled with Display Control parameters instead. However, it is just as common to use both display settings and layers to achieve desired display results.

The strategy will become clear in the chapters that follow. The first step is to understand how automatic layering works. Objects are automatically placed on a layer as they are created. The current Layer Standard and Layer Key Style determine the exact layer they use. Layering assigned to internal object components is typically a parameter of the object style. Tools on tool palettes can also contain Layer Key information that supersedes the defaults of the particular object class.

CAD Manager Note: You can customize the Layer Standard and Layer Key Styles to comply with your existing office standard layering scheme. Assign the Layer Standard in the Drawing Setup command and save it as the default. You can be sure that your latest settings are available to all users by also checking the Always import Layer Key Style when first used in drawing option. Within the file are Layer Key Styles. Layer Key Styles are used to assign AEC objects to their required layers. If a layer does not exist at the time the object is created, the Layer Key for that object will generate the layer on the fly. All AEC objects and any object or group of objects inserted from the Custom tab of the DesignCenter as well as objects created from tool palette tools can use Layer Keys.

SET THE LAYER STANDARD

ADT automatically places all new architectural objects on a predetermined layer. Layer standards which facilitate this feature can be loaded in the Drawing Setup window.

1. On the drawing status bar, click the Open Drawing Menu and choose **Drawing Setup**. (The Open Drawing Menu is depicted in Figure 1–31 in Chapter 1.)

2. Click the Layering tab.

   Take note of the file listed in the Layer Standards/Key File to Auto-Import box.

3. If the Layer Standards/Key File to Auto-Import field is blank or incorrect, click the small browse (…) button to the extreme right of the window, and locate the correct file, *AecLayerStd.dwg* (see number 1 in Figure 3–14).

   Your particular office's standards might be set to Auto-Import a file other than *AecLayerStd.dwg*. If this is the case, then leave the setting as is. Check with your CAD Manager or CAD support personnel for more information.

CAD Manager Note: ADT 2006 uses Windows XP compliant file locations for all resources. This means that the path to the *AecLayerStd.dwg* drawing file defaults to: *C:\Documents and Settings\All Users\Application Data\Autodesk\ADT 2006\enu\Layers*. This path will vary with locale. For instance, 'enu' stands for 'English United States.' If you have installed a non-English version of ADT, or non-US version, this path would vary slightly. It is also important to note that the *Application Data* folder is a hidden folder by default in Windows 2000 and XP. Therefore, you will need to set your Folder Options in Windows Explorer to show hidden folders. Bear in mind that this path and any other support file search path can be changed to any location that you find suitable, including locations on a network server. In fact, it will often be much easier to administer resources for several users if all common files (those within the *All Users* folder) are stored on the server. The option to move most content and support files to alternate locations on your system or a network server are available during installation.

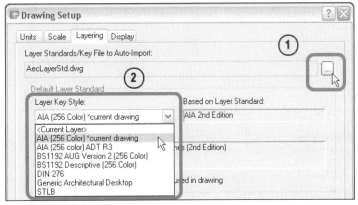

Figure 3–14 *The browse button used to load a Layer Standards file and Layer Standards included in the AECLayerStd file*

The *AecLayerStd.dwg* file includes many built-in industry standard layering schemes such the one published by The American Institute of Architects (AIA)—which is a part of the US National CAD Standard (NCS) and several other layering schemes. From the Layer Key Style list, you can choose the layer standard appropriate to your firm's office standards (see number 2 in Figure 3–14). If your firm uses a standard not included in this file, your CAD Manager can advise you regarding the proper file to load here. If you are a sole practitioner or otherwise do not have a CAD Manager or an established office standard, it is recommended that you adopt one of the standards in this file, such as AIA (256 Color) as your layering standard. This Layer Key Style is fully compliant with the NCS. Adopting an industry standard provides immediate benefit to you and your firm:

▶ Sharing drawings with other firms is easier because the layering systems are more likely to be compatible.

▶ Orientation is easier when new employees are hired.

▶ The developers of ADT have already done the work of configuring the standard and including it in the software.

▶ The structure is logical and extensible.

One additional setting on the Layers tab bears mention. The "Always import Layer Key Style when first used in drawing" check box instructs ADT to check the default Layer Standards/Key file each time a drawing is opened to see if there is a more recent version of the Layer Key Style. If so, it will update the current drawing to the latest version of the Layer Key Style. This setting is very useful for maintaining CAD Standards across the office.

4. Place a check mark in the Always import Layer Key Style when first used in drawing box.

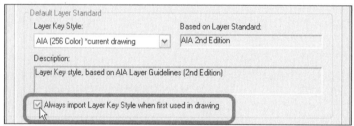

Figure 3–15 *Layer Standards included in the AECLayerStd file*

5. Click Apply and then OK to dismiss the Drawing Setup dialog box.

Once they have been established, the majority of settings in the Drawing Setup dialog box do not need to change. Typically, as project needs dictate, the Scale setting might be changed as the project progresses. However, it is important to note that changing the values in the Units, Scale and Layering tabs may impact objects already contained in the drawing. For instance, if you change the Units setting, a dialog box will appear when you apply the change. This dialog box will give you the option to rescale all objects in the drawing to the new Unit. It will be your choice, which you should consider carefully depending on what you are trying to achieve. Regardless, all objects created from that point on would use your new settings. Re-layering of objects is NOT automatic. A command to remap layers is, however, provided on the Format menu. Choose the Layer Management > Remap Object Layers command. The Display tab gives access to the System Default Display Control settings. This is an alternative to the Display Manager discussed in the previous chapter.

TEMPLATE FILES

AutoCAD has long used template files as a means to quickly apply setup information, enforce company standards and project-specific settings, and save time.

Autodesk Architectural Desktop takes full advantage of the same benefits. In much the same way as other popular Windows software packages, ADT uses template files at the time of file creation to establish a whole host of user settings and overall configuration. (Please note that template files apply only at the time of drawing creation.) A template file is an AutoCAD drawing file preconfigured for a particular type of task. Template files have a DWT extension and are available from the **File** > **New** and **Qnew** commands. In addition to the time saved when drawings are created from templates, templates help to ensure file consistency by giving all drawings the same basic starting point. ADT 2006 ships with several pre-made template files. These templates are ready to be used "as is." However, because office standards and project-specific needs vary, feel free to modify the default templates as necessary. The exact composition of the template used to create a drawing is not as important as ensuring that a template *is* used to create all new drawings. Table 3-2 shows the AEC template files included in the box with ADT 2006.

Table 3–2   *Out-of-the-Box Template Files*

Template File Name	Drawing Units	Plot Style Type
Aec Model (Metric D A CH Ctb).dwt	Meters	Color
Aec Model (Imperial Ctb).dwt	Inches	Color
Aec Model (Imperial Stb).dwt	Inches	Named
Aec Model (Metric Ctb).dwt	Millimeters	Color
Aec Model (Metric Stb).dwt	Millimeters	Named
Aec Sheet (Metric D A CH Ctb).dwt	Meters	Color
Aec Sheet (Imperial Ctb).dw	Inches	Color
Aec Sheet (Imperial Stb).dwt	Inches	Named
Aec Sheet (Metric Ctb).dwt	Millimeters	Color
Aec Sheet (Metric Stb).dwt	Millimeter	Named

Notes on Table 3–2-:
AEC Model—Template files are used for creating "Model" files. A model file is a file containing actual full-scale building data. This template is used to create files in which all of the day-to-day work is performed.
AEC Sheet—Template files are used for creating "Sheet" files. A sheet file is used *exclusively* for printing drawings.
In addition to the DWT templates listed here, the *Template* folder also contains Project templates and Sheet Set templates. (Refer to Chapter 5 for more information.)

NOTE Please refer to the "Elements, Constructs, Views and Sheets" heading in Chapter 5 for complete information on Model and Sheet files.

Caution: Please avoid creating drawings without a template. This is because a scratch drawing requires an enormous amount of user configuration before serious ADT work can begin.

CAD Manager Note: ADT saves virtually all data and configuration within the drawing file. This includes system variables, blocks, object styles, and display configurations. Therefore, template files provide an excellent tool for promoting and maintaining office standards. ADT 2006 ships with several sample template files to help you get started. These include Model and Sheet templates for both Imperial and Metric units. Refer to Table 3–2 above for more information. Certainly, you will find one of these pairs of templates suitable for your firm's needs. When you establish and configure office standards, it is highly recommended that these default templates be used as a starting point. Modify them to suit individual project or office-wide needs. Styles, blocks, and other resources can also be stored in separate drawing files stored on the network or hard drive. The resources can be accessed with custom tool palette tools. This keeps the template file size smaller by including only those items needed in all drawings, yet it provides a central repository for additional office standard items. This method also provides additional ongoing flexibility because new items can be easily added to the office standard library.

 In addition, ADT 2006 now has the ability to establish Project Standards. Project Standards allow you to establish one or more drawing or template files as the source for all project styles and display configurations, and then use them to keep all other drawings synchronized with the standard Use of this feature requires that the drawing management system be used to manage projects. ADT projects are covered throughout this book starting in Chapter 5.

START A DRAWING WITHOUT A TEMPLATE (START FROM SCRATCH)

The best way to demonstrate the importance of using template files is to compare drawings created with and without template files. A drawing begun with the Start from Scratch option has little in the way of architectural settings. (To start from scratch, choose **File > New**, and then click the small arrow on the Open button to reveal a small menu. Then choose either **Open with no Template – Imperial**, or **Open with no Template – Metric**. See Figure 3–16.) Of the three major tools covered in the previous chapter, the display system, object styles, and anchors, display configurations and object styles make excellent additions to a template file. The "scratch" drawing will have no useful settings for any of these critical items.

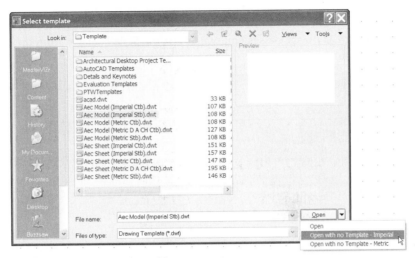

Figure 3–16 *Accessing the No Template (Start from scratch) option*

CAD Manager Note: For actual production work, using template files is the only viable option; however, using Start from Scratch can be an excellent way to troubleshoot styles or a corrupted file or to import a file from another CAD package or outside vendors.

A drawing created without a template will have the following characteristics:

▶ Scratch drawings include two Layout tabs; neither has its Page Setup configured (see Figure 3–17).

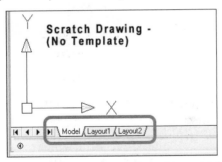

Figure 3–17 *Scratch drawings contain two non-configured Layout tabs*

▶ Only Layer 0 is present in a scratch drawing (see Figure 3–18).

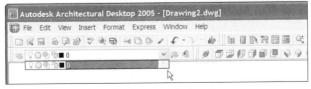

Figure 3–18 *Layer 0 is the only layer in the scratch drawing*

▶ There are no ADT styles in a scratch drawing. (Figure 3–19 shows the Style Manager with all object classes empty.) Access the Style Manager from the Format menu.

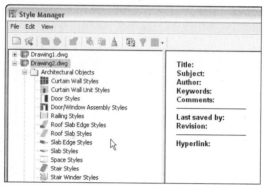

Figure 3–19 *The Style Manager showing no styles*

▶ There will be only the most basic Display Configurations and Display Control settings in a scratch drawing (see Figure 3–20).

 NOTE On the Drawing Status Bar, (Figure 1–30 in Chapter 1), none of the ADT-specific controls will appear on the right side. However, if you open the Display Manager, as covered in the "Use Display Manager" heading in Chapter 2, several very basic Display Configurations are auto-created and added to the scratch drawing (see Figure 3–20).

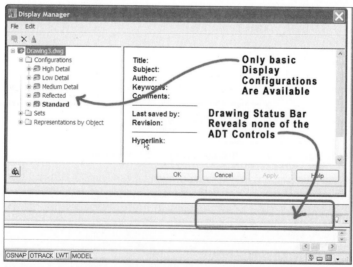

Figure 3–20 *Only basic Display Configurations are present in the scratch drawing*

▶ Scratch drawings start in model space zoomed in to a size roughly equal to a sheet of notepaper (US Letter Size or A4 Size).

CREATE A DRAWING USING AN ADT TEMPLATE

▶ Template drawings typically include purpose-built Layout tabs. These are pre-configured for purposes like Working or Plotting at a particular Sheet size (see Figure 3–21).

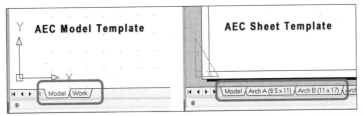

Figure 3–21 *Templates use purpose-built Layout tabs, pre-configured for specific tasks*

▶ Templates typically contain several pre-made layers (see Figure 3–22).

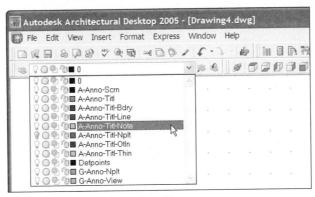

Figure 3–22 *Layer list from a typical template file*

▶ Templates often contain some ADT Styles. (Figure 3–23 shows the Style Manager with several classes populated with styles. A plus sign (+) next to a class indicates that styles are present.)

NOTE The inclusion of Styles in a Template is not as critical a factor as the others noted here due to a Tool Palette's ability to automatically import a remote Style.

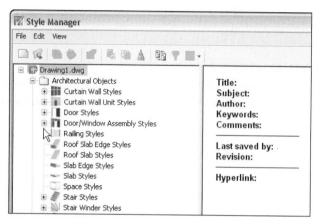

Figure 3–23 *A Template is often pre-populated with Styles in one or more categories*

> Perhaps the biggest benefit of an AEC template is the preconfigured Display Configurations and Display Control settings (see Figure 3–24).

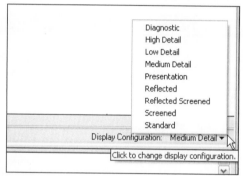

Figure 3–24 *Templates typically have several pre-defined Display Configurations designed for a variety of drawing types and tasks*

> Template drawings start in model space zoomed out to a size appropriate for building models.

CAD Manager Note: The templates shipping with ADT 2006 are ready to use straight out-of-the-box. If you wish to build your own office standard template, they are excellent starting points for developing your own template file(s). It is highly recommended that you become comfortable with the out-of-the-box offerings and then if necessary, customize them to meet your firm's specific needs. Consider including those settings and display items that people will use most frequently. Once you have a standard template in place, it is imperative that all users be required to use it, or a project-based derivative of it, for all project work. (See Chapter 5 for more information on Projects.)

Hopefully, you are beginning to see the benefits to starting new drawings with an AEC template. There is really no compelling reason to begin drawings any other way. As you work through the exercises in the coming chapters, you will certainly discover areas where the default templates could be enhanced and improved. Make note of these observations as you go. When you are ready, try your hand at creating your own template file. The basic steps are simple:

▶ Create a new drawing using the template that most closely matches the one you wish to create.

▶ Edit any settings as you see fit.

▶ Choose File > Save As.

▶ From the Files of type list, choose AutoCAD Drawing Template File (*.dwt) (see Figure 3–25).

▶ Type a name for your new template and then click the Save button.

Figure 3–25 *Saving a new template file*

 NOTE Choosing AutoCAD Drawing Template File (*.*dwt*) as the file type automatically switches to the correct folder for saving templates. This location is configured on the Files tab of the Options dialog (see the "Configure the New Profile" heading above). Check with your CAD Manager before saving templates to be sure of the correct location.

Templates will also be used in ADT by the Drawing Management system. This very important and powerful tool set will be covered extensively throughout this book starting in Chapter 5. Whether or not you choose to use this powerful ADT drawing and XREF management system, template files are *the* way to start new Architectural Desktop drawings.

SUMMARY

A profile helps manage the system level settings in ADT.

Adding a profile switch (/p) to the ADT icon on the desktop loads the profile automatically when you launch ADT.

Choose Drawing Units before all other settings in the Drawing Setup dialog box; this will set other choices to appropriate matching values.

The Layer Standard chosen determines which layers will be auto-created by objects in the drawing.

Using the Start from Scratch option to create new drawings is not recommended for typical usage.

The AEC templates come preloaded with a variety of useful sample content and settings.

Adopt templates as your office/project standards to manage standards, quality control and support issues when managing ADT drawings.

The Building Model

In this section, we will explore in detail the creation of the Building Information Model. The focus in this section is on the "*Building Model*" itself, with the "*Information*" part coming later. We will actually be working on two separate projects from start to finish throughout the course of this book; one residential and the other commercial. We begin in Chapter 4 by constructing a floor plan, which will become part of the project files later. In Chapter 5 we use the Project Management system in ADT to create and link together several files, each with its own unique focus, used together to create the complete composite building model. In Chapters 6 through 12, we undertake a thorough exploration of the major Autodesk Architectural Desktop object types such as Walls, Doors, Curtain Walls, Column Grids, Stairs, Roofs and Fixtures, each with its own special focus on how the parts fit into the complete building model.

Section II is organized as follows:

CHAPTER 4

Beginning a Floor Plan

INTRODUCTION

Any discussion about using Architectural Desktop to create floor plans needs to begin with walls. After all, walls are the major component of floor plans. The Wall object will be our first thorough examination of an ADT object. The last several chapters have tried to get you comfortable with the theoretical underpinnings of ADT. Now that you are in the correct mind-set, get ready to roll up your sleeves—it is time to produce some drawings

OBJECTIVES

Throughout the course of the following hands-on tutorials, we will lay out the existing conditions floor plan for our residential project. In this chapter, we will explore the various techniques for adding and modifying walls, doors, and windows. In addition, we will add plumbing fixtures and other elements to make the floor plan more complete.

- Understand Wall objects.
- Add and modify walls.
- Explore Wall properties.
- Add and modify Doors and Windows.
- Add plumbing fixtures.
- Work with Wall Plan Modifiers and associated tools.

 ## WORKING WITH WALLS

Basic object creation in ADT involves frequent interaction with the Tool and Properties palettes as well as a heavy use of right-click menus. An overview of the various right-click menus can be seen in Chapter 1. Although many of the command options do appear at the command line, remember that it is often quicker and more direct to interact with the Properties palette or the right-click menus. If you have Dynamic input enabled (see Chapter 1) you will also receive heads-up prompts directly on screen at your cursor location. Working with walls is very similar to drawing simple lines. You add Walls point by point just as you do lines, and like a line, each Wall segment remains a separate entity distinct from its neighbors. However, unlike lines, Walls know that they are Walls and behave accordingly. Walls will "cleanup" with intersecting Walls. Walls have height and width parameters, can have custom shapes and profiles, and can receive Doors and Windows by automatically creating openings and anchors for them. Finally, like lines, Walls can also be copied, trimmed, filleted, extended, offset and arrayed. Many of these commands will be used with Walls in this chapter; however, if you wish, you can consult an AutoCAD reference or the online help for further information on basic AutoCAD commands.

INSTALL THE CD FILES

1. If you have not already done so, install the dataset files located on the Mastering Autodesk Architectural Desktop 2006 CD ROM.

   Refer to 'Files Included on the CD ROM' in the Preface for information on installing the sample files included on the CD.

2. Launch Autodesk Architectural Desktop 2006 from the desktop icon created in Chapter 3.

If you did not create a custom icon, you might want to review "Create a New Profile" and "Create a Custom ADT Desktop Shortcut" in Chapter 3. Creating the custom desktop icon is not essential; however it makes loading a custom profile easier.

3. Create a new file using the *Aec Model (Imperial Stb).dwt* [*Aec Model (Metric Stb).dwt*] template.

   Depending on your system's settings, this can normally be done by simply clicking the QNEW icon on the Standard toolbar.

GETTING STARTED WITH WALLS

ADT has two "New" commands: the "NEW" command on the File menu and "QNEW" on the Standard toolbar. QNEW automatically uses a default template file, whereas NEW presents you with a dialog box of templates from which to choose (see Figure 4–1). To set a default template, see the "Configure the New Profile" heading in Chapter 3. It is highly recommended that you always begin ADT files from an ADT Template file. Review the "Template Files" heading in Chapter 3 for more information.

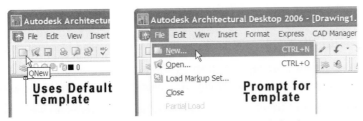

Figure 4–1 *The QNEW setting on the Files tab of the Options dialog box*

   4. On the Design palette, click the Wall tool (see Figure 4–2).

 NOTE If the Tool Palettes are not open on screen, choose **Tool Palettes** from the Window menu, or press CTRL + 3. If the Design Tool Palettes Group is not visible, right-click the Tool Palettes title bar and choose **Design**. See the "Tool Palette Groups" heading in Chapter 1 for more information.

The Properties palette will appear when the Wall tool is clicked. If it was already open on your screen, then the parameters listed within it will now be for the new Wall that you are creating. If you have your Properties palette open but set to Auto-hide, then you will have to hover your mouse over the title bar to make it pop open to view or edit the Wall creation parameters (see the "Understanding Tool Palettes" heading in Chapter 1 for more information on Auto-hide).

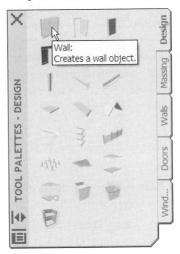

Figure 4–2 *The Wall tool on the Design palette*

   5. On the Properties palette, change the Width to **8″ [188]**, change the Base Height to **9′ [2750]**, and change the Justify to **Center**.

The following list explains the major fields and controls in the Properties palette while you are adding a Wall (see Figure 4–3).

General Properties:

- **Description**—Click to give this particular Wall a detailed Description.

- **Style**—Contains a list of all Wall styles within the current drawing file. Think of Wall styles as "wall types." (Refer to Chapter 10 for more information.)

- **Cleanup automatically**—A Yes or No value that determines whether or not this Wall will automatically clean up with other Walls. This is Yes by default, and should rarely be changed (refer to Chapter 9 for more information).

- **Cleanup group definition**—Used to limit cleanup of Walls to other Walls within the same Group. (This topic is covered in Chapter 9.)

- **Segment type**—Walls can be added with Line or Arc segments. Adding Straight segments is like adding lines. Adding Arc segments is like drawing 3 point arcs.

Dimensions Properties:

- **Width**—The thickness of the Wall. This is also referred to as the "Base Width."

- **Base height**—The Floor to Ceiling height of the wall.

- **Justify**—One of four possible points within the thickness of the Wall from which the Width is referenced. The Wall's grip points will occur at its justification. Center places the grips at the middle of the Wall width. Left and Right justification correspond to the left and right edges of the Wall width when looking from the Wall's start point toward its end. (Refer to Figure 9–1 in Chapter 9.) Baseline justification occurs at the zero point of a Wall's width as determined by its style. (Refer to Chapter 10 for information on editing and creating Wall styles.)

- **Baseline offset**—Can be used to place the Wall at a specified distance parallel to the points actually picked in the drawing.

- **Roof line offset from base height**—The top edge of Walls can be projected above the Base height in elevation and 3D.

- **Floor line offset from base height**—The bottom edge of Walls can be projected below the Base height in elevation and 3D.

All of these parameters are shown in the illustration included on the Properties palette.

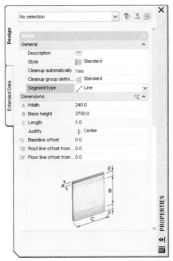

Figure 4–3 *The Properties palette for adding Walls*

6. At the "Start point" command line prompt, click a point toward the lower left corner of the screen to start the first Wall (see Figure 4–4).

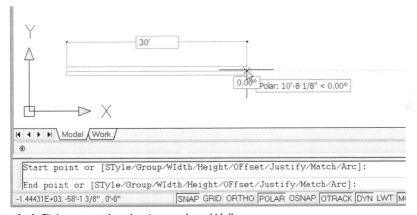

Figure 4–4 *Click start and end points to draw Wall segments*

7. Verify that Polar Tracking is active.

The button labeled POLAR at the bottom of the screen (on the Application Status Bar) should appear depressed. If it is not, click it.

8. Move the cursor directly to the right.

Use the Polar Tracking as a guide to be sure you are drawing the wall perfectly horizontal.

9. Type **30′ [9000]** and then press ENTER (see Figure 4–5).

TIP Look up Direct Distance Entry and Use Dynamic Input in the online help for information on this technique.

10. Move the cursor straight up (90°).

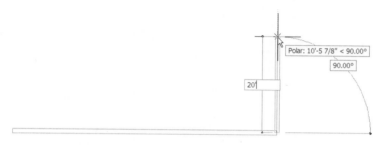

Figure 4–5 *Using Direct Distance Entry with Polar Tracking*

11. Type **20′ [6000]** and then press ENTER.

    Notice that the corner where the two Walls join has formed a clean intersection.

12. Move the cursor to the left (180°), type **15′ [4500]** and then press ENTER.

USE POLYLINE CLOSE

13. With the command still active, press the down arrow to reveal command options. (You can also right-click in the drawing window to see a menu). With either method, choose **Close** from the context menu (see Figure 4–6).

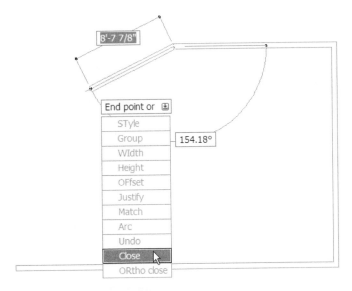

Figure 4–6 *Use the Polyline Close option*

Choosing the Close option will add a single Wall segment joining the current wall to the first Wall drawn in this sequence, and then complete the add Wall command.

ADD A CURVED WALL SEGMENT

Let's add a curved Wall segment this time.

14. On the Design palette, click the Wall tool.

15. On the Properties palette, choose Arc for the Segment type.

16. Click a point anywhere outside the existing room drawn in the last sequence.

17. At the Mid point prompt, click anywhere within the room.

    Move the mouse slowly before clicking in the next step to get a sense for the behavior of the curve.

18. At the End point prompt, click anywhere outside the room again.

19. Right-click and press ENTER to end the command (see Figure 4–7).

    Notice that the curved Wall segment also cleans up nicely with the others.

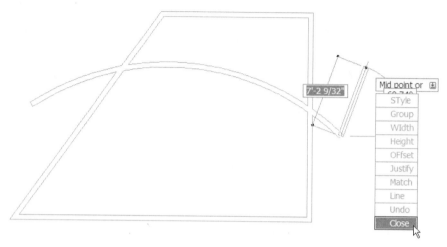

Figure 4–7 *Adding a curved wall*

20. Close the file without saving the changes (**File > Close**).

CREATE AN EXISTING CONDITIONS PLAN

Now that we have practiced adding a few Walls and seen some of the properties available while doing so, let's begin creating an actual model. We will start with the first floor existing conditions plan for our residential project.

1. Create a new file using the *Aec Model (Imperial Stb).dwt* [*Aec Model (Metric Stb).dwt*] template.

   This file will later become our first floor existing conditions for the residential project.

2. At the command line, type **z** (for zoom) and then press ENTER.

3. Type **-10',-30'** [**-3000,-9000**] and then press ENTER.

4. Type **40',10'** [**12000,3000**] and then press ENTER.

Any time that the command prompt requests a point (location, corner, etc.), you are able to either use your mouse to click a point or type in a coordinate entry. Should you choose keyboard input, there are two types of coordinate entry, absolute and relative. Absolute coordinate entry is so called since it references the absolute 0,0,0 point of the drawing—called the "origin." Relative coordinate entry, by contrast, is measured from (that is, relative to) the current point. In this case we have used absolute entry to zoom in to a region of the drawing starting from down and to the left of the origin, to up and to the right. Under normal circumstances, it is rare that you would zoom this way, but for purposes of the tutorial, it is effective

since we will begin drawing at 0,0 and it is desirable to see the layout we are drawing at a comfortable zoom level. For more information on coordinate entry, consult the "Use Coordinates and Coordinate Systems" topic in the online help.

USE ORTHO CLOSE

5. On the Design palette, click the Wall tool.

6. Set the Width to **12″ [311]**, the Height to **9′ [2750]** and the Justify to Right
    (see Figure 4–8).

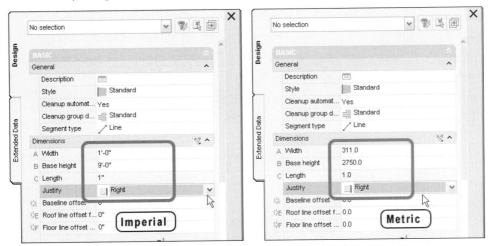

Figure 4–8 *Change the settings on the Design palette*

7. At the "Start point" command line prompt, type **0,0** and press ENTER.

8. With POLAR active (click on the Application Status Bar or press F10), move
    the cursor down (270°), type **24′ [7300]** and press ENTER.

9. Using POLAR, move the cursor to the right (0°), type **33′ [10000]** and then
    press ENTER.

10. Press the down arrow to access the Wall command options or right-click in
    the drawing window and choose **Ortho close**.

Ortho is a close command that will complete the Wall Add sequence by adding
two wall segments. The first will follow the angle indicated by the next mouse
click. The other will be placed perpendicular to the first wall segment drawn in
this sequence. If we point straight up with the mouse in this case, we get a rectangular shape.

The command line should prompt "Point on wall in direction of close:"

11. To make a rectangular space, use POLAR to click 90° straight up (see Figure 4–9).

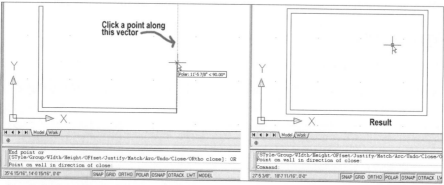

Figure 4–9 *Using the Ortho close option*

This option is very commonly used to create rectangular spaces. However, plenty of other shapes are possible depending on the angle you indicate with the mouse (see Figure 4–10).

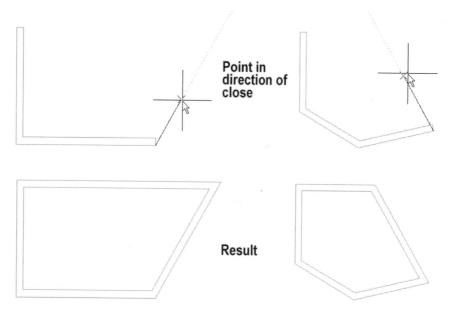

Figure 4–10 *Using the Ortho close option*

OFFSET WALLS

Most standard AutoCAD editing commands work normally on Walls. We will use offset, trim, extend and fillet to continue to lay out the interior walls of the floor plan for the residential project. All of these commands are available on the right-click menu via the basic modify tools submenu, or you may type the appropriate command line version or hot key for the command. (Refer to the online help for more information on command line versions of the commands.)

12. Right-click anywhere in the drawing and from the context menu, choose **Basic Modify Tools > Offset**.

 TIP You can also use keyboard shortcuts for many commands as well if you wish. The hot key for offset is O (followed by **ENTER**). If dynamic input is turned on, typing will appear directly at the cursor on screen, otherwise, it will appear at the command line.

13. At the "Specify offset distance" prompt, type **12'-11.5"** [**3950**] and then press ENTER.

14. At the "Select object to offset" prompt, click on the lower horizontal Wall.

15. At the "Specify point on side to offset" prompt, click a point anywhere within the building (see Figure 4–11).

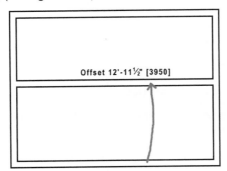

Figure 4–11 *Offsetting a new wall*

16. Press ENTER to end the offset command.

17. Press ENTER again to repeat the offset command.

 TIP Remember that except for text editing commands, the **SPACEBAR** functions like an **ENTER** key at the command line. This would also be a good time to use the "thumb on SPACEBAR" technique to repeat the previous command.

18. Type **12'-8.5"** [**3874**] for the offset distance this time.

19. Click the leftmost vertical wall, and offset it inside as well (see Figure 4–12).

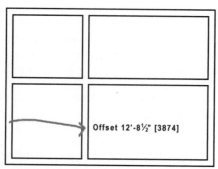

Offset 12'-8½" [3874]

Figure 4–12 *Two new walls offset inside*

20. Press ENTER to end the offset command.

21. From the File menu, choose **Save as**.

22. Save the file to the *C:\MasterADT5\Chapter04\Residential* folder, using the name *First Floor Existing.dwg*.

One of the features that makes Architectural Desktop such a powerful tool is the ability to change an object's parameters at any time, as design needs change.

Let's take a look at modifying some of the Walls as we continue with the layout of the first floor existing conditions for the residential project.

23. Select the two internal Walls created in the "Offset Walls" heading above.

24. Right-click, and choose **Properties**.

25. On the Properties palette, set the Width to **5″ [125]** and the Justify to **Center** (see Figure 4–13).

   Notice the change to the drawing, particularly the way the walls shift when changing justification.

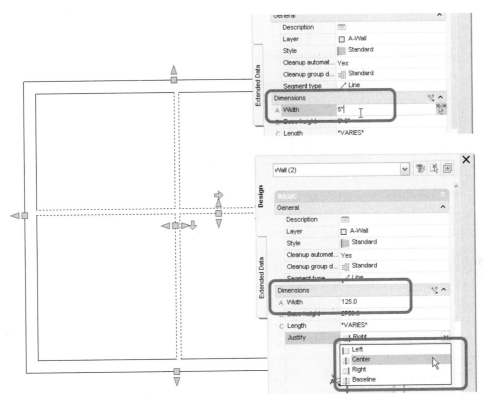

Figure 4–13 *Manipulate Wall properties in the Properties palette*

> 26. Right-click and choose **Deselect All** or press the ESC key.

Justification is an important consideration when working with walls. Again, the Properties palette allows you to receive immediate feedback on the change while the object selection is still active. Therefore, if you decide you are not happy with the change, you can simply try another on the fly. When calculating offset distances, you must take the wall's width and justification into account. Offsets performed with the AutoCAD Offset command are measured between the justification lines of the walls (or center to center), and the Trim and Extend commands use the justification line as the Cutting/Boundary edge.

LAY OUT THE REMAINING WALLS

Now that we have the interior Walls at the correct width and justification, we will continue adding Walls using the same technique.

> 27. Using the offset command, offset the interior horizontal Wall up **4'-3"** [**1295**] and the interior vertical Wall to the left **2'-11"** [**889**] and **6'-11"** [**2108**] to the right (see Figure 4–14).

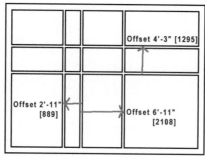

Figure 4–14 *Offset two more Walls*

At this point we have several Wall segments crossing one another. Before we add any more Walls, it is best to clean up what we have. We will use the AutoCAD Trim command to eliminate the unwanted portions of the Walls. If you are familiar with the AutoCAD Trim command, it works the same way with Walls as it does with other AutoCAD geometry. If you are not familiar with this command, pay close attention to the command prompts as you execute it.

The first prompt reads: Select cutting edges. At this prompt you can select one or more objects that will be used as trim points for other objects (pressing ENTER at this prompt will select all objects in the drawing as edges). All of the interior Walls will be used as cutting edges in this step, as indicated in Figure 4–14. You can do each one in a separate Trim command or do them all together. It is up to you.

28. Right-click anywhere in the drawing and choose **Basic Modify Tools > Trim** (or type **TR** and then press ENTER).

29. At the "Select cutting edges" prompt, select all of the Walls indicated in Figure 4–14 as Cutting Edges.

30. Press ENTER to complete the selection of cutting edges.

This is an important step. Before you can move on to the "Select object to trim" prompt, you must press ENTER at the "Select objects" prompt to stop selecting cutting edges. For more information on Trim, consult the online help.

31. At the "Select object to trim" prompt, and using Figure 4–15 as a guide, Trim all of the unneeded Wall segments.

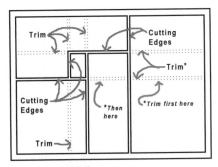

Figure 4–15 *Trim away unnecessary Wall segments*

The layout is coming along. We need to offset two more Walls to frame out a toilet room at the top of the plan. We could use the AutoCAD offset command as we did above; however, as we have pointed out, the distances must be calculated carefully since AutoCAD Offset does not give an option for the reference point. On the Wall right-click menu, there is such an option. Using the **Offset** > **Copy** function, we can offset Walls from face to face rather than center to center.

32. Select the rightmost interior vertical Wall (labeled "Wall A" in Figure 4–16), right-click and choose **Offset > Copy**.

     Move the mouse left and right and notice the red line that shifts from side to side on the Wall.

33. Click to set the red line on the left edge of the Wall (see panel I in Figure 4–16). This is the offset edge.

     Begin moving the cursor further to the left and note the dynamic dimension.

34. Type **4′-4″ [1323]** and then press ENTER. (see panel 2 in Figure 4–16). Like the AutoCAD Offset command, this offset function will repeat. Press ENTER again to complete the command.

35. Using **Offset > Copy** again, offset "Wall B" up **1′-8″ [510]** (see panel 3 in Figure 4–16). (Please note that the look of your dimensions might vary slightly from those in the figure.)

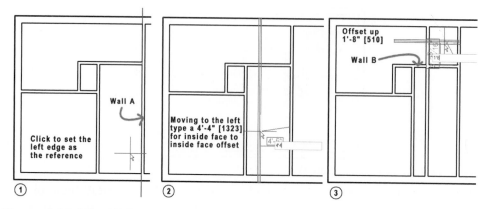

Figure 4–16 *Offset Walls for the toilet room*

36. Right-click anywhere in the drawing and choose **Basic Modify Tools > Fillet** (or type **F** and then press ENTER).

37. Using Figure 4–17 as a guide, remove the unneeded wall segments. Be sure to click on the part of the Wall that you wish to keep.

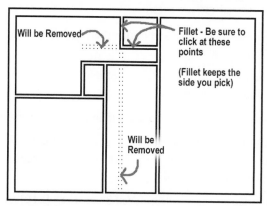

Figure 4–17 *Trim away unnecessary wall segments*

38. **Offset > Copy** two more walls to frame out the stairway in the center of the plan. Offset from the left wall of the hallway to the right **3'-1″ [942]**, and up **5″ [129]** from the horizontal wall on the left of the plan (see Figure 4–18).

   If you would rather use the AutoCAD Offset command, add 5″ [125] to each dimension to account for the thickness of the Wall. (Please note that the look of your dimensions might vary slightly from those in the figure.)

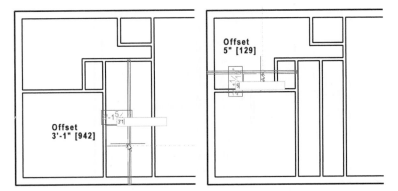

Figure 4–18 *Offset walls for the stairway*

39. Using the Extend command and Figure 4–19 as a guide, extend the required wall segments.

   If you are new to the Extend command, its prompts are nearly identical to the Trim command. Follow the command line closely as you work.

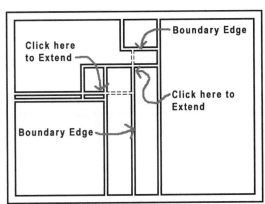

Figure 4–19 *Extend wall segments*

40. Using the Trim command, and Figure 4–20 as a guide, trim the unneeded wall segments.

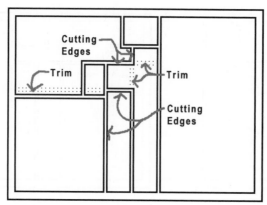

Figure 4–20 *Trim away unnecessary wall segments*

41. Select the vertical Wall bounding the right side of the stairway in the center of the plan.

42. Click the triangular shaped grip at the bottom end of the wall to make it "hot."

 NOTE The default setting will turn the grip red; choose **Format > Options** and click the Drafting Settings tab to change this color if you wish. See Chapter 1 for more information on grip shapes.

43. Click the triangular grip and drag it up (the triangular or "Lengthen" grip will keep movement constrained to the existing angle of the Wall without the need for POLAR or ORTHO). See Figure 4–21.

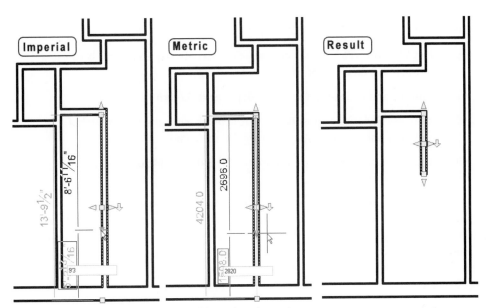

Figure 4–21 *Grip stretching the Wall bounding the stairway*

Three Dynamic Dimensions will appear: one overall, one representing the new length of the Wall and the other representing the amount added or subtracted to the Wall's original length (the delta). One of the three will be highlighted. If the one highlighted is not the delta portion, press the TAB key. Repeat until the correct one is highlighted. (Please note that the look of your dimensions might vary slightly from those in the figure.)

44. With the delta dynamic dimension active, type **9′-3″** **[2820]** and then press ENTER.

45. Deselect the Wall and save the file.

To verify the dimensions of your plan, use the Distance tool on the Drafting palette (in the Document Tool Palette Group; see the "Tool Palette Groups" heading in Chapter 1 for more information).

CREATE AND ASSIGN A WALL STYLE

We have completed the layout of existing walls on the first floor of the house. Assigning Wall styles to the walls will help us distinguish them as existing construction as the project progresses. Wall styles can be thought of as "wall types" and usually indicate the construction of the Wall and a variety of other parameters. Wall Styles will be covered in detail in Chapter 10. For now, let's simply create and assign a new Wall Style to our Walls and place them on an existing construction layer.

Continue in the *First Floor Existing.dwg* file.

46. Select all of the Walls in this drawing. (The easiest way to do this is to use a window selection or to press CTRL + A).

47. Right-click and choose **Copy Wall Style and Assign**.

48. In the Wall Style Properties dialog that appears, click the General tab if it is not already chosen, and type **Existing** for the name.

49. In the Description field, type **MasterADT Existing House** and then click OK.

You will not see any change on the screen. This very simple change merely created a new Style named "Existing" and assigned it to all of the selected Walls. If you wish to verify this, click to select any Wall and check the Style setting on the Properties palette.

CREATE A LAYER

In addition to the Style, Layers are typically used to distinguish existing construction from new. In this sequence, we will create a Layer for existing walls. ADT includes Layer Standards to simplify the creation of Layers and help to standardize naming.

50. On the Layer Properties toolbar (see Figure 1–1 in Chapter 1), click the Layer Properties Manager icon.

51. On the toolbar across the top, click the New Layer from Standard icon (see Figure 4–22).

52. In the New Layer from Standard dialog, click the small browse (...) icon next to the Status field.

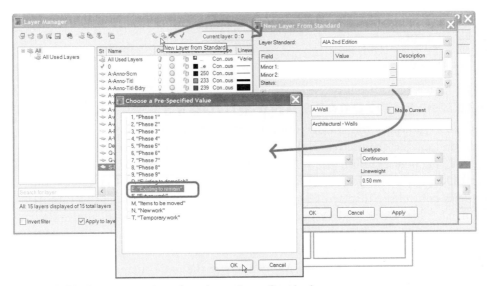

Figure 4–22 *Create a new layer based on a Layer Standard*

53. In the Choose a Pre-Specified Value dialog, choose **E, "Existing to remain" [X, "Existing"]** and then click OK.

54. Back in the New Layer from Standard dialog, from the color list, choose Select color, choose color **#64** and then click OK.

55. Change the Plot Style to **Medium** and the Lineweight to **0.35 mm** and then click OK twice.

You will now have an existing conditions wall Layer named: A-Wall-E [A-Wall-GX]. Next we will assign all of our Walls to this Layer.

CAD Manager Note: This book makes use of two out of the box Layering systems, AIA (256 Color) for Imperial files and BS1192 Descriptive (256 Color) for Metric files. Your firm may use a different Layering scheme and Standard. You can discuss this with your CAD Manager and/or CAD Support Personnel to identify what issues and standards are in place at your firm. Feel free to use whatever Layering system is in place at your firm in place of the ones suggested in this text.

APPLY A LAYER TO WALLS

56. Select all of the Walls within the drawing. (Use the same procedure as above.)

57. On the Properties palette, choose **A-Wall-E [A-Wall-GX]** from the Layer list.

    Notice that all of the Walls change color.

58. Right-click in the drawing and choose **Deselect All** to complete the operation.

59. Save the file.

WORKING WITH DOORS AND WINDOWS

Doors and Windows automatically interact with Walls when inserted to create the opening and anchor themselves to the Wall in a parametric way. (As discussed in Chapter 2, anchors establish the rules that physically link two ADT objects together.) Doors and Windows have a dedicated Elevation Display Representation that makes controlling display of Doors and Windows in sections and elevations more manageable. Doors and Windows offer a variety of routes to customization, resulting in virtually no Door or Window condition that cannot be rendered with accuracy and relative ease in ADT. Let's continue work on our floor plan by adding the required doors and windows. We will cover the basics here. For more detailed information on Door and Window advanced usage and customization, refer to Chapter 11.

ADD DOORS

For this exercise, turn off your Object Snaps (OSNAP toggle on the Application Status Bar or press F3). Object Snaps can actually hinder the proper placement of Doors and Windows by causing them to jump to the wrong Wall. Use the Location settings on the Properties palette when adding Doors and Windows. These provide greater precision than OSNAP and less possibility for error.

1. Continuing in the *First Floor Existing.dwg* file, at the bottom of the screen, click the OSNAP button to pop it up (see Figure 4–23).

   If it is already popped up, do nothing.

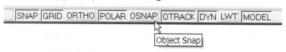

Figure 4–23 *Turn off the Object Snaps*

2. On the Design palette, click the Door tool (see Figure 4–24).

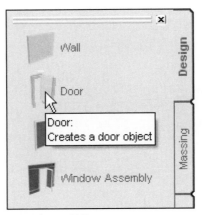

Figure 4–24 *Choose the Door tool to add Doors*

On the Properties palette, open the Style list (in the Basic Properties area under the General grouping).

Notice that there is only the Standard style currently available in this drawing. Most of the time, you will wish to use a style other than Standard for Doors and Windows. To do this, you use a tool from a different palette. Tools have the ability to reference a specific style (external to the current drawing), which will be automatically imported as the tool is executed.

Click the Doors tab to make the Doors tool palette active.

3. Click the ***Hinged – Single*** tool.

4. On the Properties palette, choose **3'-0" x 6'-8" [900 x 2200]** from the Standard sizes list.

5. In the Location grouping, choose **Unconstrained** from the Position list (see Figure 4–25).

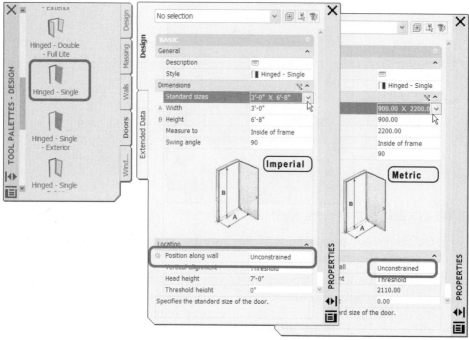

Figure 4–25 *Setting Door parameters on the Properties palette*

The following list explains the major fields and controls in the Properties palette while adding Doors.

General Properties:

> ▶ **Description**—Click to give this particular Door a detailed Description.

> ▶ **Style**—Contains a list of all Door styles within the current file. Think of Door styles as door types. (Refer to Chapter 11 for more information.)

Dimension Properties:

> ▶ **Standard Sizes**—List of standard sizes built into the Door style. Choose from this list or type in a Width and Height manually.

> ▶ **Width**—Width of the door opening. Refer to the Note below the Width and Height fields to see if this door measures to the inside or outside of the frame.

> ▶ **Height**—Height of the Door opening. Refer to the Note below the Width and Height fields to see if this door measures to the inside or outside of the frame.

> ▶ **Measure to**—Determines whether the Door size relates to the Door Leaf or the Frame size.

NOTE To change whether the dimensions measure to inside or outside of the frame, choose an option from the Measure to list.

> ▶ **Swing Angle**—Use to set the number of degrees open that the Door swing should be drawn.

Location Properties:

> ▶ **Position along Wall**—When set to Offset/Center, ADT automatically sets the position of the Door relative to Wall intersections and midpoints.

> ▶ **Automatic Offset**—When the Position along Wall is set to Offset/Center, this value determines the offset from Wall intersections.

> ▶ **Vertical Alignment**—Sets the reference point and offset within the height of the Wall to which to measure the height of the Door.

> ▶ **Head height**—Location of the Door Head as measured from the Baseline of the Wall.

> ▶ **Threshold height**—Location of the Door Threshold as measured from the Baseline (floor line) of the Wall.

6. At the "Select wall, space boundary or RETURN" command prompt, click the topmost horizontal exterior Wall.

7. Move the mouse around on screen a bit.

   Move to the left, then to the right, and then back again.

Notice how the door you are placing keeps up with the mouse movements, but at all times remains attached to a wall. Notice also the dynamic dimensions showing the Door size and location relative to the Wall (see Figure 4–26).

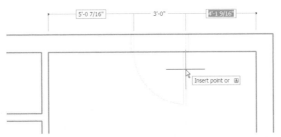

Figure 4–26 *Dynamic Dimensions interactively show Door location and size*

8. Move the mouse back to the top right side of the plan.

9. Position the Door roughly (don't try to be precise at this step) in the center of the top exterior Wall and then click the mouse (see Figure 4–27).

Notice that the Door appears in the drawing and cuts a hole in the Wall.

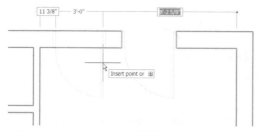

Figure 4–27 *The new Door cuts a hole in the Wall*

As you add doors, you can change settings on the fly by simply clicking back into the Properties palette, changing the desired setting(s) and then returning to the main drawing window.

> 10. On the Properties palette, change the Standard Size to **2′-6″ x 6′-8″ [760 x 2200]**.
>
> > Move the cursor to the upper left corner of the plan and position it so that the Door is being added to the topmost horizontal exterior Wall.

Dynamic Dimensions may also receive keyboard input. Use the TAB key to cycle through each editable dimension, and then type the desired value.

> > Press the TAB key. Repeat until the dimension indicating the offset to the left of the Door is highlighted.
>
> 11. Type **2′-4″ [762]** and then press ENTER (see Figure 4–28).

A door will appear offset 2′-4″ [762] from the top left exterior Wall intersection.

> 12. Press ENTER again to complete the command.

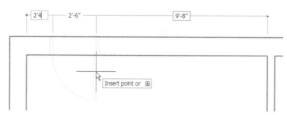

Figure 4–28 *Using Dynamic Dimensions for accurate placement*

> 13. On the Doors palette, click the Bifold – Single tool.
>
> 14. On the Properties palette, from the Standard Sizes list, choose **2′-2″ x 6′-8″ [660 x 2200]**.
>
> 15. In the drawing window, click the right vertical Wall of the closet in the middle of the plan (see Figure 4–29).

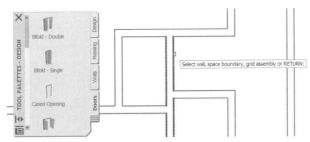

Figure 4–29 *Add a Bifold Door to the closet*

16. On the Properties palette, beneath the Location grouping, change the Position along Wall to **Offset/Center**. In the Automatic Offset field, type **4″** [**100**] in the text box (see Figure 4–30).

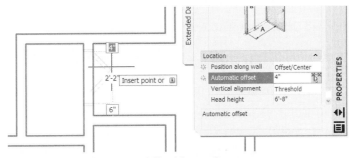

Figure 4–30 *Configure the Automatic Offset/Center feature*

17. Move the mouse around near the closet again. (If the cursor is unresponsive, click once in the drawing to shift focus away from the Properties palette.)

    Do NOT click the mouse to place the door yet.

18. Make your mouse movements slower and more deliberate.

Notice the behavior with Automatic Offset/Center active. The door automatically snaps to a set distance (4″ [100] in this case) from a nearby corner or jumps to the center of the Wall segment. (In this case, both dynamic dimensions will show 5″ [128] when it is centered.)

Notice also that slower and more deliberate movement allows you to control the direction of the swing of the door more precisely as it is being placed (see Figure 4–31).

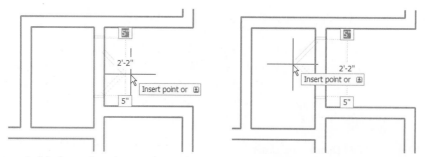

Figure 4–31 *Controlling the position and swing direction with the Automatic Offset/Center feature*

19. When you are happy with the placement of the door and its swing position, click the mouse to place the door.

20. Right-click and choose **Enter** to end the command (or press ENTER).

CHANGE DOOR SWING WITH DIRECT MANIPULATION

To manipulate or relocate a door, simply use its grips.

21. Click on the hinged Door at the top right, (the first Door placed in the previous steps above).

Notice there are several grips: two triangular shaped, one square and two arrows. Each of these grip shapes performs a different function. The arrow grips are "trigger" grips. A trigger grip will perform a designated function with a single click and then return to the previous condition with a second click. In this case, the arrow triggers are used to change the Door swing direction.

22. Hover over either one of the two arrow trigger grips.

It will turn green and a "Flip" tool tip will appear to indicate its function (see Figure 4–32).

23. Click either one of the two arrow trigger grips to flip the Door swing.

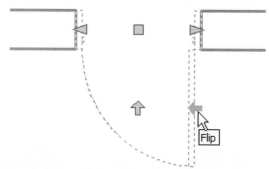

Figure 4–32 *Using trigger grips to perform an object-specific function*

CHANGE DOOR SIZE WITH DIRECT MANIPULATION

The two triangular grips allow you to interactively resize the Door object. If the Door Style contains a list of "Standard sizes" (see the definition above), then movement of these grips will snap to these Standard sizes. If there are no Standard Sizes saved in the Style, then resizing will be unconstrained. Door Styles are covered in more detail in Chapter 11.

24. Hover over one of the two triangular grips and pause for a moment.

    A tool tip will appear indicating the grip's function as well as a Dynamic Dimension indicating its current size (see Figure 4–33).

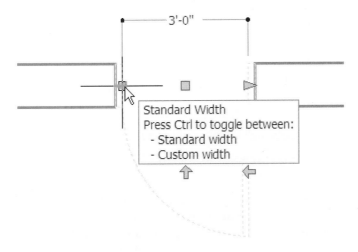

Figure 4–33 *Hover over a grip to reveal its function in a tool tip*

25. Click on one of the triangle grips and begin dragging it (see Figure 4–34).

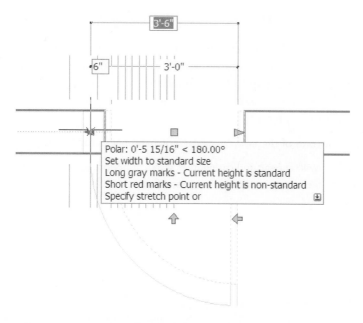

Figure 4–34 *Snap marks (gray or red) indicating standard sizes appear while grip editing*

Click on one of the gray marks to snap to a standard size.

 NOTE In some cases, some of the tick marks may appear red and a bit shorter than the gray ones. This indicates that the size has a standard width, but that the height is not standard. This cannot be seen with the Door Style that we have used here, but if you use another Style, you may encounter this behavior.

 TIP If you wish to use a non-standard size, press the **CTRL** key and then drag to the new size. You may also type a value into the Dynamic Dimensions to set the new non-standard size accurately.

MOVE A DOOR PRECISELY WITH OBJECT SNAP TRACKING

The square grip in the middle of the Door is a "location" grip. With this grip, you can change the location of the Door relative to the length, thickness or height of the Wall. (Use CTRL to cycle between modes.)

26. Right-click on the OSNAP toggle at the bottom of the screen and choose settings (see Figure 4–35).

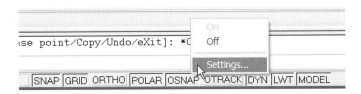

Figure 4–35 *Access the settings of OSNAP*

27. In the Drafting Settings dialog box, on the Object Snap tab, place a check mark in Object Snap On (F3), Object Snap Tracking On (F11) and Midpoint (see Figure 4–36).

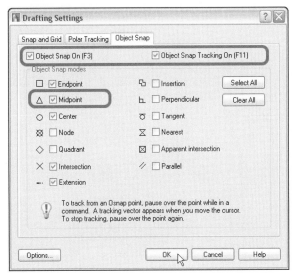

Figure 4–36 *Configure the Drafting Settings*

It is not necessary to deselect any of the other Object Snap modes that may be turned on in this dialog box; simply make certain that Midpoint is among those selected. For future reference, make note of the keyboard shortcuts F3 and F11 listed next to Object Snap and Object Snap Tracking. You can use these to quickly turn on and off these features.

28. Click OK to dismiss the Drafting Settings dialog box.

    Continue to work with the Door at the top right corner of the plan.

29. Select the square grip point in the middle of the Door to make it hot (the grip turns red when "hot").

30. Move the mouse to bottom horizontal wall of the room and hover over the Midpoint. Do NOT click yet.

A small tick mark should appear on the edge of the Wall at the midpoint. The midpoint is indicated by a triangle symbol.

31. Move the mouse all the way back straight up to the original Wall (see Figure 4–37).

A light dotted tracking vector will trace your movement and keep you lined up with the Midpoint at the bottom. If you move too far to the left or right, it will vanish, so keep it lined up.

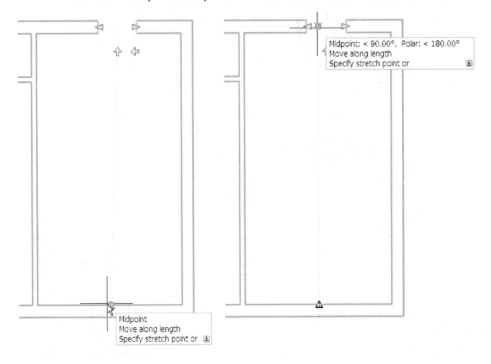

Figure 4–37 *Using Object Snap Tracking for precise alignment*

32. When your mouse is over the original Wall, click the mouse to set the new position.

 TIP Polar Tracking (**F10**) can make this process even easier, by giving additional tracking vectors in the opposite direction.

33. Right-click and choose **Deselect All**, or simply press the ESC key to deselect the Door.

ADD THE REMAINING DOORS

Using Figure 4–38 as a guide and the techniques covered above, place the remaining doors in the plan.

HINT You can select a Door, then right-click, and choose Add Selected as a way of quickly adding Doors with similar parameters.

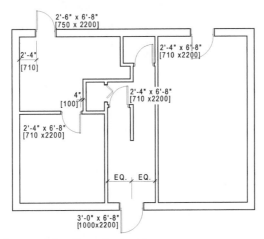

Figure 4–38 *Place the remaining Doors in the plan*

ADD WINDOWS

Working with Windows is nearly identical to working with Doors. Many of the parameters are the same and placement and manipulation of Windows works the same as with Doors.

34. On the Windows palette, click the Double Hung tool.

Refer to "Add Doors" above for a description of each of the settings on the Properties palette that are active while adding Windows, except that with Windows, "Swing Angle" becomes "Opening Percent" and "Threshold Height" becomes "Sill Height."

35. For the Width type **3'-0″** [**900**] and for the Height type **4'-8″** [**1422**].

The Standard Size will indicate that this is a Custom Size.

36. In the Location grouping, choose **Offset/Center** from the Position along Wall list.

Set the Automatic Offset value to **2'-6″** [**762**] (see Figure 4–39).

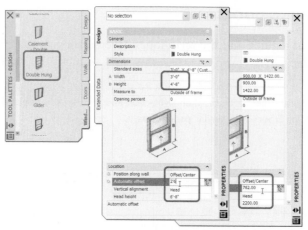

Figure 4–39 *Set parameters on the Properties palette for adding Windows*

Just as we were able to do for Doors, when adding Windows we can set the position of the Window vertically within the Wall. We will not see the effect of this setting in plan view, but you will see it later when you generate an elevation from your plan or view the model in 3D. The Vertical Alignment setting includes two options in a list used to toggle the reference point from Sill to Head and a text box for each to set the offset from these points relative to the Wall's Baseline. For this exercise, verify that the Head is chosen for Vertical Alignment and that 6'-8" [2200] appears for the offset. Using Head Height as the Vertical Alignment point measures the window height *down* from the position set for the Head, in this case 6'-8" [2200]. Using Sill would measure the window height *up* from the position set for the Sill.

37. At the "Select Wall, space boundary or RETURN" command prompt, click on the exterior wall at the bottom of the screen.

38. Position the Window in the center of the horizontal Wall in the bottom left room.

39. Using Figure 4–40 as a guide, place the remaining double-hung Windows. Add them all at the same height initially and then change to the height on the Properties palette for the two Windows at the top of the drawing. When you do this, they will seem to disappear. Please see the "Adjust Cut Plane" heading below to correct this.

40. Right-click and choose ENTER when finished.

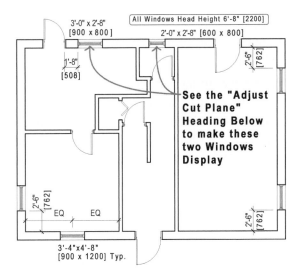

Figure 4–40 *Double-hung Window placement*

ADJUST CUT PLANE

When you add the two Windows at the top and then adjust their height, they seem to disappear. Do not repeat the steps again or undo them. Although the two windows do not appear, they are in fact still there. The problem here is a matter of the way in which the display parameters are configured in the default templates. The display parameter in question is the Cut Plane. Just as taught in traditional architectural drafting, the Cut Plane is an imaginary plane parallel to the ground plane that cuts through the entire floor plate. This plane determines the way in which each ADT object will be drawn in a Plan Display Configuration. (If you are unfamiliar with Display Configurations, please review Chapter 2). In this particular instance, the default Cut Plane height of 3'-6" [1400] is not high enough to cut through these two Windows; therefore, they are excluded from the Plan display. However, if you were to view the drawing as an elevation or in 3D display, they would appear. To fix this problem, we will simply override the Cut Plane height for the Wall in which the two problem Windows reside. This is not the only way to correct this situation, but it is effective and simple to achieve.

41. Select the top horizontal Wall, right-click and choose **Edit Object Display**.

42. On the Display Properties tab, place a check mark in the Object Override box next to Plan.

A Display Properties (Wall Override) dialog will appear.

43. Click the Cut Plan tab, place a check mark in the Override Display Configuration Cut Plane check box, and then change the Cut Plane Height value to **4′-6″** [**1400**].

44. Click OK twice to return to the drawing.

The Windows at the rear of the house should now be displayed. If you inadvertently added some extra ones, please delete them now. For a more detailed discussion of this topic, see the "Explore Display Representations" heading in Chapter 10.

45. Save the file.

TIP All other techniques covered above for Doors work the same for Windows—Automatic Offset/Center, Grip Edits, Standard Sizes, etc.

Try repeating the grip editing steps above on some of the Windows.

ADD OPENINGS

An Opening is a parametrically defined hole in a Wall. Openings do not use styles, but do use a Shape definition. All parameters of an Opening object apply directly to the Opening object. So far, when adding Doors and Windows, we have used tools on the various tool palettes. However, Doors, Windows, Openings, and Door/Window assemblies can be inserted directly into a wall from the Wall object's right-click menu. We did not use this technique for Doors and Windows because we wanted their respective tools to automatically import the required Styles. Since Openings do not use Styles, we will practice this technique now by placing a few Openings.

46. Select the left vertical Wall of the hallway and then right-click.

47. From the right-click menu, choose **Insert > Opening** (see Figure 4–41).

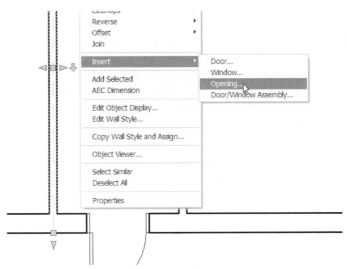

Figure 4–41 *Add Wall penetrations with the right-click Insert menu*

48. On the Properties palette, set the Shape to Rectangular, the Width to **2'-6"** [**762**], the Height to **6'-8"** [**2200**] and the Automatic Offset/Center to **6"** [**150**].

 NOTE You must type these values, as Openings do not have a Standard Sizes list.

49. Place openings in the locations shown in Figure 4–42.

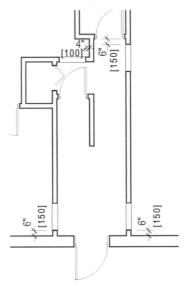

Figure 4–42 *Opening locations*

An Opening is unique in the sense that it is a "negative" object. No other object in ADT is used exclusively to represent a hole within another. In order to access the parameters of the Opening object however, it must be selectable; this is why some lines appear in the hole rather than nothing at all. The easiest way to manage this situation is with the Opening object's layer or its display properties. Opening objects are placed on the layer like all other ADT objects; "A-Wall-Open" if using the AIA layer scheme. Simply access the Layer Properties dialog box and set this layer to No Plot (click the little printer icon off). This will leave the layer visible on screen but render it invisible when printed. The other way to manage Openings is to turn them off in the current Display Set (see Chapter 2 for more information on Display Sets).

CAD Manager Note: You can make this change the default in the Layer Key Style file (if using the Layer technique) and template files (for the Display option) for the office. This will save the users from making this change manually within each drawing. Other choices would include leaving the Opening objects displayed for plotting, and change the linetype, lineweight or other display parameters to match your office graphic standards.

LOAD CUSTOM TOOL PALETTE

There is one additional opening object type that we can add to Walls: a Door/Window Assembly. These objects behave very much like Doors, Windows and Openings, but have the potential for more complex Styles and designs. Using a Door/Window Assembly, you can define opening styles containing one or more integral Doors and Windows as well as any variety of Frame and Mullion conditions. Door/Window Assembly objects are nearly identical in concept and function to Curtain Wall objects. You can learn more about Curtain Wall and Door/Window Assembly design and composition in Chapter 8. For now, we will focus on how to import and use a pre-made Style. A content file containing a custom Door/Window Assembly Style and a custom tool palette is included with the other Chapter 4 files from the Mastering Autodesk Architectural Desktop 2006 CD ROM. In this sequence, we will load the Mastering Autodesk Architectural Desktop Tool Palette and use its tools to import a custom Door/Window Assembly Style for the picture window at the front of the existing house.

 NOTE The process portrayed here specifically showcases the importing of a Door/Window Assembly Style, but can be used to import any type of ADT Style.

50. On the Navigation toolbar, click the Content Browser icon (or press CTRL + 4).

The Content Browser window will open. Refer to the "Content Browser" topic in Chapter 2 for more information.

51. In the bottom left corner of the Content Browser, click the Add Catalog icon.

52. In the Add Catalog dialog box, choose **Add an existing catalog or website** from the bottom and then click the Browse button.

53. Navigate to the *C:\MasterADT 2006\Catalog* folder, choose the *MasterADT 2006.atc* file, click Open and then click OK (see Figure 4–43).

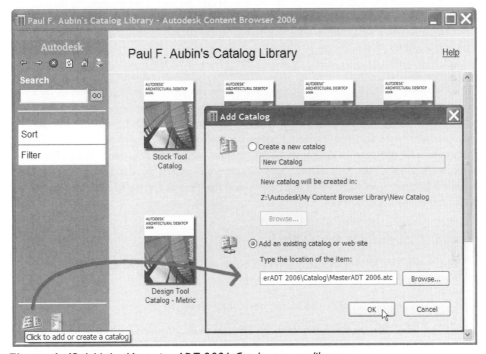

Figure 4–43 *Add the Mastering ADT 2006 Catalog to your library*

IDROP A TOOL PALETTE

54. Click on the newly loaded catalog (click on the image to open it).

Contained within the *MasterADT 2006* Tool Catalog are four tool palettes named *MADT Residential, MADT Residential - Metric, MADT Commercial* and *MADT Commercial - Metric.*

55. Click and hold down the mouse on the small eyedropper icon for the *MADT Residential [MADT Residential – Metric]* tool palette and drag it into the ADT drawing window.

A new *MADT Residential* [*MADT Residential – Metric*] tab will appear on your Tool Palettes.

56. Click on the *MADT Residential* [*MADT Residential – Metric*] tab on your Tool Palettes.

You will see three tools contained in this palette, two Wall tools and one Door/Window Assembly tool. We will use the Door/Window Assembly tool next.

 TIP You can create a new MasterADT Tool Palette Group and move the new tool palette to this group to consolidate the number of tabs on your Tool Palettes. To do this, right-click the Tool Palettes title bar and choose Customize. Create a new group and then drag the MasterADT palette to it. For more information, see the "Tool Palette Groups" heading in Chapter 1 and the online help.

CAD Manager Note: You might want to have a single central location on your system or on a network server (where all users have access) to store all catalogs and content files. This will make it easy to access and maintain the company library of content and styles. Establish a procedure for submitting new content to the library as the team creates it. It is also wise to develop a standard for locating project-specific content that may or may not become part of the firm-wide library. These project resources can then be made available to project team members via a project-specific tool palette. See the 'Setting up Standard Tools in a Project' and the 'Project Standards' topics in the online help for more information.

ADD A DOOR/WINDOW ASSEMBLY

57. On the MADT Residential [MADT Residential – Metric] tool palette, click the Existing Living Room Window tool.

58. When prompted, click the lower horizontal Wall.

    The Tool has many of the settings already preconfigured including the Style reference like other tools that we have used, as well as the basic dimensions of the assembly.

59. Verify that the length is set to **8'-0"** [**2400**], the Height to **6'-0"** [**1800**] and the Vertical alignment is set to Head with a Head height of **6'-8"** [**2200**].

60. Using the **Offset/Center** option, add a Door/Window Assembly centered in the room on the right.

    Be sure that the sill appears on the outside (see panel A in Figure 4–44).

61. Press ENTER to complete the command.

TIP If you add the Assembly with the sill to the inside, do not erase it and start over. Instead, select the Door/Window Assembly (you must click the sill to select it, do not click the nested Windows within it), right-click and choose **Wall Anchor > Flip Y.** This will reverse the direction of the Assembly relative to the Wall.

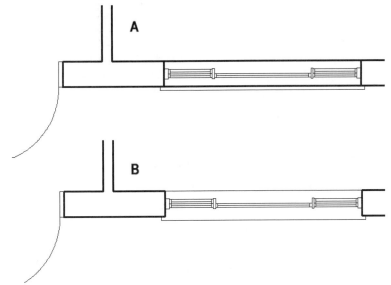

Figure 4–44 *Place a Door/Window Assembly with the sill on the outside*

Notice that a second line appears between the Assembly itself and the Assembly's Sill (see panel A in Figure 4–44). This is actually the line of the hole within the Wall below the Assembly. The graphics of the Plan display would be cleaner without this line displayed. We can make that adjustment easily by editing the display properties of the Wall. The result will look like panel B in Figure 4–44.

ADJUST WALL DISPLAY

The display characteristics for an ADT object are managed by the display system. We learned in Chapter 2 that at a global level, the display system dictates what objects display and how they display in the context of a particular type of drawing such as a Plan or Model. At the object level, Object Display (also part of the display system) controls the graphic display of a particular object and the various subcomponents contained within the object, such as Frame, Panel, Stop, and Threshold for a Door, or Components, Shrinkwrap and Cut Plane parameters for a Wall.

The Door/Window Assembly that we have just added includes an embedded Sill. This is achieved using the display parameters of the Door/Window Assembly's Sill Plan Display Representation. (The topic of Sill Plan display is covered in the "Add

Sill Extensions" heading in Chapter 11, and a related topic below, "Adjust Door Threshold Display.") We will not edit the Sills of the Assembly in this exercise; instead we will edit the display properties of the Wall to hide the portion of the Wall that occurs below the Door/Window Assembly in Plan.

62. Select the horizontal Wall at the front of the house, right-click and Choose **Edit Object Display**.

63. On the Display Properties tab, be sure that Plan is highlighted, and then click the Edit Display Properties icon to the right.

64. In the Display Properties (Drawing Default) dialog, click the Other tab and then place a check mark in the "Hide Lines Below Openings at Cut Plane" check box.

65. Click OK twice to return to the drawing (see panel B in Figure 4–44).

If you wish to have the same effect on the horizontal Wall at the top of the plan (the one we applied the Object Override to above), then you will need to select that Wall and repeat the steps. Drawing Default affects all Walls in the drawing, however, once you add an override to the display, the Drawing Default settings no longer affect it.

65. Save the file.

MODIFYING DOOR SIZE AND STYLE

Modifying Doors and Windows is easy. We can use the direct manipulation techniques already covered above, or we can also use the Properties palette. The following will show you the technique to follow when using the Properties palette.

67. Select the Single Hinged door in the top right corner of the plan (the first one we added), right-click and choose **Properties**.

This will open the Properties palette if it is not open already. On the Design tab of the Properties palette, you can change many of the basic parameters of the Door: Layer, Size, Style and Anchor point.

68. Change the Width to **6'-0"** [**1800**] and then press ENTER.

Repeat the steps in "Move a Door Precisely with Object Snap Tracking" to re-center the door in the space. (You should be able to track to the center of the Door/Window Assembly this time.)

This Door is now a bit large to remain a Single Hinged Door. However, if you keep the Door selected and look at the available styles in the Style list on the Properties palette, no double Door style is currently available. We will need to import a double Door style into this drawing. We could use the Content Browser as before;

however, it is always a good idea to first check what is available on your tool palettes. The tools on tool palettes often have built-in styles that can be applied to existing drawing objects.

69. Click the Doors tab on the Tool Palettes.

    If you do not see the Design tool palette, right-click the Tool Palettes title bar and first load the Design tool palette group. This reveals the Doors tool palette and its collection of Door tools.

70. Locate the Hinged – Double tool and right-click its icon.

71. Choose **Apply Tool Properties to Door**.

The selected Door will change to match the properties of the tool, including the application of the Hinged – Double style. Here we used a Door tool, but you can perform this process with most tool palette tools on most types of ADT objects.

    If you have already deselected the Door that we enlarged above when you right-click the tool, you will be prompted to "Select Door(s)". Select the Door in the top right corner of the plan that we enlarged and press ENTER.

  NOTE You can use this technique to convert nearly any object type into any other. For instance, a Door or Opening can be converted to a Window. A Window can be converted to a Door/Window Assembly. When we begin exploring other object types in later chapters, we will also see further examples of conversions that we can perform.

ADJUST DOOR SWING ANGLE

If you wish to use 45° Door Swings for existing Doors, perform the following steps.

72. On the Properties palette, click the Quick Select icon (the small icon with the funnel and lightning bolt).(See item A in Figure 4–45).

Quick select allows you to select objects within the drawing that have one or more properties in common.

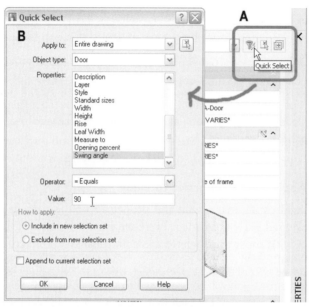

Figure 4–45 *Use the quick select command to select objects with similar properties*

73. From the Apply to list, choose **Entire Drawing**.

74. From the Object type list, choose **Door**.

75. From the Properties list, choose **Swing angle**.

76. Leave Operator set to **= Equals** and type **90** for the Value (see item B in Figure 4–45).

 NOTE In Palettes and dialog boxes such as these, the degree (°) symbol is not necessary, 90 will be correctly interpreted as 90°.

77. Click OK to apply the selection criteria to the drawing.

All of the hinged (both single and double) Doors should be selected. We could have right-clicked one of the doors here and chosen Select Similar, but since we had both Single and Double hinged door types, only the ones of the same type as the one right-clicked would have been selected. In most cases, Select Similar is much quicker and easier, however in this case we can see that Quick Select is also very useful and offers more control.

78. On the Properties palette, type **45** for the Swing angle.

The doors you selected now have 45° swings.

79. Right-click and choose **Deselect All** or simply press the ESC key.

NOTE The Bi-fold Door remains unchanged. This is because the parameter for this type of Door is actually "Open percent" rather than Swing angle.

80. Select the Bi-fold Door (in the closet at the center of the Plan).

81. On the Properties palette, change the Open percent to **75** and then press ENTER.

82. Save the file.

All of the techniques above would also work for Windows. (Windows would typically not have a "Swing Angle" but rather an "Opening percent".) Most would apply to Door/Window Assemblies and Openings as well. The overall process using both the Properties palette and the "Apply Tool Properties to" object techniques will work with all ADT objects. Try out a few variations on your own if you wish.

ADJUST DOOR THRESHOLD DISPLAY

In this sequence, we will look at adding thresholds to the exterior Doors.

83. Select the exterior double Door (top right corner).

84. On the Doors palette, right-click the Door Tool labeled Hinged – Double – Exterior and choose **Apply Tool Properties to Door.**

85. Right-click and choose **Deselect All.**

Notice that the Door now shows a Threshold on the exterior.

86. Repeat the same process on the two remaining exterior Doors. Use the Hinged – Single – Exterior this time.

This process worked fine for the Door at the top, but the Door in the center of the bottom exterior Wall (which happens to be the house's front door) has the threshold on the wrong side. Changing the swing (with the grips as explained above) would flip both the swing and the threshold. But if we want to keep the Door swinging out, and have the threshold to the outside as well, we must edit the Display Properties of the Door.

 TIP Be certain to select only one object when using the Edit Object Display command. If you select more than one object before you right-click, only the General Properties tab will be available.

87. Select only the Door in the bottom exterior Wall and right-click.

88. Choose **Edit Object Display,** and then click the Display Properties tab if not already active.

Here are listed all of the Display Representations for Doors. As stated in Chapter 2, a Display Representation is a specific set of graphics used to convey a particular type of drawing such as Plan or Elevation. The items that are in bold are active in the current view of the drawing. In this case, both "Plan" and "Threshold Plan" are active.

The subcomponents for interior and exterior thresholds are contained in the Threshold Plan Display Representation. (For more information on the display system and Display Representations, refer to Chapter 2.)

TIP In most cases you will want to work with only Display Representations that are bold. Otherwise, you will not see the effects of your edits in the current view.

Notice that next to Plan, the Display Property Source reads Drawing Default, and next to the Threshold Plan, the Display Source reads Door Style Override - Hinged - Single – Exterior. This means that for the Plan components, which include the Door Panel, Swing, Frame and so forth, the settings for this Door use the default for the entire drawing. In other words, there is a global setting for all Doors, regardless of style, and this Door style follows that default. In contrast, the Hinged - Single – Exterior Door style controls the display of components contained in the Threshold Plan Rep. The Property Source can also be overridden at the object level. (We applied this type of override above for the Wall when we adjusted the Cut Plane.) This gives us three levels of control for Display Property Source. Let's define each level of contribution in descending order.

- ▶ **Drawing Default**—Applies to all objects of a particular type (in this case all Doors) regardless of style.

- ▶ **Style**—Applies to all objects of the same style (in this case all Hinged – Double Doors). All Drawing Default settings are ignored at this level. Objects of other styles are not affected.

- ▶ **Object**—Applies only to the selected object (in this case the main entrance Door of the existing house). All Drawing Default and Style based settings are ignored at this level. No other objects in the drawing are affected in any way.

These three property sources apply hierarchically. When you apply a style-level override, for instance, the Drawing Default settings are copied to the style. You can then edit those settings at the style level, which would then apply to all Doors of that style.

Tip As a general rule of thumb, apply Display Properties first at the Drawing Default level, then at the Style level, and finally, only if necessary, at the Object level. Try to avoid object level overrides if you can.

89. Returning attention to the Object Display dialog box, place a check mark in the box in the Object Override column next to Threshold Plan.

A Display Properties (Door Override) dialog box will appear (see Figure 4–46).

The Display Properties dialog box will appear, with a list of Components available in the Threshold Plan display representation. Each Display Representation has its own list of Components; the Threshold Plan representation has six: Threshold A, Threshold B and four "Cut Plane" Threshold Components. For now we will focus only on Threshold A and B. Threshold A is the threshold for the swing side of the door, Threshold B is on the opposite side.

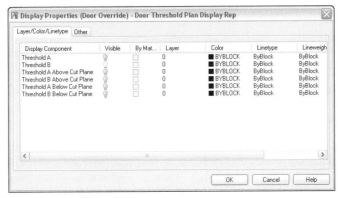

Figure 4–46 *Attaching the Object level override*

90. Click the light bulb next to Threshold A to turn it on.

91. Click the light bulb next to Threshold B to turn it off.

92. Click the Other tab to control the size and offset of the threshold.

Values A and B apply to Threshold A, while C and D apply to Threshold B.

93. Set A- Extension to **2″ [50]** and B- Depth to **4″ [100]** (see Figure 4–47).

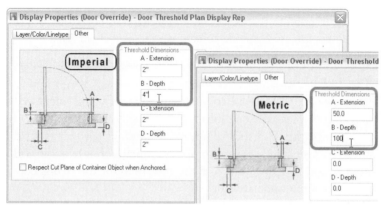

Figure 4–47 *Configure Threshold settings*

94. Click OK twice to return to the drawing.

    Notice the change.

95. Save the file.

CAD Manager Note: Object Display Properties are very powerful tools for all ADT object types. They can also be very complex. Having guidelines and procedures firmly established in your office for their proper use will yield enormous benefit. For instance, Door styles with various typical size thresholds can be included in the office library, and made into tools on all users' tool palettes, thereby preventing the need (as in this example) to attach Object level overrides.

ADDING PLUMBING FIXTURES

A vast library of drawing components has been provided with Architectural Desktop. Items such as furniture, toilets, trees, parking lot layouts, equipment, electrical fixture symbols, targets, tags, and much more have been included. Most of these items are created from an ADT object called a Multi-View Block. A Multi-View Block (MVB) is similar to an AutoCAD block; however, as implied by its name, it has the addition of display control intelligence (or "multiple viewing") built in. This allows completely separate AutoCAD blocks within the same MVB to be displayed under different viewing conditions. Items that have not been programmed as true parametric objects such as toilets, furniture, and many others mentioned above can be created as MVBs. The two major benefits in doing this are the ability to take advantage of ADT display control (Chapter 2) with custom drawn graphics and the ability to anchor design items such as furniture and fixtures to other ADT objects (also Chapter 2). AutoCAD entities such as lines, arcs and circles cannot use display control or anchors. In addition, property data and keynotes can also be applied to MVBs.

ADD PLUMBING CONTENT FROM THE CONTENT BROWSER

MVB content provided with ADT can be accessed through the Content Browser and tool palettes. (In Chapter 13, we will also use the AutoCAD DesignCenter to access AEC Content). If you use the same content frequently, you can add it to a custom tool on a tool palette as we did in the "Add a Palette" heading in Chapter 2.

1. On the Navigation toolbar, click the Content Browser icon (or press CTRL + 4).

2. Click the Design Tool Catalog - Imperial [Design Tool Catalog - Metric] to access its contents (see Figure 4–48).

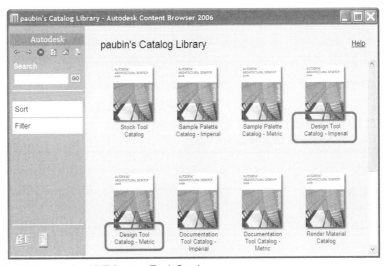

Figure 4–48 *Access the ADT Design Tool Catalogs*

This will reveal a text page describing the contents of this catalog.

As we have seen, the Content Browser has several catalogs. (See "Work with the Content Browser" in Chapter 2 for a complete list with descriptions.) *The Design Tool Catalog - Imperial* and the *Design Tool Catalog - Metric* both contain a mixture of Style definitions and AEC Content. We will browse within these catalogs to locate the plumbing fixtures that we need.

3. On the left side on the Navigation pane, scroll to and choose the desired category.

If you are using Imperial units, hover over *Mechanical* and then choose *Plumbing Fixtures*. A collection of categories will appear on the right; click the Toilet category (see the left pane of Figure 4–49).

If you are using Metric units – hover over *Bathroom* and then choose *Toilet* (see the right panel of Figure 4–49).

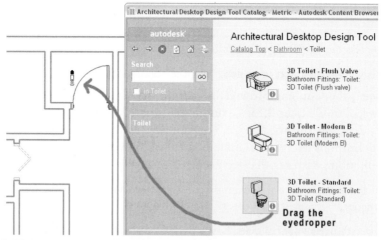

Figure 4–49 *In Content Browser, browse to the Toilet Fixtures location*

A collection of toilet symbols will appear.

4. Click and hold down over the eyedropper for the toilet named **Tank 1** [**3D Toilet – Standard**] (wait for it to fill up) and then drag it to the drawing window and release (see Figure 4–50).

5. Click a point to place the toilet in the drawing.

The exact position is not important, because we will move it precisely in the next step.

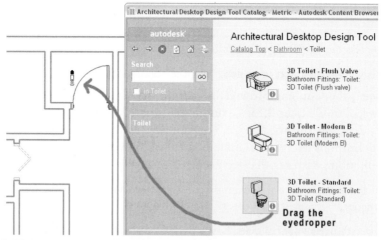

Figure 4–50 *Drag the eyedropper icon into the drawing to insert the symbol*

6. Press ENTER to complete the insertion.

Notice that although the icon in the Content Browser was in 3D, once dragged into the current drawing, the symbol came in "flat – 2D." This is typical behavior of a Multi-View Block.

The MVB contains one or more view blocks used to portray the object under different viewing conditions. In the Content Browser, you will often see a 3D icon used; however, it is automatically adjusted to show a simple 2D symbol when viewed from top (Plan) view; appropriate to a floor plan in this case. As a matter of convention, 3D icons typically convey symbols that are multi-view, while a 2D icon typically indicates a 2D only symbol.

7. On the left side of the Content Browser, click the Back icon. (Content Browser should still be open on your Windows Task Bar; simply click it to make it active again.)

 NOTE You can also use any common Web browser technique to go back as well.

8. Click the Lavatory [Basin] link.

9. Locate the symbol called **Wall I [Oval]** and drag it (also using the eyedropper icon) into the drawing and release.

 NOTE In the Imperial catalog, Wall I is on screen 3, so you will need click Next a few times to locate it.

10. Pick your insertion point and then press ENTER to complete the insertion.

11. Move (and Rotate if necessary) the toilet and the Lavatory [Basin] to a proper location in the small half bath at the top of the plan, as shown in Figure 4–51.

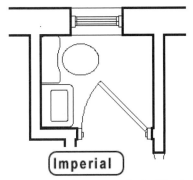 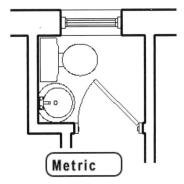

Figure 4–51 *Adding fixtures to the half bath*

VIEWERS

Being able to visualize a design change quickly from many angles and under a variety of display settings is critical to the design process. Throughout ADT a series of viewers provide this functionality. We will consider the Object Viewer here. With the Object Viewer, you can quickly study any selection of objects from all angles, in 2D or 3D, in a separate floating window. In addition, the viewers provide full access to display configurations. Access the Object Viewer from the right-click menu.

USE THE OBJECT VIEWER

1. Using a crossing selection window, select all of the objects in and around the small half bath (see Figure 4–52).

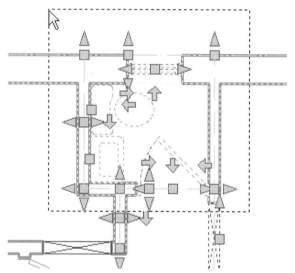

Figure 4–52 *Select the half bath and surrounding walls*

TIP To make a crossing selection, click below and to the right of the room, move the cursor up and to the left, click again outside and above the room. More information on object selection can be found in the AutoCAD online help.

2. Right-click anywhere in the drawing window and choose **Object Viewer**.

The Object Viewer is a separate resizable window used for viewing portions of your model at any angle. Only the current selection will appear within the Object Viewer, making it easy to isolate portions of the model for quick study. At the top of the object viewer is a collection of icons and menus.

3. From the Display Configuration list, choose Medium Detail.

4. From the View Control list (located at the top right), choose Back (see Figure 4–53).

Figure 4–53 *Change the viewing angle to Back and display configuration to Medium Detail*

5. If you have a wheel mouse, roll the wheel up a click or two to zoom in on the image. If you don't have a wheel mouse, choose one of the Zoom icons and zoom in on the bathroom.

 TIP When you use your mouse wheel in the Viewer, be sure to click in the Viewer window first before rolling the wheel or the focus will remain on the View Direction menu and change the view direction rather than zoom.

Notice the height of the window sill. Also notice the diagonal swings for the doors and the direction arrows for the double-hung windows. These are features rendered by the display system (see the "View Direction Dependent Configurations" heading in Chapter 2 for more information).

6. Choose the Orbit icon (looks like the planet Saturn).

7. Drag within the viewer to orbit the model dynamically into 3D view.

Drag slowly and watch the toilet and sink as you drag. They will appear as 2D elevation symbols until you release the mouse. This experiment showcases the qualities of the Multi-View Blocks. With "Back" view active, a 2D elevation view block is displayed. When you orbit to a 3D isometric view, the MVB dynamically swaps out the 2D symbols with 3D models.

Experiment with dragging inside and outside the green circle. Each orbits differently. If the Orbit shifts the image off screen, switch to Pan (If you have a wheel mouse, simply drag with your wheel held down) to bring it back into view, and then return to Orbit. If you don't have a wheel mouse, hold down the SHIFT key and drag in the Object Viewer; this will temporarily pan. Hold down the CTRL key and drag to zoom temporarily. For more information, search for "Object Viewer" and "3D Orbit" in the online help.

8. Click the Flat Shaded icon (see Figure 4–54).

Notice that the lavatory [basin] is sitting on the floor. Many of these symbols have an insertion point that places them this way. The practical application of such an insertion point is that it allows this content item to be mounted at different heights. We will explore how in the next section.

9. Close the Object Viewer to return to the main drawing window.

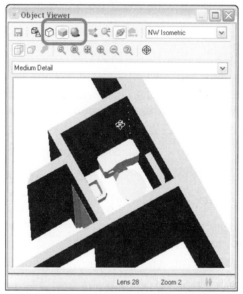

Figure 4–54 *Orbiting and shading the model in 3D within the Object Viewer*

Spend a little time trying each of the controls in the Object Viewer, such as Hidden, Perspective, and Set View. Refer to the online help for complete descriptions of each tool. When you are finished, reset the drawing to Top view and 2D Wireframe if you changed it (**View > 3D Views > Top** and **View > Shade > 2D Wireframe** menu).

USING EDIT IN VIEW

As useful as the Object Viewer is to quickly study an isolated selection of drawing objects, it does not allow any edits to be made. As its name implies, objects may only be "viewed" in the Object Viewer. If you wish to study an isolated selection of objects and edit them in place, use the Edit in View commands.

1. Right-click in the drawing and from the context menu choose **Isolate Objects > Edit In Section** (see Figure 4–55).

Figure 4–55 *Choose Edit in Section from the Isolate Objects menu*

2. At the "Select Objects" prompt, click beneath the door to the half-bath and then move the mouse up and to the left crossing through the left vertical wall and the topmost horizontal wall (see leftmost panel of Figure 4–56).

3. Click again and then press ENTER to complete selection.

 NOTE You can optionally select objects first, before right-clicking and choosing the Edit in Section command.

4. At the "Specify first point of section line or ENTER to change UCS" prompt, click beneath the door move the mouse straight up (using POLAR or ORTHO) and click again outside the window and then press ENTER.

5. At the "Specify section extents" prompt, move the mouse to the left past the left vertical wall and then click (see middle panel of Figure 4–56).

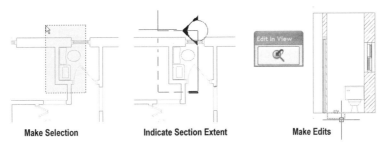

Make Selection   **Indicate Section Extent**   **Make Edits**

Figure 4–56 *Following Edit in Section prompts, designate the portion of the model to edit*

The drawing will enter the Edit In View mode. The selected objects will be isolated and the drawing will automatically zoom to the selected region (see rightmost panel of Figure 4–56). As you have seen, the Edit In View commands are located on the Isolate Objects menu. Isolate objects allows you to temporarily hide a selection of objects in the drawing thereby isolating other objects and making them easier to view and edit. When one of the Isolate commands is used, you can continue editing the visible objects normally. In this case, we can now see that the sink is not mounted at the proper height and we can edit it directly in this view.

There are two ways that we can adjust the height of the sink. We can simply move it in this view, or we can edit the Multi-View Block's insertion offsets. The first method seems obvious and is quick and easy, however, it can cause undesirable results (that are not obvious) in the plan view. As we have said already, a Multi-View Block is an ADT object that contains one or more AutoCAD Blocks within it. It is basically a "smart" container object that will only display one of its view blocks in any given viewing angle. Try this experiment: select the toilet MVB, right click and choose Object Viewer. Slowly switch the view direction through several of the options: Top, Left, Right, Front, etc. Notice that there are at least four separate blocks swapped in and out in this MVB. Next, while still in the Object Viewer, switch to Top, Left or Front view, and then slowly drag the mouse in the green circle while holding down the button. Notice that the view block used for these views is 2D. Release the mouse. Notice that the 3D block is immediately swapped in when this is done. Continue to experiment until you have a good feel for the MVB behavior.

6. Click the sink MVB, and move it up to **3'-0″ [900]**. (You can use the Move command or the Properties palette).

7. Use the Object Viewer (or the commands on the **View > 3D Views** menu) and view the half bath from all directions.

So far, everything appears fine. However, if you use the Object Viewer starting from a Top view and slowly rotate the view into 3D, you can see the problem. The sink will show as 2D in plan, but will appear to "float" above the plane of the rest of the floor plan (see Figure 4–57). This situation can prove undesirable as the project progresses. Problems can include issues with object snaps and snapping to the wrong Z height, and problems when you export the drawing to a 2D background for consultants. For these reasons, it is recommended that you use Insertion Offsets rather than the simple move technique. This process is covered next.

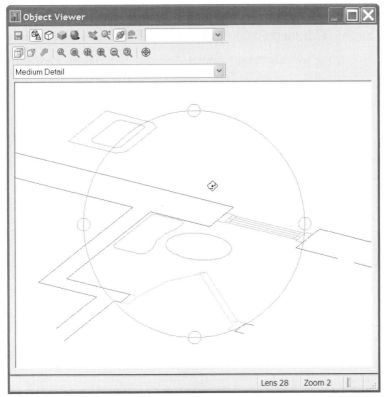

Figure 4–57 *Orbit from Top to a 3D view without releasing the mouse to see the 2D symbol "float"*

8. Move the sink back to Z = 0. You can do this on the Properties palette by setting the Location parameter to **0**.

9. On the Edit In View toolbar, click the Exit Edit In View icon.

Notice that once you exit the Edit In View mode, all of the hidden objects return to view and the drawing's viewpoint before the Edit In View is restored.

ADJUST THE LAVATORY [BASIN] HEIGHT

As we noted in the last sequence, we have two options on how to change the "Z" location (height) of the lavatory [basin]. We can move the entire symbol, or we can offset some of its internal view blocks. Moving the entire symbol is quicker, but not completely correct. Display Control allows ADT objects to have both a 2D (flat) Plan display component and a 3D Model component. If we move the Multi-View Block up in 3D, we move all of its views, including its Plan view. This would mean that the 2D symbol used to represent the Plan would be floating above the Plan. If instead, we move (change the offsets) just the appropriate view blocks within the MVB, the Plan will stay "flat" while the 3D and elevation views display at the correct height. This method takes a bit more effort, but is more accurate.

10. Select only the lavatory [basin] symbol, right-click and choose **Properties**.

11. Click the Insertion Offsets icon in the Advanced Properties area of the Properties palette.

This worksheet lists each of the view blocks contained within the MVB. The names of the view blocks are the same except for the suffix at the end. The one that ends in "_P" is for Plan, "_M" for Model, "_F" for Front and so on. Using this interface, we can shift any of the view blocks in any direction: X, Y or Z.

12. In the Z offset column, type **3'-0″ [900]** for each View Block except the one ending in "_P" (the "_P" block is indicated in Figure 4–58 by a small dot).

13. Click OK to dismiss the worksheet.

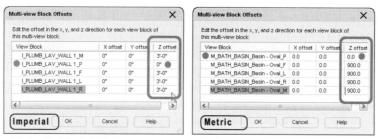

● Leave the "Plan" Block set to 0

Figure 4–58 *Using Insertion offsets to move just the 3D and Elevation blocks along the Z axis to the correct height*

14. Reselect the same objects as before, and return to the Object Viewer to view the change.

Notice that this time, when you use the Object Viewer and orbit from a Top view to a 3D view that the sink remains "flat" and on the floor. Even though a bit more effort to achieve, this is the recommended approach to dealing with mounting heights of Multi-View Block symbols.

CAD Manager Note: Please note that the insertion offsets are applied at the object level. You cannot make these changes at the style level for the Multi-View Block. However, if you always wish to insert a particular symbol with the same offsets, there are two options. In the first, open the content item within the library and edit the actual view blocks contained within. You can move the graphics up relative to their respective insertion points. In this way, even though the insertion point remains at Z=0, the block itself would appear at the correct height. The problem with this approach is that your newly edited MVB will work only at one height. If you have no intention of inserting the item at a different height, then this approach is acceptable. However, in the example of the Lavatory [Basin] showcased here, some design scenarios might require mounting it at a height other than 3'-0″

[900]. In that case, if you use this technique, you would need to make, potentially, several copies of the same piece of Content, one for each height. In the second approach, leave the MVB Definition unchanged (with all of its view blocks inserted at Z=0), and instead, create a custom tool (on a tool palette) referencing this Content item, and adjust the Insertion Offsets of this tool. The process to create such a tool is a bit more advanced, but the advantage is greater flexibility with a single piece of Content. For more information, search for 'Creating a Multi-View Block Tool' in the online help. Which ever approach you choose, you should document your choice clearly so that team members will know how to use the content items properly.

You can create your own Multi-View Blocks to meet specialized needs. Refer to Chapter 11 for a complete tutorial on the process.

CREATING WALL PLAN MODIFIERS

The first floor existing conditions plan is nearly finished. We still need to add the fireplace in the living room. We will use wall Plan Modifiers to create this type of condition. A wall Plan Modifier is a variation in the surface condition of the wall. Wall Plan Modifiers can represent any protrusion or indentation in the surface of the wall, such as pilasters, piers, column enclosures, or niches.

CONFIGURE OSNAP SETTINGS

1. At the bottom of the screen, right-click on the OSNAP button and choose **Settings**.

2. In the Drafting Settings dialog box, click Clear All.

3. Put a check mark in Endpoint and Midpoint.

4. Put a check mark in Object Snap On (F3) and Object Snap Tracking On (F11), if they are not already on, and then click OK.

ADD A WALL MODIFIER

5. Select the right exterior vertical Wall and right-click.

6. Choose **Plan Modifiers > Add**.

7. At the "Select start point" command line prompt, place the cursor over the lower right outside corner of the building and hover for a second or two.

 Caution: Do Not click yet.

A small tick mark will appear at the corner.

8. Using POLAR or ORTHO, slowly move the cursor up 90° (see Figure 4–59).

This will activate OSNAP Tracking.

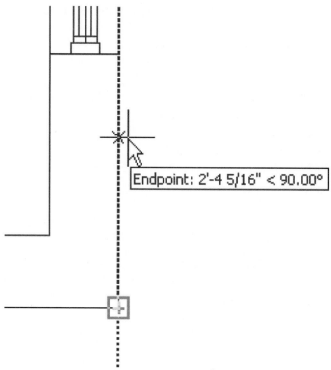

Figure 4–59 *Tracking the first point of the wall modifier.*

Note the small tick mark at the corner point. This is referred to as an "acquired" point. By thus acquiring a point, we are able to reference it with a dimension at the command line. For more information on Object Snap Tracking, refer to the online help.

9. Type **7'-9″** [**2370**] at the command line and press ENTER.

This sets the first point of the Wall Modifier.

10. At the "Select end point" command line prompt, continue tracking up at 90°, type **6'-2″** [**1880**] and then press ENTER.

NOTE It will appear as though nothing has happened with the second dimension; however, if the command line prompts "Select the side to draw the modifier:" then everything is working correctly. Always remember to read the command line! Review the "Prerequisite AutoCAD Skills" heading in Chapter 1 if you have not done so already. If you have Dynamic Input enabled (the DYN toggle on the status bar is pressed) then the prompts will appear on screen as well.

11. At the "Select the side to draw the modifier" prompt, move the cursor outside the plan (to the right) and click anywhere.

    This indicates whether the modifier is a "protrusion" or an "indentation" of the Wall's surface.

12. At the "Enter wall modifier depth" prompt, type **10″ [250]** and press ENTER.

    The Add Wall Modifier dialog box will appear.

The following options are available in the Add Wall Modifier dialog box:

▶ **Modifier Style**—Choose from an existing list of styles. Modifier styles determine the shape of the modifier in plan. The default Standard Style is rectangular in shape; custom shapes can be described by drawing polylines (see below).

▶ **Wall Component**—Assigns the modifier to a specific component of the Wall. Wall components are determined by the Wall style. The Walls in this file have only a single component called "unnamed." (Refer to Chapter 10 for more information on multi-component Wall styles.)

▶ **Offset Opposite Face**—Shapes the opposite side of the Wall component to the same shape as the side with the modifier. This is useful for showing Gyp. Board wrapping around a column as shown in Figure 4–62.

▶ **Start and End Elevation Offsets**—Use these to make a modifier that is not the full height of the Wall. Useful for niches and buttress type protrusions. We will make use of this below to shape the Hearth and Mantel.

13. Accept all the defaults by clicking OK (see Figure 4–60).

This creates a modifier representing the exterior shape of the chimney.

Figure 4–60 *The completed wall modifier for the protrusion of the chimney on the exterior*

IMPORT OBJECTS WITH THE CLIPBOARD

1. Open the file named *Fireplace.dwg* [*Fireplace-Metric.dwg*].

This file is among the files installed with the Chapter 4 sample files. You will find it in the *C:\MasterADT 2006\Chapter04* folder.

2. Select both polylines on screen and right-click.

3. Choose **Clipboard > Cut** from the right-click menu.

4. Hold down the CTRL key and press the TAB key to switch back to the *First Floor Existing* file.

TIP Use this CTRL + TAB technique to quickly cycle between open drawing files.

5. Right-click and choose **Clipboard > Paste to Original Coordinates**.

CAD Manager Note: If you have shunned the use of clipboard commands in AutoCAD due to past functionality, please note that the clipboard commands no longer add anonymous blocks by default to the drawing when pasting (unless you deliberately choose Paste as Block). Therefore, please do not hesitate to use this excellent functionality, particularly the "Paste to Original Coordinates" feature showcased here. By contrast, you will typically want to avoid "Paste as Block."

CONVERT POLYLINES TO WALL MODIFIERS

1. Again select the right exterior vertical Wall and right-click.

2. Choose **Plan Modifiers > Convert Polyline to Wall Modifier.**

3. Click on the hearth Polyline shape (see Figure 4–61).

Figure 4–61 *Creating a hearth from a polyline*

4. At the "Erase layout geometry" command line prompt, right-click and choose **Yes** (or type **Y** and press ENTER).

5. In the New Wall Modifier Style Name box, type **Hearth** and then click OK.

6. In the Add Wall Modifier dialog box, verify that the Start Elevation Offset is set to **0″** and choose **Wall Baseline** in the "from" list.

7. Set the End Elevation Offset to **2″** [**50**] and choose **Wall Baseline** in the "from" list (see Figure 4–62).

The Wall Baseline is equivalent to the floor line. Therefore we are making this modifier 2″ [50] thick and setting it at the floor.

8. Click OK.

Figure 4–62 *Setting the vertical limits of the Wall Modifier 2" [50] from the floor*

One obvious effect of this new modifier in plan view is the change in color of the hearth component. We will study the total effect in the Object Viewer after the next sequence.

9. Select the same Wall once more and right-click.

10. Choose Plan **Modifiers > Convert Polyline to Wall Modifier** again.

11. Click the firebox Polyline shape this time.

12. Again, right-click and choose **Yes** at the "erase the layout geometry" prompt.

13. Type **Firebox** for the New Wall Modifier Style Name and then click OK.

14. In the Add Wall Modifier dialog box, change the Start Elevation Offset to **2″** [**50**] and choose **Wall Baseline** in the "from" list.

    This will start the firebox just above the hearth.

15. Set the End Elevation Offset is set to **3′-9″** [**1140**] and choose **Wall Baseline** in the "from" list.

16. Click OK (see Figure 4–63).

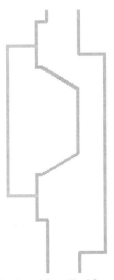

Figure 4–63 *The results of adding both polyline Modifiers*

EDIT IN PLACE FOR PLAN MODIFIERS

It is perhaps now obvious that the chimney is not centered on the firebox and hearth. Although it is possible that this could be the true configuration, in this case, this represents an error in our original numbers used to create the first Plan Modifier. This is easy to rectify with the Edit in Place feature.

1. Select the same Wall again, right-click and choose **Plan Modifiers > Edit In Place**.

 NOTE If a message stating: "The Wall has a modifier that is not drawn to size. In order to edit it, it must be converted" appears, simply click Yes.

Several grips will appear around the Wall modifier.

2. Click the cyan triangle grip in the center of the modifier and begin dragging it up (see Figure 4–64).

Figure 4–64 *Move a Wall modifier while in edit in place*

    3. Snap it to the Midpoint of the firebox (or the hearth, either one will suffice).

 NOTE The triangular grips will always move in an angle constrained to the original. This is why you can snap to either Midpoint.

The other grips around the modifier can be used to reshape the chimney vertex by vertex if you like. Notice that they are magenta while the one that we just edited was cyan. When you edit shapes in place like this, cyan indicates an edit unique to the selected object while magenta indicates that the edit will be applied to the style. In this case the style would be the rectangular shaped Plan Modifier style. If you edited its shape and then added another Modifier using this same Modifier Style somewhere else, it would match the same shape.

    4. Click the rectangular grip at the midpoint of the right edge of the chimney modifier.

    5. Drag it to the right **5″ [130]**.

    6. On the In-Place Edit toolbar, click the Save All Changes icon (see Figure 4–65).

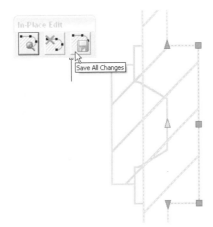

Figure 4–65 *Exit the Edit in Place mode and save all changes*

    7. Select this Wall and bring it into the Object Viewer (see above).

       Have a look at it in 3D.

The fireplace could use a mantel and possibly some more articulation on the chimney. However, because there will be no new work done in the living room of this project and therefore no sections or elevations of the fireplace, that extra level of detail is unnecessary for this tutorial. What we have created works well for plan. If you wish to try it anyway for the practice, feel free. Use the add wall modifier command as in the first sequence, with a 3'-9" [1140] Start Offset, and perhaps a 4'-0" [1220] End Offset, both measured from the Wall Baseline, to make the mantel. Experiment with other modifiers by drawing more polylines to make it fancier.

Another point to note as we move further into ADT functionality: the process that was used here to create a fireplace is one of several possible approaches. Since there is no dedicated parametric ADT fireplace object, you can model the fireplace with whatever technique you see fit. Other options include modeling it with Mass Element objects, creating a Multi-View Block or using Wall Body Modifiers. Many of these objects will be covered in the chapters that follow. If you wish to apply materials to the fireplace created here (like bricks), then you would need to build it differently. In this example, the fireplace is integral to the Wall. Therefore, without some modification, we could not assign a brick material to it without applying that material to the entire Wall. Materials will be covered in more detail in several upcoming chapters.

    8. Save the file.

CAD Manager Note: To best take advantage of wall Plan Modifiers, create a 'graphics standards' section in your office CAD manual. Show a few project examples of the best graphic conditions in which to take advantage of wall Plan Modifiers. Figure 4–66 provides an example.

…

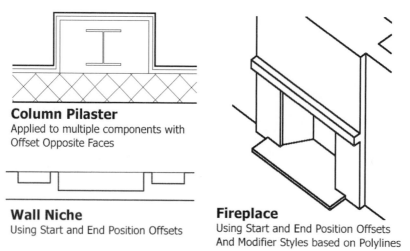

Column Pilaster
Applied to multiple components with
Offset Opposite Faces

Wall Niche
Using Start and End Position Offsets

Fireplace
Using Start and End Position Offsets
And Modifier Styles based on Polylines

Figure 4–66 *Examples of Wall modifiers*

FINISHING TOUCHES

The last item needed in our first floor existing conditions file is the stairs. (Stairs are covered in detail in Chapter 7.) For now, we will copy and paste some pre-built stairs in the file to finish it. Later, when we have explored Stairs in depth, we will revisit the Stairs in this file and build them from scratch.

IMPORT STAIRS

1. Open the file called *First Floor Stairs.dwg* [*First Floor Stairs - Metric.dwg*].

This file is among the files installed with the Chapter 4 sample files. You will find it in the *C:\MasterADT 2006\Chapter04* folder.

2. Press CTRL + A to select all objects.

3. Right-click and choose **Clipboard > Cut**.

4. Close the file without saving.

5. In the *First Floor Existing* file, right-click and choose **Clipboard > Paste to Original Coordinates**.

This will add the Stairs to the floor plan. Congratulations, the first floor existing conditions file is finished (see Figure 4–67).

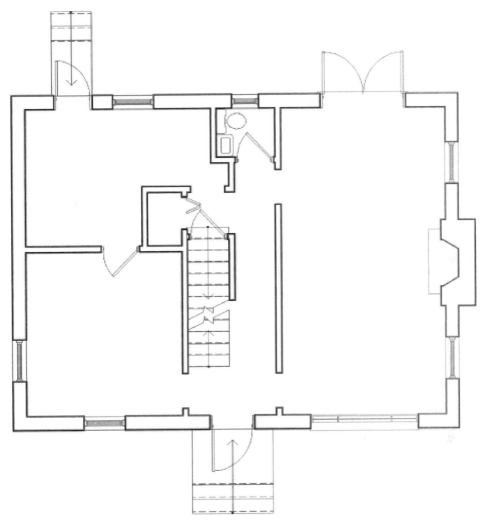

Figure 4–67 *The final first floor existing conditions*

6. Save and Close the file.

ADDITIONAL EXERCISES

Additional exercises have been provided in Appendix A. In Appendix A, you will find exercises for the Basement and Second Floors of the Existing Conditions as well as an exercise for completing a commercial Floor Plan. It is recommended that you complete these exercises for practice with techniques covered in this chapter before continuing to the next chapter. Completed projects for each of the exercises have been provided in the *Chapter06/Complete* folder.

SUMMARY

When adding objects in ADT, remember that the Properties palette remains open and interactive as you work.

Walls can be added one segment at a time (or converted from sketch lines as shown in Chapter 1).

Doors, Windows, Openings and Door/Window Assemblies automatically "cut" a hole in, and remain attached to, the receiving Wall.

Add Walls, Doors and Windows quickly, and modify their properties to add detail later.

Use normal AutoCAD commands to erase, move, copy, trim, extend, fillet or offset Walls.

Wall, Door and Window styles give access to a variety of Wall, Door and Window types.

Ortho close and Close offer convenient ways to close Wall geometry while drawing.

Direct manipulation of objects via their grips makes shifting Door and Window placement and orientation simple.

Through the Content Browser, you can access the vast ADT content library.

Multi-View Blocks are like AutoCAD blocks with display control added for different views in elevations or 3D.

Use Object Viewer to quickly study design changes of selected objects in 2D or 3D.

Wall Plan Modifiers offer a simple way to articulate the surface of Walls.

CHAPTER 5

Setting Up the Building Model

INTRODUCTION

Autodesk Architectural Desktop can be used successfully in all types of architectural projects and within all phases. This is due in part to the versatility of the architectural tools provided and in part simply to ADT's being built on the AutoCAD foundation. This book pays most attention to the design development and construction documentation phases. The tutorial exercises in this book will explore two building types concurrently, starting at different points in the project cycle, to give some sense of the variety of ways you can approach the design process. Please don't be limited by the techniques covered here. The aim of the tutorials is to get you thinking in the right direction. Exploration and play time are highly encouraged.

Programmatic and preliminary design information might be received by the project team in a variety of forms, such as hand-drawn sketches, SketchUp Files, AutoCAD files, or other CAD files. If data is hand drawn but dimensionally accurate, it can be scanned into any popular graphics file format (such as TIF) and attached to an ADT file using the built-in support for raster images. Once you have the scan in AutoCAD, scale it to the proper size and trace over it with ADT objects. If preliminary design data includes field notes or non–dimensionally accurate sketches, transpose the information into ADT using objects and common AutoCAD drafting techniques (as shown in Chapter 4).

Getting started is sometimes the most difficult part. If SketchUp was used to generate preliminary designs, these models can be imported into ADT with the SketchUp plugin. Visit www.sketchup.com for more information. Regardless of the source of preliminary design data, gather all project data together in ADT early in the project cycle and consider developing a digital "cartoon set." A cartoon set will allow you to quickly assess the quantity and composition of each of the drawings and files required in the final document package. Remember, like everything else in ADT, the cartoon set will evolve and become more refined as the project progresses. The goal is to simply build a rough road map and gain a jump start on production.

OBJECTIVES

In this chapter, we will explore the Drawing Management features of ADT 2006 to create several files that will provide the basis for the projects explored throughout the rest of the book. One of the projects will be a commercial office building, while the other is a residential addition to a single family home. The main goal of this chapter is to help you understand the procedure used to set up a project using the Drawing Management tools in ADT. The techniques covered here can be used on all types and sizes of projects. The following list summarizes the goals of this chapter:

- Set up preliminary files for ongoing tutorial projects.
- Build comfort with Drawing Management.
- Set up all preliminary Model and Sheet files.
- Set up callouts, elevations, and sections.
- Work with Sheets and Sheet Sets.
- Print a digital cartoon set.

BUILDING A DIGITAL CARTOON SET

When the time comes in a project cycle to begin thinking about how many sheets of drawings will be required and what those sheets will contain, it is time to build a digital cartoon set. Just like the traditional cartoon set, the digital version will help make good decisions about project documentation requirements and the impact on budget and personnel considerations. One extra advantage of the digital cartoon set is that it will evolve into the actual CAD files of the project, which means that the layout of a cartoon set is actually the layout of the real building model and document set! Don't be concerned with the finality that this seems to imply. The documents remain completely flexible and editable, making this approach consistent with the goal of progressive refinement as defined in Chapter 2.

Caution: Please do not skip this step when setting up your own projects. Establishing the Building Information Model structure and the digital cartoon set at the beginning of a project is critical to maximizing the potential of the ADT tools and methods with that project and will go a long way toward ensuring success.

In this chapter, we will set up the project structure for two different types of projects. Both project structures involve a heavy dose of external references (XREFs). However, using the Drawing Management system in ADT 2006, the use of XREFs is made easy and nearly transparent. If you are uncomfortable with XREFs, do not shy away from this task—complete the exercises in this chapter anyway. You will find that most of the work associated with the attachment and maintenance of the required XREFs is handled automatically and intelligently by the software. If you want to learn more about XREFs outside the context of ADT Drawing Management, consult an AutoCAD resource or the online help. File naming strategies and organization used herein are based on accepted industry standard practices. Although practices vary from company to company, (and region to region) the recommendations made by the *United States National CAD Standard – Second Edition* (NCS) are the most prolific (in the US). Therefore, these guidelines will be suggested and utilized throughout this chapter and the rest of the book. More specific information can be found in that publication. Every attempt has been made to follow NCS recommendations wherever possible and appropriate. We will see however, that certain modifications to NCS naming recommendations are necessary to accommodate the specific needs of building information modeling when working with the ADT drawing management system. It is hoped that the intent is still discernable if not directly applied.

While NCS naming is utilized throughout the tutorials in this book, specific best practice guidelines and recommendations will also be made as appropriate (for example, see the "File Naming Guidelines" sidebar later in this chapter). Therefore, if your firm has specific naming guidelines in place that differ from the US National CAD Standard, feel free to utilize your firm's naming scheme for files created in this text instead. The exact file name that you choose should not negatively impact the intent or function of the file assuming that the naming scheme used is logical and understandable and that naming guidelines mentioned herein have been considered.

PART I – SETTING UP A COMMERCIAL PROJECT

The first project is a 30,000 SF [2,800 SM] commercial office building. The project is mostly core and shell with some build out occurring on one of the tenant floors. The project will run through construction documents and will include plans, sections, and elevations. The tutorials that follow walk through the setup of the commercial project in several stages: setting up the project, creating Model files (Constructs and Elements), configuring plans, elevations, and sections (Views), and then generating sheet files and printing the cartoon set (Sheets). The completed files for the project are installed with the files from the CD in the *C:\MasterADT 2006\Chapter05\Complete* folder on your hard drive. There you find the completed ADT project files and a DWF (Design Web Format) file of the complete set of printed sheets as they appear at the end of the chapter.

INSTALL THE CD FILES AND CREATE A NEW PROJECT

In this first exercise, we use the Project Browser, a dialog box used to navigate through and manage all of your Architectural Desktop projects, to set up a new project.

1. If you have not already done so, install the dataset files located on the Mastering Autodesk Architectural Desktop 2006 CD ROM.

   Refer to "Files Included on the CD ROM" in the Preface for information on installing the sample files included on the CD.

2. Launch Autodesk Architectural Desktop 2006 from the desktop icon created in Chapter 3.

If you did not create a custom icon, you might want to review "Create a New Profile" and "Create a Custom ADT Desktop Shortcut" in Chapter 3. Creating the custom desktop icon is not essential; however, it makes loading the custom profile easier.

3. On the Navigation toolbar, click the Project Browser icon (or from the File menu, choose **Project Browser**—see Figure 5–1).

By default, when you install Architectural Desktop, a folder named *Autodesk* that contains a subfolder named *My Projects* will be created in the current user's *My Documents* folder. When you open the Project Browser for the first time, it will be set to this location. However, if you completed Chapter 1 or 2, it will remember the location of the last project that you had loaded there instead.

It is not necessary that you create projects in the *My Documents* location and in fact, in a team environment it is preferable to work from a network server location instead. Files installed from the Mastering Autodesk Architectural Desktop 2006 CD ROM are installed in the *C:\MasterADT 2006* folder (see the "Files included on the CD ROM" section of the Preface for more information). Even though you will typically work from a server location on your real projects, for the tutorials in this book, it is required that you work in the *C:\MasterADT 2006* folder. Several of the project files provided rely on their being in this location.

UNDERSTANDING THE PROJECT BROWSER WINDOW

The Project Browser is a file browser mechanism that allows you locate project files (.APJ files) anywhere on your computer (local or network). Using the Project Browser, you can locate projects and make them current, and move, copy, and create projects. Figure 5–1 illustrates the various icons and controls within this window and includes brief descriptions of each in the caption below.

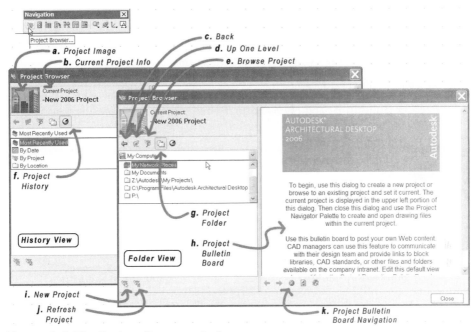

Figure 5–1 *The Project Browser window*

a. **Project Image**—A custom-defined BMP image can be assigned to the project and displayed here. For example, load the client's logo here.

b. **Current Project Info**—The Name, Number and Description of the current project will display here.

c. **Back**—Click to go back to the previous folder.

d. **Up One Level**—Click to go to parent folder.

e. **Browse Project**—Click to open a standard Browse Window to locate and load project files (projects have an APJ extension).

f. **Project History**—Click to browse for projects that were previously active (see image inset for additional history view options). While in the History view, you can right click and remove items from the history list as well as reset the list.

g. **Project Folder**—Click to browse for projects within the folder tree. This option gives access to My Computer, My Documents, My Network Places and any additional locations that you add to the AEC Project Location Search Path (see Figure 5–2 below).

h. **Project Bulletin Board**—The user-defined Project Bulletin Board Web Page, a fully customizable project-specific HTML Web page. ADT starts with a simple generic page, you can load your own custom one in the Project properties.

i. **New Project**—Creates a new project within the current folder.

j. **Refresh Project**—Refreshes the current folder.

k. **Project Bulletin Board Navigation Tools**—Typical browser functions for the bulletin board page.

CAD **Manager Note:** You can add to the default search path locations used by the Project Browser in the Options dialog box. From the Tools menu choose **Options**, and then click the AEC Project Defaults tab. There you can add paths to the AEC Project Location Search Path. You can also edit the default Project Templates, Project Bulletin Boards, and Project Images on this screen. It is recommended that at the time of installation, you reset these paths for users to a location on the server where project templates and files are typically stored (see Figure 5–2).

Please note that new in ADT 2006 is the ability to create a project from a template project. In this instance, the settings for Default Project Template Files will be ignored and the settings saved with the template project will be used instead.

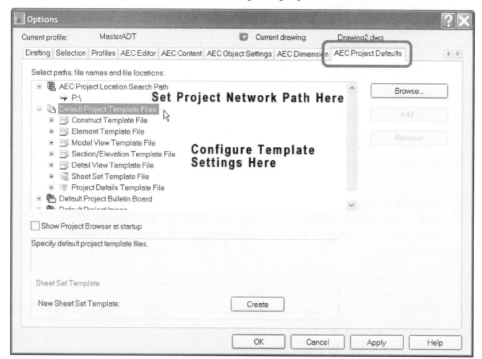

Figure 5–2 *Changing the default path settings for AEC Projects*

4. In the Project Browser, be sure that the Project Folder icon (see item 7 in Figure 5–1) is active and choose *My Computer* from the list.

5. Double click the C Drive, then the *MasterADT 2006* folder and finally the *Chapter05* folder.

6. With the *Chapter05* folder showing in the drop down list, click the New Project icon (see item 9 in Figure 5–1) at the bottom of the Project Browser window.

 NOTE Be sure to browse to the folder first, then click the New Project icon. Review items 3 through 7 in Figure 5–1 for the tools used to navigate within Project Browser.

 The Add Project worksheet will appear.

7. In the Project Number field, type **2006.1**.

8. In the Project Name field, type **MADT Commercial**.

9. Click in the Project Description field, and in the small dialog box that appears, type **Mastering Autodesk Architectural Desktop 2006 Commercial Project**.

10. Make sure the "Create from template project" checkbox is selected and then click the small browse button (**...**).

11. In the dialog that appears, navigate to your template folder.

    By default, the project template folder is located at: *C:\Documents and Settings\All Users\Application Data\Autodesk\ADT 2006\enu\Template\ADT Template Project (Imperial)* [*C:\Documents and Settings\All Users\Application Data\Autodesk\ADT 2006\enu\Template\ADT Template Project (Metric)*].

12. Select *ADT Template Project (Imperial).apj* [*ADT Template Project (Metric).apj*] and then click Open.

 NOTE This is the default template project and should already be selected unless you have previously created a new project and changed the setting.

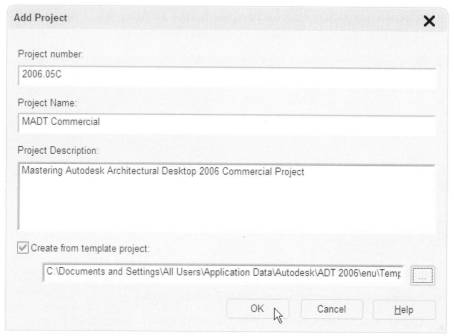

Figure 5–3 *Create the Project from a template in the Add Project worksheet*

13. In the Add Project worksheet, click the OK button to create the project (see Figure 5–3).

CAD Manager Note: The project template that we used to create the project included a bulletin board. You can replace the default one with something more specific to the project. Any HTM or HTML file can be used as a project Bulletin Board. You can use virtually any word processor, text editor, or HTML editor to create or edit project Bulletin Board files. A project Bulletin Board can be used by a Project Coordinator or CAD Manager to keep project team members informed on project news, or as a way to provide links to project standards, project tool palettes, or even a DWF of the project. (A good example of this is seen in the ADT Sample Project included with the software—the ADT Sample Project is located in the *C:\Program Files\Autodesk Architectural Desktop 2006* folder). The Project Web page displays in its own integrated Web Browser window within the Project Browser, complete with its own Back, Forward, Home, and Refresh icons (see item 11 in Figure 5–1 above). In this way, the project Bulletin Board can actually reference a home page to an entire intranet project Web site. If you prefer not to use the Bulletin Board feature in projects, simply leave the reference to the default Bulletin Board page unchanged.

The Project Image file is a logo for the project. Any image file can be used for this image, but it must be saved in BMP format. Use Windows Paint or any other image editor to save the image file. If your project has its own logo, you can load it here. Otherwise, you can use your company logo, the client's logo, or this can be left set to the default ADT image. If you prefer not to use the Project Image feature, simply leave the reference to the default Image.

Some firms like to include the job number as a prefix to drawing file names. This can be accomplished automatically with the "Prefix Filenames with Project Number" option. To set this, right-click the project in Project Browser and choose **Properties**. Then simply choose **Yes** from the drop down next to this feature to enable it. For simplicity in this book, the Project Number Prefix has not been used. However, feel free to use this feature if you wish. Finally, you can also choose different default template files for each of the various files managed by the ADT Drawing Management system, such as Constructs, Elements, Views, Sheets, and Sheet Sets. (Each of these file types is explained below.)

> The new MADT Commercial Project will appear in the Project Browser (highlighted in bold to indicate that it is current), and the chosen Bulletin Board and Project Image will appear (see Figure 5–4).

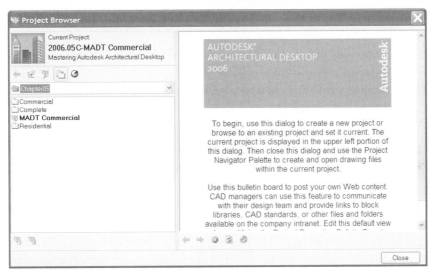

Figure 5–4 *The Project Browser with the newly created project set current*

Some additional tools can be found on the right-click menu within the Project Browser; right-click on the project name to access this menu. Use these tools to set the current project, refresh, copy, and browse project files. You can also close the current project and rename and move inactive (not current) projects. You will not find an option to delete projects in the Project Browser. To delete a project, exit ADT, navigate to the project location in Windows Explorer, and delete the folder containing the project files. Please exercise caution when doing this, however, as this will completely and permanently remove the project and *all* of its drawing files.

TIP If you wish to have the Project Browser appear automatically each time you open ADT, choose **Options** from the Format menu, and click the AEC Project Defaults tab. There, place a check mark in the Show Project Browser at startup box and then click OK. With this active, the Project Browser will launch each time you start ADT.

14. Click Close to close the Project Browser and return to ADT.

    The Project Navigator palette will appear on screen.

 The tool palettes will also change, automatically loading the MADT Commercial tool palette group.

When you create a new project, you have the option to have a tool palette group associated with the project and have this group automatically load in project team members' workspaces. This is typically accomplished by creating the project from a template such as we have done here. However, as you will notice, the tool palette group that loaded automatically contains a single palette that contains no tools. At first, this can seem a bit troubling when tool palettes seem to have disappeared. However, as we saw in Chapter 1, tool palettes can be grouped. To return to your previous (or any other) group, you simply need to right-click the tool palettes title bar. In coming chapters, we will work with the Project Tools tool palette. For now, let us return to the Design tool palette group to continue our project setup.

15. Right-click the Tool Palettes title bar and choose **Design** to re-load the Design tool palette group.

PROJECT NAVIGATOR TERMINOLOGY

Projects in Architectural Desktop consist of a collection of drawing files and project information files saved together in a common location. Taken together, this collection of graphical and non-graphical data is used to assemble a complete Building Information Model (BIM). The graphical drawing data associated with the project fall into four types of ADT drawing (DWG) files: *Constructs*, *Elements*, *Views*, and *Sheets*; each defined below. The non-graphical project information files include a single Autodesk Project Information file (APJ) that contains the basic framework of a project, a Sheet Set template file (DST) that determines the organization and configuration of the printed Sheets within the project, and several individual project data (XML) files (one per drawing) describing how each individual drawing fits into the overall project structure.

When you create a project, you enter the basic descriptive information (as we did above), such as Name, Description and Project Number. The next task is to determine how the building will be subdivided into Levels and Divisions. Levels are the floor Levels and divide the building horizontally. You can also subdivide the building laterally into Divisions. Although you can edit Levels and Divisions at any time, it is typical to establish the Levels and Divisions that will describe the basic framework of our Project at the onset. These tasks are performed in the Project Navigator.

 TIP If Project Navigator did not appear automatically on screen when you created the project, choose **Project Navigator Palette** from the Window menu (or press CTRL + 5).

Levels and Divisions (Project Framework)

▶ **Level**—A horizontal separation of building model data. A Level is typically an actual floor level in a building. Levels can be established for actual building stories, and also for mezzanines, basements, and other partial levels. You also use Levels to establish Grade level, Roofs and Datum levels (see Figure 5–5).

▶ **Division**—A vertical separation of the building model data. Divisions are typically used to articulate a physical separation such as a Wing, an Annex, or an Addition to a building. Divisions can be used to subdivide large floor plates into pieces that are more manageable in size (See Figure 5–5).

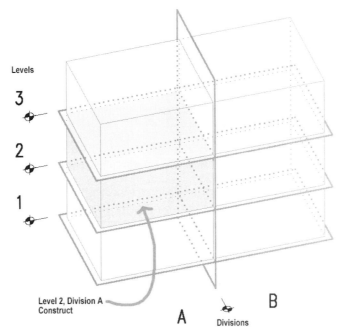

Figure 5–5 *Divide your building into Levels and Divisions to create your project structure*

 NOTE See next topic for a definition of Construct.

Model and Sheet Files

Standard industry practice and the US National CAD Standard (NCS) recommend the creation and maintenance of two types of file: Model files and Sheet files. We will set up both Model and Sheet files in this book. This practice is widely used in the industry and offers many benefits.

- ▶ **Model File**—A file containing actual building data drawn at full size (1 to 1 scale.) This is a file in which all of the *day-to-day* work is typically performed. In ADT 2006, Constructs, Elements, and Views can all be considered Models in the traditional sense; see definitions below.

- ▶ **Sheet File**—A file that is used *exclusively* for printing drawings. No data is saved in this file. It typically contains only a title block and external references to the project's various Model files.

Model files are referenced to Sheet files for printing (or when appropriate, other Model files). Most daily work is performed in Model files. In contrast, Sheet files exist solely for printing final documentation sets for distribution. One or more Model files are "gathered" by the Sheet file, composed on a title block sheet, scaled properly with the proper Display Configuration active, layers and objects visible, and then printed. The Sheet is saved in this state so that documents can be printed again any time, at a moment's notice. To perform physical edits and design changes, return to Model files and perform the changes there. Those changes will appear in the Sheet file the next time the XREFs in that Sheet are reloaded, which happens automatically when the Sheet is opened.

 Caution: The Sheet file's "ready-to-print" status is maintained only if all project team members agree to work **only** in Model files and **not** in the Sheet files.

Elements, Constructs, Views, and Sheets (Project Drawing Files)

Autodesk Architectural Desktop formalizes the creation of project files based on the Model/Sheet concept in the Drawing Management system and accompanying procedures. To fully realize the goals of the Model/Sheet file concept, ADT provides the project navigation system that we have begun to see in the preceding passages (as well as in tutorials in the previous chapters). This system incorporates the industry-standard use of Model and Sheet files and introduces an additional layer of granularity (Elements, Constructs, and Views) to help formalize the process. This is necessary simply because Model/Sheet file recommendations as published in the *NCS* and its predecessor the *AIA CAD Layer Guidelines* are not written specifically for ADT or even for AutoCAD. Rather they are general CAD guidelines applicable to any software package. Therefore, when we apply these recommendations to the specific toolset offered by ADT, we find that the Element, Construct,

View, and Sheet framework greatly enhances our ability to fully achieve the Model/Sheet file intent and push the functionality and usefulness much further.

 NOTE Some of the recommendations made in the UDS Module 01 (part of NCS 2.0) refer to the earlier "AIA CAD Layer Guidelines - Second Edition" document. The overall intent of these documents has been summarized here. However, for the complete explanation of these recommendations and all supporting materials, you are encouraged to refer to the above referenced documents directly.

ADT 2006 Model Files

▶ **Element**—A discrete piece of a design *without* explicit physical location within the building. Often an Element represents components that are repeated (typicals) more than once in the design. Elements are also ideal for storage of project libraries and resources. Elements are drawing files that are XREFed to other files (Constructs, Views, Sheets, or even other Elements) as project needs dictate.

▶ **Construct**—A discrete piece of a design *with* explicit physical location within the building. More specifically, it is a unique piece of the building occurring within a particular zone (Division) on a specific floor (Level) of the building. Constructs typically contain only items representing "real" building components, and typically do not contain any annotation, notes, or dimensions. Constructs are not specific to any particular type of drawing. They are not "Plans" or "Sections" but rather Models that can be used to generate plans, sections, or any other type of drawing. A Construct is distinguished from an Element by its unique identifiable physical location within the building. Constructs are drawing files that are XREFed to other files (other Constructs, Views, and Sheets) as project needs dictate.

▶ **View**—A "working" report of the building model. A View gathers all of the Constructs (and their nested Elements) required to correctly represent a specific *slice* (or *view*) of the building. Views are akin to a particular type of drawing such as a "plan" or "section" and will contain those project annotations like notes, dimensions, and tags appropriate to the drawing type in question. Views are drawing files that are XREFed into Sheets as project needs dictate.

For example, imagine a three-story commercial building. You might create Elements to represent a typical egress stair, typical restroom layouts, and even furniture grouping configurations. You would have at least one Construct for each floor, although there could be, and often are, several. For instance, in many cases it is advantageous to separate the interior from exterior construction. In this case, you would have a "First Floor Interior" Construct and a "First Floor Exterior" Construct. Nested within the interior Construct would be Elements for stairs and toilet rooms. A similar structure would be established for each of the other floors. If some unique element occurred on one or more of the upper floors, such as a Curtain Wall that spans from second to third on three sides of the building, it would be built in its own "Spanning" Construct. This Construct would then be referenced to both (or all) the floors to which it applies. When you were ready to begin creating construction documents to convey your design, you would start creating drawing specific Views. A View allows you to create a unique snapshot of a portion of the building. For instance, if you wished to work on the Third Floor East Wing in plan, you would create and work in a View that would gather and correctly represent all of the Constructs (and their nested Elements) that are required by that *physical* portion of the building. Another similar View could be made of the same physical slice of the building but configured for a reflected ceiling plan, and yet another for furniture or finishes. Views are not limited to just plans. Views can be made to accommodate the creation of Sections, Elevation, Schedules, Details, and even full 3D Models. Be careful, however, to distinguish the "working" nature of Views from the "output" or plotting nature of Sheets.

ADT 2006 Sheet Files

▶ **Sheet**—A "just for printing" report of the building model. While Elements, Constructs, and Views can all to be considered *Models* as defined by NCS/AIA, the ADT 2006 Sheet exactly emulates the purpose and intent of the NCS/AIA recommended Sheet file as noted above. A Sheet file will gather all required building model components (Views with their annotations and all nested Constructs and Elements) and compose them on a title block sheet, at a particular scale ready to print.

There are those who argue that annotation and dimensions ought to be placed in the Sheets. Some go further to promote that these items be placed in Layout space on top of viewport images of the project files. While both these approaches are certainly possible, the ADT Drawing Management toolset supports the approach championed by this text as indicated in the definition of "Sheet" above. It is the position of this text that Sheets should be set up once and maintained from then on as "for plotting only" files. The goal is to provide a set of files (one for each physical paper sheet in a document set) that is always ready to be opened and printed with no advance notice or *tweaking* required.

When project team members are allowed (or encouraged) to *work* in Sheet files, it is possible or even likely that they will leave the drawings in a state that is less than ideal for "ready printing." For instance, one might close the drawing with model space active, or change the LtScale, or accidentally forget to freeze or thaw the correct layers. These are just some examples of the types of small mundane settings that if set incorrectly at the time of plotting can force a drawing to require reprinting. Not only is this frustrating to the person making the plots, but it needlessly wastes time, paper, and money. It is therefore *strongly* recommended that Sheets be used for plotting only and all work be done in Elements, Constructs, and Views.

ESTABLISH THE PROJECT FRAMEWORK

Now that we have defined the terms that we will be using throughout the rest of this chapter (and the entire book), we are ready to add some structure to our commercial project.

SET UP PROJECT LEVELS

1. If the Project Navigator palette is not showing on screen, choose **Project Navigator Palette** from the Window menu (or press CTRL + 5).

There are four tabs on this palette. The first one, labeled Project, should already be active. On the Project tab is listed the Current Project information, Divisions, and Levels. If you wish to edit any of these, click the small Edit icon (a small pencil icon as seen in Figure 5–6) next to the appropriate item.

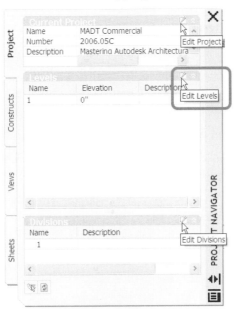

Figure 5–6 *The Project tab of the Project Navigator and the accompanying Edit icons*

This project will have four stories, site conditions, and a roof. We will establish each of these as a Level.

2. In the Levels area, click the Edit Levels icon (see Figure 5–6).

The default Level is named "1". We need to make a few edits to its ID, Floor Elevation and the Floor to Floor Height.

3. Click in the Floor Elevation column, and edit the value to **3'-0"** [**900**].

4. Change the Floor to Floor Height to **12'-0"** [**3650**].

5. Change the ID to **G** and edit the Description to read **First Floor** (see Figure 5–7).

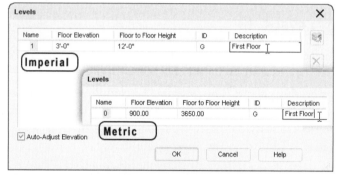

Figure 5–7 *Edit the parameters of the First Floor Level*

6. At the top right corner, click the Add Level icon.

Notice that the new Level has automatically been named "2" and begins at 15'-0" [4550]. This is due to the Auto-Adjust Elevation setting, which is on by default (see the bottom left corner of Figure 5–7).

7. Edit the Description to read **Second Floor**.

8. Click the Add Level icon twice more and edit the Descriptions for **Third Floor** and **Fourth Floor** (see Figure 5–8).

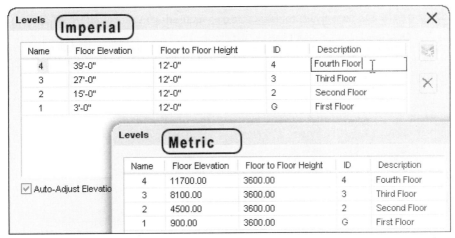

Figure 5–8 *Add and edit each of the four Floors*

As stated above, we will also require a Level for the Site and another for the Roof.

9. Be sure that Level 4 is still selected, and then click Add Level again.

10. Click in the ID column and change the Name to **R**.

11. Change the Floor to Floor Height to **3'-0" [900]**(the parapet height) and type **Roof**for the Description (see Figure 5–9).

Figure 5–9 *Add the Roof Level*

12. In the Name column, select I, right-click, and choose **Add Level Below** (see Figure 5–10).

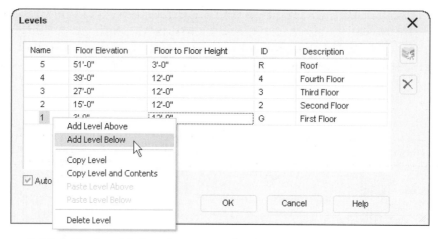

Figure 5–10 *Add a Level below the current level*

A Level is added below the current level; it is named "6." This name and the Floor Elevation of this Level require adjustment.

13. Click in the Name column (on the number "6"), pause a moment, and then click again.

    This should activate the rename mode for Level 6. You can also right-click and choose **Rename Level**.

14. Change the Name to **0**.

15. Click in the Floor Elevation column and edit the value to **0**.

    This will distort all of the other Floor Elevation values. Don't worry, with Auto-Adjust on, it is easy to fix.

16. Change the Floor to Floor Height to **3'-0″** [**900**].

17. Change the ID to **S**(for "Street") and the Description to **Street Level**(see Figure 5–11).

Name	Floor Elevation	Floor to Floor Height	ID	Description
5	15300.00	900.00	R	Roof
4	11700.00	3600.00	4	Fourth Floor
3	8100.00	3600.00	3	Third Floor
2	4500.00	3600.00	2	Second Floor
1	900.00	3600.00	G	First Floor
0	0.00	900.00	S	Street Level

Figure 5–11 *Completing the edits to the Project Level Structure*

18. Click OK to accept the values and dismiss the Levels worksheet.

19. A prompt will appear asking you to "Update all Project Views." Click the Yes button (see Figure 5–12).

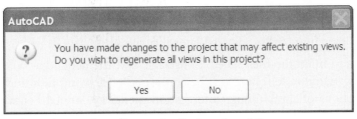

Figure 5–12 *Completing the edits to the Project Level Structure*

Answering "yes" to this prompt will update all drawings that reference the Project Levels to incorporate the new values just entered. There are not yet any drawings in this project that would require such updating, but you should always answer "Yes" to this prompt regardless. This will ensure the integrity of your project files and their relationships to one another.

This project has a small and simple footprint that requires only a single Division. (All projects must have at least one Division.) By default, this single Division is named simply "1." We can, if we wish, change that name, but for this project we will leave it set to the default.

Your Project Navigator should look like Figure 5–13.

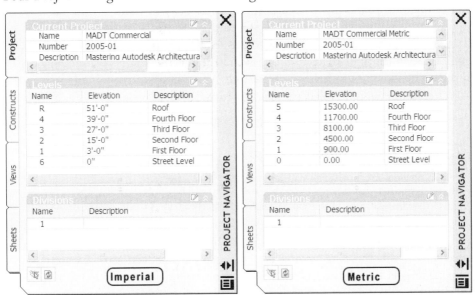

Figure 5–13 *The Project Navigator complete with Divisions and Levels*

At this point, we have completed the general project parameters and established that the building will be divided into six Levels (including the Street and Roof

Levels) and a single Architectural Division. Keep in mind that we can revise this structure later if necessary and adjust the Level and Division structure as changing project needs dictate. Naturally we would want to try to avoid drastic changes to the Level and Division structure wherever possible. The important point of note here is that this structure *can* be adjusted later if required.

The next step is to build the files called Constructs and Elements that will populate this building framework. These will be comprised of drawing files that represent various portions of the building information model as defined below in the next topic.

CREATE CONSTRUCTS

Now that we have created the project database, the first files that you need to create in your new project are the Constructs. These files will make up the pieces of the building model and therefore are required before you can create Views and Sheets. It is not necessary that you create every Construct that your project will eventually have at the early cartoon set phase, but you typically will create each of the major building components at this stage. For instance, in this sequence, we will create a Construct for each level of the building as well as the site.

OPEN AN EXISTING SITE CONDITIONS FILE AND CREATE A CONSTRUCT

The first Construct that we will create is the Site. We will build a Site Terrain Model utilizing some existing data. For this tutorial, we will assume that we received a sketch of the site plan as a drawing file. This drawing is in rough form, with only basic outlines of roads and alleys that define the site and some simple contours, but it is enough to help us get started. We will first save this file within the project structure as the project's site Construct.

1. Click the Open icon on the Standard toolbar (or choose **Open** from the File menu).

2. Navigate to the *Commercial* folder in the *C:\MasterADT 2006\Chapter05* folder and locate the file named *Site.dwg* [*Site-Metric.dwg*].

   Even though the site data is preliminary, it gives us the most important information required at this point in the project: the extent of the building footprint.

As stated above, the Building Model is composed of a collection of Constructs (and as appropriate, their nested Elements). To incorporate the provided Site data contained in this file into our project structure, we will create a Construct from it and assign it to the Street level defined above.

3. On the Project Navigator palette, click the Constructs tab.

Notice that there are two folders shown on this tab: *Constructs* and *Elements*. (Later, we can add additional sub-folders to these as project needs dictate).

CAD Manager Note: Creating a well-planned Folder (Category) structure for your projects can prove extremely beneficial. Category sub-folders can be added to each of the root folders: *Constructs*, *Elements*, and *Views*. While you can also add folders to Sheets, the Sheets typically use Sheet Set Subsets for organization. These folders can be added within the Project Navigator or in Windows using standard folder creation methods. Once you have established a suitable folder structure, you can even re-use it in future projects by the **Copy Project Structure** command (available as a right-click option in the Project Browser), which will copy all of the sub-folders in the project to a new Project name and location that you specify. This will help you maintain consistency in project setup. Please note that **Copy Project Structure** does not copy any of the files. To create a project from another including all of its folders, Constructs, Views and Sheets, use it as a project template when creating new projects.

4. Right-click on the Constructs folder and choose **Save Current Dwg As Construct**.

   The Add Construct dialog box will appear.

5. Type **Site** in the Name field.

6. Click in the Description field and type **Site Information**.

7. In the Assignments area, place a check mark next to Street Level (see Figure 5–14).

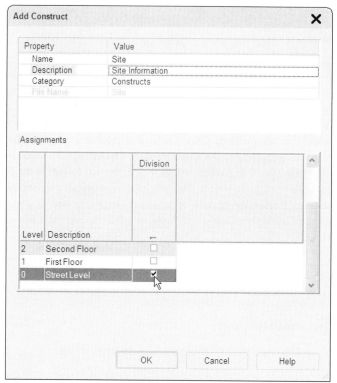

Figure 5–14 *Name the new Construct and assign it to the Street Level*

The Assignments area of the Add Construct dialog box is very powerful. It is here that you tell ADT and the Project Management system which portion of the building this particular Construct represents. Since the Project Management system is aware of all the Levels and Divisions within a project, ADT will be able to correctly XREF and locate this Site Construct drawing relative to all other drawings in the project.

8. Click **OK** to accept all values and create the new Construct.

Notice that there is now a new Construct in the Project Navigator (Constructs tab) named *Site* (see Figure 5–15).

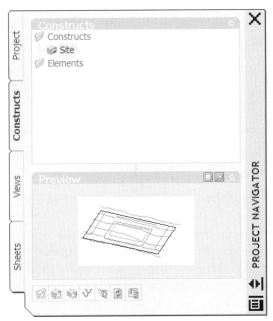

Figure 5–15 *The new Site Construct appears on the Project Navigator palette in the Constructs folder*

BUILD A TERRAIN MODEL FROM THE SITE DATA

The *Site* Construct should still be open. If it is not, double click it on the Constructs tab of the Project Navigator. Looking carefully at the geometry in the file, you will notice that there are polyline contours that have been set at their respective heights in the Z direction. To see this, click on one of the polylines, right-click and choose **Properties** and on the Properties palette, check the value of the Elevation field. Note that each polyline you do this to has a different elevation. We will use these elevated polylines to generate a terrain Mass Element.

9. Select one of the gray polyline contours, right click and choose **Select Similar**.

10. With the gray polylines selected, on the Massing tool palette, click the Drape tool.

   If the Massing palette is not available, right click the tool palettes title bar and choose **Design** to load the Design tool palette group.

11. At the "Erase selected contours" prompt, choose **Yes** from the dynamic prompt (or type Y and then press ENTER).

12. At the "Generate regular mesh" prompt, choose **No** from the dynamic prompt (or type N and then press ENTER).

13. At the "Generate rectangular mesh" prompt, choose **No** from the dynamic prompt (or type N and then press ENTER) (see Figure 5–16).

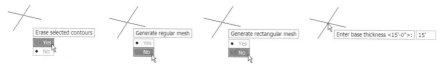

Figure 5–16 *Establish the rectangular bounds of the Drape object*

14. At the "Enter base thickness" prompt, type **15'-0"** **[4500]** and then press ENTER.

   Upon completion of the Drape, you will have a 3D Mass Element terrain model.

We erased the contour lines in the first prompt because they were only needed to generate the terrain model. When creating your own terrain models, if you wish to keep the contours to display in a site plan, you can instead choose **Yes**. In this example, we actually have a second set of contours on a frozen layer for that purpose which we will thaw in the exercise below. The "regular" mesh prompt determines whether a uniform number of mesh segments is created in the X and Y directions. If you choose yes for this prompt, you are then prompted to indicate how many mesh segments you wish in each direction. By choosing **No** here, the Drape instead generated a varying number of mesh segments determined by the shape of the contours. This type of mesh will give more segments only where they are needed. The "rectangular" mesh prompt determines the shape of the plan footprint of the mesh. It can be either rectangular in shape (in which case you indicate the extent of the rectangle on screen) or irregularly shaped based on the extent of the selected contours. This is the option we used by choosing **No** for a rectangular mesh. If you like, you can undo the command and try it again with different answers to the prompts. Each of the options chosen in this sequence is new to ADT 2006.

Next we will restore a hidden 3D Mass Element shaped like the roads. We will use this Mass Element to cut the shape of the roads away from the Drape. We would not have been able to achieve the same thing with the linework that we erased above.

15. On the Drawing Status Bar click the red light bulb icon and choose **End Object Isolation** from the popup menu (see Figure 5–17).

TIP If the light bulb icon does not appear red, right-click in the drawing and choose **Isolate Objects > End Object Isolation** instead. This will achieve the same result.

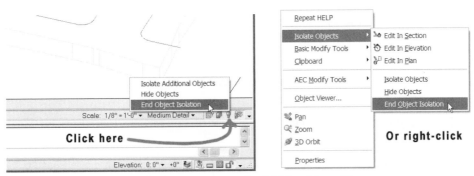

Figure 5–17 *End Object Isolation to reveal a hidden Mass Element*

In the original Site file that contained the contours, there was also a hidden Mass Element object. The Isolate Objects tool (small light bulb on the Drawing Status Bar) allows you to hide and isolate selected objects in a file. If objects in the file are hidden, then the small light bulb appears red. As mentioned in the tip above, sometimes the icon fails to appear in red. Use the right-click menu instead in this case.

16. Select the Drape (Mass Element) that you just created, right-click and choose **Boolean > Subtract**.

17. When prompted, select the gray Mass Element representing the roads (the one that was hidden) and then press ENTER.

18. At the "Erase layout geometry" prompt, choose **Yes**.

    The result is subtle, but the shape of the roads is now carved out of the terrain Mass Element. To get a better look, select the terrain, right-click and choose **Object Viewer** and then orbit the model around a bit. Try the shade icons in the viewer as you orbit.

19. Using the Layer drop down list, thaw the **A-Site-Clin** Layer.

This reveals some other contour lines that are more accurately depicted than the ones we erased above. They are set at 1'-0" [300] intervals, rather than the 6" [150] ones used to generate the terrain. Although these contours could have been used to generate the terrain, the Drape tool uses a smoothing algorithm that would have smoothed all the sharp edges at the curb lines. This is why we did the two-step technique shown here.

20. Save the file.

BEGIN THE FIRST FLOOR CONSTRUCT

Each floor of our project will consist of one or more Constructs. In fact, there will often be several Constructs per floor. The next several steps continue from the previous exercise (building the site) and will walk through the process of beginning a Construct for the first floor. This will in turn be used when we create first floor plan View and Sheet files.

If you closed the *Site* Construct, double click it in Project Navigator to re-open it now.

1. Select the light blue rectangle in the middle of the *Site* Construct. (Click on the edge.)

The rectangle in the center should highlight with four grip points displaying at its corners. It is on the Layer: G-Anno-Nplt.

2. With the CTRL key held down, click and hold down on the edge (not a grip) of the selected rectangle and drag it on top of the *Constructs* folder of the Project Navigator.

3. Release the mouse button to create the Construct (see Figure 5–18).

 NOTE Holding down the CTRL key while dragging makes a copy of the rectangle and leaves the original rectangle in the Site file. Had we simply dragged, it would have moved the rectangle (deleting it from the original file).

 NOTE If you have the auto-hide feature turned on for Project Navigator, drag the rectangle over the title bar of the Project Navigator until it pops open and then continue to drag the item to the *Constructs* folder.

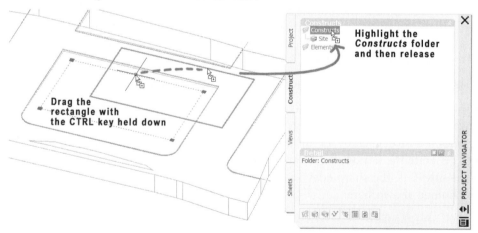

Figure 5–18 *Drag drawing geometry to the Project Navigator to create a Construct*

4. In the Add Construct worksheet, type **01 Shell and Core** for the Name.

5. In the Description field, type **First Floor Core and Exterior Shell**.

6. In the Assignments area, place a check mark in the First Floor Level check box and then click OK to finish (see Figure 5–19).

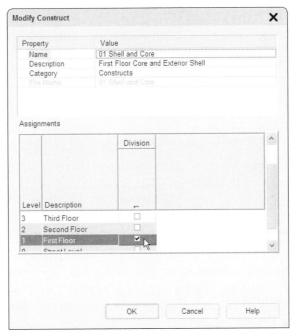

Figure 5–19 *Name the Construct and assign it to a Level 1*

7. Save and close the *Site* Construct file.

8. On the Project Navigator, double click *01 Core and Shell* (or right-click the *01 Core and Shell* Construct, and then choose **Open**).

9. Zoom in on the rectangle.

10. On the Design tool palette, right-click the *Wall* tool and choose **Apply Tool Properties to > Linework** (see Figure 5–20).

NOTE If the Tool Palettes are not open, choose **Tool Palettes** from the Window menu (or press CTRL + 3).

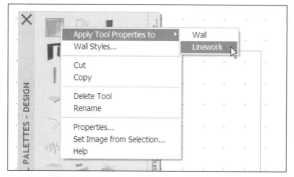

Figure 5–20 *Access the **Wall**tool's right-click options*

11. At the "Select lines, arcs, circles, or polylines to convert into walls" prompt, click on the edge of the rectangle.

12. Press ENTER to accept the selection.

13. At the "Erase layout geometry" prompt, choose **Yes** (or type Y and press ENTER).

We are erasing the rectangle, because we needed it only to provide the overall shape of our building footprint. The rectangle has now been replaced with four Walls. They are still highlighted, and the Properties palette has appeared (or has become active if it was already on screen) with the properties of the selected Walls showing.

14. On the Properties palette, change the Width to **1'-0"** [**300**] and the Height to **12'-0"** [**3600**] and then right-click and choose **Deselect All** (see Figure 5–21).

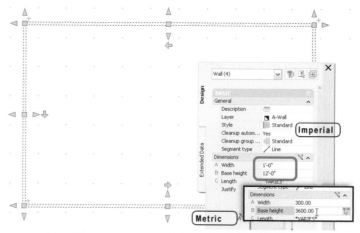

Figure 5–21 *Change the dimensions of the selected Walls*

The Walls generated in this file are temporary *"stand-ins"* for the real Walls that we will add later. For purposes of the cartoon set, it is necessary to have some objects in each file as *placeholders* and to provide the rough size and shape of the building. Think of them as a sketch similar to what you might create by hand in the traditional cartoon set.

PREPARING FOR PROJECT STANDARDS

In the coming chapters, we will progressively add refinements to the simple sketch models that we are building here. New in Architectural Desktop 2006 is the Project Standards feature. With it, we will be able to establish a master library file for our Wall styles (types) and other styles and then synchronize changes across the entire project. In this way, you can quickly make an edit to a style in one file and then apply it to the entire project. At this stage of the project, we do not want to concern ourselves with developing the specifics of a Wall style. However, we can simply name the style being utilized here so that later when we do get more specific, it will be a simple task to update the stand-in Walls that we have here.

15. With the four Walls still selected, right click and choose **Copy Wall Style and Assign**.

16. In the Wall Style Properties – Standard (2) dialog, choose the General tab.

17. In the Name field, type: **Exterior Shell**.

18. In the Description field, type: **Mastering Autodesk Architectural Desktop Commercial Project Exterior Building Shell Wall** and then click OK.

19. Save and Close the file.

CREATE THE UPPER FLOOR CONSTRUCTS

Once you have built one floor, you can easily copy its Construct to create additional floors via the Project Navigator where we are able to copy a Construct to other levels.

1. On the Project Navigator, right-click on the Construct *01 Core and Shell*.

2. From the context menu, choose **Copy Construct to Levels** (see the left side of Figure 5–22).

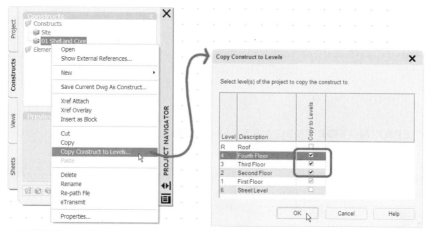

Figure 5–22 *Copy Construct to Levels command*

3. In the Copy Construct to Levels dialog box, place a check mark in the Second, Third, and Fourth Floor boxes, and then click OK (see the right side of Figure 5–22).

The first floor Construct has been duplicated to each of the other floors. However, notice that the names read "*01 Core and Shell (2)*" and so on.

CAD Manager Note: This method is useful to maintain drawing consistency when starting a project. It ensures that every project file uses the same template files and maintains a consistent global origin for XREFs. Also, when setting up a project in anticipation of using project standards as noted above, this method in copying the first floor "Exterior Shell" Walls also will make it easy to synchronize our styles later.

4. Right-click on *01 Core and Shell (2)* and choose **Rename**.

5. Type **02 Core and Shell** and then press ENTER.

6. Repeat the process for the third making it **03 Core and Shell**, and the fourth as **04 Core and Shell**.

7. Right-click on *02 Core and Shell* and choose **Properties**.

Notice that the check mark for the level is correctly at the Second Floor but that the description incorrectly refers to the first floor.

8. Click in the Description field, change it to **Second Floor Core and Exterior Shell**, and then click OK.

9. Repeat these steps for the Third and Fourth floors as well (see Figure 5–23).

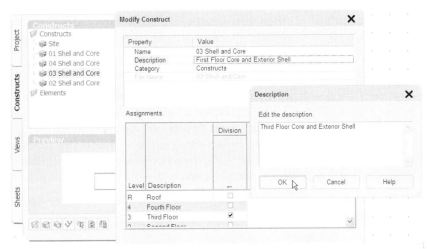

Figure 5–23 *Editing the Descriptions of the copied Constructs*

10. At the bottom of the Project Navigator, click the Re-path icon (all the way to the right, see Figure 5–24).

 Tip: If you get an error about this function's not being able to run in Zero Doc State, click the Qnew icon on the main toolbar to open an empty drawing and try again. Some operations are not available in Project Navigator when in "zero doc state" (no drawings open). Always keep at least one drawing open while working in Project Navigator to avoid this error message.

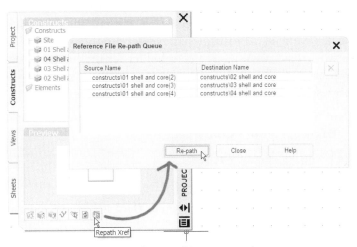

Figure 5–24 *Re-path the project after making name changes.*

11. In the Reference File Re-path Queue dialog box, click the Re-path button.

Because we renamed each of our new Constructs, it is important to let ADT re-path the project. This will ensure that all external references are properly maintained. Do this any time you rename a file in Project Navigator.

 Caution: If you close Autodesk Architectural Desktop without updating the file paths in your project, all queued path information is lost. The next time you open your project, it will not have current and correct file path information. Therefore, it is highly recommended that you update your project during each work session. This note is excerpted from the ADT Help file; for more information please refer to the "Updating (Re-pathing) the Project" topic in the online help.

To display the Constructs in the correct numerical order, click the Refresh icon (immediately to the left of the Repath Xref icon).

CREATE A SPANNING CONSTRUCT

In Chapter 8, we will build a Curtain Wall element that spans the Second through Fourth floors of the front façade of the building. Building components that occupy more than one level in an ADT Model are referred to in Project Navigator as "spanning" Constructs. Spanning Constructs will automatically be added to Views that reference them for each floor in which they span. In this case, we will build a spanning Construct for a future front façade condition that has yet to be designed at this stage of the project. However, as is often the case in the early stages of a project, the designers usually have some notion of the types of design elements that they hope to incorporate. Remembering our aim in this chapter, to build a mock-up set to be fleshed out as the project progresses, we will not allow ourselves to get carried away with the particulars of this design element; rather we will simply add a Construct for it with a simple Wall as a placeholder for the future Curtain Wall design. This approach is consistent with the rest of the steps taken here so far and the notion of progressive refinement promoted throughout this book.

1. In Project Navigator, on the Constructs tab, right-click on the *Constructs* folder and choose **New > Construct**.

2. For the Name type **Front Façade**.

3. For the Description type **Front Fa\c¢cade Spanning Curtain Wall**.

4. In the Assignments area, check the Second, Third and Fourth floors and then click OK (see Figure 5–25).

   Take note of the message that appears at the bottom of the Add Construct worksheet indicating that by selecting more than one Level, we have made this a "Spanning" Construct (see Figure 5–25).

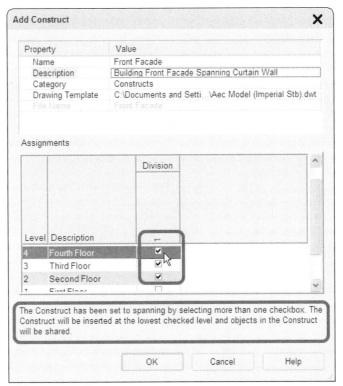

Figure 5–25 *Assigning the Construct to multiple levels makes it a "spanning" Construct*

5. Double-click on *Front Fa\c¢cade* in the Project Navigator to open it.

6. From the Insert menu, choose **Block**.

7. Click the Browse button, and in the *C:\MasterADT 2006\Chapter05\Commercial* folder, locate the file named *Base Curve.dwg* [*Base Curve-Metric.dwg*] and then click Open.

8. In the Insert dialog box, in the Insertion point area, deselect the Specify On-screen check box and leave the values for X, Y and Z set to **0**.

9. Place a check mark in the Explode check box and then click OK.

The *BaseCurve.dwg* file contains a single arc segment. This segment is a sketch of the curved Curtain Wall footprint that we will build in Chapter 8. We will now convert this arc to a Wall.

10. On the Design tool palette, right-click the *Wall* tool and choose **Apply Tool Properties to > Linework**.

11. At the "Select lines, arcs, circles, or polylines to convert into walls" prompt, select the arc segment and then press ENTER.

12. At the "Erase layout geometry" prompt, choose **Yes**.

13. On the Properties palette with the Wall still selected, change the height to **36'-0"** [**10,950**].

 NOTE The metric value is shown here with a comma separating the thousands. This is done for clarity. Please type the value without the comma in ADT.

14. From the View menu, choose **3D Views > SE Isometric** (or click the SE Isometric icon on the Views toolbar).

15. Save and Close the file.

We now have four *Core and Shell* Constructs, a single spanning *Fa\c¢cade* Construct, and a *Site* Construct. We will create additional Constructs for this project as needs dictate in later chapters. But for now, we have all of the Constructs we will require for our cartoon set.

CREATE VIEWS

The next step in project setup will bring all of these floor plate Constructs together to form a single composite model of the whole building in its current (albeit very schematic) form. This composite model will be our first *View* file as defined earlier in the "Elements, Constructs, Views, and Sheets (Project Drawing Files)" topic under the "Project Navigator Terminology" heading. There can be many reasons to create such a composite model early on. The most common reason is simply to use a check to verify that all of the pieces (Constructs) are fitting together properly as expected.

BUILD A COMPOSITE MODEL VIEW

A composite model View will help us visualize what we have so far. The tools in Project Navigator make creating a View easy. By simply designating which portions of the building we wish to include in the View, we can have the system gather all of the required XREFs for us and assemble them at the correct relative heights.

1. In Project Navigator, click on the Views tab to activate it.

   There are currently no files on this tab.

2. Right-click on the *Views* folder and choose **New View Dwg > General**.

Remember: You must have at least one drawing open to perform certain Project Navigator operations. If the function is unavailable, click the Qnew icon on the main toolbar to open an empty drawing and try again.

3. For the Name, type **A-CM00** and for the Description, type **Composite Building Model** (see Figure 5–26).

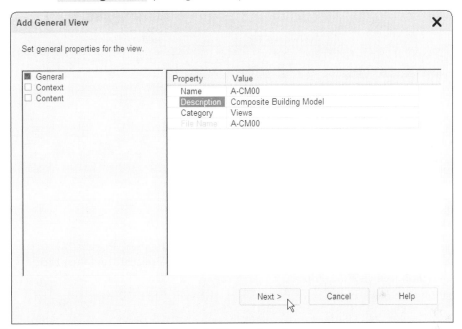

Figure 5–26 *The first page of the Add View wizard*

File naming conventions vary widely from one firm to the next. While it is possible to adapt existing file naming conventions to Project Navigator files when migrating to ADT, it is often necessary to make some adjustments. In this book, we will use simple descriptive names for Constructs and both descriptive names and US National CAD Standard (NCS) names for Views. See the "File Naming Guidelines" sidebar below for recommendations on Project file naming. Feel free to use your firm's file naming strategies rather than those recommended in this text. Changing the names of the files will not alter the tutorials in any way. In this case, for our Composite Model, "A" stands for "Architectural," "CM" for "Composite Model" and "00" is simply a placeholder for enumeration. For a floor plan this would be a floor number designation and for sections and elevations it is typically a simple sequential number.

4. Click Next to move to the Context page.

5. Right-click anywhere in the Levels area and choose **Select All** (see Figure 5–27).

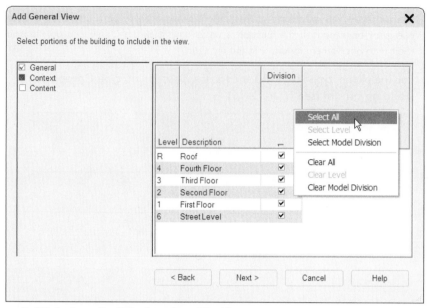

Figure 5–27 *Assign all of the Levels to the View*

6. Click Next to move to the Content page.

7. On the Content page, verify that all Constructs are selected and then click Finish (see Figure 5–28).

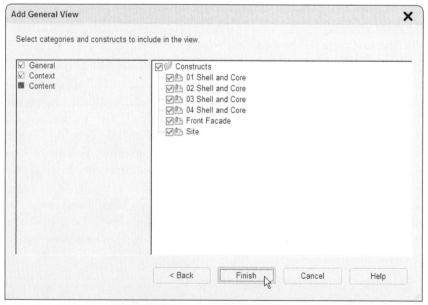

Figure 5–28 *Complete the Add View wizard by verifying that all Constructs are included*

8. On the Project Navigator palette, Views tab, double-click the new *A-CM00* View file to open it.

9. From the View menu, choose **3D Views > SE Isometric**.

10. From the View menu, choose **Shade > Hidden** (See Figure 5–29).

Note that all of the Construct files have been XREFed into this View file and inserted at their correct heights respectively relative to the Levels settings on the Project tab of the Project Navigator. This is one of the major benefits of the Project Management system in ADT. Once the basic parameters have been established, all of the Level and height information as well as which files are required to assemble a particular View are handled automatically by the system.

Naturally we still have quite a bit of work to do. For instance, there is no roof, but at this early stage of design, this model begins to give a good idea as to how this proposed project will sit on the site and a sense of the overall proportion of the building. You can also get to this point by using Mass Modeling techniques. The building's overall form can be massed out using tools on the Massing palette in ADT or in a program like SketchUp. These models can be then brought into the current project by saving the files as Elements and Constructs on the Constructs tab (using the technique covered above in the "Open an Existing Site Conditions File and Create a Construct" heading) and then sliced to generate floor plates. Although this is a perfectly valid schematic design process, it will not be elaborated on here.

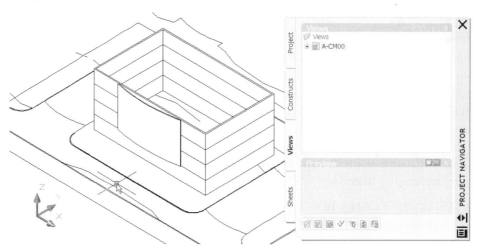

Figure 5–29 *The composite model viewed from the South West*

11. Save and Close the file.

FILE NAMING GUIDELINES

In general, try to name all files as descriptively as possible. Whether this is with names that include the actual contents of the file, like: "Architectural First Floor Plan," or with well-established abbreviations like: "A-FP01," the name should evoke the file's contents in some way.

Naming Element files: Use names that describe the contents such as "Typical Toilet Room" or "Temporary Building Outline." Elements don't belong to specific levels and should not include Level or Division descriptors in the names.

Naming Construct files: Use names that describe the contents of the file such as "01 Partitions" or "First Floor New" or "North Stair Tower." Do *not* name Constructs with names evocative of specific drawing types. For example, Constructs should *not* be named as "Plans," "Sections," or "Elevations." Constructs are models and are used to generate all of these types of drawings. Name them for what they contain. Drawing "type" names should be used for View files instead.

Naming View files: Views are created with a particular type of drawing in mind. Views can be named descriptively with a name that evokes the type of drawing that they will spawn, such as "First Floor Plan," "Exterior Building Elevations," "Third Floor Interior Elevations," or "Door and Frame Details." Views can also be named using existing naming conventions or industry-standard naming conventions such as the U.S. National CAD Standard. For instance, using NCS names, you might end up with View names like "A-FP01 (for Architectural Floor Plan First Floor)," "A-EL01 (for Architectural Elevations number one)" and "A-SP00 (for Architectural Site Plan)." NCS style names are used here for the Commercial project View files, while descriptive names (not based on NCS) are used in the Residential project to follow.

Naming Sheet files: Sheets are typically named after their Sheet number in many architectural firms. The AutoCAD Sheet Set functionality is incorporated directly into Project Navigator. The default behavior of Sheet Sets is for the file name to combine both the Sheet number and the Sheet title in the drawing file name. This is done automatically and is very difficult to override. For example, Sheet "A-101" titled "Floor Plans" is automatically named "A-101 Floor Plans" by Project Navigator (and its integral Sheet Set).

CREATE FLOOR PLAN VIEWS

The next task we have is to create a series of Floor Plan View files. Like the Composite Model, these will also be "General" Views; however, they will each include only one level of the project rather than all levels as in the composite model. Remember, when creating a View, that the software will automatically gather all of the correct Construct files required to make the View at a particular Level and Division combination. The View file thus created will be ready to receive notes, dimensions, and other annotation appropriate to floor plans (or whatever specific type of drawing is intended). Later we can then compose one or more Views (including their drawing specific annotations) onto a Sheet for printing.

CREATE A SITE PLAN VIEW

We will start at the bottom of the model and work our way up. The first plan View that we will create is a Site Plan.

1. On the Project Navigator palette, on the Views tab, right-click the *Views* folder and choose **New View Dwg > General**.

2. Name the file **A-SP00**, give it a Description of **Building Site Plan**, and then click Next.

3. On the Context page, place a check mark in the "Street" Level box and then click Next.

4. On the Content page, verify that only the *Site* Construct is checked and then click Finish.

   The new View file has been created.

5. Double-click on *A-SP00* to open it, and zoom to see the complete Site Plan file.

Try switching to 3D as we did above and note that this time only the Site Construct is included. Return to plan view (**View > 3D Views > Top**) before closing and saving the file.

6. When you are satisfied that the file has been created correctly, save and close the file.

CREATE THE FIRST FLOOR PLAN VIEW

The process for the first floor is nearly identical. However, we will take the floor plans a bit further.

7. On the Project Navigator palette, on the Views tab, right-click the *Views* folder and choose **New View Dwg > General**.

8. Name the file **A-FP01**, give it a Description of **Architectural First Floor Plan**, and then click Next.

9. On the Context page, place a check mark in the First Floor box and then click Next.

10. On the Content page, verify that only the *01 Shell and Core* Construct is checked and then click Finish.

The new First Floor Plan View file has been created.

11. Double-click on *A-FP01* to open it, and zoom to see the complete file.

Notice that only the four Walls that make up the First Floor Construct have been XREFed into this file. Neither the front fa\c¢cade, which begins on the Second Floor, nor the Site Plan has appeared here. This is correct based on the levels that we specified for this View.

FIRST FLOOR PLAN MODEL SPACE VIEW

Because View files are created with specific architectural drawing types in mind, like plans in this case, we can go further than simply gathering the correct Constructs as we have here. Let's make some decisions about how we would like this particular View to appear on a Sheet for plotting. For instance, we can designate the portion of the model that we wish to appear within the viewport of our Sheet Layout even before creating the Sheet. We can assign a display configuration, a title, and plotting scale ahead of time as well. To aid us in this process, we will insert a guideline from another file that represents the correct size and scale that we need for the First Floor Plan viewport. This step is optional in your own projects, but it is being shown here because even though this adds an extra step, it will be much easier to later compose your Sheets if you make this extra bit of pre-planning effort up front.

12. On the Insert menu, choose **Block**.

NOTE To save time in this tutorial, the 2D Layout Grid is provided in a separate file for quick retrieval. See the CAD Manager Note below for how this object was created.

13. Click the Browse button, and in the *C:\MasterADT 2006\Chapter05\Commercial* folder, locate the file named *Live Area Guide.dwg* [*Live Area Guide-Metric.dwg*] and then click Open.

14. In the Insert dialog box, in the Insertion point area, be sure that the Specify On-screen check box is not checked and leave the values for X, Y and Z set to **0**.

15. Place a check mark in the Explode check box and then click OK.

The *Live Area Guide.dwg* file contains an AEC Layout Grid 2D (available from the Stock Tool Catalog in Content Browser—CTRL + 4). It will appear as a dashed purple rectangle surrounding the plan (see the left side of Figure 5–32). We will use this rectangular gridline to create a Model Space View that we wish to use for the creation of our Sheet viewport. AEC Layout Grid 2D objects automatically appear on a non-plotting layer by default.

CAD Manager Note: The best way to plan these items is to know on what size title block the Sheets will be composed. For this project, we will be using the default Sheet size 42" x 30" [1189 x 841] and the default title block provided out of the box. In order to plan how our plan will fit on the final Sheet, we must know the "Live Area" of the title block. The Live Area is the portion of the title block minus all borders, margins and title strips available to place drawings. This title block has a Live Area measuring 36" x 28¾" [1040 x 780]. We can create a rectangular boundary using this Live Area dimension within our View file. This will give a visual reference for how much of our View file's geometry can fit on the title block at a given scale. In order for this guide to be valid, we must insert it into the View at a scale equal to that which we intend to plot. A simple rectangle would do for this purpose, but consider using an ADT Layout Grid 2D (available in the Stock Tool Catalog in Content Browser—CTRL + 4). Using such a grid makes it easy to compose Sheet layouts where multiple viewports are involved. This is because it has parameters for the number of cells in both the X and Y directions. To do this, you set both X and Y directions to "Space Evenly" and then input the number of bays you desire. In this case, both were set to 1 bay. But for an elevation View, you might do 2 or 4 vertically (in Y) and 1 in X, while a detail sheet might be 5 high by 6 wide for example.

16. On the Project Navigator palette, select the Views tab, right-click the *A-FP01* file, and choose **New Model Space View** (see Figure 5–30).

 NOTE A Model Space View is simply the AutoCAD Named View that is essentially a "saved zoom." You could achieve almost the same result by choosing Named Views from the View menu.

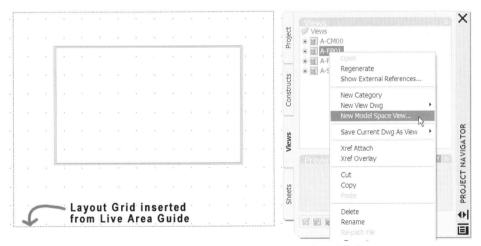

Layout Grid inserted from Live Area Guide

Figure 5–30 *Create a New Model Space View in the A-FP01 file*

17. In the Name field, type **First Floor Plan**.

18. Verify that the Display Configuration is set to Medium Detail and that the Scale is set to **1/8"=1'-0"** [**1:100**].

It is not necessary to type a Description for this exercise.

19. Click the Define View Window icon on the right (see Figure 5–31).

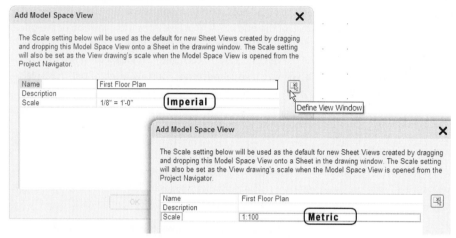

Figure 5–31 *Name the View and then designate its boundaries*

20. At the "Specify first corner" prompt, snap to the lower left corner of the dashed purple rectangle (the one just inserted).

21. At the "Specify opposite corner" prompt, snap to the upper right corner of the dashed purple rectangle (the one just inserted), and then click OK to dismiss the Add Model Space view worksheet.

In Project Navigator, indented beneath *A-FP01*, an icon labeled "First Floor Plan" should appear. AutoCAD Named Views (Model Space Views) appear in Project Navigator on both the Views and Sheets tabs. You can select these in Project Navigator to view detail information or a preview at the bottom just as you can the actual drawing files. You can toggle between the Detail and Preview views with the icons on the preview pane (see bottom right corner of Figure 5–32). To open a file and zoom right to the Model Space View, double-click it in Project Navigator.

22. In Project Navigator, double-click First Floor Plan beneath *A-FP01* (see Figure 5–32).

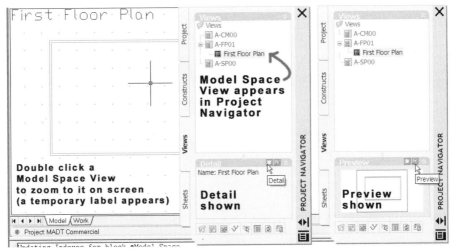

Figure 5–32 *Name the View and then designate its boundaries*

A temporary label will appear on screen with the Model Space View name (First Floor Plan in this case). This will vanish the next time you zoom or pan.

New in ADT
2006

ADD A TITLE MARK

In addition to providing a convenient way to zoom to a particular portion of a View file directly from Project Navigator and giving us a way to pre-assign the extents of a Sheet Viewport (as we will see below), the name of the Model Space View is referenced automatically into drawing title marks.

23. Right-click the title bar of the Tool Palettes and choose **Document** to load the Documentation tools group.

24. Click the Callouts tab, and then click the first *Title Mark* tool, (the one with both title and drawing number bubble).

    Move the mouse around on screen. You will notice that the First Floor Plan Model Space View shows a temporary label and border and that whenever you move your mouse within it, the border highlights in red—this indicates that it will be associated to the Model Space View name.

25. At the "Specify location of symbol" prompt, click a point within the highlighted border and beneath the plan (see Figure 5–33).

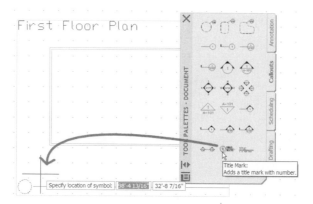

Figure 5–33 *Add a title mark beneath the Plan, within the First Floor Plan Model Space View boundary*

> 26. At the "Specify endpoint of line" prompt, drag to the right and click to designate the length of the title bar.

Notice that the label of the Title Bar has automatically picked up the name of the Model Space View. There is gray shading surrounding this label. This indicated that this is a field code. Field codes can be added to any piece of text including within Block Attributes (as is the case here) and then set to reference data from some other location. In this case, the field is configured to read the name of the Model Space View in which it is contained. Had we inserted it outside the Model Space View boundaries, it would have read the name of the drawing file instead. This is why it was important to insert the Title Mark within the boundaries of the Model Space View. Notice that the Scale has also been inserted as a field code and correctly reads the values that we assigned above.

Finally, there is a third field code within this title mark: the number within the round bubble. Currently it is displaying a question mark (?). This is because it is tied to the actual drawing number for this plan from the Sheet. Since we have not yet built the Sheet, the field cannot yet display the correct number. Later when we build our Sheets, this question mark will automatically be replaced with the correct designation (see Figure 5–34).

Figure 5–34 *The title mark contains field codes that reference the drawing name and scale*

27. When you are satisfied that the file has been created correctly, save and close the file.

CREATE THE SECOND FLOOR PLAN VIEW

To create the remaining floor plans, we could repeat all of the steps above. In this example, we will copy them from the First Floor Plan View and then edit them.

28. On the Project Navigator palette, choose the Views tab, right-click on *A-FP01*, and then choose **Copy**.

29. Right-click again and choose **Paste**.

    The new file will be named *A-FP01 (2)*.

30. Right-click *A-FP01 (2)*, and choose **Rename**. Change the name to **A-FP02** and press ENTER to accept the new name.

31. Right-click on the newly renamed *A-FP02* and choose **Properties**.

32. On the General page on the left, change the Description to **Architectural Second Floor Plan**.

33. Click on the Context page on the left, and then clear the check mark next to Level 1 and place a check mark in Level 2 (see Figure 5–35).

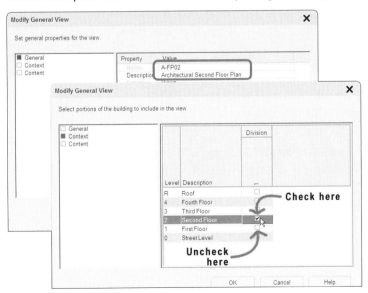

Figure 5–35 *Changing the Properties to reference Level 2*

34. On the left, click the Content page and verify that the Constructs checked are *02 Shell and Core* and *Front Fa\c¢cade*. Click OK to dismiss the Modify View dialog box.

35. Double-click on *A-FP02* to open it.

POST LINKING FIELD CODES

Notice that the title bar under the plan still reads "First Floor Plan." The reason for is that the field codes within the title mark still reference the First Floor Plan Model Space View from the *A-FP01* View file.

36. On the Project Navigator palette, right-click the Model Space View indented beneath the *A-FP02* file (currently named "First Floor Plan") and choose **Properties**.

37. Change the name to **Second Floor Plan** and then click OK.

The Model Space View name is now changed. Let's redirect the link of the fields now from the original copy of the file (*A-FP01*) to *A-FP02*.

38. Using a crossing window selection, select all pieces of the title mark.

 Tip: To make a crossing selection, click to the right of the title mark and then move the mouse to the left until it is near the other side of and surrounding the bubble and click again. New to ADT 2006, the crossing selection will appear in color—green by default.

There should be three objects selected. You can verify this on the Properties palette at the top: "All (3)" should appear in the selection list.

39. Click and hold down on any of the highlighted items and drag to the Project Navigator. (Do not click on any grips, drag from the highlighted edge.)

40. Drop the selection directly on top of the Model Space View indented beneath *A-FP02* (the one we just renamed to "Second Floor Plan," see Figure 5–36).

This procedure is called "post linking" and is used to link up the values in existing field codes with particular nodes within the Project Navigator. The title mark should now correctly reference the new name (Second Floor Plan) assigned to the Model Space View in *A-FP02*.

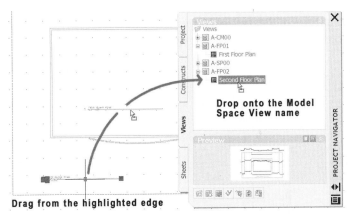

Drag from the highlighted edge

Figure 5–36 *Drag and drop the title mark to the Second Floor file to "post link" the field codes*

41. When you are satisfied that the file has been created correctly, save and close the file.

CREATE THE THIRD AND FOURTH FLOOR PLAN VIEWS

The process for creating the upper floors is the same. You can repeat either procedure.

42. Repeat the steps in the "Create the Second Floor Plan View" and "Post Linking Field Codes" topics to create the Architectural Third and Fourth Floor Plans.

   If you prefer, you can repeat the steps in "Create the First Floor Plan View," "First Floor Plan Model Space View" and "Add a Title Mark" instead.

43. Name them **A-FP03** and **A-FP04** respectively.

 Tip: Be sure to open the new file before renaming its Model Space View and post linking. There can be problems with the applying the new name if you do not open the file first.

44. Save and close all the Floor Plan View files. **Caution:** Do not close all files. Be sure at least one empty drawing remains open. The easiest way to do this is to close and save all project files, and then click the Qnew icon to create a new empty drawing.

We have copied and renamed several files. To make all changes permanent and avoid broken XREFs later, we must instruct Project Navigator to re-path all XREF files.

45. At the bottom of the Project Navigator palette, click the Re-path icon and then click Re-path in the dialog box that appears (see Figure 5–37).

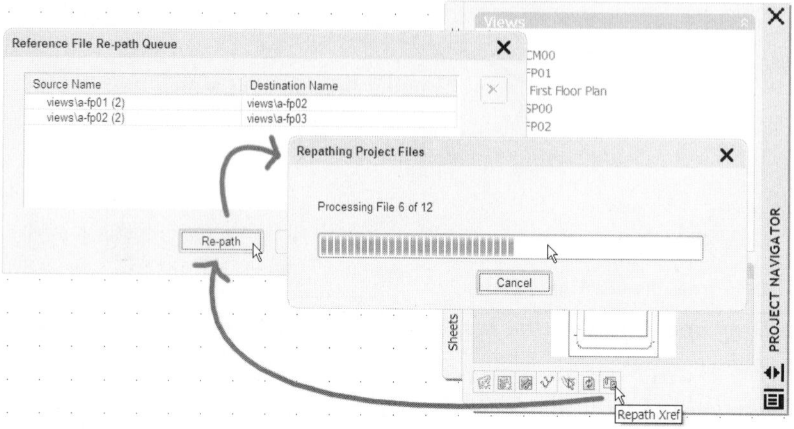

Figure 5–37 Re-path the Project to ensure that XREF links are not lost next time ADT is loaded

ELEVATIONS AND SECTIONS

Architectural Desktop provides several tools to view and document building sections and elevations. For a quick elevation view, you can view the model from one of the orthographic views such as Left or Front. However, these views show the model as is, without the level of abstraction typically required for a printed elevation or section. With a bit more effort, you can cut a "live section" view through the model that crops away the portion behind the Section line and reveals the sectioned portion in a live view. Depending on how these are cut and configured, live sections can even be printed under certain circumstances.

For most design development and construction document needs, the ADT 2D Section/Elevation object gives us the required level of control by creating a separate two-dimensional drawing that remains linked to the original building information model and can be updated when the original changes. This linked drawing functions like a graphic report of the data within a building model. To get just the right section or elevation requires a bit of careful configuration. (In Chapter 16, we will explore the 2D Section/Elevation object in detail.)

In this exercise, we carry on from the previous exercise and generate simple elevations that will serve as placeholders in our cartoon set, in the same way that the basic walls we added earlier serve as stand-ins for our plans.

CREATE THE BUILDING ELEVATION VIEW

We can generate sections and elevations from any ADT model. For instance, for quickly assessing design edits and changes, Live Sections can be enabled in any Construct or View file. These are used typically for this purpose and rarely for printed documentation. However, Live Sections can be printed if desired and can make very interesting presentation drawings. In these cases, they ought to be generated in separate View files and then dragged to Sheets.

2D Section/Elevation objects can also be created in any ADT file, but often should be created in separate Elevation/Section Views. The benefit of this approach is that each Section/Elevation View can be uniquely configured for its specific purpose, and it enables flexibility in work flow by enabling different team members to work simultaneously on different parts of the project. Following this rationale, we will start by creating a new Section/Elevation View file for our building elevations. We will use the Callout routines on the Tool Palettes to assist us with this task.

1. In Project Navigator, on the Views tab, double-click *A-FP01* to open it.

 Note: If you left the First Floor Plan View open above, then this action will simply make that file active.

2. On the Callouts tool palette, click the **Exterior Elevation Mark A3**tool (see the right side of Figure 5–38).

   If you do not see this palette or tool, right-click the Tool Palettes title bar and choose **Document**, and then click the Callouts tab.

You may be wondering, if we intend to create a separate Section/Elevation View file, why we have begun the process in the First Floor Plan file. The callout routine will add four elevation callouts, one on each side of the building, and then generate four corresponding elevations. The callouts will be placed in the current file (in this case *A-FP01*), and when prompted, we will generate the elevations in a new Section/Elevation View file.

3. At the "Specify first corner of elevation region" prompt, click a point below and to the left of the building.

4. At the "Specify opposite corner of elevation region" prompt, click a point above and to the right the building (see Figure 5–38).

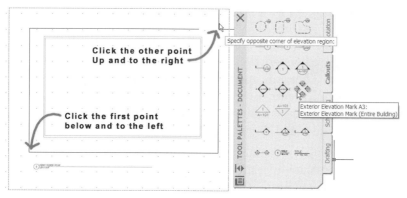

Figure 5–38 *Designate the extent of the building elevation region*

The Place Callout worksheet will appear.

5. In the Place Callout worksheet, verify that a check mark does appear in the Generate Section/Elevation and the Place Titlemark check boxes.

6. For Scale, verify that **1/8″ = 1′-0″ [1:100]** is chosen from the list.

These are the default settings for the exterior elevation callout routine. The Generate Section/Elevation check box instructs ADT to create a 2D Section/Elevation object from each of the four sides of the rectangular region designated in the previous steps. The Place Titlemark setting is used to add a title mark beneath each 2D Section/Elevation like the ones we added manually to the plans the plans above.

7. In the New Model Space View Name field, change the text to read **North Building Elevation;East Building Elevation;South Building Elevation;West Building Elevation**.

When the 2D Section/Elevation object is created, a Model Space View (like the ones we created above for our Plans) will be created around each one. These Model Space View names must be unique within the project, so plan carefully. You can use any names that you find suitable. Remember, these are the names of the Model Space Views being created, which in turn will become the names automatically referenced by the title marks. The Scale setting is applied to this Model Space View and is also used to scale annotation applied to the 2D Section/Elevation object such as the automatically created title mark.

8. Verify that your settings match Figure 5–39.

Figure 5–39 *The Place Callout worksheet*

There are four behaviors in this dialog box. If you choose **Callout Only** at the top, you will simply get the section/elevation Callouts, and nothing else will be generated. This is useful if the section or elevation already exists somewhere and you simply wish to have a callout reference it; such as on an upper floor. The other three icons allow you three destinations for the generated section or elevation. The choices are a new View drawing, and existing View drawing or the current drawing. In this case, we are creating the elevations and a new View drawing in the same operation.

> 9. In the middle of the Place Callout worksheet, in the "Create in" area, click the New View Drawing icon.

The Add Section/Elevation View wizard will appear. This is the same wizard that we have worked through already to create the composite model and plan View files.

> 10. For the name, type **A-EL01**, type **Architectural Building Elevations** for the Description, and then click Next (see item 1 in Figure 5–40).

> 11. Verify that all levels are selected and then click Next (see item 2 in Figure 5–40).

A Section/Elevation View differs from a General View only slightly. Basically, the Section/Elevation View assumes that you wish to create your section or elevation from all levels and has them all pre-selected, while the General View, as we have seen, does not start with any items selected. Also if you look carefully on the Content screen, you will note that the XREFs for a Section/Elevation View use Overlay, unlike those of the General View, which use Attach.

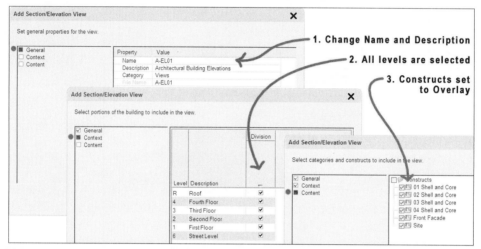

Figure 5–40 *The Add Section/Elevation View wizard*

12. Verify that all Constructs are selected and then click Finish (see item 3 in Figure 5–40).

 Caution: Do not press the ENTER or ESC keys. You are not done with the Callout routine yet!

Look at the command line and notice the message that has appeared. It will read:

** You are being prompted for a point in a different view drawing **

If you turned off your command line, you will still see a prompt at your cursor if dynamic input is on. (Refer to Chapter 1 for more information on these options). When you create elevations to an existing or new View file, you still must indicate within the current drawing where you would like the elevations to be created in that file. This prompt serves to inform you of that. Since the Section/Elevation View file being created will XREF the same Constructs that appear in the current file, it is usually best to pick a point off to the side of the plan.

13. At the "Specify insertion point for the 2D elevation result" prompt, click a point in the drawing to the right of the floor plan beyond the purple rectangle (see item 1 in Figure 5–41).

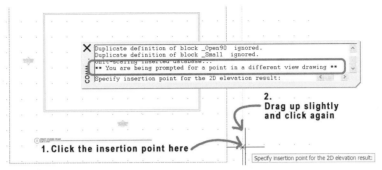

Figure 5–41 *Click two points to the side of the drawing to indicate the insertion point and direction to generate elevations*

14. At the "Pick a point to specify the spacing and direction of elevations" prompt, move the mouse up slightly and then click again (see item 2 in Figure 5–41).

The first point is the insertion point of the elevations and will become the lower left corner of the elevation object. The distance and direction between the second and first points will be used to determine the direction in which to draw the elevations and how far apart to space them. At this point, the routine is complete. You should have four Callouts in your First Floor Plan–*A-FP01* file. There should also be a new View file named *A-EL01* on your Project Navigator.

> If any of your Callouts are overlapping other geometry, feel free to move them.

15. Save the *A-FP01* file.

16. On the Project Navigator palette, click to expand the small plus (+) sign next to *A-EL01*.

Notice that there are four Model Space Views with the names that we assigned above already configured within this View file (see Figure 5–42).

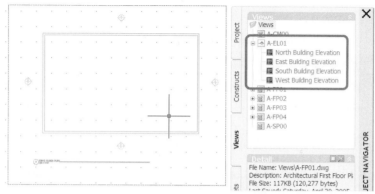

Figure 5–42 *The First Floor Plan now contains Callouts, and the new A-EL01 View appears in Project Navigator complete with four Model Space Views*

17. On the Project Navigator palette, double-click *A-EL01* to open it and then Zoom Extents.

Examine the Elevation objects in this file. Notice that there are four of them, spaced one on top of the other as we indicated above. If you zoom and pan around a bit, you will also notice that the building appears to "float" above the ground in each of the elevations. Don't worry; we will address this in later chapters. Also notice that the title mark beneath each one references the title of the associated elevation. These titles come from the Model Space View names that were created with this file. Finally, depending on where you clicked your elevation insertion point in the other file, the elevations may be overlapping the *Site* Construct. This is because the *Site* file is included here, but was not in the First Floor Plan View. We can simply move the elevations and their title marks a bit to the right if necessary.

18. Click on one of the Elevation objects.

Notice the dashed outline that appears around the elevation and the magenta grips on each edge. This is the boundary of the associated Model Space View. The magenta grips allow you to resize the Model Space View (see Figure 5–43). If you move the 2D Section/Elevation object, the Model Space View will move as well. This is very handy, as we will see. You will also notice that the Bldg Elevation Line associated with this elevation also highlights in red when the associated elevation is selected. The Bldg Elevation line is an object that determines what portion of the plan is to be included in the section or elevation. We will look more at this object in Chapter 16.

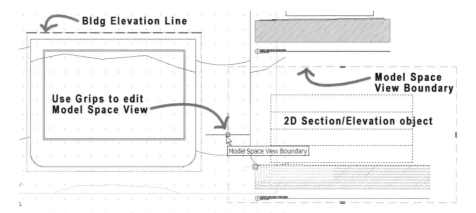

Figure 5–43 *The Model Space View associated with each elevation moves with the elevation and can be grip edited*

19. Select all four elevations and all four title marks.

Tip: Be sure to select all three pieces of each title mark. Use a window selection to select everything easily. Make a window selection by clicking above and just to the left of the top elevation, and then clicking again below and to the right of the bottom one. Your window must surround all of the items to select them. New to 2006, the Window selection will shade in color—blue by default. It will not select items not completely surrounded like the *Site* Construct in this case.

All four elevations and their associated Model Space View boundaries and title marks should all be highlighted.

20. Move all selected items to the right until they no longer overlap any of the XREFs.

These elevations are very simplistic; however, they are more than adequate for us to use in planning an elevation Sheet file. This is our main goal at this time: to set up the required Model and Sheet files and to print a cartoon set. As the design evolves in the upcoming chapters, these elevations remain linked to the model, and we will update them periodically.

21. Save and Close the *A-EL01* file.

CREATE A SECTION VIEW

We can create sections using a nearly identical process. On the Callouts tool palette are several variations of Callout symbols, each with slightly different behaviors tailored to their intended functions. There are two types of section that we can create in ADT: the 2D Section/Elevation object (like the ones created in the previous sequence) is typically used for Design Development and Construction Documentation purposes, and the Live Section is more of a design and presentation tool. We will begin by creating a 2D Section/Elevation object through the building following a process nearly identical to the steps covered in the previous topic. Then we will create a Live Section from the same Bldg Section Line. This will give us two ways to consider how the design is shaping up throughout the course of the design.

 Note: Please note that both types of section use the same Bldg Section or Elevation Line object. Therefore, any Section or Elevation Line object can be used to generate both 2D and Live Sections and Elevations.

22. In Project Navigator, on the Views tab, double-click *A-FP01* to open it.

   If you left the First Floor Plan View open above, then this action will simply make that file active.

We are going to run a section line vertically through the middle of the plan.

23. On the Callouts tool palette, click the **Section Mark A2T** tool.

   If you do not see this palette or tool, right-click the Tool Palettes title bar and choose **Document**, and then click the Callouts tab.

24. At the "Specify first point of section line" prompt, click a point on the Construction Line near the top of the plan just beneath the Elevation Callout (see Figure 5–44).

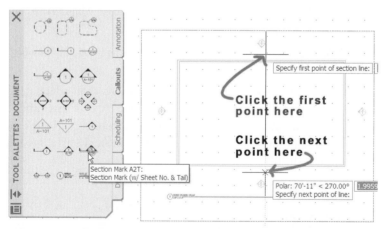

Figure 5–44 *Draw a Construction Line through the middle of the plan*

25. At the "Specify next point of line" prompt, move the mouse down to just beneath the lower horizontal Wall and slightly above the elevation callout and click (see Figure 5–44).

26. At the subsequent "Specify next point of line" prompt, press ENTER to complete the operation.

27. At the "Specify section extents" prompt, move the mouse to the left side of the section line and then click outside the building Walls.

    Use the interactive dashed section line boundary as a guide to click in the right spot (see Figure 5–45).

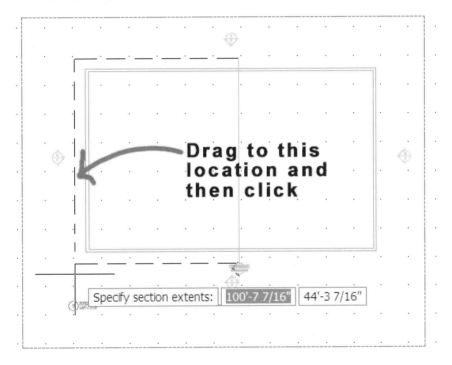

Figure 5–45 *Drag the dashed section line boundary to enclose the left of the building*

28. In the Place Callout worksheet, type **Transverse Building Section** for the Model Space View name and accept all other defaults.

29. In the "Create in" area, click the New View Drawing icon.

30. For the name, type **A-SC01** and **Architectural Building Sections** for the Description, and then click Next.

31. On the Context page, verify that all levels are selected and then click Next.

32. On the Content page, verify that all Constructs are selected and then click Finish.

Caution: Do not press ENTER or the ESC key. You are not done with the Callout routine yet!

33. Pan over to the left side of the plan a distance about equal to the plan itself.

Remember to pay close attention to the command line here. Refer to the "Create the Building Elevation View" topic and Figure 5–41 above for the command line warning during this routine. We are placing the section on the opposite side of the plan this time. However, the insertion point of the section is still the lower left corner; this is why we need to pan to the left to allow enough room for the section in the new View drawing. If you pick too close to the plan, it is not a big concern. You can always move the generated section later, as we saw with the elevations above.

34. At the "Specify insertion point for the 2D elevation result" prompt, click a point in the drawing to the far left of the floor plan.

35. Save the *A-FP01* floor plan file.

The routine is now complete, and there should be a new *A-SC01* View drawing on the Project Navigator and a new section Callout in the current plan file. Notice that the section line is continuous the full height of the plan. In some cases, it is desirable to break this line in the middle to make it easier for the plan drawing to read. You can do this now, using the AutoCAD Break command or wait until later when the plan is more fleshed out. Either way, breaking this polyline in the plan file, will have no impact on the actual section object in the *A-SC01* file.

36. On the Project Navigator palette, double-click *A-SC01* to open it.

37. Zoom Extents.

    If the section overlaps the XREFs at all, use the process above in the "Create the Building Elevation View" topic to move the section, its Model Space View, and title mark over to the left until it no longer overlaps.

38. Save the file.

ENABLE A LIVE SECTION

Working in the same file, *A-SC01*, we can enable a Live Section from the Bldg Section Line created by the Callout routine. In this way, we can use the Live Section to assist in making design decisions, and then once changes have been made, refresh the 2D Section/Elevation object to incorporate those changes into the printed Sheets that we will set up below.

39. On the Project Navigator palette, double-click *A-SC01* to open it.

Note: If you left the Section View open above, then this action will simply make that file active.

40. Change the active Display Configuration to **High Detail**.

The active Display Configuration can be chosen from a list at the bottom right corner of the screen on the Drawing status bar. (See the "Loading a Display Configuration" topic in Chapter 2 for more information). Live Sections apply to the active Display Configuration. Since the one we are enabling here will be used as a design tool, it is useful to enable it in the High Detail Display mode.

41. Select the Section line object in the middle of the plan, right-click and choose **Enable Live Section**.

42. From the View menu, choose **3D Views > SE Isometric** (see Figure 5–46).

Notice that the entire model has been sectioned transversely through the middle. Hatching appears at the section line. Also notice that the 2D Section/Elevation object has disappeared. There is no cause for alarm. Recall the lessons of Chapter 2 in the "Working with the Display System" topic. There we explored the way ADT objects change their display behavior under different viewing conditions. Here, because we are viewing the model in 3D, the 2D Section/Elevation object has simply been turned off. If you return to Top view, it will re-appear.

> To make the model easier to see, zoom in a bit, choose **Hidden** from the **View > Shade** menu, and then turn off the GRID using the toggle button on the Application status bar. View the model from other views or using 3D Orbit if you wish.

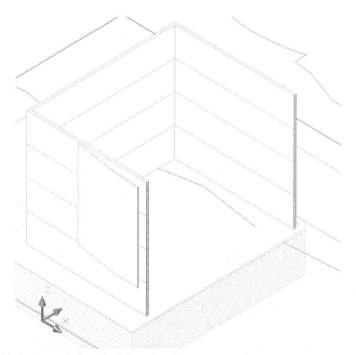

Figure 5–46 *A view of the Model with a live section enabled*

43. Save and Close the file.

We will look more at this Live Section in later chapters.

ADD ELEVATION AND SECTION CALLOUTS TO UPPER FLOOR PLANS

When we drag the elevations to a Sheet below, the field codes embedded within the Callouts in the First Floor Plan file will update to reference the correct elevation number and Sheet number. We would also like to see Callouts referenced to the same information appearing on the upper floors of the project. We could repeat the same Callout routine in the Second and other upper floors and simply choose the "Callout Only" option this time. (The Callout Only option creates only the Callout symbols and does not create a 2D Section/Elevation object or a View file.) The only problem with that approach is that we would then need to use the technique covered above in the "Post Linking Field Codes" topic to properly reference the fields in the Callouts to the correct elevations. It is much easier to simply copy and paste the required Callouts from the First Floor Plan to the other floors. Recall the behavior above before we post-linked the title mark copied from the first floor to properly reference the second floor. At first, it simply continued to link to the original reference. In this case, this is exactly the behavior that we want.

44. On the Project Navigator palette, click the Views tab and then double-click *A-FP01* to open it.

 Note: If you left the First Floor Plan open above, then this action will simply make that file active.

45. Select all four elevation Callouts, the section Callout, and both ends of section line, right-click, and choose **Clipboard > Copy**.

 Note: Each Callout is comprised of two separate objects. Be sure to select both in each case, or eight objects total.

 Tip: You can select one elevation callout, one section callout and the section line and then right click and choose **Select Similar** to select the rest.

46. On the Project Navigator palette, double-click *A-FP02* to open it.

47. Right-click and choose **Clipboard > Paste to Original Coordinates**.

48. Repeat the **Paste to Original Coordinates** in both *A-FP03* and *A-FP04*.

 Tip: If you prefer, you can drag the items with the CTRL key held down to each of the floor plan View files.

49. Save and Close all Floor Plan files.

SHEET FILES AND THE CARTOON SET

In this exercise, we will create Sheet files for each of the floor plans as well as an elevation and section Sheet file. Finally, we will print the cartoon set. The Auto-CAD Sheet Set functionality is fully incorporated into Project Navigator. When you click the Sheets tab, you will see a hierarchical series of items that look similar to a folder tree. These are the Sheet Set and its nested Subsets. Sheet Sets are used to organize drawing files and their Layout tabs for plotting. Coupled with all of the other functions of Project Navigator, Sheets Sets provide the final component of the ADT Drawing Management system.

When you create a new project, a Sheet Set will automatically be created. A Sheet Set Template is used to create all of the initial Subsets. If you click on the Sheets tab of Project Navigator, you will see that the default Sheet Set template creates a General and an Architectural Subset. The Architectural Subset contains several additional Subsets that further help to organize a large document set.

CAD Manager Note: The entire collection of Subsets is completely customizable. If your firm uses Sheet categories different from those included here, they can easily be modified to suit your firm's needs. To do this, open a project and then modify the Sheet Set by adding, modifying or deleting Sub Sets. Right click each Sub Set and choose its Properties, such as template file to use. Once you are satisfied with the Sub Set organization, copy the DST file from Windows Explorer to your ADT templates folder. You can choose **Format > Options** and click the AEC Project Defaults tab to set this DST as the default for new projects that are created without template projects. If you are planning to implement the project template feature (used at the start of this chapter to create the Commercial Project) then you will want to load that template project (using Project Browser) and then modify its Sheet Set instead. This will give all future projects created from that template project a standard and consistent Sheet Set with office standard Sub Sets and Sheet templates.

CREATE THE SITE PLAN SHEET FILE

Let's begin with the Site Plan. Let's create this new Sheet within the Architectural – General Subset.

1. On the Project Navigator palette, click the Sheets tab (see Figure 5–47).

   Be sure that the Sheet Set view is active. If it isn't, click the Sheet Set icon at the top right corner of the Sheets tab. If Sheet Set View is active, but only MADT Commercial shows in the list, right-click on *MADT Commercial* and choose **Expand**.

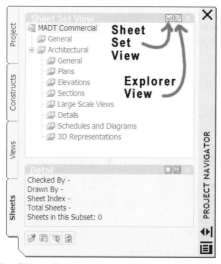

Figure 5–47 *Be sure the Sheet Set view is active*

2. Right-click on the *General* Subset beneath the *Architectural* Subset and choose **New > Sheet**.

3. For the Number, type **A-100** and for the Sheet Title, type **Site Plan**. Click OK to dismiss the New Sheet dialog box (see Figure 5–48).

Notice that the File Name is created automatically by concatenating these two fields. You are able to edit the File name field manually before clicking OK.

New in ADT 2006, you can now rename a Sheet file and choose whether the associated drawing file renames as well. When you do this, you will have the option to use the Name only or the Name and Number in the new file name. To see these options, select a Sheet, right-click and choose **Rename and Renumber**.

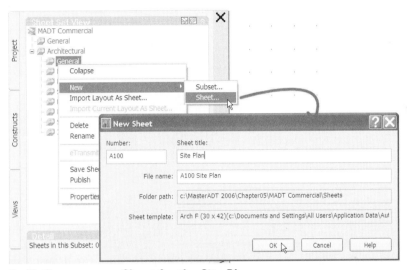

Figure 5–48 *Create a new Sheet for the Site Plan*

4. In the *Architectural – General* Subset, double-click *A-100 Site Plan* to open it.

   A Layout tab named the same as the file is active. Examine the fields in the title block. Notice that many of them have filled in automatically with project data field codes.

CAD Manager Note: You will notice that the template used for Sheets is different from the one used by the other file types. When you first set up a project, there are settings for this. You can assign the template used directly to the Sheet Set, or each individual Sub Set can have its own template—this can be very useful in multi-discipline firms. We are using the defaults here (based on the template project which we used to create MADT Commercial), which load the *Aec Sheet (Imperial Stb).dwt* [*Aec Sheet (Metric Stb).dwt*] template file to create Sheets. You can change these settings for projects begun without template projects on the AEC Project Defaults tab of the Options dialog box (Format menu) or within the Sheet Set of the template project that you use to create a project. To do this, make the template project current using Project Browser and then right-click on the Sheet Set at the top of the tree (the top node of the Sheet Set will have the same name as the project) and choose **Properties**.

5. On the Project Navigator palette, click the Views tab.

6. Drag and drop the *A-SP00* View from the Project Navigator palette directly onto the Sheet Layout (see Figure 5–49).

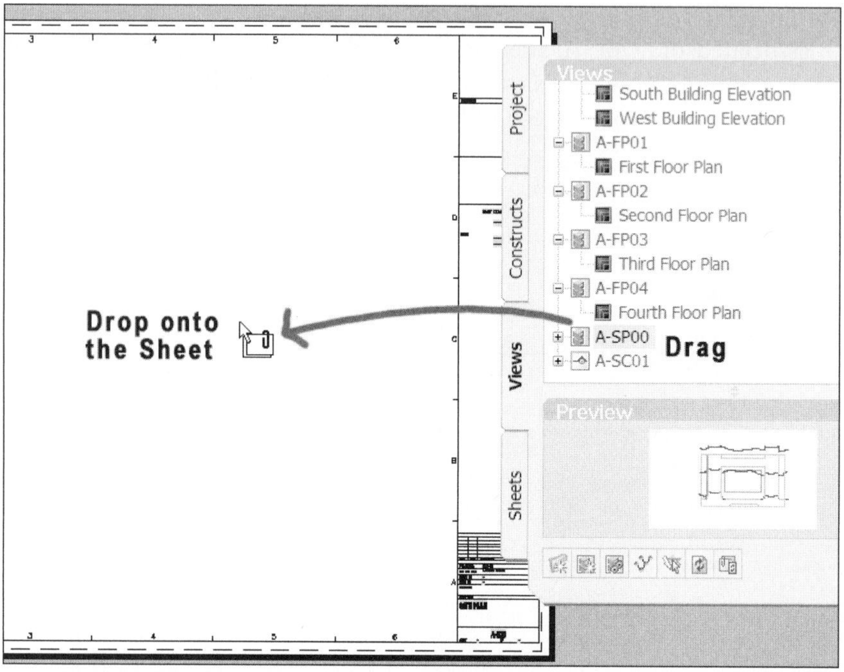

Figure 5–49 Drag the Site Plan from the Project Navigator onto the drawing sheet

The image of the file will appear on screen with the lower left corner attached to the cursor.

7. Move the Viewport to position the Site Plan to an optimal position on screen.

You may need to fine tune the position or size of the viewport after placement. You can move it around and resize it with the grips.

8. Save and close the Site Plan Sheet.

Dragging the View onto the Sheet in this way has created a single Viewport scaled at 1/8″=1′-0″ [1:100]. (We can verify the scale by selecting the Viewport object and right-clicking and choosing **Properties**). Notice that the viewport has "Display Lock" set to **Yes**. This prevents the viewport scale from being changed accidentally when someone has the viewport active and then zooms. Notice also that the viewport is automatically placed on a non-plotting layer.

CREATE THE REMAINING FLOOR PLAN SHEET FILES

Let's now create another Sheet and add the remaining floor plans to it.

9. In Project Navigator, click the Sheets tab.

10. Right-click on the *Plans* Subset beneath the *Architectural* Subset and choose **New > Sheet**.

11. For the Number, type **A-101** and for the Sheet Title, type **Floor Plans**. Click OK to dismiss the New Sheet dialog box.

    In the numbering scheme used here, "1" stands for Plans (Horizontal Views), and 01 is the floor. Above, "1" was also used for the Site Plan, because it is a horizontal view, but "00" was used to denote the Street level.

12. In the *Architectural – Plans* Subset, double-click *A-101 Floor Plans* to open it.

13. On the Views tab, expand the plus sign (+) next to *A-FP01* and then drag and drop the "First Floor Plan" Model Space View onto the Sheet (see Figure 5–50).

14. Snap the lower left corner of the viewport to the midpoint of the left vertical border of the Sheet (immediately adjacent to the letter C on the border—see the insert in Figure 5–50).

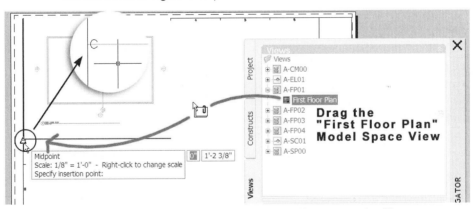

Figure 5–50 *Drag the First Floor Plan Model Space View onto the Sheet to create a viewport from the pre-defined boundary*

Notice that the bubble next to the title mark for the First Floor Plan has filled in automatically with the number 1. This is because this is the first drawing on this Sheet. When we drag in the Second Floor Plan next, it will become number 2, and so on. The elevation Callouts have not yet updated. This will happen after we drag the elevations to their own Sheet.

15. Expand the plus (+) sign next to *A-FP02* and then Drag and drop the Second Floor Plan Model Space View onto the Sheet.

16. Snap the lower left corner of the viewport to the lower right corner of the First Floor Plan viewport.

Again note that the title mark updates immediately to reflect this change.

17. Repeat this process to drop the Third and Fourth Floor Plan Model Space Views onto the Sheet (see Figure 5–51).

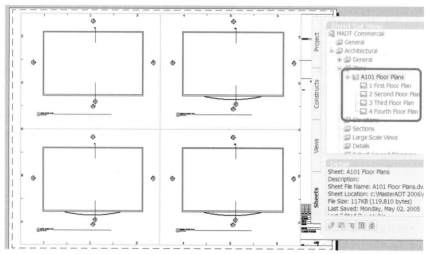

Figure 5–51 *The Floor Plans Sheet with all of the Plans added*

Take a look at the Sheet Set in Project Navigator. Notice that once we have dragged several plans onto the *A-101 Floor Plans* Sheet, there are now four Views indented beneath the Sheet name (see Figure 5–51). These are more AutoCAD Named Views. (We have been calling them Model Space Views till now; however, that name is not applicable in Paper Space Layouts.) These Named Views represent the area of the Sheet that is associated to a particular number. Try double-clicking on one of these Views. It will simply zoom to that location on the Sheet.

18. Save and close the *A-101 Floor Plans* Sheet.

CREATE THE ELEVATION SHEET FILE

We create the Elevation Sheet in the same fashion.

19. On the Sheets tab of the Project Navigator, create a new Sheet in the *Architectural – Elevations* Subset numbered **A-201** and with the Sheet Title set to **Building Elevations**.

"2" is the code for elevations, and "01" makes it the first elevation sheet.

20. In the *Architectural – Elevations* Subset, double-click *A-201 Building Elevations* to open it.

21. On the Views tab of the Project Navigator, drag and drop *A-EL01* onto the Sheet.

NOTE Drag the entire View file this time—not the individual Model Space Views. It will successively insert each Model Space View and number them accordingly.

22. As each Model Space View is inserted, place it on the Sheet until all four elevations are placed.

    Make any necessary adjustments to the position and size of the viewports. You can use the grips to stretch the viewports if needed.

Examine all the elevation numbers. Notice that they have been automatically numbered in the sequence in which they were added to the Sheet (see Figure 5–52).

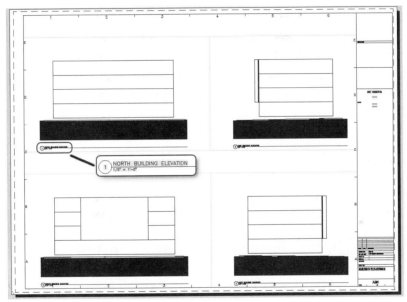

Figure 5–52 *Complete the setup of the Elevation Sheet*

23. Save and Close the file.

CREATE THE SECTION SHEET FILE

Although we generated both a 2D Section/Elevation object and a Live Section above, here we will place only the 2D Section/Elevation object on the Sheet.

24. On the Sheets tab of the Project Navigator, create a new Sheet in the *Architectural – Sections* Subset numbered **A-301** and with the Sheet Title set to **Building Sections**.

    "3" is the code for sections (Vertical Views), and "01" makes it the first section sheet.

25. In the *Architectural – Sections* Subset, double-click *A-301 Building Sections* to open it.

26. On the Views tab of the Project Navigator, drag and drop *A-SC01* onto the Sheet—do not click to place it yet.

27. Before placing the viewport, right-click.

Notice that a menu of standard scales appears. Suppose that at the time you created the View file, you chose a scale, or perhaps forgot to set a scale that you now realize is incorrect in relation to the Sheet. Using this menu, you can set a different scale for this viewport on the fly.

28. From the list of scales, choose **1/4"=1'-0"** [**1:50**], and then place the viewport on the Sheet (see Figure 5–53).

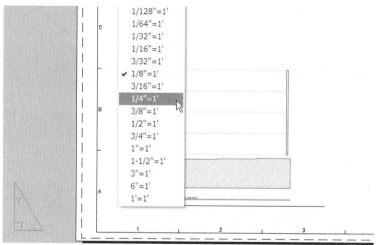

Figure 5–53 *Right-click while placing a viewport to change the scale*

Even though this setting can be very handy at this early stage of the project, take a look at the title mark. It is now the wrong scale for the viewport. You will need to make adjustments to the View file anyhow. Let's take a look at that now.

29. Save and Close the file.

30. On the Project Navigator palette, click the Views tab and then double-click Transverse Building Section beneath *A-SC01* to open it.

Notice that this action opens the *A-SC01* file and then automatically restores the Transverse Building Section Model Space View. A temporary label will appear with this name on screen until you change the zoom or pan.

31. On the Project Navigator palette, right-click on Transverse Building Section beneath *A-SC01* and choose **Properties**.

32. In the Modify Model Space View worksheet, change the scale to **1/8"=1'-0"** **[1:100]** and then click OK.

33. In the drawing window, right-click and choose **Basic Modify Tools > Scale**.

34. At the "Select Objects" prompt, select all pieces of the title mark and then press ENTER.

35. At the "Specify Base Point" prompt, choose the center of the bubble and then press ENTER.

36. At the "Specify scale factor" prompt, type **.5** and then press ENTER.

37. Save and Close the file.

38. On the Sheets tab of the Project Navigator palette, double-click A-301 Building Sections to open it.

    Notice that the Title mark is the correct size. If you left this file open, a balloon should appear alerting you that the XREF file has changed. Simply click the link in this balloon to reload the *A-SC01* XREF file.

39. Save and close the *A-301 Building Sections* file.

40. On the Sheets tab of the Project Navigator palette, double-click *A-101 Floor Plans* to open it.

Zoom and pan round the file and notice that now that we have created the elevations and sections Sheets, all of the Callouts now correctly display the associated elevation or section to which they reference on all four floors. If you left the file open before, then choose **Regen All** from the View menu to force the fields to update with the new information. You do *not* need to reload the XREFs for this update to occur.

41. Save and Close the file.

CREATE A COVER SHEET

Let's create one more Sheet before printing our cartoon set. A Cover Sheet template has been provided with the Chapter 5 dataset files. We will use it now to give our set a Cover Sheet.

1. On the Sheets tab of the Project Navigator, right-click the *General* Subset and choose **Properties**.

    Be sure to choose the *General* category above *Architectural* and not the one indented beneath *Architectural*.

2. Place a check mark in the Prompt for template check box and then click OK (see Figure 5–54).

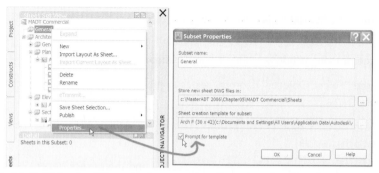

Figure 5–54 *Enable the "Prompt for template" option for the General Sub Set*

Each Subset of a Sheet Set can reference its own template file. This is very useful for multi-discipline firms where, for instance, the MEP Consultant uses a different template than the Architects do. In this case we are specifying that we wish to be prompted for a template. You will not typically want this setting in most Subsets; however, it seems appropriate here in the General category, which will have a Cover, perhaps a Legend, and other types of Sheets that can use slightly different templates.

3. Right-click the *General* Subset and choose **New > Sheet**.

Notice that a dialog box appears to prompt you to select a template file for this Sheet.

4. Click the Browse (…) icon next to the Drawing template file name field.

The default template folder on your system will be displayed. This could be in the *Documents and Settings* folder of your system, on your office server, or any other location. The template that has been provided is in the same location as the other dataset files.

5. In the Select Drawing dialog box, on the left side, click the Desktop icon, double-click *My Computer*, and then browse to the *C:\MasterADT 2006\Chapter05* folder.

6. Choose *Cover Sheet.dwt* [*Cover Sheet-Metric.dwt*] and then click Open.

In the bottom half of the dialog box, any Layouts saved in the template will appear. You choose a Layout here, and it will be used to create the new Sheet.

7. Choose Cover Sheet (30x42) [Cover Sheet (841 x 1189)] and then click OK to continue to the New Sheet dialog box.

8. For the Number, type **G-100** and for the Sheet Title type **Cover Sheet**.

"G" indicates "General," where "A" indicated "Architectural on all of the other Sheets". "100" is used here for consistency with the naming of the other Sheets and simply indicates that it is the first Sheet.

9. Double-click *G-100 Cover Sheet* to open it (see Figure 5–55).

It is a pretty simple Cover Sheet. Naturally, you can customize this template or use your own in real projects. Notice that several of the fields have empty values. These are retrieved from the project database. At this time, we have not yet entered those values. We can edit these values now, and the Cover Sheet will update.

10. In Project Navigator, click the Project tab, and then click the Edit Project icon at the top (see Figure 5–6 above).

11. At the bottom of the dialog box, click the Edit button next to Project Details.

12. Edit any fields that you wish, and then click OK twice to return to the drawing. From the View menu, choose **Regen** to update the fields.

The first field below the Project title is the "Project Address." The Date field is actually the "Project Start Date." "Owner" and "Architect" fields are clearly noted in their respective categories. One final touch that our Cover Sheet could use is a Sheet Index.

13. On the Sheets tab of the Project Navigator, right-click the top node of the Sheet Set labeled *MADT Commercial* and choose **Insert Sheet List**.

14. In the Insert Sheet List Table dialog, place a checkmark in the "Show subheader" check box and then click OK.

The Sheet List Table dialog will appear and warn you that if you make manual edit to the table, that they will be lost the next time the table is updated.

15. Click OK to dismiss the Sheet List Table dialog.

16. Place the Table in the lower right corner of the Cover Sheet in the space labeled for it (see Figure 5–55).

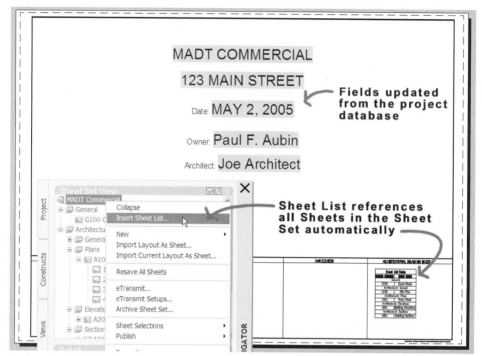

Figure 5–55 *Create a Cover Sheet from a different template update its fields and insert a Sheet List*

    17. Save and Close the file.

PUBLISH THE CARTOON SET

Now that all of the preliminary files have been created, we can publish our Cartoon Set.

    18. In Project Navigator, click the Sheets tab.

    19. At the top of the Sheet Set list, right-click the *MADT Commercial* node and choose **Publish > Publish to DWF**.

    20. In the Select DWF File dialog box, accept the default name (*MADT Commercial.dwf*), browse to the *C:\MasterADT 2006\Chapter05\Commercial* folder, and then click Select.

Sit back and watch it process the entire Sheet Set. You can access this same command from each node of the Sheet Set. Therefore, you can Publish the entire drawing set as we have done here, or you can publish smaller Subsets. You can even create custom Sheet Selections and publish only those. We will explore this technique in Chapter 17. When the DWF is complete, a balloon will appear in the status bar (see left side of Figure 5–56). Click on it to see a report.

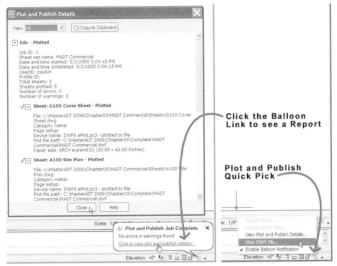

Figure 5–56 *A notification appears when a Plot and Publish job is complete, right-click for options*

The Publish routine can create hard-copy plotted Sheets using the settings saved in the Page Setup of each file, or it can also create a multi-page DWF (Design Web Format) file. This single file will contain several pages, one for each Sheet, and can be distributed electronically. This is what we have created here. To view and print DWF files, use the free Autodesk DWF Viewer software available from Autodesk.com and installed automatically with Autodesk Architectural Desktop. If the recipients wish to mark up the DWF file, they can purchase Autodesk DWF Composer instead. Markups generated in Composer can be loaded back into ADT for reference while picking up changes (see Chapter 19 for more information). Let's view the DWF file now.

21. On the Application status bar, at the right side, right-click the small Plot and Publish icon and choose **View DWF File** (see right side of Figure 5–56).

    Depending on whether you have DWF Composer or DWF Viewer installed, the DWF will display in one of these applications.

22. Explore the file in DWF Composer or DWF Viewer (see Figure 5–57).

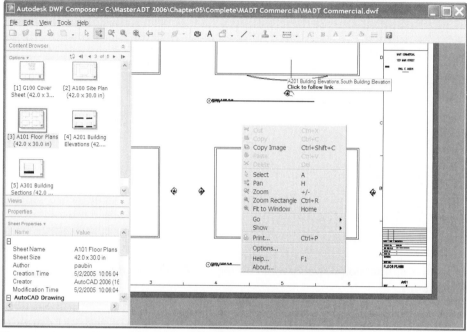

Figure 5–57 *Viewing multi-sheet DWF files in Autodesk DWF Composer or Viewer*

All Sheets of the set will appear in DWF Composer or Viewer in a panel on the left as icons, or in a list. Click the icon to view a particular Sheet. All Callouts from the original files will contain hyperlinks in the DWF file. Simply hold down the CTRL key and click a link to jump directly to the referenced Sheet in the DWF. If you have Composer, you can make markups directly on top of the DWF file, save them, and then load them back into ADT. Right-click for additional options and navigation controls.

23. Close DWF Composer or DWF Viewer when you are finished viewing the DWF file.

If you prefer paper plots, and you are already familiar with AutoCAD plotting, choose the Plotters named in page setups option instead, and proceed to generate "real paper" plots. (For more information on plotting in ADT, refer to Chapter 19 of this book or consult the online help.)

24. Save and Close all Commercial Project files.

CAD Manager Note: Although your office standards may vary considerably from the file naming and XREF structure presented here, the more important issue is the strategy of using consistent file naming and XREF structures from one project to the next. It is also critical to set up the project files as early in the project life cycle as possible. Building the set early allows for easier setup and maintenance and allows the project team members to follow an established standard.

New in ADT 2006 If you are also planning to utilize the new Project Standards features of Architectural Desktop 2006, then you will also want to configure project standards files, and potentially project Content Browser libraries, Project Tool Catalogs, and Project Tool Palette Groups as well. All of these items can be included in your own custom Project Template that can then be used to create future projects. Examples of some of these items can be found throughout future chapters such as Chapter 8, 10 and 11.

PART 2 – SETTING UP A RESIDENTIAL PROJECT

The second project is an 800 SF [75 SM] residential addition. This project will require a little bit of demolition and new construction and will require plans, sections, and elevations. The residential project has similar ADT file requirements. However, there are a few differences. We will incorporate the files built in the last chapter into this project. The following tutorial covers the setup of the Residential project. The completed files for the Residential project cartoon set are available in the *Chapter05\Complete* folder with the files installed from the CD.

RESIDENTIAL PROJECT FRAMEWORK

In the Commercial project above, we created several levels, but left the single default Division. In the Residential project, there is a clear line separating the existing house from the new addition. Therefore, we will create two Divisions, one for the existing construction (which will include both existing to remain and demolition) and one for new construction. In addition, Demolition, Existing, and New Construction will be articulated with Layers and Cleanup Groups (an ADT Wall property). This will give us the required separation between construction phases and portions of the project.

CREATE THE RESIDENTIAL PROJECT

1. From the File menu, choose **Project Browser** (see Figure 5–1 at the beginning of this chapter).

2. In the *C:\MasterADT 2006\Chapter05* folder, create a New Project using the default template project as we did above and name it **MADT Residential**.

3. Input the required project information: Project Number: **2006.2**, Description: **Mastering Autodesk Architectural Desktop 2006 Residential Project** and then click OK.

SET UP THE PROJECT STRUCTURE

Now let's edit the Divisions and Levels for the Residential Project.

4. In the Project Navigator, click the Edit Divisions icon and use Table 5–1 as a guide and set up the two Divisions (see Figure 5–58).

Table 5–1 *Residential Project Divisions*

Name	ID	Description
Existing	E	Existing and Demolition
New	N	New Construction

Figure 5–58 Adding new Divisions

5. In Project Navigator, click the Edit Levels icon and use Table 5–2 as a guide to set up the project's Levels (see Figure 5–59).

Table 5–2 *Residential Project Levels*

Name	Floor Elevation	Floor to Floor Height	ID	Description
Roof	18'-0" [5500]	0" [0]	R	Roof
Second	9'-0" [2750]	9'-0" [2750]	2	Second Floor
First	0" [0]	9'-0" [2750]	1	First Floor
Grade	-3'-1" [-940]	3'-1" [940]	G	Grade Level
Basement	-8'-9" [-2600]	5'-8" [1660]	B	Basement Floor

Figure 5–59 Set up the Levels of the Residential project

BUILD THE CONSTRUCTS

We now have the project structure established. It is time to build some Constructs for the building model. (Refer to the Commercial project above for complete information on Constructs.) If you completed the last chapter, you already have the first floor existing conditions. If you also did the additional exercises in Appendix

A, you have the second floor existing conditions and possibly the Basement Existing Conditions as well. If you did not complete those exercises, these files have been provided in the *Chapter05\Residential* folder with the files inserted from the CD. The steps that follow will refer to the files installed from the CD for Chapter 5. Feel free to use the ones you created in Chapter 4 instead if you wish.

6. From the *C:\MasterADT5\Chapter05\Residential* folder, open the file named *First Floor Existing.dwg [First Floor Existing-Metric.dwg*

7. On the Project Navigator palette, click the Constructs tab. </Step>

8. Right-click the Constructs folder and choose **Save Current Dwg As Construct.**

   In the Add Construct worksheet, place a check mark in the "First" Level row and the "Existing" Division column (see left side of Figure 5–60).

This assigns the Construct to the "First" floor Level and the "Existing" Division.

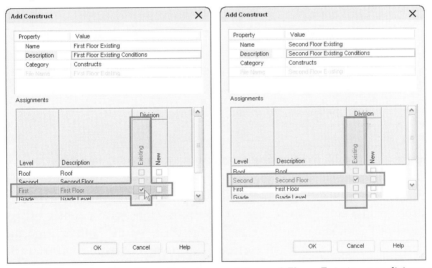

Figure 5–60 *Save the First Floor Existing and Second Floor Existing conditions as Constructs*

   A new Construct will appear in Project Navigator.

9. From the *C:\MasterADT 2006\Chapter05\Residential* folder, open the file named *Second Floor Existing.dwg [Second Floor Existing-Metric.dwg].*

10. Repeat the same process to create the *Second Floor Existing* Construct (see right side of Figure 5–60).

11. Repeat the same process once more for the *Roof Existing.dwg [Roof Existing-Metric.dwg]* file and assign it to the "Roof" Level and the "Existing" Division.

NOTE The Roof file has been provided with the files from the CD (in the *Chapter05\Residential* folder) for Chapter 5.

12. On the Project Navigator palette, right-click on the *Constructs* folder and choose **New > Category**.

13. Name the new Category **2D Drawings**.

14. From the *C:\MasterADT5\Chapter05\Residential* folder, open the file named *Site Plan.dwg* [*Site Plan-Metric.dwg*].

15. Right-click the *2D Drawings* folder and choose **Save Current Dwg as Construct**.

16. In the Add Construct worksheet, place a check mark in the "Grade" Level row and both the "Existing" and "New" Division columns (see Figure 5–61).

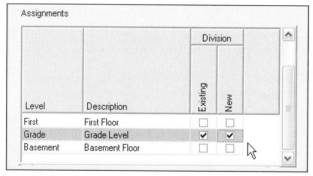

Figure 5–61 *Assign the 2D Existing Site Plan to both the Existing and New Divisions*

When you check more than one Level or Division for a Construct Assignment, the Construct is said to be "spanning." A spanning Construct occupies two or more portions of the building. In this case, the Site Plan "spans" both existing and new construction. We created a vertically spanning Construct (the *Front Fa\c¢cade*) for the Commercial project above. The site plan is just a 2D line drawing drawn in AutoCAD. It is important to include it in the project structure; however, since it is 2D and will not contribute anything to the Sections, Elevations or 3D model, we gave it its own category. This will make it easier later to isolate this drawing (and any others like it) from the project Views later. For the Sections, Elevations or 3D model, we will build a 3D terrain model based on this Site Plan the way we did above for the Commercial project. In some cases, you may also choose to place the 2D Drawings in the Elements folder instead. When doing so, they will get no Level or Division information and will need to be dragged manually to Constructs when needed.

17. From the *C:\MasterADT5\Chapter05\Residential* folder, open the file named *Terrain.dwg [Terrain-Metric.dwg]*.

18. Follow the steps above in the "Build a Terrain Model from the Site Data" topic to create a terrain model from these contours.

Caution: Be careful not to select the rectangle as one of the "Select objects representing contours." The rectangle is provided so that you can answer "Yes" to the rectangular mesh question and then use its corners to set the extent of the mesh.

19. Select just the new terrain (drape) Mass Element object.

20. Drag the terrain from the drawing window and drop it on the *Constructs* folder in the Project Navigator palette.

    An Add Construct worksheet will appear as before.

21. In the Add Construct worksheet, name the Construct **Terrain Model**, give it a similar description, and give it the same assignments as the Site Plan (see Figure 5–62).

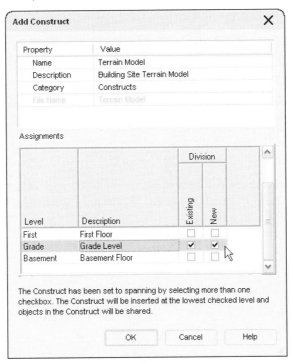

Figure 5–62 *Adding the Terrain Model Construct*

22. Close the original *Terrain.dwg [Terrain-Metric.dwg]* drawing without saving. (Leave the Terrain Model Construct open.)

We still need a few more Constructs to complete the building components of our Residential Project. We need to create the Existing conditions Basement, and the new construction for both First and Second floors.

23. On the Project Navigator palette, double-click the *First Floor Existing Construct*.

 NOTE Double-clicking a file in Project Navigator opens the file if it is not open already. If it is open already, as this file was, double-clicking it in Project Navigator will simply bring that file to the front (make it active).

24. Select the four exterior Walls in this file. With CTRL held down, drag these four Walls to the *Constructs* folder of the Project Navigator.

25. Following steps similar to the other Constructs, create **Basement Existing** and assign it to the "Basement" Level and "Existing" Division.

If your Walls in the First Floor disappeared, undo and try it again. Holding CTRL while dragging will make a copy of the Walls in the new Construct. Simply dragging (as we did with the Terrain Model) will move the objects. We will need to make adjustments to these Walls in the Basement file, such as editing their Height, Style, and the Fireplace. But we will save those edits for later in the project.

Continue in the *First Floor Existing* Construct.

26. On the Design tool palette, click the *Wall* tool, and set the Width to **10″** [**240**], Height to **9′** [**2750**] and the Justify to **Right**.

If you do not see this palette or tool, right-click the Tool Palettes title bar and choose **Design**, and then click the Design tab.

27. Starting at the outside top right corner of the building, draw a rectangular addition on the back of the house **15′** [**4500**] deep and the full width of the house (see Figure 5–63).

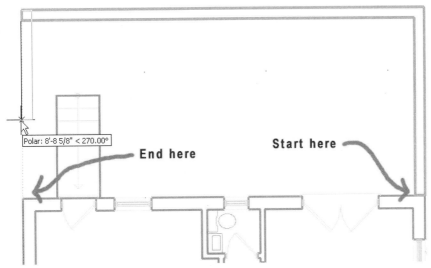

Figure 5–63 *Add new Walls for the Addition*

28. Select all three Walls, right-click and choose **Copy Wall Style and Assign**.

29. On the General tab, rename the new Style to: **New Construction Exterior** and then click OK.

30. Using the techniques already covered, hold down the CTRL key and drag these three new Walls to the *Constructs* folder to create the **First Floor New** Construct.

31. Repeat this step (using the CTRL key drag method) to create the **Basement New** Construct.

32. Using the techniques already covered, drag these three new Walls to the *Constructs* folder to create the **Second Floor New** Construct. (Do not use the CTRL key this time.)

We first copied the Walls to the *First Floor New* and the *Basement New* Constructs (with the CTRL key), and then moved them (without the CTRL key) to the *Second Floor New* Construct. This technique is the quickest way to reuse the same Walls in both Constructs. We could also have dragged the Walls without the CTRL key first to create one of the Constructs, and then used the Copy Construct to Levels technique covered above in the "Create the Upper Floors" topic. The final result would have been the same. At this stage of the project, we know only the overall extent of the addition, but not its form or material. Therefore, we are using a simple Wall style copied from the Standard Wall Style, with approximate dimensions. In later chapters, we will begin to flesh out the design, and therefore, we will make edits to these Walls at that time. That completes all of the required Constructs for the Residential project. We are now ready to build Views and Sheets.

SET UP THE RESIDENTIAL PROJECT GENERAL VIEWS

The process for setting up the Views and Sheets for the Residential project is identical to the process outlined above for the Commercial project. Let's begin with the Plan and Model Views, which will be General Views. Refer to the steps in the "Create Floor Plan Views" topic for reference as you create the Views shown in Table 5–3.

Table 5–3 *Residential Project View Files*

View Name	Description	Context (Level – Division)	Content
A-CM00	Composite Building Model	All Levels All Divisions	Include all Constructs Deselect *2D Drawings* category
A-FP00	Basement Floor Plan	Basement – Existing, Basement – New	Deselect *2D Drawings* category
A-FP01	First Floor Plan	First – Existing, First – New	Deselect *2D Drawings* category
A-FP02	Second Floor Plan	Second – Existing, Second – New	Deselect *2D Drawings* category
A-FP03	Roof Plan	Roof – Existing, Roof – New	Deselect *2D Drawings* category

Important: When creating all Views, on the Content page, deselect the 2D Drawings *folder. This will entirely remove this category from the View, thereby preventing future files in the 2D Drawings category from being automatically added to the View.*

To create any of the Views listed in Table 5–3, right click the *Views* folder on Project Navigator, and choose **New View DWG > General**. Type the name indicated, input a Description and then click Next. On the Context page, check the Levels and Divisions indicated in the table. On the Content page, verify that the correct Constructs are chosen and deselect the box next to the *2D Drawings* folder.

Now let's open the Composite Model and see how everything came together. You will see that all of the files have been XREFed into the Composite Model at the correct location and height as with the Commercial project above.

1. In Project Navigator, on the Views tab, double-click the *A-CM00* View to open it.

2. On the View toolbar, click the SE Isometric icon, and then on the Shade toolbar, click the Gouraud Shaded icon. Further adjust the view with the 3D Orbit icon if you wish (see Figure 5–64).

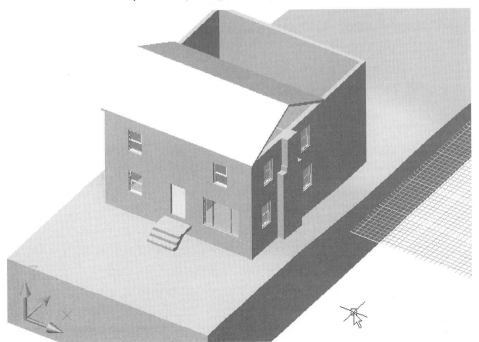

Figure 5–64 Open the Composite Model and view it in 3D

3. Save and Close the *A-CM00* file.

MODEL SPACE VIEWS AND TITLE MARKS

Following the process that was used above in the "First Floor Plan Model Space View" and "Add a Title Mark" topics, we can add a Live Area Guide to our plans, create a Model Space View with the desired drawing title, and then add title mark Callouts as we did above to label these drawings and make them ready to drop onto Sheets. As with the Commercial project, there is a *Live Area Guide.dwg* [*Live Area Guide-Metric.dwg*] file included with the Chapter 5 residential dataset. Within this file is an ADT Layout Grid 2D that matches the proportions of the Live Area of the Sheet files we will be creating below. Using this Grid, we can create Model Space Views that fit two plans snugly onto the Sheet files without any trial and error on the Sheets. Again, the use of the Live Area Grid technique is optional, but is presented as a way to eliminate guess work when creating Views and Sheets.

4. In Project Navigator, on the Views tab, double-click *A-FP01* to open it.

5. At the bottom of the screen, on the Drawing Status Bar (see Figure 1–2 in Chapter 1), choose **1/4"=1'-0" [1:50]** from the Scale pop-up list.

6. Follow the process outlined above to insert the *Live Area Guide.dwg* [*Live Area Guide-Metric.dwg*] file.

   Just as we did in the "First Floor Plan Model Space View" topic above for the Commercial Project, insert the file at 0,0,0 and be sure to check the "Explode" check box.

7. Zoom extents to see the whole grid.

This Layout Grid represents the area available on one half of the Sheet when using 1/4"=1'-0" [1:50] scale.

8. In Project Navigator, right-click *A-FP01* and choose **New Model Space View**.

   For the Name, type **First Floor Plan**. Verify that the Display Configuration is set to **Medium Detail** and change the Scale to **1/4"=1'-0" [1:50]**. It is not necessary to type a Description.

9. Click the Define View Window icon on the right (shown in Figure 5–32 above).

10. Snap to the corners of the Layout Grid.

11. Following the same process as outlined above in the "Add a Title Mark" topic, add a Title Mark from the Callouts tool palette within the First Floor Plan Model Space View beneath the house.

The title (First Floor Plan) and the scale (1/4"=1'-0" [1:50]) should automatically fill in based on the values designated in the Model Space View.

12. Open each of the other plans and repeat the entire process (including setting the scale to **1/4"=1'-0" [1:50]** from the Drawing Status Bar). Save and close each file when finished.

 Tip: Select the Grid in *A-FP01*, Copy, and then Paste to Original Coordinates in the next file. Create the Model Space View, and then add the Titlemark Callout. Repeat in the next file.

SET UP THE RESIDENTIAL SECTION/ELEVATION VIEW

Following a process very similar to that of the Commercial project above, we will add Callouts to the Plans and create the associated Elevations and Sections.

13. In Project Navigator, on the Views tab, double-click *A-FP01* to open it.

Refer to the steps in the "Elevations and Sections" topic above for reference as you add Callouts and their associated elevations, sections, and View file to the project.

14. Using the **Exterior Elevation Mark A3**Callout tool and the process outlined above in the "Create the Building Elevation View" topic, create Callouts and generate elevations in a New View Drawing.

    In the Place Callout worksheet, accept the default Model Space View names, be sure that both the "Generate Section/Elevation" and "Place Titlemark" boxes are checked, change the Scale to **1/4"=1'-0"** [**1:50**], and create the results in a New View Drawing.

    Name the new Section/Elevation View *A-CM01*, Description **Section and Elevation Composite Model**, verify that all Levels and Divisions are selected on the Context page, and clear the *2D Drawings* category on the Content page.

15. For the "insertion point for the 2D elevation result" click a point next to the plan in *A-FP01* to the right (see Figure 5–65).

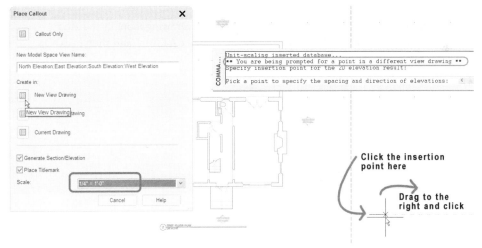

Figure 5–65 Add Callouts and Elevations

CREATE SECTIONS IN AN EXISTING VIEW

To create a couple of Sections for the Residential project, we will follow a process very similar to that of the Commercial project, but this time, rather than create a new View Drawing, we will add the Sections to the *A-CM01* View created for the elevations. It is not necessary to create them in separate Views. You typically make the decision on whether to use one or several Views based on project size, scope, and team size. However, if the scale changes, or the drawing type is different (plan vs. elevation or section) or there is a different discipline (such as Mechanical or Electrical) you should always make a new View file. In this case, the Residential project is small enough and the sections and elevations are similar enough (and will later use the same styles) that it is practical to put sections and elevations together in the same View file in this project.

16. In Project Navigator, on the Views tab, double-click *A-FP01* to open it.

NOTE If you left the First Floor Plan View open above, this action will simply make that file active.

17. Using the **Section Mark A2T** Callout, cut vertically through only the new addition. Cut through the double door on the right and look to the left.

    Using the process outlined in the "Create a Section View" topic above as a guide, use a Construction Line as necessary and include a break in the Section Line.

18. In the Add Callout worksheet, type **Transverse Building Section** for the name and choose **1/4″=1′-0″ [1:50]** for the Scale.

19. In the "Create in" area, click the Existing View Drawing icon. In the Add Model Space View dialog box that appears, select *A-CM01* and then click OK (see Figure 5–66).

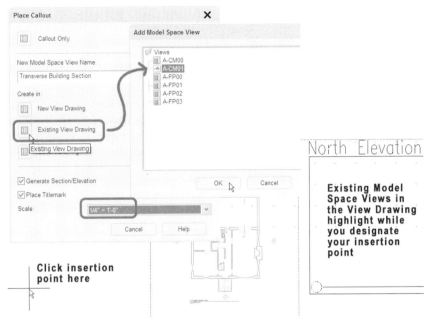

Figure 5–66 Add Callouts and Section to the same existing A-CM01 View file

20. Repeat this process to create a **Longitudinal Building Section**, cutting horizontally through the new addition and looking down.

    Again, use a Scale of **1/4"=1'-0" [1:50]**, create it in the *A-CM01* Existing View Drawing, and place the Section next to the Transverse Building Section.

21. In Project Navigator, on the Views tab, double-click *A-CM01* to open it. Make any adjustments to positions of the elevations and sections as required.

22. Follow the process outlined in the "Add Elevation and Section Callouts to Upper Floor Plans" topic above to copy and paste the elevation and section Callouts from the *A-FP01* First Floor Plan to the other plans.

23. Save and Close all project files.

CHANGE THE SHEET TEMPLATE FOR THE SHEET SET

Residential projects typically use smaller Sheet sizes and larger scales than commercial projects. We have used 1/8"=1'-0" [1:100] for all our commercial files and 1/4"=1'-0" [1:50] for all the residential. The Commercial project used the default title block and Sheet size of Arch F (30 x 42) [ISO A0 (841 x 1189)]. If you opened up the default Sheet file template named *Aec Sheet (Imperial Stb).dwt* [*Aec Sheet (Metric Stb).dwt*], you would find that it contains several layout tabs at different common sheet sizes. When you create a Sheet file, the size that is pre-assigned in the Sheet Set is copied to the new Sheet file and renamed to the same name as the file. Therefore, changing sheet size is a simple matter of changing the Sheet Set default.

24. In Project Navigator, on the Sheets tab, right-click the *MADT Residential* Sheet Set node at the top and choose **Properties**.

25. In the Sheet Set Properties dialog box, click the Browse icon next to Sheet creation template and in the dialog box that appears, choose **Arch D (24 x 36) [ISO A1 (594 x 841)]** and then click OK twice (see Figure 5–67).

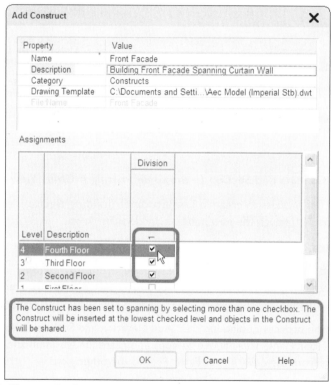

Figure 5–67 Change the Layout for new Sheets

When you perform the change, you will be asked: "Do you want to apply these changes to all nested subsets?"

26. Click the Yes button to accept this change.

SET UP THE RESIDENTIAL PROJECT SHEETS

27. Following similar steps to those in the "Sheet Files and the Cartoon Set" topic above, create a Sheet in the *Architectural – Plans* Subset numbered **A-101** with the Sheet Title **Floor Plans**n.

28. Drag the Basement Floor Plan and the First Floor Plan Views onto the Sheet.

   As we saw above, Viewports will be created automatically as you drag the files. Also, since we built the Model Space Views to match the Layout Grid, the plans fit perfectly onto the Sheet.

   If it appears that the title marks are showing in both viewports, choose **Regenall** from the View menu. This will correct the problem.

29. Create the remaining Sheets as indicated in Table 5–4.

Table 5–4 *Residential Project Sheet Files*

Sheet Number	Sheet Title	Model Space Views to drag and drop
A101	Floor Plans	Basement Floor Plan and First Floor Plan
A102	Floor Plans	Second Floor Plan and Roof Plan
A201	Elevations	North Elevation and South Elevation
A202	Elevations	East Elevation
A203	Elevations	West Elevation
A301	Sections	Transverse Building Section and Longitudinal Building Section

Notes for Table 5–4:

1. You will probably need to adjust the viewports a bit to make them narrower for the elevations and sections.

30. Following the steps in the "Create a Cover Sheet" topic, create a Cover Sheet from the same template file used above. This time choose the **Cover Sheet (24x36) [Cover Sheet (594 x 841)]** for the Layout.

   If you wish, edit the Project data fields as we did for the Commercial Project.

PUBLISH THE RESIDENTIAL CARTOON SET

All that is left to do now is publish the set.

31. Following the steps similar to those in the "Publish the Cartoon Set" topic above, print out your Residential Project Cartoon Set.

32. Save and Close all project files.

CONGRATULATIONS!

You have set up your first ADT Projects and completed your first ADT cartoon sets. The files are now ready to receive project design data. At any time, you can open the Sheet files to assess your progress, compare with your schedule and budget projections, and re-plot as necessary. The completed set of files has been provided with the files from the CD. You will find them in a folder named *Chapter05\Complete*. Launch the Project Browser, and load the completed versions to compare them to the ones created here if you wish.

 Note: When you wish to view the completed versions of the Projects from the CD, be sure to use Project Browser on the File menu to make the project current. If you are prompted to "Re-path" the project, always answer yes to this query. This will ensure that all required XREF paths are properly configured for your machine.

 New in ADT 2006 is the ability to use Relative XREF Paths for your projects. The default project template that we used here is configured this way; therefore, all of the projects included with Mastering Autodesk Architectural Desktop now make use of relative paths. This means that even if you move a project and choose not to re-path the XREFs, it should still function. However, you should still always answer "Yes" when the project location dialog appears (see Figure 5–68).If you wish to change the type of XREF paths that are stored in your project files, click the Project tab of the Project Navigator, and then click the Edit Project icon. In the Modify Project worksheet, choose "No" for the "Use Relative Xref Paths" setting.

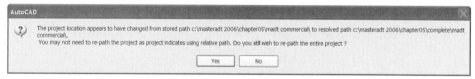

Figure 5–68 Repath dialog for projects with Relative Paths

ADDITIONAL EXERCISES

Additional exercises have been provided in Appendix A. In Appendix A you will find an exercise for adding a Furniture Element file to the Project dataset and then using it to build a Furniture Plan. It is not necessary that you complete this exercise to begin the next chapter, it is provided to enhance your learning experience. Completed projects for each of the exercises have been provided in the *Chapter06/Complete* folder.

SUMMARY

Thorough project setup can help give a good sense of project drawing requirements early in the project cycle.

Using the Project Browser and Project Navigator tools makes setting up a project quick and easy.

ADT Drawing Management tools make use of XREFs to relate files to one another.

XREF Overlay is used when you want the XREF to go only one level deep.

XREF Attach creates nested references, which create a hierarchical reference structure.

Model files are full-scale drawings used to generate actual project data on a daily basis.

Constructs and Elements are Model files representing individual pieces of a complete building model.

Constructs have a unique physical location (an address) within the Building Model; Elements do not.

Views are used to gather a collection of Constructs (and any nested Elements that they may contain) for a specific viewing purpose.

Views make an excellent location for adding annotation.

Several provided Callout routines make the process of creating sections and elevations with linked annotation as simple as following a wizard.

Sheet files are used for setting up "ready to plot" sheets for printing document sets.

Demolition, Existing to Remain, and New Construction can be separated into XREFs and/or managed with layers and cleanup groups.

Sheet Sets can be quickly plotted to Multi-Sheet DWF files that can be opened, viewed, plotted and redlines in Autodesk DWF Viewer of Composer applications.

CHAPTER 6

Column Grids and
Structural Layout

INTRODUCTION

In this chapter, we will explore the layout of structural components for the commercial project begun in the last chapter. As we have seen, the design is a four-story structure of modest footprint. We will begin with the layout of the column bay grid. We will add framing members and explore how to incorporate these items into appropriate files within our project structure and revisit the Residential Project to create a foundation plan.

OBJECTIVES

We will begin adding a Column Grid and Columns to a file that will be used as a typical layout on all floors. This grid layout will be added to each level of the project including bubbles and dimensions. We will then add Beams and Joists to complete the framing. The main tools covered in this chapter include the following:

- Explore the Structural Member Catalog.
- Explore the Structural Member Wizard.
- Work with the Column Grid tools.
- Annotate a Column Grid with bubbles and dimensions.
- Create Structural Beams and Braces using automated layout routines
- Create a foundation Wall with integral footings.

 ## STRUCTURAL MEMBERS

A structural member is an AEC object that is used to represent Columns, Beams or Braces. Each structural member belongs to a style, which in turn must reference one or more "Structural Member Shapes." The shape is simply the cross section of the structural member at a given point along its length. A structural member must contain at least one shape, but may have several. A shape is extruded along the path of the structural member. (A Structural Member may optionally have more than one shape along its path). Structural Member shapes are retrieved from an extensive catalog of industry-standard structural shapes. This catalog includes typical sizes of concrete, timber, and steel. A simple wizard can also be used to select shapes. Custom shapes can also be defined, thus, the potential of structural members virtually limitless. In order to be used in a design, shapes located in the catalog (or custom defined shapes) must be referenced in styles in the current drawing. In this tutorial, we will import a few steel shapes from the catalog and wizard and define them as styles for our commercial project.

INSTALL THE CD FILES AND LOAD THE CURRENT PROJECT

If you have already installed all of the files from the CD, simply skip down to step 3 below to make the project active. If you need to install the CD files, start at step 1.

1. If you have not already done so, install the dataset files located on the Mastering Autodesk Architectural Desktop 2006 CD ROM.

   Refer to "Files Included on the CD ROM" in the Preface for information on installing the sample files included on the CD.

2. Launch Autodesk Architectural Desktop 2006 from the desktop icon created in Chapter 3.

If you did not create a custom icon, you might want to review "Create a New Profile" and "Create a Custom ADT Desktop Shortcut" in Chapter 3. Creating the custom desktop icon is not essential; however, it makes loading the custom profile easier.

3. From the File menu, choose **Project Browser**.

4. Click to open the folder list and choose your *C:* drive.

5. Double-click on the *MasterADT 2006* folder, then the *Chapter06* folder.

   One or two commercial Projects will be listed: *06 Commercial* and/or *06 Commercial Metric*.

6. Double-click *06 Commercial* if you wish to work in Imperial units. Double-click *06 Commercial Metric* if you wish to work in Metric units (you can also right-click on it and choose **Set Current Project**). Then click Close in the Project Browser.

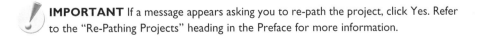 **IMPORTANT** If a message appears asking you to re-path the project, click Yes. Refer to the "Re-Pathing Projects" heading in the Preface for more information.

If you want to learn more about ADT Projects, refer to the previous chapter.

CREATE A NEW ELEMENT FILE IN THE PROJECT

The same Column Grid layout will occur on all four floors of the project. We could build the grid layout, and then copy it to each of our existing floor plates, but an easier approach will be to take advantage of Project *Elements*, as described in the last chapter. By making the Column Grid an Element, we can use it in each floor level *Construct*, while maintaining a link to the original Element file. If the Column Grid layout needs to change, we simply edit the Element file, and it will update in all of the Constructs. Naturally this approach would be less effective on buildings where the column layout varies from floor to floor. However, anytime that you have a repetitive portion of your building design—such as a typical Stair, Toilet Room or Column Grid as in this case—you can use Element files in Project Navigator to manage them. Refer back to the "Elements, Constructs, Views and Sheets (Project Drawing Files)" heading in Chapter 5 for more information.

7. On the Project Navigator, click the Constructs tab. (If the Project Navigator did not open automatically when you closed Project Browser, press CTRL + 5 to open it now.)

8. Right-click on the *Elements* folder and choose **New > Element**.

    Type **Column Grid** for the Name and **Typical Column Grid** for the Description and then click OK (see Figure 6–1).

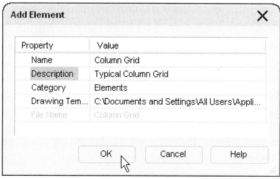

Figure 6–1 *Create a new Column Grid Element file*

9. Double-click on the *Column Grid* Element file to open it.

ACCESS THE STRUCTURAL MEMBER CATALOG

Let's now import some Structural Member Shapes to use in our Grid layout.

10. From the Format menu, choose **Structural Members > Catalog** (see Figure 6–2).

Disclaimer:

The shapes used in this book are chosen only for illustration purposes and are not presented as a design solution or to be construed as a recommendation of structural integrity. No structural analysis of any kind has been performed on the designs in this book.

Navigate the Structural Member Catalog in the same way as you would Windows Explorer. On the left are two main categories: *Metric* and *Imperial*.

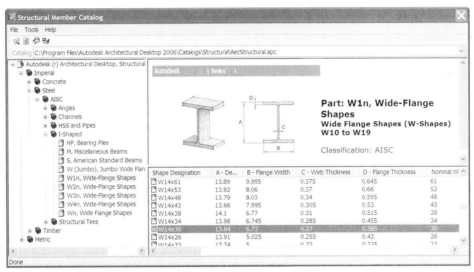

Figure 6–2 *The Structural Member Catalog*

11. Click the Plus (+) sign next to each category to expand the tree.

    Both *Imperial* and *Metric* are divided into three sections based on material: *Concrete*, *Steel*, and *Timber*. However, the divisions within each material vary with regional differences (see Figure 6–2).

12. Expand the *Steel* entry, then the *AISC* entry, and finally *I-Shaped* [for Metric, expand one level deeper to *Wide Flanges*].

 NOTE Nearly every imaginable industry standard shape is included in this hierarchy. However, as new shapes become available, the catalog's XML format makes it very easy to update.

13. Select the *W1n, Wide-Flange Shapes* [*W3nn, Wide Flanges*] category.

14. Scroll through the list, locate the **W12x87 [W310x97]** shape, and select it (see Figure 6–3).

    Scroll horizontally and notice the complete list of properties associated with each shape.

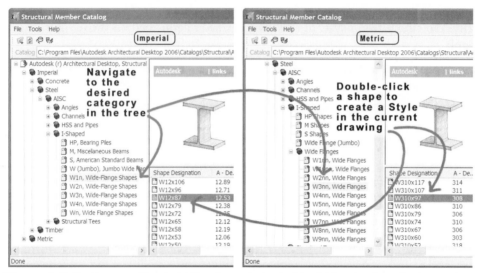

Figure 6–3 *Create a new Structural Member style by double-clicking the shape in the catalog*

15. Double-click **W12x87 [W310x97]**.

    The Structural Member Style dialog box will appear. Double-clicking a shape name imports the shape into the current drawing and creates a style that references it.

There are three types of structural members available in ADT: Columns, Beams and Braces. The shapes available in the Structural Member Catalog are used for all three forms of Structural Members. When you double-click a shape, ADT will suggest that the name of the shape being imported be used as the name for the style being created. Nothing in particular about the shape names themselves indicates your intent to use them as columns, beams or braces in your project. Therefore, you might want to add a descriptive suffix when creating Structural Member styles, like "Main Columns," or "First Floor Beams." This is optional, and if you prefer, you can simply accept the name that is offered instead.

16. Type **W12x87 (Main Columns) [W310x97 (Main Columns)]** and then click OK (see Figure 6–4).

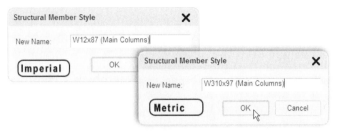

Figure 6–4 *Name the new Structural Member style with a descriptive suffix*

> 17. Continue browsing the Structural Member Catalog if you wish. When finished, close the Structural Member Catalog.

USE THE STRUCTURAL MEMBER STYLE WIZARD

If you wish, you can bypass the Structural Member Catalog when creating Structural Member styles. This is useful if you have not yet consulted your structural engineer and you wish to create a shape based on overall size. To do this, we use the Structural Member Style Wizard. Later, when accurate sizes have been calculated by your structural engineer, you can swap those styles in to replace the ones created with the Wizard.

> 18. On the Format menu, choose the **Structural Members > Wizard**.
>
> 19. From the list under the *Steel* category, choose **Wide Flange (I)** and then click Next.
>
> 20. For both the Sectional Depth and Sectional Width type **8″ [210]** and then click Next.
>
> 21. For the Style Name, type **Front Skin Columns** and then click Finish (see Figure 6–5).

Figure 6–5 *The Structural Member Style Wizard*

> 22. Save the file.

VIEWING STRUCTURAL MEMBER SHAPES IN STYLE MANAGER

Structural Members, as we have mentioned, contain one or more Structural Member Shapes as cross sections. In Architectural Desktop 2006 you can now view and edit these shapes in the Style Manager. Structural Member Shapes can contain up to three shapes: one each for High, Medium and Low Detail Display Configurations (see Chapter 2 for more information on Display Reps and Display Configurations). It is not necessary to utilize all three Display Reps, but doing so gives

Structural Members based on the Shapes a great deal of flexibility. To understand this better, let's take a look in the Style Manager and preview the Shapes that we just added to the drawing.

23. From the Format menu, choose **Structural Members > Member Shapes**.

    The Style Manager will appear filtered to the Structural Member Shape Definitions node.

24. On the left side, click on the **W12x87 [W130x97]** Shape.

Note that here we are looking at the Structural Member Shape Definitions, not the Structural Member Styles that reference the Shapes. Therefore, the name listed here is simply the Shape name, not the longer name (including "Main Columns") that we assigned to the Structural Member Style above.

25. On the right side of the Style Manager, click the Design Rules tab.

26. Click through each level of detail and note that the corresponding Shape will highlight in green (see Figure 6–6).

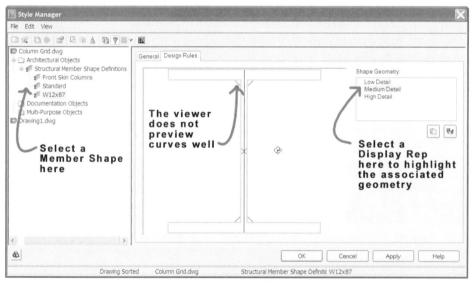

Figure 6–6 *Viewing Structural Member Shape Definitions in the Style Manager*

The chamfered corners of the High Detail Shape are actually rounded. The embedded viewer within Style Manager does not have the resolution required to display the curves smoothly. Despite that limitation, the ability to view and edit Structural Member Shapes directly from Style Manager is a useful feature. There are two icons beneath the Display Reps. The one on the right (Create Style) allows you to create a Structural Member Style from the Shape that you are viewing. The icon on the left (Specify Rings) will return you to the drawing editor and allow you to select linework from which to create the selected Shape. If you wish to create your own Shape, you can start with the polylines embedded in an existing Definition or draw new ones from scratch. To start with the existing ones, you would exit the Style Manager and use the **Structural Members > Insert Member Shape** command on the Format menu. This command will allow you to select one or more of the rings embedded in the Shape to insert into the drawing. Once there, you can edit them and then return to the Style Manager to reassign them to the Shape Definition with the "Specify Rings" icon. Feel free to try this on your own later. Remaining in Style Manager, let's take a look the Structural Member Style that references the Shape that we have been viewing here.

> The Structural Member Style Properties dialog box will appear.

27. On the toolbar across the top of the Style Manager, locate the Filter Style Type icon and then click it (see the top of Figure 6–7).

    The tree at left will expand to include all ADT Style types organized into three folders. Structural Member Styles appear directly below Structural Member Shape Definitions.

28. On the left side in the tree view, expand Structural Member Styles and then select **W12x87 (Main Columns)** [**W310x97 (Main Columns)**].

29. On the right side, on the General tab, type **Main Structural Columns All Bays** in the Description field.

30. Click the Design Rules tab.

Any shapes referenced by the Structural Member Style are listed here. This is where the Structural Member Shapes that we just explored are assigned to the Structural Member Style. A basic Structural Member Style will have only one Shape. However, it is possible to design a very complex and intricate Member style incorporating several Shapes at various cross sections along the length of the Structural Member. (The Add and Copy buttons at the bottom right of the dialog would be used to accomplish this). Member styles can be tapered, and a variety of parameters such as rotations, mirroring, and offsets can be built into the style. To access these advanced functions, click the Show Details button (see the bottom of Figure 6–7). We will skip over the remaining tabs at this time.

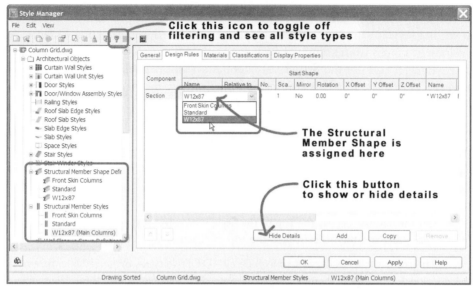

Figure 6–7 *Understanding how Structural Member Shapes are assigned to Structural Member Styles*

31. Click OK to dismiss the Style Manager and return to the drawing.

WORK WITH STRUCTURAL MEMBERS

Now that we have seen how to create Structural Styles and how their components fit together, let's use them in the drawing.

1. On the Design tool palette, click the **Structural Column** tool.

   If you do not see this palette or tool, right-click the Tool Palettes title bar and choose **Design** (to load the Design Tool Palette Group) and then click the Design tab.

2. On the Properties palette, choose **W12x87 (Main Columns)** [**W310x97 (Main Columns)**] from the Style drop-down list.

3. At the "Insert point" prompt, click a point anywhere on screen and then press ENTER to complete the command. Press ENTER again to complete the routine.

   Zoom in as required to see the Column.

Structural members, like most ADT objects, have three levels of detail: Low Detail, Medium Detail and High Detail. As we noted above in the Style Manger, it was a bit difficult to see each of the Shapes clearly and the filleted corners were showing chamfered in High Detail. Let's have a look at the Structural member in the drawing now in each Display Configuration to see this more clearly. The scale of the drawing will typically determine which level of detail is appropriate (see Figure 6–8).

▶ **Low Detail**—Single-line diagram, good for small-scale drawings.

▶ **Medium Detail**—Basic double-line display with square corners; good for 1/8″=1′-0″ [1:100] and 1/4″=1′-0″ [1:50] scales.

▶ **High Detail**—Shows a high level of detail including filleted corners; good for large-scale details, 1/2″=1′-0″ [1:25] and larger.

Figure 6–8 *Three levels of display detail*

If you are not zoomed in on the Column, zoom in on it now.

4. Change the current Display Configuration to **Low Detail**.

Note the change to the structural member; a simple line sketch.

5. Change the current Display Configuration to **High Detail**.

Again note the change to the display of the structural member; In High Detail, the fillets and hatching are shown. Hatching is a parameter of the Structural Member Style, not the Shape. Therefore, we did not see hatching in the Style Manager preview above.

6. Change the current Display Configuration back to **Medium Detail**.

7. Erase the Column on screen before proceeding to the next sequence.

8. Save the file.

COLUMN GRIDS

An overview of layout tools and Anchors appears in the "Anchors" topic of Chapter 2. A Column Grid object is a type of 2D Layout Grid. There are two shapes, rectangular and radial. Grid spacing, orientation and location are controlled parametrically. Spacing can be set to equal spacing, repeat spacing or manual spacing. A Node Anchor point exists at each grid intersection. Column objects can be anchored to these points. Anchored column line labels generate automatically, and AEC dimensions contribute to form a complete assembly of components. It is also possible to create a manual grid from linework. We will explore these features in detail by building the Column Grid layout for our commercial building.

ADD A COLUMN GRID WITH ANCHORED COLUMNS

1. Zoom back out a bit; on the Design palette, click the **Structural Column Grid** tool.

2. On the Properties palette, choose **Rectangular** for the Shape.

The overall size of the Column Grid is set in the Width and Depth fields. This should correspond to the size of the overall building footprint. In buildings with a non-rectilinear footprint, you can use more than one grid and set the Width and Depth to the size of the section of the building for which you're designing the grid.

3. In the Dimensions grouping, set the X - Width to **80'-0"** [**24,000**] and set the Y - Depth to **60'-0"** [**18,000**].

There are two ways to set the initial bay spacing within the grid. The total Width or Depth dimension can be divided by a certain quantity of bays. To do this, choose **Space Evenly** from the Layout type list. A fixed bay dimension can also be used. To do this, choose **Repeat** for the Layout type instead. In this example, we will do one of each to compare and contrast.

4. Be sure the X Axis Layout type is set to **Repeat**, and set the X - Baysize to **20'-0"** [**6000**].

5. For the Y Axis Layout type choose **Space evenly**, and set the Number of bays at **3**.

6. In the Column grouping, choose **W12x87 (Main Columns)** [**W310x97 (Main Columns)**] from the Style drop-down list.

7. Set the Logical Length to **11'-5"** [**3475**] (see Figure 6–9).

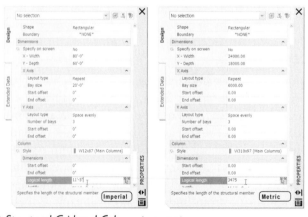

Figure 6–9 *Set Structural Grid and Column parameters*

8. At the "Insertion point" command line prompt, click a point anywhere on screen to place the grid.

   Slowly move the mouse and notice that the grid's insertion point is its lower left corner and that you are able to rotate the grid relative to this point.

9. Press ENTER to accept the default rotation of **0°**. Press ENTER again to complete the routine.

   Zoom out as necessary and notice the appearance of the Grid with a Column at each intersection (see Figure 6–10).

Figure 6–10 *A Column Grid with anchored columns*

MOVE COLUMN GRIDS AND COLUMNS

10. Right-click in the drawing window and choose **Basic Modify Tools > Move**.

11. Move the grid.

    Be sure to select only the Column Grid object and not the Columns. It doesn't matter where or how much you move it.

Notice that all of the columns move with the Column Grid and remain attached to the grid intersections. This is the effect of the anchor used to link the Columns to the Grid. Now we will move a Column.

12. Repeat the Move command.

13. Select any Column (just the Column, not the Grid).

14. Click any Base point and move the column just a little (approximately **3'-0"** **[900]**) in any direction.

Notice that the Column did not move at all! Again, this is the effect of the Node Anchor, which establishes the rules that are used to link these two objects (the Column and the Grid) together. (Refer to the "Anchors" heading in Chapter 2 for more information.)

15. Try Move one more time, but this time move the Column farther, about **20'-0" [6000]** in the direction of another bay.

Notice that the Column appears to have been deleted. In fact, what has happened is that it has moved to a different Node.

The rules of the Node Anchor relationship state that the middle center point at the base of the column must be attached to a grid intersection—it does not matter which one. Therefore, if you move too far, you force it to "jump" to the next Node. Make a window selection around the Node you moved toward and you will find that there are now actually two Columns located there. (Look at the Properties palette to see this. It will read "Structural Member (2)" at the top of the palette in the selection list). Be careful with this in your projects. In general, you would not want two Columns on the same node.

16. Use the Undo command (Edit menu or type **U** and press ENTER) until the Column shifts back to its original position.

  TIP New to ADT 2006, you can now undo a series of Zoom and Pan commands with a single Undo. To enable this feature, choose **Options** from the Format menu and then click the User Preferences tab. Place a checkmark in the "Combine zoom and pan commands" checkbox and then click OK.

IDENTIFY ANCHORS

It is not always obvious when objects are anchored to one another. Over time, with practice and experience it does become more obvious, but initially it can be a little confusing. The next few steps are an exercise designed to help you identify when objects are anchored. There are a few simple techniques.

17. Select any Column and right-click (see Figure 6–11).

Notice the Node Anchor item on the right-click menu. The presence of this submenu indicates that the object is anchored. (The type of Anchor varies, such as Wall Anchor, Cell Anchor, Leader Anchor or Node Anchor, as is the case here.)

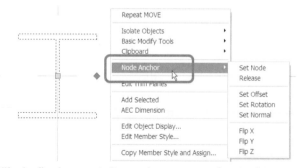

Figure 6–11 An "Anchor" sub-menu indicates the presence of an anchor

18. Choose **Properties** from the right-click menu.

   In the Location grouping, notice that there is a "Location on node" grouping and that an Anchor worksheet icon is present beneath it. An Anchor worksheet icon will always be present in the Properties palette for any object with an anchor. Clicking this icon will open a worksheet with additional anchor parameters (see Figure 6–12).

19. Click the Anchor worksheet icon (see Figure 6–12).

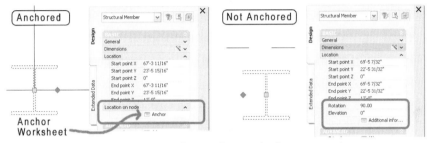

Figure 6–12 *Identifying the presence of an anchor on the Properties palette*

20. In the Insertion Offset X field, type **3′-0″** [**900**] and then click OK.

   Note the shift in Column position. You may have to zoom out if you are too close to the Column.

21. Undo this change.

   There is another way to perform this same sort of change that is a bit easier.

22. Select the same Column, right-click and choose **Node Anchor > Set Offset.**

23. Use your mouse to set the offset. You can also use POLAR or ORTHO and type values in for precision.

24. Undo this change to return the Column to the node.

MOVE THE GRID TO THE CORRECT COORDINATES

When we added the Column Grid object, we placed it randomly. In order for this Column Grid to be useful in our Commercial Project, we need to locate it correctly relative to the rest of the building. With all the Columns being anchored to the grid, relocating it is as simple as moving the grid itself. All of the anchored Columns will automatically follow.

1. Select the Column Grid and right-click.

Review the right-click menus. Notice the absence of an Anchor menu. The Column Grid is the "parent" object in this relationship; it is not anchored to anything and can be freely moved.

Because this object is not anchored to another, it can move and rotate freely. If you view the coordinates in the "Location" grouping on the Properties palette for the Column Grid, the coordinates in the X and Y boxes are currently random because we simply used a mouse click when creating the Column Grid. It would be helpful to have some geometry from the project to help us place the grid. To do this, we will XREF one of the floors.

2. Choose **Deselect All** to dismiss the right-click menu.

3. Open the Project Navigator palette if it is not already open (Window menu, or CTRL + 5).

4. On the Constructs tab, drag the *01 Shell and Core* file (in the Constructs folder) and drop it anywhere on screen (see Figure 6–13).

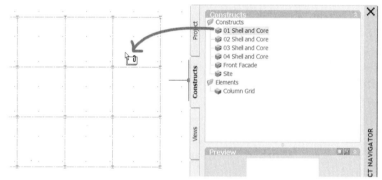

Figure 6–13 *XREF the First Floor Shell for reference when placing the grid*

This gives us the First Floor Walls to use for reference so that we can place the Grid in the correct location. You can use this technique anytime you want to use another floor's geometry for reference. To save you a bit of trouble, however, the exact coordinates are provided in the next few steps.

5. Select the Column Grid (just the Grid, not any Columns) and then right-click and choose **Properties**.

6. In the Location grouping, click the Additional Information worksheet icon.

7. In the Insertion Point X field, type **105′-5″** [**32,130**], in the Y field, type **53′-0″** [**16,154**], and then click OK (see Figure 6–14).

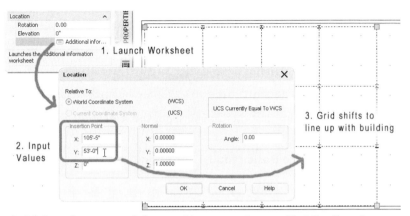

Figure 6–14 *Open Location worksheet and input coordinates of building footprint*

Notice the grid and its anchored Columns shift to its new location.

MODIFY COLUMN GRIDS

The parameters established for the grid upon creation continue to control its dimensions as modifications are made.

8. Select the Column Grid object.

   Notice the four grip points, one at each corner of the grid.

9. Click the upper right grip point to make it hot.

10. Slowly drag the grip up and to the right. Do NOT click yet.

11. Move about half a bay in each direction.

    Notice that the bays in the X direction do not appear to be affected; however, in the Y direction, the spacing is changing dynamically as the mouse moves.

12. Continue moving the mouse slowly up and to the right.

    Notice that at approximately one bay width to the right, a new bay begins to appear.

The behavior in both the Y and the X directions is simply based on the original parameters set at the time of creation. We built this grid to use a fixed bay size of 20″-0″ [6000] in the X direction and to evenly space the entire Depth (Y direction) by 3 bays.

13. Click the mouse anywhere to finalize the change and add one bay to the right.

    Notice that new Columns were *not* added in the X direction.

When you add a Column Grid, you are able to simultaneously add Columns, but these are separate and distinct objects. To add Columns to the new bay, we would need to add them manually. This can be accomplished with the *Structural Column* tool on the Design tool palette or simply by copying one of the existing Columns. When copying an object with anchors, the copy will *also* be anchored. Try it out.

14. Select all four Columns (just the Columns, not the grid) at the extreme right of the grid.

    Right-click and choose **Basic Modify Commands > Copy** (*not* Clipboard > Copy).

 Caution: If you use *Clipboard > Copy* on an anchored object, it will also copy the parent object when pasted, thereby anchoring the newly copied objects to this newly copied parent object. This is not what we want in this case.

15. Click a "Base point" near the original Columns and a "Second point of Displacement" near the new bay.

16. Be sure to press ENTER to complete the Copy command.

 NOTE Unlike previous versions of AutoCAD, the "Multiple" Copy option is the default behavior of the Copy command in AutoCAD and ADT 2006.

Because of the anchor, we do not need to use Object Snaps on this operation. The rules built into the anchor have higher priority to ADT than Object Snaps. As a rule, however, using Object Snaps is always a good idea; particularly when Anchors are not involved. Many objects do not use anchors, and for them, Object Snaps are often the only way to guarantee accuracy.

You should now have a five-bay-wide by three-bay-high Column grid with Columns anchored to all points.

17. Save the file.

MANIPULATE COLUMN LINE SPACING

It is a rare building that has perfectly regular Column grid bay spacing. The automatic spacing we used in the beginning of this exercise is a good way to get started, but the grid will likely need to be manipulated manually to achieve the proper spacing of bays in your design.

1. Continuing in the *Column Grid* file, select the Column Grid, right-click, and choose **Properties**.

In the Dimensions grouping, notice that the X Width and the Y Depth parameters have both changed because of the grip editing in the last sequence.

2. In the X - Axis grouping change the Layout type to **Manual**.

   Notice the appearance of grips at each of the grid lines in the X direction. Notice also that all parameters on the Properties palette have been replaced with a worksheet icon.

3. Do the same in the Y - Axis grouping making the Layout type **Manual** as well.

4. Near the bottom left corner of the grid, select the second grip point from the left to make it hot (see Figure 6–15).

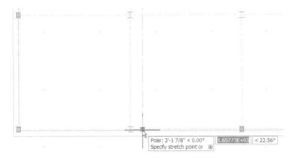

Figure 6–15 *Select the grip point of a single grid line*

5. Make sure that either POLAR or ORTHO is on, move the mouse directly to the right, type **2′-7″** [**860**], and then press ENTER.

   Notice that only the selected grid line has shifted and that each of the anchored Columns along its length has shifted with it.

This method is effective for simple modifications where immediate feedback is needed. However, if several bays need to shift, it is easier to perform the change in the Column Grid Properties worksheet.

6. On the Properties palette, in the Dimensions > X Axis grouping, click the worksheet icon next to Bays.

   Notice that each Column line in the X direction is listed with two sets of dimensions.

   ▶ **Distance to Line**—Is measured from the origin (lower left corner) of the grid.

   ▶ **Spacing**—Is measured from the previous grid line.

7. Type **14′-4″** [**4400**] in the Spacing Column for Column line Number 2 and press ENTER.

Notice that the difference between the new value and the old value has been applied to the next Column line (Number 3) and that each Column line in the X direction is listed with two sets of dimensions.

 TIP For this reason, you should always begin at the top of the list and work your way down.

8. Work your way down the list and set the following values: Column Line 3 = **28'-0"** [**8500**], Column Line 4 = **14'-4"** [**4400**], Column Line 5 = **22'-7"** [**6860**] (see Figure 6–16).

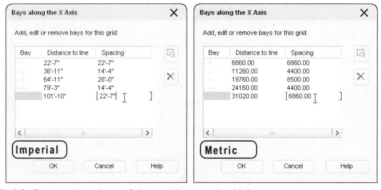

Figure 6–16 *Change the values of the grid lines in the X Axis*

Notice that as you change the Spacing values, the Distance to Line values also change.

9. Click OK to close the Bays along the X Axis worksheet.

10. Click the worksheet icon next to Bays in the Dimensions > Y Axis grouping.

11. Set the Spacing for each grid line as follows: Column Line 1 = **20'-0"** [**6100**], Column Line 2 = **21'-9"** [**6600**], Column Line 3 = **20'-0"** [**6100**].

12. Click OK to close Bays along the Y Axis worksheet.

13. Right-click in the drawing and choose **Deselect All**.

14. Save the file.

ROTATE AN ANCHORED COLUMN

Column orientation can be adjusted to suit design needs. When objects are anchored to the grid as these Columns are, the standard AutoCAD Rotate command will not do the trick. We must change the parameters governing the anchor of the Columns we wish to rotate. Fortunately this is easy to do with grips.

15. Select one of the Columns on the left edge of the grid.

16. Click the diamond-shaped grip and move the mouse.

    Notice the dynamic angular dimensions that appear.

17. Type **90°** and then press ENTER.

 NOTE The degree symbol (°) is used here for clarity. You need only type "90" into the dynamic dimension to rotate. ADT is already expecting input in degrees for this value.

You can use this method to rotate all of the Columns on the left and right edges one at a time, or it is possible to perform the rotation on all of them at once. The method to do so, however, is not quite as intuitive.

18. Select all of the Columns on both the extreme left bay and the extreme right bay. (Do NOT select any Columns in the middle.)

19. On the Properties palette, click the Anchor worksheet in the Location on node grouping.

20. In the Orientation area, change the Rotation Y to **90°** and then click OK.

 TIP If you prefer, you can perform this rotation on all selected Columns via the right-click menu instead. With them selected, right-click and choose **Node Anchor > Set Rotation**, and then input **90°** for each of the X, Y and Z rotations.

WHY ROTATE "Y"?

A Structural Member can be a Column, a Beam or a Brace. Each of these items has a different default orientation. If all of the rotation values were set to 0°, we would essentially have a Beam. Therefore, although not very intuitive, in this case rotating the Y value was appropriate to achieve the desired effect. Had we been rotating any other anchored object in the plan (such as a piece of furniture, a light fixture, etc) we would have actually been rotating around the Z axis instead.

REMOVE UNNECESSARY COLUMNS

The front of the building (at the bottom of the screen) will receive a curtain wall in an upcoming chapter. A secondary Column Grid will be placed here with a different type of Column. Therefore, we will remove the unnecessary Columns at the bottom row of the grid.

21. Select the two middle Columns along the bottom row.

22. Right-click and choose Basic **Modify Tools > Erase** (or just press DELETE).

23. Save the file.

CONVERT LINEWORK TO A COLUMN GRID

Along the front of the building will be a gently curved curtain wall. To provide support for this object, we will create a second Column Grid object. Sometimes it is easier to simply draw the grid we want with lines and then convert those lines into a Column Grid.

1. Click on the Manage XREFs quick pick icon located in the lower right of the Drawing Status Bar (looks like a small binder clip—or type XR and then press ENTER).

2. Select the *01 Shell and Core* XREF, click the Detach button and then click OK.

   This will make it easier to see the grid as we work.

3. From the second grid intersection of the bottom left corner of the grid, draw a Line **6′ [1800**] long straight down (270°).

   TIP Zoom in as required to snap accurately to the Endpoint of the Column Grid, and not to the points on the Column. This is where a wheel mouse comes in very handy for zooming.

4. Repeat for the next three grid intersections moving to the right or just copy this line three times (see Figure 6–17).

Figure 6–17 *Draw the vertical grid lines for the second Column Grid*

5. Offset the two inner lines **9′-0″ [2700**] toward the middle (see Figure 6–18).

Figure 6–18 *Offset two additional vertical lines*

6. Draw a line from the bottom endpoint of the line at the left to the bottom endpoint of the line at the right.

7. Offset this line up **8″ [200**].

8. Move both horizontal lines up **2′-0″ [610**] (see Figure 6–19).

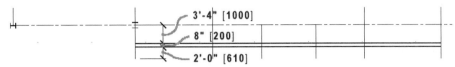

Figure 6–19 *Complete the grid lines with two horizontal lines*

9. On the right side, extend each of the horizontal lines about **12'-0" [3600]** using the grips.

10. On the Design palette, right-click the **Structural Column Grid** tool and choose **Apply Tool Properties to > Linework**.

11. Select the eight lines just drawn and then press ENTER.

12. At the "Erase selected linework?" prompt, choose **Yes**.

You should now have a small custom Column Grid in place of the lines.

ADD COLUMNS TO THE NEW GRID

13. On the Design tool palette, click the Structural Column tool and on the Properties palette, change the Style to **Front Skin Columns** and the Logical Length to **11'-5" [3475]**.

    The prompt will read: "Insert Point."

14. On the Application Status Bar turn off the Osnaps by clicking the OSNAP toggle (or press F3 on the keyboard).

Columns will automatically Anchor to a Grid if you click on one. As you move the mouse around on screen, if you hover over a Grid, you will see this behavior and options will appear at the cursor. This is why we will temporarily turn off the OSNAPs.

15. Move the mouse over the new Grid that we just created and pause for a moment—do not click yet (see Figure 6–20).

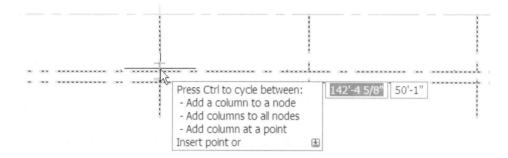

Figure 6–20 *Hover over a Grid to see automatic Node Anchor options*

> ▶ **Add a column to a node**—This is the default option and will add a single column to a single node on which you click.

> ▶ **Add a column to all nodes**—Press CTRL once to toggle to this option. A column will be attached to all nodes with a single click.

> ▶ **Add column at a point**—This simply adds a freestanding, non-anchored column.

If you wish to try the "all nodes" option, feel free to do so. Just Undo before proceeding to the next step.

16. At the "Insert point" prompt, click on each of the intersections indicated as "Add New Columns" in Figure 6–21 and then press ENTER.

Figure 6–21 *Add Front Skin Columns to new Column Grid by picking the nodes*

17. Select the two Columns (on the original grid) indicated as "Swap Style" in Figure 6–21 and change the Style on the Properties palette to **Front Skin Columns**.

18. Right-click and choose **Deselect All**, and then Save the file.

Add and Delete Grid Lines

*Column Grids (both parametric and manual) can be further manipulated. While in the Column Grid Bays worksheet, in either the X or Y direction, click in the blank space below to add a new grid line. Select a grid line by clicking in the Number column. Delete a grid line by selecting it by number and pressing DELETE. Grid lines can also be added and deleted in the drawing using the right-click menu. To do this, select the grid, right-click and then choose **X Axis > Add Grid Line**, **Remove Grid Line** or **Y Axis > Add Grid Line**, etc. To add grid lines to a manual grid, draw the line in place where you want it, then right-click the manual grid and choose **Add Grid Line**.*

Right-click and choose **Remove Grid Line** *to delete a line. Feel free to experiment with some of these techniques before continuing. Either Undo to return the file to the point saved here, or save the file before you begin experimenting and then simply close the file without saving when you are done and then reopen it to continue.*

COLUMN GRID LABELS AND DIMENSIONS

Once you have a Column Grid, grid labeling and dimensioning can be accomplished easily. Grid labels are Multi-View Blocks with Leader Anchors to attach them to the Grid. The dimensioning routine provided quickly adds AEC Dimensions in two strings. These dimensions stay linked to the Grid and update as the Grid changes.

SET DRAWING SCALE

1. From the Open Drawing Menu (the little arrow in the lower left corner on the Drawing Status Bar as shown in Figure 1–2 in Chapter 1), choose **Drawing Setup**.

2. Click the Scale tab. In the Drawing Scale area, choose **1/8″=1′-0″** [**1:100**].

    ADT annotation objects and symbols will automatically scale relative to this setting.

TIP You can also choose the scale directly from list on the Drawing Status Bar.

3. In the Annotation Plot Size box, type **5/32″** [**4**].

    This value works with the Drawing Scale to determine the final scale factor used for scaling annotation symbols. The value typed here represents the final plotted height of the text portion of annotation symbols. The default is **3/32″** [**2.5**] but for Column Bubbles you will typically want a larger value.

4. Click OK.

CAD Manager Note: In general, if you refer back to the "Elements, Constructs, Views, and Sheets (Project Drawing Files)" heading in Chapter 5, you will note that you are intended to build your model geometry in Constructs and that annotation should be added to Views. However, Column Grids present a bit of a limitation with regard to the annotation in this approach. The limitation is a function of the Anchors that are used with Grids. Technically, the Columns which are part of the model should be in Constructs (or Elements as we are doing here) and the Grid which is actually technically annotation should be in the View file. The same would be true of the bubbles and the dimensions. However, Anchors unfortunately do not function across XREFs. Therefore, in order to keep all of the parts anchored, the Columns, Grid, Bubbles and Dimensions are all typically kept together in the Construct or Element file. Theoretically you could choose not to Anchor these objects and thereby place them in the proper file. However, this is a situation where "breaking the rule" is considered to have more benefit overall than following it. Therefore, it is recommended that you keep all of these components together in the Construct or Element file as we are doing here in your own projects.

ADD COLUMN GRID LABELS

5. Select the main grid, right-click and choose **Label**.

6. In the Column Grid Labeling dialog box, on the X - Labeling tab, put a check mark in the "Automatically Calculate Values for Labels" check box.

This will automatically calculate the values of the labels based on the first value you enter.

7. In the Automatic Labeling Rules area, choose **Ascending**, place a check mark in the Never Use Characters check box, and type **I,O** in the text field. (Be sure to type this uppercase.)

This will allow you to type certain characters that you wish to have the software skip when calculating the automatic numbering values (in this case, capital "I" because it looks very similar to "1" and capital "O" because it looks very similar to "0").

8. In the Bubble Parameters area, place a check mark in both "Top" and "Bottom" check boxes and set the Extension to **12'-0"** [**4000**].

These settings will add bubbles to both the top and bottom of the plan and will place them 12'-0" [4000] away from the edge of the grid. Choose this value based on the size of your project and the scale at which it will print.

9. Make sure "Generate New Bubbles On Exit" is checked.

   This will replace any labels already in the drawing and add new ones. This is valuable if you renumber the column lines during design.

10. On the left side of the dialog box, type **H** (uppercase) in the first Number box and then press ENTER.

    Notice how the remaining values fill in automatically based on the parameters we set, including skipping the letter "I."

11. Click the Y - Labeling tab.

12. Set everything the same as on the X - Labeling tab, except type **I** for the first value this time (see Figure 6–22)

TIP It is not necessary to choose "Never Use Characters" on the Y Labeling because we are labeling with numerals here rather than letters.

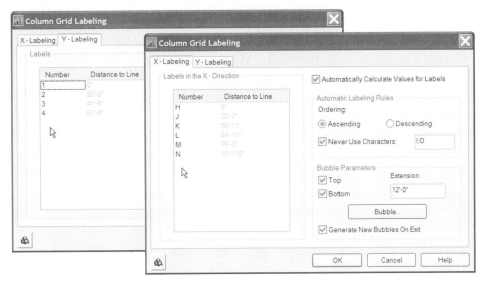

Figure 6–22 *Configure the Column Grid Labels*

13. Click OK to add the labels.

 TIP If you make a mistake while labeling, just right-click and select Labels again, adjust the settings, and click OK. The column bubbles will be adjusted to the proper settings, provided you place a check mark in the "Generate New Bubbles on Exit" check box.

CREATE GRID LINE EXTENSIONS AND PROJECT TOOL PALETTES

In many cases, the labels are desired on only one side of the plan. This is easy enough to accomplish by simply not checking one of the choices of Top, Bottom, Left, or Right as the bubbles are being added. However, it is also often desirable to have the column grid lines extend past the end of the grid. There is no Grid parameter to do this, which is why we added bubbles to both sides. The following steps illustrate a workaround that will resolve this situation and convert the bubble on one side to grid extensions.

In order to understand the logic, it is necessary to understand exactly what the column bubbles are in ADT. As covered in Chapter 4, a Multi-View Block (MVB) is an ADT object containing one or more AutoCAD blocks. The Grid bubbles use an MVB called "StandardGridBubble". This MVB is anchored to the Column Grid using a Column Grid Anchor. The Column Grid Anchor behaves similarly to the Node Anchor, except that a visible leader attaches it to the Grid Node. (See the discussion on Anchors in Chapter 2 and in the headings above.) If you click on one of the bubbles, you will note that the leader is highlighted with the bubble. If you delete one of the bubbles, you will note that the leader is deleted as well—the two are connected and cannot be separated. What we need to do is make one set of bubbles in each direction where the actual bubble is *invisible*. We can make the bubble invisible without losing the leader. When we do this, it will "appear" as though the grid lines extend past the limit of the grid.

When you load a Project in ADT, a Tool Palette Group for the current Project is also loaded. The Commercial Project that we have current has such a Tool Palette Group associated with it. We will load this Group now to access a custom Multi-View Block tool that we will use to swap out the bubble on one side to create extensions.

14. Right click on the Tool Palettes title bar and choose **06 Commercial [06 Commercial Metric]** to load the Project Tool Palette Group (see Figure 6–23).

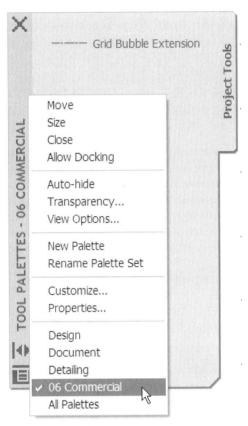

Figure 6–23 *Load the Project Tool Palette Group*

When the 06 Commercial [06 Commercial Metric] Tool Palette Group loads, a single tool palette tab named "Project Tools" will appear.

15. Carefully select all of the bubbles at the bottom and right sides of the plan.

16. Right-click on the **Column Bubble Extension** tool on the 06 Commercial [06 Commercial – Metric] tool palette and choose **Apply Tool Properties to Multi-View Block Reference**.

Notice the change to the drawing. It now appears as though there are grid line extensions.

If you wish to see what makes this Multi-View Block display this way, right-click one of the Column Grid Extensions and choose **Edit Multi-View Block Definition**. Then click the View Blocks tab and notice if you click the General or Model Display Representation on the left, all the View Direction checkboxes are cleared on the right. Even though a View Block named *BubbleDef* is loaded, by having no boxes checked, it is effectively *invisible*. Therefore, all we see in the drawing is the extension line that is the Column Bubble Anchor (a type of Leader Anchor).

If you prefer, you can create this Multi-View Block yourself rather than use the provided tool. To do this, select all of the Bubbles on the bottom and the right, right-click and choose **Copy Multi-View Block Definition and Assign**. On the General tab, rename it to: **Grid Bubble Extension** and on the View Blocks tab, deselect all checkboxes in both the General and Model Display Reps. Click OK to see the results.

CAD Manager Note: This method might seem a bit complicated, and it might be tempting to add labels to only two sides and create the extension by drawing a line manually. This practice is NOT recommended. The line workaround is not properly anchored to the grid. AutoCAD objects such as lines cannot use anchors. The workaround suggested above, although not perfect, keeps the Grid intact and can be modified globally by editing the Multi-View Block to modify the bubbles back to the original or to a new graphic standard. In addition, if Column Grid lines move, the extensions created in the technique above will move as well since they are anchored. Give it a try.

ADD LABELS AND EXTENSIONS TO THE SECOND GRID

The second Grid is a manual Grid created from linework. Manual Grids do not use the automatic labeling routine. Instead we will use the Column Bubble tool on the Annotation tool palette.

17. Right-click on the title bar of the Tool Palettes, load the **Document** group and then click the Annotation tool palette. On the Annotation tool palette, click the **Column Bubble** tool.

18. At the "Select node to label" prompt, click the vertical line just to the right of Column Line K.

19. In the Create Grid Bubble dialog box, type **K.3** for the Label, and set the Extension to **8'-0"** [**2400**] (see Figure 6–24).

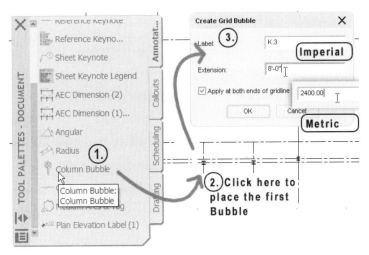

Figure 6–24 *Add a manual Grid Bubble*

The prompt will repeat.

20. At the "Select node to label" prompt, click the vertical line just to the left of Column Line L (and the right of the one you just labeled).

21. In the Create Grid Bubble dialog box, type **K.7** for the Label this time, leave the Extension set as is.

22. Repeat the process again for the two horizontal grid lines. Add the bubbles only to the left this time by clearing the "Apply at both ends of gridline" check box.

    Use **0** and **0.1** for the Labels and **12'-0"** [**4000**] for the Extension.

23. Click OK to complete the routine.

ADJUST THE BUBBLES WITH GRIPS

24. Select the two bubbles (K.3 and K.7) at the top.

25. On the Properties palette, change the Definition to **GridBubbleExtension**. (Or repeat the steps above to apply the *Grid Bubble Extension* tool properties).

26. Click on bubble 0 at the left.

27. Click the grip point in the middle of the leader to make it hot.

28. Move the mouse straight down (using POLAR or ORTHO), type **6'-0"** [**1800**], and press ENTER (see Figure 6–25).

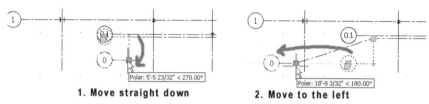

1. Move straight down   **2. Move to the left**

Figure 6–25 *Grip editing bubble extensions*

29. Click the same grip point again and move it directly to the left **22'-7"** [**6860**].

    This will make it line up with the bubbles from the other grid.

30. Repeat with Gridline 0.1 moving it down only **2'-0"** [**600**] this time. Move it to the left by the same amount.

TIP If you have trouble grip editing the Gridlines, temporarily turn off the Dynamic (DYN) toggle on the Application Status Bar.

TIP Grid bubble values can be modified manually without regenerating all of the bubbles by right-clicking the bubble you wish to change and choosing Properties. Click the Attribute worksheet icon in the Advanced properties grouping.

ADD AEC DIMENSIONS

31. On the Annotation tool palette, click the **AEC Dimension (2)**tool.

    If you do not see this palette or tool, right-click the Tool Palettes title bar and choose **Document** (to load the Document Tool Palette Group), and then click the Annotation tab.

32. Select the top horizontal edge of the main Column Grid and then press ENTER.

33. Click a point above the grid to place the Dimensions.

TIP Tweak dimension placement by using the small triangular shaped grip.

34. Repeat the Dimensions steps on the left side of the grid (see Figure 6–26).

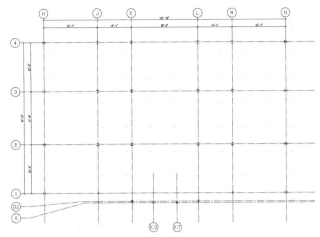

Figure 6–26 *The complete Column Grid*

35. Save the file.

 CAD Manager Note: Create a Dimension Style for Column Grids with both extension lines suppressed. Use this Dimension Style for Column Grids to avoid having the extension lines of the dimensions cover up the dashed lines of the Column Grid.

 NOTE Both the parametric Column Grid and the manual (converted from linework) grid were showcased here. Both techniques are suitable for creating Column Grids for use in real projects. The choice of which technique to use is largely a matter of personal preference. Parametric grids work well for repetitive spacing while manual grids work well for irregular spacing and orientation. Sometimes using a little of both (as we have done here) is the best choice.

STRUCTURAL FRAMING

So far we have limited our exploration of Structural Members to Columns only. While Beams and Braces have been part of Architectural Desktop for several releases, ADT 2006 introduces many new enhancements to the existing toolset that we will explore now.

ADD BEAMS

Let's start by revisiting the Structural Member Catalog and importing some more shapes.

1. Following the steps above in the "Access the Structural Member Catalog" heading, create two new Structural Member Styles: **W18x40** [**W460x52**] and **L6x6x3_8** [**L152x152x9.5**].

   Name them **W18x40 (Beams)** [**W460x52 (Beams)**] and **L6x6x3 8 (Braces)** [**L152x152x9.5 (Braces)**]

We'll use the "W" shape for the Beam in this exercise. Below we will use the "L" shape for the Braces.

2. Close the Structural Member catalog when finished.

3. On the View menu, choose **3D Views > SE Isometric**.

4. Right-click on the title bar of the Tool Palettes, load the **Design** group and then click the Design tool palette. On the Design tool palette, click the **Structural Beam** tool.

5. On the Properties palette, choose **W18x40 (Beams)** [**W460x52 (Beams)**] for the Style.

6. In the Dimensions Grouping, choose **Edge** for Layout Type.

The Edge option will place beams along grid lines between two columns. The Fill option will array the Structural Members within one or more bays.

7. Slowly move your cursor around on screen hovering over the Column Grid—do not click yet.

A tool tip similar to the one we saw above for the columns will appear indicating the various CTRL key options for placing beams. Notice also that this tip changes as you move off of the Grid object. Furthermore, notice that when the Grid highlights, that the Beam will automatically match the length of the nearest Grid bay (see Figure 6–27).

 TIP Turn off Object snaps to facilitate the automatic placement of Structural Members. The easiest way to toggle them off is to press the F3 key or click the OSNAP toggle (to pop it up) on the Application Status Bar.

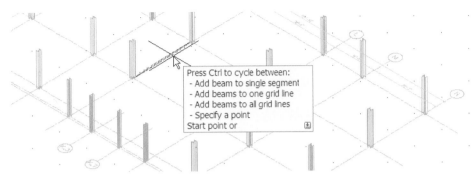

Press Ctrl to cycle between:
- Add beam to single segment
- Add beams to one grid line
- Add beams to all grid lines
- Specify a point
Start point or

Figure 6–27 *Beams have many* **CTRL** *key options and automatically match the size of Grid cells*

8. Click the mouse to create a Beam.

The only trouble with the Beam that we just placed is that it attached to the same height as the Grid. We have a tool we can use to override this behavior.

9. At the bottom of the screen, on the Application Status Bar, click the number next to the word Elevation (see item I in Figure 6–28).

10. In the Elevation Offset worksheet that appears, click the Pick point icon (see item 2 in Figure 6–28).

11. Using your Object Snaps (toggle them back on with F3 if they are still turned off from above) snap a point to the top of one of the columns and then click OK when the Elevation Offset worksheet returns (see item 3 in Figure 6–28).

12. Next to the Elevation value on the Application Status Bar, click the small Replace Z toggle icon—it will stay pushed in to indicate that it is active (see item 4 in Figure 6–28).

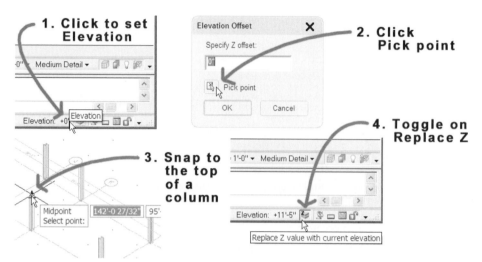

Figure 6–28 *Set the Z Elevation snap*

This procedure has now moved the Z Elevation up to the top of the columns. If we repeat the Beam placement from above, the Beams will now appear at the top of the columns. If you still have a stray Beam down at Z=0, delete it now.

13. On the Design tool palette, click the **Structural Beam** tool.

14. On the Properties palette, choose **W18x40 (Beams) [W460x52 (Beams)]** for the Style.

15. From the "Trim automatically" list, choose **Yes**.

16. From the "Layout type" list, choose **Edge**.

    Verify that "Justify" is set to Top Center.

17. Hover over one of the outer edges of the Column Grid—do not click yet.

18. Press the CTRL key once (see middle of Figure 6–29).

    Note that beams will be added to all bays along a single edge.

19. Press the CTRL key again (see right side of Figure 6–29).

    Notice that now Beams will be added to all edges.

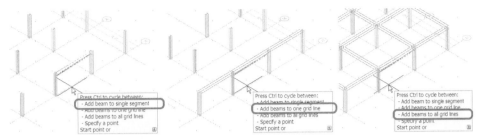

Figure 6–29 *Cycling through the Beam placement options*

20. Click the mouse to place the Beams.

At this point we have the makings of a nice framing model. However, if you take a look at the front of our building where we added the small manual Grid, the beams don't quite match the columns. This is easy to fix. It is important to realize, however, that all of our Structural Members are currently "connected" to one another in logical ways. Try selecting any Beam or column and lengthening or shortening it with the grips. Notice that when you do so, the other "connected" beams and columns adjust as well. While this behavior looks similar to anchoring, it is not a result of using Anchors. This is special behavior by which Structural Members relate to one another. Undo any grip edit that you made.

21. Delete the Beam on column line I between lines K and L.

22. Using the lengthen grip (pause over each grip to find lengthen), begin to lengthen the Beam on column line K.

23. Press the ctrl key once to toggle off the connected behavior and then snap to the Columns on column line 0.

TIP Hold down the SHIFT key, right-click and then choose **Node Osnap**. The Node snap will allow you to easily snap to the middle center of the column.

24. Repeat this process for the Beam on column line L.

25. On the Design tool palette, click the **Structural Beam** tool.

    Verify that all of the previous settings are still configured: "Style" is **W18x40 (Beams)** [**W460x52 (Beams)**], "Trim automatically" is **Yes**, "Layout type" is **Edge** and "Justify" is Top Center.

26. Move the mouse near the Column at column line K0.1. When the Column highlights, click the mouse (no Osnap is necessary).

27. Highlight the free-standing Column at column line K0 and click.

A Beam will appear between these two columns trimmed neatly to fit between them (see Figure 6–30).

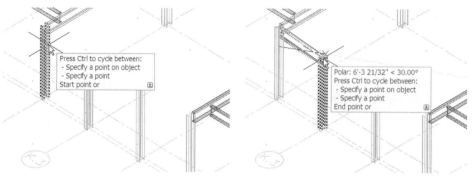

Figure 6–30 *Create a Beam that automatically connects to neighboring Columns*

28. Repeat the process to add two more Beams (see Figure 6–31).

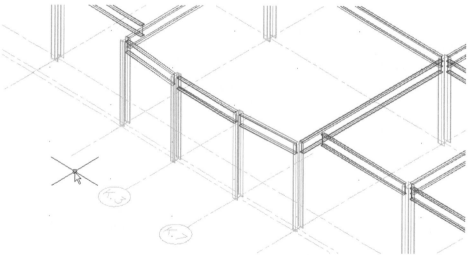

Figure 6–31 *Add three Beams to complete the front fa\c¢cade bay*

29. Save the file.

ADD BRACES

We can add braces using similar methods. Let's add a few Braces where the building core will be. We will add our first brace along column line 3 between lines L and M.

30. Zoom into the area near column lines K through L and 3 and 4.

31. On the Design tool palette, click the **Structural Brace** tool.

32. On the Properties palette, choose **L6x6x3_8 (Braces)** [**L152x152x9.5 (Braces)**] for the Style.

33. From the "Trim automatically" list, choose **Yes**.

34. From the "Specify rise on screen" list, choose **No**.

35. For the "Method" choose **Distance** and for both "Distance along first member" and "Distance along second member" type **6'-0"** [**1800**].

36. At the "Start point (Pick Beam or Column)" prompt, hover over the Column at grid intersection L3 and then click.

37. At the "End point (Pick Beam or Column)" prompt, hover over the horizontal Beam at grid line 3 and then click.

38. Repeat the same process to add an additional Brace between the same Beam and Column M3.

TRY EDIT IN VIEW

If you are having difficulty getting the Braces to link up to the Columns and Beams properly, try using the Edit in Section command on the Isolate objects right-click menu to get a better view and indicate the relationship with more precision. The Edit in View commands allow you to isolate a small selection of objects and zoom directly to them in either Plan, Elevation or Section. In this case, we can add the Braces using Edit in Section.

Select just the Columns and Beams on and between column lines L3 and M3. Right-click in the drawing, and choose Isolate Objects > Edit in Section. Following the prompts, click two points in front of the Column object and parallel to column line 3 and then press ENTER. (This is sometimes easier to do if you go to a Top view first). When prompted, drag the temporary section line back large enough to surround the Columns and the press ENTER again. You will be automatically zoomed to a section view of the selected objects and all other objects will temporarily disappear. Add your Braces following the steps here, and then click the Exit Edit in View icon when finished. This will return you to the drawing, restore the previous view and all the hidden objects will return.

Place one more Brace.

39. Click near the base of Column K3 and near the top of Column K4 (see Figure 6–32).

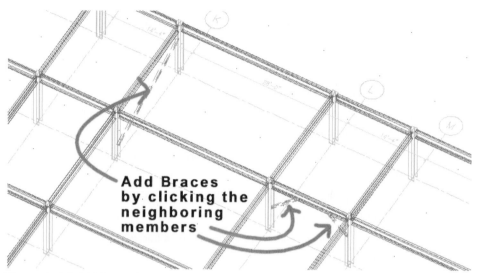

Figure 6–32 *Add Braces between the existing Columns and Beams*

There are several methods of placement for Braces. Feel free to experiment with others if you like. Most of the methods require that a Brace be placed between two perpendicular members such a Column and Beam. However, as you can see from our last example, you can achieve results between parallel members like two Columns as well.

ADD JOISTS

To complete our structural layout of the commercial project, let's add some joists. For this task, we could return to the Structural Catalog or Wizard to create a new joist Beam Style. However, there are some pre-made steel bar joists provided in the library with the software. These are accessed via the Content Browser. Recall in Chapter 4 that we utilized the Content Browser and even loaded a custom catalog. In this exercise, we will simply access one of the out-of-the-box catalogs. Please note that the bar joists are provided only in the Imperial catalog.

40. On the Navigation toolbar, click the Content Browser icon (or press CTRL + 4).

41. Click on the *Design Tool Catalog – Imperial* catalog to open it.

42. On the left side, choose the *Structural > Bar Joist* category.

43. Click and hold down the mouse on the small eyedropper icon next to the Steel Joist 16 tool.

44. Wait for the eyedropper to "fill up" and then drag it and drop it into the ADT drawing window (see Figure 6–33).

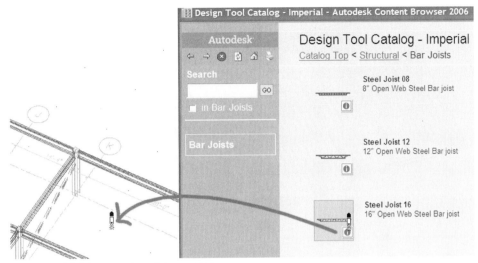

Figure 6–33 *The complete Column Grid*

The Structural Beam command will execute and the Steel Joist 16 Style will automatically import and become active in the Properties palette.

45. On the Properties palette, change the "Layout type" to **Fill** and the "Justify" to **Baseline**.

46. In the Layout grouping, set "Array" to Yes, "Layout method" to Repeat and set the "Bay size" to **2′-0″ [600]**.

47. Hover the mouse over a grid line—do not click yet.

Notice that the entire Grid cell will be filled with members. Try moving the mouse to hover over both horizontal and vertical Grid edges. Notice that the orientation of the joists flips to match the highlighted edge.

48. Press the CTRL key once.

Notice that now all cells are highlighted with joists.

49. Orient the joists parallel to the lettered grid lines and then click the mouse to create joists in all bays.

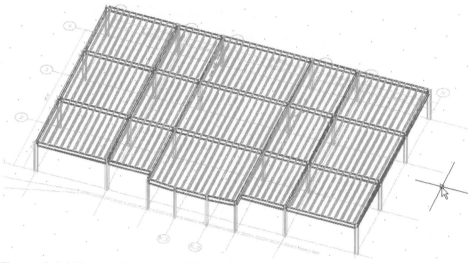

Figure 6–34 *The complete structural framing system*

    50. Save and close the file.

ADDING A STRUCTURAL CATEGORY TO THE PROJECT

Despite the effort expended in Chapter 5 to get the project set up, you will often identify additional drawings and Sheets that are required as the project progresses. This in no way invalidates the goals of Chapter 5. Rather it exposes a situation that occurs very frequently in *real* projects. In this case, we could simply take the Element file (that we currently have open) and attach it to each of the *Shell and Core* Constructs. However, if your firm is multidisciplinary, or if you would like greater flexibility in project structure, then it is useful to keep the disciplines separate and create a Column Grid Construct for each floor.

ADD A STRUCTURAL CATEGORY

1. On the Project Navigator (Window menu or press CTRL + 5) click the Constructs tab.

2. Right-click on the *Constructs* folder and choose **New > Category**. Name the new Category **Structural**.

   A Category will appear as a folder in the Project Navigator.

3. Repeat this process and make a new Category in *Constructs* named **Architectural**.

4. Repeat this process again in the *Elements* folder, creating the same two Categories, **Architectural** and **Structural**.

5. In the *Elements* folder, drag the *Column Grid* Element file onto the *Elements\Structural* folder to move it to that folder.

6. In the *Constructs* folder, drag and drop each of the existing Constructs onto the *Architectural* folder.

   Unfortunately, this must be done one Construct at a time.

CREATE STRUCTURAL GRID CONSTRUCT FILES

Now that we have extended our project folder structure to accommodate a Structural discipline, let's build a Column Grid Construct for each floor of the building.

7. Right-click on Structural Grids and choose **New > Construct**.

8. For the Name type **01 Grid** , and for the Description type **First Floor Structural Grid**.

9. In the Assignments area, place a check mark next to First Floor and then click OK (see Figure 6–35).

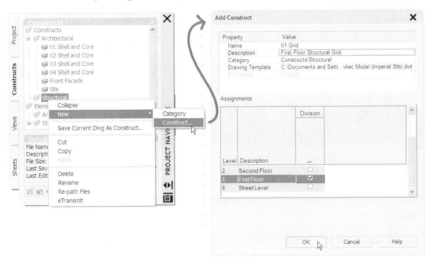

Figure 6–35 *Create a First Floor Column Grid Construct within the Structural Category*

10. Repeat these steps to create **02 Grid** , **03 Grid** and **04 Grid**. Edit the Descriptions and Level assignments accordingly.

11. In the *Elements\Structural* folder, right-click the *Column Grid* Element file and choose **Attach Element to Constructs**.

12. Place a checkmark in the checkbox next the *Structural* folder (see Figure 6–36).

This will XREF Attach the *Column Grid* Element to each of the Grid Constructs.

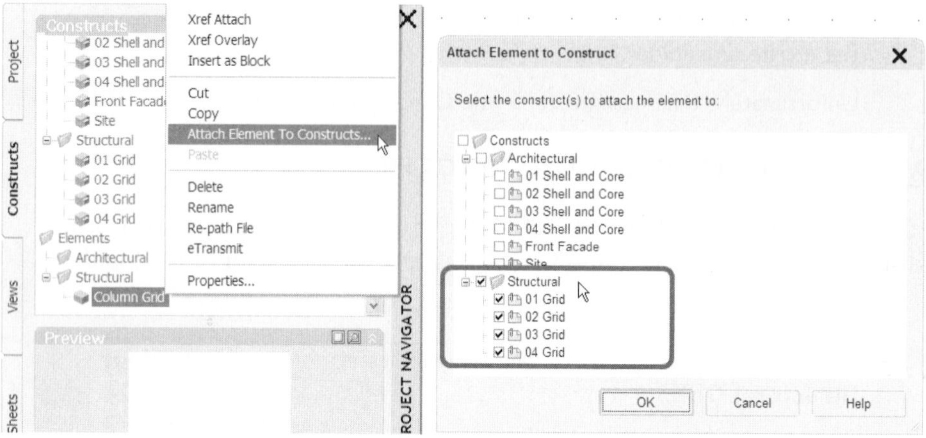

Figure 6–36 *Attach the Column Grid Element to each of the Grid Constructs*

13. On the Constructs tab, in the *Constructs\Structural* folder, double-click *01 Grid* to open it.

    Note that the Grid has been XREFed to this file. Feel free to open others as well.

Since moving Constructs to different categories literally moves the drawing files to a different folder, the XREF paths will fail if you do not re-path the project. Also, when you rename files you can experience broken XREF links. For these reasons, we must re-path all of the XREFs in the project. Refer to the "Create the Upper Floor Constructs" heading in Chapter 5 for more information on these issues.

14. At the bottom of the Project Navigator palette, click the Re-path icon, and then in the Reference File Re-path Queue dialog box, click the Re-path button.

15. That completes the setup of the Structural Column Grid and all associated Project files.

16. Save and Close all project files.

CAD Manager Note: Adding subfolders (Categories) to the Constructs and Elements folders is a very common way to manage multidiscipline firms. You could add additional folders for MEP and other disciplines as well. It would also be fairly common to add subfolders to the Views tab. In this project we will leave the Views tab uncategorized for simplicity. However, feel free to experiment with this. Always remember to re-path after moving and/or renaming files. Also, files cannot be moved or re-pathed while they are open for editing, so be sure that all files are closed first.

UPDATE PROJECT VIEWS

The Constructs that we have created here will need to be added to the Project Views (refer to the previous chapter for complete details on Views). In particular, we would typically wish to see Column Grids in the Floor Plan Views and in the overall Composite Model and Section Model. It would not be useful to have the Column Grids appear in the Composite Elevation Model.

17. On the Project Navigator, click the Views tab.

18. Right-click on *A-CM00* and choose **Properties**.

19. Click on the Content page and verify that all of the Grid files are checked, and then click OK.

    Notice that simply clicking on this tab is sufficient to make the View re-query the project and gather the new Constructs that meet its selection criteria.

20. Repeat the same steps on *A-SC01*, and all of the Plan Views.

    Naturally in each Plan View, only one Structural Grid file will be selected for that particular level.

 Caution: *Be sure to check the Content page after each change or the updates may not Re-generate properly.*

21. Right-click on the *A-EL01* View and choose **Properties**.

22. Click on the Content page place a check mark in the Structural folder, and then remove it.

    This action will deselect the entire Category and all future files from this View. It is not necessary to load Structural Grids for the exterior Building Elevations.

23. Right-click on the Views folder and choose **Regenerate**.

This will update all of the View files with the changes that we just made. Go ahead and open any of the Views to verify the addition of the Column Grid to each of them.

If you wish, open the *A-SC01* View file, select the Section object, right-click and choose **Refresh** to see all the framing added to the section. Publish an updated DWF of the entire set on the Sheets tab as well if you wish.

24. Save and Close all commercial project files before continuing.

CREATING A FOUNDATION PLAN

In addition to Columns and Grids, the structural tools in ADT include Beams, Braces, Slabs and certain Wall styles. In this exercise, we will look at a concrete Wall style with integral footing. We will do this in the context of the Residential Project. However any of the techniques we will cover would work equally well in the Commercial or any other Project.

LOAD THE RESIDENTIAL PROJECT

Be sure that all drawings from the Commercial Project are saved and closed. However, you must have at least one drawing open in ADT for the Project Browser to work. Therefore, if you don't have any drawings open; click the Qnew icon on the Standard toolbar to create a new blank drawing. (If a dialog opens prompting you to select a file, review Chapter 3 for information on how to correct this).

1. On the File menu, choose **Project Browser**.

2. Navigate to the *C:\MasterADT 2006\Chapter06* folder and make the *06 Residential [06 Residential Metric]* project current.

   If the Re-Path dialog appears, simply click the Re-path icon to continue. There may still remain some queued XREF path changes in the Commercial Project. This operation will ensure that none of them are missed before you change the current project. *Always* click the Re-path button when asked! It is possible you could get this alert twice here, once for closing the commercial project and then again for the residential project. Always click Re-path.

3. On the Constructs tab, double-click the *Basement New* file to open it.

4. On the Walls palette, right-click the **Concrete-8 Concrete-16x8-footing** [*Concrete-200 Concrete-400 x 200-footing*] tool and choose **Apply Tool Properties to > Wall**.

   If you do not see this palette or tool, right-click the Tool Palettes title bar and choose **Design** (to load the Design Tool Palette Group), and then click the Design tab.

5. At the "Select Wall(s)" prompt, select all three Walls in this drawing and then press ENTER.

   Notice the change in the Walls and the appearance of the footing line below. The Walls, however, still need some adjustment.

6. On the Project Navigator palette, drag and drop the *Basement Existing* Construct into the drawing.

NOTE When you drag one Construct into another Construct, the file will be XREFed as an Overlay. Overlaid XREFs apply only to the current drawing and do not carry forward if the current drawing is XREFed to another.

Adding the Overlaid XREF of the Existing Basement Conditions reveals that the foundation Walls on the addition do not properly align with the existing construction.

7. With the same three Walls still selected, right-click and choose **Properties**.

8. On the Properties palette, change the Justify to **Baseline**.

    Notice that the face of the concrete Wall is now aligned with the existing construction (rather than the face of the footing as before), but the wrong face is aligned.

9. With the Walls still selected, right-click and choose **Reverse > Baseline** (see Figure 6–37).

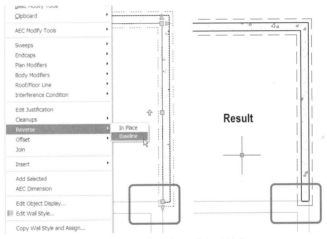

Figure 6–37 *Use the grips to reverse the direction of the Walls*

10. Select all three Walls again and on the Properties palette, change the Base Height to **8′-9″** [**2600**].

11. Right-click and choose **Deselect All**.

    There will be a small galley and new basement stair on the left side of the new addition.

12. Offset the left vertical Wall into the plan (toward the right).

    For the Offset Distance type **5′-0″** [**1500**].

13. Using both of the vertical Walls at the left as cutting edges, trim off the piece of horizontal Wall between them (see Figure 6–38).

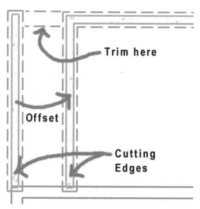

Figure 6–38 *Adding a Basement access gangway*

A few more things need attention in this file. For instance, the foundation Wall appears to encroach on the existing Wall [and does not touch in the Metric file]. This is easily remedied with Wall Endcaps. However, we will save those edits for later chapters.

14. Save and Close the file.

ADDITIONAL CONTENT

In addition to the Wall tool that we accessed here, there are many other Wall Styles in the Content Browser that can be used for footings of various kinds. Some are concrete Wall Styles with integral footings like the one showcased here, others are styles of just the footing that you add separately from any Walls. One of the advantages of using Wall Styles for footings is that like other Walls, you get automatic cleanup (see Chapter 9) and the ability to control how they display below the cut plane (see Chapter 10). To view the many styles provided, return to the Content Browser (as covered above) and browse to the *Design Tool Catalog – Imperial* [*Design Tool Catalog – Metric*] and then the *Walls > Concrete* category. There you find several concrete footing and wall with footing styles ready to iDrop into your projects (see Figure 6–39).

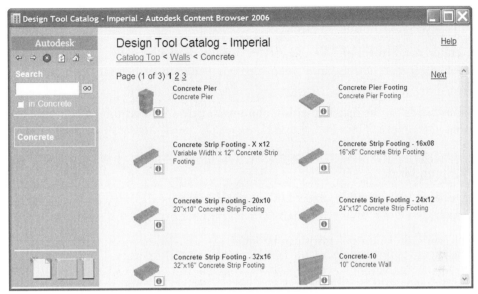

Figure 6–39 Concrete *Footing Styles in the Content Browser library*

ADDITIONAL EXERCISES

Additional exercises have been provided in Appendix A. In Appendix A you will find an exercise for updating the Basement Existing Construct of the Residential Project to include the Beams and Columns. It is not necessary that you complete this exercise to begin the next chapter, it is provided to enhance your learning experience. Completed projects for each of the exercises have been provided in the *Chapter06/Complete* folder.

SUMMARY

Use either the Structural Member Style Wizard or the Structural Member Catalog to import shapes into your file and generate Structural Member styles.

Column Grids can be inserted with Columns already anchored to each intersection.

Anchors provide powerful parametric relationships between the Columns and the Column Grid.

A parametric Column Grid object can have grid spacing set to a fixed bay size, or equally divide the total grid size by a fixed number of bays.

Grid spacing of parametric grids can be manually adjusted to meet any design or existing conditions requirement.

You can draw a grid exactly the way you want it using lines and then convert those lines to a Column Grid object.

Automatic grid labeling and dimensioning are fast and flexible tools for annotating a Column Grid.

Grid Extensions can be achieved with a custom Multi-View Block used in place of the standard bubble.

Beams, Braces and Joists can be added to your parametric Column Grid using many powerful automated placement routines.

Categories can be added to the Project folders to organize the project files by discipline.

Wall Styles with integral footings can be used to create foundation plans.

Many footing styles both integral to Walls and independent can be found in the Content Browser library.

CHAPTER 7

Vertical Circulation

INTRODUCTION

In this chapter, we will look at Stairs and Railings. The core plan of the commercial building will include Stairs, elevators, and toilet room layouts. The exterior entrance plaza leading up to the commercial building calls for Stairs, a ramp, and Railings. The Residential Project contains existing Stairs on the interior and an existing exterior Stair at the front entrance.

OBJECTIVES

We will add Stairs for the basement and first floor of the residential project and lay out the core file for the commercial project. Our exploration will include coverage of the Stairs, Stair styles, and Railings. We will also look at tools available to represent elevators and ramps. The main tools covered in this chapter include the following:

- Add and modify Stairs.
- Add and modify Railings.
- Add and modify ramps.
- Lay out elevators and restrooms.
- Working with Stair Tower Generate.
- Stairs and Railings.

 ## STAIRS AND RAILINGS

When we work with Stairs, most parameters are configured in the style (refer to Chapter 2 for an overview of Object styles in ADT). Style-based settings include riser and tread relationships, stringer settings, landing rules, and display settings. Width and height parameters and clearances belong directly to the Stair object (and are therefore not part of the style). Stairs do *not* automatically include Railings; however, rules can be assigned to the Stair objects that control the placement of Railings. Railings are then added as a separate object, anchored to the Stair (refer to Chapter 2 for an overview of Anchors). Multi-View Blocks are used to generate elevators, and ramps are created from specially configured Stair styles included with the out-of-the-box ADT content library.

Adding and modifying Stairs is much like adding and modifying any other ADT object. Choose the appropriate tool from the tool palettes and set the initial parameters in the Properties palette, then you place the object into the drawing. Finally you progressively refine the Stair over time as project needs dictate. To begin our exploration of Stairs, we will revisit our first floor existing conditions plan for the Residential Project. In this file, we will build the existing Stairs in the main house.

INSTALL THE CD FILES AND LOAD THE CURRENT PROJECT

If you have already installed all of the files from the CD, simply skip down to step 3 below to make the project active. If you need to install the CD files, start at step 1.

1. If you have not already done so, install the dataset files located on the Mastering Autodesk Architectural Desktop 2006 CD ROM.

   Refer to "Files Included on the CD ROM" in the Preface for information on installing the sample files included on the CD.

2. Launch Autodesk Architectural Desktop 2006 from the desktop icon created in Chapter 3.

If you did not create a custom icon, you might want to review "Create a New Profile" and "Create a Custom ADT Desktop Shortcut" in Chapter 3. Creating the custom desktop icon is not essential; however, it makes loading the custom profile easier.

3. From the File menu, choose **Project Browser**. (You can also click the Project Browser icon on the Navigation toolbar).

4. Click to open the folder list and choose your *C:* drive.

5. Double-click on the *MasterADT 2006* folder, then the *Chapter07* folder.

   One of two residential Projects will be listed: *07 Residential* and/or *07 Residential Metric*.

6. Double-click *07 Residential* if you wish to work in Imperial units. Double-click *07 Residential Metric* if you wish to work in Metric units (You can also right-click on it and choose **Set Project Current.**) Then click Close in the Project Browser.

IMPORTANT If a message appears asking you to re-path the project, click Yes. Refer to the "Re-Pathing Projects" heading in the Preface for more information.

ADD A STAIR TO THE RESIDENTIAL PLAN

1. On the Project Navigator, double-click *First Floor Existing* in the *Constructs* folder.

This is the first floor existing conditions file from Chapter 4. At the end of that lesson, we inserted some pre-built Stair objects into our Residential First Floor Existing Conditions file. The interior Stair in the main hallway has been removed here so that we can learn the techniques needed to build it ourselves.

2. Open the Content Browser (click the icon on the Navigation toolbar or press CTRL + 4).

3. Click on *Design Tool Catalog - Imperial* [*Design Tool Catalog - Metric*]

4. Navigate to the Stairs and Railings category and then the Stairs category.

5. Using the eyedropper icon, drag the **Stair** [**Wood-Saddle**] tool into the drawing window.

   The Add Stair command will begin and on the Properties palette, Wood-Saddle will be set as the Stair Style.

6. On the Properties palette, choose **Straight** for the Shape and choose **Left** for Justify (see Figure 7–1).

7. Set the Width to **3'-1"** [**942**], the Height to **9'-0"** [**2750**] and the Tread to **11"** [**280**].

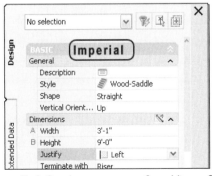 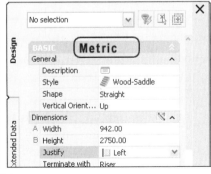

Figure 7–1 *Set the parameters for adding a Stair*

8. At the "Flight Start Point" prompt, use OSNAP and OTRACK and track **3'-6"** [**1075**] from the corner to the left of the front door (see Figure 7–2).

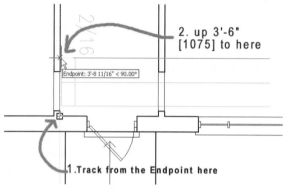

2. up 3'-6"
[1075] to here

Endpoint: 3'-8 11/16" < 90.00°

1.Track from the Endpoint here

Figure 7–2 *Track from the front door 3'-6"*

9. From that point, move the mouse straight up (90° using POLAR or ORTHO), click the mouse to place the Stair. Press ENTER to complete the command (see Figure 7–3).

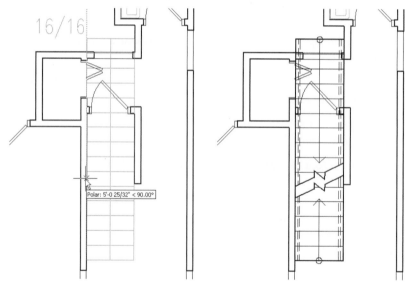

16/16

Polar: 5'-0 25/32" < 90.00°

Figure 7–3 *The Stair is too long*

Notice the overall box defining the footprint of the Stair as it is being placed. There is also feedback on the number of risers placed (in this case 16 out of 16). This is based on the riser to tread relationship and the 9'-0" [2750] height we assigned. In this case, the box is too long. By default, the Stair is placing too many risers. Furthermore, there are rules that govern the relationship of treads to risers in the Stair style. To correct this, we will first edit the Stair style, and then the Stair properties.

WORK WITH STAIR STYLES

In this sequence, we will edit the parameters of the Stair style and make some minor modifications to it to help it conform to the needs of the existing structure. This will involve removing the constraints enforced by Design Rules. Since the stairs existing in this house were built before current code requirements were in place, an "Existing Stair" style needs to be created allowing more freedom.

10. Select the Stair that you just drew, right-click and choose **Copy Stair Style and Assign**.

11. In the Stair Styles dialog, on the General tab, change the name to **Existing Stairs.**

12. Change the Description to: **Existing Conditions Stair – Residential Project**.

13. Click the Floating Viewer icon in the lower left corner of the dialog box and set it to **SE Isometric** view and **Flat Shaded** (see Figure 7–4).

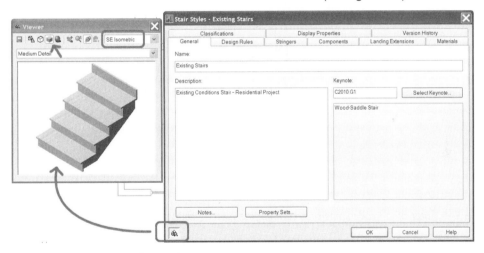

Figure 7–4 *Rename the Stair style*

This is an existing Stair, and therefore we will draw what is actually there rather than what the building code states *should* be there.

14. Click the Design Rules tab.

Notice the parameters.

Here you can enter a range of values for minimum and maximum tread and riser. You can also enter building code values for your jurisdiction in a rule-based calculator. Values assigned to these rules will constrain the parameters of the Stair as it is being created. Again, because the Stair we are building is an existing Stair, we will set the limits very broadly so that we can enter actual values without restriction.

15. In the Maximum Slope fields, type **12″ [300]** for Riser Height and **8″ [200]** for Tread Depth.

    The Optimal Slope and Minimum Slope values can remain as they are for this Stair.

16. Clear the Use Rule Based Calculator check box (see Figure 7–5).

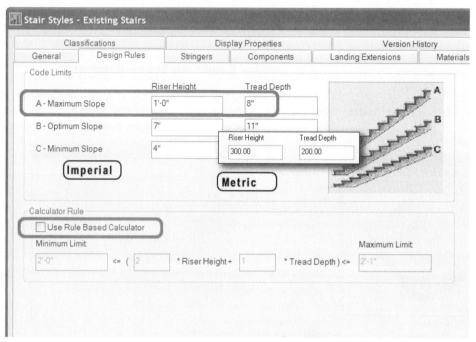

Figure 7–5 *Enter broad code limits to accommodate existing conditions*

These settings will give us the freedom we need to create the existing Stair. (We hope we never run across a 12″ [300] riser with 8″ [200] tread, but you never know.)

17. Click the Stringers tab.

ADT Stairs can have four different types of stringer. The choices are Saddled, Housed, Slab, and Ramp. A Saddled stringer is notched in the shape of the treads and risers and supports the treads from underneath. The Housed stringer occurs at the edges of the Stair with the treads spanning in between. A typical wooden Stair would use Saddled, while a steel pan Stair would be Housed. Slab creates a solid mass of material across the full width of the Stair underneath the treads and risers. Solid Slab is used primarily for poured concrete Stairs. In ADT, there is no dedicated Ramp object. To create a ramp, you use a Stair style with Ramp stringers. Take a quick look at the other settings and match them up with the letters in the diagram at the left of the dialog box. We do not need to change anything here for our existing Stair.

18. Click the Components tab.

This tab contains the tread and riser settings. Match up the settings with the letters in the diagram. Deselect tread or riser if you do not wish to display either of these components. For instance, by deselecting the Riser checkbox, you can create an open riser Stair. Check Sloping Riser for steel pan Stairs; leave it unchecked for our Wood Stairs.

19. Place a check mark in the Allow Each Stair to Vary check box (see Figure 7–6).

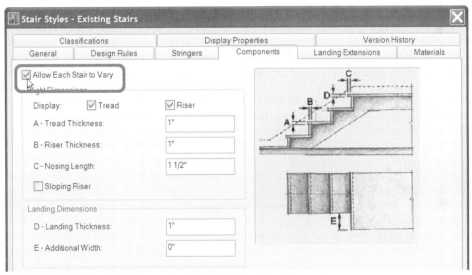

Figure 7–6 *The Components tab controls the Tread and Riser dimensions*

This is another helpful setting for an existing Stair. Each Stair you create will take its original parameters from the style. However, when you allow each Stair to vary, all style-based parameters can be edited on each individual Stair object without affecting the others. This setting is recommended only if your project has few Stairs that share the same constraints. The alternative is to simply make a style for each unique situation.

20. Click the Landing Extensions tab.

Study the settings. For this Stair, we will not set any extensions, but when we return to the commercial core plan, we will look further at this feature. Don't worry about the other tabs yet. We will edit those parameters later.

21. Click OK to accept the changes and return to the drawing.

MODIFY STAIR PROPERTIES

22. Select the Stair, right-click and choose **Properties**.

23. In the Dimensions grouping, click the Calculation rules worksheet icon, labeled Tread (see Figure 7–7).

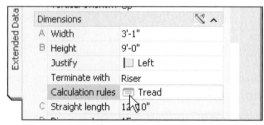

Figure 7–7 *Access Calculation rules worksheet*

24. Change D - Tread to **10″ [250]**.

25. Click the Automatic (lightning bolt) icon next to Riser Count (see Figure 7–8).

This will make the Riser Count field editable.

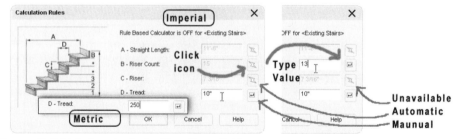

Figure 7–8 *Unlock and edit the Riser Count field*

26. Type **13** in the B - Riser Count field and then click OK.

Notice the dynamic reduction in Stair length. This is due to the now shallower treads and taller risers.

27. Save the file.

MODIFY STAIR DISPLAY

The default display characteristics for Stairs in plan is to represent them as two flights, one up and the other down. This is helpful in many plans, but it does not work well in all plans. If the Stairs on each floor are different heights or shifted slightly from one another, this graphic representation will not be accurate. At the early stages of a project, it might be acceptable to leave the Stair representation as is, but it will likely need to be changed as the design enters a more refined state. Editing the Display Properties of the Stair object will allow us to produce whatever graphic display our specific design scenario requires. In this residential plan, the floor-to-floor height of the first to second floor is 9'-0" [2750]; however, it is only 8'-9" [2675] from basement to first floor. To resolve this situation, we will edit the display of the Stair we created in the last step to show only the Stairs going up. Then we will copy this Stair and modify it to show only the basement Stairs going down.

28. Select the Stair, right-click and choose **Edit Object Display**.

29. Click the Display Properties tab.

Ordinarily we would prefer to edit display properties at the Style level. However, in a project such as this where each Stair is unique, we will find it easier to edit at the Object level.

30. In the Display Representation column highlight Plan (it should be bold).

31. In the Object Override column, click the check box next to Plan.

32. On the Layer/Color/Linetype tab, click the light bulb icon next to Stringer Up and each of the Down components (see Figure 7–9).

The light bulbs will turn dim, indicating that the display of these components is off.

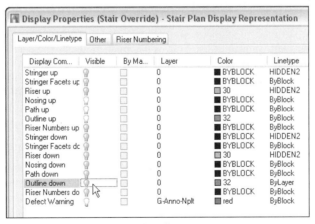

Figure 7–9 *Turn off the Stringers and all "down" components*

33. Click the Other tab and change the Elevation in the Cut Plane area to **6'-0"** **[1800]**.

This will move the diagonal break line and show more of the Stair in plan.

34. Click OK twice to return to the drawing.

    Notice that the down flight has disappeared, and the stringers going up have disappeared as well. These were not required and made the drawing look too busy. In addition, the Cut Plane moved, showing us more of the Stair in plan.

CREATE THE "DOWN" STAIR

35. Select the Stair, right-click and choose **Add Selected**.

36. On the Properties palette, change the Vertical Orientation to Down.

37. Change the Height to **8'-9"** [**2675**] and the Justify to **Right**.

38. Click the point shown in Figure 7–10 as the Start Point of the Stair.

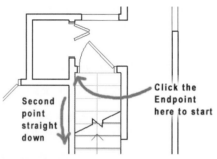

Figure 7–10 *Pick the Endpoint at near the Basement Door as a start point*

39. Pull the mouse straight down 270° (using POLAR or ORTHO) and then click. Press ENTER to end the command.

    A new Stair, very similar to the first, will appear in nearly the same spot as the first (see Figure 7–11).

40. Select the newly created Stair, right-click and choose **Edit Object Display**.

 TIP To be certain that you selected the correct one, check the Vertical orientation on the Properties palette; the newly created Stair should have a "Down" orientation. If the one you selected is "Up", deselect it and try again.

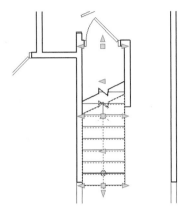

Figure 7–11 *Select the newly created overlapping Stair*

41. Return to the Display Properties tab.

   The Stair object level override is already attached because it was copied from the original Stair.

42. Click the Edit Display Properties button in the top right corner of the dialog box (or double-click Plan).

43. On the Layer/Color/Linetype tab, click the light bulb icons next to Nosing down, Path down and Outline down to turn them on.

44. Click the light bulb icon next to each of the Up components to turn them off (see Figure 7–12).

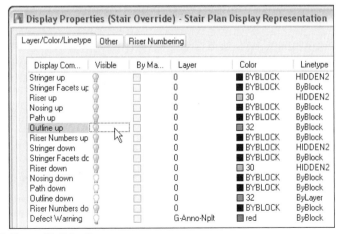

Figure 7–12 *Turn off all "up" components*

45. On the Other tab, change the Elevation in the Cut Plane area to **6'-6″** [**1600**] and then click OK twice to return to the drawing.

Notice that the Stairs now display correctly. The change in Vertical orientation is not yet evident because we are viewing the plan. (We will see the effect of this change in the "Adjust Wall Roof Line" sequence below.) If you like, switch to an isometric display and have a look. Be sure to return to Top View before continuing to the next exercise.

PROJECT A STAIR EDGE

The edge of the Stairs can be customized to conform to neighboring objects or po-lylines. In this sequence, we will widen the Stair as it passes the short Wall in the middle of the hallway.

46. Draw a polyline (Shapes toolbar) from the outside corner of the half wall in the center of the plan straight down past the end of the Stair (see Figure 7–13).

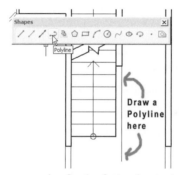

Figure 7–13 *Draw a polyline as an edge for the Stair edge projection*

47. Select the first floor Stair, right-click and choose **Customize Edge > Project**.

48. At the "Select an edge of a Stair" prompt, select the right edge of the Stair.

49. At the "Select a polyline or connected AEC entities to project to" prompt, select the polyline just drawn and press ENTER.

The edge of the Stair projects to the line defined by the polyline.

50. Erase the polyline (see Figure 7–14).

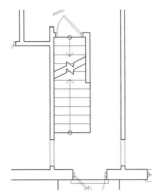

Figure 7–14 *The final existing residential Stair corridor in plan Wall Roof Line*

ADJUST WALL ROOF LINE

Although our Stairs look fine in plan, 3D View would reveal that the Wall at the Basement Door is too tall and intersects the Stair.

51. Using a crossing window, select both Stairs and the surrounding walls and doors of the hallway.

52. Right-click and choose **Object Viewer**.

53. Use the Orbit (click and drag in the green circle) to dynamically change the view to 3D (or choose a preset view from the View Control list like **SE Isometric**).

We can now see the effect of the Down orientation of the basement Stair. However, if you look at the top of the first floor Stair, it is passing right through the wall. If you will use this drawing only for plans, you could ignore this situation. However, it will be noticeable in 3D or Sections. Let's see the steps to fix it in 3D View.

54. Close the Object Viewer.

55. Click on the small horizontal wall with the door leading to the basement (the one that passes though the Stair); select just the Wall.

56. Right-click and choose **Roof/Floor Line > Modify Roof Line** (see Figure 7–15).

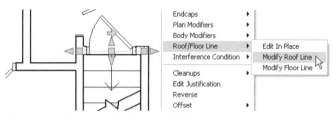

Figure 7–15 *Choose Roof Line to edit the top of the Wall*

57. If you have DYN turned on, choose **Auto project** from the list of options, otherwise, right-click and choose it from the list of options.

58. At the "Select objects" prompt, click the first floor Stair (not the one going to the basement) and then press ENTER.

    The command line in response should read something like "[1] Wall cut line(s) converted". Please note that if you have turned off the Command Line, this prompt will not appear.

59. Press ENTER again to end the command.

60. Select the Stairs, walls and doors of the hallway again, right-click and return to Object Viewer to view the results (see Figure 7–16).

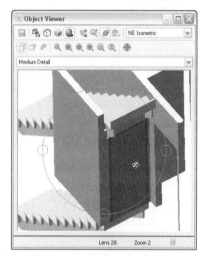

Figure 7–16 *The top of the wall now stops below the Stair*

61. Save the file.

ADD A RAILING

Railings can be added to Stairs or drawn free-standing as guardrails. We will look at both types in this chapter. In this sequence, we will add a Railing to the existing Stair.

62. Open the Content Browser (CTRL + 4).

63. In the *Design Tool Catalog - Imperial* [*Design Tool Catalog - Metric*], browse to the *Railings* category. Click Next to go to the second page of Railings.

64. Drag the eyedropper icon for **Guardrail - Wood Balusters 01** and drop it into the drawing.

65. On the Properties palette, choose **Stair** from the Attached to list, and from the Automatic placement list choose **No** (see Figure 7–17).

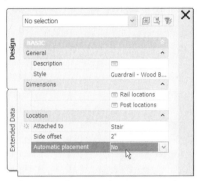

Figure 7–17 *Set the parameters to attach the Railing to the Stair*

When you turn off automatic placement, you will indicate the extent of the Railing with two clicks of the mouse. When you turn on automatic placement, the Railing will automatically extend to the full length of the Stair.

66. At the "Select a Stair" prompt, click the right side of the first floor Stair.

67. At the "Railing start point" prompt, click near the bottom of the Stair.

68. At the "Railing end point" prompt, click the endpoint of the Wall where the Stair jogs (see Figure 7–18).

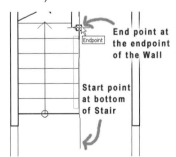

Figure 7–18 *Manually click the start and end points of the Railing*

69. Right-click and choose **Enter** (or press ENTER) to end the routine.

There is a slight gap between the end of the Railing and the wall. We can adjust this with grips.

70. Click on the Railing.

There are two triangular shaped grips at each end: The one that points away from the Railing (End Offset) will lengthen the Railing, and the one that points into the Railing (Fixed Post Position) will move the post.

71. Click the Fixed Post Position grip at the top of the Railing (the one pointing down in this case) and drag it slightly up till it snaps to the Wall.

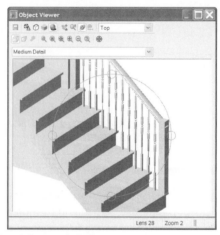

Figure 7–19 *The completed Stair with Railing*

72. Save and Close the file.

COMMERCIAL CORE PLAN

The Commercial Project files for Chapter 7 include a new Element file named *Core*. This file, included in files installed from the Mastering Autodesk Architectural Desktop 2006 CD ROM already includes the core walls. This will allow us to focus exclusively on the vertical circulation elements.

LOAD THE COMMERCIAL PROJECT

Be sure that all drawings from the Residential Project are saved and closed. However, you must have at least one drawing open in ADT for the Project Browser to work. Therefore, if you don't have any drawings open; click the QNew icon on the Standard toolbar to create a new blank drawing.

1. On the File menu, choose **Project Browser**.

2. Navigate to the *C:\MasterADT 2006\Chapter07* folder and make the *07 Commercial* [*07 Commercial Metric*] project current.

3. On the Constructs tab, double-click the *Core* file in the *Elements\Architectural* folder to open it.

   As stated above, you will find core Walls already laid out in this file.

ADD A NEW STAIR TO THE COMMERCIAL PLAN CORE

The Column Grid created in the last chapter is already XREFed to this file as an overlay. With Overlay as the reference type, the Column Grid will not be included as a nested XREF whenever the Core file is attached to other drawing files.

4. Open the Content Browser (CTRL + 4).

5. In the *Design Tool Catalog - Imperial* [*Design Tool Catalog - Metric*] catalog, navigate to the *Stairs and Railings* category and then the *Stairs* category.

6. Using the eyedropper icon, drag the "Steel" tool into the drawing window.

   The Add Stair command will begin and on the Properties palette, "Steel" will be set as the Stair Style.

7. On the Properties palette, choose **U-shaped** for the Shape.

8. Set the Width to **3'-8"** [**1100**], the Height to **12'-0"** [**3650**] and the Tread to **11"** [**280**].

9. Choose **1/2 Landing** for the Turn type, choose **Clockwise** for the Horizontal Orientation, choose **Outside** for Justify, and choose **Riser** for Terminate with (see Figure 7–20).

Figure 7–20 *Set the parameters for adding a U-shaped Stair*

Shapes include U-shaped, Multi-landing, Spiral, and Straight. The Width is measured from stringer to stringer across the flight of stairs. The Height is measured floor to floor. Justification assists with placement of the Stair as it is being added, but it has little impact after placement. Horizontal Orientation determines which leg of the U-Shaped Stair is the starting leg. Terminate with determines what type of component; riser, tread or landing will be placed at the end of the Stair.

The Stair tower is the upper left space in the *Core* file. Make sure that both OS-NAP and OTRACK are turned on in the Application Status Bar at the bottom of the screen. These will both aid in placement of the Stair.

10. Hover the mouse over the lower right corner of the Stair tower space to acquire the inside wall intersection.

11. Track the mouse directly to the left, type **5′ 8″ [1800]** at the command line, and press ENTER (see the left side of Figure 7–21).

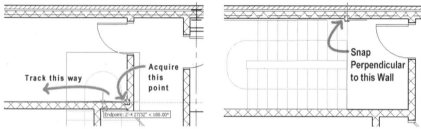

Figure 7–21 *Acquire the lower right corner with Object Snap Tracking. Set the second point perpendicular to the opposite Wall*

12. Hold down the SHIFT key, right-click, and choose **Perpendicular**.

13. Move the mouse straight up to the opposite side of the Stair tower and click when the Perpendicular icon indicates the inside face of the Stair tower wall (see the right side of Figure 7–21).

14. Press ENTER to complete the routine (see Figure 7–22).

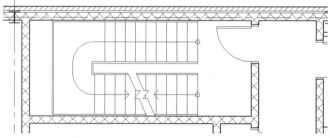

Figure 7–22 *Completed Stair object*

NOTE If the two flights are not even, be sure that the Stair is still selected and on the Properties palette, in the Advanced > Constraints grouping, choose **Lower flight** from the "Uneven tread on" list.

EXTEND THE LANDING

Like most ADT objects, we can use grips to fine tune the shape of the Stair after it is placed. In this case we will enlarge the landing.

15. Click on the Stair to activate the grips.

    As we have seen in previous exercises, if you pause your mouse over a grip, a grip tip will appear indicating the function of that grip point and shape.

16. Locate the "Edit Edges" Trigger Grip (it is a small gray circle) for the outer edge and click it (see Figure 7–23).

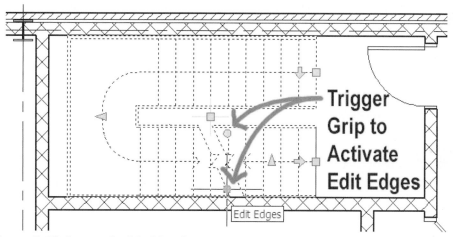

Figure 7–23 *Activate the Edit Edge Grips with the small gray Trigger Grip*

    Several more Grips will now appear around the outer edge of the Stair. Hover over each one to see their function.

17. On the left side of the Stair, in the middle of the landing are two grips, one square shaped, the other triangular. Click the triangular-shaped grip.

    This grip shape manipulates the width of the landing.

18. Drag this grip point and snap it perpendicular to the core Wall to the left of the Stair (see Figure 7–24).

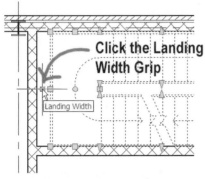 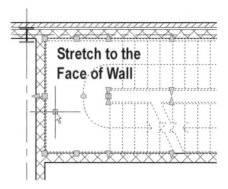

Figure 7–24 *Grip editing the Stair shape*

19. Experiment with additional grip edits. Undo when finished.

 TIP Grip editing offers an excellent way to get the shape of the Stair exactly the way you want it. Using the **CTRL** key while grip editing a Midpoint will extend just that edge without moving the ones connected to it. Try it out on one of the horizontal edges of the landing. Undo when finished experimenting.

COPY AND ASSIGN THE STAIR STYLE

Now let's make some modifications to the Stair Style for the Commercial Building Core.

20. Select the Stair, right-click and choose **Object Viewer**.

   In Object Viewer, view the Stair in 3D. You will notice that this particular Stair style has open risers. Let's change this to a Steel Pan configuration.

21. Select the Stair, right-click and choose **Copy Stair Style and Assign**.

22. On the General tab, change the name to *MADT Steel Pan*, and type **Mastering Autodesk Architectural Desktop Commercial Project Steel Pan Stair** for the Description.

Unlike the Residential Project, the Commercial Project is all new construction. Therefore, we will leave the Rule-based calculator on for the Commercial Stair. View the settings on the Design Rules tab if you wish. If you change anything, be sure to Undo the change before you proceed.

23. Click the Stringers tab and change the A - Width for both Stringers to **2″** [**50**].

24. Change the E - Waist (for the Landing) for both Stringers to **8″** [**200**].

   This will drop Stringers surrounding the Landing platform down relative to the landing.

25. Click the Components tab.

   We need to make some adjustments here to transform this open riser Stair into a Steel Pan configuration.

26. In the Flight Dimensions area, place a check mark in the Riser check box.

27. Change the A - Tread Thickness to **2″ [50]**, the B - Riser Thickness to ½″ **[12]** and place a check mark in the Sloping Riser check box.

28. In the Landing Dimensions area, set the D - Landing Thickness to **2″ [50]** (see Figure 7–25).

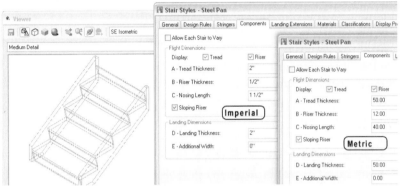

Figure 7–25 *Set the parameters for a Steel Pan Stair on the Components tab*

   We do not need to change any of the other settings on the other tabs at this time.

29. Click OK to see the results.

   Use the Object Viewer to see the effect of all the changes in 3D.

ADD RAILINGS

As mentioned earlier, Railing objects do not appear automatically when a Stair is placed. They are separate objects. However, they can be anchored to Stairs. In this way, the Railing will respond to changes made to the Stair. Railing rules applied to the Stair object help control the exact placement of the Railing relative to Stair as well as any extensions required by building codes. As with the Stair (and any other object type), we can use the Content Browser to locate an appropriate style.

30. Open the Content Browser (CTRL + 4).

31. In the *Design Tool Catalog - Imperial* [*Design Tool Catalog - Metric*], navigate to the *Stairs and Railings* category and then to the *Railings* category.

32. Using the eyedropper icon, drag the **Guardrail - Pipe + Rod Balusters** tool into the drawing window.

    The Add Railing command will begin and on the Properties palette, Guardrail - Pipe + Rod Balusters will be set as the Railing Style.

33. From the Attached to list, choose **Stair**. Set the Side Offset to 1″ [**25**], and choose **Yes** for Automatic placement.

34. At the "Select a Stair" command line prompt, click the inside edge of the Stair and then press ENTER to complete the routine.

    Notice that the Railings follow the inside edge of the Stair. They are also displaying posts in plan. Let's simplify the Plan display.

35. Select the Railing, right-click and choose **Edit Railing Style**.

36. On the Display Properties tab, place a checkmark in the Style Override checkbox next to Plan.

37. On the Layer\Color\Linetype tab, turn off (click the light bulb) Posts Up and Posts Down and then click OK twice to return to the drawing.

    The railing posts no longer display in Plan.

38. Return to the Content Browser and click Next twice to go to last page of Railing Styles.

39. Drag the eyedropper of the **Handrail - Round + Return (Escutcheon)** Style into the drawing.

40. On the Properties palette, change the Side offset to **-1 1/2″** [**-50**] and then choose **Stair Flight** from the Attached To list.

41. At the "Select a Stair" prompt, click the bottom edge of the Stair, and then click to the top edge when the prompt repeats. Press ENTER to complete the routine.

42. Using a crossing window, select the Stair and the Railings that were added, right-click and choose the **Object Viewer**, and set to **SW Isometric** to view the results of the added Railings (see the left side of Figure 7–26).

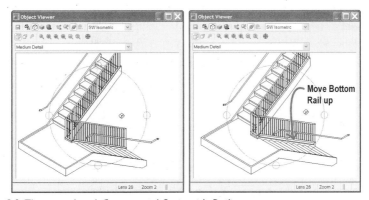

Figure 7–26 *The completed Commercial Stair with Railings*

Notice that the bottom rail of the inner railing nearly sits directly on top of the Stair Stringer. We can adjust this in the Railing Style.

43. Close the Object Viewer, select the inner Railing, right-click and choose **Copy Railing Style and Assign**.

44. On the General tab, change the name to *MADT Guardrail - Pipe + Rod Balusters*, and type **Mastering Autodesk Architectural Desktop Commercial Project Guardrail - Pipe + Rod Balusters** for the Description.

45. On the Rail Locations tab, change both C - Horizontal Height and C - Sloping Height to **6″ [150]**.

    Feel free to explore the other tabs in this dialog and experiment with additional changes.

46. Click OK to close the dialog and accept the changes. Select the Stair and Railings again, and return to Object Viewer to see the results (see the right side of Figure 7–26).

47. Save the file.

GENERATE A STAIR TOWER

In this topic, we will explore the Stair Tower Generate routine. Using this tool, we can create a single "spanning" Stair Tower for use in all floors of the project. (As we saw in Chapter 5, in Project Navigator the term "Spanning" refers to a Construct that occupies more than one Level or Division). The routine *must* be used within the context of the ADT Drawing Management system. If you do not have a Project active in Project Browser, the routine will fail. The routine will automatically copy a Stair, and optionally any Railings and Slabs associated with it, to one or more levels of the project. Stair heights will automatically be adjusted to match the level floor-to-floor heights.

Since we are going to use this routine to create our Stair Tower, it might be useful to create a Slab for the landing between the Stair and the Door to the Stair Tower. Slab objects are ADT objects used to represent floor slabs, or nearly any horizontal surface. They can be simple flat slabs, sloped or embellished with custom edge conditions. Slabs will be explored in more detail in Chapter 12. For now, a tool with an associated pre-built custom Slab Style has been included on the Project Tools tool palette.

USE ISOLATE OBJECTS

Let's begin by hiding the Railings so we can work with the Stair more easily.

1. On the Drawing Status Bar, click the Isolate Objects icon, and then choose **Hide Objects**.

    It appears as a small light bulb icon in the bottom right corner of the screen (shown in Figure 1–2 in Chapter 1).

2. At the "Select objects" prompt, select all three Railing objects (zoom in as required) and then press ENTER.

 NOTE The Isolate Object icon will turn red, indicating that Object Isolation Mode is active. In other words, there are currently some invisible objects in the drawing. To display all objects, simply click the Isolate Objects icon again and choose **End Object Isolation**.

ADD A LANDING SLAB

3. On the Project Tools tool palette, click the *MADT Stair Landing* tool.

This tool palette was loaded in the "Create Grid Line Extensions" heading in Chapter 6. If you did not complete Chapter 6, please right click the tool palettes title bar and choose **07 Commercial [07 Commercial Metric]** to load the project tool palette group and access this palette.

4. At the "Specify start point" prompt, snap to the inside right corner of the Stair tower (the same point indicated as "Acquire this point" in Figure 7–21 above).

5. At the "Specify next point" prompt, snap to the inside upper corner of the room (directly above the first point and the Door).

6. Continue adding points as shown in Figure 7–27, right-click and choose **Close** to finish the Slab, and then press ENTER to end the command.

Point 3 and 4 are the top Endpoints of the Stringer, and Points 5 and 6 are the bottom Endpoints of the Stringer.

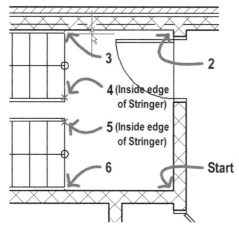

Figure 7–27 *Snap to the Stringer Endpoints to complete the Slab*

EDIT THE LANDING SLAB

A second line will appear around the entire Slab. This is the Slab Edge condition, which has been designed to emulate the Stringer of the Stair. We of course need to make a few adjustments.

7. Click on the Slab and note the Grips. Click the top Edge Grip (rectangular shape on the top horizontal edge) and move it down **2″ [50]**.

8. Click on the bottom Edge Grip (rectangular shape on the bottom horizontal edge) and move it up **2″ [50]**.

9. With the Slab selected, right-click and choose **Edit Slab Edges**.

10. At the "Select edges of one slab" prompt, click the vertical edge to the right (at the Door) and the two vertical edges to the left that abut to the Stair flights.

    The three edges should highlight. Press ENTER to complete edge selection. In the Edit Edges dialog, Edge 1, 3 and 5 should be listed.

11. In the Edit Slab Edges dialog, change the Edge Style of all three edges to ***None*** and then click OK.

12. Click on the Slab and click the Edge Grip between the two Stair flights (see item 1 in Figure 7–28).

13. Drag the Grip to the right with POLAR or ORTHO on and then press CTRL once (see item 2 in Figure 7–28).

This will toggle to the "Add New Edges" mode, which will move the selected edge while adding new perpendicular edges to connect it to the existing vertices.

14. Type **2″ [50]** and then press ENTER. Right-click and choose **Deselect All** (see item 3 in Figure 7–28).

TIP To make this step work properly and have the edge move exactly 2″ [50], you may need to temporarily disable Dynamic Input – click the DYN button on the Drawing Status Bar.

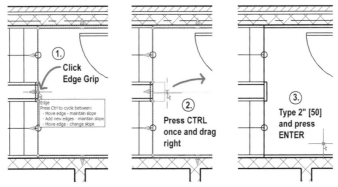

Figure 7–28 *Toggle Edge Grip to "Add New Edges"*

15. On the Drawing Status Bar, click the Isolate Objects icon and then choose **End Object Isolation**.

ANCHOR A RAILING TO THE SLAB

The last item we need to create for the Stair core is a small piece of Railing between the two Stair flights. We can anchor this Railing to the Slab.

16. Select the inner Railing (zoom in as necessary), right-click and choose **Add Selected**.

17. On the Properties palette, from the Attached to list, choose: ***None***.

18. Following the prompts, snap from Midpoint to Midpoint as indicated in Figure 7–29.

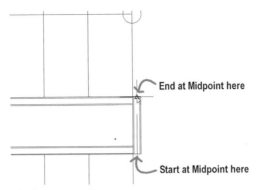

End at Midpoint here

Start at Midpoint here

Figure 7–29 *Draw a Railing manually*

Remain zoomed in on the Railing.

19. Select the new Railing, right-click and choose **Railing Anchor > Anchor To Object**.

20. At the "Select AEC Objects" prompt, select the Slab and press ENTER. Press ENTER twice more to accept the other defaults.

21. Select both the Slab and the new Railing, right-click and choose **Object Viewer** to examine the results.

22. Close the Object Viewer and the Save the file.

EDIT THE SLAB DISPLAY

One last finishing touch for our Slab: let's get rid of that line in plan at the Door.

23. Select the Slab, right-click and choose **Edit Object Display**.

24. On the Display Properties tab, place a check mark in the Object Override check box next to Plan.

25. In the Display Properties dialog that appears, on the Layer/Color/Linetype tab, turn off (dim the light bulb) both the Below Cut Plane Body and Baseline components. Turn on the Below Cut Plane Outline and the Above Cut Plane Outline components and then click OK twice.

26. Select the Slab, right-click and choose **Properties**.

27. On the Properties palette, scroll down to the Location grouping and change the Elevation to **12'-0"** [**3600**].

MOVE STAIR AND RAILINGS TO THE SPANNING STAIR TOWER CONSTRUCT

Now that we have all of the pieces of our Stair – the Stair itself, the Railings and the Landing Slab – we are ready to generate a Stair Tower. To do this, we must work in a Spanning Construct (see the "Create a Spanning Construct" heading in Chapter 5 for more information). A Construct named: Stair Tower has been included in the Chapter 7 dataset. We will open this file, move the Stair, Railings and Slab to it, and then we will generate the Stair Tower.

1. On the Project Navigator palette, click the Constructs tab, and expand the *Architectural* folder.

2. Right-click the *Stair Tower* Construct and choose **Properties**.

In the Assignments area, notice that all levels are selected and at the bottom notice the message indicating that the Construct is "Spanning" (this message is illustrated in Figure 5–25 in Chapter 5).

3. Click OK to return to the drawing.

4. In the *Core* file, Select the Stair, all four Railings and the Landing Slab. Drag them onto the Project Navigator palette, and release them on the *Stair Tower* Construct (in the *Architectural* folder).

   The Stair, Railings and Landing Slab will disappear from the *Core* file.

5. Save and Close the *Core* Element file.

GENERATE THE STAIR TOWER

The Stair Tower Generate routine copies a selected Stair to one or more levels in the project. It will automatically adjust the height of the copied Stairs to match the levels in your project.

6. Double-click the *Stair Tower* Construct to open it.

7. Zoom in on the Stair.

8. On the Design tool palette, click the Stair Tower Generate tool.

   If you do not see this palette or tool, right-click the Tool Palettes title bar and choose **Design** (to load the Design Tool Palettes Group) and then click the Design tab.

9. At the "Select a stair" prompt, click the Stair. (Do NOT press ENTER).

10. At the "Select railings and slabs" prompt, select the Railings and Slab and then press ENTER.

11. In the Select Levels dialog, verify that all available levels are selected, and then place check marks in both the "Include Anchored Railings" and the "Keep Landing Location when Adjusting U-Shaped Stair" check boxes (see Figure 7–30).

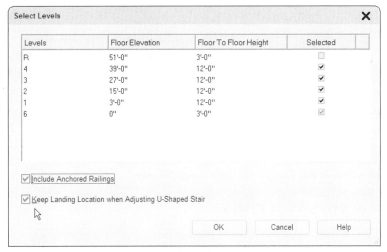

Figure 7–30 *Select Levels and Options for Stair Tower Generate*

The "Include Anchored Railings" option will copy the Railings to each level with the Stairs. The "Keep Landing Location when Adjusting U-Shaped Stair" will stack all copied Stairs relative to the landing position regardless of the height of the level. With this option turned off, the start of the Stair flight will be maintained on each floor and if the floor heights vary, the landings will not be aligned in the plan.

12. Click OK to complete the operation.

You may see an error marker when you click OK. This is caused in this case by the same option that allowed the Landings to maintain alignment. The Stair Tower Generate routine copied the Stairs to all levels including the Ground level, which is only 3'-0" [900] tall in this project. Therefore, when it tried to insert a Landing to comply with the "Keep Landing Location when Adjusting U-Shaped Stair" option, the Stair failed. This is not an issue, since we will actually delete this unneeded Stair at this level.

13. From the View menu, choose **3D Views > Front**.

14. Select the bottom Stair (the short one) and the bottom Landing and erase them (see Figure 7–31).

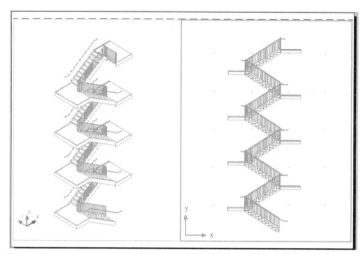

Figure 7–31 *Select Levels and Options for Stair Tower Generate*

ADJUST STAIR DISPLAY SETTINGS

We will make one final modification to the this file, and then we will examine how our Stair appears in the View files of this project.

15. Select the Stair, right-click and choose **Edit Object Display**.

16. On the Display Properties tab, double-click Plan (or select Plan and then click the Properties icon – do not attach an override).

17. Click the Other tab, and clear the check mark from the "Override Display Configuration Cut Plane" check box.

18. Click OK twice to return to the file.

If you return to Top view, the Stairs now display the up direction only. With the Display Configuration Override removed, the first floor will show Stairs only going up, middle floors will show both up and down, and the top floor will show Stairs only going down. Let's see the effect of this change in the Plan View files for the project.

19. Save and Close the file.

ATTACH CORE TO CONSTRUCTS

Before we open the Plan View files and view the Stair display, let's add the *Core* Element from above to each of the *Shell and Core* Constructs. This will give our Stair Tower its enclosure on each floor.

20. On the Project Navigator palette, on the Constructs tab, right-click the *Core* Element in the *Elements\Architectural* folder and choose **Attach Element to Constructs**.

21. In the Attach Element to Construct dialog, select all of the *Shell and Core* Constructs and then click OK (see Figure 7–32).

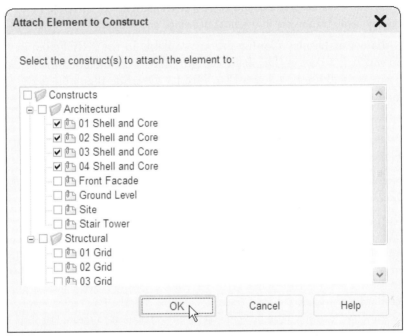

Figure 7–32 *Attach Element to all Shell and Core Constructs*

22. On Project Navigator, click the Views tab and then double-click *A-FP01* to open it.

    Zoom in on the Stairs and note that only the up direction is currently showing.

23. On Project Navigator, click the Views tab and then double-click *A-FP02* to open it.

    Zoom in on the Stairs and note that both up and down are currently showing.

24. Close *A-FP02*.

OVERRIDE XREF DISPLAY

If we had a Roof Plan for this project, we could open that file and see that the Stairs there would only show going down. We will build a Roof Plan later in Chapter 12. Meanwhile, in the First Floor Plan, even though the Stairs correctly display up only, we may wish to see more of the Stairs than is currently displayed.

It is also often desirable to see the above portion that is cropped away displayed in dashed line. This can be accomplished with an XREF Display Override.

25. Click on the Stair in *A-FP01*, right-click and choose **Edit Object Display**.

    Since this is an XREF, you will get a slightly different dialog.

26. Click the XREF Display tab, and place a checkmark in the "Override the display configuration set in the host drawing" check box.

This reveals the list of Display Configurations available in the XREF file, in this case Stair Tower. A few custom Display Configurations were included in the dataset file. The one we will use here has a Plan Display Override that adjusts the Cut Plane to a higher height and displays in dashed lines the outline of the Stairs above the Cut Plane.

27. From the list of Display Configurations, choose **Medium Detail Bottom Floor** and then click OK.

    Note the change to the Stairs in the First Floor Plan. Feel free to re-open the Stair Tower file and explore the settings more carefully.

CAD Manager Note: The custom Display Configurations provided in this file, were created in the Display Manager. The Plan Display Representation of the Stair object was duplicated twice (once for the bottom level and once for the top level). These were then assigned to custom Sets which were in turn applied to custom Configurations. To do this, open Display Manager, expand Representations by Object and then expand the Stair object node. Right click the Plan representation and choose Duplicate and give it a name (Plan Bottom was used here). Repeat to create Plan Top. Edit the properties of these two new Display Reps. Both have the Cut Plane Override removed. Plan Bottom has all down components turned off except Outline Up which is set to a dashed line. Plan Top has all up components turned off. Adjust the Cut Plane heights for each as desired. In the Sets node of Display Manager, Duplicate the Plan Set twice and name them Plan Top Floor and Plan Bottom Floor. In Plan Top Floor, uncheck Plan and check Plan Top for Stair objects. Do the same for Plan Bottom checking Plan Bottom instead. Finally, on the Configurations node, duplicate Medium Detail twice, name them and change the Set selection for Top view to the two custom Sets created (Medium Detail Top Floor uses Plan Top Floor and Medium Detail Bottom Floor uses Plan Bottom Floor).

28. Save and Close all Plan View files.

RAMPS AND ELEVATORS

In this section, we will add some ramps and elevators. Ramps and elevators are handled quite differently from one another. To create a ramp in ADT, use a Stair style defined for ramps. There are a few examples provided in the Content Browser. Elevators are created from Multi-View Blocks. There are a few sample Elevator Multi-View Blocks in the Content Browser as well.

ADD A RAMP

A ramp is made by configuring a Stair style to "look like" a ramp. This is accomplished by two tricks: First, the ramp is actually a very wide Housed stringer. Second, on the Components tab, both Tread and Riser are removed. In this way, a wide stringer is drawn without any Treads or Risers. This gives us a ramp. In all other ways, it has the same behavior and parameters as any other Stair.

1. On the Project Navigator, double-click the file named *Ground Level* in the *Constructs\Architectural* folder.

This file has been added to the project structure for Chapter 7.

2. Open the Content Browser again, and return to the Stairs category.

3. Drag (using the eyedropper icon) the Ramp - Concrete Curb tool into the drawing.

 NOTE Don't worry if it looks like a Stair while attached to your cursor. It won't when we are finished.

4. On the Properties palette, change the Shape to **Multi-landing** and set the Turn type to **1/2 Landing**.

5. Set the Width to **3'-0"** [**900**], the Height to **2'-8"** [**810**] and the Justify to **Right**.

Three sets of polylines have been included in this file. We will use the magenta one to draw the ramp.

6. At the "Flight start point" prompt, snap to the endpoint of the magenta polyline. (Snap to the lower one along the angled leg of the polyline.)

7. At the "Flight end point" prompt, snap to the next endpoint along the angled magenta polyline (see Figure 7–33).

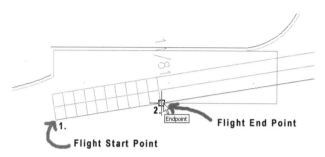

Figure 7–33 *Begin drawing the ramp by tracing points on the magenta polyline*

8. With either POLAR or ORTHO on, move directly to the right (0°), type **3'-0"** **[900]** and then press ENTER.

The prompt changes back to "Flight start point". When you have a multi-landing Stair, you draw a flight of Stairs (or section of ramp in this case), then indicate some space for a landing, and then draw the next flight. The distance between the last "Flight end point" (or the top of the last flight) and the next "Flight start point" is the size of the landing.

9. The next point ("Flight end point") is the bottom right corner endpoint of the magenta polyline.

10. Click the top right corner endpoint of the magenta polyline next—this is the end of the landing and the start of the next run of ramp (see Figure 7–34).

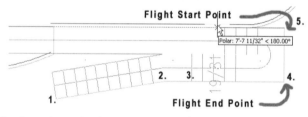

Figure 7–34 *Continue alternating between ramp and landing*

11. With either POLAR or ORTHO on, move directly to the left (180°), type **12'-0"** **[3650]** and then press ENTER.

12. With either POLAR or ORTHO on, continue moving directly to the left (180°), type **3'-0"** **[900]** and then press ENTER.

13. Snap to the final endpoint (top left) of the magenta polyline to complete the ramp (see Figure 7–35).

IMPORTANT If you still have lots of leftover ramp, or if the ramp completely disappears, check your Properties palette, and be sure that the Calculation Rules are set to Height. To do this click the worksheet icon next to Calculation Rules and lock all settings by clicking the icons to change them to lightning bolts.

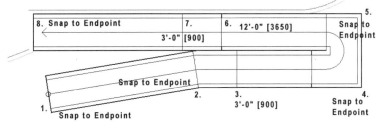

Figure 7–35 *Completed ramp with each point labeled*

14. Using the steps above in the "Add Railings" topic, add a Railing to the ramp.

15. Select the new ramp and Railing, right-click and choose **Object Viewer** (see Figure 7–36).

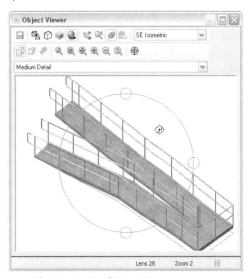

Figure 7–36 *The ramp and Railing in the Object Viewer*

16. Mirror the ramp and Railings to the other side.

COMPLETE ENTRANCE TERRACE

17. On the Design tool palette, right-click the Slab tool and choose **Apply Tool Properties To > Linework and Walls**.

18. At the "Select walls or polylines" prompt, click on the green polyline and then press ENTER.

19. At the "Erase layout geometry" command line prompt, right-click and choose **Yes** (or type Y and press ENTER).

20. At the "Creation mode" prompt, right-click and choose **Direct**. At the "Specify slab justification" prompt, press ENTER to accept the default.

21. With the Slab still selected, change the Thickness on the Properties palette to **2′-8″ [810]**.

22. With the Slab still selected, right-click and choose **Copy Slab Style and Assign**.

23. On the General tab, change the name to **MADT Entrance Terrace**. For the Description, type **Mastering Autodesk Architectural Desktop 2006 Commercial Project Entrance Terrace**.

24. Click the Materials tab and change the Slab material to **Concrete.Cast-in-Place.Flat.Grey** and then click OK.

25. On the Design tool palette, right-click the Railing tool and choose **Apply Tool Properties to > Polyline**.

26. Follow the prompts to convert the three blue polylines to Railings and delete the polylines. On the Properties palette, choose the same Railing Style that you used for the Ramp.

    Feel free to experiment further in this file. Use the technique above to anchor the three railings to the terrace Slab.

27. Save and Close the file.

ADD ELEVATORS

1. On the Project Navigator palette, right-click the *Elements\Architectural* folder and choose New Element. Name the new Element file *Elevators*. Double-click *Elevators* to open it.

2. On the Project Navigator, right-click the *Core* Element file and choose XREF Overlay. Zoom in on the Core.

3. Open the Content Browser (CTRL + 4) and in the Design Tool Catalog - Imperial, navigate to the *Conveying* category and then the *Elevators* category.

 NOTE The Metric Catalog does not contain Elevator symbols. However, the Imperial ones will automatically scale when inserted.

    There are three Elevator Multi-View Blocks from which to choose.

4. Drag the eyedropper icon of the Elevator named Square into the drawing file.

It does not matter where you drop it; we will move and rotate it in the next step.

5. Rotate the elevator 270° and then move it into position in the space below the Stair tower.

6. Copy the elevator next to the first to create the second one (see Figure 7–37).

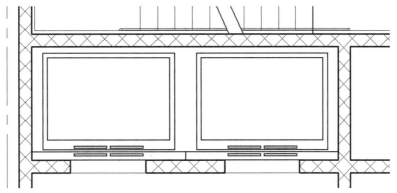

Figure 7–37 *Elevators added to the Core file*

7. Select both elevators, right-click, and choose **Object Viewer**.

8. View from all sides, and close the Viewer when finished.

 NOTE These Elevators are Multi-View Blocks. You can edit them to suit your needs. If you receive blocks from an Elevator manufacturer, you can build your own, or customize this Multi-View Block from those files. See Chapter 11 for more information on creating and editing Multi-View Blocks.

We could have simply added the Elevator Multi-View Blocks directly to the *Core* file rather than create a new and separate *Elevators* file. However, by taking this extra step, we can later add the Elevators just once to the composite model from which we will cut our sections. This way, we will only see one set of elevator cabs in the section cut rather than one on every floor. If you are primarily concerned with plans, you can skip the step of creating the *Elevators* file above and add the elevators directly to the *Core* Element file.

TOILET ROOMS

The last item needed in our core plan is the restrooms. There are pre-made fixture layouts in the Content Browser, as well as a collection of individual fixtures.

ADD TOILET LAYOUTS

1. Open the Content Browser (CTRL + 4).

2. In the *Design Tool Catalog - Imperial* [*Design Tool Catalog - Metric*]catalog, navigate to the *Mechanical > Plumbing Fixtures > Layouts* [*Bathroom > Layouts*] category.

3. Drag the eyedropper for the **Rest Room (Women)** [**Toilet (Women)**] layout into the drawing.

4. The Insert Block dialog box will appear; accept all defaults and click OK. (However, if Explode is checked, uncheck it.)

This object is actually an AutoCAD block containing several ADT objects. Leave it as a block to position it in the room. Explode it to manipulate the actual pieces of the layout. Be sure to explode it only once! If you explode a second time, you will destroy the nested ADT Multi-View Blocks within.

5. Use the single grip point at its insertion to position it in one of the rooms at the right.

6. Move, rotate, and mirror as necessary.

7. Once it is positioned correctly, explode the block.

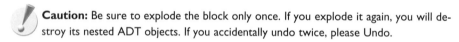 **Caution:** Be sure to explode the block only once. If you explode it again, you will destroy its nested ADT objects. If you accidentally undo twice, please Undo.

8. Delete one stall, and move the counter and lavatories to fit the room.

9. Repeat the steps above for the **Rest Room (Men)** [**Toilet (Men)**] layout (see Figure 7–38).

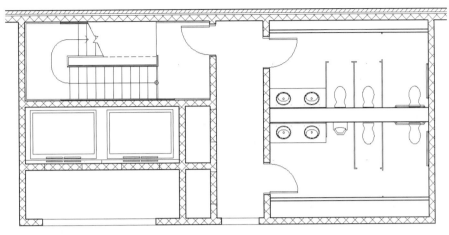

Figure 7–38 *The final core plan layout*

10. Save and Close all project files.

ADDITIONAL EXERCISES

Additional exercises have been provided in Appendix A. In Appendix A you will find an exercise for adding a new Stair to the *Basement New* Construction file of the Residential Project. There is also an exercise to complete the front entry plaza for the Commercial Project. It is not necessary that you complete these exercises to begin the next chapter, they are provided to enhance your learning experience. Completed projects for each of the exercises have been provided in the *Chapter06/ Complete* folder.

SUMMARY

Stairs and Railings offer flexible configuration with style-based parameters and optional object-based variations.

A single Stair object can be configured to represent one or two flights in the Display Properties.

Stair shapes can be manipulated with grips.

Slabs can be used with Stairs to create Landings.

Stair Tower Generate can be used with ADT Project Navigator to create Spanning Stair Constructs.

Custom Display Properties can be configured to show Stairs correctly on the bottom, middle and upper levels.

Display Overrides can be attached to XREF files to force them to display in a different Display Configuration than the host drawing.

Railings can be anchored to Stairs or drawn free form as guardrails.

Ramps are created from Stair styles.

Pre-made elevators and toilet layouts are included in the Content Browser.

CHAPTER 8

The Building Shell

INTRODUCTION

In this chapter, we will enclose our Commercial Project with a building skin. The skin will be comprised of masonry enclosure on three sides, with a Curtain Wall on the front of the building. The Curtain Wall is a "spanning" element beginning on the second floor and spanning the third and fourth floors. Our project setup (completed in Chapter 5) will handle the correct display of the spanning Curtain Wall on each floor respectively.

OBJECTIVES

In order to complete the shell of the Commercial Project we will apply a Wall style to the Shell files created in Chapter 5 and use the Project Standards feature to synchronize the change to all floors. We will also build a custom Curtain Wall for the front façade. We will take a comprehensive look at ADT's Curtain Wall object to build this front façade—this will be the major focus of the chapter. The main tools covered in this chapter include the following:

- Work with project-based Tool Palettes.
- Use Project Standards (new in 2006) to synchronize styles.
- Convert Walls to Curtain Walls.
- Build a Curtain Wall.
- Create a Custom Curtain Wall.
- Convert an Elevation Sketch to a Curtain Wall.
- Work with Curtain Wall In-Place Edit.

 ## CREATING THE MASONRY SHELL

In this exercise, we will refine the *Shell and Core* Construct files (created in Chapter 5) with an appropriate Wall style from the Content Library and then synchronize the change across the project.

INSTALL THE CD FILES AND LOAD THE CURRENT PROJECT

If you have already installed all of the files from the CD, simply skip down to step 3 below to make the project active. If you need to install the CD files, start at step 1.

1. If you have not already done so, install the dataset files located on the Mastering Autodesk Architectural Desktop 2006 CD ROM.

   Refer to "Files Included on the CD ROM" in the Preface for information on installing the sample files included on the CD.

2. Launch Autodesk Architectural Desktop 2006 from the desktop icon created in Chapter 3.

If you did not create a custom icon, you might want to review "Create a New Profile" and "Create a Custom ADT Desktop Shortcut" in Chapter 3. Creating the custom desktop icon is not essential; however, it makes loading the custom profile easier.

3. From the File menu, choose **Project Browser**.

4. Click to open the folder list and choose your *C:* drive.

5. Double-click on the *MasterADT 2006* folder, and then the *Chapter08* folder.

   One or two commercial Projects will be listed: *08 Commercial* and/or *08 Commercial Metric*.

6. Double-click *08 Commercial* if you wish to work in Imperial units. Double-click *08 Commercial Metric* if you wish to work in Metric units. (You can also right-click on it and choose **Set Current Project**.) Then click Close in the Project Browser.

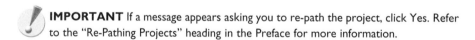 **IMPORTANT** If a message appears asking you to re-path the project, click Yes. Refer to the "Re-Pathing Projects" heading in the Preface for more information.

COPY WALLS BETWEEN FILES

In Chapter 5 we built four Shell files containing four Walls in the basic shape of the building's footprint. This was useful for the early stage of the project. It is now time to begin adding some more detail to these files and refine the Wall layout.

1. On the Project Navigator palette, in the *Constructs\Architectural* folder, double-click the *Ground Level* Construct to open it.

The Walls in this file more accurately portray the desired Wall layout for the upper floors. We will copy and paste a couple of them to assist with the upper floors.

2. Select the two short vertical Walls on either side of the entry terrace (see Figure 8–1).

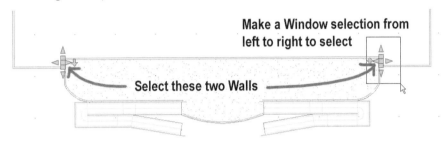

Figure 8–1 *Select the two short vertical Walls*

3. Right-click in the drawing and choose **Clipboard > Copy** (or press CTRL + C).

4. Close the *Ground Level* Construct (it is not necessary to save) and then on the Project Navigator palette, double-click the *01 Shell and Core* Construct to open it.

5. Right-click in the drawing and choose **Clipboard > Paste to Original Coordinates.**

6. Right-click in the drawing and choose **Basic Modify Tools > Trim** (or type TR at the command line and then press ENTER).

7. At the "Select cutting edges" prompt, select both of the small Walls just pasted and then press ENTER.

 TIP Remember, to receive on-screen prompts, turn on Dynamic Input by pushing in the DYN toggle icon on the Application Status Bar. Refer to Chapter 1 for more information.

8. At the "Select object to trim" prompt, click in the middle of the bottom horizontal Wall and then press ENTER to complete the command (see Figure 8–2).

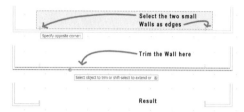

Figure 8–2 *Trim the middle segment of the Wall away using the two pasted Walls as edges*

ADD A TOOL TO THE PROJECT TOOL PALETTE GROUP

When you load a Project in ADT, a Tool Palette Group for the current Project is also loaded. Recall that in Chapter 6 we discussed the presence and usage of these "Project-based Tool Palette Groups." If you did not complete Chapter 6, review the "Create Grid Line Extensions and Project Tool Palettes" heading now for more information. The 08 Commercial [08 Commercial Metric] Tool Palette Group (and the Tool Palette which it contains) can be loaded giving us access to any project-based tools which it contains.

9. From the File menu choose **Open**.

10. Navigate to the *C:\MasterADT 2006\Chapter08\MADT Commercial\ Standards\Content [C:\MasterADT 2006\Chapter08\MADT Commercial Metric\Standards\Content]* folder and open the *Commercial Styles.dwg [Commercial Styles - Metric.dwg]* drawing file.

NOTE This file must be opened manually; it is not accessible from Project Navigator.

This file is the "Style Library" for the project. We will create a Wall Style in this file for the exterior shell of our building. To do this, we will start with one of the out-of-the-box Wall styles provided in the default Content Browser library. We'll bring it into our Style Library file, rename it and then make a tool for it.

11. On the Navigation toolbar, click the Content Browser icon (or press CTRL + 4).

12. Click the *Design Tool Catalog - Imperial [Design Tool Catalog - Metric]* catalog, navigate to the *Walls* category and then the *CMU* category. (Navigation is on the left).

In the *Imperial* Catalog, browse to the fourth page of CMU Styles, in the *Metric* Catalog, browse to the second page.

13. Locate the **CMU-8 Air-2 Brick-4 [CMU-190 Air-050 Brick-090]** tool, click and hold down the mouse on the small eye-dropper icon and then drag the tool into the drawing window in ADT.

Release the mouse button anywhere in the ADT drawing window to begin drawing a Wall with the tool (see Figure 8–3).

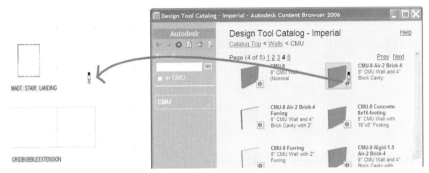

Figure 8–3 *Use the iDrop icon to drag the tool from Content Browser to the drawing window*

14. Draw a small horizontal segment of Wall in the file next to the other objects onscreen and then press ENTER to complete the command.

15. Select this Wall on screen, right-click and choose **Edit Wall Style**.

16. On the General tab, rename the style to: **Exterior Shell** add: **Mastering Autodesk Architectural Desktop Commercial Project Exterior Building Shell Wall** to the Description and then click OK.

By renaming the style this way, we make it easy to later edit the style and synchronize the change across the entire project. Adding the description is optional, but usually a good idea. If you wish, you can explore the other tabs in the Wall Style Properties dialog however, these will be covered in detail in Chapter 10.

17. Save the *Commercial Styles.dwg* [*Commercial Styles - Metric.dwg*] drawing file.

CAD Manager Note: You will notice the text object beneath each of the two objects already in the *Commercial Styles* file. When you create a Style Library, consider adding a piece of text beneath each object and then insert a field within the text. The field should reference "Objects" from the Field Category and then click the "Select Object" icon to pick the object onscreen. Finally, choose "Style" from the Property list to have the field automatically read the Style name of the selected object. This makes it easy to see the names of the items within your library at a quick glance. Feel free to add such a note to the new Wall Style in this file.

18. Right click on the Tool Palettes title bar and choose **08 Commercial [08 Commercial Metric]** to load the Project Tool Palette Group (see Figure 6–23 in Chapter 6 for an illustration—note however that the figure will show "06" rather than "08").

There are two tools on the Project Tools palette; the one we used in Chapter 6 and the one from Chapter 7.

19. Select the Wall in the drawing (the one we just renamed). Drag it from the drawing and drop it on the Project Tools tool palette (see Figure 8–4).

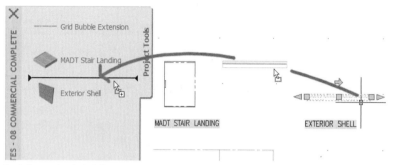

Figure 8–4 *Drag the Wall from the drawing to the Project Tools palette to make a new tool*

To do this properly, hold the mouse down for a moment over the highlighted edge of the Wall (not a grip) and then drag.

A Wall tool should now appear on the Project Tools tool palette in your workspace. Since we added this new tool to the Project Tools palette, it will only be available (as with the other tools on this palette) when this project is current.

20. Right-click the new tool and choose **Properties**.

21. In the Tool Properties worksheet, change the Base Height to **12′-0″ [3650]** and then click OK.

22. Save the *Commercial Styles.dwg [Commercial Styles - Metric.dwg]* drawing file.

CAD Manager Note: When preparing to implement customized Tool Palettes and Tool Palette Groups in Architectural Desktop, there are a few important considerations. Tool Palettes can be deployed on a network, from the project or locally. When you deploy office standard or project-based Tool Palettes, they can be deployed as "shared" or "per-user." Shared Tool Palettes are accessed directly by all users from a centralized location on the network server. Per-user Palettes are deployed from a common location but are copied to the local work station when accessed. Each user ends up with a local copy of the Tool Palettes which is accessed by ADT on each subsequent load. To keep local copies synchronized with a common copy on the server, you must enable "linked" Tool Palettes. This is a setting within the Content Browser and will create a link back to the source from which users may periodically refresh their local copies.

If you wish to use office standard and/or project palettes, you should consider using shared or linked. One person should be put in charge of maintaining the server version of the palette and its published catalog located in a read-only folder or using limited Windows security permissions. This is very important to prevent users from accidentally overwriting the standard tools. If you use shared, you will need to include the appropriate paths in each user's ADT Profile (see Chapter 3 for more on Profiles). If you use linked, users will need to load the Content Browser catalog (as we did in Chapter 4), drag the palette into their workspace and use its tools. If edits are made to the catalog, they can click Refresh to retrieve these changes. Auto-Refresh can also be enabled so that the tool palette automatically refreshes each time ADT is launched.

This is a brief explanation of the various options. It is important to read the online help on the topic of Tool palettes to fully understand all of the issues and options available. When you create a new blank Tool Palette, a direct link to the appropriate section of the help is provided on the palette. You can also navigate there directly by browsing the help to: *Architectural Desktop User's Guide > The Workspace > Working with Tool Palettes.*

RE-IMPORT WALL STYLE AND APPLY TOOL PROPERTIES

You may recall that in Chapter 5, we built the shell walls using a style named: Exterior Shell—the same name that we just assigned to a more detailed Wall Style in the library file. We will now update our project files with the new version of the style.

23. On Project Navigator, double-click *01 Shell and Core* to open it.

 NOTE If you left the *First Floor Shell* and *Core* open above, then this action will simply make that file active.

24. On the Project Tools palette, right-click your new ***Exterior Shell*** tool and choose **Re-import 'Exterior Shell' Wall Style**.

The exterior Wall will now change to show all the correct poché and detail of the Wall. The two small Walls pasted at the start of the chapter did not change with the rest. This is because they use a different Style than the Walls that were already in the file. Let's address those now.

25. At the bottom of the drawing, select the two small vertical Walls (the ones pasted at the start of this chapter).

26. On the Project Tools palette, right-click your new ***Exterior Shell*** tool and choose **Apply tool properties to > Wall**.

27. Right-click and choose **Deselect All**, and then **Save** the file.

The strategy just employed would be referred to as "Progressive Refinement". This concept has been mentioned in previous chapters. In ADT, your goal is often to start by simply populating your drawings with very basic geometry and then throughout the course of the project, you slowly and progressively refine the detail and data contained within those files. The parametric nature of ADT objects makes this approach not only practical, but very desirable.

WORKING WITH PROJECT STANDARDS

We could open each of the Shell and Core files and manually apply this technique to each one. While that approach would certainly work, let's use the new Project Standards feature to synchronize the new version of the style to the other project drawings. To use Project Standards, we must designate one or more drawing files to be "Standards Drawings." In our case, we will use the *Commercial Styles.dwg* [*Commercial Styles - Metric.dwg*] drawing file for this purpose. We must then apply a "Version ID" to the Styles within this file. Next we enable Project Standards, and then instruct our project files to synchronize with the Standards Drawing. When you synchronize, ADT will compare Style names and Version IDs looking for matches and Styles that are out of date.

SET UP PROJECT STANDARDS

Before we can synchronize, we must configure the Project Standards.

28. On the Project Navigator palette, click the Project tab.

29. Click the Configure Project Standards icon at the bottom of the palette (see item 1 in Figure 8–5).

TIP You can also click the quick pick icon in the Drawing Status Bar shown in Figure 1-2 in Chapter 1 and choose **Configure Standards.**

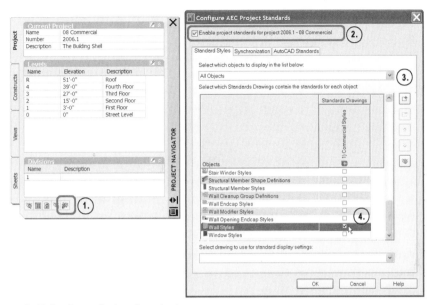

Figure 8–5 *Configure Project Standards*

30. At the top of the Configure AEC Project Standards dialog, place a checkmark in the "Enable project standards for project 2006.1 - 08 Commercial" checkbox (see item 2 in Figure 8–5).

31. On the Standard Styles tab, click the Add Drawing icon on the left (see item 3 in Figure 8–5).

32. Navigate to the *C:\MasterADT 2006\Chapter08\MADT Commercial\ Standards\Content [C:\MasterADT 2006\Chapter08\MADT Commercial Metric\Standards\Content]*folder and open the *Commercial Styles.dwg [Commercial Styles - Metric.dwg]* drawing file.

33. In the Objects list, scroll down and place a checkmark next to Wall Styles (see item 4 in Figure 8–5).

34. On the Synchronization tab, verify that "Manual" is selected and then click OK.

    A Version Comment dialog will appear.

35. In the Version Comment dialog, type: "Initial Standards Configuration" and then click OK.

Project Standards compares Style names and Version IDs. In order to assist you in deciding which version is most current, comments are requested (and highly encouraged) when you configure standards.

36. On the Project Navigator palette, on the Project tab, click the Synchronize Project icon (see Figure 8–6).

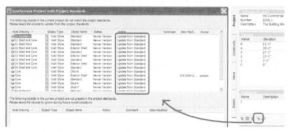

Figure 8–6 *Synchronize the Project to the Standards*

After the progress bar completes the scan of all project files, the Synchronize Project with Project Standards dialog will appear.

37. Verify that **Update from Standard** is chosen in the Action column for all entries and then click OK.

38. On the Project Navigator palette, double click to open each of the other *Shell and Core* Constructs (*02 Shell and Core*, *03 Shell and Core* and *04 Shell and Core*).

Notice the more detailed brick and block wall in place of the previous simple two-line wall.

39. In the *01 Shell and Core* Construct file, copy the two small vertical Walls to the clipboard.

40. Paste them into each of the other three *Shell and Core* Constructs and repeat the trim steps from above.

CAD Manager Note: This has been a very brief introduction to the powerful Project Standards feature. A complete set of tools is available on the **Window > Pulldowns > Cad Manager** menu. You can use the basic process followed here to configure additional Style types to synchronize. It is also possible to load more than one Standards Drawing and organize them hierarchically for synchronization. Using the Project Standards, you are also able to synchronize Display Configurations, Sets and Representations. We can also update Standards files with the styles and display settings from project drawings. We will see additional examples later in this book. Also please refer to the online help for more detailed descriptions of all the various Standards functions.

41. When you are finished, Save and Close all of the *Shell and Core* Constructs.

ADDING AND MODIFYING CURTAIN WALLS

The Curtain Wall object is similar to the Wall object in use and function. Add Curtain Walls to the drawing in the same way that you add Walls or even convert existing Walls to Curtain Wall objects. What makes the Curtain Wall object special is its ability to represent complex grid patterns and designs within its mass. A series of interwoven horizontal and vertical members defines cells, which are infilled by other nested grids or infill objects such as Door/Window Assemblies or panels. A common use for the Curtain Wall object will be to represent the exterior skin of a building. Whether the design is expressed with steel and glass skin, or heavy masonry piers and infill panels, the Curtain Wall object offers tremendous flexibility and design potential. In fact, Curtain Walls can be used to model all sorts of objects that would not necessarily be thought of as Curtain Wall. Casework, Seating configurations and Wrought Iron fences are just a few examples of objects that have been modeled with the Curtain Wall object.

To begin our exploration of Curtain Wall objects, we will open the first floor Shell file modified earlier in this chapter, and begin adding Curtain Wall objects to the front entry Wall.

ADD A CURTAIN WALL

1. On the Project Navigator, double-click the *01 Shell and Core* file in the *Constructs\Architectural* folder to open it.

Some column pier enclosures at the front entrance of the building have been provided in a separate file. For convenience, this file has been saved to the *Elements* folder. The *Elements* folder is a handy location to store temporary project files such as this one until they are given a permanent location in the project structure. Let's insert these piers from the *Element* file to assist us with building the Curtain Wall along the first floor front façade.

2. In the *Elements* folder of the Project Navigator, right-click the *01 Shell Piers* and choose **Insert as Block** (see Figure 8–7).

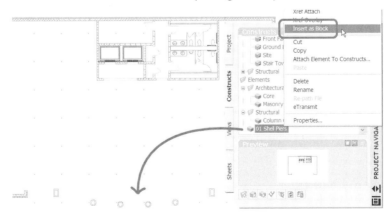

Figure 8–7 *Insert the First Floor Shell Piers into the 01 Shell and Core Construct*

3. In the Insert dialog, clear the "Specify on Screen" check mark for Insertion Point, place a check mark in the "Explode" check box and then click OK.

Several column pier enclosures will appear in the void at the bottom of the plan.

4. On the Design tool palette, click the **Curtain Wall** tool.

If you do not see this palette or tool, right-click the Tool Palettes title bar and choose **Design** (to load the Design Tool Palette Group) and then click the Design tab.

5. On the Properties palette, set the Base Height to **12'-0"** [**3650**].

We will draw the Curtain Wall across the space at the bottom of the plan, just above where we trimmed away the Wall at the start of the chapter.

6. Using tracking set the first point of the Curtain Wall **6"** [**150**] below the end of the short vertical masonry Wall on the left (see Figure 8–8).

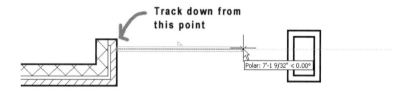

Figure 8–8 *Start point of Curtain Wall at the small vertical Wall on the left*

7. Using tracking again (or Perpendicular), set the end point **6"** [**150**] below the end of the masonry Wall on the other side (see Figure 8–9).

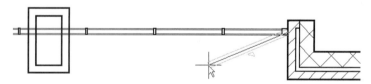

Figure 8–9 *End point of the Curtain Wall at the small vertical Wall on the right*

8. Press ENTER to complete the command.

TRIM THE CURTAIN WALL

Like Walls, Curtain Walls can be offset, trimmed and extended.

9. Zoom in to the left end of the Curtain Wall just added.

10. Trim the piece of the Curtain Wall that passes through the column enclosure, using the two Walls as cutting edges (see Figure 8–10).

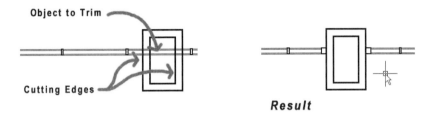

Figure 8–10 *Trim the Curtain Wall*

11. Repeat the same steps on the other side.

 NOTE The Trim action in this case yielded the precise result we needed because the position of the justification of the Walls is Left (outside edge relative to the column enclosure). Trim and Extend use the justification line of the Wall or Curtain Wall as the cutting or boundary edge. Keep this is mind as you work, and plan ahead accordingly.

MERGE CELLS

Examine the Curtain Wall on screen. The small rectangles are "frames" and "mullions" and the spaces in between are "cells." The cells of a Curtain Wall style are "infilled" parametrically, in this case with a simple infill panel expressed here as two parallel lines in plan. There are times, however, when the parametrically defined infill isn't appropriate for a particular cell, such as the front entry bay of the building. Rather than redesign the entire Curtain Wall style, we can apply an override to a particular cell.

12. Select all three Curtain Wall objects, right-click, and choose **Copy Curtain Wall Style and Assign**.

 TIP You can select all three at once using this method: select one Curtain Wall object, right-click and choose **Select Similar**.

13. On the General tab, name the Style **MADT Front Entry**. For the Description type **Mastering Autodesk Architectural Desktop 2006 Commercial Project Front Entrance**.

14. Click OK to dismiss the Curtain Wall Style Properties dialog.

15. Zoom in to the middle of the Curtain Wall (between the round columns).

16. Select the Curtain Wall object, right-click, and choose **Infill > Merge**.

Between each pair of round columns are three cells. We will use the Infill > Merge command to merge these three cells into one for each pair of columns (three total).

17. At the "Select cell A" prompt, click one of the cells between a pair of round columns (see "First Merge" in Figure 8–11).

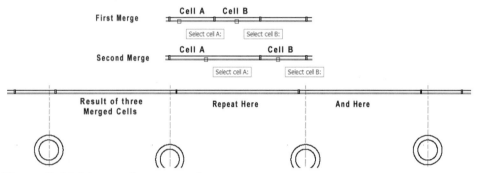

Figure 8–11 *Select a cell to merge cells*

18. At the "Select cell B" prompt, click the cell adjacent to the one selected for Cell A.

    Notice how the two cells have fused into one larger cell.

19. Repeat the process to merge the double cell with the third cell adjacent to it (see "Second Merge" in Figure 8–11).

20. Repeat the entire process (to make three large cells total) for each of the additional pairs of round columns.

TIP This is an excellent time to use the "Rule of thumb" to repeat the last command. Just press the space bar to repeat the **Infill > Merge** command.

ADD THE FRONT ENTRY INFILL OVERRIDE

When all merges are complete, the Curtain Wall should look like the lower portion of Figure 8–13. We will now add a revolving door entry to each of these large cells.

21. Open the Content Browser (CTRL + 4).

22. In the *Design Tool Catalog - Imperial* [*Design Tool Catalog - Metric*] catalog, navigate to the *Doors and Windows > Door and Window Assemblies* category.

23. Browse to the sixth (6) page of the *Door and Window Assemblies* category.

24. Using the eyedropper icon, drag the **Revolving 6-0×6-8 Ctr (R) + Sidelights + Transom** [**Revolving 1800×2050 + Sidelights + Transom**] tool into the drawing window.

25. At the "Select wall, grid assembly or RETURN" prompt, click the Curtain Wall.

26. At the "Select grid assembly cell to add door/window assembly" prompt, click one of the large merged cells.

27. In the Add Infill worksheet, choose the "Add as Cell Override" radio button.

    "New Infill" will be the only available option in the Infill section.

28. Type **Front Entry Infill** for the name.

29. In the Override Frame Removal area, place a check mark in the Left, Right and Bottom boxes and then click OK (see the right side of Figure 8–12).

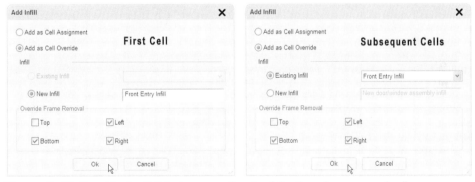

Figure 8–12 *Override the Cell Assignment of each of the three large merged cells*

The "Select grid assembly cell to add door/window assembly" prompt will repeat.

30. Select the next large merged cell and this time, choose "Existing Infill" but all other settings should match the first (see the left side of Figure 8–12).

31. Repeat again for the third cell.

32. From the View menu, choose **3D Views > SE Isometric** (or choose the SE Isometric icon on the Views toolbar).

33. Select the Curtain Wall, right-click, and choose **Edit Curtain Wall Style**. (Be careful not to select the Front Entry Infills.)

34. On the Materials tab, for the Default Infill component choose **Doors & Windows.Glazing.Glass.Clear** and for both the Default Frame and Default Mullion components choose **Doors & Windows.Metal Doors & Frames.Aluminum Frame.Anodized.Dark Bronze.Satin**.

    These Material Definitions were imported automatically as part of the Door/ Window Assembly tool used above.

35. Click OK when finished to return to the drawing and then click the Gouraud Shaded icon on the Shading toolbar (see Figure 8–13).

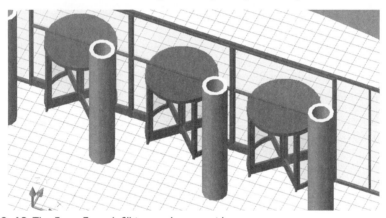

Figure 8–13 *The Front Entry Infill inserted as overrides*

36. Save the *01 Shell and Core* file, but leave it open for the next sequence.

CURTAIN WALL STYLES

After that quick primer on the Curtain Wall object, it is time to get more advanced. The true power and utility of the Curtain Wall object comes in its endless customization potential. It is perhaps the most complex and powerful object offered by Architectural Desktop. For this reason, developing good procedures is critical to success. In this lesson, we will walk through the process of designing and building a custom Curtain Wall style. The process involves detailed planning and good procedure. To begin, we will open a sample file and add some Curtain Wall objects using the Convert from Walls feature. This technique allows you to place Walls in your design in the early stages as "stand ins" for the Curtain Walls you will add later. This way, you are not hindered with the specifics of a complex Curtain Wall design before the project warrants it. The Curtain Wall style, like those of many ADT objects, can start very simply and then be slowly refined and embellished with detail as the design evolves.

UNDERSTANDING CURTAIN WALL TERMINOLOGY

The basic structure of a Curtain Wall is actually quite simple. One or more grid structures are nested together to form a complex design. Each grid can be horizontal or vertical, spaced evenly or repetitively, and might have other grids nested within it. In order to work successfully with Curtain Walls, you must first understand some basic terminology (see Figures 8–14 and 8–15).

▶ **Grid**—Establishes the quantity and orientation of the *cells* in a particular design. A grid references a *division* to determine its orientation and spacing. A division is NOT a grid, but rather the rules used to create one. Each grid is a distinct item, but several grids can reference the same division.

▶ **Cell**—The space formed by the intersecting *grids*. This space can be filled with the contents of an Infill or another Nested Grid.

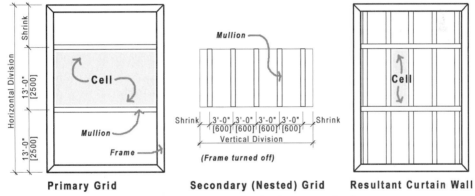

Primary Grid   **Secondary (Nested) Grid**   **Resultant Curtain Wall**

Figure 8–14 *Understanding the relationship of grids, divisions and cells*

In order to create the two basic structural components (grid and cell), a Curtain Wall design makes use of four "Element Definitions": Divisions, Infills, Frames and Mullions. Each of these is defined as follows:

▶ **Division**—Sets the spacing and orientation of the Curtain Wall bays. Orientation can be either horizontal or vertical. Spacing can be configured to repeat (use a specified bay dimension), space evenly (divide the total equally into a specified number of bays), or space manually (specify each mullion explicitly). A horizontal division can also be set to reference the baseline and base height of the Curtain Wall object. A vertical division can be configured to use a polyline for the spacing. In this configuration, each vertex of the polyline will become a mullion division in the grid.

▶ **Infill**—Determines what each grid cell contains. Each cell of each grid is filled with either an Infill definition or a Nested Grid. Infills can use a simple panel (solid slab of material) or they can reference a Door, Window, Curtain Wall Unit, Window Assembly, or AEC Polygon style as the infill.

- ▶ **Frame**—The outer edge of each grid and nested grid. The Frame can be the same on all four sides, or each one can be different. Frames can also be turned off on internal nested grids if not required by the design. Each Frame component is defined by its width and depth and can also optionally use a custom profile shape.

- ▶ **Mullion**—Has the same parameters as frames, but is used for the internal grid divisions to separate each cell from one another. Mullions can also use a custom profile shape.

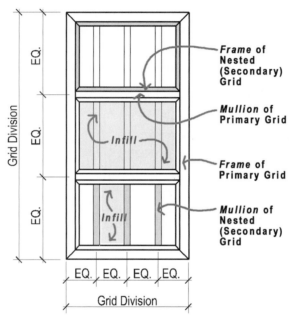

Figure 8–15 *Curtain Wall Element Definitions*

PUTTING IT ALL TOGETHER

In summary, a *grid* is a collection of one or more cells defined by a division. A *division* defines the spacing of each grid of the Curtain Wall. A division establishes the number of cells in a single direction. To define a basic rectangular grid pattern requires two grids, one horizontal and one vertical. The *frame* is the outermost edge of a grid on all sides. Each nested grid can have its own frame definition. The edge between each cell is a *mullion*. The *cell* is the space defined by frames and mullions. Each cell can contain another nested grid, or an infill. An *infill* can be comprised of solid material or it can reference one of several object style types.

EXPLORING EXISTING CURTAIN WALL STYLES

To further understand some of these elements and their relationships, let's dissect some existing Curtain Wall styles in order to understand the hierarchy and function of elements. We will explore the Curtain Wall interface in the *01 Shell and*

Core file by examining the composition of the Style used here. This Style was copied from the Standard style, so with the exception of the Infill we added above, it is
identical in composition.

EXPLORE THE INTERFACE

1. Select the main segment of the Curtain Wall, right-click, and choose **Edit Curtain Wall Style**.

2. Click the Design Rules tab.

3. Click the Viewer icon at the lower left corner and position the windows on screen to be next to one another, enlarging them as much as your screen will allow.

4. In the Viewer, change the View Direction to **SE Isometric** (see Figure 8–16).

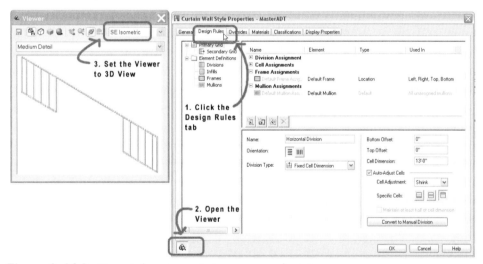

Figure 8–16 *Position windows to maximize screen real estate*

The Design Rules tab of the Curtain Wall Style Properties dialog box can be challenging to master. However, once you get the hang of it, you will see that it is organized quite logically (see Figure 8–17). The left side of the dialog box is a tree
view divided into two sections. At the top of the tree, one or more grids are organized hierarchically. The main grid is listed at the top (in this case, it is called "Primary Grid"). If there are additional nested grids, there will be a minus (-) sign in
front of main grid and they will listed indented below. Click on the minus sign to
collapse the grid tree. In this case "Secondary Grid" is the name of the only nested
grid. Also in the tree view is the Element Definitions node (depicted by a folder
icon). Here you gain quick access to the definitions for each of the four Curtain

Wall Element types: Divisions, Infills, Frames, and Mullions (see definitions above). The right side of the dialog box is divided into two areas. The top right section lists (or previews) whatever item is selected at the left tree view. At the bottom right, detailed parameters will be available for the item selected at the top right (where appropriate).

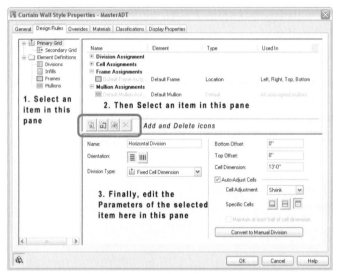

Figure 8–17 *Understanding the Curtain Wall Style Properties dialog box interface*

The basic flow of movement through the dialog box is as follows:

▶ Choose an item on the left tree view, either a Grid (at the top) or an Element Definition (at the bottom).

▶ On the right at the top, select an item to edit. (Depending on the selection at left, there could be several choices.)

▶ Edit the parameters at the bottom half of the window.

All changes occur immediately. Buttons to add and delete elements are located in a strip across the middle of the right pane (see Figure 8–17). Hover your mouse and pause over each icon to see a tool tip indicating its function.

EXPLORE THE ELEMENT DEFINITIONS

5. On the left hand tree view, under Element Definitions, choose **Divisions**.

   On the right side at the top, two divisions will appear, Horizontal Division and Vertical Division.

6. Choose **Horizontal Division** and review its settings in the bottom half of the window (see Figure 8–18).

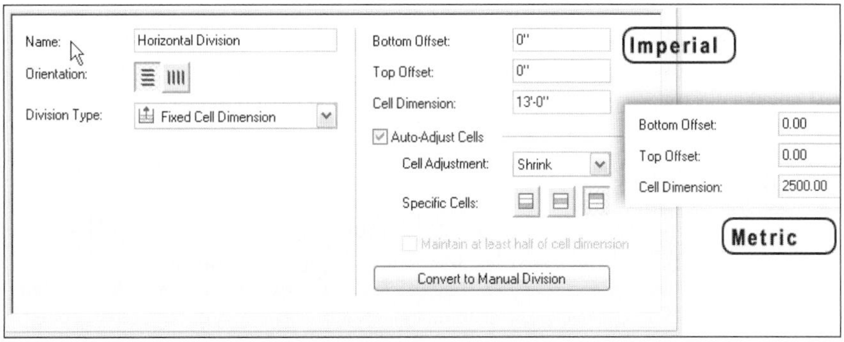

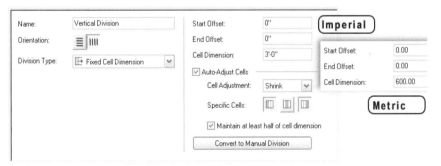

Figure 8–18 *Study the parameters of the Horizontal Division*

Next to Orientation, the Horizontal button is depressed, thus the name Horizontal Division. Below that is a list of Division Types. This particular division uses the Fixed Cell Dimension type. To the right of these settings, the Cell Dimension is 13'-0" [2500] with zero unit offsets. Finally, if the grid does not work out to be an exact multiple of 13' [2500], it will reduce (Shrink) the size of the Top cell. In summary, this division will repeat a 13' [2500] bay until it runs out of space; the final bay at the Top will be allowed to be smaller than 13' [2500].

7. Choose **Vertical Division** and review its settings (see Figure 8–19).

Figure 8–19 *Study the parameters of the Vertical Division*

Next to Orientation, this time the Vertical button is depressed. This one is also a Fixed Cell Dimension, with a Cell Dimension of 3'-0" [600] this time. Should the grid not multiply evenly by 3' [600], both the Start and the End will equally shrink to accommodate the variance. However, we also have a setting to instruct the Curtain Wall style to maintain at least half the Cell dimension size when dividing the difference between two or more Cells. In summary, this Division will repeat 3' [600] as many times as possible and split the leftover between both ends as long as it can do so while maintaining half the Cell size.

8. On the left hand tree view, under Element Definitions, select Infills.

There is only a single Infill Definition named Default Infill. This Infill is a two-inch [10 millimeter] thick Simple Panel. A Simple Panel is drawn graphically as a solid slab of material (two parallel lines in plan). Infill panels can be aligned to left, right or center.

9. Select Frames next.

There is only one Frame Definition, called Default Frame. It is 3″ [50] square. Frames may have offsets and use profiles for their shapes (see below).

10. Select Mullions.

There is also one Mullion Definition, Default Mullion, 1″×3″ [30×30]. Mullions may have offsets and use profiles for their shapes (see below). Now that we have explored the elements, let's look at how they are put together.

EXPLORE THE GRID STRUCTURE

11. On the left hand tree view, at the top, select the Primary Grid item.

    On the right side the parameters of the Primary Grid will appear.

12. On the right, click the minus (-) sign next to each entry (see Figure 8–20).

13. This will help you to focus on just the component you are working on at any given moment.

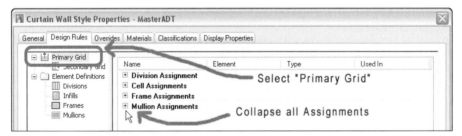

Figure 8–20 *Collapse all components*

14. Expand *only* the Division Assignment (click the plus (+) sign).

Reading in the Element column, the Division Assignment for the Primary Grid is Horizontal Division. As shown above, the Horizontal Division establishes a 13′ [2500] horizontal spacing. We also saw that the spacing begins at the bottom, divides the total height by 13′ [2500] cells and leaves any uneven space at the Top. Each of the Cells resulting from this Grid will have an Assignment as well. To see what component is loaded into each Cell, expand the Cell Assignments item.

15. Expand the Cell Assignments.

    Here the Element is a Nested Grid. (This means that these Cells will be subdivided further.)

16. Open the list of choices by clicking on *Nested Grid* (see Figure 8–21).

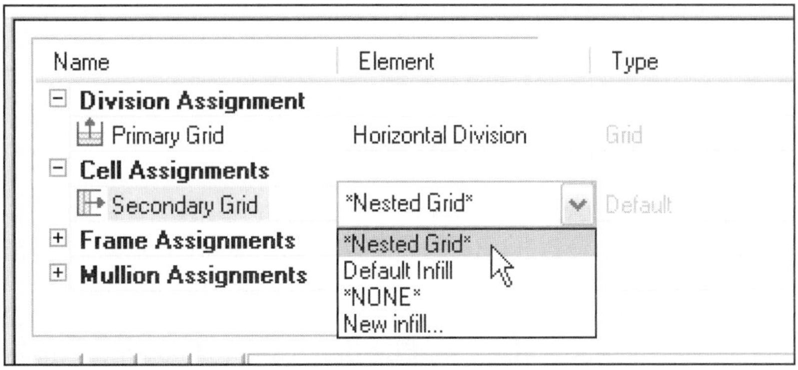

Figure 8–21 *Infill elements available in the current style*

Two choices will always be available, ***Nested Grid*** and ***NONE***. The asterisks indicate that these are hard coded into the software. In addition, **Default Infill** and **New infill** are available. **Default Infill** is the simple two-inch [50 millimeter] thick panel we looked at above in the Infill Definitions item. Choosing **New infill** allows a new component to be created on the fly. Additional infills will appear on this list as they are defined (see later in this exercise). In this case, each of the 13′ [2500] horizontal Cells created by the Horizontal Division are "filled" with another Grid called Secondary Grid.

17. Expand Frame Assignments.

There is only one Frame Assignment as well; it uses the Default Frame Definition looked at above, and it is applied on all four sides of the Primary Grid as seen in the Used In column.

18. Expand Mullion Assignments.

There is also only one Mullion Assignment. It occurs between any two cells (of the Primary Grid in this case) and is expressed with the 1″×3″ [30×30] Default Mullion explored above.

19. On the left hand tree view, at the top, select the Secondary Grid item.

This is the Nested Grid referenced by the Cell Assignment of the Primary Grid. It has similar parameters to the Primary Grid, with a few notable differences. First, its Division Assignment uses the Vertical Division. This creates the 3′ [600] mullion spacing as seen above. The Cell Assignment for the Secondary Grid is the **Default Infill**. This means that the grid does *not* subdivide any further and each 3′ [600] Cell is simply infilled with the Simple Panel designated by the Default Infill definition. The last item to note is the Frame Assignment. The Default Frame *is* assigned here as well; however, in the Used In column it specifies "None." Therefore, the Frame although assigned, is not *expressed* at this level of the design. The

Mullion Assignment is however the same as the Primary Grid, forming a flush appearance between Primary and Secondary Grid Mullions in the final design (see Figure 8–14).

20. Click Cancel to return to the drawing.

OUT OF THE BOX CURTAIN WALL STYLES

Until now, we have used the Content Browser to access the styles and content provided out of the box with ADT. There is another tool that we can use to access styles, in this case Curtain Wall styles. You can use the following procedure to access and work with any kind of style, Curtain Wall or other objects, those provided with ADT in the box, and others that your firm may have in library files on a network server. This is simply an alternative method. This method is also useful when Styles exist, but tools have not yet been created from them.

21. On the Design tool palette, right-click the Curtain Walls tool and choose **Curtain Wall Styles**.

    The Style Manager dialog box will appear. The Style Manager is used to create and edit ADT Styles.

22. Across the top of the Style Manager dialog box is a row of icons. Click the Open Drawing icon.

23. On the left side of the dialog box that appears is a series of icons. Click the Content icon on this toolbar.

24. This shortcut opens a folder with several more folders (see Figure 8–22).

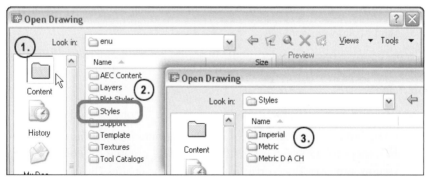

Figure 8–22 *Use the Content icon to access the Out of the Box Style Library*

25. Double-click the *Styles* folder, and then if you wish to use Imperial Styles, double-click the *Imperial* folder [if you wish to use Metric, double-click the *Metric* folder instead].

26. Finally, double-click the *Curtain Wall & Curtain Wall Unit Styles (Imperial).dwg* [*Curtain Wall & Curtain Wall Unit Styles (Metric).dwg*].

   This will return you to the Style Manager with the *Curtain Wall & Curtain Wall Unit Styles (Imperial).dwg* [*Curtain Wall & Curtain Wall Unit Styles (Metric).dwg*] file loaded on the left in the tree view. Beneath this drawing are three categories of Style represented by folder icons. Curtain Walls belong to the Architectural Objects category.

27. Expand the *Architectural Objects* category and then the Curtain Wall Styles item.

New in ADT

2006

28. At the top of the Style Manager, click the "Inline Edit Toggle" icon.

   This will enable the interactive viewer directly in the Style Manager.

New in ADT 2006 is the ability to edit a style inline within the Style Manager. The toggle icon used here toggles the traditional viewer behavior with the new edit inline behavior. If you double-click a style in the list, it also toggles to the Edit inline mode. Notice also, that if Project Standards are enabled (as we did above) that any Standards Drawing(s) in use in your project will be listed at the top of the Style Manager tree. You can expand the Standards Drawing node(s) and view and edit the styles like any other node in the tree. Right-click to get additional Standards options.

29. Select any Curtain Wall Style on the left to preview it in an interactive Viewer on the right. Be sure not to double-click (see Figure 8–23).

   TIP The interactive viewer and the ability to edit inline are the primary features that set the Style Manager apart from the Content Browser. If you wish to preview and edit styles interactively, use Style Manager.

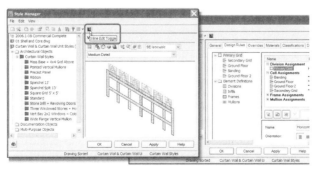

Figure 8–23 *Select styles on the left to preview or edit them on the right*

NOTE One limitation of the Viewer within the Style Manager is that it does not preview any nested components. Some of the Curtain Wall styles reference other AEC styles for Infills. This was mentioned above. Those nested styles will not preview in the Viewer. You would need to add a Curtain Wall of that style to the drawing to properly see those infills.

30. In the tree drag the **Mass Base + 4×4 Grid Above** Style from the *Curtain Wall & Curtain Wall Unit Styles (Imperial).dwg* [*Curtain Wall & Curtain Wall Unit Styles (Metric).dwg*] file and drop it on top of the *01 Shell and Core.dwg* file. (You can also copy and paste instead.)

    This action will import the style into your current drawing.

31. Double-click on the style that you just imported in the list.

    This will toggle the inline Edit in the right pane.

32. Repeat the procedure that we performed above in the "Explore the Element Definitions" topic.

    You will notice several differences from the **Standard** Style explored previously.

33. Continue this process on as many styles as you wish. Click OK to close the Style Manager when you are finished.

 NOTE if a dialog box asking you to save the Content file appears when you click OK, choose **No**. This is asking you if you wish to save the library file, not the drawing you have on screen. You should not save any changes to the Library file at this time.

The exercise just completed will help you to get familiar with strategies and techniques to planning and building a Curtain Wall style. In the next exercise, we will build a custom Curtain Wall Style from scratch.

BUILDING A CUSTOM CURTAIN WALL STYLE

The Standard Curtain Wall style defines a single Frame, Mullion, and Infill component, each called Default. For example, the mullion defined in the Standard Curtain Wall style is called Default Mullion. You can name the elements anything you wish. There are two predefined Division elements: one for the default horizontal Grid, the other for the default vertical Grid. Each of the default elements can be renamed and redefined, and additional elements can be created. When creating your own style, careful naming of each component is critical to successful implementation of your design, as it helps to avoid confusion and keep you organized. Outlined below is the recommended procedure for creating a new Curtain Wall style.

PROCESS FOR DESIGNING A CURTAIN WALL STYLE

Outlined next are three steps to follow when designing a custom Curtain Wall style. Curtain Walls can be complex, but following good procedure when you lay them out can mean the difference between successful implementation and frustration.

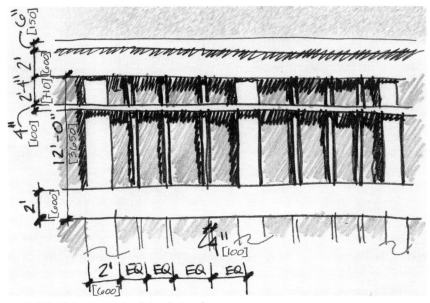

Figure 8–24 *Make a sketch of the design first*

Step 1 - Plan

Make a sketch of your design ideas on paper first (see Figure 8–24). A simple sketch is all you need in order to help you determine the basic grid structure and orientation. This sketch will also reveal how many Division, Frame, Mullion, and Infill elements you will need. The purpose of the sketch is to establish the proper organization and structure of the Curtain Wall design. The dialog box interface of the Curtain Wall object is complex, and having a clear "road map" before beginning will help keep you on track.

Don't worry about making the sketch too perfect or refined, because it will likely change as the design develops.

Step 2 - Create the Element Definitions (Build a Kit of Parts)

Referring to the Curtain Wall terminology covered above, you can begin to understand the primary goal of the sketch. The purpose of the sketch is to help you determine what Element Definitions (Divisions, Frames, Infills, and Mullions) will be required by your design. Once you have your rough sketch, create all of the elements you will need in your design. You can create additional elements on the fly later, but it will be easier to choose from a ready-made kit of parts as you work. First, refer to your sketch, and determine how many Divisions you will need; each level of your grid structure will be defined by a Division (see Figure 8–25).

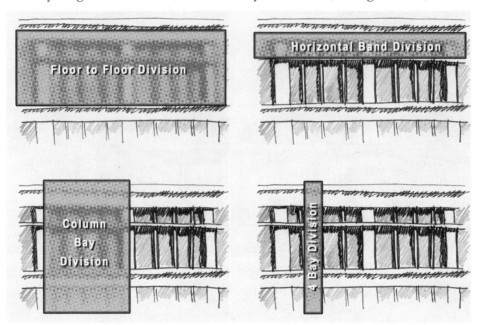

Figure 8–25 *Determine the required Divisions*

Next determine what will go inside each Cell (a Nested Grid or Infill). Finally, consider how each Grid level will be "edged" with Frames and Mullions (see Figure 8–26).

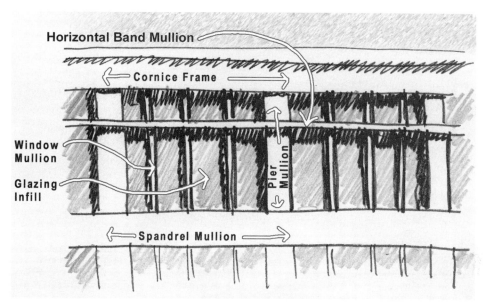

Figure 8–26 *Determining the required Element Definitions*

Following the recommendation of Step 1, a rough design has been sketched out in Figure 8–24. Figure 8–25 reveals that four divisions will be required in the Curtain Wall style. Finally, according to the sketch in Figure 8–26, several frames and mullions will also be needed. Table 8–1 summarizes the required Element Definitions.

Step 3 - Build the Grid Structure

Once you have your kit of parts built, it is short work to put all of the pieces together in a grid structure that will yield the results you desire. You can add as many grids as necessary to achieve the desired effect. However, as a general rule of thumb, if your design requires more than five nested grid levels, you might want to consider creating Infills that reference Curtain Wall Unit styles or Door/Window Assembly styles (see the note below). If you add these types of components to your design, the possibilities are limitless, and the grid structure of the design remains manageable. The most important consideration when establishing the grid structure is careful planning of the hierarchy. Using descriptive naming throughout will aid tremendously in this endeavor. This tip cannot be stressed enough: although it is tempting to accept the default names while creating elements, it is highly recommended that you take the time to consider your naming scheme carefully and pick names that make sense for the design. More important still, pick names that will still make sense two or three months after the design is complete. You never know when you will need to revisit a design scheme and when that happens, good naming will pay back tenfold. For all of these reasons, you should also avoid the default names (such as "Primary Grid," "Horizontal Division" and "Default Mullion").

These are simply not descriptive enough in most cases. The strategy can be summarized as follows:

▶ Limit your design to three to five nested grids.

▶ Create Infill Definitions that refer to Curtain Wall Units or Door/Window Assemblies in complex designs.

▶ Have variations of styles, both Curtain Wall and Infill, for quick swapping and "what if" scenarios.

▶ Plan your grid hierarchy carefully

▶ Use clear, descriptive, and simple naming

▶ Avoid complex naming schemes with cryptic abbreviations or acronyms.

Table 8–1 *MADT Front Façade Curtain Wall Style Element Definitions List*

Element Name	Type	Dimensions	Other
Divisions			
Floor to Floor Division	Fixed Cell Dimension	12'-0" [3650]	Shrink Top
Column Bay Division	Fixed Cell Dimension	10'-0" [3000]	Shrink Left & Right Don't Maintain half Cell
Horizontal Band Division	Manual	3'-8" [1100] from Top	
4 Bay Division	Space Evenly	4 equal	
Infills			
Glazing Infill	Simple Panel	1" [25] thick	
Frames			
Vertical End Frame	Basic (No profile)	6" [150] wide × 8" [200] deep	
Cornice Frame	Basic (No profile)	24" [600] wide × 12" [300] deep	
Mullions			
Pier Mullion	Basic (No profile)	12" [300] wide × 12" [300] deep	
Spandrel Mullion	Basic (No profile)	24" [600] wide × 12" [300] deep	
Horizontal Band Mullion	Basic (No profile)	4" [100] wide × 18"[450] deep	

Window Mullion	Basic (No profile)	4″ [100] wide × 6″ [150] deep

CURTAIN WALL UNITS AND DOOR/WINDOW ASSEMBLIES

Curtain Wall Units and Door/Window Assemblies are very similar to Curtain Walls in form and function, with a few minor differences. Curtain Wall Units are meant specifically to be smaller components of a larger Curtain Wall design. Use them as nested infills rather than building endless nested Grids and Divisions directly within the Curtain Wall Style. They are intended to be nested within Curtain Wall cells, and because they are style based, they offer a powerful way to consider alternate schemes and contingencies while designing. A Door\Window Assembly can be used in similar fashion as an infill in a Curtain Wall, and it can be inserted in a Wall just like a door or window. In this way, Window Assemblies are ideal for doors with sidelights, complex grouping of windows, transoms, and storefronts. Refer to the exercise under the "Add the Front Entry Infill Override" heading above, and to the "Add Window Assembly" topic in Chapter 4 for some specific examples. Although we will not specifically build Curtain Wall Units and Door/Window Assemblies in this lesson, the interface to these items is nearly identical to that of the Curtain Wall and all of the same techniques apply.

CREATING A CUSTOM CURTAIN WALL STYLE

We now have a strategy to follow when contemplating a Curtain Wall. Let's put it into practice and get started building a custom Curtain Style Wall for the Mastering Autodesk Architectural Desktop Commercial Project.

CONVERT THE WALLS TO CURTAIN WALLS

1. On the Constructs tab of the Project Navigator in the *Constructs\Architectural* folder, double-click the *Front Façade* Construct file to open it.

You may recall from Chapter 5 that we built a file for the front façade of the building and placed a stand-in Wall in that file as a temporary place holder. Some additional Wall segments have been added to this file to help complete the Curtain Wall shape. The first step will be to convert these Walls to Curtain Walls, and then we will adjust the basic height parameters. Once we have a Curtain Wall in place, we can begin developing our custom Curtain Wall design.

2. On the Design tool palette, right-click the Curtain Wall tool and choose **Apply Tool Properties to > Walls**.

3. At the "Select Walls" prompt, select all of the Walls and then press ENTER.

4. At the "Curtain Wall baseline alignment" prompt, choose **Center** (if you don't have DYN toggled on, right-click and choose **Center**).

5. At the "Erase layout geometry" prompt, choose **Yes** (or type Y and then press ENTER).

We now have Curtain Wall objects in place of the Wall objects. Let's configure the basic height parameters.

6. With the Curtain Walls all selected, right-click and choose **Properties**.

7. Set the Base height to **35'-0"** **[10,650]** (see Figure 8–27).

Figure 8–27 *Setting the height parameters and Floor Line offsets*

8. With the Curtain Walls still selected, right-click in the drawing and choose **Roof Line/Floor Line > Modify Floor Line**.

9. At the "Floorline" prompt, choose **Offset** (if you don't have DYN toggled on, right-click and choose **Offset**).

10. At the "Enter offset" prompt, type **-12'-0"** **[-300]** and then press ENTER twice.

With the Floor Line projected down slightly, the Curtain Wall will cover the floor slab of the second floor in the building model. Likewise the Base Height that we assigned will allow enough room for a Roof Slab above.

11. From the View menu, choose **3D Views > SE Isometric** (or click the SE Isometric icon on the Views toolbar).

CREATE A NEW CURTAIN WALL STYLE

12. Select all of the Curtain Walls on screen (5 total), right-click and choose **Copy Curtain Wall Style and Assign**.

13. On the General tab, name the New Style **MADT Front Façade**.

On the General tab, it is usually a good idea to put a description. Typical descriptions include a reference to the project, a detailed description of components in the design, or reference to the manufacturer if it is prefabricated.

14. In the Description field, type **Mastering Autodesk Architectural Desktop 2006 Commercial Project Front Façade Curtain Wall**.

As you make edits, it will be handy to see some visual feedback. For this reason, it is always useful to have the floating viewer open as you work.

15. If the Viewer is not already open, click the Floating Viewer icon at the bottom left corner of the dialog box to open the Viewer window, and choose **SE Isometric** from the View Control list.

Notice that even though the viewer appears to be showing us the correct style, its image does not look like the Curtain Wall we have in the drawing. This occurs in the floating viewer when you select multiple objects before editing Styles as we did here. It is easily remedied.

16. Click OK to dismiss the Curtain Wall Style Properties dialog, select the curved front Curtain Wall, right-click and choose **Edit Curtain Wall Style**.

Notice that the viewer now displays only the curved front Curtain Wall segment. This will be much more useful as we progress.

CREATE DIVISIONS

As we work through the Design Rules, they will at first be very familiar to those explored above. This is because the Curtain Walls here were created from the generic Curtain Wall tool on the Design tool palette. This tool uses the Standard style, making the "MADT Front Façade" style a copy of Standard. We have not yet configured its Design Rules to make it vary from Standard.

17. Under Element Definitions (on the left in the tree), click Divisions.

18. Choose **Horizontal Division** at the top right.

19. In the bottom right area, in the Name field, rename it to **Floor to Floor Division**.

20. Change the Cell Dimension to **12'-0"** [**3650**].

Leave the Orientation set to **Horizontal** and the Division Type set to **Fixed Cell Dimension**. Leave the Auto-Adjust Cells section configured as is. This will reduce the Top cell if the Curtain Wall's height does not multiply cleanly. Experiment with the alternative setting of Grow and the other choices under Specific Cells. Check the results in the Viewer as you experiment, and change it back to **Shrink** and **Top** when you are done.

21. In the top right pane of the dialog box, choose "Vertical Division" and rename it **Column Bay Division**.

22. Change the Cell Dimension to **10'-0"** [**3000**] and leave the remaining settings as they are.

Notice the effect of the settings in the Viewer. With the current settings, the curved Curtain Wall ends up divided into six bays. The two at the end are a bit smaller. For this design, we would actually prefer a center bay with an equal number of bays on either side of it. To do this, we need to experiment with the Maintain at least half of cell dimension setting. When this is turned on, cells smaller than a half Cell are not permitted, so rather, all Cells will be shifted. If this setting is not checked, then the two Cells at the ends can be any size, thereby not requiring a shift in Cells. Experiment with the Maintain at least half of cell dimension setting as well as some of the other Auto-Adjust Cells settings.

23. When finished experimenting, be sure that Auto-Adjust Cells is set to **Shrink Left** and **Right**, and that Maintain at least half of cell dimension is *not* checked (see Figure 8–28).

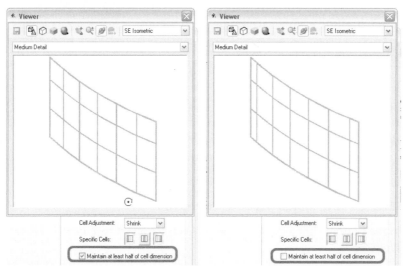

Figure 8–28 *Comparing the effect of the Maintain at least half of cell dimension setting*

These changes occur immediately because the two Divisions are already assigned to items in the grid structure. Review the steps in "Explore the Hierarchy" above for more information.

In the middle of the window, there are a couple of icons similar to those highlighted in Figure 8–17. These icons will change function when you select a different node on the tree at the left. Pause your mouse over each icon to see a tool tip indicating its function.

24. Click the New Division icon.

25. Name the New Division **Horizontal Band Division**.

26. Click the Horizontal icon next to Orientation.

27. Open the Division Type list and choose **Manual** for the type.

28. To the right of the window, click the Add Gridline icon.

29. Change the Offset of Gridline 1 to **3'-8" [1100]**, and the From point to **Grid Top** (see Figure 8–29).

> We will use this to form a single horizontal band across the top of each main horizontal cell.

Figure 8–29 *Add a manual grid line*

 NOTE There will not be any feedback on this change in the viewer. This is a new Division and until it is added to the Curtain Wall design, the effect will not be apparent.

30. Click the New Division icon again.

31. Name it **4 Bay Division** and change the Orientation to **Vertical**.

32. Choose **Fixed Number of Cells** from the Division Type list and set the Number of Cells field to **4**.

CREATE INFILLS, FRAMES, AND MULLIONS

33. On the left in the tree, select Infills under the Element Definitions heading.

34. Select the Default Infill and rename it to **Glazing Infill**.

35. Change the Panel Thickness to **1" [25]**.

36. On the left in the tree, select **Frames** under the Element Definitions heading.

37. Select the Default Frame and rename it **Vertical End Frame**.

38. Set the Width to **6" [150]**, and the Depth to **8" [200]**.

39. Click the New Frame icon (where the New Division icon was).

40. Name the new Frame **Cornice Frame** and make its Width **24" [600]** and its Depth **12" [300]**.

41. Following the same procedure, refer to Table 8–1 and create the Mullions.

42. Create all new Mullion Definitions. Do *not* rename and reuse Default Mullion this time.

 TIP You may want to click OK here, save the drawing and then return to the Curtain Wall Style to continue. It is a good habit to save after you have completed a procedure, and unfortunately, you cannot save while in a dialog.

BUILD THE GRID STRUCTURE

1. Select the Primary Grid node at the top of the tree. Pause for a second and then click again.

   This will allow the item to be renamed. This is the same technique used to rename items in Windows Explorer.

2. Type **Level 1 Grid** and then press ENTER.

On the right, note that the Level 1 Grid is already referencing the Floor to Floor Division, because we simply renamed the existing division in the steps above. This is a good time to pause and make certain that you understand this behavior. It is important to grasp this interrelationship between the parts. Here again, good naming will help you keep the hierarchy and relationships straight.

3. In the top right pane, expand Frame Assignments, click in the Used In field and then click the small browse (...) button within the field (see Figure 8–30).

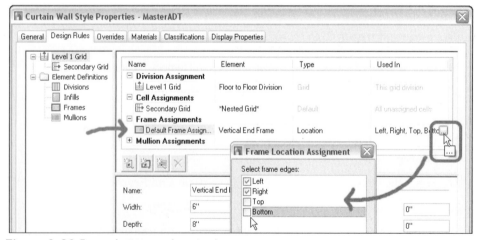

Figure 8–30 *Frame Assignment Location button*

The Element currently referenced by the Default Frame Assignment is the Vertical End Frame. We only need the Vertical End Frame to show at the left and right edges of the Curtain Wall.

4. In the Frame Location Assignment dialog box, deselect Top and Bottom and then click OK (see Figure 8–30).

   Notice that the frame has disappeared in the top and bottom edges in the Viewer window.

5. In the middle of the dialog box, click the New Frame Assignment icon.

   A new entry labeled "New Frame Assignment" will appear in the Frame Assignments grouping in the top pane.

6. Click directly on this new entry and pause for a second.

   This activates the Rename mode.

7. Type **Top and Bottom** for the new name and press ENTER to accept it.

 Caution: Be sure to rename Assignments directly in the Name column at the top half of the dialog box. Do not type in the Name box in the lower half. This is because the name below is the name of the Element being referenced, not the Assignment (see Figure 8–31).

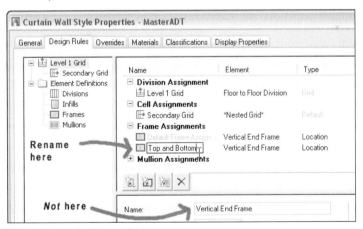

Figure 8–31 *Renaming the Frame Assignment*

Element assignments and element definitions are not the same thing. An element assignment, such as the frame assignment we are discussing here, is part of the actual grid structure of the Curtain Wall design. Element definitions are members of our "kit of parts," which are available for *use* in element assignments. Another way to look at it: an element assignment uses an element definition, not the other way around. Please note that unlike all of the other components we have seen in this dialog box so far, the Default Assignments cannot be renamed.

For example, a wall is often framed with a top and bottom plate, and studs at 16″ on center. 16″ OC would be a Division assignment, as would be the specification of a top and bottom plate. The inclusion of both together as a specification for wall construction would be a Grid in the Curtain Wall. Furthermore, you could use 2×4s or 2×6s to do the actual framing. A 2×4 is an element definition, as is a 2×6, whereas their use for all of the members spaced at 16″ OC is the element assignment.

Please note that unlike all of the other components we have seen in this dialog box so far, the Default Assignments cannot be renamed.

8. Click in the Element column (currently reading "Vertical End Frame").

   A small menu will appear.

9. Choose **Cornice Frame** (see Figure 8–32).

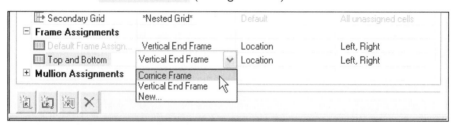

Figure 8–32 *Choose the Frame component*

Notice that the list includes the Frame elements we took the time to define in the previous sequence. It is much easier to simply choose from a premade list of elements than to create them as you build the grid hierarchy.

10. Click in the Used In field, click the browse (…) button, and place a check mark in Top and Bottom; clear Left and Right this time.

    Notice the change in the Viewer. A much "heavier" frame now occurs at the top and bottom edges.

11. Under the Mullion Assignments grouping, next to Default Mullion Assignment, choose **Spandrel Mullion** from the Element list.

    Notice the change in the Viewer. Again, having predefined the list makes the process of assigning them much simpler.

That completes the first grid in the structure. Now we can begin nesting other grids within the cells of this main grid to build a more complex design.

12. On the left in the tree, click and rename Secondary Grid as we did above for Primary Grid. Call it **Level 2 Grid**.

Here we are going to make changes that are slightly more dramatic than those for the Level 1 Grid.

13. In the Division Assignment area for Level 2 Grid, choose **Horizontal Band Division** from the Element list.

    Notice the change in the Viewer. The design has become entirely horizontal. This will prove temporary as we continue (see Figure 8–33).

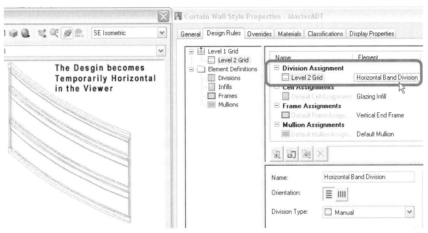

Figure 8–33 *Adding the Horizontal Band Mullion grid*

14. Skip down to Mullion Assignments and choose **Horizontal Band Mullion** from the Element list for the Default Mullion Assignment.

    Notice the change in the Viewer. The Level 2 Grid Mullion is now a bit deeper.

15. Move up a bit to Cell Assignments, open the list in the Element column, and choose ***Nested Grid*** (see Figure 8–34).

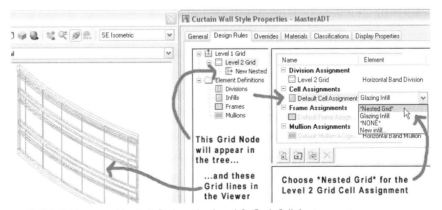

Figure 8–34 *Add a new Nested Grid as the Level 2 Grid Cell Assignment*

Notice the appearance of a new node in the tree called New Nested Grid and the appearance of a new grid in the Viewer.

16. Select New Nested Grid in the tree on the left, and rename it **Level 3 Grid**.

17. With Level 3 Grid still highlighted in the tree on the left, in the Division Assignment area, ensure that **Column Bay Division** is selected from the Element list (it may automatically default to this choice).

18. In the Mullion Assignments area, choose **Pier Mullion** from the list.

19. In the Cell Assignments area, choose ***Nested Grid*** once more.

20. In the tree on the left, rename the New Nested Grid to **Level 4 Grid**.

21. Change the Division Assignment to **4 Bay Division** and the Mullion Assignment to **Window Mullion**.

We don't need to change the Cell Assignment this time, because we do not need any more Grids and it is already referencing the Glazing Infill by default. As with the other nested grids above, the Frame Assignment is irrelevant because it is not assigned to any edges (in the Used in column).

Take a good look at the design in the Viewer (see Figure 8–35). Click the Shade button to see the details even better. The design is coming very close to the intention of the sketch, but it could use some improvements. For instance, the entire thing is currently a dull gray with no material articulation. To make this design read better, we need to assign some materials.

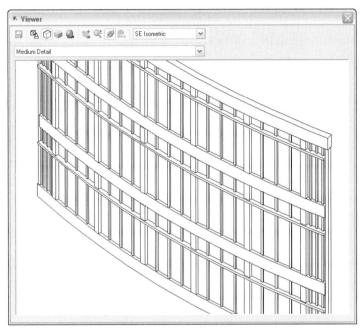

Figure 8–35 *Completed Grid Assignments*

 New in ADT 2006

ASSIGN MATERIALS USING A MATERIAL TOOL

Material Definitions offer a powerful way to articulate the plan, surface, section and 3D parameters of any component in any ADT style. Materials define linework properties such as layer, color and linetype. They also designate hatching and rendering material parameters for all components in all Display Configurations.

1. In the Curtain Wall Style Properties dialog box, click the Materials tab.

There are two columns on this tab. On the left are listed all the Components of the Curtain Wall design that we just added (each Element Definition is listed). The right column shows which Material has been assigned to that component. Currently all components are assigned to the Standard material. The Standard Material simply uses the Layer color in plan and usually yields a dull gray in 3D. Let's assign materials to each of our components.

2. Click on Standard in the Material Definition column next to Glazing Infill.

    This will reveal a drop-down list. Notice that the only choice is Standard. This will limit our ability to enhance the Materials of this design.

3. Click OK to close the Style Properties dialog and Save the file.

There are three ways to import Material Definitions into a drawing. You can use the Material Definitions command on the Format menu, which loads the Style Manager and filters it to Materials only. You must then use the technique covered in the "Out of the Box Curtain Wall Styles" heading (above) to load a drawing with Materials and drag and drop or copy and paste them into the current drawing. However, in many cases, you can also import the desired Materials simply by using a Style that references them. In other words, since Materials are not stand-alone objects, they are always assigned to other Styles. Therefore, if you know a Style that already uses the Material you want, simply use that Style and the Material will automatically be imported with the Style. For instance, if you wanted a basic Brick Material, you could simply use one of the Wall tools on the Walls palette to import it. Try it out if you like. Click on any Wall tool on the Walls palette and then return to the Materials tab of the Curtain Wall Style. You will now have the Material(s) from that Wall in addition to Standard.

The third and final way to import and assign materials is to use a Material Tool on a tool palette. This method allows the direct import and assignment of materials from a library file. On the Design Tool Palette is a basic Material Tool that references the out-of-the-box Material Definition library file. This file, located in the *C:\Documents and Settings\All Users\Application Data\Autodesk\ADT 2006\enu\Styles\Imperial* [*C:\Documents and Settings\All Users\Application Data\Autodesk\ADT 2006\enu\Styles\Metric*] folder by default, contains nearly 200 Material Definitions to choose from.

4. From the View menu, choose **Shade > Gouraud Shaded**.

5. On the Design tool palette, click the Material tool.

6. At the "Select an object" prompt, click one of the Curtain Wall objects.

   The Apply Materials to Components worksheet will appear.

7. From the drop-down list at the top, choose ***Doors & Windows. Glazing.Glass.Clear***.

8. Next to the Glazing Infill Component choose **Style** from the "Apply to" list and then click OK.

   The Glazing Infills should now be transparent.

   If your glazing is not transparent, please visit the Autodesk Knowledge Base at: http://usa.autodesk.com/adsk/servlet/ps/item?id=5503469&linkID=2476059&siteID=123112 for a solution to the issue.

9. Click the Material tool again and then select the Curtain Wall again when prompted.

10. Choose ***Doors & Windows.Metal Doors & Frames.Aluminum Frame.Anodized.Dark Bronze.Satin*** from the Material list at the top.

11. Hold down the CTRL key and select Vertical End Frame, Horizontal Band Mullion and Window Mullion.

12. From the "Apply to" list (next to any of the selected components) choose **Style** (see Figure 8–36).

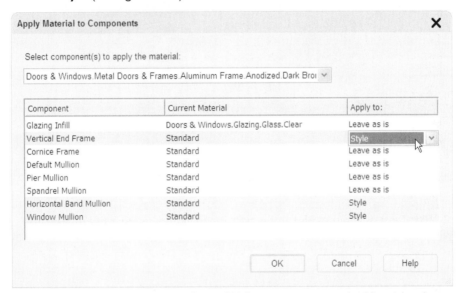

Figure 8–36 *Assign Materials to the Curtain Wall components using the Material tool*

13. Repeat one more time and assign ***Masonry.Stone.Marble. Square.Stacked.Polished.White-Brown-Black*** to the remaining components excluding the Default Mullion.

The results should look something like Figure 8–37.

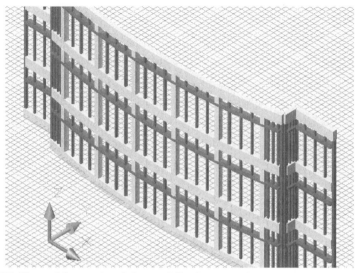

Figure 8–37 *Material Assigned and Gouraud Shading turned on*

14. Save the file. (Do not close the file.)

TIP There is no rule of ADT that is more important than remembering to Save often!

CHECK DIMENSIONS

At an early stage of design, the Curtain Wall style we have built would be sufficient. However, closer examination would reveal some issues that need to be resolved, particularly the way the Frame and Mullion Elements are placed relative to the Curtain Wall dimensions and Grid Divisions. For convenience, let's explore these issues in another file.

15. In the *Elements* folder (on the Constructs tab) of the Project Navigator, double-click the file named *Curtain Wall Offsets* to open it.

Note the two dimensions on the left side of the screen. 35'-0" [10,650] is the height parameter and 1'-0" [300] is the projection of the bottom edge. These were applied to the Curtain Wall at the beginning sequence. These two values total: 36'-0" [10,950], or exactly three vertical bays according to the 12'-0" [10,650] dimension used in the Level 1 Grid. The important issue to note is the difference in the way that frames and mullions are applied relative to these overall dimensions. The frame falls *completely* within the limits of the Curtain Wall on all sides. However, mullions are always *centered* relative to the cell divisions by default. The problem this presents is fairly obvious—the space within the bays at the ends will not be equal to those in the middle. To resolve this situation, we will explore the Offset parameters of the Frame and Mullion Elements.

16. Select the Curtain Wall, right-click, and choose **Edit Curtain Wall Style**.

17. On the Design Rules tab, choose Frames from the Element Definitions listing in the tree, and then select the "Cornice Frame" on the right.

Make sure the Viewer window is open and set the View to Front.

On the bottom right of the dialog box are four fields in the Offsets area. The X and Y offsets adjust the position of the frame component relative to the Curtain Wall. The Start and End Offsets adjust the length of the Frame material (see Figure 8–38).

▶ **A positive X**—Moves the frame *out* away from the Curtain Wall center.

▶ **A negative X**—Moves the frame *in* toward the Curtain Wall center.

▶ **A positive Y**—Shifts the frame in plan view above the baseline (relative to a Curtain Wall drawn from left to right). When viewed in elevation as we have here, again with the start point on the left and the end point on the right, a positive Y will move the component away from us.

▶ **A negative Y**—Shifts the frame in plan view below the baseline (relative to a Curtain Wall drawn from left to right). When viewed in elevation as we have here, again with the start point on the left and the end point on the right, a negative Y will move the component toward us.

▶ **Positive Start or End offsets**—Shorten the length of the frame component.

▶ **Negative Start or End offsets**—Extend the length of the frame component.

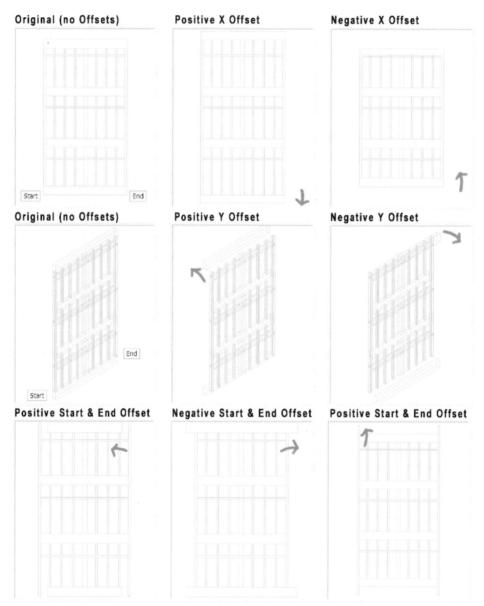

Figure 8–38 *The various effects of offsetting Frames and Mullions*

18. Type **2'-0"** [**600**] in the X Offset box and then press ENTER.

   Note the change in the Viewer. The top Cornice frame will move up and the bottom one will move down.

19. Change the X Offset to **-2'-0"** [**-600**].

   The Cornice frames will move in the opposite direction.

20. Change the Viewer to an isometric view and then experiment with the Y Offsets.

    Note that the movements of the Cornice frames are now relative to the depth of the Curtain Wall.

21. Type **2′-0″ [600]** in the Start Offset box and then press ENTER.

    Note the gap on the left side of the Cornice frame. The Cornice frame length has been shortened by 2′-0″ [600].

22. Continue to experiment with the various offsets and compare the results to Figure 8–38.

23. Return all Offsets to **0** (zero) before continuing.

24. Beneath Element Definitions, select the Cornice Frame component again.

25. Type **1′-0″ [300]** in the X Offset field and then click OK.

    Notice the shift of the Frame component on both the top and bottom (relative to the dimensions).

A check of the dimensions will now yield evenly spaced horizontal bays. We will perform a similar technique on the mullion spacing.

26. Return to the Design Rules tab of the Curtain Wall Style Properties dialog box.

27. Under Element Definitions, select Divisions.

Divisions can also have offsets just like Frames and Mullions.

28. Select the 4 Bay Division, type **4″ [100]** in both the Start and End Offsets, and click OK to see the change.

The Start and End Offsets will move the points used to divide the cell closer to the center (reduce the width of the cell). Notice that the distance between the face of the mullions and the face of the frames is now the same as the distance between the face of the mullions in relation to each other (see Figure 8–39). You can check this spacing using either the AutoCAD Distance (type DI) command or by grip editing the existing dimensions to the new points.

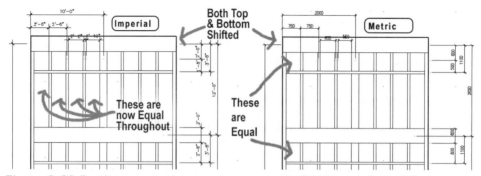

Figure 8–39 *Equalizing spacing between components with Division and Frame Offsets*

29. Save the file.

ADD ANOTHER STANDARDS DRAWING TO THE PROJECT

When using Project Standards, you can optionally have more than one Standards Drawing. In some cases it will make sense to this. For instance, if all projects used the same Schedule Table Styles and Property Sets you could have an office-wide Standards Drawing assigned to all projects to keep those items in synch. Each project could then reference a separate Standards Drawing that contained the project-specific Wall and Door Styles. In another example you may want one type of Style to be maintained by one individual while another type is handled by someone else. In this case, both Standards Drawings might be project-specific. There could be several other reasons to save standard styles in more than one Standards Drawing. In the next topic, let's look at adding the *Curtain Wall Offsets* file as an additional Standards Drawing.

SYNCHRONIZE THE STYLE

We are now ready to apply the changes to this Style to the version in the *Front Façade* Construct. There are actually several ways to do this in Architectural Desktop. We could use the Style Manager to manually copy the new version and overwrite the one in the Front Façade file; we could create a Tool on our Project Tools palette and use it to "Re-import" the Style into the Front Façade file (as we did for the Exterior Shell Wall above; or we could use Project Standards. (We could also do all three or any combination of them if we wished—see the CAD Manager Note below.) All methods would achieve the same initial result, but over the life of the project, the Project Standards approach is the most useful.

1. On the Project Navigator, click the Project tab and then click the Configure Project Standards icon.

2. Click the Add Drawing icon on the right.

The Select Standards Drawing dialog should have defaulted to the previous folder.

3. Browse to the *Elements* folder of the current project, choose *Curtain Wall Offsets* and then click OK.

TIP You should start off in the *Standards\Content* folder of your current project. Click the "Up one Level" icon twice and then double click the *Elements* folder.

4. Place a check mark next to Curtain Wall Styles beneath the newly added *Curtain Wall Offsets* drawing column and then click OK.

5. When prompted to apply a Version, type **Modified Frame and Division Offsets** for the comment and then click OK (see Figure 8–40).

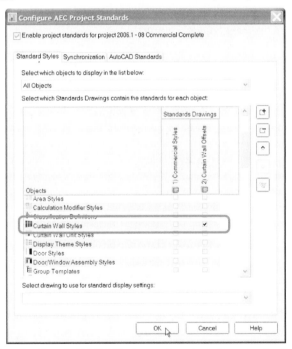

Figure 8–40 *Assign the Curtain Wall Offsets as a Standards Drawing for Curtain Wall Styles*

6. Save and close the *Curtain Wall Offsets* file.

SYNCHRONIZE A DRAWING TO THE NEW STANDARDS FILE

Now let's synchronize the *Front Façade* file to the new Standards Drawing.

7. On the Project Navigator, open *Front Façade* from the *Constructs\Architectural* folder.

8. At the bottom right corner of the drawing, right click the AEC Project Standards quick pick and choose **Synchronize Drawing** (see Figure 8–41).

Figure 8–41 *Synchronize the Front Façade Construct*

9. Be sure that the Action reads "Update from Standard" and then click OK.

 Caution: Remember, once you designate a drawing as Standards Drawing, it controls the composition of that Style from then on. If you should make a change to the Style in any other project drawing, it can be overwritten the next time you synchronize the standards. This is a very powerful tool in ADT. However, make certain that you really want to overwrite the existing style before you synchronize. Once it has been over-written, the original style is gone. It is always a good idea to save often and perform regular backups. Changes made in project drawings can be "pushed" to the Standards Drawing where appropriate. The process is called "Updating." Refer to the "Updating Project Standards Drawings" topic in the online Help for more information.

As noted above, there are other ways to update the Style. We could make a custom tool from our Curtain Wall style on our Project Tools palette. Even if you use Standards to synchronize the Styles, it is often useful to create tools for the most common project styles as well. We have seen examples of this already. Objects from the drawing can be dragged to Tool Palettes to create tools. You can also drag styles from the Style Manager to create tools. When you create a new tool, it will write the full path to the host drawing file as the Style source location. An icon is also generated automatically for the new tool. It is best to work in Standards Drawings when you make tools. This way, the tool's style reference will point to the master version of the Style and there will be no ambiguity as to which version is most current. To try this out, open one of the Standards Drawings, make the Project Tools palette active and then drag an object (or Style from Style Manager) to the Project Tools palette. After the tool is created, right-click the new tool and choose **Properties,** and then look at the Style Location field. It will reference the Standards Drawing name and path.

Please note, that it is very important to save the file *before* you drag to create the tool. This is the only way that the file reference in the Style Location field will be correct. The style must *exist* in the saved version of the file before a tool that references it can be properly made.

If you are unhappy with the icon image that was automatically created, you can scroll to the bottom of the Tool Properties worksheet and adjust the image in the Viewer to your liking. When you have finished, right-click the icon image at the top of the worksheet and choose **Refresh image.** Specify image allows you to choose an image file from your hard drive (these should be small, 64 pixels square). Finally, if you close the Tool Properties worksheet, you can right-click on the tool and choose **Set Image from Selection,** and then have an icon generated from an object selection in the drawing (see Figure 8–42).

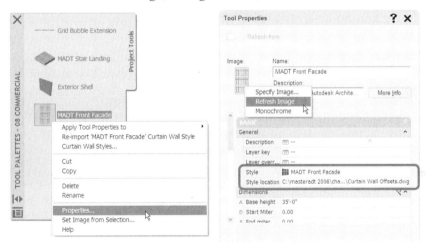

Figure 8–42 *Create a Project Tool for the Curtain Wall and customize the image*

Once you have built a tool of this sort, it can be applied to objects in other drawings in the project by simply right-clicking the tool and choosing "Apply to." You can also "Re-import" style to manually update a drawing without performing a Standards Synchronization.

Finally, if you create tools in your workspace, they must still be made available to other team members on a project. This can be done via the Content Browser. Copy and paste the tool to a common catalog in Content Browser to which all team members have access.

REFINING THE MADT FRONT FAÇADE CURTAIN WALL DESIGN

Editing the offsets to help equalize the frame and mullion spacing was helpful, but there are a few additional refinements that we can make to the design to make it more visually appealing.

Due to the "Shrink" rule of the Grid Division, the cells on the ends are smaller than the ones in the middle. Subdividing them into four cells, like the other full-sized bays, does not work well in this circumstance. In addition, the corner conditions need some refinement as well. We will redefine the style so that the Level 4 Grid (4 Bay Division in this case) does not apply to the first and last cells. Then we will apply overrides to the corners to get nicer transitions and cleanup. Finally, because the two segments parallel to the front of the building do *not* need the Start and End cell override that the others do, we will make a copy of the original Curtain Wall style to apply to them.

ADDRESS THE END BAYS

1. Make sure that *Front Façade* is the current drawing.

    If it isn't, choose *Front Façade* from the Window menu, press CTRL + TAB to cycle it into view, or simply double-click it on the Project Navigator palette again.

2. Select the two straight horizontal Curtain Wall segments on the right and left (the ones that will intersect the masonry Walls), right-click and choose **Copy Curtain Wall Style and Assign**.

3. In the Curtain Wall Style Properties dialog box, click the General tab and type **MADT Front Façade 3 Bay** for the name and then click OK.

We now have two styles that will work independently of one another. This will enable us to edit them separately and create slight variations to the design between the main Curtain Wall and the two smaller pieces that frame it.

4. Select the curved Curtain Wall segment, right-click and choose **Edit Curtain Wall Style**.

5. On the Design Rules tab, click the Level 3 Grid node in the tree.

6. Click the New Cell Assignment icon. Rename the New Cell Assignment to **End Bays**.

    Verify that in the Element column, Glazing Infill appears and that "Used In" displays Start, End (see Figure 8–43).

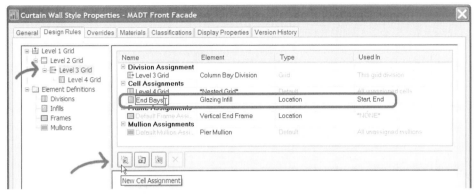

Figure 8–43 *Add a new Cell Assignment for the end bays*

7. Click OK to accept the changes and close the dialog box.

Notice the change to the drawing from the adjustment of the Level 3 Grid. The Curved Curtain Wall and the small perpendicular segments now look better. Since we first copied and applied a new style to the two horizontal segments that will join up with the masonry Walls they are unaffected. However, these might look better with only three bays instead of four.

EDIT DIVISIONS IN PLACE

8. From the View menu, choose **3D Views > Front**.

   If you have been working in shaded mode, you might want to choose **2D Wireframe** (Views > Shading menu, or the Shading toolbar) for the next operation. This will make it easier to see while you work.

9. Select the straight horizontal Curtain Wall segment on the right.

10. Right-click and choose **Design Rules > Transfer to Object**.

11. Click again on the Curtain Wall and then click the small round gray (Edit Grid) grip at the bottom (see item 1 in Figure 8–44).

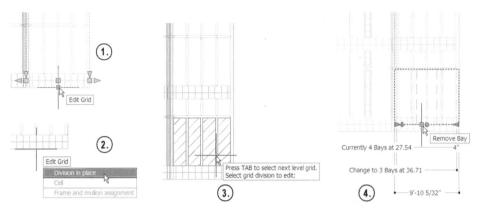

Figure 8–44 *Edit Grid of the Curtain Wall and edit the Division in Place*

12. At the "Edit Grid" prompt, choose **Division in Place** (see item 2 in Figure 8–44).

    Move the cursor over the Curtain Wall and notice the way the grid cells highlight.

13. Press the tab key as required to highlight one of the 4-Bay Divisions as shown in item 3 in Figure 8–44 and then click to select it.

    A series of magenta grips will appear. Magenta indicates that you are editing a style—in this case the 4 Bay Division.

14. Hover over each grip to see a tool tip indicating its function.

15. Click the minus sign (-) grip to remove one bay.

    Notice that this applies to the entire Curtain Wall not just the selected bay.

16. On the In-Place Edit toolbar, click the Save all Changes icon.

17. Select the same Curtain Wall, right click and choose **Design Rules > Save to Style**.

18. In the Save Changes worksheet, click OK.

This action will apply the edit made to the first Curtain Wall to the one on the other side as well. Let's return to an isometric view and study the changes.

19. From the View menu, choose **3D Views > SE isometric**.

The design is starting to shape up nicely. However, corners still don't cleanup very nicely. Let's resolve that next.

OVERRIDE CORNER CONDITIONS

20. From the View menu, choose **3D Views > Top** (or click the Top icon on the Views toolbar).

    If you have been working in shaded mode, you might want to choose **2D Wireframe** (Views > Shading menu, or the Shading toolbar) for the next operation. This will make it easier to see while you work.

TIP In general, when viewing Top, you should always choose 2D Wireframe.

21. Zoom in on the right side of the plan where the three segments of Curtain Wall come together.

22. Select the horizontal Curtain Wall segment, right-click, and choose **Edit Curtain Wall Style**.

23. On the Design Rules tab, in the tree, select the Frames item beneath Element Definitions.

24. Click the New Frame icon.

25. Name the new Frame **Corner Frame**.

26. Set both the Width and Depth to **8″ [200]**, and the X Offset to **4″ [100]**.

27. Click the Materials tab and set the material for Corner Frame to **Doors & Windows.Metal Doors & Frames.Aluminum Frame.Anodized.Dark Bronze.Satin** and then click OK to return to the drawing.

    Now let's apply this new Element to the corners that need it.

28. Select the horizontal Curtain Wall segment, right-click and choose **Frame/ Mullion > Override Assignment**.

29. At the "Select an edge" prompt, click the Frame to the left (see Figure 8–45).

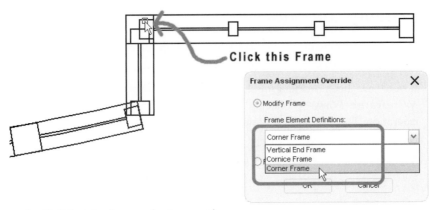

Figure 8–45 *Select the Frame edge to override.*

30. In the Frame Assignment Override dialog box that appears, choose **Corner Frame** from the Frame Element Definitions list and then click OK.

31. Repeat the steps 22 through 30 on the small perpendicular Curtain Wall segment (see Figure 8–46).

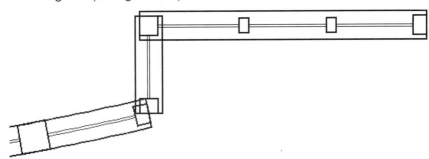

Figure 8–46 *Result of the Square Corner Override applied to both segments*

It was necessary to repeat all of the steps including the definition of the square Frame component, since these two segments are now two different styles. You will need to repeat the Override steps again on the other side of the plan as well. However, in this case, since the Corner Frame is already defined for both styles, you can simply repeat the Override Frame steps. If you wish, you can delete the two segments on the left side and mirror them over instead.

32. Select the small perpendicular Curtain Wall segment or the curved one, right-click, and choose **Edit Curtain Wall Style**.

33. On the Design Rules tab, select Level 1 Grid.

34. In the Frame Assignments area, next to the Default Frame Assignment item, click in the Used In box, then click the (...) button.

35. Deselect the Right and Left check boxes, and then click OK.

The Used In will now display *NONE*.

36. Click OK again to view the change (see Figure 8–47).

The Frame edges of the Curtain Wall will disappear. However, the square edge of the small perpendicular Curtain Wall retained its override.

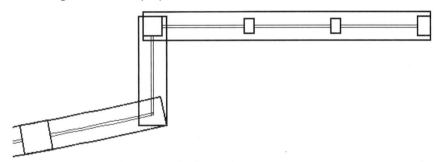

Figure 8–47 *The result of removing the Frame Assignment.*

APPLY A MITER TO THE CORNERS

37. Select the curved Curtain Wall segment and the small perpendicular segment, right-click, and choose **Set Miter Angles**.

Figure 8–48 *Miter the corners*

38. Repeat the Miter on the other side and on the intersection between the horizontal and perpendicular straight Curtain Wall segments.

 NOTE If the angles calculated by the Set Miter Angles command are not correct, you can manually adjust them in the Properties palette.

39. From the View menu, choose **3D Views > SE Isometric**, zoom in on the corner, turn on Gouraud Shading and study the result.

40. Save the file.

UPDATE PROJECT STANDARDS

Recall that we mentioned above that if we were to synchronize the project right now with the Standards, we would loose most of the changes that we have just made. This is because the Standards Drawing still contains the old Version of the Style. While it is typical that the workflow will be for changes to be made in the Standards Drawing and then synchronized to the project files, it is possible to take changes made in the project file and push them back to the Standards file. This is called "Updating" the Standards Drawing.

UPDATE THE STANDARDS DRAWING

We will use this process to update the *MADT Front Façade* style and to copy the new *MADT Front Façade 3 Bay* style to the Standards Drawing.

1. From the Format menu, choose **Style Manager**.

2. Expand the *Architectural Objects* category and then highlight the *Curtain Walls* category.

3. Right-click the **MADT Front Façade** style and choose **Version Style**.

4. In the Version Object dialog, type: **Modified End Bays and Added Corner Frame** for the Version Comment and then click OK.

5. Repeat this process for the **MADT Front Façade 3 Bay** style using the comment: **Created 3 Bay Variation**.

6. Right click the Curtain Wall Styles node on the left and choose **Update Standards from Drawing**.

7. Verify that the action next to **MADT Front Façade** is **Update Project Standards** and that the action next to **MADT Front Façade 3 Bay** is **Add to Project Standards** and then click OK (see Figure 8–49).

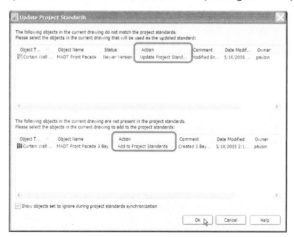

Figure 8–49 *Update Project Standards from the current Drawing*

8. Click OK to close the Style Manager. In any "Save Changes" dialogs that appear, click Yes.

If you wish, you can open the Curtain Wall Offset file (in the Elements folder) select the Curtain Wall onscreen, right click and choose Edit Curtain Wall Style. Click the Version History tab to see that the new Version with the "Modified End Bays and Added Corner Frame" comment is the current Version.

DIRECT MANIPULATION OF CURTAIN WALL COMPONENTS

In many respects, the preceding tutorial could be considered "the hard way" of designing and building Curtain Walls, Curtain Wall Units and Door/Window Assemblies. The overall goal of the exercise was to convey a complete understanding of Curtain Wall terminology and procedure. The result of what we created was a fully parametric Curtain Wall design. However, there are easier ways to build Curtain Walls. Perhaps the simplest method is to convert linework to a Curtain Wall, Curtain Wall Units or Door/Window Assembly object. In addition, we have several in place edit and direct manipulation functions as well. Let's look at a few here.

CONVERT LINEWORK TO A DOOR/WINDOW ASSEMBLY

1. On the Project Navigator palette, click the Constructs tab.

2. In the *Elements* folder, double-click *Window Assembly Sketch* to open it.

    This file was added to the dataset for this chapter.

3. On the Design tool palette, right-click the Door/Window Assembly tool and choose **Apply Tool Properties to > Elevation Sketch**.

4. At the "Select elevation linework" prompt, select all of the lines in the drawing and then press ENTER.

5. At the "Select baseline or RETURN for default" prompt, press ENTER.

6. At the "Erase layout geometry" prompt, choose **Yes** (if you do not have the DYN toggle turned on, right click and choose **Yes**).

    A new Door/Window Assembly will be made from the linework. Let's add a Door to the main Cell.

7. On the Doors tool palette, click the ***Hinged - Double - Full Lite*** tool.

 NOTE For convenience, this Door Style has been provided in this file. In your own projects, you may have to use any of the techniques covered in this chapter to import the Door Style.

8. At the "Select wall, space boundary, grid assembly or RETURN" prompt, click the Door/Window Assembly object.

    This will turn on the cell markers (an icon in the center of each cell). These icons are used to select cells when you wish to override their contents.

9. Hover over the cell marker in the center cell.

Hatching will appear highlighting the cell.

10. Click to select the cell (see Figure 8–50).

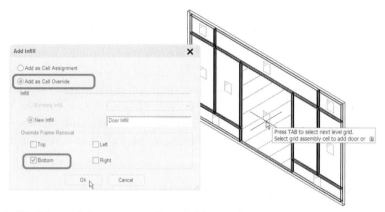

Figure 8–50 *Select a Cell to receive a Door Infill override*

11. In the Add Infill worksheet, choose the "Add as Cell Override" option.

12. Type **Door Infill** for the New Infill name and then place a check mark in the Bottom box for Frame Removal.

13. Click OK to see the results and then press ENTER to complete the command.

14. Right-click the Door/Window Assembly and choose **Design Rules > Save to Style**.

15. Click the New icon, type **MADT Retail Store Front** for the name, and then click OK.

16. Place a check mark in the "Transfer Infill Overrides to Style" check box and then click OK (see Figure 8–51).

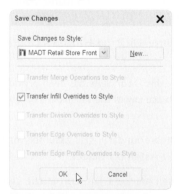

Figure 8–51 *Saving the custom Door/Window Assembly to a new Style*

17. On the Design palette, click the Material tool.

18. Apply the **Doors & Windows.Glazing.Glass.Clear** Material to the Default Infill component at the Style level.

19. Apply the **Doors & Windows.Metal Doors & Frames.Aluminum Frame.Anodized.Dark Bronze.Satin** Material to the Default Frame and the Default Mullion also at the Style level.

20. Save the file.

For more detailed instruction on using the Material tool, refer to the "Assign Materials using a Material Tool" heading above.

CREATE A CUSTOM DOOR/WINDOW ASSEMBLY TOOL

21. Right click the tool palettes title bar and choose **08 Commercial [08 Commercial Metric]** to load the Project Tool Palette Group.

22. Select the Door/Window Assembly and the nested Door, right-click and choose **Clipboard > Copy**.

23. From the File menu, choose **Open**. Browse to the Standards\Content folder for the current project and open the *Commercial Styles [Commercial Styles Metric]* drawing file.

24. Right-click anywhere in the drawing, choose **Clipboard > Paste** and then click a point on screen to locate the pasted object.

25. Save the *Commercial Styles [Commercial Styles Metric]* drawing file.

26. Select the Door/Window Assembly object and (being careful not to select a grip) drag it and drop it on the Project Tools palette.

27. Right click the new tool and choose **Set Image from Selection** and then when prompted, select both the Door/Window Assembly and the nested Door and then press ENTER.

28. Save and Close the *Commercial Styles [Commercial Styles Metric]* drawing file.

ADD THE DOOR/WINDOW ASSEMBLY TO THE 01 PARTITIONS CONSTRUCT

29. On the Project Navigator, double-click *01 Partitions* (in the *Constructs\Architectural* folder) to open it.

    This file was added to the dataset for this chapter. It contains a few Walls defining a lobby and an XREF Overlay of the *01 Shell and Core* file for reference.

30. On the Project Tools tool palette, click the **MADT Retail Store Front** tool.

31. Add one to each of the two vertical Walls in this file and then press ENTER when finished (see Figure 8–52).

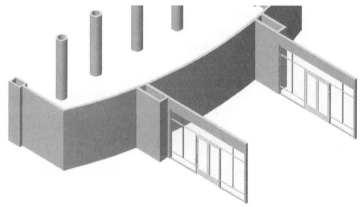

Figure 8–52 *Adding the Door/Window Assemblies to the 01 Partitions file*

32. Save and Close both the *Window Assembly Sketch* and the *01 Partitions* files.

CUSTOMIZE A MULLION PROFILE

The shape of all of our Frames and Mullions are rectangular at this point. But they can be any shape we like. For the next sequence, we do a bit of experimentation. Be sure that the *Front Façade* file is open on screen.

Open the *Curtain Wall Offsets* file. (If it was closed, double-click it on the Project Navigator to open.)

1. From the View menu, choose **3D Views > SE Isometric** (or choose the SE Isometric icon on the Views toolbar).

2. Zoom in close on one of the square Pier Mullions in the center bay of the Curtain Wall.

3. Select the Curtain Wall, right-click, and choose **Frame/Mullion > Add Profile**.

4. At the "Select a frame or mullion of the grid assembly to add a profile" prompt, click on one of the Pier Mullions in the center.

 TIP If you have trouble selecting one of the Pier Mullions, change to Top view first and then try again.

5. In the Add Mullion Profile dialog box, be sure that Profile Definition is set to **Start from scratch**.

6. Type **Pier Mullion Profile 1** for the name, and make sure that "To Shared Mullion Element Definition" is chosen (see Figure 8–53).

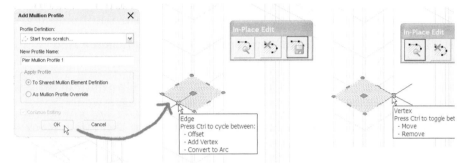

Figure 8–53 *Enter Edit in Place for a new Mullion profile*

An Edit in Place profile will appear with grips at each corner, one at each midpoint and one at the center. You can edit the shape of the profile with these grips, or you can right-click for more options. Like most ADT grips, each shape has a different function. Hover your mouse over a shape to see a tip indicating its function and any CTRL key options. For instance, with the thin rectangular grips, you can add vertices and convert the edge to an arc. On the small circular grips, you can move or remove vertices. You can also add and remove vertices from a right-click menu, as well as add, remove and replace rings. A ring is a shape that you draw with a closed polyline first, and then incorporate into the Edit In Place profile. Feel free to experiment with these grips before proceeding. You can stretch and reshape the profile any way you wish. The Curtain Wall will respond interactively. When you are finished experimenting, be sure to undo the edits until you return the Edit In Place profile to its original square shape. For our purposes here, we will make a very simple edit.

7. Click the small rectangular grip on the edge facing out and then press CTRL once to add a vertex.

8. Drag the new point using POLAR or ORTHO perpendicular to the Curtain Wall **2″ [50]** and then press ENTER (see Figure 8–54).

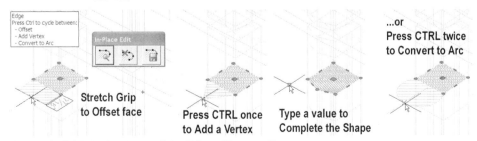

Figure 8–54 *Edit the shape of the Edit in Place profile*

9. Click the Save icon on the In-Place Edit toolbar.

   Note the change to all of the Pier Mullions in the design.

10. Zoom in on the middle bay of the top row of the Curtain Wall.

11. Select the Curtain Wall, right-click and choose **Frame/Mullion > Add Profile**.

12. When prompted to "Select a frame or mullion of the grid assembly to add a profile" click the third Pier Mullion from the left (in the top row).

13. In the Add Mullion Profile dialog box, be sure that Profile Definition is set to **Start from scratch** again.

14. Type **Pier Mullion Profile 2** for the name, and choose "As Mullion Profile Override" from the Apply Profile group this time.

   By choosing this option, the edit will be applied to only the selected Mullion.

15. Repeat the Add Vertex steps exactly as before. (Or press CTRL twice and add an arc segment this time.)

16. Click the same grip again, and drag it back in the opposite direction **2″ [50]** this time and then click the Save icon on the In-Place Edit toolbar.

   Where the first profile produced a pointed Mullion profile, this edit produces a concave shape.

17. Select the Curtain Wall, right-click and choose **Frame/Mullion > Add Profile** once again.

18. This time, select the third Pier Mullion from the right.

19. Instead of "Start from Scratch," this time choose **Pier Mullion Profile 2** from the Profile Definition list and be sure that As Mullion Profile Override is chosen from the Apply Profile group again.

   You may repeat these steps to add the Pier Mullion Profile 2 to as many of the top row Mullions as you wish.

   If you like, open the *A-CM00* Composite Model file on the Views tab of the Project Navigator to see your Curtain Wall in the context of the entire building.

20. Save and Close all project files.

Here are a couple of additional tools to explore:

▶ **Roof Line/Floor Line**—Includes options such as Project, which will match the top edge (Roof Line) or bottom edge (Floor Line) to the shape of an open polyline. Change to an elevation view; draw a polyline above or below the Curtain Wall in the shape that you want it to project. Right-click the Curtain Wall and choose **Roof Line/Floor Line > Edit Roof Line** or **Edit Floor Line**. Choose the **Project** option, and when prompted, select the polyline. There is also an Edit in Place option.

▶ **Reference Shape**—Allows the baseline in plan of the Curtain Wall to follow another shape. Draw a polyline, spline, or ellipse in the plan in the drawing. On the Design palette, right-click the Curtain Wall tool and choose **Apply Tool Properties to > Referenced Base Curve**. Follow the prompts. The Curtain Wall will remain linked to the shape as it changes.

ADDITIONAL EXERCISES

Additional exercises have been provided in Appendix A. In Appendix A you will find an exercise for building a screen porch in the Residential Project using a Curtain Wall (see Figure 8–55). In it you will review topics covered here, and explore the Edit in Place functionality further. It is not necessary that you complete this exercise to begin the next chapter, it is provided to enhance your learning experience. Completed projects for each of the exercises have been provided in the *Chapter08/Complete* folder.

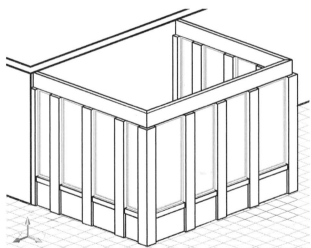

Figure 8–55 *Add a Sun Porch to the rear of the Residential Project*

SUMMARY

Project Standards offers a way to keep all Styles (and Display settings) in a project synchronized with one or more master Standards Drawing files.

Create Project Tools that point to the styles in your Project Standards files for additional functionality and convenience.

The Curtain Wall object is a powerful and flexible tool for designing a building skin.

Proper planning and good naming are critical to successful Curtain Wall designs.

Sketch out your design intent first.

Try to determine and build as many of the components as you will need ahead of time—Kit of Parts.

Assemble the style from your Kit of Parts.

Establish proper offsets on the Frame and Mullion elements to ensure proper cell spacing.

The Edit in Place mode allows for flexible "what if" scenarios that can be saved back to the original style or to a new style.

CHAPTER 9

Mastering Wall Cleanup

INTRODUCTION

The extent to which you understand and can control wall cleanup is a direct determinant of your success or failure in using Architectural Desktop—A strong statement to be sure, but a true one nonetheless. If you skip any chapter in this book, *this is not the one!* That being said, let's discuss why wall cleanup is so critical. No one can dispute that walls are the major component of any floor plan or building model. It is also fair to say that the majority of production time on a typical architectural project is often consumed while working on floor plans, and by extension, editing walls. It therefore stands to reason that being able to master the usage of walls is paramount to success. Proper implementation of wall cleanup will play a huge role in this scenario. Wall cleanup determines the interaction between ADT Wall objects and controls all aspects of their successful intersection. Master wall cleanup and you will gain an immediate increase in your efficiency and productivity. This will translate into less time drafting and more time designing. In this chapter, you will learn the basic steps necessary to achieve successful wall cleanup in *all* circumstances.

OBJECTIVES

In this chapter, we will take a break from our dual projects and instead explore the concepts of wall cleanup in several small files prepared specifically to focus on individual aspects of the topic. As we work through the lessons, we will be building a list of rules and guidelines to follow when performing and troubleshooting wall cleanup. These rules and guidelines are summarized in Appendix B—Wall Cleanup Checklist. Keep it handy for ongoing reference. The objectives of this chapter are:

- Understand the "rules" of automatic wall cleanup.
- Learn to manipulate wall cleanup parameters.
- Understand the Graph Display Representation.
- Learn when to use "manual" wall cleanup (Wall Merge).
- Understand Wall component cleanup priorities.

WHAT IS WALL CLEANUP?

Let's begin with some terminology associated with wall cleanup.

- **Wall cleanup**—The interaction between two or more Wall segments that causes them to respond to one another and form "correct" and/or "clean" corners and intersections.

- **Baseline**—The zero point of a Wall object's width as defined by its style. The parallel edges of a Wall are measured from this point. The Baseline and its significance in Wall Styles is explored in detail in the next chapter.

- **Justification line**—The point within the Width of the Wall that is used for reference as the Wall is drawn. This is the line where the Wall's grips will appear in the drawing. The justification can be Right, Left, Center, or Baseline. Right and Left are relative to the direction of the Wall as determined when looking toward the end point from the start point (see Figure 9–1). You can easily see which end is the start and which is the end by hovering the cursor over the grips.

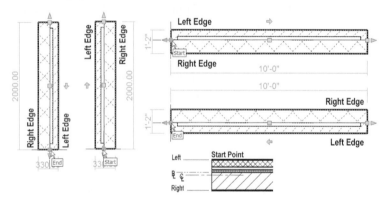

Figure 9–1 *Wall justification and its relationship to Wall orientation*

- **Graph line**—A line coincident with the Wall's length that is used to determine cleanup. The graph line can be at the Wall's justification line, or forced to the center of the Width (see Figure 9–2).

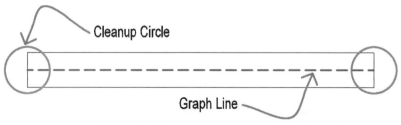

Figure 9–2 *Graph line and cleanup circles*

- **Cleanup Circle**—A tolerance applied at each end of a Wall segment and each intersection between Walls to force wall cleanup to occur. Also referred to as cleanup radius (see Figure 9–2).

- **Defect Warning**—A small circular marker indicating display calculation errors in the drawing. These usually appear when wall cleanup cannot be calculated properly (see Figure 9–3).

Figure 9–3 *Wall Defect Warning symbol*

CAD Manager Note: In the Drawing Default Wall Display Properties, Defect Warnings are assigned the color Red and placed on the A-Anno-Nplt layer. This is good practice, because in most cases you would not want the Defect Markers to print. (The Defpoints layer, AutoCAD's default "no plot" layer, was used in previous releases of ADT.) The color Red serves as a "red flag" on screen and helps these warnings to be noticed. It is recommended that you reserve color Red for this purpose and use it consistently throughout your standards.

PART I – AUTOMATIC WALL CLEANUP

By default in ADT, whenever walls intersect one another, an automatic cleanup is performed.. When automatic cleanup is successful, it requires the least amount of interaction by the user. It simply works! However, as powerful and sophisticated as automatic cleanup is, it is not always successful. Therefore, we must first understand exactly what factors determine if automatic cleanup will occur, and then what to do about it if it does not.

Install the CD Files and Open a Sample File

If you have already installed all of the files from the CD, skip step 1. If you need to install the CD files, start at step 1.

1. Install the files for Chapter 9 located on the *Mastering Autodesk Architectural Desktop 2006* CD ROM.

   Refer to "Files Included on the CD ROM" in the Preface for information on installing the sample files included on the CD.

2. Launch Autodesk Architectural Desktop 2006 from the desktop icon created in Chapter 3.

If you did not create a custom icon, you might want to review "Create a New Profile" and "Create a Custom ADT Desktop Shortcut" in Chapter 3. Creating the custom desktop icon is not essential; however, it makes loading the custom profile easier.

3. From the File menu, choose **Open** and browse to the *C:\MasterADT 2006\Chapter09* folder. Open the file named *Wall Cleanup1.dwg* [*Wall Cleanup1-Metric.dwg*].

RULE I – USE WALL GRAPH DISPLAY

Wall cleanup is calculated automatically by the software using a graph line coincident with the length of the Wall and an optional cleanup circle occurring at each end of the Wall. Using these components, the software constructs a "graph" of all Walls and then applies all of the other Wall parameters like justification, width, endcap conditions, and openings to draw the actual plan display on screen and in print. (For more information on the specifics of this process, search for the topic "Wall Cleanups and Priorities" in the online help.) The position of graph lines within the width of the Wall and the radius of cleanup circles are configured individually for each Wall object. Wall graphs can be displayed on screen to assist in achieving proper wall cleanup. To display Wall graphs, use the following steps:

TOGGLE THE WALL GRAPH DISPLAY

Several Walls appear on screen in the *Wall Cleanup1.dwg* [*Wall Cleanup1-Metric.dwg*] file.

1. Select any Wall on screen, right-click and choose **Cleanups > Toggle Wall Graph Display** (see Figure 9–4).

Figure 9–4 *Toggle Wall Graph Display on the Wall Object right-click menu*

When the Wall Graph Display is toggled on, it reveals the graph lines and radii for all Walls in the drawing. Each of these components is usually color coded to help them stand out. (If your drawing contains Walls within XREFs, you may need to Regen the drawing before the graph will display for the XREFs). Examine each condition in this file. Notice that two intersecting radii (the rightmost condition) do not cause cleanup (see Figure 9–5). Cleanup will occur if any of the following conditions exist:

▶ The graph line of one Wall intersects the graph line of another Wall.

▶ The graph line of one Wall intersects the Cleanup radius of another Wall.

▶ The Cleanup radius of one Wall intersects the graph line of another Wall.

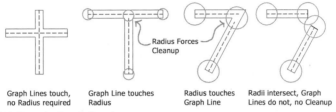

| Graph Lines touch, no Radius required | Graph Line touches Radius | Radius touches Graph Line | Radii intersect, Graph Lines do not, no Cleanup |

Figure 9–5 *Walls with Graph Display on*

TIP Toggle Wall Graph Display on whenever you are adding or modifying Walls. Toggle it off before printing.

CAD Manager Note: The graph lines and cleanup circles are color coded by default in the *Aec Model (Imperial Stb).dwt* [*Aec Model (Metric Stb).dwt*], template files. This is very helpful when using Graph Display mode. Choose unique colors and/or linetypes that are not being used anywhere else in your company's standards. To change the Graph Display settings, edit the Drawing Default property source of the Graph Display Representation in your template file(s). In addition, assign these components to a non-plotting layer such as A-Anno-Nplt.

Graph Display is your best diagnostic tool to assess Wall Cleanup issues. Toggling Graph Display on is akin to the doctor's taking an X-Ray to assess whether you have broken a bone. Just as it would be very difficult to assess many medical conditions without an X-Ray, it is nearly impossible to correctly assess a Wall Cleanup situation without the Graph Display.

RULE 2 – START WITH A CLEANUP RADIUS OF 0 (ZERO)

The larger the Cleanup radius, the farther away from a true intersection the Walls can actually be. In some cases, this is desirable to achieve a particular cleanup effect, but in general, large Cleanup radii should be avoided, and any Cleanup radii should be as small as possible. Also, notice from the image at the left in Figure 9–5 that the Cleanup radius can be set to a value of 0 (zero, no radius). As a rule of thumb, all Cleanup radii should initially be set to 0 and only increased when proper cleanup requires it. With the radius set to 0, you are forced to close your geometry. The benefits of good, cleanly drafted floor plans and models go far beyond simple wall cleanup, as any veteran AutoCAD user can attest (more on this in Rule 3 below).

SET THE DRAWING DEFAULT CLEANUP RADIUS

When you draw Walls with the Wall tool on the Design tool palette (or any tool that does not have its own Cleanup radius value), you can designate a cleanup radius value on the Properties palette. ADT will remember the last value set for Cleanup radius. This setting becomes the Drawing Default. This is true even after you have quit and restarted ADT.

1. On the Design tool palette, click the Wall tool.

2. On the Properties palette, scroll down to the Cleanups grouping in the Advanced category.

3. Verify the value of Cleanup radius and set to **0″ [0]** as necessary and then press ENTER (see Figure 9–6).

Ultimately, regardless of whether the value is preset in the Properties palette or preset in a Wall tool, the Cleanup radius can always be changed directly on the Wall object (see Rule 4 below). Having the default in place however, helps to avoid any inadvertent errors. A zero cleanup radius is the ADT default out of the box. This default should be maintained.

4. Save the file.

CAD Manager Note: The default Wall Cleanup radius set to 0 (zero) is another good value to have set in the office template files.

Figure 9–6 *Set the Wall Cleanup radius on the Properties palette*

RULE 3 – PRACTICE GOOD, CLEAN DRAFTING

This is perhaps the most important rule of all. A striking majority of all cleanup problems can be avoided by simply practicing good, clean drafting technique. All

AutoCAD editing commands can be performed on Walls. Commands such as Move, Copy, Array, Rotate, Break, Fillet, Chamfer, Trim, and Extend are among the many commands useful for editing Walls. There are several Wall-specific command options as well. Always use Object Snaps! Always type in dimensions when they are known. Use Offset, Trim, and Extend and the 'L' and 'T' Cleanup tools (see below) often. They offer an excellent means of performing layout tasks quickly while nearly always facilitating proper cleanup. Review the tutorials in Chapter 4 for proof of this statement. In that chapter, we laid out an entire floor plan without discussing the rules of wall cleanup. This was possible because the first three rules of wall cleanup, as covered here, were anticipated and built into the exercise. Let's look at a few good drafting techniques in particular.

AVOID "DOUBLES"

1. Select any Wall in the drawing, right-click, and choose **Basic Modify Tools > Copy**.

2. For the base point, type **0,0** and then press ENTER.

3. For the second point, type **0,0** and then press ENTER again.

   Notice the red Defect Warning appearing on the Wall.

Avoid deliberately or inadvertently creating double Walls. The problem is that the Wall is trying to cleanup with itself. As you can see, it is not having an easy time of it.

4. Erase the duplicate Wall and then save and close the file.

 TIP Click the Defect Marker to easily select the duplicate Wall. If you have several duplicates, try using a crossing selection to select all of them (note the quantity on the properties palette) and then hold down the **SHIFT** key and deselect just the one on top.

USE EXTEND AND TRIM

5. In the *C:\MasterADT 2006\Chapter09* folder, open the file named *Wall Cleanup2.dwg* [*Wall Cleanup2-Metric.dwg*].

   Notice that the Wall graph lines are already displayed in this file. There are also text labels to help you work through this tutorial.

6. Right-click in the drawing and choose **Basic Modify Tools > Extend**.

7. On the upper left side of the drawing (labeled Extend), click the leftmost horizontal Wall (at the top) as the "boundary edge" and then press ENTER.

8. At the "Select object to extend" prompt, click near the top of the leftmost vertical Wall (at the top) and then press ENTER (see Figure 9–7).

Notice that the Wall has actually extended from its centerline to the centerline of the boundary Wall.

Figure 9–7 *The justification line of the Wall is used with Extend*

9. Right-click in the drawing and choose **Basic Modify Tools > Trim**.

10. On the upper middle section of the drawing (labeled Trim), click the middle horizontal Wall (at the top) as the "cutting edge" and then press ENTER.

11. At the "Select object to trim" prompt, click near the top of the middle vertical Wall, and then press ENTER (see Figure 9–8).

Notice that the Wall has actually trimmed from its centerline to the right edge (the justification) of the cutting Wall.

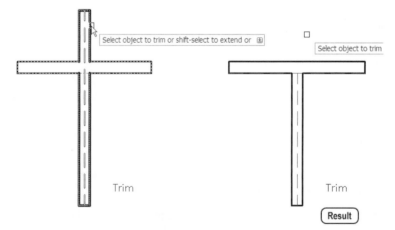

Figure 9–8 *The justification line of the Wall is used with Trim*

In the next set of Walls in the top right corner of the drawing (labeled Fillet/Chamfer); try performing a Fillet and Chamfer. Undo after each operation.

Change the Fillet Radius and the Chamfer Distances and try them again.

AutoCAD editing commands such as Trim, Extend, Fillet and Offset reference the justification line of Wall segments. Because the graph line is almost always positioned at the justification line (see below), having the Graph Display mode on while editing is very helpful. It is important to consider this fact when using these functions with Walls and other similar ADT objects, such as Curtain Walls. With practice, you begin to anticipate this behavior and incorporate it into your process as you work.

'L' AND 'T' CLEANUPS

In addition to the standard AutoCAD commands, there are two very useful tools on the Wall Cleanups right-click menu. 'L' Cleanup is similar to Fillet, and 'T' Cleanup is like Extend or Trim. Although the results are similar to these well known AutoCAD counterparts, the usage varies a bit.

12. On the bottom left side of the drawing ('L' Cleanup), select the horizontal and the leftmost vertical Walls.

13. Right-click and choose **Cleanups > Apply 'L' Cleanup**.

14. Repeat on the same horizontal Wall and the other vertical Wall that intersects it (see Figure 9–9).

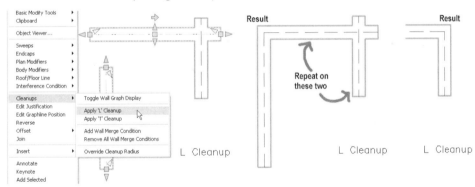

Figure 9–9 *Using the 'L' Cleanup routine*

The 'T' Cleanup routine works in a similar fashion, but requires you to select the Wall that you wish to trim or extend first, right-click it and choose **Apply 'T' Cleanup**, and then select the bounding Wall when prompted.

15. In the bottom middle section of the drawing ('T' Cleanup), select the vertical Wall.

16. Right-click and choose **Cleanups > Apply 'T' Cleanup**.

17. At the "Select boundary Wall" prompt, click the lower horizontal Wall.

18. Repeat for the upper Wall.

EXPLORE AUTOSNAP

When grip-editing Walls, there is a special snap setting that facilitates the automatic intersection of Wall baselines. This helps force good drafting and ensure proper cleanup.

19. In the bottom right section of the drawing, (labeled Grip Edit), click on the rightmost vertical Wall (at the top) to highlight it.

20. Select the top grip point (either the square or the triangular one) and begin stretching the Wall up.

21. Use the Midpoint Object Snap to snap to the bottom face of the horizontal Wall (see Figure 9–10).

    Notice that the grip point jumps to the top edge (the justification line) of the horizontal Wall.

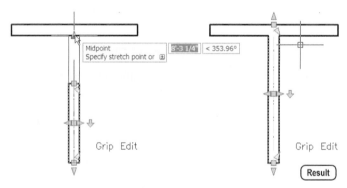

Figure 9–10 *"Autosnap Grip Edited Wall Baselines" feature while grip editing*

With the Autosnap Grip Edited Wall Baselines setting enabled, if a grip edit falls within the drawing's designated Autosnap radius setting, the grip point will automatically snap to the justification line of the Wall. This behavior can be toggled on or off. As a general rule of thumb, keep this setting turned on, with a modest Autosnap radius (the default is 6″ [75]). Turn the setting off in situations where you deliberately need to grip edit within the allowable tolerance.

22. Click the rightmost vertical Wall to highlight it again.

23. Right-click and choose **Add Selected.**

TIP Add Selected is a great way to add a Wall (or any other object) with the same parameters as an existing one.

24. Set the start point to the left of this Wall.

25. Move straight up and snap Perpendicular to the same horizontal Wall as in the last sequence. (Hold the SHIFT key and right-click to get perpendicular.)

26. Press ENTER to end the command.

Notice that the behavior is the same as with the grip editing; the Wall has automatically "snapped" to the justification line of the horizontal Wall (see Figure 9–11).

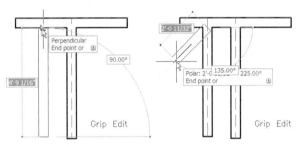

Figure 9–11 *New Walls also Autosnap*

The Autosnap New Wall Baselines feature uses the same Autosnap radius as the grip editing Autosnap feature. Both features are configured in the Options dialog box, on the AEC Object Settings tab. Both features can be turned on or off independently and the snap strength can be adjusted as well.

27. Undo both the new Wall and the grip edit.

28. From the Format menu, choose **Options** (you can also right-click in the command line and choose **Options**).

29. Click on the AEC Object Settings tab (see Figure 9–12).

The Autosnap options are in the Wall Settings area. Clear the check box next to Autosnap Grip Edited Wall Baselines to turn it off. Clear the check box next to Autosnap New Wall Baselines to turn it off. Place a check mark in either box to turn them on.

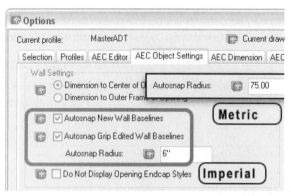

Figure 9–12 *Autosnap features on the AEC Object Settings tab of Options*

When either of these Autosnap features is active, the Autosnap Radius field controls the tolerance level. Enter a larger value if you want the effect to be stronger; enter a smaller value if you want the effect to be weaker.

30. Clear both boxes, and click OK to accept your changes.

31. Repeat the steps above for grip editing and creating a new Wall.

    Notice the behavior in contrast to the first time. This time, the Walls did not automatically snap to the neighboring Wall's justification line.

32. Return to Options, turn both settings back on, and then click OK.

In the Options dialog box, the Autosnap features display Wall Baseline when in fact the Autosnap is actually to the graph line position, which is almost always the same as the justification line. (See the definitions above for more clarification). Keep this "typo" in mind as you use the feature. Also note that both of these settings in Options are saved in the current drawing as indicated by the drawing icon (refer to Chapter 3 for more information).

Graph Line Position

As mentioned in the last example, the graph line position ultimately controls the behavior of the Autosnap feature. The graph line can be set to coincide with the Wall justification line, or forced to remain in the Center of the Wall. If you follow the zero Cleanup radius recommendation in Rule 2 above, you should leave the graph line set to the Wall justification line. Moving the graph line to the Center will cause certain Wall Cleanups to fail even though good drafting techniques were followed. This is because most editing commands like Trim and Extend use the justification line, but Autosnap and Wall Cleanup use the graph line. If the two are set to different points within the Wall's width, you will be forced to increase the Cleanup radius to compensate. It is highly recommended that you leave the graph line set to the Wall justification line, which is its default setting. There is little good reason to change it.

If you do wish to change it, the setting can be changed as the default for the entire drawing on the Properties palette (in the Advanced – Cleanups grouping).

RULE 4 – TWEAK THE CLEANUP RADIUS

Sometimes, despite graph line position and good drafting technique, cleanup might still prove troublesome. In many cases, tweaking the Cleanup radius can solve the problem. It is usually easier to solve problems by increasing the radius than by decreasing; therefore start with low initial values, and slowly increase until the optimal setting is reached. When it becomes necessary to increase the initial value, Cleanup radii can be set in three ways:

▶ A default can be set for all Walls in the entire drawing. (However, as stated above, this should be avoided and the default should remain 0″ [0].)

▶ A default can be set in the Properties for each Wall tool (which also should be zero as a rule).

▶ A radius can be applied individually to each Wall segment, and even each end of each Wall separately as needed. In general, if a radius is required, this is the recommended approach.

A common example of a troublesome situation occurs when two Walls of varying thickness touch end to end. Consider the condition on the left of Figure 9–13. The problem is that no matter which of the above measures we employ, the graph lines are parallel. Therefore, they will never intersect, thwarting our efforts to achieve cleanup. This is a perfect example of when it is appropriate to increase the Cleanup radius.

As you can see in middle and right of Figure 9–13, using Extend does not help. In fact, in some cases (image on the right) it can actually make matters worse. The only solution in this scenario is to adjust the Cleanup radius.

Troublesome Intersection   Extend does not help...   ...and can even make it worse

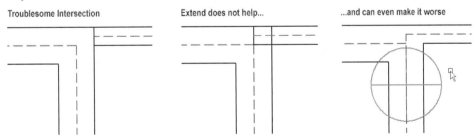

Figure 9–13 *Using Extend on this troublesome intersection does not help*

EXPLORE COMMON CLEANUP RADIUS SITUATIONS

1. In the *C:\MasterADT 2006\Chapter09* folder, open the file named *Wall Cleanup3.dwg* [*Wall Cleanup3-Metric.dwg*].

There are four cleanup issues illustrated in this file. The numbers 1 through 4 below each situation are referred to in the following steps. Each situation requires an adjustment of the Cleanup radius. Changing the value of the Cleanup radius for a particular Wall segment can be accomplished in a couple of ways.

Examine Situation 1

2. Select the thinner Wall at the right, right-click and choose **Properties** (see Figure 9–14).

3. Scroll down to the Advanced category, and locate the Cleanups grouping.

When necessary to increase the Cleanup radius beyond 0, use the following guidelines to determine how large a radius to use:

▶ If the justification is **Center** or **Baseline**, choose a value between 1/2 and 3/4 the width of the largest segment in the intersection.

▶ If the justification is **Left** or **Right**, choose a value between 1/2 and 1 times the width of the largest segment in the intersection.

In this case, the Walls in Situation 1 are 12″ [300] and 6″ [150] in thickness. The 12″ [300] Wall therefore is larger; the value ought to be between 6″ [150] (1/2 × 12″ = 6″ [1/2 × 300 = 150]) and 9″ [225] (3/4 × 12″ = 9″ [3/4 × 300 = 225]). However, this is not an exact science. The goal is simply to get the cleanup circle of the Wall to touch the Graphline of the neighboring Wall.

4. In the Cleanup radius field type a value in between the range like **8″ [200]** and then press ENTER (see Figure 9–14).

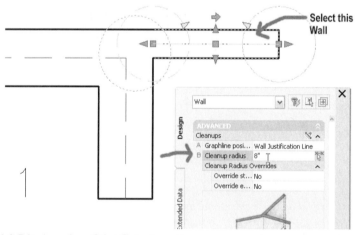

Figure 9–14 *Edit the value of the Cleanup radius*

If you want to see the results of the intersection without the graph lines, toggle the Graph Display off temporarily. (Select the Wall, right-click and choose **Cleanups > Toggle Wall Graph Display**.) Be sure to toggle it back on to complete the rest of the exercise.

Examine Situation 2

Situation 2 is nearly identical to the first. There are just two Walls of varying thickness, but again, the graph lines are parallel and will never touch. This situation will give us the opportunity to look at an alternative (and easier) way to set Cleanup radius for a particular Wall.

5. Be sure that Graph lines are displayed; if they are not, select the Wall, right-click and choose **Cleanups > Toggle Wall Graph Display**.

6. In situation 2, select the thinner Wall at the right.

Notice that there are two triangular shaped grips (one at each end of the Wall) that are at a slight angle to the Wall's centerline. These are the Cleanup radius grips (see Figure 9–15).

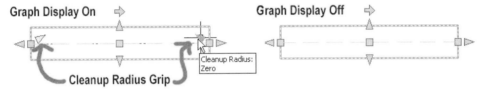

Figure 9–15 *Cleanup radius grips show only when the graph lines are displayed*

7. Click the Cleanup radius grip (on the left end of the Wall) and drag it outward.

Notice the Radius enlarges interactively. There is also a dynamic dimension into which a value may be typed if you wish, or you can simply click the mouse to set the size of the radius just large enough to touch the neighboring graph line. In most cases, this is much easier than performing the calculations detailed above.

8. Drag the grip until the Cleanup Circle just crosses the graph line of the thick Wall, and then click to set the radius (see Figure 9–16).

If you prefer, using the formula in the bullet points above, you could type **6″ [150]** and then press ENTER.

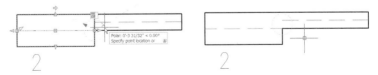

Figure 9–16 *Assigning the Cleanup radius using grips*

9. Deselect the Wall segment.

NOTE The first technique covered here works on selections of multiple Walls. The grip edit method must be done on one Wall at a time. However, in most cases, the grip edit technique is much quicker.

Examine Situation 3

Situation 3 is a bit trickier. First, experiment with the Cleanup radius. A value of 6″ [150] is at the low end of the recommended range; however, it might not be quite enough. Or it might work today, but fail the next time the drawing is opened. One important thing to realize about wall cleanup is that it is part of the display system and is calculated parametrically each time the drawing opens or is regenerated. If the value set for the radius is "border line," it might work one time and fail another for no apparent reason. One way to help prevent this is to always select a value in the middle of the range. As you experiment with different values, you can get a better sense for this situation. It takes some practice to get the hang of it.

As you can see, adjusting the Cleanup radius can solve this problem. However, there is a better solution. Always remember there are several interconnected "rules" being outlined here. Also remember that the rules are meant to be followed in order. The correct solution to this problem is Rule 3 – Practice Good, Clean Drafting. Jumping too quickly to Rule 4 – Tweak the Cleanup radius, can lead to problematic drawings. Use Trim in this case to eliminate the small piece of Wall that extends past the graph line. This will solve the cleanup and satisfy the first two rules; the Cleanup radius can remain 0″ [0] and the geometry is cleanly closed. No need to move to Rule 4 at all

Examine Situation 4

Situation 4 is also a tricky situation and represents what happens when the Cleanup radii are too large. The "one-half to three-quarters" rule should not be used if the radius ends up being larger than the length of the Wall segment itself. If this happens, the start of the Wall attempts to cleanup with the end of the Wall and defects occur. Change all of the radii back to 0″ [0] and these Walls should cleanup up properly. This will certainly be true in this case. However, if you encounter a similar situation in your own files and they don't cleanup after taking these measures, set the value as close as possible to the allowable Cleanup radius range without exceeding the length of any of the Wall segments involved. If none of these techniques works, refer to the "Part 2 – Manual Wall Cleanup (Using Wall Merge)" section below.

OTHER AUTOMATIC CLEANUP TIPS

In addition to the rules listed in the steps above, keep the following guidelines in mind:

Justification

▶ Wall justification will have an impact on cleanup. Typically you will want to use Center justification for most interior partitions. Baseline is often used for structural bearing Walls and exterior Walls. Left and Right can in some cases solve problems that required radii with Center or Baseline, such as situations I and 2 above.

▶ Left and Right justification will work well for exterior Walls or interior conditions where alignment with some existing feature (existing construction or building setback line, for instance) is needed.

▶ Use Baseline justification for multi-component exterior Walls where drafting based on the line of demarcation between structural and non-structural components is desirable.

In the *C:\MasterADT 2006\Chapter09* folder, open the file named *Wall Cleanup4.dwg* [*Wall Cleanup4-Metric.dwg*]. There you will find several combinations of Wall justifications and graph line positions (see Figure 9–17). Experiment with additional variations in this file. Use a combination of the techniques covered above in your explorations.

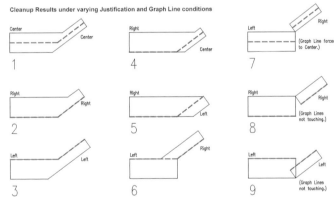

Figure 9–17 *Justification can have an impact on wall cleanup*

Cleanup Groups

▶ Cleanup Groups can also be used to control cleanup. A Cleanup Group limits cleanup to Walls sharing the same group. Walls will not cleanup outside their group. This is often used to prevent one type of Wall, such as new construction, from cleaning up with another type, such as demolition. The next chapter covers some examples of this in more detail. Another common use is to isolate Wall Styles used to create countertops from the Walls to which they are attached.

▶ In order for Cleanup to occur between an XREF and its Host, the "Allow Wall Cleanup between host and XREF drawings" setting in the Cleanup

Group must be enabled. Wall Cleanup Group Definitions are accessed from the Style Manager, as described below.

XREF Cleanup is controlled with Wall Cleanup Groups. To configure a Cleanup Group for XREF cleanup, choose **Style Manager** from the Format menu. Expand the *Architectural Objects* folder, and then the Wall Cleanup Groups category. Create a new or edit an existing Wall Cleanup Group. Edit the new Group and on the Design Rules tab of the Wall Cleanup Group Definition Properties dialog box, place a check mark in the "Allow Wall Cleanup between host and XREF drawings" box (see Figure 9–18). Be sure to do this in at least the XREF file. You can do it in both drawings if you wish, but it is not required in the host drawing.

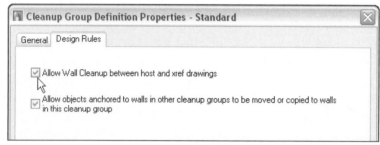

Figure 9–18 *Turn on XREF Cleanup in the Wall Cleanup Group Definition Properties dialog box*

CAD Manager Note: By default this setting is disabled in the out-of-the-box drawing template files (DWT). It is recommended that you consider changing this setting and enabling XREF Cleanup in your office template files. It is more likely that this make the best default setting for typical project files. Users can always turn it off for special cases.

Z Elevation

▶ Wall cleanup will *not* occur between Walls that are not in the same X,Y plane. If Walls are at different Z heights or rotated relative to the ground plane, they will not cleanup. *Never* move Walls in the Z direction. Use the Roofline/Floorline options instead to modify the top and bottom edges of Walls relative to Z=0.

Component Priorities

▶ Complex Wall styles use Component Priorities to determine how their individual subcomponents cleanup; refer to the "Wall Component Priorities" topic below.

Wall Merge

▶ If all attempts at automatic cleanup fail, use manual cleanup, Wall Merge; refer to the "Part 2 – Manual Wall Cleanup (Using Wall Merge)" topic next.

PART 2 – MANUAL WALL CLEANUP (USING WALL MERGE)

So, you followed all of the rules and guidelines listed here in an attempt to get some stubborn Walls to cleanup to no avail. Frustration is the natural reaction, and in some cases people are tempted to give up. Fortunately, if the automatic cleanup tools and procedures fail to generate the desired cleanup condition, we have the ability to manually force the cleanup we need. The command to do this is called "Wall Merge." In this lesson, you will learn how to perform manual wall cleanup in situations where automatic cleanup fails to achieve the desired effect. Please bear in mind that we should try all of the techniques covered in the last lesson before resorting to Wall Merge. This is because turning on manual cleanup requires that automatic cleanup be turned off.

TRY TO TRIGGER AUTOMATIC CLEANUP

1. In the *C:\MasterADT 2006\Chapter09* folder, open the file named *Wall Cleanup5.dwg* [*Wall Cleanup5-Metric.dwg*].

This file contains Wall situations that will not cleanup properly using the automatic cleanup techniques covered earlier. The various recommendations are at odds with one another due to the complexity of the intersection (see Figure 9–19). On the left is the file as it appears when you open it, and on the right is the desired result. Remembering that the rules outlined in the last lesson are meant to be applied in order, let's see if we can achieve the required results.

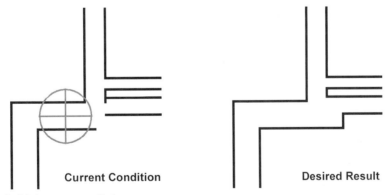

Current Condition   **Desired Result**

Figure 9–19 *Improper wall cleanup*

2. Beginning with Rule 1: Select any Wall on screen, right-click and choose **Cleanups > Toggle Wall Graph Display** (see Figure 9–20).

You can see that the bottom horizontal Wall on the right already has a Cleanup radius greater than zero applied, to no avail in this circumstance.

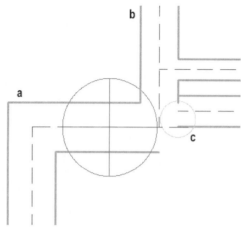

Figure 9–20 *Toggle Graph Display on*

Refer to the labels in Figure 9–20 in the following steps.

3. Following Rule 2, change the Cleanup radius of the lower horizontal Wall (Wall "c") to **0″ [0]**.

 TIP To set the radius to zero using the grips, begin dragging the Cleanup radius grip and then type 0 and press **ENTER.**

Notice that it has eliminated the defect warning, but it hasn't really helped (see left panel of Figure 9–21).

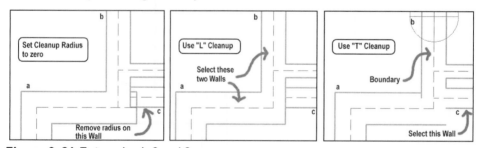

Figure 9–21 *Trying rules 1, 2 and 3*

4. Use 'L' Cleanup on the Wall "a" and Wall "b" (see middle panel of Figure 9–21).

We were making some progress overall, but it is still not right.

5. Use 'T' Cleanup on the Wall "c," and use Wall "b" as the Boundary (see right panel of Figure 9–21).

> We have now made matters worse. The defect warning is back. (It may be time to rethink the rules.)

Perhaps we need only to adjust the Cleanup radius. This is after all one of the rules, and we have exhausted the possibilities with Rules 1 through 3. Let's give it a try.

6. Using the grip, enlarge the Cleanup radius of Wall "a."

7. Undo, and try the grip of Wall "c" instead (see left panel of Figure 9–22).

This eliminates the defect warning, but it is still not quite right. Perhaps if we just extend Wall "a" slightly to the right it might do the trick. If we were looking for a clean corner between Walls "a" and "b" (as we have in the middle of Figure 9–21), then we could simple drag the Lengthen grip (the triangular shaped one at the end) of Wall "a" and we would be finished (see the middle of Figure 9–22). However, if you refer back to Figure 9–19, we are actually looking for a slightly different condition.

8. Click the triangular Lengthen grip at the right end of Wall "a," and snap it to a point on the right face of Wall "b" (see middle panel of Figure 9–22).

 NOTE At this point you may no longer have cleanup errors, but the results are still not exactly what we are trying to achieve in this condition.

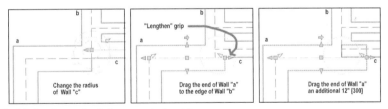

Figure 9–22 *Appling rule 4 does not help either*

9. Drag the Lengthen grip again to the right **12″ [300]** (see right panel of Figure 9–22).

Don't forget to run through the additional considerations as well: Are all the Walls in the same X,Y plane (in other words, Z=0)? Do they all belong to the same Cleanup Group Definition? Would modifying the justification help at all? If you experiment with these items, be sure to undo your changes before continuing.

APPLY A WALL MERGE

It appears that we have exhausted our options in this situation. We will need to apply a Wall Merge to this intersection to force cleanup manually. With Wall Merge, we can "show" ADT the exact cleanup behavior we require. The disadvantage is that we must maintain this intersection manually from this point on. For this reason, use Wall Merge sparingly and only after you have exhausted the other possibilities, as we have here. Remember the rules

10. Select the Wall "a," right-click, and choose **Cleanups > Add Wall Merge Condition**.

11. At the "Select Walls to merge with" command line prompt, select the Wall "c" (see the left side of Figure 9–23).

12. The Wall will be highlighted as it is selected, the command prompt will acknowledge with "I found," and the "Select Walls to merge with" prompt will be repeated.

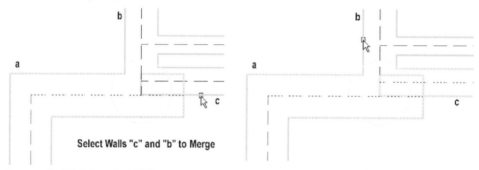

Select Walls "c" and "b" to Merge

Figure 9–23 *Select the Walls to merge with*

13. Select the Wall "b," and then press ENTER (see the right side of Figure 9–23).

The Wall Merge is complete, and the intersection should now display properly. Compare the final result with Figure 9–24. As you can see, the Wall Merge provides a critical bit of functionality without which we would face considerable frustration in situations like the one above. With complete understanding of both automatic and manual cleanup at your disposal, there is virtually no cleanup situation you will be unprepared to handle.

If you select a Wall that has a Merge Condition while the Graph Display is active, curved "ghost" lines will appear to indicate where the Merge Conditions occur. Each of these ghost lines also has a small "minus" (-) grip. You can use this grip to remove the Wall Merge Condition if required.

14. Select Wall "a" to reveal the Merge Condition grips.

15. Click either one of the Remove Wall Merge Condition grips (see Figure 9–24).

16. Undo the last command (to restore the Wall Merge).

You may want to toggle the Graph Display off to see the final result.

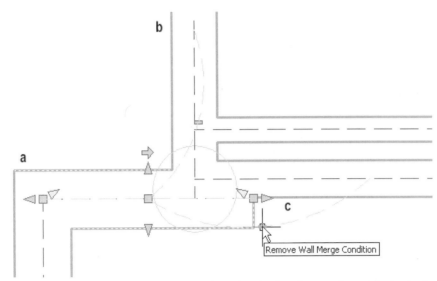

Figure 9–24 *Wall Merge Conditions are indicated with ghost lines and special grips in the Graph Display*

17. Select any Wall on screen, right-click and choose **Cleanups > Toggle Wall Graph Display** (see Figure 9–25).

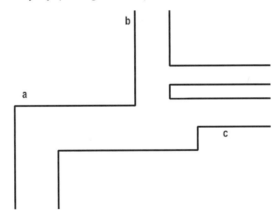

Figure 9–25 *The final result of the Wall Merge condition*

PART 3 – WALL COMPONENT PRIORITIES

Up to this point in this book, we have worked mostly with the Standard Wall style. The Standard Wall style is a simple two-line Wall (that contains a single component), centered relative to its width. Walls however, can include up to 20 subcomponents in their style. Although it is rare that a style will have that many, it is common for Walls to have more than one component. These "multi-component" Walls have an additional factor to consider when calculating cleanup. The

Walls must determine how each component within the Wall will cleanup with each of the components of the neighboring Wall. The parameter used to control this behavior is a Wall component's "priority." In this lesson you will learn how cleanup priorities assigned to Wall components interact and influence overall wall cleanup behavior.

EXPLORE WALL COMPONENT PRIORITIES

1. In the *C:\MasterADT 2006\Chapter09* folder, open the file named *Wall Cleanup6.dwg* [*Wall Cleanup6-Metric.dwg*].

Study the intersection between the two Walls. Each Wall is of a different Wall style. The horizontal Wall at the top uses the Style *CMU-12 Air-2 Brick-4 Furring* [*CMU-300 Air-050 Brick-090 Furring*], while the Wall at the bottom uses the Style *CMU-8 Furring Both Sides* [*CMU-190 Furring Both Sides*]. Notice that all the internal component materials are cleaning up properly with the corresponding materials in the other Wall. This is due to the component priorities assigned to each component within the respective Wall styles.

 NOTE A complete discussion on editing and building Wall styles follows in the next chapter.

2. Select the horizontal Wall (brick veneer), right-click, and choose **Edit Wall Style**.

3. Click the Components tab (see Figure 9–26).

Study the list of components. Notice that there are five components in this Wall style, and that each component has a number in the Priority column. The numbers are as follows: Brick = 810, Air Gap = 700, CMU = 300, Stud = 500 and GWB = 1200.

Index	Name	Priority	Width	Edge Offset	Bottom Offset
1	Brick	810	4"	2"	0"
2	Air Gap	700	2"	0"	0"
3	CMU	300	1'-0"	-1'-0"	0"
4	Stud	500	7/8"	-1'-0 7/8"	0"
5	GWB	1200	5/8"	-1'-1 1/2"	0"

Figure 9–26 *Component Priorities of the CMU-12 Air-2 Brick-4 Furring [CMU-300 Air-050 Brick-090 Furring] style*

4. Click Cancel to dismiss the dialog box.

5. Select the vertical Wall (Gyp Bd. Both sides), right-click, and choose **Edit Wall Style**.

6. Click the Components tab (see Figure 9–27).

Notice that this style also has five components, many of them the same as the previous style. Note that the Priority numbers for the like components match. This is critical to successful cleanup between the two styles.

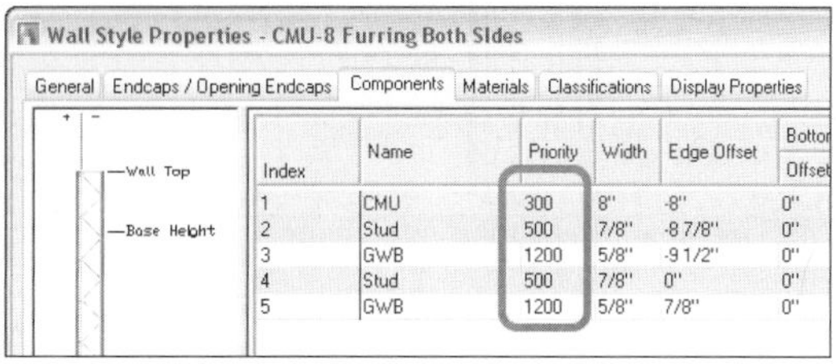

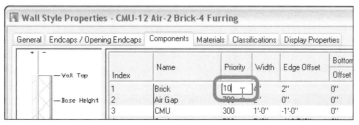

Figure 9–27 *Component Priorities of the CMU-8 Furring Both Sides [CMU-190 Furring Both Sides] style*

CHANGE COMPONENT PRIORITIES

Let's perform the following experiment:

7. Select the horizontal Wall (brick veneer), right-click, and choose **Edit Wall Style**.

8. Click the Components tab.

To see the full effect of priorities, let's mess these Walls up a bit.

9. Select Index 1 (Brick) and change the Priority to **10** (see Figure 9–28) and click OK.

Figure 9–28 *Change the Brick component Priority to 10*

No obvious changes yet—perform the next step.

10. Select the vertical Wall (Gyp Bd. Both sides), right-click, and choose **Edit Wall Style**.

11. Click the Components tab.

12. Select Index 1 (CMU) and change the Priority to **10** and click OK.

Notice how the CMU of the vertical Wall now passes through the CMU of the horizontal and attempts to cleanup with the Brick beyond (see Figure 9–29).

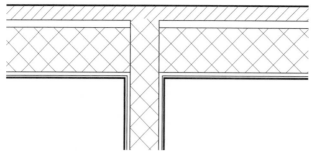

Figure 9–29 *Brick and CMU are now attempting to cleanup*

Let's keep going.

13. Select the vertical Wall (Gyp Bd. Both sides), right-click, and choose **Edit Wall Style**.

14. Click the Components tab.

15. Select Index 2 (Stud) and change the Priority to **20**.

16. Select Index 4 (Stud) and change the Priority to **20** and click OK.

17. Select the horizontal Wall (brick veneer), right-click, and choose **Edit Wall Style**.

18. On the Components tab, select Index 2 (Air Gap), change the Priority to **20**, and then click OK (see Figure 9–30).

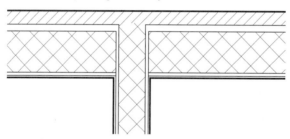

Figure 9–30 *Interesting, but it is not exactly a construction-worthy detail*

Experiment

Do a few more experiments; see how badly you can mess it up. Then try to fix it.

UNDERSTANDING WALL COMPONENT PRIORITIES

Having completed the previous tutorial, you can probably begin to draw some conclusions about the behavior of component priorities. Priorities determine which materials will cleanup with one another and which will be interrupted. Priorities function like this:

▶ Components with the same priority number *will* cleanup.

> ❱ Lower-numbered components will interrupt (pass through) higher-numbered components.

> ❱ Wall priorities will take precedence over drawing order. (The order in which Walls were drawn has no bearing.)

Establishing a consistent and logical material and priority list is critical to successful cleanup and having a useable library of Wall styles. All sample Wall styles provided with ADT use the same list of priorities. The default list is based on the order of construction. Materials installed first, like poured concrete, get low numbers, while items installed last like stucco and toilet partitions get high numbers. Numbers in the chart deliberately do not start at one (1). The reason for this is that number 1 is the highest Priority. To allow for flexibility and future expansion, numbers start in the hundreds. For all practical purposes, there is no lower limit. (In reality, it is 2147483647, but it is unlikely that you will ever set actual priorities that low.) The following chart is a complete list of all the priorities used in the sample content provided with ADT. You are not required to use this list, but it is *highly* recommended that you do. Check with your CAD Manager for the actual list used at your firm. If you are your own CAD Manager, use this list! It works, it is done already and some 200 or so Wall Styles that ship with the product are configured with these values.

 NOTE The first two items, "Existing to Remain" and "Demolition," are not part of the default ADT offerings. They are included in the chart here as recommended additions to the default list, and they have been incorporated into files used in the tutorials of this book.

CAD Manager Note: It is recommended that you adopt the chart in Table 9–1 as your office standard. Consistency in assigning wall cleanup priorities is perhaps as important as an office standard Layer scheme. Having consistency will make it easy to add Wall styles created on specific projects to the office library without having to rework them first. A consistent list will also save hours wasted in needless cleanup troubleshooting related to incompatible priority lists. Note that plenty of space has been left between each material. (They jump in increments of 100.) It is recommended that you follow this guideline and maintain an adequate increment between numbers if you develop your own list. This will give you flexibility and room for future expansion in between materials without needing to redefine your entire collection of office standard styles each time the priority list changes. Following this method, a new material component can be slipped in wherever it fits best.

Table 9-1 *Wall Component Cleanup Priorities*

Component Name	Priority Number
Existing to Remain (Mastering ADT Recommendation)	50
Demolition (Mastering ADT Recommendation)	100
Concrete	200
Concrete (Footing)	200
CMU	300
Air Gap (CMU/CMU)	305
CMU Veneer	350
Precast Panel	400
Rigid Insulation (Brick)	404
Stud	500
Air Gap (Stud/Stud)	505
Insulation (CMU/Brick, Stud/Brick)	600
Air Gap	700
Brick	800
Air Gap (Brick/Brick)	805
Brick Veneer	810
Siding	900
Metal Panel	1000
Stucco	1100
Glass	1200
GWB	1200
GWB (X) - First Layer	1200
GWB (X) - Second Layer	1210
GWB (X) - Third Layer	1220
GWB (X) - Fourth Layer	1230
Bulkhead	1800
Casework - Upper	2000
Casework - Base	2010
Casework - Counter	2020
Casework - Backsplash	2030
Toilet Partition	3000

ADDITIONAL EXERCISES

Additional cleanup exercises have been provided in Appendix A. In Appendix A you will find the Third Floor Partitions (*03 Partitions*) file of the Commercial Project. Load the project (using Project Browser on the File menu) and open the *03 Partitions* Construct. Follow the guidelines outlined in this chapter as you move around the plan. Fix all of the cleanup problems you encounter. It is not necessary that you complete this exercise to begin the next chapter, it is provided to enhance your learning experience. However, it is highly recommended that you do complete this exercise for practice with techniques covered in this chapter before continuing to the next chapter. Completed projects for each of the exercises have been provided in the *Chapter09/Complete* folder. The solution to the wall cleanup exercise is provided in Appendix E.

SUMMARY

At the beginning of this chapter, the importance of mastering wall cleanup was stressed. As you can see, although critically important, mastering wall cleanup can be achieved easily by following the systematic approach outlined here. Like all things in ADT, wall cleanup is simply the interaction between a series of user-defined parameters. Learn to master those parameters, and you will have mastered the tool. A summarization of all of the points covered in this chapter is included in Appendix B.

Intersecting Wall graph lines or Wall graph lines intersecting Wall Cleanup radii cause Walls to automatically cleanup.

Rule-1: Toggle the Graph Display of Walls on while working with cleanup.

Rule-2: Start with the Cleanup radius 0.

Rule-3: Practice good clean drafting using normal AutoCAD editing commands such as Offset, Trim, Extend, and Object Snaps.

Rule-4: When necessary, tweak the Cleanup radius to between 1/2x and 1x the Wall width.

Justification, Cleanup Groups, and the Z Elevation of Walls all affect cleanup success or failure.

Use Wall Merge when automatic wall cleanup fails to produce desired results.

Matching component priority numbers cleanup; lower priority numbers interrupt components with higher priority numbers.

It is recommended that you adopt the out-of-the-box list for your firm.

CHAPTER 10

Progressive Refinement – Part I

INTRODUCTION

As a design scheme progresses, it is often desirable to begin including more detail in the articulation of Walls, Doors, and Windows. We may want to begin indicating construction materials of Walls with poché and other articulation. We have stressed throughout the book that ADT is designed to allow us to add objects with little detail early in the design cycle and progressively refine our design as more information becomes known. The key to being able to achieve this is working with the various Object styles. In this chapter – part I in our exploration of progressive refinement, we will focus on Wall styles, Wall Endcaps and Opening Endcaps.

OBJECTIVES

In this chapter, we will refine the first and second floor plans of the residential project. First, we will assign styles to all of the Walls. Then we will look at the steps involved in editing existing and creating a new Wall style. In addition, we will explore the following topics:

- Understand Wall styles.
- Import Wall styles.
- Edit Wall styles.
- Understand Wall and Opening Endcaps.
- Create custom tools.
- Build a Custom Wall style.

WALL STYLES

A Wall style is analogous to a Wall type in the Construction Document set. One Wall style might represent a stud Wall, while another represents a 2-hour fire rated masonry Wall. As you develop your floor plan, think in terms of having a Wall style for every Wall type you will have in your Wall type legend. Wall styles control the component makeup of a Wall type, and they control the display characteristics of the Walls as well. Wall styles control virtually every aspect of Wall configuration. In this lesson, you will learn how to import Wall styles from other drawing files and use them in an existing floor plan, and we will build some custom Wall styles from scratch.

INSTALL THE CD FILES AND LOAD THE CURRENT PROJECT

If you have already installed all of the files from the CD, simply skip down to step 3 below to make the project active. If you need to install the CD files, start at step 1.

1. If you have not already done so, install the dataset files located on the Mastering Autodesk Architectural Desktop 2006 CD ROM.

   Refer to "Files Included on the CD ROM" in the Preface for information on installing the sample files included on the CD.

2. Launch Autodesk Architectural Desktop 2006 from the desktop icon created in Chapter 3.

If you did not create a custom icon, you might want to review "Create a New Profile" and "Create a Custom ADT Desktop Shortcut" in Chapter 3. Creating the custom desktop icon is not essential; however, it makes loading the custom profile easier.

3. From the File menu, choose **Project Browser**.

4. Click to open the folder list and choose your *C:* drive.

5. Double-click on the *MasterADT 2006* folder, then the *Chapter10* folder.

   One or two residential Projects will be listed: *10 Residential* and/or *10 Residential Metric*.

6. Double-click *10 Residential* if you wish to work in Imperial units. Double-click *10 Residential Metric* if you wish to work in Metric units. (You can also right-click on it and choose **Set Current Project**.) Then click Close in the Project Browser.

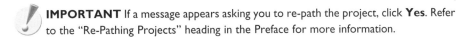 **IMPORTANT** If a message appears asking you to re-path the project, click **Yes**. Refer to the "Re-Pathing Projects" heading in the Preface for more information.

WALL STYLE BASICS

Architectural Desktop ships with a vast library of pre-made styles. Before we embark on the process of creating our own styles, let's explore some of the out-of-the-box offerings. There are two ways to access this content. We could use the Content Browser to access the style that we need, draw a Wall with it, and then copy and Edit the Style. The other method is to use the Style Manager to access a remote Style library drawing, and then either copy or edit it. Both methods are valid and have already been covered in previous chapters. As a general rule of thumb, we have been using Content Browser to access and *use* styles, reserving Style Manager for creation and editing. We will continue that practice in this chapter and work mostly with Style Manager for the following tutorials.

IMPORT STYLES FROM A CONTENT FILE

The default templates shipped with ADT 2006 contain very few embedded styles. Therefore, we must typically open a remote content file to import styles for use in the current drawing.

1. On the Project Navigator palette, click the Constructs tab.

2. Double-click the *First Floor New* file in the *Constructs* folder to open it.

The plan has been refined a bit since we last opened it. The new construction has been laid out with the Standard Wall style. Doors and windows have been added and a screen porch on the back of the house (top of the plan) has also been added. (The screen porch is built using Curtain Walls and was the subject of an additional exercise for Chapter 8. Details on its construction can be found in Appendix A.) Throughout this chapter, we will work with this file and refine it. By now, you should also be getting comfortable with the Project Navigator and the process of working in ADT projects. When the *First Floor New* file opens, the Existing Conditions file is not showing. However, as you work on new construction, it will be helpful to have the existing conditions visible as you work. To do this is an easy drag-and-drop process from Project Navigator.

3. On the Project Navigator, drag the *First Floor Existing* file and drop it into the drawing area work space.

   You will note that it appears in the correct location at the bottom of the plan.

It is important to note that when you drag a Construct into another Construct (as we have here); the XREF is overlaid *not* attached. This is very important. If the file were attached, you would receive "circular reference" errors later when you tried to work with your project's Views and Sheets. If you wish to verify that the *First Floor Existing* file is in fact overlaid, click the Manage XREF quick pick in the Drawing Status Bar to open the XREF Manager and have a look.

4. Right-click the title bar of the tool palettes and choose **Design** to load the Design tool palette group.

5. On the Design tool palette, right-click the Wall tool and choose **Wall Styles** (see Figure 10–1).

Figure 10–1 *Access Wall Styles*

6. On the row of icons appearing across the top of the Style Manager, click the Open Drawing icon (see Figure 10–2).

Figure 10–2 *Click the Open Drawing icon*

7. In the Open Drawing dialog box, click the Content icon in the Outlook bar at the left, and then double-click the *Styles* folder (see Figure 10–3).

Style content is saved in ADT drawing files. These can be referred to as library files. The default ADT Style Content folder is located at *C:\Documents and Settings\All Users\Application Data\Autodesk\ADT 2006\enu\Styles*. If you work with a non-English version of ADT, this path might be slightly different. Also, your firm may have located this folder elsewhere on your company's network. Check with your CAD manager in this case for the correct location. The Content icon on the Outlook bar is a shortcut to this location, which makes it much easier to navigate there. In the default content folder, there are several Wall Style (and other style type) library files provided. Each file contains styles of a particular construction type. Be sure to explore each of these files at some point to get a good sense of what has been provided with the software.

Figure 10–3 *Use the Content shortcut on the Outlook bar to access the Content folder on your system*

CAD Manager Note: Please note that the *Application Data* folder is hidden by default. You can turn on all hidden files in Windows XP using **Folder Options** from the Tools menu in Windows Explorer. Some CAD Managers may be uncomfortable at the prospect of users' turning on hidden files on their systems. This is required if they want to view the files in Windows Explorer or browse to this folder manually; however, it is *not* required to access them from ADT when using the Content shortcut provided on the Outlook bar. Another way to avoid the issue is to relocate the Styles folder to a Server location. If you do this, you will need to edit the tools in the Content Browser to point to the new location as well. Options to do this are offered at the time of installation. If you installed ADT with the defaults, you will need to uninstall it, and then reinstall and use the options in the installation wizard to locate Content on the server.

8. If you are working in Imperial units, double-click the *Imperial* folder, select the file named *Wall Styles – Stud (Imperial).dwg* and then click Open. If you are working in Metric units, double-click the *Metric* folder, select the file named *Wall Styles – Stud (Metric).dwg* and then click Open.

This will return you to the Style Manager, and this content file will now be listed among any other files you have open in ADT. Your tree view in Style Manager should look something like Figure 10–4.

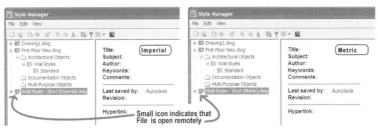

Figure 10–4 *The library file appears in the tree view of Style Manager with a small icon indicating that it is open remotely*

 NOTE The special icon (similar to a Windows Shortcut icon) indicates that the file is open remotely in the Style Manager, and not directly in ADT.

9. Expand the *Architectural Objects* folder and then the list of Wall styles beneath the remote file *(Wall Styles – Stud (Imperial).dwg* [*Wall Styles – Stud (Metric).dwg*]) entry.

10. On the Style Manager toolbar, click the Inline Edit Toggle icon (farthest to the right).

11. Click on each entry, one at a time, in the list in the tree.

As you click on a Style name in the tree, a preview will appear to the right. The preview window is interactive like the others in ADT. Feel free to use Zoom, Orbit or to change shading.

12. Locate the style named **Stud-5.5 Air-1 Brick-4** [**Stud-140 Air-025 Brick-090**].

13. Right-click it and choose **Copy** (see Figure 10–5).

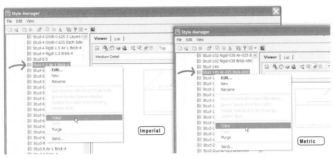

Figure 10–5 *Copy a style from the remote file*

14. Scroll the list back to the top, right-click on *First Floor New.dwg*, and choose **Paste**.

You have now imported the style into the current drawing file. When you copy a style in Style Manager, you can right-click on the file name, the *Architectural Objects* folder or the *Wall Styles* category to paste. Regardless of the specific node you select, ADT will paste it to the proper category. When you copy, if you right-click the file name, the folder or category, you will be copying all styles from that drawing, folder or category. If you want just one, be sure to right-click directly on the style you want.

 TIP Multiple styles can be copied at the same time by selecting with the SHIFT and CTRL keys on the right side of the Style Manager. Select the heading *Wall Styles* at left to see the list appear at right. Multiples cannot be selected at the left.

15. Click OK to close the Style Manager.

If prompted to save the content file, answer no.

CAD Manager Note: When Content files are accessed remotely as we have done here, ADT will often prompt for you to save the remote file even if all you did was copy something from it. In general, you will not want users to have the ability to save the library content files, particularly if you moved them to the server. To prevent this, simply make the *Styles* folder read only in Windows for typical users. If the folder is read only, users will still be able to remotely open files and copy styles as we have done here, but they will not be able to edit the styles directly in those remote files. The message to save the file will also cease to appear, since the file can no longer be saved.

SWAP STYLES

16. Click to select the three exterior Walls of the addition (right, left, and top, not the screen porch).

17. If the Properties palette is not open on screen, right-click and choose **Properties**.

18. From the list of styles, choose *Stud-5.5 Air-1 Brick-4* [*Stud-140 Air-025 Brick-090*].

    Notice the change to the drawing, and notice that the brick is on the wrong side (see Figure 10–6).

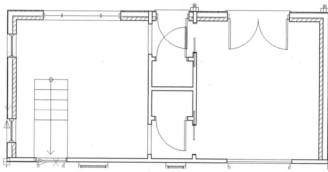

Figure 10–6 *The new style applied to the Walls*

19. With the same three Walls selected, right-click, and choose **Reverse > Baseline**.

 NOTE You could also use the Reverse Direction (arrow shaped) grip to reverse each Wall directly one at a time.

That solved the problem of the brick, but now the justification line (in this case, Right) of the Walls is on the inside of the plan and the Walls are sticking out. (Click any Wall to see this; the justification line is the edge with the grips.) Right justification was chosen when these Walls were added so that they would remain flush with the existing house. Now that we have reversed the Walls, we need to switch the justification to Left to maintain the flush relationship and shift the Walls back where they belong. We can do this for all three Walls at once on the Properties palette. We can also change the justification of the Walls directly with grips, one at a time, using a right-click option. Let's look at both techniques.

20. Select the vertical Wall on the right, right-click and choose **Edit Justification**.

Four diamond shaped grips will appear in the middle of the Wall. Each one corresponds to a different Wall justification. The grey one is the current justification (in this case Right).

21. Hover your mouse over each grip point to see the grip tip indicating its function.

Notice that the grip tips indicate that holding down the CTRL key will maintain the baseline. This means that the Wall footprint will not move when the justification is repositioned. If you do not hold the CTRL key, then the Wall will shift. You can see this indicated before you click by the small gray ghost line that appears when you hover over the grip. In this case, we need the Wall to shift back to its original position before we reversed it, so we will *not* use the CTRL key.

22. Click the Left justification grip (see Figure 10–7).

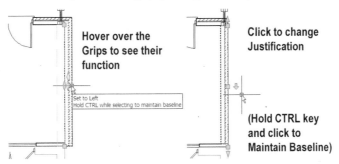

Figure 10–7 *Left justify positions the Walls correctly*

The Wall is now oriented, positioned and justified correctly.

23. Deselect the first Wall and then select the other two Walls, right-click and choose **Properties**.

24. On the Properties palette, in the Dimensions grouping, change the Justify to **Left**.

 NOTE In some situations, when you perform these types of edits, the Walls will no longer cleanup. The Apply 'L' Cleanup routine covered in the last chapter can easily fix this.

An alternative to the approach that we took here would be the **Reverse > In Place** option instead of **Reverse > Baseline**. Reverse in place would have flipped the brick without shifting the Walls, which would initially have appeared to be correct, but the justification line would then be the inside edge rather than the outside edge. Therefore, we would still have to shift the justification lines to Left and potentially apply "L" Cleanups. Either approach is valid so long as the end result is achieved.

Despite the fact that the exterior Walls are now displaying properly and in the correct orientation, the cleanup between them and the interior Walls is incorrect. It appears as though the interior Walls penetrate through the brick veneer on the exterior. This clearly needs to be addressed. This is actually an example of a simple cleanup problem like those we explored in the last chapter. The issue is that the interior Walls use the Standard Wall style, which uses a Cleanup Priority of 1 and therefore does not interact nicely with other Wall Styles. The priorities of the components in the exterior Wall style that we just applied are all much higher than this (refer to Chapter 9 for detailed information on Wall Component Priorities). However, this situation is easily resolved when we apply a Wall style other than Standard to the interior Walls as well. As a general rule of thumb, you should build your plan entirely from the Standard Wall style, or not use it at all.

25. Deselect all Walls.

26. Select one of the interior Walls, right-click and choose **Select Similar** (see Figure 10–8).

Select similar is quick and powerful. With a simple right-click, you can select all objects in the drawing that match (Style and Layer) the one selected.

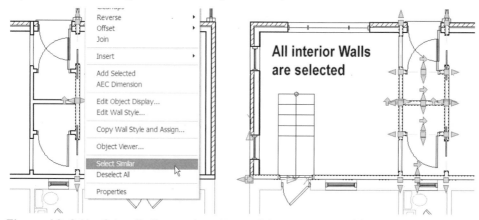

Figure 10–8 *Use Select Similar to select objects of the same type and layer*

All of the interior Walls in the new addition will be highlighted. This is because they were all similar to the one you originally selected.

27. On the Walls tool palette, right-click the *Stud-4 GWB-0.625 Each Side [Stud-102 GWB-018 Each Side]* tool and choose **Apply Tool Properties to Wall**.

    If you do not see this palette or tool, right-click the Tool Palettes title bar and choose **Design** (to load the Design Tool Palette Group) and then click the Design tab.

All of the selected Walls now use the style that was built into the tool.

28. Zoom in to the top middle of the plan where the interior Walls meet the exterior.

Study the Walls on all sides of the closet, including the way they join and cleanup with the exterior. Note that the stud Walls are made from a stud component with a layer of gypsum wallboard (GWB) on each side, (represented here by the extra set of internal Wall lines). The masonry Wall is composed of a stud structure with a brick veneer. The stud component of the interior walls and the stud back-up of the masonry walls now cleanup properly with each other. Recall the Wall Component Priority lesson in the last chapter and note the way the interior Walls behave before and after, as shown in Figure 10–9. All of these features are built into the respective Wall styles of each Wall object.

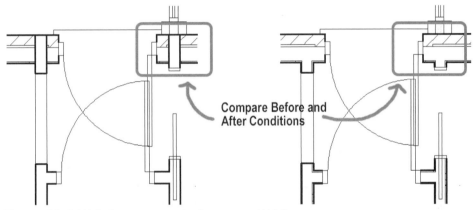

Figure 10–9 *Wall cleanup in standard versus stud Wall*

BUILDING AND EDITING WALL STYLES

It has been stated already that there are hundreds of pre-built Wall styles provided with the software. They are stored in library files like the one we accessed in the steps above. When you begin contemplating the creation of your own custom Wall styles, you should first thoroughly explore the ones that ship in the box. Even if the out-of-the-box offerings do not meet your needs, it will usually be much easier to start with one of the existing styles and edit it, rather than build one completely from scratch. Reverse engineering existing styles also provides an excellent means to learn about Wall style composition.

EXPLORE THE WALL STYLE PROPERTIES

1. Select the exterior masonry Wall (the one at the top of the screen), right-click, and choose **Edit Wall Style**.

The Wall Style Properties box has six tabs: General, with name and description fields, and five others, most of which will be covered in the upcoming exercises (see Figure 10–10). On the General tab, there are also some other buttons: Notes, Property Sets and the Keynote assignment area. Notes simply calls a dialog box with a single, large text field. You can enter any information here you wish: notes to the drafter, information on the style, etc. Property Sets are covered in Chapter 15, "Generating Schedules." Keynotes provide a means to assign construction notes to the Style from a central database. These notes can then be referenced with field codes on the printed sheets and even pass through to elevations, sections and details (see Chapter 17 for more information).

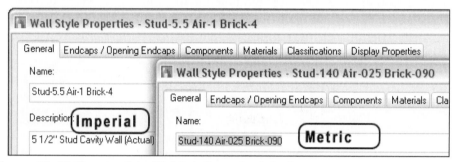

Figure 10–10 *The tabs of the Wall Style Properties dialog box*

2. Click the Components tab (see Figure 10–11).

Components in the Wall Style represent the major elements of the Wall Style's construction. A Wall style contains a minimum of one component and can contain up to 20. The Components tab is used to define each component's width, height, and position within the Wall, both horizontally and vertically. At the left is an interactive Viewer that defaults to a vertical end (Left) view of the Wall style. You can use Zoom and Pan in this Viewer as well as access a standard right-click menu as needed. Your wheel mouse will work in this Viewer as well. The main center section contains an interactive list of all the Wall components. On the right is a bank of icons to add, remove and shuffle components within the list.

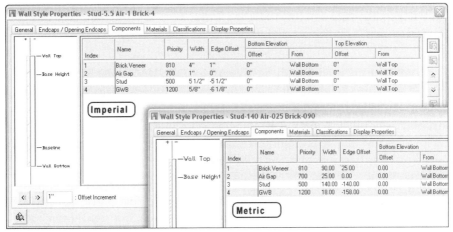

Figure 10–11 *The Components tab*

To edit the properties of a particular component, select it in the list. Set the horizontal dimensions in the left columns; configure the vertical dimensions in the columns at the right. Let's define the relevant terminology on the Components tab.

▶ **Index**—A number assigned by ADT from 1 to 20. Indexes are always in numeric order, however, you can use the icons at the right of the dialog box to reorganize (and thereby reindex) any of your components.

▶ **Name**—User defined designation; usually refers to the material that the component represents. Naming schemes should be standardized throughout the office. A complete list of common components (with their associated priorities) appears in Table 9-1 in Chapter 9.

▶ **Priority**—Numeric value used to determine how the component will clean-up with components of other Walls. Refer to the last chapter for complete information on priorities.

Geometrically, each component is basically a long thin box. In plan, this is some portion of the total width of the Wall. The Edge Offset and the Width parameters define the two lines that are used to draw a component relative to the Wall Baseline.

▶ **Baseline**—The "zero point" of both the Wall's width (in plan) and height (in elevation). Its location relative to the Wall section (and plan) is indicated by the long red vertical line in the Viewer. A Baseline text label indicates its location relative to elevation.

▶ **Edge Offset**—Locates first edge of the component, it is measured from the Baseline. This can be a positive or negative value.

▶ **Width**—The *true* width of the component, measured from the Edge Offset, *not* the Baseline. This can be positive or negative.

TIP When trying to understand or build a Wall style, remember that the Edge Offset is measured from the Baseline, while the Width is measured from the Edge Offset.

Positive and negative are used to indicate direction in the Wall style. Small plus (+) and minus (-) signs are shown at the top of the Viewer for reference. When a Wall is drawn from left to right in plan, the positive direction is up, and the negative is down. Another way to think of it is to imagine standing at the start point of the Wall and looking toward the end point. In this orientation, positive offsets would be to your left, while negative offsets would be to your right.

- ▶ **Top and Bottom Elevation Offset**—Establish the default top and bottom heights of a component in elevation relative to one of four possible points along the Wall height. Typically, components have no offset (zero) from both Wall Bottom and Wall Top. This makes them "full height" components. Change these values for components that occur in elevation such as horizontal bands, footings, and even countertops, moldings or soffits. Vertical components can be referenced from the following four points within the height of the Wall (see Figure 10–12).

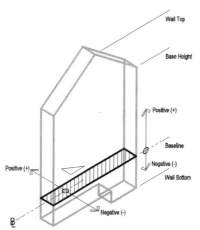

Figure 10–12 *Key datum points within a Wall object*

- ▶ **Wall Bottom**—In elevation or 3D, the absolute lowest point on the Wall's height when measured normal to the ground plane. Wall Bottom defaults to Baseline (typically at Z = zero). Use the Floor Line command (on the right-click menu) or the Roof/floor Line worksheet on the properties palette to change it on an individual Wall object. (Recall the use of Floor Line and Roof Line Offsets in the Curtain Wall style in Chapter 6, and the Wall at the residential Stair in Chapter 7.)

▶ **Baseline**—The datum line of the Wall in elevation or 3D. This point corresponds to the Wall's Z location (on the properties palette) and is typically Z=0. This point corresponds to the finish floor line in most cases.

▶ **Base Height**—The basic height parameter of the Wall on the properties palette and as seen in elevation or 3D. It is best to think of this point as the finish ceiling line of the Wall in most cases.

▶ **Wall Top**—In elevation or 3D, the absolute highest point on the Wall's height when measured normal to the ground plane. Wall Top defaults to Base Height. Use the Roof Line command (on the right-click menu) or the Roof/ floor Line worksheet on the properties palette to change it on an individual Wall object.

 NOTE The Wall Bottom in some cases may be above the Baseline. The Wall Top in some cases may be below the Base Height although neither of these scenarios is common. Typically, Wall Bottom will be equal or lower than Baseline while Wall Top will typically be equal or higher than Base Height.

CHANGE THE AIR GAP WIDTH

Now that we have an understanding of the basic terminology used in Wall styles, let's explore the specific settings of the Wall style that we are currently editing (see Figure 10–11 and 10–13).

Look closely at the settings for the Air Gap and the Stud. The Air Gap begins at zero and is 1″ [25] wide. The Stud is offset negative 5 1/2″ [140] from zero, with a width of 5 1/2″ [140] (see Figure 10–13). This puts the left edge of the Stud component directly at zero. This means that this Wall style is built with a Baseline at the edge between the Stud and the Air Gap. This is the edge of the structural material (in this case the Stud). Placing the Baseline at the edge of structure is a common approach seen in nearly all out-of-the-box styles. This enables you to draw this Wall with Baseline justification and trace along the slab edge or a beam, and have the Wall properly positioned as you work. It also allows you the freedom to adjust the width of the structure, or the veneer, at a later stage in the design process without needing to move the Walls. It is not required that the Baseline be positioned this way; it is done for the convenience of the user. Also remember that regardless of Baseline position within the Wall Style, you can always choose to draw with Left, Right or Center justification instead.

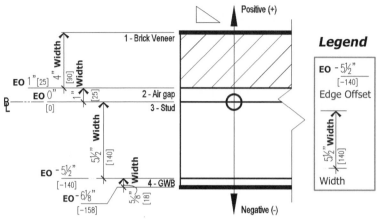

Figure 10–13 *Note each component's Edge Offset and Width and their relationship to the Baseline*

Let's assume we wanted to change the width of one of the components. For instance, suppose we wanted to have a 2″ [50] Air Gap.

3. Select the Air Gap component (Index number 2).

   Notice it is highlighted in green in the Viewer.

4. Click in the Width column, change the value to **2″ [50]** and then press ENTER (see Figure 10–14).

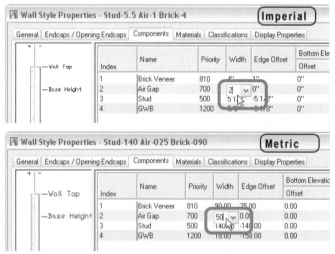

Figure 10–14 *Change the Width of the component*

Now that we have made the width of the Air Gap larger, the Brick no longer has the correct Edge Offset. (At an Edge Offset of 1″ [25], it overlaps the Air Gap, which is not recommended.)

5. Click on the Brick Veneer component (Index 1).

6. At the bottom left corner of the dialog box, click the Increment Wall Component Offset icon (see Figure 10–15).

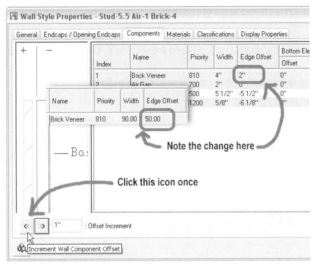

Figure 10–15 *The Brick Veneer is still flush, but the Endcaps need work*

Notice the change to the Edge Offset for the Brick component. It is now 2″ [50]. (If it went to zero, you clicked the Decrement icon instead. If so, click the Increment icon twice to fix it.)

With this tool, you can interactively adjust the position of components within a Wall style without the need to calculate the values. Adjust the offset Increment as appropriate before clicking the Increment and Decrement icons.

7. On the General tab, rename the Style to **Stud 5.5 Air-2 Brick-4 [Stud-140 Air-050 Brick-090]** and then click OK to return to the drawing and see the change.

 NOTE We are keeping the component-based naming convention in this example. If you choose, you could use a more descriptive name such as "Exterior Wall" like we did in the commercial project. The exact naming convention is less important than being certain that you apply it consistently across the project. This is especially true if you intend to use Project Standards to synchronize styles as we did in the commercial project.

Note that the exterior Walls have become wider. However, the exterior edge has remained flush with the exterior of the existing building just as we required. This is because the justification of the Walls is Left. Had we used Baseline instead, the brick veneer would have shifted 1″ [25] all the way around the building and we would have been forced to move the Walls to compensate. This is an example of how understanding the interaction of the various Style parameters (component dimensions) and Object-based parameters (justification in this case) provides insights valuable to forming good strategy.

8. Save the file.

GUIDELINES TO SUCCESSFUL COMPONENT COMPOSITION

There are a few issues to consider when setting up the components of a new Wall style.

▶ Will you use Nominal or Actual dimensions?

▶ How much detail is appropriate and necessary (quantity and composition of components, hatching choices, etc.)?

▶ Where should you position the Baseline (relative to the edge of the structural component or elsewhere)?

▶ Will the Wall style use a fixed (hard coded) Width or a variable (user specified on the properties palette) Width?

▶ If you use a variable Width, will the variable (Base Width on the properties palette) represent the overall width of the entire Wall or a single component within the Wall such as just the CMU?

▶ Will the Wall style require fixed heights for some or all of its components in elevations and 3D?

The answers to these questions depend somewhat on your personal philosophy or project needs. For example; should a brick component be 3 5/8" or 4"? Compelling arguments can be made to support either approach. Will you design the Wall style to draw lines representing each layer of drywall or will you simply represent the major structural components or overall wall width? Be sure to consider the scale at which the drawings will be printed when deciding on an appropriate level of detail and quantity of components. Keep in mind that the default template includes three levels of detail: Low, Medium and High (see Chapter 2). You will need to consider these points at each level of detail that you plan to use. On most default out-of-the-box ADT Wall styles, the Baseline is at the face of the structural component. This is typically a good location, but it is not critical. Once you have determined your own personal leanings concerning these criteria, the following three guidelines will help guarantee the successful creation of your new Wall style:

A. Regardless of nominal or actual, make sure that your math works. Red Defect Markers are a virtual certainty if any of your Edge Offsets and Widths does not add up properly.

B. Leave no gaps within the thickness of your Wall. If your Wall is 12″ [300] in total Width, make sure that your entire component Widths total to exactly 12″ [300] and that each component is touching one another without overlap. Air Gaps (in construction such as between brick and CMU) should be defined as Air Gap components with Display Properties turned off. Otherwise, you will experience display, cleanup and area/volume calculation problems.

C. In any decision, be consistent! This is the golden rule. No matter what you decide, actual or nominal, do it consistently. If you like to use negative Edge Offsets and positive Widths, do it consistently. If you change the component priority for brick in a new Wall style, assign it consistently. Making good decisions that support the way your firm works is half the battle. Following those decisions with consistency will make your job and those who work with you *much* easier.

CAD Manager Note: The key to solid and consistent building models and drawings involves a combination of technical and drafting standards. Do not simply address file naming, layer naming, XREF/block creation, and plotting methodologies. Your firm may already have a drafting standard, perhaps created in the days prior to CAD. Management needs to be aware of the opportunities and options the CAD software offers, because production time will be affected. A project methodologies guideline and CAD/Drafting Guideline are essential in order to take advantage of many ADT features. It is not necessary, as the CAD Manager, to create every possible Wall style anyone in the firm could ever need. Rather, if good guidelines based on time-tested drafting standards and object modeling best practices are put in place (incorporating the suggestions noted here); it will become much easier for all members of the team to be involved in the process of populating and using the library.

ENDCAPS

While components are the major aspect of a Wall Style, there are several other settings and characteristics of a Wall Style to consider; for example, how they terminate at the ends.

1. Zoom in on left side of the plan, where the masonry Wall meets the existing house.

Notice that the ends of the Walls at both the connection to the existing house and penetration of the windows show a special condition for the brick, yet appear a bit disjointed because of the change to the Air Gap component. Let's explore Endcaps to learn how to solve this problem.

Where the Edge Offset and Width define the two parallel lines that represent the component, an "Endcap" terminates the component on both its ends and at each opening (penetrations such as windows and doors). An Endcap is simply the condition at the end of a Wall. Specifically, it is the termination of each component within the Wall. The simplest form of Endcap is a straight line. The default Endcap named *Standard* is simply a straight line. Graphically this will terminate the Wall as if it were cut straight across all components. However, Endcaps can be virtually any shape; look at Figure 10–16 for some examples of Endcap conditions. Shown are four Wall styles contained in the out-of-the-box Content files. The name of the style is at the top; beneath them are sketches of the polylines used to create the Endcap. Next are the description of the Endcap condition, and finally an image of the resultant Wall condition.

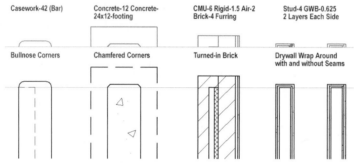

Figure 10–16 *A selection of sample Endcap conditions*

ASSIGN A DIFFERENT ENDCAP

Let's start by assigning a different Endcap.

2. Select the vertical masonry Wall at the left, right-click, and choose **Edit Wall Style**.

3. Click the Endcaps / Opening Endcaps tab (see Figure 10–17).

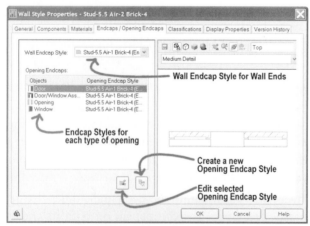

Figure 10–17 *Endcaps can be assigned to the Wall ends and each type of Opening*

There are three areas on this tab. In the top right corner, you can assign from a list the Endcap to use for the Wall's ends. Beneath this are the Opening Endcaps assignments; a different Opening Endcap Style may be assigned for each type of Wall Opening. The third area of this tab is the familiar Viewer. If the Wall style is complex and requires a more articulated termination, a custom Endcap can be devised. To create an Endcap, simply draw the way that each material should end with a polyline. We will build a new Endcap for this style below, but first, let's assign a different Wall Endcap style from the predefined list. Currently the Wall uses an Endcap named *Stud-5.5 Air-1 Brick-4 (End 1)* [*Stud-140 Air-025 Brick-090 (End 1)*]. This Endcap turns the brick veneer in to close the Air Gap. However, because it is rare to have a Wall of this type in a freestanding situation, the brick really only needs to turn in at the openings and not at the ends. Therefore, let's assign the Standard Endcap (which as mentioned above is a simple straight line) to the Wall Endcap Style condition.

4. From the Wall Endcap Style list, choose **Standard** (see Figure 10–18).

Figure 10–18 *Choose Standard from the list*

5. Click OK to return to the drawing and view the change (see Figure 10–19).

Notice the condition where the new walls meet the existing walls. The brick now simply terminates into the existing wall rather than turning in. Notice also that this change occurs on both sides of the addition. This is because the change was made in the Wall Style.

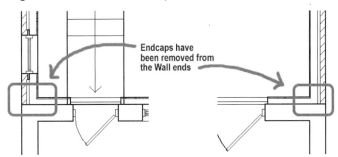

Figure 10–19 *Standard Endcap applied to the ends*

EDIT AN ENDCAP IN PLACE

We have solved the condition where the new and existing come together nicely. However, if you observe the condition at the Windows and other Wall penetrations, you will note that the problem caused by increasing the Air Gap is still evident. To fix this, let's edit the existing Endcap style.

6. Select the vertical masonry Wall at the left, right-click, and choose **Endcaps > Edit in Place**.

7. At the "Select a point near Endcap" prompt, click near one of the Window frames (see the left side of Figure 10–20).

   This will activate the Edit in Place mode. The end of the Wall will shade and grip editable shapes will appear at the end of each component. The In-Place Edit toolbar will also appear.

8. Click on the hatching inside the turned in Brick shape.

   Grips will appear on this shape (see the right side of Figure 10–20).

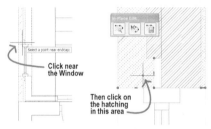

Figure 10–20 *Click the hatching in the Brick component to reveal its In-Place Edit Grips*

You can use any of the grip points to edit the shape of the Endcap. Corner grips (round shape) edit just that vertex. The midpoint edge grips (long thin rectangle) move the entire edge parallel to itself.

9. Click the edge grip on the vertical edge of the selected profile.

10. Snap this point the perpendicular edge across the Air Gap (see Figure 10–21).

11. On the In-Place Edit toolbar, click the Save all Changes icon.

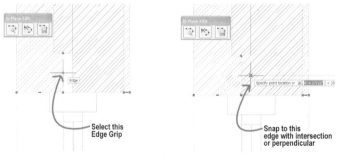

Figure 10–21 *Draw polylines for a custom Endcap*

12. Zoom in to one of the windows in the plan.

Notice the change to the Endcap. Because we redefined the existing one, the change was immediate and applied to all Window and Door locations. However, even though we have successfully redefined this Endcap, it is still bears its original name. It is recommended that you rename the Endcap Style to reflect its new configuration. To do so, choose **Style Manager** from the Format menu, expand *Architectural Objects* and highlight *Wall Endcap Styles*. Right-click *Stud-5.5 Air-1 Brick-4 (End 1)* [*Stud-140 Air-025 Brick-090 (End 1)*] and choose **Rename**. Change the name to *Stud-5.5 Air-2 Brick-4 (End 1)* [*Stud-140 Air-050 Brick-090 (End 1)*], and repeat for *Wall Opening Endcap Styles*, then click OK. An Opening Endcap Style is used at Doors, Windows, Openings and Door/ Window Assemblies. It is a style that can apply one or more Wall Endcap Styles to each of the four sides of an opening: Sill, Head, and both Jambs. This is why we had to rename two different styles. One was the Opening Endcap Style, the other was the nested Wall Endcap Style within it.

13. Save the file.

CORRECT ENDCAPS AT STUD WALL OPENINGS

14. Zoom in on the small foyer at the top middle of the plan.

Take a look at the Endcap used for the interior partition Walls in the center of the plan. Everything appears correct where Doors penetrate the Walls. However, there are two Opening objects in the vertical Wall that bounds the left side of the closets. In this case the drywall should wrap around the end of the Wall; but the way it shows now, the internal stud would be exposed. If you are working in Imperial units, this does not occur automatically.

15. Select the vertical Wall on the left side of the closets in the middle of the plan, right-click, and choose **Edit Wall Style**.

16. Click the Endcaps/ Opening Endcaps tab.

 NOTE if you are using Metric units, you do not need to perform these steps, because the style is already correct. However, you may want to follow along to learn the technique nonetheless.

This problem is very easy to fix. An Opening Endcap style named *Stud-4 GWB-0.625 Each Side (End 1)(2-Sided)*has been assigned to the Door\Window Assembly condition, and the Standard style has been assigned to all others. This requires an adjustment. *Standard* should be used for Doors, Door\Window Assemblies and Windows. *Stud-4 GWB-0.625 Each Side (End 1)(2-Sided)*should be used for Openings. Let's correct this.

17. From the list next to Door\Window Assembly in the Objects column, choose **Standard**.

18. From the list next to Opening, choose **Stud-4 GWB-0.625 Each Side (End 1)(2-Sided)**.

19. Click OK to view the change (see Figure 10–22).

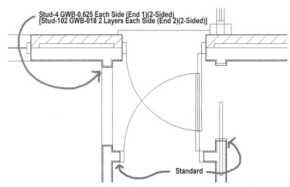

Figure 10–22 *Assign a "wrap around" drywall Endcap to Openings in Stud Wall styles*

Unfortunately, this change will need to be made in most of the out-of-the-box Imperial Stud Wall styles. However, it is a very easy process, and once you have made the change once, you can resave the styles back to the Content file in the library for future use.

CAD Manager Note: To fix this problem globally, open the Stud Wall Content library file named *Wall Styles - Stud (Imperial).dwg* located by default in the *C:\Documents and Settings\All Users\Application Data\Autodesk\ADT 2006\enu\Styles\Imperial* folder. Fix all of the styles in that file and then save it. If you have not moved your styles and content to a server location, copy the updated file to each user's machine.

ENDCAPS WITH INVISIBLE SEGMENTS

When an Endcap is built, all components in the Wall style must be terminated individually. However, take a close look at the Endcap we just assigned to the Opening in this style. The gypsum board appears to wrap continuously around the opening. It appears as though a single shape in the Endcap is being used to terminate two gypsum board components. In fact, what is occurring is that the shapes used in the Endcap contain some invisible edges. In other words, because each layer of gypsum *must* be terminated separately, an edge is made invisible where the two Endcap pieces "touch." To see this more clearly, we can insert the original polylines used to create the Endcap and study them.

20. If you have Dynamic Input enabled, with the mouse anywhere on screen, type **AECWALLENDCAP** and then press ENTER.

    If you have disabled Dynamic Input, you can type this command at the command line instead.

21. At the "Wall Endcap" prompt, choose **as Pline**. (if you have toggled off DYN, right-click and choose **as Pline** instead).

22. In the Endcap Styles dialog box, choose *Stud-4 GWB-0.625 Each Side (End 1)(2-Sided)* [*Stud-102 GWB-018 2 Layers Each Side (End 2)(2-Sided)*] and then click OK.

23. Pick a point on screen to insert the resulting polylines and then press ENTER to end the command.

24. Zoom in on the polylines and move them apart from one another.

There are two polylines here. Move them apart so that you can see each one separately. Click on either polyline to highlight the grips. Notice that there is a single segment of the polyline that has a *non-zero* width (it is thicker than the other edges). This is the secret to an invisible segment within an Endcap. While you are drawing a polyline to use for an Endcap, choose the **Width** option from the right-click menu (or the command line) and key in any non-zero width (any width greater than zero will work). The example on screen here uses a width of 1 unit (1 inch for Imperial and 1 millimeter for Metric). Each segment drawn in this way will render invisible in the final Endcap style (see Figure 10–23).

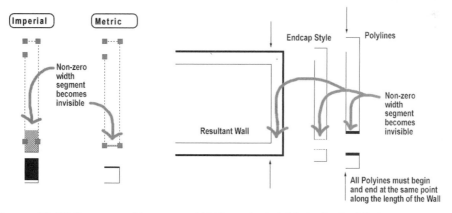

Figure 10–23 *Segments with non-zero Width render invisible in the final Endcap*

25. Erase the polylines and zoom back out.

If you don't like to type in commands

The AECWALLENDCAP command is available from a pull-down menu as well. This menu is the Design menu, which is hidden by default. To access it, choose **Pull-downs > Design pulldown** from the Window menu. A Design pull-down menu will appear. This menu was provided for compatibility with ADT R3.3. However, in most cases, you will not need this menu to execute commands. If you prefer the pull-down menu interface over tools and palettes, you can leave this menu loaded. A Desktop pull-down is available as well. Once you have loaded the Design pull-down menu, the equivalent command for inserting Endcaps is **Walls > Insert Endcap Style as Polyline** from the Design pull-down menu.

There is a third way to access this (or any) command. Open the Content Browser and open the Stock Tool Catalog. In the Helper Tools category, there is a tool labeled "Command Tool." Drag this tool onto one of your tool palettes (such as the Project Tools palette). Right-click the tool just added and choose **Properties**. Type in a name for the tool, and then click the Command worksheet icon. Type **AECWALLENDCAP P** in the Edit the command field. Be sure to include a space between the command name and the "P." Click OK to complete the tool. Click the tool to execute the command. For commands that have more parameters, use the Command Tool with Properties tool from the catalog instead.

CUSTOMIZE A WALL STYLE

Along the Wall between the existing house and the addition, a new Stud Wall is needed to abut the Masonry Wall of the existing house. Currently this Wall uses the same Stud Wall style as the other interior Walls. However, it is not likely that a layer of drywall would be installed between the existing brick and the new stud. Therefore, it would be good for us to create a new Wall style that has drywall on only one side. This will also require a custom Endcap for the Openings that penetrate this new Wall.

26. Select the horizontal Wall between the existing house and the addition, right-click and choose **Copy Wall Style and Assign**.

27. The Wall Style Properties dialog will appear ready to receive edits for the new style.

28. On the General tab, change the name to **Stud-4 GWB-0.625 One Side [Stud-102 GWB-018 One Side]**. Edit the description as appropriate.

29. Click the Components tab.

30. Select component number 3 – GWB and then click the Remove Component icon (see Figure 10–24).

In some cases making such an edit would require you to change the dimensions of the components remaining in the style. Refer back the "Guidelines to Successful Component Composition" listed above. Be certain that all criteria are met before leaving the components tab. In this case, the removal of the second GWB component simply makes the Wall narrower. This is fine in this case, since this is what would really occur.

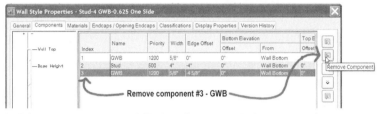

Figure 10–24 *Remove one layer of GWB – Component Index number 3*

31. Click the Endcaps/ Opening Endcaps tab.

32. Change all Endcap entries to **Standard** (see Figure 10–25).

Figure 10–25 *Change all Endcap entries to Standard for the new Wall style*

33. Click OK to return to the drawing.

Notice that the Wall got a bit narrower moving the remaining layer of drywall closer to the existing house. If you click the Wall you can see why. The justification of this Wall is Right. This keeps that bottom face of the Wall touching the existing house no matter what we do to the component composition. This is another good example of choosing your justification carefully (see Figure 10–26).

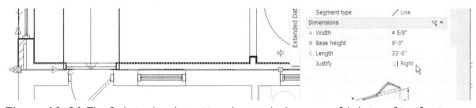

Figure 10–26 *The Style is already positioned properly due to careful choice of justification*

CREATE A CUSTOM ENDCAP

This process so far has taken care of removing the unnecessary layer of drywall. However, we now need to build a custom Endcap style for the Openings that penetrate this Wall.

34. Select the horizontal Wall between the existing house and the addition, right-click, and choose **Add Selected**.

    This will call the Add Wall command and match all of the parameters to the selected wall.

35. Click a random point outside the house toward the bottom of the screen

36. Move the mouse straight up (90° using ORTHO or POLAR) and click again anywhere, and then press ENTER to end the command (see Figure 10–27).

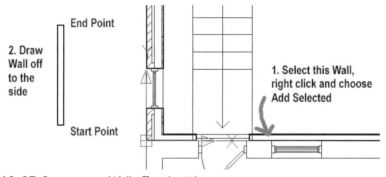

Figure 10–27 *Create a new Wall off to the side*

You should have a single segment freestanding Wall off to the side of your plan. The length is not important. We will use this Wall as a guide to build the new Endcap. Always start this way when building Endcaps because it is often easier to draw them separately from the surrounding Walls.

37. Zoom in on the top end of the Wall segment just drawn.

38. Click the Polyline icon on the Shapes toolbar (or type **PL** and press ENTER at the dynamic prompt or command line).

Use Figure 10–28 as a guide for the next several steps.

39. Snap the first point to Endpoint of the line between GWB and Stud components (see point 1 in Figure 10–28).

40. Move directly to the right and snap at the other side of the Stud (see point 2 in Figure 10–28).

    Make sure POLAR (or ORTHO) is on.

41. Move directly up (90°), type **5/8″ [18]**, and then press ENTER (see point 3 in Figure 10–28).

42. Move directly to the left (180°), type **4-5/8″ [120]**, and then press ENTER (or use OSNAP Tracking to line up with the corner) (see point 4 in Figure 10–28).

43. Snap to the left corner (outside corner) of the GWB and then press ENTER to end the command (see point 5 in Figure 10–28).

    Do *NOT* close the polyline.

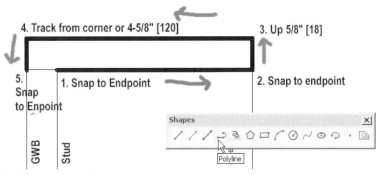

Figure 10–28 *Draw a polyline for the custom Endcap*

With this polyline, we are "showing" the Wall Style the way we wish the drywall to terminate; it will wrap around the end to cover the Stud component.

44. Select the Wall, right-click, and choose **Endcaps > Calculate Automatically**.

45. At the "Select polylines" prompt, click the polyline that you just drew and then press ENTER.

46. At the "Erase selected polyline(s)" prompt, choose **Yes**. (If you have Dynamic input turned off, right-click and choose **Yes**).

47. At the "Apply the new Wall endcap style to this end as" prompt, choose **Wallstyledefault**. (If you have Dynamic input turned off, right-click and choose **Wallstyledefault**).

48. In the New Endcap Style dialog box, type *Stud-4 GWB-0.625 One Side [Stud-102 GWB-018 One Side]*.

    The polyline will be converted to an Endcap directly on this Wall.

49. Select the Wall, right-click and choose **Edit Wall Style**.

50. Click the Endcaps/Opening Endcaps tab.

Notice that at the top left corner of the dialog, the new Endcap Style is applied to this Wall Style. This is because we chose the *Wallstyledefault* (Wall Style Default) option when creating it.

CREATE AN OPENING ENDCAP STYLE

So far we have focused on simple Wall Endcap styles. Openings do not use Wall Endcap styles directly; they actually use an Opening Endcap style, which then references one or more Wall Endcap styles. This is done because you can build an Opening Endcap Style that has a different Wall Endcap at its Sill, Head and each Jamb. Therefore, before we can assign the new Endcap that we have built to the Opening condition on the Wall style, we first need to create a new Opening Endcap Style for it. We could open the Style Manager to do this, but it turns out that there is a shortcut button on the Endcaps/Opening Endcaps tab of the Wall Style dialog box which we already have open on screen.

1. At the bottom of the dialog box, click the "Add a New Opening Endcap" icon (see Figure 10–29).

Figure 10–29 *Click the Add a New Opening Endcap Style icon to create a new Opening Endcap Style*

2. On the General tab, name the new style **Stud-4 GWB-0.625 One Side Openings [Stud-102 GWB-018 One Side Openings]** and then click the Design Rules tab.

3. Change all Positions, Jamb Start, Jamb End, Head and Sill to the Endcap style we created above: **Stud-4 GWB-0.625 One Side [Stud-102 GWB-018 One Side]**.

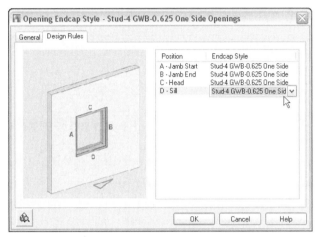

Figure 10–30 *Assign the Endcap style created above to all conditions*

4. Click OK to return to the Wall Style Properties dialog box.

5. Change the Opening Endcap for all Opening Objects to **Stud-4 GWB-0.625 One Side Openings** [**Stud-102 GWB-018 One Side Openings**] and then click OK (see Figure 10–31).

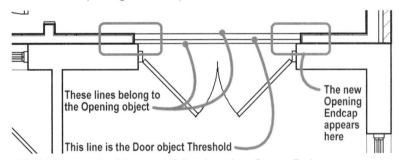

Figure 10–31 *The result of the new Wall style and its Custom Endcap*

6. Erase the stray Wall and zoom back to the drawing to see the result.

Some extra lines show at the door between the existing house and the new addition. This is because there are actually two Walls here: the existing Wall in the *First Floor Existing* Construct XREF and the new Wall in the current file. Opening objects have been added to the new Wall at each of the Door and Window locations between the existing and new house. The two outer lines are this ADT Opening object, and are on their own layer: A-Wall-Open [A-Opening-G]. You can make this layer non-plotting in the Layer Properties Manager by clicking the small Plot\NoPlot icon next to this layer. The third line in the center is the threshold from the existing Door (see Figure 10–31).

ADD A BRICK SILL

The possibilities with Opening Endcap Styles are extensive. Using an Endcap Style, we can create Sills for our Windows that penetrate the masonry Walls. To save time, some of the work has been done for you and provided in a file in the *Elements* folder of Project Navigator.

> 7. On Project Navigator, on the Constructs tab, double-click the *Opening Endcaps* file in the *Elements* folder.

In this file is a segment of the Wall style used in our exterior Wall. A Wall Endcap Style that incorporates a Brick Sill shape has already been defined (using the steps above) and applied as an override to the end of one of the Wall segments.

> 8. Select the Wall on the right with the Endcap applied to it, right-click, and choose **Clipboard > Copy**.
>
> 9. Close the file, and then in the *First Floor New.dwg* file (which should still be open), right-click, and choose **Clipboard > Paste**.
>
>   This was done as a quick way of importing the Endcap style to the current drawing.
>
> 10. Select any exterior masonry Wall, right-click, choose **Edit Wall Style** and then click the Endcaps/ Opening Endcaps tab.
>
> 11. Repeat the steps in the "Create an Opening Endcap Style" heading above and name this new style: *Stud 5.5 Air-2 Brick-4 Openings* [*Stud-140 Air-050 Brick-090 Openings*].
>
> 12. Change Jamb Start, Jamb End and Head to *Stud-5.5 Air-2 Brick-4 (End 1)* [*Stud-140 Air-050 Brick-090 (End 1)*].
>
> 13. Change Sill to *Stud-5.5 Air-2 Brick-4 Sill* [*Stud-140 Air-050 Brick-090 Sill*] and then click OK.
>
> 14. Assign this new style to Windows and then click OK

 NOTE Do not add this style to the other Opening types, only Windows.

> 15. Erase the stray Wall and then Save your file.

You will not see the effect of this change in Plan. Furthermore, even if you were to switch to 3D, this change would not yet be readily apparent. This is because by default, this sort of Endcap will only show in High Detail. We need to make a slight modification to the Display Properties of this Wall style to see it in Medium Detail. Wall style display properties are covered next. In the "View Opening Endcaps in 3D" heading below, we will look at our new Opening Endcap Style in 3D.

WALL STYLE DISPLAY PROPERTIES

In Chapter 2, it was stated that the display system determines how Autodesk Architectural Desktop objects are displayed under different viewing conditions and circumstances. In the context of Wall styles, the display properties that will concern us are the individual components of the Wall in plan, section, elevation, and 3D Model. Within the Display Properties of a Wall, Layer, Color, Lineweight, Hatching, Cut Plane, and level of detail are among the myriad of settings available.

Explore Display Representations

1. Select any masonry Wall, right-click, and choose **Edit Wall Style**.

2. Click the Display Properties tab.

You may recall that in Chapter 2 we learned that the display system contained Configurations, Sets and Representations. In Chapter 2 we explored the display system as a global set of controls that affected the display of the entire drawing and as such, focused mostly on Configurations and Sets. In this exercise, we will approach display from the other end of the spectrum and consider very specific display characteristics at the object level (of Walls in particular). Therefore, we will be focused entirely on Display Representations in this sequence. Looking at the Display Properties tab, you can see that Walls have several Display Representations (see Figure 10–32).

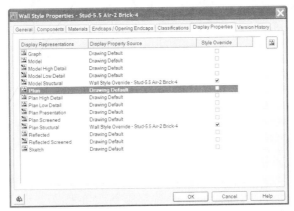

Figure 10–32 *The Display Properties tab of the Wall Style Properties dialog box*

The list can be a bit daunting due to the quantity of Representations available. (The quantity of Representations will vary depending on the type of object selected—Walls have more Representations than most objects as a rule.) In most cases, you will want to focus your attention on the Representation that is highlighted in bold. Bold indicates that this Representation is currently active in the drawing window. Since it is easier to edit parameters that you can see immediately affected, in most circumstances, you will want to edit only the Display Representation that is bold. This will make your task more manageable.

In general, the various Representations serve the purpose of representing the object in a fashion appropriate to their names. For instance, "Plan High Detail," "Plan" and "Plan Low Detail" are used to represent the object with 2D Plan graphics in three levels of detail. There are three similar Representations for 3D Models: "Model High Detail," "Model" and "Model Low Detail." We also see Representations for Reflected Ceiling Plans, Presentation Plans and Screened Plans. The "Graph" Display Representation is unique to Walls and was covered in detail in the last chapter (this becomes active when you toggle Wall Graph Display). "Sketch" is a diagnostic Display Representation for Walls that shows the Center line, Justification line and the Direction (from start to end) of the Walls.

Three columns are available on this tab. The Display Representations available for Walls are listed at the left. In the middle is listed the "Display Property Source." At the right are check boxes used to apply and remove display overrides. The Property Source can be one of three possible values:

- **Drawing Default**—Affects all objects of a particular object class globally. In this case all Walls regardless of style.

- **Style Override**—Affects all objects of a particular object class that belong to the same style. In this case, all Walls of the type *Stud-5.5 Air-2 Brick-4 [Stud-140 Air-050 Brick-090]*. No other Walls would be affected. (The Style Override column is only available when you select an object, right-click and choose **Edit Style**, and then click the Display Properties tab.)

- **Object Override**—Affects *only* the selected object. No other objects are affected regardless of object class or style. (The Object Override column is only available when you select an object, right-click and choose **Edit Object Display**.)

As you can see in this case, most of the Display Representations for the currently selected Wall style use the Drawing Default Display Properties. This makes managing those properties simple, since virtually anything that we do while editing this Wall will simultaneously affect all other Walls in the drawing.

3. Be sure that Plan is selected and then click the Edit Display Properties icon in the upper right corner (see Figure 10–32).

This calls another multi-tabbed dialog box.

EDIT DISPLAY PROPERTIES

> 4. Click the Layer/Color/Linetype tab.

There is a long list of components available to edit. Scroll through the list and look at the names of the components available (see Figure 10–33).

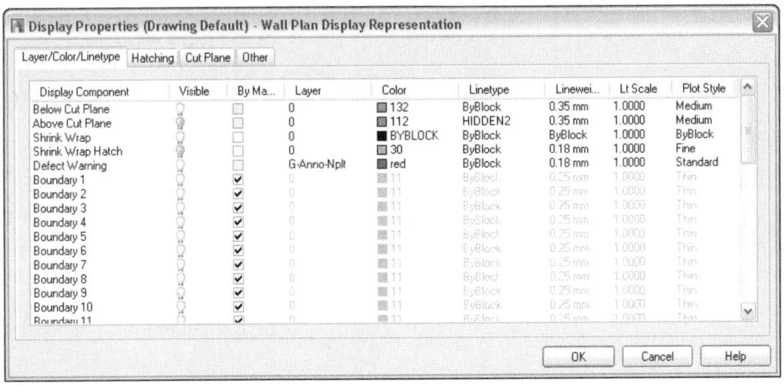

Display Properties (Drawing Default) - Wall Plan Display Representation

Layer/Color/Linetype | Hatching | Cut Plane | Other

Display Component	Visible	By Ma...	Layer	Color	Linetype	Linewei...	Lt Scale	Plot Style
Below Cut Plane		☐	0	☐ 132	ByBlock	0.35 mm	1.0000	Medium
Above Cut Plane		☐	0	☐ 112	HIDDEN2	0.35 mm	1.0000	Medium
Shrink Wrap		☐	0	■ BYBLOCK	ByBlock	ByBlock	1.0000	ByBlock
Shrink Wrap Hatch		☐	0	☐ 30	ByBlock	0.18 mm	1.0000	Fine
Defect Warning		☐	G-Anno-Nplt	■ red	ByBlock	0.18 mm	1.0000	Standard
Boundary 1		✔	0	■ 11	ByBlock	0.25 mm	1.0000	Thin
Boundary 2		✔	0	■ 11	ByBlock	0.25 mm	1.0000	Thin
Boundary 3		✔	0	■ 11	ByBlock	0.25 mm	1.0000	Thin
Boundary 4		✔	0	■ 11	ByBlock	0.25 mm	1.0000	Thin
Boundary 5		✔	0	■ 11	ByBlock	0.25 mm	1.0000	Thin
Boundary 6		✔	0	■ 11	ByBlock	0.25 mm	1.0000	Thin
Boundary 7		✔	0	■ 11	ByBlock	0.25 mm	1.0000	Thin
Boundary 8		✔	0	■ 11	ByBlock	0.25 mm	1.0000	Thin
Boundary 9		✔	0	■ 11	ByBlock	0.25 mm	1.0000	Thin
Boundary 10		✔	0	■ 11	ByBlock	0.25 mm	1.0000	Thin
Boundary 11		✔	0	■ 11	ByBlock	0.25 mm	1.0000	Thin

[OK] [Cancel] [Help]

Figure 10–33 *Common Wall components*

The length of this list and its contents vary considerably from object to object. Walls have perhaps the longest list of components of any ADT object. The first four components are related to the Cut Plane of the Wall. The Cut Plane is a user-definable distance above the floor line from which the Floor Plan graphics are derived. This Floor Plan information will then be used to generate the two-dimensional Plan Display Representations. Each 2D Plan Representation has its own Cut Plane parameters.

> ▶ **Below Cut Plane**—Any edges visible below the Cut Plane height will be rendered to this component (see Figure 10–34).

> ▶ **Above Cut Plane**—Any edges visible above the Cut Plane height will be rendered to this component (see Figure 10–34).

> ▶ **Shrink Wrap**—A continuous outline drawn around the outermost edge of all the Walls after cleanup has occurred and derived at the height of the Cut Plane.

> ▶ **Shrink Wrap Hatch**—An infill within the shape of the shrink wrap using any standard AutoCAD hatch pattern.

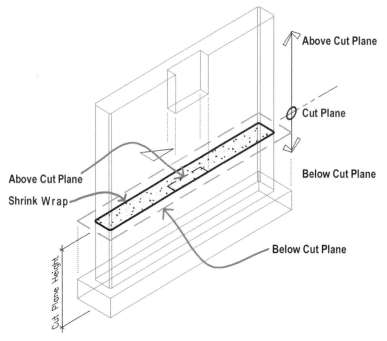

Figure 10–34 *How Above and Below Cut Plane components are determined*

▶ **Defect Warning**—A small "bull's-eye" marker that appears at the midpoint of the Wall segment when unresolved cleanup errors occur. (The marker is shown in Figure 9-3 in the last chapter.) This component is an important diagnostic tool. It is on a non-plotting layer by default and should be left on in the Display Properties.

 NOTE Keep in mind that even though the content of these components is determined from the three-dimensional Wall body, the resulting Plan graphics are drawn two-dimensionally on the X, Y plane at Z=0.

5. Scroll down to the lower portion of the component list.

Notice that there are several "Boundary" components and several corresponding "Hatch" components.

6. Select any Boundary or Hatch component.

The light bulb in the Visible column indicates that the component is turned on in the drawing. However, all of the other columns are grayed out. This is because there is a check mark in the "By Material" column. In this particular Wall Style, the Brick is showing as hatched in plan, but the Stud shows no hatching. This is because both components are turned on here; however, since they are both configured By Material, their respective hatching (or lack of it) comes from the Material assigned to that component. Try deselecting the By Material box. Notice that all of the other columns become available again. Be sure to reselect the By Material box when you are finished exploring. If you were to leave this unselected, it would not be possible to differentiate Boundary and Hatch 1 in one Wall Style from any other Style without applying a Style Override to each Style. The By Material setting allows the overall Drawing Default Wall properties to govern all Styles, while also allowing each Style to display its own unique hatching and display parameters.

It is important to understand what is occurring here. Every ADT object has one or more Display Representations. Every Display Representation has one or more components. In this case, we have a Wall selected, which has 14 Display Representations on the previous screen, and 45 components on this screen. Each one of these components can have its own Layer, Color, Linetype, Lineweight and Plot Style settings. Or we can defer all of these settings to the Material Definition. A Material Definition is an ADT style that represents a real life material like brick or wood or concrete. By controlling an object's Display Properties by Materials, we get a much more realistic set of plans and models, and the management of the display parameters is much simpler. It is simpler because we only need to configure the parameters of brick once in the brick Material Definition, rather than having to duplicate those settings in each Wall style that uses brick. Materials are covered in more detail in later chapters.

    7. Select the Shrink Wrap component.

Notice that the settings of this component are a little different from those of the Hatch component. Specifically, it is assigned to Layer 0, and the other properties are assigned "ByBlock." Every object in AutoCAD is assigned to a Layer, Color, Linetype, Lineweight and Plot Style. Walls are no exception. The layer of Walls in this drawing is A-Wall [A-Wall-G] (the default for ADT). By assigning the Shrink Wrap to Layer 0, it defers to the layer of the Wall itself (in this case A-Wall [A-Wall-G]) for its display parameters. Likewise, the ByBlock property is a special AutoCAD property that allows the components within the object to likewise defer to the Color, Linetype, Lineweight or Plot style of the object itself. Think of ByBlock as "By Object." Using this property allows individual Walls to hold overrides for these properties if required. If no override is assigned to a specific Wall, then they will use the drawing default, which is typically ByLayer. The net result is to have the shrink wrap, which outlines the entire Wall (and all adjoining Walls) to appear bolder than the components within (assuming the Lineweight and/or Plot Style of Layer A-Wall [A-Wall-G] is set properly for this occur). This may all be a bit confusing at first. The most important thing to remember is that the Wall Layer is controlling the properties of the shrinkwrap and everything else at the cut plane comes from the Materials assigned to each component.

    8. Click the Hatching tab (see Figure 10–35).

Here each of the Hatch components is listed again, with the AutoCAD hatching parameters such as pattern and scale available. Here again, since most of the components are being rendered By Material, most of the parameters on this tab are unavailable for edit. Only the Shrink Wrap Hatch component can be edited. However, for any changes to Shrink Wrap Hatch to be visible in the drawing, you would need to return to Layer/Color/Linetype and turn on the Shrink Wrap Hatch component. Shrinkwrap Hatch can be used to shade in the entire thickness of the Wall across all components with a single hatch pattern.

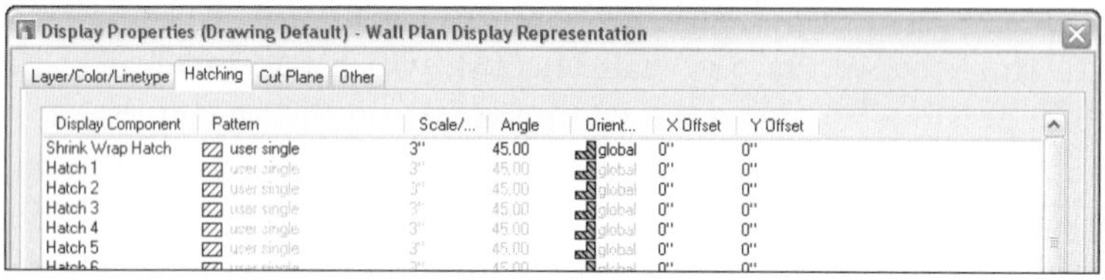

Figure 10–35 *Hatching parameters on the Hatching tab*

    9. Click the Cut Plane tab (see Figure 10–36).

The Cut Plane for all objects is an imaginary plane that sits at a fixed height above Z=0. All objects are cut at this height to determine their outlines and which graphics to draw in 2D plans. The height of the Cut Plane is 3′-6″ [1400] by default. This value is configured on the current Display Configuration (see "Use the Display Manager" topic in Chapter 2 for more details) and applies to all ADT objects. (This is sometimes referred to as "Global Cut Plane.") If you should need to override the Global Cut Plane, you can place a check mark in the Override Display Configuration Cut Plane check box. This will make the Cut Plane Height field available for edit (see exercise below). Normally you will want to leave the check mark in the Automatically Choose Above and Below Cut Plane Heights check box. Again, with this checked, the Global Display Configuration Cut Plane parameters will apply. However, if you have a particular feature above or below that you want to show explicitly, clear this and click the Add button to set the heights manually.

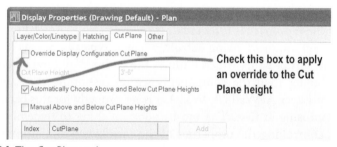

Figure 10–36 *The Cut Plane tab*

    10. Click the Other tab.

Miscellaneous settings are available here to fine-tune the behavior of components in the plan display. Many of these will be addressed in later lessons.

11. Click Cancel twice to return to the drawing.

APPLY AN OBJECT-LEVEL DISPLAY OVERRIDE

As we saw above, there is a new stud bearing Wall sistered to the existing masonry Wall for support of the new addition. Since this gives us a double Wall at this location, Openings have been added to the stud Wall where the existing Doors occur. The same has been done for the Window above the kitchen sink on the left side of the existing plan. However, it currently appears as though both Windows are covered over, and it is only intended that the Window in the existing bathroom be covered over. The reason the Opening is not displaying is that it is inserted above the Display Configuration Cut Plane (3'-6" [1400] by default). To see this, view the Wall in the Object Viewer in 3D. Therefore we need to raise the Cut Plane height of this Wall.

1. Select the horizontal stud Wall separating the addition from the existing house.

2. Right-click and choose **Edit Object Display**.

3. Click the Display Properties tab.

The Plan Display Representation should already be highlighted and bold. Notice that the third column is now labeled Object Override. This is because the Edit Object Display command allows us to edit at the currently assigned level (Drawing Default in this case) or at the object level. The last time we were in this dialog box, we were actually editing the Wall style, and therefore this column was labeled "Style Override."

4. Place a check mark in the Object Override column next to Plan.

   The Display Properties (Wall Override) - Wall Plan Display Representation dialog box will appear.

5. Click the Cut Plane tab.

6. Place a check mark in the Override Display Configuration Cut Plane check box (see Figure 10–37).

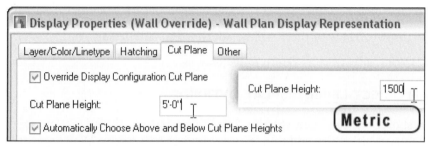

Figure 10–37 *Override the Global Cut Plane*

7. Change the Cut Plane Height to **5′-0″** [**1500**] and then click OK twice (see Figure 10–38).

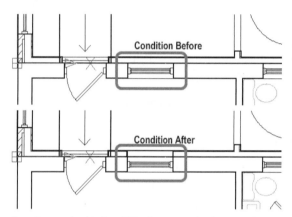

Figure 10–38 *The Opening at the kitchen Window now displays in plan*

Notice that the custom Endcap Style that we applied above is also now visible at this opening.

VIEW OPENING ENDCAPS IN 3D

Recall above that we added Sills to the Windows in the masonry Wall, yet we have yet to see them. Displaying Endcaps in 3D is within the province of Display Properties.

8. On the Views toolbar, click the NW Isometric icon (or choose **3D Views > NW Isometric** from the View menu).

Notice that all of the Walls of the addition are displaying materials.

9. Zoom in on one of the Windows.

10. Notice that the Sills are not displaying (see Figure 10–39).

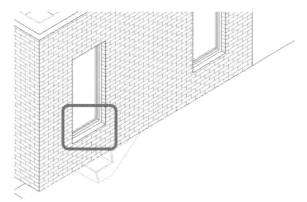

Figure 10–39 *When viewing the Windows in 3D the Sills do not yet display*

11. On the Drawing Status Bar, click the Display Configuration pop-up menu (depicted in Figure 1–2 in Chapter 1) and choose **High Detail**.

You should immediately see a change in the brick hatching. It will now be denser and show the mortar as well as the bricks. However, the sill will likely still not display. The model must be regenerated to force a display update. To do this, use the tool on the Standard toolbar or the View menu.

12. On the Standard toolbar, click the Regenerate Model icon (or choose **Regen Model** from the View menu) and then press ENTER.

The brick sills should display when regeneration is complete. Keep this "Regenerate Model" tool in mind. It is very helpful to force the Display System to recalculate all (or selected) objects in the drawing. This can often correct most display abnormalities. Switching to High Detail has enabled the display of the custom Endcaps, but now everything is more detailed. We can instead modify the Medium Detail display to show Opening Endcaps in 3D if we wish.

13. On the Drawing Status Bar, click the Display Configuration pop-up menu and choose **Medium Detail**.

14. Select the masonry Wall, right-click and choose **Edit Object Display**.

    Notice that Model is now bold, indicating that in this view, it is active.

Recall that in Chapter 2, we discussed the way that the display of the drawing changed with a change in viewing direction. If you are unsure about this change please review the "The Display System" topic in Chapter 2.

15. Click the Edit Display Properties button.

    A list of Components similar to the one shown above appears on the Layer/Color/Linetype tab.

16. Click the Other tab.

17. Place a check mark in the "Display Opening Endcaps" check box, and then click OK twice to return to the drawing (see Figure 10–40).

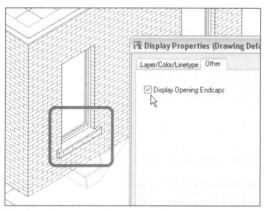

Figure 10–40 *With Display Opening Endcaps turned on, Endcaps display in 3D*

Again use Regenerate Model if necessary to force Medium Detail to display the new settings.

Note that the large picture window on the adjacent wall already shows a sill before this change. This is because the object used for the picture window is actually a Door/Window Assembly object. Door/Window Assembly objects are nearly identical to Curtain Wall objects and as such you can define custom frame components for one or more sides of the style. (Curtain Wall Styles were covered in detail in Chapter 8). In this case, a custom Frame component of the Door/Window Assembly forms the "sill" for this style.

18. On the Views toolbar, click the Top icon (or choose **3D Views > Top** from the View menu).

If you turned on shading, go back to 2D wireframe.

 NOTE The Display Opening Endcaps setting is a 3D Model setting only. It will not appear in the Plan view. To make sills appear in Plan use the Sill Plan Display Rep. More information on the Sill Plan Display Rep is found in the next chapter.

19. Save the file.

DEMOLITION

There are just a few final changes left to the first floor plan. We need to show the items that need to be demolished. The simplest way to achieve this is to move the "demo" items to another layer. The layer A-Demo has been included in this file. It is assigned a gray color and dashed linetype.

DEMOLISH EXISTING ITEMS

1. Select the *First Floor Existing* XREF, right-click, and choose **Edit XREF in Place**.

2. In the Reference Edit dialog box, click OK.

   The *First Floor New* file will be gray, and a toolbar will appear while you are in the XREF Edit mode.

3. Select the Door and Window in the kitchen (on the left), the Window in the bathroom and the Stair along the Wall between the new work and the existing.

4. From the Layer list on the Properties palette, choose **A-Demo** (see Figure 10–41).

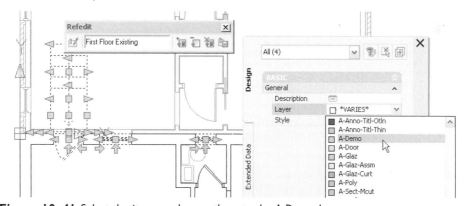

Figure 10–41 *Select the items and move them to the A-Demo layer*

5. Right-click and choose **Deselect All**.

6. On the Refedit toolbar, click the Save back changes to Reference icon.

7. Click OK in the confirmation dialog that appears.

8. Save the *First Floor New.dwg* file.

DEMOLITION WALL STYLE

These items are now properly showing as demolition. The second floor plan has a little more demolition. On this plan, we will need to remove some of the Walls. We could use the same technique we just covered, but it can also be worthwhile to use a Wall style and tool palette tool specifically for demolition. Like the commercial project, this project has a project-based tool palette named: "Project Tools." That tool palette contains the Existing Conditions Wall style and the Door/Window Assembly tool that we used in Chapter 4. It also has a Demolition Wall style on it that we will use now.

REVIEW THE DEMO WALL TOOL PROPERTIES

1. On the Project Navigator, in the *Constructs* folder, double-click the *Second Floor New* file to Open it.

2. Drag and drop the *Second Floor Existing* file from the Project Navigator into the *Second Floor New* drawing window to overlay the XREF.

3. Right click on the tool palettes title bar and choose **10 Residential [10 Residential Metric]** to load the project tool palette group.

   The Project Tools tool palette should appear.

4. On the Project Tools palette, right-click the ***Demo Wall*** tool and choose **Properties**.

All of the salient features of this tool are evident in the first few fields (see Figure 10–42). In the Basic parameters grouping, we find the Wall Style, the Cleanup Group and a Layer Key Override. The Wall style itself is actually quite simple, so simple in fact that this tool could just as easily reference the Standard Wall style to similar effect. Were you to edit the Demo Wall style, you would find a simple single component Wall style with variable width. The only difference between its parameters and the Standard Wall style's is the Cleanup Priority of 100 as recommended in Chapter 9. Also evident in Figure 10–42 is the use of a "Demo" Cleanup Group. You may recall the mention of Cleanup Groups in Chapter 9. The function of a Cleanup Group is very simple. Walls will only cleanup with other Walls in the same Cleanup Group. If they belong to different groups, they will not cleanup. Therefore, using a Demo Cleanup Group allows this Wall style to be used freely in a drawing that also includes existing or new construction without concern.

Perhaps the most significant setting of this tool is the Layer Override. This setting references the rules set by the current Layer Standard to automatically place Walls drawn with this tool on a Wall Demolition layer.

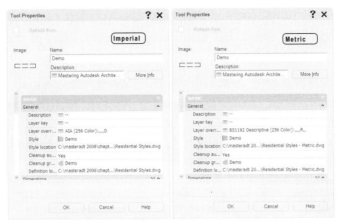

Figure 10–42 *The Properties of the Demo Wall tool*

> 5. Click Cancel when you are finished viewing the settings.

Now that we have seen the way this tool will behave, there is one small issue that we must address. This tool simply overrides the normal Wall Layer name by adding a suffix to it to indicate that it is Demo. However, it does this using the same parameters that are assigned to the existing Wall Layer. Typically we would want different settings for a demolition Wall Layer. Therefore, we will create the Wall Demo Layer first, and then use the tool. If the Wall Demo Layer already exists in the drawing, the tool will use the existing one rather than create one based on the Wall Layer.

CREATE A NEW LAYER BASED ON A LAYER STANDARD

Layer Standards are used to ensure that layers created in ADT drawings use the proper layer naming conventions. It is easy to create a layer based on the current Layer Standard.

> 6. Select the *Second Floor Existing* file in the drawing, right-click, and choose **Open XREF.**
>
>   This is an alternative to the in-place edit method used above. This technique actually opens the file in a separate window. Either approach is perfectly valid. It is a matter of personal preference.
>
> 7. On the Layer Properties toolbar, click the Layer Properties Manager icon.
>
> 8. At the top of the dialog, locate the "New Layer from Standard" icon (see the top of Figure 10–43).
>
> 9. In the New Layer form Standard dialog, choose **AIA 2nd Edition** [**BS1192 Descriptive**] from the Layer Standard list.

The layer name defaults to A-Wall. We will start with this and simply add a Status field to the end of the name, and then change the parameters of the layer to display as desired for demolition.

10. Click the small browser icon (…) next to the Status field.

11. From the Status list, choose: **D**, "Existing to demolish" [**R**, "To be removed"] and then click OK.

In the bottom of the dialog, assign Color to **9**, Plot Style to **Thin Screened**, Linetype to **Hidden2**, and Lineweight to **.25mm** (see Figure 10–43).

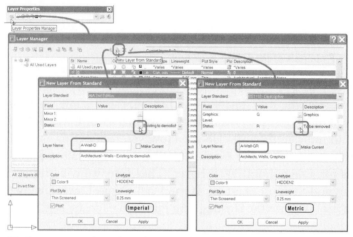

Figure 10–43 *Create a new layer based upon a Layer Standard*

12. Click OK twice to accept the new layer and return to the drawing.

CAD Manager Note: To implement the Demo Wall tool properly, you should edit the Layer Key style for your firm. A Layer Key style is a collection of instructions on which layers should be auto-created by ADT and how they should be configured. The reason that we had to build this layer is because it is not included in this default Layer Key style. The default Layer Key style is imported into a drawing automatically from a static location on your user's hard drive or a network server. In most cases you will want to move it to the server so that there is only one file to keep up to date (see the "Layer Standards" topic in Chapter 3 for more information). Regardless of the location of the default Layer Key style, if you edit it to include a Layer Key for the custom tools that you create, you can then reference those Layer Keys and/or Overrides in the tool Properties. This would prevent your having to include the Layers in your template or project files. For instance, in this case, if a WallDemo Layer Key had been established in the default Layer Key style, then it would have been unnecessary to include the A-Wall-D Layer in the current file.

TEST THE DEMOLITION WALL TOOL

Now that we have our Wall Demo Layer defined, we are ready to test the Demo Wall tool.

13. On the Project Tools tool palette, click the **Demo Wall** tool.

14. Click a random point outside the building to the left.

15. Click another point on the right side of the building.

    Allow the Wall to pass diagonally through the building and cross through several of the other Walls.

16. Add another segment or two in any direction (see Figure 10–44).

17. Press ENTER to end the Add Wall command.

View the result. Notice that the segments of Demo Wall have properly cleaned up with one another but have completely ignored the other Walls. The Demolition Cleanup Group defined with the Demo Wall style governs this behavior. Also, notice that the Demo Walls are properly displaying on the Demolition layer with a dashed linetype.

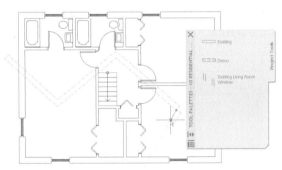

Figure 10–44 *Draw a few random Demo Walls*

18. Erase the Demo Walls just drawn.

DEMOLISH EXISTING WALLS BY APPLYING TOOL PROPERTIES

19. Turn on the Graph Display as covered in Chapter 9 (select any Wall, right-click, and choose **Cleanups > Toggle Wall Graph Display**).

    The Graph lines will make the next several steps easier.

20. Right-click in the drawing and choose **Basic Modify Tools > Break** (or type **BR** and press ENTER at the command line).

21. Select the existing horizontal Wall at the bathrooms at the top of the plan where the new and existing construction meet (see Figure 10–45).

22. Right-click in the drawing and choose **First point** (or type **F** and press ENTER at the command line).

23. At the "Specify first break point" prompt, snap to the point where the closet Wall and outside Wall meet (see Figure 10–45).

 NOTE Notice how it is easy to pick the desired point by snapping to the Graph line endpoint.

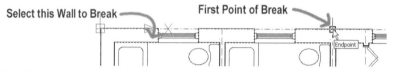

Figure 10–45 *Select the Wall to Break and use the "Specify the first point" option*

24. Click the same point again for the second point.

The Wall segment has now been broken into two segments.

Repeat the Break command on the same horizontal Wall and break out a segment of the horizontal bathroom Wall equal to the width of the abutting vertical Wall (see Figure 10–46). Choose your first and second point at endpoints of the vertical Wall separating the two bathrooms.

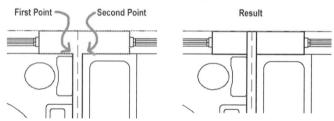

Figure 10–46 *Use Break to remove a small segment of Wall*

25. Select the two horizontal exterior bathroom Walls.

26. On the Project Tools tool palette, right-click the **Demo Wall** tool and choose **Apply Tool Properties to > Wall**.

Notice that the Demo Walls now overlap the existing ones.

27. Use the Wall Lengthen grips (the triangular ones) to adjust the Demo Walls as shown in Figure 10–47.

28. Follow the steps above to change the layer of the two Windows and the bathtub on the left to the A-Demo layer.

Toggle the Wall Graph Display off.

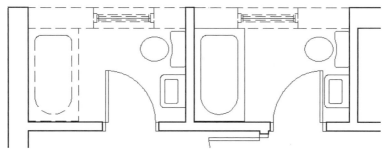

Figure 10–47 *Fine-tune how the Existing and Demo Walls meet*

29. Save and Close the file.

    When you close the file, you will be returned to the *Second Floor New* file, which was left open in the background. A message should appear in the bottom right corner of the screen alerting you that the XREF has changed.

30. Click on this message to reload the *Second Floor Existing* XREF (see Figure 10–48).

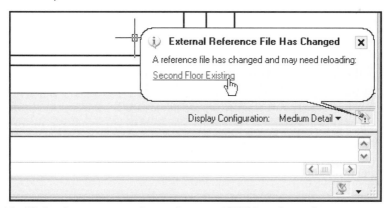

Figure 10–48 *An alert pops up to let you know an XREF has changed*

COMPLETE THE SECOND FLOOR PLAN REFINEMENTS

The second floor plan looks much like the first did at the start of this chapter. All of the Walls currently use the *Standard* style. Rather than repeat all of the steps that we performed above on the first floor, we can save time by reusing the styles that we developed there. To do this, we will make tools from the styles that we used on the first floor. This will help us maintain consistency throughout the project and streamline workflow. We will also need to create a new Wall Style for the second floor patio condition that does not occur on the first floor.

New in ADT
2006

PROJECT TOOL PALETTES

Up to this point in the book, our interaction with project-based tool palettes has been limited to the "Project Tools" palette that is created as part of the out-of-the-box template project. This template project (*ADT Template Project (Imperial)* [*ADT Template Project (Metric)*]), you will recall, was used to create both the commercial and the residential projects in Chapter 5.

While options do exist for the way that project tool palettes can be configured, the scope of this book will limit us to discussing the specific implementation utilized in the provided template projects. With that in mind, there are three critical behaviors to remember. First, when switching projects in the Project Browser, project-based tool palettes will be loaded and the previously loaded project's palettes will be hidden. Second, the "Project Tools" tool palette seen in your workspace is actually copied to your local machine from the project folder the first time you load the project. From then on, when you load and reload the project, it is the local copy of Project Tools (*not* the one in the project folder) that you will see and interact with in your workspace. For one-person firms and project teams, this behavior will be perfectly suitable. For project teams with two or more team members, you will need to implement "linked project based palettes" in order to share tools and keep them in synch with all team members (this is covered below). Third, tools can be built from styles in *any* drawing file. However, when creating project-based tools, it is very important to save all source Styles within the folder designated as the "Tool Content Root Path" saved with the project's properties. This will ensure that tools function properly on all team members' systems.

CREATE CUSTOM TOOLS

The styles from which we will create our tools currently reside in the *First Floor New* Construct file. As was mentioned in the previous passage, the project tools we are about to create will function better if they reference a drawing within the project's Tool Content Root Path.

1. On the Project Navigator, click the Project tab.

2. At the top of the palette, click the Edit Project icon.

Take note of the path listed next to Tool Content Root Path. By default it will be: *C:\MasterADT 2006\Chapter10\MADT Residential\Standards\Content* [*C:\MasterADT 2006\Chapter10\MADT Residential Metric\Standards\Content*]. We will not change this path. The important point here is that any tools that we build for the project should be created from drawings that live in the path listed here, or a sub-folder within it (see Figure 10–49).

Figure 10–49 *Edit the Project to note the Tool Content Root Path*

3. Click OK to dismiss the Modify Project worksheet.

4. From the File menu, choose **Open**.

5. Browse to the *C:\MasterADT 2006\Chapter10\MADT Residential\ Standards\Content [C:\MasterADT 2006\Chapter10\MADT Residential Metric\Standards\Content]* folder, select *Residential Styles.dwg [Residential Styles - Metric.dwg]* and then click Open.

This file contains one of each of the objects that are currently included on the Project Tools palette. We need to copy the additional styles from the *First Floor New* file over to this file before making tools from them.

6. On the Project Navigator, on the Constructs tab, double click the *First Floor New* Construct to open it.

If you left *First Floor New* open above, this action will simply switch to this file.

There are several ways to copy styles from one file to another. The two most common are using the Style Manager, or Copy and Paste. Let's use the Copy and Paste method.

7. Select one the exterior brick Walls, one interior stud partition (with GWB both sides) and the single stud wall with GWB on only one side (between the new and existing construction).

8. With all three Walls selected, right click and choose **Clipboard > Copy**.

9. From the Window menu, choose **Residential Styles [Residential Styles – Metric]** to make that file active.

10. Right click and choose **Clipboard > Paste** and click a point anywhere on screen to paste the Walls.

Zoom as necessary to place them in a convenient spot. The exact location is not important.

If you wish, you can adjust the lengths of the Walls to more closely match those of the Walls already in the file. You can also add text labels to them like the other objects. These steps are not required, but do make for a cleaner library file.

11. Save the *Residential Styles* [*Residential Styles – Metric*] file.

12. Right-click the tool palettes title bar and choose **10 Residential** [**10 Residential Metric**] to load the project tool palette group.

13. Click on the first pasted Wall and drag it to the Project Tools palette (see Figure 10–50)

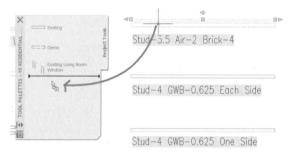

Figure 10–50 *Drag a Wall to the palette to make a tool*

A new Wall tool should appear. It has been noted in previous chapters that we can make tools in two ways, we can drag a Style from Style Manager and drop it on a palette, or we can drag an object from the drawing and drop it on a palette. In both cases the result is nearly the same. They vary slightly in the defaults that will be written to the new tool.

14. Right-click on the new tool and choose **Properties**.

Notice that the Wall Style and Style location correctly reference the current file and Wall Style. Scroll down a bit and notice that both the Base height and the Justify parameters have automatically been set to match the Wall that was dragged (see Figure 10–51). This will be useful later when we apply the tool properties to the Walls in the *Second Floor New* file.

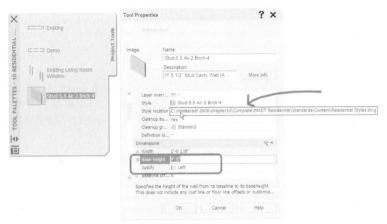

Figure 10–51 *Properties of a tool dragged from an object, match the properties of the original object*

15. Click OK to dismiss the Tool Properties worksheet.

New to Architectural Desktop 2006 is the ability to edit several tool's common properties at the same time. To do this, hold down the CTRL key, select two or more tools of the same type, such as two or more Wall tools, and then right-click and choose **Properties**. You can then edit common parameters as appropriate, such as Width, or Base height or Justify. This makes creating several tools with custom properties quick and easy. Furthermore, if you wish to create several tools from the styles in a file, you can use the Tool Catalog Generator. This tool, located on the CAD Manager pulldown (use the command on the Window menu to load the CAD Manager pull-down if you don't have it loaded) allows you to create tools automatically from all of the Styles in drawing or folder. Once the tools are created, you can edit them using the techniques outlined here and then publish the catalog to your team. Look for more information on this tool in the online help.

There are other Catalog enhancements as well. Categories in Content Browser now show an iDrop icon and can be dropped in their entirety to the ADT workspace. When you iDrop a Category from Content Browser, it creates a Tool Palette Group in the ADT Workspace. Be careful when doing this however. If you drag items while a project is active, they become part of the project and will disappear when another project is loaded. If you want to add a Palette or Tool Palette Group to your default workspace that is always available regardless of the currently active project, first go to Project Browser, right-click the current project and choose Close Current Project. Then iDrop the items you want into your workspace. When finished, you can reload the current project and the new palettes and groups will persist. Finally, Catalogs in Content Browser can be exported via the right-click menu to create XML files or Registry files. You can use these features to assist in deploying catalogs to a large number of users in the office. More information on these items is also in the online help.

16. Repeat the drag and drop on the other two Wall objects to create two more tools.

17. Save the *Residential Styles* [*Residential Styles – Metric*] file.

    You should have three new Wall tools: **Stud-5.5 Air-2 Brick-4** [**Stud-140 Air-050 Brick-090**], **Stud-4 GWB-0.625 Each Side** [**Stud-102 GWB-018 Each Side**] and **Stud-4 GWB-0.625 One Side** [**Stud-102 GWB-018 One Side**].

We now have our tools that correctly reference styles in the *Residential Styles* [*Residential Styles – Metric*] file. If you are a single-person firm, or have a single-person project team, you do not need to do anything else. You can use the tools as configured directly from the Project Tools palette as is. However, if you wish to share the same collection of tools among several members of a project team, proceed to the next topic to learn how to create a linked-shared Project Tools palette.

SET UP A SHARED PROJECT TOOLS PALETTE

Now that we have tools created we want to make them accessible to all project team members. One additional feature of the out-of-the-box project template (upon which our current project is based) is the inclusion of a project-specific Content Browser Library. As you recall from previous chapters, the Content Browser provides access to a much broader collection of tools than is typically active in the current workspace. ADT loads a default Content Browser library when you click the icon on the Navigation toolbar. This is the way we used the Content Browser in Chapters 4 and 6. We can also create a library that is project specific. We can populate such a library with just the catalogs, palettes and tools that are needed for a project. This makes it easier for project team members to locate the content and tools that they need quickly. Our first task will be to launch our Project Content Browser Catalog and then create a "Shared Project Tools" palette within it. Like the Tool Content Root Path setting above (see Figure 10–49), the Tool Catalog Library location for a project is also configured in the Modify Project worksheet and is also part of the template project used to create the current project. Therefore, the library is already configured for us and ready to go.

18. On the Project Navigator palette, click the Content Browser icon (see Figure 10–52).

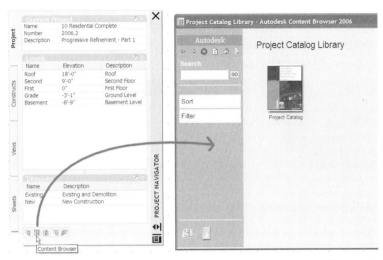

Figure 10–52 *Access the project based Content Browser Library from the Project Navigator*

19. Click on the Project Catalog.

20. At the bottom left corner, click the New Palette (rightmost) icon.

21. Name the palette **Shared Project Tools** and give it an appropriate description (see Figure 10–53).

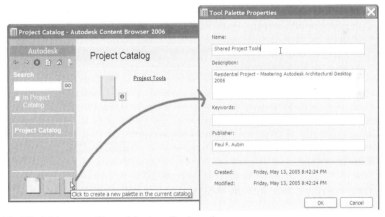

Figure 10–53 *Add a new Shared Project Tools palette*

22. Click on the new Shared Project Tools palette.

Since we just created this palette from scratch it is currently empty. We will now add the tools we made above to this palette.

23. Leave Content Browser running and return to ADT. (You can click the Architectural Desktop icon on the Windows task bar, or press ALT + TAB to cycle to it).

24. Right-click on one of the Wall tools created above and choose **Cut**.

25. Switch back to Content Browser (press ALT + TAB again) and then right-click and choose **Paste** (see Figure 10–54).

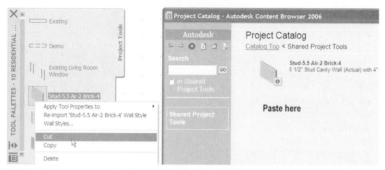

Figure 10–54 *Cut the new tools from the palette in ADT and Paste to Content Browser*

26. Click the "Catalog Top" link at the top of the page.

This will return you to the root of the catalog where both palette icons are located.

27. Click and hold down on the eyedropper icon next to the Shared Project Tools palette and wait for it to "fill up."

28. Continue to keep the mouse button depressed and drag and drop into the ADT drawing window (see Figure 10–55).

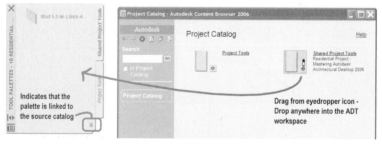

Figure 10–55 *Drop the new Palette into the Workspace*

When you drop the palette into the workspace, it will appear as a new tab in the active tool palette group—which is the 10 Residential [10 Residential Metric] group in this case. Also note the small refresh icon that appears at the lower right corner of the palette (see the left side of Figure 10–55). This icon indicates that the palette is linked to the source catalog. Clicking this link icon will force the palette to refresh and download the latest version of all tools on the palette from the source catalog. Since the source catalog is common to all project team members, use this technique to keep all team members in synch with the latest project tools. Now let's see this in action. While it is difficult in a book exercise to exactly replicate a project team dynamic, we can come pretty close. Let's add the remaining two Wall tools to the catalog and then refresh the palette to synchronize our workspace.

29. Make the Project Tools palette active.

30. Hold down the CTRL key and click each of the remaining two Wall tools created above.

31. Right-click and choose **Cut**.

32. Return to the project Content Browser (which should still be open on the Windows Task Bar) and paste the tools into the Shared Project Tools palette.

33. Return to ADT, make the Shared Project Tools palette active and then click the refresh icon to reload the palette.

Once the palette is refreshed, all three Wall tools that we created above should now appear on the Shared Project Tools palette. If another person joined the project team, they would get a copy of Project Tools (with its default tools) automatically when loading the project. They would then need to click the Content Browser icon on the Project Navigator to get the project specific library. They could then iDrop the Shared Project Tools palette to their own workspace to get access to all of the project tools and maintain a link back to the catalog to quickly capture edits, deletions and additions to the palette. New to ADT 2006, you can now add text (labels) and separators to your tool palettes. If you would like to add these, return to the Project Tools palette, right-click anywhere in the palette (except directly on a tool) and choose either **Add Text** or **Add Separator**. You can then Cut and Paste these to the catalog and then refresh the link to make them appear on the Shared Project Tools palette on each workstation. Also new in 2006 is the ability to sort a palette by name or type. Remember, "when in doubt, right-click."

Before we move on to the next topic, you might also want to experiment with loading a different project in the Project Browser. When you do so, the two palettes that are associated with the current project should disappear and be replaced by a Project Tools palette for the new project that you load.

Palettes dragged from a catalog default to "linked." This setting can be modified for the entire catalog. If you would rather that tool palettes dragged to the workspace not be linked to the source catalog, you can right-click the catalog in Content Browser, choose Properties and then clear the checkmark from the "Link items when added to workspace" checkbox. When doing this, palettes dragged to the workspace will then be unlinked copies on the local machine. It is recommended that you leave catalog set to "linked."

APPLY TOOL PROPERTIES TO ANOTHER FILE

We now have our tools and their palettes ready to be used.

34. If the *Second Floor New* file is not open, double-click it on the Project Navigator to open it now.

35. Using the technique covered above in the "Demolish Existing Walls by Applying Tool Properties" topic, break the two Walls (top horizontal and left vertical) as indicated by the "Break here" notes on screen.

TIP Remember to toggle the Graph Display to make it easier to break at the correct point.

Erase the two "Break here" notes when finished.

36. From the Shared Project Tools palette, right-click the **Stud-5.5 Air-2 Brick-4 [Stud-140 Air-025 Brick-090]** tool and choose **Apply Tool Properties to > Wall**.

37. At the "Select Wall(s)" prompt, click the three exterior Walls shaded dark in Figure 10–56 and then press ENTER.

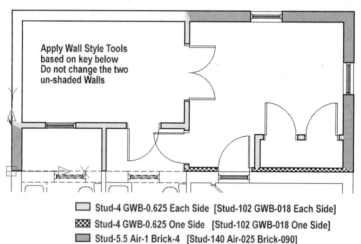

Figure 10–56 *Apply the new Wall tools to the existing Walls*

38. Following the key in Figure 10–56, repeat these steps for the remaining Walls.

OVERRIDE AN ENDCAP

39. Zoom in on and select the new section of Wall between the two bathrooms.

40. Right-click and choose **Endcaps > Override Endcap Style**.

41. At the "Select a Point" prompt, click a point at the end of the Wall where the new and old Walls meet.

42. In the Select an Endcap Style dialog box, choose **Standard**, and then click OK.

Notice the change. This technique quickly removes the unnecessary layer of drywall between new and existing construction.

43. Adjust the justification of any of the Walls as required.

BUILDING A CUSTOM WALL STYLE FROM SCRATCH

To complete the second floor and this chapter, we will build a complete Wall style from scratch. It is rare that you will need to build a Wall style from scratch, because in most cases you can simply modify an existing one more quickly. Nonetheless, it is beneficial to chronicle the steps needed to build your own style. In this exercise, we will build a parapet Wall style for the patio on the second floor. We will then apply it to the two remaining Standard Walls (at the top and left of the plan).

CREATE A NEW STYLE

To be consistent with our approach above, let's build the new Wall Style for the patio parapet in the Style Library file for the project. This will make it easy to create a tool for it following the same procedure outlined above. It will also keep all of our project's styles in a nice tidy location. Remember the "rules" above: Be consistent

1. From the Window menu, choose *Residential Styles.dwg* [*Residential Styles - Metric.dwg*].

   If you closed the file above, choose Open from the File menu and then browse to: the *C:\MasterADT 2006\Chapter10\MADT Residential\Standards\Content* [*C:\MasterADT 2006\Chapter10\MADT Residential Metric\Standards\Content*] folder, select *Residential Styles.dwg* [*Residential Styles - Metric.dwg*] and then click Open.

2. Click the Project Tools tab on the tool palettes to make it active and then right-click any Wall tool and choose **Wall Styles**.

3. On the right side of the Style Manager, right-click and choose **New**. (There is also a New Style icon on the Style Manager toolbar).

4. Type **Parapet Wall** for the Name of the new style.

5. Double-click the *Parapet Wall* style to edit it. (Double click the name).

6. On the General tab, type **Mastering Autodesk Architectural Desktop Residential Project Patio Parapet Wall** for the Description.

USE THE WALL STYLE COMPONENTS BROWSER

7. Click the Components tab.

This Wall style will have three components: two wythes of brick and a stone cap. We could define each of these components completely from scratch, but it will be easier to use the Wall Style Browser to locate existing components that have already been defined and copy them to our new style. We can then modify them as required for the current design.

8. On the right side of the Component tab, click the Wall Style Components Browser icon (the one at the bottom).

The Wall Style Components Browser is organized much like the Style Manager. It has a tree view at the left, and a list view and viewer at the right. By expanding the Wall styles listed at the left, you can drag and drop components from existing styles to use in the one you are creating. Currently the left tree view is showing us the styles that reside in the current drawing. We need to create two brick components. One of the existing Wall styles has brick in it, so let's start by copying this component to our new style.

9. In the tree of the Wall Style Components Browser, select **Stud-5.5 Air-1 Brick-4** [**Stud-140 Air-025 Brick-090**].

This will list all of its components on the right (see Figure 10–57).

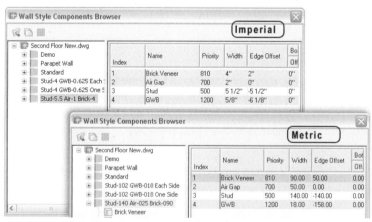

Figure 10–57 *Browsing Wall style components*

10. Drag Brick Veneer from the Wall Style Components Browser and drop it in the Wall Style Editor (anywhere on the Components list).

11. At the top of the Wall Style Components Browser, click the Open Drawing icon.

    The dialog box should default to the last location that you were browsing, which should be the location of the Style Content library files. If it has opened to some other location, click the Content icon in the Outlook bar at left and then navigate to your *Styles* folder, and then either Imperial or Metric depending on your preference.

12. Open the *Wall Styles – Brick (Imperial).dwg* [*Wall Styles – Brick (Metric).dwg*] file.

13. In the tree of the Wall Style Components Browser, select **Brick-4 Brick 4 [Brick-090 Brick 090]**.

14. Drag the Brick Veneer (Structural) component from the Wall Style Components Browser and drop it in the Wall Style Editor anywhere on the Components list.

15. Repeat the process and open the *Wall Styles - Concrete (Imperial).dwg* [*Wall Styles - Concrete (Metric).dwg*] file.

16. In the tree of the Wall Style Components Browser, select **Concrete-10 [Concrete-250]**.

17. Drag the Concrete component from the Wall Style Components Browser and drop it in the Wall Style Editor anywhere on the Components list.

    That is all of the "raw" components that we will need.

18. Close the Wall Style Components Browser.

19. Click the Materials tab.

Notice that all of the Materials are already assigned to each component. When you use the Wall Style Components Browser to create your components you gain many benefits:

▶ The basic sizes and parameters are already set.

▶ The names are always consistent.

▶ The Cleanup Priorities are properly configured.

▶ All Material Definitions are already assigned.

20. Click back to the Components tab.

    Now we need to make some adjustments to the dimensions, because currently all of these components are on top of one another.

21. Select the Unnamed component and then click the Remove Component icon at the right.

Table 10–1 *Parapet Wall Style Component Parameters*

| Edge Offset | Offset | Bottom Elevation | | Top Elevation | |
		From	Offset	From	
1-Brick Veneer	-4″ [-90]	0	Wall Bottom	2′-6″ [850]	Baseline
2-Brick Veneer (Structural)	-8″ [-180]	0	Wall Bottom	2′-6″ [850]	Baseline
3-Concrete	-9″ [-215]	2′-6″ [850]	Baseline	3′-0″ [1000]	Baseline

The Unnamed component is the default component added to all Wall Styles when created from scratch. It cannot be deleted if it is the only component in the style, this is why we add the new components first, and then delete it here. We can adjust the order of the components in any way that suits us.

22. Using the Up and Down arrow buttons on the right, organize the components so that the Brick Veneer component comes first (index 1), Brick Veneer (Structural) is next (Index 2) and the Concrete is last (index 3).

Earlier in this chapter, we outlined several points to consider when building a Wall style. One that we must decide on before we can go further is where we wish our new Wall Style's Baseline to be. Remember that the Baseline is the "zero point" of the Wall. For this Wall style, it will be most useful to have the Baseline at the face of the brick. This way, we can easily flush the brick from this Wall with the adjacent Walls.

23. Change the values of each component as shown in Table 10–1 and Figure 10–58.

TIP You should maximize the width of the Style Manager as wide as your screen will allow so it is easy to work with. You can either type the values for the Edge Offsets directly into the fields or use the Offset Increment ions beneath the viewer. The Top and Bottom Elevation offsets must be typed into the fields.

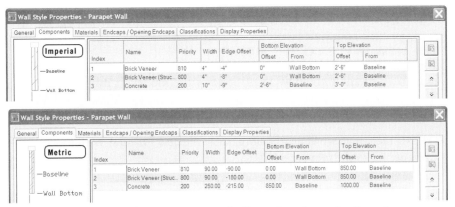

Figure 10–58 *Reorganize the components in the list and configure the dimensions*

This Wall style will butt into the other surrounding Walls and will not have any penetrations; therefore, we do not need to configure anything on the Endcaps/Opening Endcaps tab. We will not configure anything on Classifications or Display Properties for now either.

24. Click OK to complete the Wall Style edit and close the Style Manager.

 NOTE While it is possible to drag the style directly from the Style Manager to a palette to make a tool, we must save the drawing first before the tool can be created properly. Unfortunately, we cannot save the drawing while Style Manager is open.

25. Select any Wall on screen (and its text label if it has one) make a copy of it to any available location on screen. (The exact location is unimportant).

26. Select the new Wall and on the Properties palette, change the Style to **Parapet Wall** and change the Justify to **Baseline**.

The new Wall should appear as two lines in plan. This is because all of the components that we defined occur below the cut plane. Therefore we are simply looking down on the topmost component (the Concrete cap) only.

27. Save the file.

28. Drag the Parapet Wall from the drawing onto the Project Tools tool palette as before.

We are consistently using our Project Tools palette here as a kind of "scratch pad." Tools must be created in the ADT workspace. Once created, they can be added to a catalog to share with the project team. Again, if you are a one-person shop, you can stop at creating the tool on the Project Tools palette—that is all you will need. However, if you work with a larger team, repeat the steps above to Cut the tool from the Project Tools palette and paste it to the catalog. Then refresh the Shared Project Tools palette to see it update.

29. Close and save the *Residential Styles.dwg* [*Residential Styles - Metric.dwg*] file.

APPLY THE NEW STYLE TO THE MODEL

All that remains is to apply the new Wall Style to the model. It might be nice to do this in a view other than plan. NW Isometric view on the View > 3D Views menu will give you the best vantage point.

30. Using the same technique as before, apply the new ***Parapet Wall*** tool's Properties to the two patio Walls.

    You will need to do some grip editing and other fine-tuning to resolve the intersections where the Parapet meets the house (see Figure 10–59). The specifics are left to the reader as an exercise. You may want to consider some of the following items that we have already covered:

▶ Grip editing the Parapet Wall ends.

▶ Endcap overrides for the exterior masonry Walls of the addition.

▶ Wall Cleanup rules.

▶ Wall Cleanup issues if any.

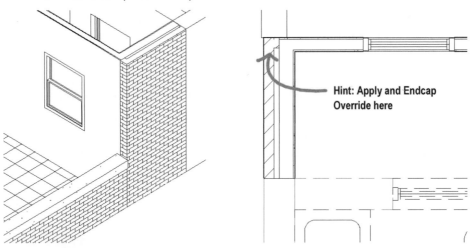

Figure 10–59 *The completed patio parapet Wall*

ADDITIONAL EXERCISES

Additional exercises have been provided in Appendix A. In Appendix A you will find suggestions for adding Casework and Equipment to the Residential and Commercial Projects. There are special Wall styles defined to behave as countertops (look in the *Wall Styles - Casework (Imperial).dwg* [*Wall Styles - Casework (Metric).dwg*] file), with and without cabinets (see Figure 10–60). There is also a collection of pre-built cabinet Multi-View Blocks (MVBs) available in the Content Browser. Equipment, Appliances, and Furniture are also available as MVBs in the Content Browser. It is not necessary that you complete this exercise to begin the next chapter, it is provided to enhance your learning experience. Completed projects for each of the exercises have been provided in the *Chapter10/Complete* folder.

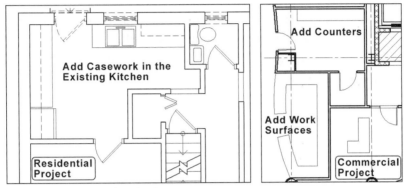

Figure 10–60 *Add Casework to both projects*

SUMMARY

Wall styles control default size and shape, internal component makeup, and graphical display settings of Walls.

Default settings such as Wall width, justification and cleanup behavior can be preset in Wall tools.

A Wall style can contain between 1 and 20 internal components, each comprised of a boundary (two parallel edges) and a hatch infill.

The termination of each component of a Wall is controlled by Endcap styles at both the ends of the Walls and at all Wall penetrations.

Cleanup Groups limit Wall cleanup to other Walls belonging to the same group.

Display Properties can be assigned at the style level to control how a specific Wall style is displayed graphically on screen and in print.

Most Wall styles use the Drawing Default Display Properties and reference the By Material setting instead.

You can build tool palette tools that reference specific styles in your project and then use those tools to apply those parameters throughout any drawing in the project.

Building custom Wall Styles is easy when you use the Wall Style Component Browser to start with pre-made components.

CHAPTER 11

Progressive Refinement – Part 2

INTRODUCTION

As a design scheme progresses, it is often desirable to begin including more detail in the articulation of Walls, Doors, and Windows. We have stressed throughout the book that ADT is designed to allow us to add objects with little detail early in the design cycle and progressively refine our design as more information becomes known. In the last chapter, we took an exhaustive look at Wall styles and Wall Endcaps. In this chapter, we will focus on Door and Window styles. In addition, we will fine-tune the placement of Doors and Windows, both along the Walls they are anchored to and within the thickness of the Walls as well. We'll also take a look at Multi-View Block content.

OBJECTIVES

In this chapter, we will refine the masonry shell of the Commercial Project and continue to work on first and second floors of the Residential Project. Now that styles have been applied to all of the Walls, we will shift our focus to fenestration. In the commercial shell file, we will adjust the placement of the Windows within the walls and look at the Sill Plan Display Representation. Returning to the Residential Project and some ancillary sample files, we will explore the features of the Door and Window styles. The chapter will conclude by finishing the residential floor plans with the creation of a custom Multi-View Block for the second floor bathroom.

- Understand Door and Window styles.
- Manipulate Door and Window Anchors.
- Customize Door and Window display.
- Build a Multi-View Block.

WINDOW STYLES

The simple Door and Window placement provided by the Add Door and Add Window commands is easy to achieve and perfectly acceptable for early design phases. However, as the design scheme becomes more refined, so too must the graphics used to represent the building fenestration. In this chapter, we will begin with a return to the Commercial Project. The shell wall on three sides of the building includes punched Window openings. We will add sills to these Windows on their exterior and reposition them within the thickness of the Wall.

INSTALL THE CD FILES AND LOAD THE CURRENT PROJECT

If you have already installed all of the files from the CD, simply skip down to step 3 below to make the project active. If you need to install the CD files, start at step 1.

1. If you have not already done so, install the dataset files located on the Mastering Autodesk Architectural Desktop 2006 CD ROM.

   Refer to "Files Included on the CD ROM" in the Preface for information on installing the sample files included on the CD.

2. Launch Autodesk Architectural Desktop 2006 from the desktop icon created in Chapter 3.

If you did not create a custom icon, you might want to review "Create a New Profile" and "Create a Custom ADT Desktop Shortcut" in Chapter 3. Creating the custom desktop icon is not essential; however, it makes loading the custom profile easier.

3. From the File menu, choose **Project Browser**.

4. Click to open the folder list and choose your *C:* drive.

5. Double-click on the *MasterADT 2006* folder, then the *Chapter 11* folder.

   One or two commercial projects will be listed: *11 Commercial* and/or *11 Commercial Metric*.

6. Double-click *11 Commercial* if you wish to work in Imperial units. Double-click *11 Commercial Metric* if you wish to work in Metric units. (You can also right-click on it and choose **Set Current Project**.) Then click Close in the Project Browser.

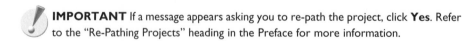 **IMPORTANT** If a message appears asking you to re-path the project, click **Yes**. Refer to the "Re-Pathing Projects" heading in the Preface for more information.

COPY AND ASSIGN A NEW WINDOW STYLE

Let's get started right away and build a new Window style for the commercial project masonry shell.

NOTE Nearly every technique that we will explore here is interchangeable for Windows or Doors. Feel free to experiment with either or both.

1. On the Project Navigator, in the *Constructs\Architectural* folder (Constructs tab), double-click to Open the *01 Shell and Core* file.

Some of the required Windows have been added to this file. The process to add Windows was covered in Chapter 4. We will start with these Windows and create a new style, adjust their Anchor parameters and their Sills. Later we will copy several of them to complete the layout.

2. Select one of the Windows, right-click and choose **Select Similar**.

    All of the Windows will highlight. (Please note that the front façade Curtain Wall will not highlight).

3. With all of the Windows selected, right-click and choose **Copy Window Style and Assign**.

4. On the General tab, type **Shell Window** for the name.

5. In the Description field, type **Mastering Autodesk Architectural Desktop Commercial Project - Window with fixed glazing and Sills** (see Figure 11–1).

 TIP It is a good habit to put a description for new styles. This will help you and others recognize the intention of the style at a future date.

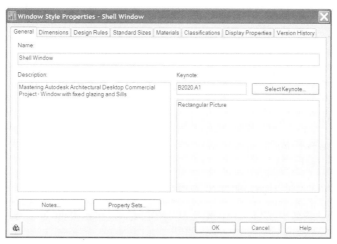

Figure 11–1 *Create and Edit the new style*

In addition to the Name and Description, notice that we also have Property Sets, Notes and a Keynote assignment area. Property Sets are covered in Chapter 15, "Generating Schedules." Notes simply calls up a dialog box with a single, large text field. You can enter any information here you wish: notes to the drafter, information on the style, etc. Keynotes provide a means to assign construction notes to the Style from a central database. These notes can then be referenced with field codes on the printed sheets and even pass through to elevations, sections and details.

EDIT THE NEW STYLE

6. Click the Dimensions tab.

The settings on this tab allow you to configure the basic sizes of each of the major Window components. Use the diagram at the left of the dialog box to help understand what each setting controls.

7. In the Frame area, set the Width to **3″ [75]** and the Depth to **5″ [125]**.

8. In the Sash area, set the Width to **3″ [75]** and the Depth to **4″ [100]** (see Figure 11–2).

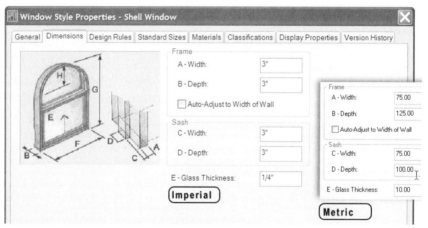

Figure 11–2 *Experiment with the settings on the Dimensions tab*

9. Click the Floating Viewer icon in the lower left corner of the dialog box (see Figure 11–3).

10. From the list of Views, choose **Top** (see Figure 11–3).

This will show the effect in plan of the settings we just changed. The size of the Frame and Sash are now exaggerated.

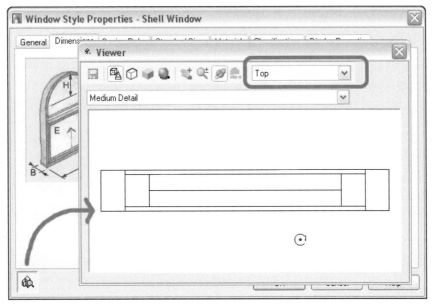

Figure 11–3 *Open the Floating Viewer*

Leave the Viewer open but position it next to the main dialog box on your screen so that it will remain visible.

11. Click the title bar of the Window Style Properties dialog box.

This will shift focus away from the Floating Viewer and back to the Window Style Properties dialog box while leaving the Viewer open.

12. Change the Width of the Frame to **2″ [50]** set the Depth of the Frame to **5″ [125]**.

Note the change in the Viewer. The Frame is now a more reasonable size.

For the Window style that we are designing, we are looking for a very simple look graphically. For instance, showing the Sash component for this style is not desired. Removing it can be achieved in a number of ways. It seems that the logical place to remove the display of Sashes would be in the Display Properties. Let's have a look.

13. Click the Display Properties tab.

As we saw with Walls in the last chapter, there are several Display Representations available for Windows. Notice that there are two bold Display Representations in this case: Plan and Sill Plan. This indicates that both Plan and Sill Plan are currently active for Windows in this drawing. Notice also that the Drawing Default setting controls all Display Representations for Windows. This is fine for our purposes—this Window style does not need to be displayed differently in plan than other Window styles.

14. Select Plan from the list.

15. Click the Edit Display Properties icon at the top right corner.

16. On the Layer/Color/Linetype tab, click the small light bulb icon next to the Sash component to turn it off (it should go "dim," see Figure 11–4).

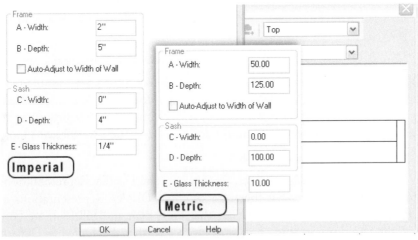

Display Properties (Drawing Default) - Window Plan Display Representation

Layer/Color/Linetype | Other

Display Component	Visible	By Ma...	Layer	Color	Linetype	Linewei...	Lt Scale	Plot Style
Frame			0	■ BYBLOCK	ByBlock	ByBlock	1.0000	ByBlock
Sash			0	■ BYBLOCK	ByBlock	ByBlock	1.0000	ByBlock
Glass			0	■ BYBLOCK	ByBlock	ByBlock	1.0000	ByBlock
Frame Above Cut Plane			0	□ 51	ByBlock	0.25 mm	1.0000	Thin

Figure 11–4 *Turn off the Sash component in the Display Properties*

17. Click OK once to return to the Display Properties tab.

Notice the result in the Viewer. The Sash is no longer displayed, but there is now a gap between the Frame and the Glazing. This is obviously not the best technique to achieve our goal.

18. Click the Edit Display Properties icon again.

19. Click the small light bulb icon next to the Sash component again to turn it back on.

20. Return to the Dimensions tab again.

21. Set the Width of the Sash to **0″ [0]** (see Figure 11–5).

Frame
A - Width: 2″
B - Depth: 5″
☐ Auto-Adjust to Width of Wall

Sash
C - Width: 0″
D - Depth: 4″

E - Glass Thickness: 1/4″

Imperial

Top

Frame
A - Width: 50.00
B - Depth: 125.00
☐ Auto-Adjust to Width of Wall

Sash
C - Width: 0.00
D - Depth: 100.00

E - Glass Thickness: 10.00

Metric

OK | Cancel | Help

Figure 11–5 *Adjusting the Sash Dimensions to Zero*

This will remove the sash more effectively. As you can see, this is different from turning off the Sash component in the Display Properties tab. There is no gap between components this time. There is one more way that our goal could be achieved. We could use the Nominal Display Representation for Windows in the current Display Configuration. However, this technique would affect all Windows in the drawing, which may not be desirable either. Zeroing out the Sash is the best compromise approach for our purposes here.

ADD SILL EXTENSIONS

In the previous chapter, we saw that we could use Opening Endcaps to add Sills in the 3D and Elevation views of the Model. However, you may recall that those Sills did not appear in plan. To make the sills appear in plan, we need to edit the Display Properties.

    22. Return to the Display Properties tab.

Sill components belong to a special Display Representation called Sill Plan (Threshold Plan for Doors). We want to adjust the way the sills are displayed on the current style only; therefore, we will apply the override for Window Style.

    23. In the list of Display Representations, next to Sill Plan place a check mark in the Style Override check box (see Figure 11–6).

        Checking the box will make the Display Properties dialog immediately appear.

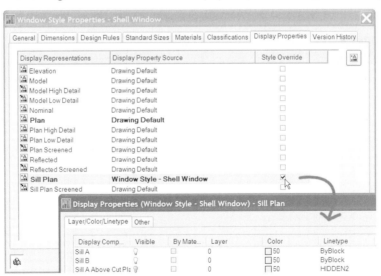

Figure 11–6 *Attach a style override to the Sill Plan Display Representation*

Notice that there are two components: Sill A and Sill B. Sill A occurs on the "swing" side of the window (the side with the arrow-shaped "Flip" grips) and Sill B occurs on the opposite side. Our Window is fixed, and therefore does not have a swing at all. In this case we will treat Sill A as the outside and Sill B as the inside.

We are going to leave Sill B flush to the inside face of the Wall, and project Sill A outward.

24. Click the Other tab.

There are four dimensions here. A - Extension and B - Depth apply to Sill A, while C - Extension and D - Depth apply to Sill B.

25. In the Sill Dimensions area, change A - Extension to **2"** [**50**], and change B - Depth to **4"** [**100**] (see Figure 11–7).

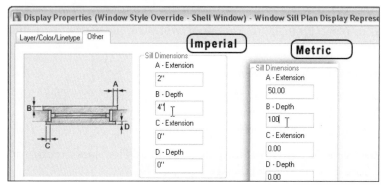

Figure 11–7 *Edit the dimensions of the sills*

26. Click OK to return to the Window Style Properties dialog box.

Note the change in the Viewer (see Figure 11–8).

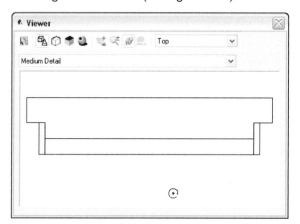

Figure 11–8 *The Sills now extend past the Window Frames*

 NOTE Sill Plan display is available for Door\Window Assemblies and Openings as well. All parameters are configured the same as shown here. A very similar Display Representation exists for Doors called Threshold Plan. It is configured and behaves the same way as Sill Plan except that the components are named Threshold A and Threshold B.

COMPLETE THE CUSTOM WINDOW STYLE

There are a few other tabs in this dialog box. Let's take a quick look at them.

27. Click the Design Rules tab.

    On this tab, the basic shape and type of Window are configured.

28. Change the Viewer to an isometric view, and experiment with the various Predefined Window Shapes and Window Types.

29. When you are finished experimenting, reset the Shape to **Rectangular**, and the Window Type to **Picture**.

30. Click the Standard Sizes tab.

Standard Sizes appear in a list on the Properties palette when Windows of this type are being added to the drawing. You are also able to snap to Standard Sizes while grip editing the size of a Door or Window that uses them. For this reason, Standard Sizes can be very useful. We will be using this Window in two situations, so let's add two Standard Sizes.

31. Click the Add button, and in the Add Standard Size dialog box, type **Shell Windows** for the Description, **5′-0″ [1500]** for Width and **6′-0″ [1800]** for Height and then click OK (see the left side of Figure 11–9).

32. Select the Standard Size: **3′-0″ [900]** Wide by **5′-0″ [1500]** High. Click the Edit button and input **Core Windows** for the Description and then click OK (see the right side of Figure 11–9).

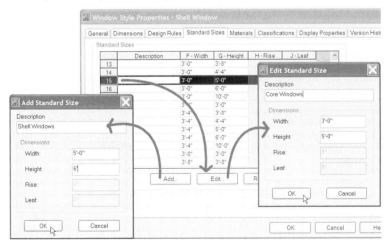

Figure 11–9 *Add some Standard Sizes to the Window style*

TIP If you wish to have only the two standard sizes that we are adding and remove all of the others, you can select the entire list by clicking on the small cell in the first column next to the description at the top left corner. With all cells highlighted, click the Remove button. Then add back the two sizes indicated above.

33. Click the Materials tab.

We have seen plenty of examples of Materials so far. Material Definitions control all hatching and textures applied to various ADT objects. They control the lineweights in plans, sections and elevations and determine special parameters such as transparency for glass. This drawing already contains a few basic Material Definitions. We will simply verify the settings. More information about accessing, creating and editing Materials can be found in Chapters 16. <StepsText> Change the Material for Muntins (currently Standard) to *Doors & Windows.Metal Doors & Frames.Aluminum Windows.Painted.White* (see Figure 11–10).

Figure 11–10 *Assign Material Definitions to the Window components*

34. Verify that **Doors & Windows.Metal Doors & Frames.Aluminum Windows.Painted.White** is assigned to both the Frame and Sash components

35. Verify that **Doors & Windows.Glazing.Glass.Clear** is assigned to the Glass component.

36. In the Viewer window choose **SE Isometric** and click the Gouraud Shaded icon.

37. Click OK to complete the configuration of the Shell Window parameters.

The new Window Style will appear applied to all Windows. Zoom in and study the results.

EDIT WALL DISPLAY PROPERTIES AT WINDOWS

Notice the addition of sills on the Windows (see Figure 11–11). Some of them, however, are oriented the wrong way, with the sill on the inside. In the next sequence, we will correct this. In addition if you are working in Imperial units, there is an extra line on the Sill side of the Window that is undesirable. This line is the line of the Wall beneath the Window. We can turn this off by editing the Object Display Properties of the Walls in this file. If you are working in Metric, review the next several steps and follow along if you wish.

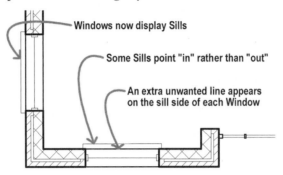

Figure 11–11 *The Windows now have sills and some extra linework*

38. Select any Wall, right-click, and choose **Edit Object Display**.

39. With Plan selected, click the Edit Display Properties button.

40. Click the Other tab.

   Place a check mark in the Hide Lines Below Openings at Cut Plane check box (see Figure 11–12).

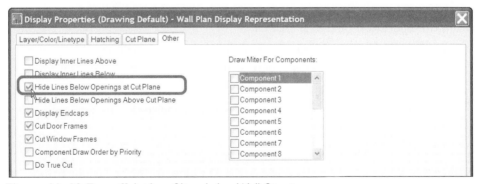

Figure 11–12 *Turn off display of lines below Wall Openings*

41. Click OK twice to return to the drawing.

   The extra line should disappear, as we instructed Walls to no longer display anything beneath openings.

42. Save the file.

**New in ADT
2006**

SYNCHRONIZE PROJECT STANDARDS

This took care of the issue in the current file, but we would need to now repeat this process in the other Constructs in this project to make the display in all files consistent. This is an ideal place to employ Project Standards. We explored the use of Project Standards to synchronize object styles back in Chapter 8. We can also use Project Standards to synchronize Display Settings. To save a bit of time, a file has been provided in the project's *Standards\Content* folder named *Commercial Displays.dwg* [*Commercial Displays - Metric.dwg*]. In this file, the change made in the previous sequence has already been applied to the Drawing Default Display Settings for Walls. We can assign this file as our Standard Display Settings drawing. When we synchronize the project, all of the files will be updated to match.

1. On the Project Navigator palette, on the Project tab, click the Configure Project Standards icon.

2. On the Standard Styles tab at the bottom, choose **Browse** from the "Select drawing to use for standard display settings" drop-down list.

   The Select Standards Drawing dialog should open to the root of the project.

3. Browse to the *Standards\Content* folder of the current project, choose the *Commercial Displays.dwg* [*Commercial Displays - Metric.dwg*] and click Open, then click OK.

4. In the Version Comment dialog, type **"Initial Display Standards Configuration"** and then click OK.

Some of the Windows have already been placed in the upper floors. The name "Shell Windows" has been applied to these Windows, but none of the style edits made so far have been applied. Let's take a look.

5. On the Constructs tab of the Project Navigator, double click *02 Shell and Core* in the *Constructs\Architectural* folder.

6. Zoom in on one of the Windows.

Notice that this Window looks the same as the Window in the first floor before we made any modifications above. It would be quite inefficient to make those modifications again. We can use Project Standards to update the current drawing to the latest version of this style. The style that we configured above has already been copied to the *Commercial Styles* [*Commercial Styles – Metric*] file. This file has also been configured to synchronize Window Styles for the project. Refer to the "Set up Project Standards" heading in Chapter 8 for more information on these configurations.

7. At the bottom right corner of the drawing on the Drawing Status Bar, click the AEC Project Standards quick pick and then choose **Synchronize Drawing**.

8. For now, verify that the only item that has the action of "Update from Standard" is the ***Shell Window*** style and then click OK (see Figure 11–13).

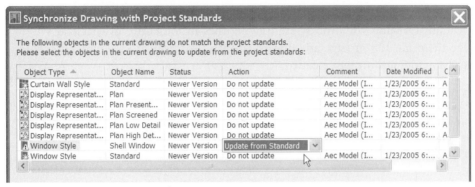

Figure 11–13 *Update only the Shell Window Style for now*

Notice that the line beneath the Sill is still showing. This is because we chose to update only the Window Style. Let's repeat the action and update everything this time.

9. From the AEC Project Standards quick pick choose **Synchronize Drawing** again.

10. In the Object Type list (in the top pane), select the first item, hold down the SHIFT key and then click the last item.

This will highlight all items in the list.

11. Choose **Update from Standard** from the Action column and then click OK.

The sill should now display properly. Project Standards must be configured carefully to achieve desirable results, but as you can see, they provide a very powerful means of propagating changes across the entire project. Now that we are certain that synchronization of the standards will give us the desired results, we can instruct ADT to synchronize the entire project all at once rather than open each file individually.

12. On the Project tab of the Project Navigator, at the bottom, click the Synchronize Project icon.

Wait for the progress bar to complete.

The list of objects in the top pane will be quite long. If you slowly scroll through them all you will note that nearly every item on the list is a Display setting of some kind. Since this is the first time that we are synchronizing Displays, nearly every Display setting in all project files is out of synch. Once we synchronize it this time, this list will be *much* shorter in future synchronizations.

13. Select the entire list of objects in the top pane (use the SHIFT key technique from above), choose **Update from Standard** for the action and then click OK.

Architectural Desktop will process all drawing files in the entire project and synchronize them all. If you have any drawings open on screen, they might indicate that XREFs now need to be reloaded.

CAD Manager Note: Only the basic "user level" procedures of the Project Standards features are being showcased in this text. If you wish to use Project Standards in you firm, you will want to read the online documentation for complete details on all its features. For instance, you can generate log files of synchronizations, perform standards audits and check AutoCAD Standards via the AEC Project Standards functions. A project can use a combination of project-based standards files (as showcased here) and office-wide standards files at the same time. For instance, you might choose to have all Display Setting synchronization come from the office standard drawing template (DWT) file that contains all of the accepted Display Settings while the Style might come from both office standard and project-based files. With most procedures, careful consideration of the options available and consistent implementation of chosen options are paramount to achieving desired results.

MANIPULATING WINDOW ANCHORS

Let's return to the first floor and continue working on our Windows. We now have sills applied to our punched Windows, but many are pointing in different directions. We can flip the Windows manually with grips, which works well for a small selection of Windows, but when many Windows are involved, a more efficient approach is needed. Among the many parameters controlled by anchors is their orientation relative to their parent Wall. We can access Anchor parameters through the Properties palette on the Anchor worksheet or on the right-click menu.

CHANGE THE ORIENTATION OF ALL WINDOWS

To flip the orientation of a single Window is easy. You simply click the arrow-shaped grip.

1. Select any two Windows.

2. Click the arrow-shaped grip pointing perpendicular to one of the selected Windows.

   Each time you click, the Sill will flip from inside to outside and back again.

This method is quick and easy, but notice that even though more than one Window is selected, only the one whose grip you clicked will actually flip. Therefore, although quick and easy, this technique works best for a single Window or a small selection of Windows, but it could get tedious for a large selection such as we have here.

3. With the Windows still selected, right-click and choose **Select Similar**.

4. Right-click and choose **Wall Anchor > Flip Y**.

> Notice that this action *did* flip all of the Windows; however, they all flipped opposite relative to their existing positions, so we are no better off than we were before. To flip only those Windows that point to the inside and make them point to the outside, we must use the Anchor worksheet.

5. Reselect the windows by using the Select Similar method shown above. In the Location grouping on the Properties palette, click the Anchor worksheet icon (see Figure 11–14).

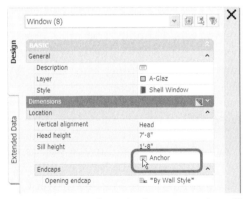

Figure 11–14 *Access Anchor worksheet from the Properties palette (the figure shows the Dimensions grouping collapsed)*

The direction of the Windows is controlled by the three check boxes at the bottom right corner of the dialog box, labeled "Flip X," "Flip Y" and "Flip Z." The check marks in Flip X and Flip Y are currently grayed over because some of the Windows are flipped and others are not. The X direction is parallel to the parent wall; the Y direction is perpendicular to it. Therefore, changing the Y setting will flip the Windows either in or out of the building. Unlike the right-click technique, this worksheet will orient all of the Windows in the same way relative to the Wall. The trick is figuring out which way they will flip, in or out. (In other words, we are not yet certain if we need the check mark in the Flip Y box or need it cleared.)

6. Clear the check mark in the Flip Y box (see Figure 11–15).

The initial orientation of each Window is determined at the time of insertion by the mouse movements. There is no *simple* way to know whether they must be flipped, but there is a 50 percent chance we will guess right.

7. Click OK to complete the change.

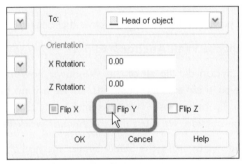

Figure 11–15 *Clear the check mark in the Flip Y box*

Notice that all of the Windows have flipped to the same direction relative to their parent Walls. In this case, we got lucky and the odds were with us. However, had chance gone the other way, we would simply repeat the steps and place a check mark in the Flip Y box, or right-click and choose **Wall Anchor > Flip Y**, which would have given the desired results this time, since all the Windows are now pointing the same way. Give it a try if you like.

ADJUST THE WINDOW POSITION

An anchor controls the Window's position within the wall as well as its orientation. (We began to explore this in Chapter 2.) By manipulating the properties of the anchor, you can shift the position of the Windows. Every ADT object has its own X, Y, and Z directions (see Figure 11–16). Typically, object direction is determined as follows:

- The *Width* of the object corresponds to X.
- The *Depth* of the object corresponds to Y.
- The *Height* of the object corresponds to Z.

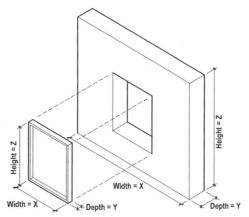

Figure 11–16 *Typical Orientation of ADT objects*

Feel free to experiment with the settings in the Anchor worksheet and see their effect on the Windows relative to these orientations. However, there is a much simpler way to manipulate the position of Doors and Windows relative to their parent Walls. We can use the Location grip in conjunction with the CTRL key for individual manipulations (see the "Explore Objects with Anchors" heading in Chapter 2 for an example of this), and we can use the reposition commands on the right-click menu.

> Make sure that all of the Windows are selected using the Select Similar technique.

8. With all Windows selected, right-click and choose **Reposition within Wall**.

A red line will appear that can be positioned relative to the faces or the center of the Window object. It is best to zoom in with your wheel on a single Window as you perform the next few steps.

9. At the "Select position on the opening" prompt, click the inside face of the Window (see Number 1 in Figure 11–17).

10. At the "Select a reference point" prompt, click the adjacent corner on the inside face of the Wall (see Number 2 in Figure 11–17).

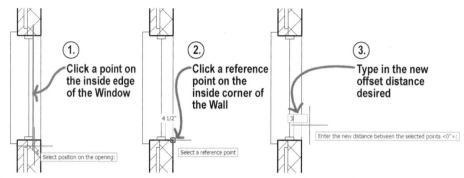

Figure 11–17 *Select the position on the Window and the Wall and indicate the new offset*

11. At the "Enter the new distance between the selected points" prompt, type
3″ [65] and then press ENTER (see Number 3 in Figure 11–17).

The change is subtle, but all of the Windows have shifted the same relative amount and now are a bit closer to the inside edge of the Wall. You can repeat the process and try other offsets and reference points if you wish. Once you have clicked your two reference points (one on the Window, the other on the Wall), a small number will move the Window closer to the point on the Wall and a larger number will move it away from the point on the Wall. If you like, you can select the Window, right-click and choose **Edit Window Style**, and edit the dimensions of the Frame on the Dimensions tab to further manipulate the overall relationship of the Window to the Wall.

12. Save the file.

COPYING EXISTING WINDOWS

We now have several Windows with a nicely configured Window Style, which are carefully positioned relative to the width of the Walls and have sills pointing to the outside. However, our left and right vertical Walls could use a few more Windows. Naturally if we copy one of the ones inserted here, the new copy will use the same Style. Also, as we saw in Chapter 2, when you copy an anchored object, all of the Anchor parameters (such as position within the Wall, Sill height and orientation to the outside) will also be copied. The two routines that we will utilize for this exercise are both new to ADT 2006: Edit in View and the Array AEC Modify Tool.

USING EDIT IN ELEVATION

We can perform the copying of Windows in any view and using any ADT or AutoCAD tool. In this sequence, we showcase a few new additions to our ADT arsenal—starting with Edit in Elevation.

1. Select both the right and left vertical exterior masonry Walls and the two Windows within them.

2. Right-click and choose **Isolate Objects > Edit in Elevation**.

3. At the "Select linework or face under the cursor or specify reference point for view direction" prompt, move the cursor close to the outside edge of the right vertical Wall. When the blue construction line appears, click the mouse (see Figure 11–18).

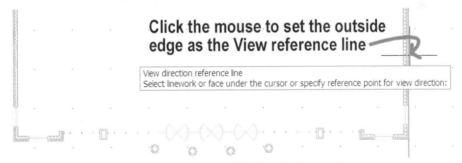

Click the mouse to set the outside edge as the View reference line

View direction reference line
Select linework or face under the cursor or specify reference point for view direction:

Figure 11–18 *Select the face of the right vertical Wall as the View reference line*

4. At the "Specify elevation extents" prompt, drag the mouse to the opposite side of the drawing and click outside the left vertical Wall.

This action will take the drawing into an elevation view and hide all objects except the two Walls and Windows that we selected. Working this way gives us a nice vantage point in which to edit and removes other objects that might prove distracting while we work. Note that an Edit in View toolbar appears floating on screen. This toolbar has a single icon on it that is used to return to the previous viewpoint and restore all hidden objects once we are finished editing.

USING AEC MODIFY ARRAY

We made certain to include both vertical Walls and their Windows above in the elevation's isolated selection so that we could array them together in the next step. Even though from the current vantage point it appears like there is only one Window, you can still select them both and array them together.

5. Click above and to the left of the Windows on screen, moving down and to the right, surround the Windows and click again.

The Properties palette should read Window (2) at the top. If it says "All (#)" instead, right-click and choose **Deselect All** and then try again.

6. With the two Windows selected, right-click and choose **AEC Modify Tools > Array** (see Number 1 in Figure 11–19).

7. At the "Select an edge to array from or ENTER to pick two points" prompt, hover over the left side of the Window and when the blue construction line appears on the outside of the frame, click the mouse to set the reference edge (see Number 2 in Figure 11–19).

Whatever edge you select, the array will move parallel to it. So be certain to highlight a vertical edge. At this point, if you move the mouse left or right, you will see the array indicated interactively. Since we are all the way to the left of our Wall, we need to move our arrayed objects to the right. A default distance between objects will appear in the dynamic dimension. Make sure the DYN icon on the Application Status Bar is toggled on, or you will not get the necessary feedback for this part of the command.

Figure 11–19 *Use the Array AEC Modify Tool to copy Windows along the Wall*

8. In the dynamic dimension that appears on screen, type **10'-0" [3000]** and then press ENTER (see Number 3 in Figure 11–17).

9. At the "Drag out array elements" prompt, drag the mouse to the right all the way to the other end of the Wall. When six Windows are indicated in the Array Count, click the mouse to complete the routine.

You should now have 6 Windows along the Wall in elevation. (There are actually 12 total, with 6 directly in front of the other 6). Notice that despite working in elevation, the Windows behave exactly as they would in plan and cut holes in the parent Wall.

10. Click the Exit Edit in View icon on the Edit in View toolbar to exit the Edit in Elevation mode and return to the previous plan view.

Notice that all of the Windows have their sills correctly pointing outside of the building.

ADD THE NEW WINDOWS TO THE UPPER FLOORS

Now that we have all of our Windows in the first floor, we need to copy them to the upper floor. Initially we will use the same Window layout on all floors. Later, if the design warrants, we can move Windows around on each floor as required. Our first instinct would likely be to simply copy and paste to original coordinates the Windows from the first floor to the other floors. While this would work, the only difficulty would be that when you copy and paste Anchored objects, they include their parent object when pasting. Therefore, we would end up with a duplicate Wall in the other files. While it would be easy enough to erase the duplicate Wall, let's explore an alternative approach. After performing the following steps to copy the Windows to the second floor, you can decide if you would like to adopt this process, or use the copy and paste method instead.

11. Repeat the steps in the "Using Edit in Elevation" heading above to return to Edit in Elevation.

    Select the same two Walls, but be sure to select all 12 attached Windows this time.

12 On the Project Navigator palette, click the Constructs tab and then drag the *02 Shell and Core* file from the *Constructs\Architectural* folder and drop it anywhere in the drawing window.

    This will XREF the *02 Shell and Core* file into the current file and stack it at the correct vertical height based on the settings in the Constructs and the Project's levels.

13. Double-click the *02 Shell and Core* XREF.

    This will active a "RefEdit" session allowing us to edit the XREF in place.

14. In the Reference Edit dialog, be sure that *02 Shell and Core* (and not *Core*) is selected and then click OK.

A Refedit toolbar will appear and the first floor will screen back a bit. When you activate a RefEdit session, you are actually editing the geometry in the XREF file. Any changes you make will apply to that file when you save the session. The XREF file is locked for editing while you are in RefEdit. Therefore, no one else on the project team can open the file and edit it while you are in RefEdit. It is exactly like you had opened the file directly except that you can see the XREF file in context of the host file (the first floor in this case). A "Working Selection Set" is automatically applied to the reference file at the start of RefEdit. This selection set includes all objects in the XREF file and excludes all objects in the host file. However, you can modify this selection set if you need to. This is what we need to do here. By adding the Windows to the working set, we can easily copy them to the second floor.

15. On the refedit toolbar, click the "Add objects to working set" icon.

16. Using a Window selection (left to right and completely surround the objects) select all 12 Windows on the first floor and then press ENTER.

    The 12 Windows should no longer be screened indicating that they are now part of the working set. If you have the command line active, you should see the message: "12 Added to working set."

17. Select all 12 Windows, and copy them up **12′-0″ [3650]**.

You can use the **Basic Modify Tools > Copy** command on the right-click menu or the **AEC Modify Tools > Array** command again to make this copy. Do not use copy and paste. As was indicated above, copy and paste would copy the Walls as well.

18. On the Refedit toolbar, click the "Remove objects from the working set" icon.

19. Select all 12 Windows on the first floor again and then press ENTER.

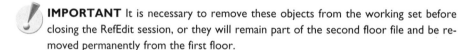 **IMPORTANT** It is necessary to remove these objects from the working set before closing the RefEdit session, or they will remain part of the second floor file and be removed permanently from the first floor.

20. On the Refedit toolbar, click the "Save changes back to reference" icon (see Figure 11–20).

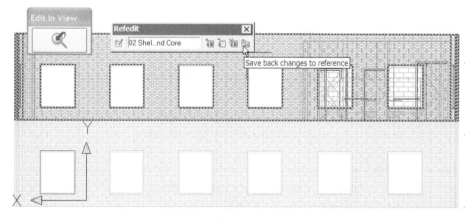

Figure 11–20 *Return the Windows to the first floor file by removing them from the Working Set and then Save the edits back to the Reference*

21. In the confirmation dialog, click OK.

22. On the Drawing Status Bar, click the Manage XREFs quick pick, select the *02 Shell and Core* file and then click the Detach button. Click OK to close the XREF Manager.

23. On the Edit in View toolbar, click the Exit Edit in View icon.

24. Save the file.

25. On the Project Navigator palette, double-click the *02 Shell and Core* file to open it.

    Notice that all the Windows have been added to this file in the correct locations. Close the file when satisfied.

26. Using the RefEdit procedure again, repeat the process to copy the Windows to the third and fourth floors.

As indicated above you can copy the Windows to the clipboard, open the third floor, and use **Paste to Original Coordinates** if you prefer. Remember that if you choose this option, you will want to delete the duplicated Walls *before* pasting.

CONVERT A WINDOW TO A DOOR

One small modification remains for the first floor shell file. There is a Window in the Core area where there ought to be an egress door. We can easily convert this Window to a Door.

27. If you closed *01 Shell and Core*, reopen it now from Project Navigator.

28. On the tool palettes, right-click the title bar and choose **Design** to load the Design tool palette group if it is not already loaded.

29. On the Doors tool palette, right click the *Hinged - Single – Exterior* tool and choose **Apply Tool Properties to > Door/Window Assembly, Opening, Window**.

30. At the "Select door/window assemblies, openings, and/or windows to convert" prompt, click the Window in the Core at the end of the small corridor and then press ENTER.

When you use this feature, the new Door will have exactly the same dimensions as the Window from which it was converted. If you wish to see this, select the Wall and Window, right-click and choose **Object Viewer** to view the Window and its Wall in 3D.

31. Select the new Door, and on the Properties palette, change the Height to **7'-0"** [**2200**], change the Vertical alignment to **Threshold** and the Threshold height to **0"** [**0**].

At the start of the chapter, we noted that most settings and procedures for Windows worked the same for Doors and other openings. Notice that this Door has a threshold line on the side opposite the swing. We need this door to swing out, and also want the threshold on the outside as well. This "Threshold Plan" Display Representation works exactly the same as the Sill Plan for Windows above. We could simply flip the Door with the grip. However, then the swing would point in, and this would likely not meet the fire codes. Instead, let's change the side of the threshold.

> Make sure the Door swings out. If it does not, click the "Flip" grip.

32. Select the Door, right click and choose **Edit Door Style**.

33. Click the Display Properties tab, highlight the Threshold Plan Display Representation and then click the Properties icon.

34. Turn on Threshold A and turn off Threshold B. Click the Other tab and in the Sill Dimensions area, change both A - Extension and B - Depth to **2″** [**50**].

35. Click OK twice to return to the drawing.

Let's make one more edit to complete the first floor. Recall the "Set up a Shared Project Tools Palette" heading in Chapter 10. A similar tool palette has been included in the commercial project. On the commercial project's Shared Project Tools palette, a single Wall tool is provided. Let's apply that tool's properties to the shell Wall of just the first floor. This will give our building a rusticated base on the lowest level.

36. On the tool palettes, right-click the title bar and choose **11 Commercial** [**11 Commercial Metric**] to load the Project tool palette group.

37. Select one of the exterior Walls, right click and choose **Select Similar**.

38. On the Shared Project Tools tool palette, right-click the *Exterior Shell Lower Level* tool and choose **Apply Tool Properties to > Wall**.

39. Save and Close the file.

MANIPULATING DOOR AND WINDOW DISPLAY

There is much more to Door and Window display properties than simply turning on and off components and manipulating Sill and Threshold dimensions. To customize the graphics of Doors and Windows, you can manipulate the display properties of Door and Window styles. You can change the Layer, Color, Linetype, Lineweight and Plot Style properties of any component, and you can add custom components. Adding custom components to Doors and Windows is, however, a little different than it was for Walls in the last chapter. Custom components are added to Doors and Windows using AutoCAD blocks and are applied at the display level. Let's have a look at the potential contained in the Display Parameters for Doors and Windows.

ENABLE A LIVE PLOT PREVIEW

1. On the Project Navigator, in the *Elements\Architectural* folder, double-click the file named *Core* to open it.

   We last had this file open in Chapter 7. No additions have been made to this file since then however, the redundant Wall along the back of the building has been removed.

2. At the bottom of the drawing window, click the Work layout tab.

This Layout tab is intended to provide a convenient way to view your model as you work in both 2D (on the right) and 3D (on the left). This can be very useful indeed. The trickiest part of using it is paying attention to which space—Model Space or Paper (Layout) Space—you are in. For the task that we are about to complete, we will want to preview the model the way it will be plotted. The only way this can be done in Model Space is by using the Plot Preview command. However, in a Layout such as this, we can turn on the "Display Plot Styles" and the "Display Lineweights" options and preview the model in a kind of *live* plot preview. Elements and Constructs use a Model template and are not intended for plotting. Therefore, we must make a few changes to the Work Layout settings to view the model as it will plot. The critical aspects of a plot preview are white paper with black linework, lineweights displayed and the proper scale.

 NOTE We should stress that we are not intending to print from the Element or Construct. Rather we are simply looking for a convenient way to preview how it will look when plotted.

With the Work Layout active, you should see a white background (this is the ADT default). If you do not, it is possible that your settings on the Display tab in the Options dialog box (**Format > Options**) have been modified from the defaults (see Chapter 3). Although it is not critical that you use a white background in Paper Space Layouts, it is recommended. To see all the linework in black, we need to display the Plot Styles.

3. On the File menu, choose **Page Setup Manager**.

4. There will be only one item in the list: *Work*. With it selected, click the Modify button.

5. In the Page Setup – Work dialog, in the top right corner, place a check mark in the Display Plot Styles check box, click OK and then click Close.

The drawing should now be displayed with all black linework. (If you have a black background the linework will display all white instead.) If your lines are still displaying in color, rather than black and white, this indicates that either the default *Plot Style Table - AIA Standard.stb* is missing or that the layers in this file are assigned to the "Normal" Plot Style, which prints as it is seen on screen. Check with your System Administrator or CAD Support person for assistance.

 NOTE Neither the proper display of Plot Style nor the absence of a white background in your Work Layout will hamper you from completing the Door Display exercise below; however, it is much easier to see the results accurately when everything is displayed in true black and white.

Let us now address Scale and Lineweight.

6. Make sure that you are working in floating model space and that the right-hand viewport (the one displaying in plan) is active.

You can tell this by the shape of the UCS Icon (L-shaped vs. triangular) and the Model/Paper indicator on the Application Status Bar. Also, the Plan viewport edge will be bold (see Figure 11–21).

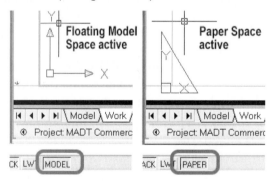

Figure 11–21 *The UCS Icon and the Model/Paper toggle indicate whether model or paper space is active*

If the plan viewport is not active, double-click inside of it to activate it. (Its edge will become bold.)

7. On the Drawing Status Bar, on the right side, click the VP Scale pop-up menu and choose **1/4"=1'-0" [1:50]** (see the left side of Figure 11–22).

Pan the view as necessary. Do *not* zoom or roll your mouse wheel! This will change the scale.

8. On the Application status bar, click the LWT button (push it in) to turn on Lineweight display (see the middle of Figure 11–22).

9. Finally, toggle to Paper Space (double-click outside the viewports, or click the Model/Paper toggle on the Application Status Bar). Select the plan viewport edge and on the Properties Palette, within the Misc Grouping, choose **Yes** from the Display locked list (see the right side of Figure 11-22).

 TIP As an alternative, you can lock a Viewport using the right-click menu. With the viewport selected, right-click and choose Display Locked > Yes.

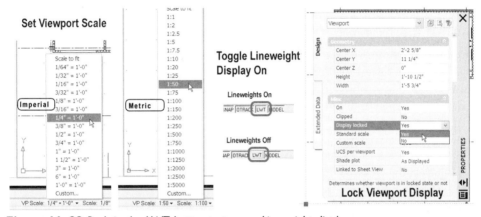

Figure 11–22 *Push in the LWT button to turn on Lineweight display*

You should now have a display that is nearly identical to the way the drawing will appear when plotted. With the viewport display locked, you can freely zoom and pan in both Paper Space and Floating Model Space. Try it out to see.

WORK WITH DOOR DISPLAY PROPERTIES

Like all ADT Objects, a Door has several components in each Display Representation. Each Display Representation has components appropriate to that Representation. In many cases, the same component will show in one or more Display Representations (such as Door Panel and Frame); in other cases, they will show in some and not others (like Glass and Swing). Each component has its own layer, color, linetype, lineweight and plot style. Each can be turned on or off. Additional subcomponents in the form of custom display blocks can also be added; we'll cover these in the next exercise. Figure 11–23 shows the basic components of a Door object in the Model, Plan and Elevation Display Representations.

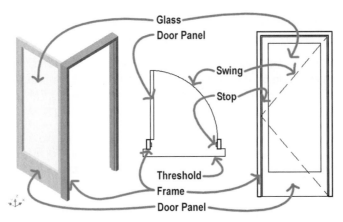

Figure 11–23 *Basic components of the Door object in three Display Representations*

Now that the drawing is giving us the proper visual feedback (as configured in the previous "Enable a Live Plot Preview" heading), we can begin fine-tuning the adjustments to the Door object display. For instance, it might be desirable to have the door swing display in a lighter lineweight than the rest of the door in plan.

10. Zoom in on one of the Doors in the Women's rest room (top right of the plan).

11. Select the Door to the Women's rest room (top right), right-click, and choose **Edit Door Style**.

12. Click the Display Properties tab, be sure that Plan is selected and then click the Edit Display Properties button.

13. Verify that the Layer/Color/Linetype tab is active, and then select the Component labeled "Swing."

14. Click the ByBlock entry in the Lineweight column, choose **0.18mm** from the Lineweight dialog box, and then click OK.

15. Click the ByBlock entry in the Plot Style column, choose Fine from the Select Plot Style dialog box, and then click OK (see Figure 11–24).

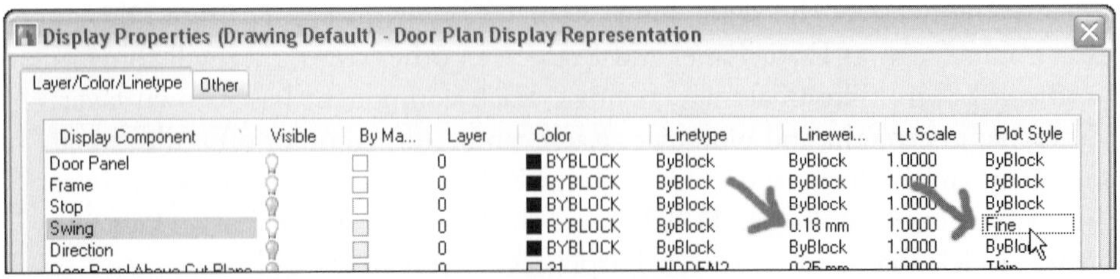

Display Component	Visible	By Ma...	Layer	Color	Linetype	Linewei...	Lt Scale	Plot Style
Door Panel		☐	0	■ BYBLOCK	ByBlock	ByBlock	1.0000	ByBlock
Frame		☐	0	■ BYBLOCK	ByBlock	ByBlock	1.0000	ByBlock
Stop		☐	0	■ BYBLOCK	ByBlock	ByBlock	1.0000	ByBlock
Swing		☐	0	■ BYBLOCK	ByBlock	0.18 mm	1.0000	Fine
Direction		☐	0	■ BYBLOCK	ByBlock	ByBlock	1.0000	ByBlock
Door Panel Above Cut Plane		☐	0	☐ 31	HIDDEN2	0.25 mm	1.0000	Thin

Display Properties (Drawing Default) - Door Plan Display Representation

Layer/Color/Linetype | Other

Figure 11–24 *Change the Lineweight of the Swing component*

16. Click OK twice more to return to the drawing and view the change (see Figure 11–25).

 NOTE Due to variances in video card displays, you may need to zoom in or out a bit for the difference between the swing and panel Lineweight to become apparent on screen.

CAD Manager Note: Traditionally, many firms would have achieved the same result by assigning different layers to each component. You may need to coordinate your choices with your specific plotting standards and/or service bureau. The approach showcased here takes better advantage of the benefits of the ADT Display System. Although the multilayer approach is possible within the Display Properties window, it is really not necessary. Also note that the change we made affects *only* the Plan Display Representation; other Representations remain unaffected. Layers do not offer this level of control.

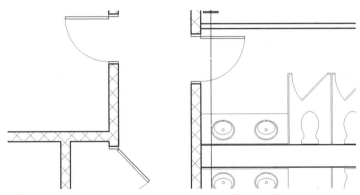

Figure 11–25 *Paper space layouts give a true preview of lineweights*

Using lineweight display will give you a good idea of the final look before committing to a plot.

Notice that if you zoom around the plan, this is a global change that affects all Doors. This is because the Door Plan Representation edits that we made were part of the *Drawing Default*. (Drawing Default affects all objects, Doors in this case, that do not have a Style or Object override.) However, the toilet stalls appear to be unaffected. This results from their having been inserted on a different layer than the Doors. The toilet stall layer uses the same lineweight settings that we assigned to the Swing component, so the entire stall door (including the Swing) is lighter.

 NOTE All of the same techniques apply to editing the Display Properties of Windows. Try the previous steps on some Window objects if you'd like to see for yourself. For instance, you could add variance between the lineweight of the Frame and the glass.

17. Return to model space by clicking the Model tab.

18. Turn off the LWT toggle.

19. Save the file.

CAD Manager Note: Note that we have just made a Display System change in a single drawing file only. If you wanted this change to apply to all Doors in all files, you could use Project Standards to synchronize the project. First, open either Style Manager or Display Manager, right-click the Core file and choose **Update Standards from Drawing**. This will update the Standards Drawing with the changes that were made here. Then you can Synchronize the project to update all the other files.

ADDING CUSTOM DISPLAY BLOCKS IN PLAN

The default graphics used to display ADT objects are suitable for most situations. Occasionally however, the default graphical display of a particular object or a sub-component within an object fails to convey the information required by the particular architectural drawing. Even with the flexibility of available object parameters, it is often necessary to add custom components to Doors and Windows to make them look precisely as required by the project's needs. Any AutoCAD block can be added to the Door or Window style in any of its Display Representations as a custom display component. This allows you to build complex Doors like revolving and overhead doors. You can also use this functionality to add hardware and custom frames, or to create bay windows.

VIEW A DOOR WITH CUSTOM BLOCKS

As you recall, in Chapter 8 we built a Curtain Wall storefront for the first floor and added some revolving Doors for entries. These revolving Doors use custom display blocks in lieu of the default graphics for the Revolving Door type.

1. On the Project Navigator palette, in the *Constructs\Architectural* folder, open the *01 Shell and Core* file.

2. On the Doors palette, click the **Revolving - Simple** Tool.

3. In the drawing, press ENTER to dismiss the "Select wall" prompt and add a free-standing Door.

4. Click a point in front of the building as an insertion point and then press ENTER twice to accept the default rotation and dismiss the command (see Figure 11–26)

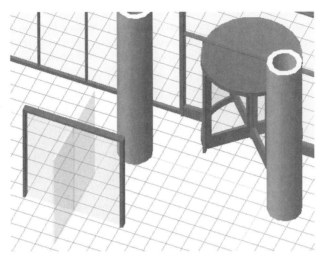

Figure 11–26 *Insert a free-standing Revolving - Simple Door next to the others to compare*

You will naturally see a big difference between the two Revolving Door styles. If you right-click either Revolving Door and choose **Edit Door Style**, you will find that there are many similarities between both styles. You will note only minor differences on the Dimensions, Design Rules and Materials tabs. However none of these differences is enough to account for the presence of the enclosure surround in the *Revolving - Custom* Style nor the absence of it in the *Revolving - Simple*. These differences occur in the Display Properties of these styles.

5. Right-click the **Revolving - Simple** Door, which you just inserted, and choose **Edit Door Style.**

6. Click the Display Properties tab.

   Note that all Display Representations for this style use the Drawing Default settings. These settings are more than adequate for most Doors, but do not suffice for a Revolving Door.

7. Cancel the dialog box and repeat the steps on the **Revolving - Custom**.

   This time you will note that Door Style Override occurs on nearly every Display Representation.

8. With Model selected, click the Edit Display Properties button.

Notice that in addition to the typical components that we are accustomed to seeing for a Door on the Layer\Color\Linetype tab, there are two additional ones: Aec_Door_Rev3D and Aec_Door_Rev3D_Glass. These are AutoCAD blocks that have been added to this Display Representation on the "Other" tab.

9. Click the Other tab.

You will see both of these blocks listed under Custom Block Display. We will go through detailed steps to add a custom display block in the next sequence, and will therefore forego the specifics of the currently selected block. However, if you wish, select one of them and click the Edit button to view its parameters. This is a great way to learn how to use this feature. Please feel free to repeat the exploration on other Display Representations as well. For instance, you will find that in this particular Door style, there are different blocks used for Plans than for Model.

10. Click Cancel to dismiss all dialog boxes when you are finished exploring.

11. Close the *01 Shell and Core* file without saving it.

ADD A CUSTOM DISPLAY BLOCK

In the next sequence, we will add a hollow metal frame to a Door style in our *Core* file. If you closed the *Core* file, please reopen it now.

1. In the *Core* file, change the current Display Configuration to **High Detail** (Drawing status bar, see Chapter 2).

2. Select the Door to the stairwell, right-click and choose **Edit Door Style**.

3. Click the Display Properties tab.

   Notice that Plan High Detail is now active (bold). This is because we changed the Display Configuration.

4. Next to Plan High Detail, place a check mark in the Style Override check box.

   In the dialog that appears, click the Layer/Color/Linetype tab.

The hollow metal frame block will completely replace the default Frame and Stop. Therefore, we have to turn off the default Frame and Stop.

5. Click the light bulb next to Frame to turn it off (Stop is already off).

6. Click the Other tab.

Here we can add custom display blocks. A block named *5HMFrame* [*150HMFrame*] has already been drawn and included in this file. The base point has been located on the left middle point of the frame. Two scaling points have been included on the Defpoints layer to force the block to scale properly when attached to the Door (see Figure 11–27).

 NOTE If you are unfamiliar with creating AutoCAD Blocks, consult a resource on creating Blocks in AutoCAD such as the online help.

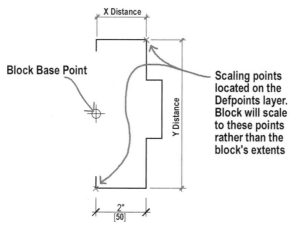

Figure 11–27 *Adding custom scaling points to a block*

A custom display block can be drawn at the specific size that you want it to be or it can be scaled dynamically when added to the Door style. If you choose scaling, by default, a custom display block will scale to the extents of the component that it is replacing. For instance, if the Door Frame is 2″ x 5″ [50 x 250], the custom block will scale to fit within a 2″ x 5″ [50 x 250] rectangle. However, it you add two scaling points to the block, then the scaling will be calculated in such a way as to fit the X Distance between the scaling points to the Width of the component (2″ [50] in the current example) and fit the Y Distance to the Depth (5″ [250] in the current example). A scaling point is simply an AutoCAD point entity that is placed on the Defpoints layer and included inside the block. If ADT finds point entities on Defpoints within a custom display block, it will treat them like scaling points.

7. In the Custom Block Display area, click the Add button (see Figure 11–28).

Figure 11–28 *Click Add to add a custom display block*

On the right side of the dialog box is an embedded Viewer window.

8. Click the Zoom button to activate the zoom mode and drag down slightly within the viewer to zoom the image out a bit (or you can use your wheel mouse instead).

This will make the changes easier to see.

9. Click the Select Block button at the top left corner of the dialog box.

10. Choose **5HMFrame** [**150HMFrame**] from the list and click OK.

11. Put a check mark in the Depth check box in the Scale To Fit area.

   This will match the 5HMFrame block depth (the "Y Distance" shown in Figure 11–27) to the frame depth set in the Door style.

12. In the Insertion Point area, change the Y Insertion Point to **Center** (see Figure 11–29).

   Notice the frame block shift in the Viewer.

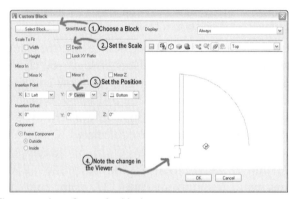

Figure 11–29 *Choose and configure the block*

This completes the setup for the frame on the left, but we need to repeat the steps for the frame on the right.

13. Click OK.

   Notice that **5HMFrame** has been added to the list field at the right.

14. Click the Add button again.

15. Click Select Block, choose the same block, **5HMFrame** [**150HMFrame**], and then click OK.

16. Adjust the zoom in the Viewer as before.

17. In the Scale to Fit area, again check Depth

18. In the Mirror In area, put a check mark in Mirror X.

   This flips the block for use on the other side.

TIP If the Viewer does not update, click OK, select the second **5HMFrame** [**150HMFrame**] in the list and then click Edit. This will return you to the same dialog box with the Viewer updated.

19. In the Insertion Point area, choose **Right** for X and **Center** for Y.

In the Viewer, notice that the frame is shifted a bit. This is because the base point of the block (as shown in Figure 11–27) is on the left side of the block, and we have shifted the block to the right (see Figure 11–30). Adding an offset equal to the width of the block will correct the problem.

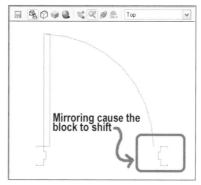

Figure 11–30 *The block is shifted*

20. In the Insertion Offset area, type **-2″** [**-50**] for the X offset and press ENTER (see Figure 11–31).

Figure 11–31 *Add an offset in the X direction to compensate*

This will correct the offset of the block.

21. Click OK to return to the Other tab.

22. Click the Layer/Color/Linetype tab.

Notice that the two custom blocks have been added as custom components. You could now change their layer and other properties if required. In this case, the default settings will be fine.

23. Repeat the edits (Lineweight and Plot Style) made to the Swing above.

24. Click OK twice to return to the drawing and notice the change to all of the Doors (see Figure 11–32).

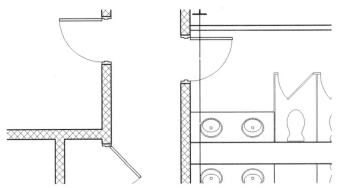

Figure 11–32 *All Doors now have a hollow metal frame in the High Detail Display Configuration*

25. Change the current Display Configuration back to **Medium Detail**.

Notice how all of the Door Frame revert back to the simple rectangle. This is another example of the power of the Display System. By adding the Custom Display Block only to the High Detail Display Configuration, we can easily swap between simple and high detail frames.

26. Save and Close all Commercial Project files.

CREATING CUSTOM-SHAPED WINDOWS AND DOORS

Several predefined shapes are available for Window and Door styles. However, as the need arises, custom shapes can be defined. Any Profile Definition (a type of ADT Style used to form the shape of several object types) in the drawing can be used to customize the elevation shape of a Window or Door style. Profile definitions appear in a list of Custom Shapes on the Design Rules tab of the Door and Window Style Properties dialog boxes. A profile is simply one or more closed polylines that have been named and saved as profiles. With the powerful edit in place functionality, customizing the shape of Doors and Windows is simple. We will begin our exploration in a separate file and then import the styles created back to the project files.

LOAD THE RESIDENTIAL PROJECT

We will continue our exploration of Doors and Windows with a return to the Residential Project. Be sure that all files from the Commercial Project have been closed and saved. However, you must make sure that a single empty file remains open. The Project Browser will not run in "zero-doc" state. The simplest way to do this is to close all files, and then use the Qnew icon to create a new empty file.

1. From the File menu, choose **Project Browser.**

2. Click to open the folder list and choose your *C:* drive.

3. Double-click on the *MasterADT 2006* folder, then the *Chapter 11* folder.

   One or two residential Projects are listed: *11 Residential* and/or *11 Residential Metric.*

4. Double-click *11 Residential* if you wish to work in Imperial units. Double-click *11 Residential Metric* if you wish to work in Metric units. (You can also right-click on it and choose **Set Current Project**.) Then click Close in the Project Browser.

 IMPORTANT If a message appears asking you to re-path the project, click **Yes**. Refer to the "Re-Pathing Projects" heading in the Preface for more information.

MODIFY THE WINDOW SHAPE

Two small files have been added to the Residential Project since the previous chapter. They are both within the *Elements* folder. Elements are typically used for project components that repeat in one or more Constructs. The *Core* and *Column Grid* files in the Commercial Project are good examples of this. However, you can also use Elements as a convenient place to store project resources that are not specifically tied to the building model in the way Constructs are. We have seen a few examples of this in the previous chapters. For more information on Projects and Project terminology, see Chapter 5.

1. On the Project Navigator, in the *Elements* folder, double-click the *Door and Window* file to open it.

The simplest way to customize the shape of a Window or Door is to use the edit in-place functionality. Let's add a new profile to the Windows in this file using edit in place.

 NOTE Profiles are not just for Windows and Doors. Many ADT objects (Slabs Edges, Roof Edges, Railings, Curtain Walls, etc.) can have their shape customized by referencing Profiles. Similar techniques will apply to those objects. Examples occur in Chapter 8 and Chapter 12.

2. Select one of the Windows on screen, right-click and choose **Add Profile**.

3. In the Add Window Profile dialog box, be sure that Start from Scratch is chosen for the Profile Definition, type **Window – Chamfered Corners** for the name, and then click OK (see Figure 11–33).

Figure 11–33 *Start a new Window Profile from Scratch*

The in-place edit profile will appear, as will the In-Place Edit toolbar. Try some random grip editing to get hang of the In-Place Edit mode. Undo when done experimenting.

4. Select the thin rectangular grip at the top edge of the Window and then press CTRL once to toggle to Add Vertex mode

There are several edit modes once a grip is activated. A tip will appear on screen to indicate them. Use the CTRL key to cycle through the choices.

5. Move the mouse directly to the right with either POLAR or ORTHO on, and then type **8″** [**200**] (see Figure 11–34).

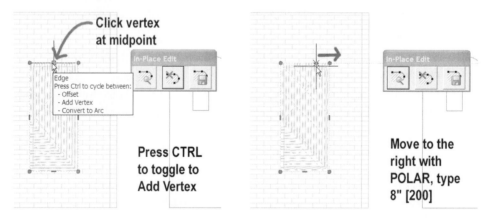

Figure 11–34 *Use the CTRLkey to toggle to Add Vertex mode*

You should now have two segments across the top.

6. Repeat the process on the longer segment on the left at the top. Drag the new vertex to the left **5″** [**125**] this time (see Figure 11–35).

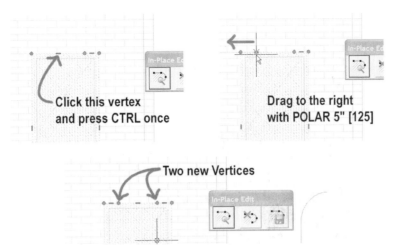

Figure 11–35 *Add another vertex to the opposite side*

7. Click the top right corner grip (small circle) to make it hot, and move it straight down (with ORTHO or POLAR).

8. Type: **8″ [200]** and press ENTER.

   Repeat on the other side (see Figure 11–36).

TIP Sometimes edits like this are easier with DYN toggled off temporarily.

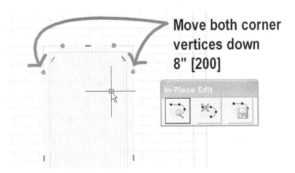

Figure 11–36 *Grip edit the corners of the shape*

9. Click the Save all Changes icon on the In-Place Edit toolbar.

   Note the results. The shape has been applied to both Windows. This was a Style-level edit.

10. Select either Window, right-click and choose **Edit Window Style**.

11. Click the Design Rules tab.

Notice that Custom is now chosen for the Shape of this Window style and that *Window – Chamfered Corners* is chosen for the Profile.

12. Click Cancel to dismiss the dialog box.

CREATE A CUSTOM-SHAPED DOOR

Creating profiles for Doors is slightly different than for Windows. If you will only be changing the outer shape of the Door's elevation, then use the same process covered in the previous exercise. If you wish your Door to have vision panels or lites, you can add another shape to the Profile. Also note that you can draw polylines for the shapes of both inner and outer rings while in the In-Place Edit routine.

13. Off to the side of the drawing are two polylines. Move them directly on top of the Door.

    You can snap the bottom midpoint of the large outer polyline to the bottom midpoint of the Door Panel.

14. Select the Door, right-click and choose **Add profile**.

15. In the Add Door Profile dialog box, be sure that Start from Scratch is chosen for the Profile Definition and Type **Door Round Corners – Small Vision Panel** for the name and then click OK.

16. Right-click and choose **Replace Ring**.

17. At the "Select a closed polyline, spline, ellipse, or circle" prompt, click the outer polyline shape (the one with the rounded corners).

18. At the "Erase layout geometry" prompt, right-click and choose **Yes** (see Figure 11–37).

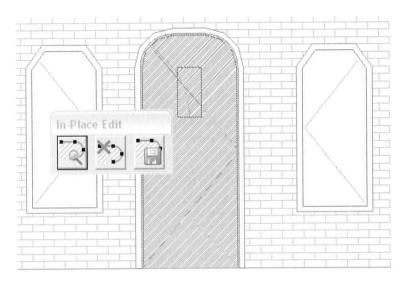

Figure 11–37 *Click the outer polyline to replace the ring*

19. Select the In-Place Edit profile (click the blue hatching), right-click and choose **Add Ring**.

20. When prompted, select the inner rectangle, and answer **Yes** to erase it.

21. Click the Save all Changes icon and view the results (see Figure 11–38).

Figure 11–38 *Add the inner ring as a void*

As before, if you were to edit the Door style, you would find that this new profile has been added on the Design Rules as a custom shape. Go ahead and select the Door, right-click and bring it into the Object Viewer and have a look at it in 3D.

Notice that the lite correctly renders as a void in the Door. However, the glass is not transparent. Let's make that correction.

22. Select the Door, right-click and choose **Edit Door Style**.

23. Click the Materials tab and assign **Doors & Windows.Glazing.Glass.Clear** to the Glass component.

24. Assign **Doors & Windows.Metal Doors & Frames.Aluminum Frame.Anodized.Dark Bronze.Satin** to the Muntins component and then click OK.

    You may notice that the curves on the Door appear faceted after the edit is complete. There is an ADT setting that controls this.

25. Type **aecfacetdev** and then press ENTER.

26. At the "Set new Facet Deviation" prompt, type **.001** and then press ENTER.

The smaller the number, the smoother the curves will render, but performance can suffer. If your drawings become too slow at the **.001** setting, try **.01** instead.

27. Save the file.

USING THE STYLES IN OTHER DRAWINGS

Once you have built a custom style in one drawing, you will often wish to use it in other drawings within the project, or even on future projects. The process was covered extensively in the last chapter in the "Project Tool Palettes" topic. To apply these styles to the Doors and Windows in the project files, we will add them to our Shared Project Tools tool palette as we did with the Walls in Chapter 10.

CREATE THE TOOLS

1. On the tool palettes, click the Project Tools tab to make it active.

    If this tab is not available, right click the tool palettes title bar and choose **11 Residential [11 Residential Metric]** to make the project tool palette group active.

In the last chapter you recall that we used "Project Tools" as a scratch pad to create the tools, and then we copied them to the "Shared Project Tools" palette via the project's Content Browser Library. If you did not do the tutorials in Chapter 10 then please review the "Project Tool Palettes" topic of Chapter 10 before proceeding.

    Be sure that the file has been saved.

2. Select the Door, drag it and drop it onto the Project Tools palette.

A ***New Addition Rear Entry*** tool will appear on the palette.

3. Select one of the Windows, drag it and drop it onto the Project Tools palette (see Figure 11–39).

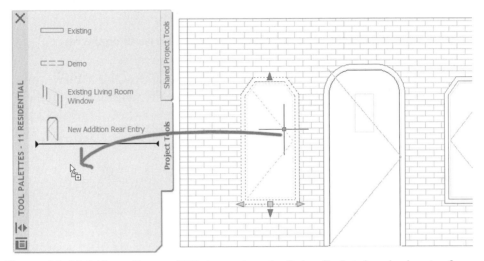

Figure 11–39 *Add new Door and Window tools to the Project Tools palette by dragging from the drawing*

As noted in Chapter 10, if you are a one person firm, you can leave the tools on the Project Tools palette. If you need to share the tools with other team members, then Copy or Cut them to the clipboard. Click the Content Browser icon on the Project Navigator to open the project's Content Brower Library and browse to the Shared Project Tools palette. Right-click in this palette and choose **Paste**. Finally return to ADT and refresh the Shared Project Tools palette.

4. Save the *Door and Window* file.

CAD Manager Note: Please note that the process that we are using here and the one outlined in Chapter 10 do not take into account the use of Project Standards. Up to this point, we have only enabled Project Standards for the commercial project. We certainly could enable them in the residential project as well. The reason that we have not done so is to showcase varying but similar techniques. In your own projects, big or small; commercial, residential or industrial; you can make use of either Project Tool Palettes, Project Standards or both. If you decide to use both Standards and Tool Palettes, it is recommended that you create your tools from styles in Standards Drawings. Otherwise, all of the other steps in the procedure would apply.

APPLY CUSTOM STYLES TO RESIDENTIAL PLAN

Now let's apply the tool properties to some objects in the project.

5. In the Project Navigator, open the *First Floor New* file.

6. Select the two Windows on the left exterior wall of the addition.

7. Right-click the **New Addition Single Casement** tool and choose **Apply Tool Properties to > Window**.

8. Repeat the process for the **New Addition Rear Entry** tool on the rear exterior Door in the addition.

There will be little evidence of the change in plan. Try using the Object Viewer or one of the preset 3D Views in the drawing window to see the styles applied to the model (see Figure 11–40).

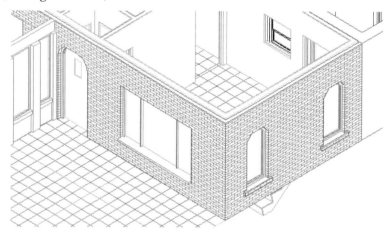

Figure 11–40 *Applying the custom styles to the model*

9. Save the *First Floor New* file.

ADDING WINDOW MUNTINS

Window muntins can be added to the Display Representations of Windows and Doors parametrically. The interface and procedure are similar to adding custom display blocks. In the steps below we will add muntins to a Window; however, the procedure is identical for Doors.

ACCESS WINDOW MUNTIN DISPLAY PROPS

1. Switch to the *Door and Window.dwg* file. If you have closed the file, reopen it from the *Elements* folder now.

The first thing we need to determine is in which Display Representations we would like the muntins to be displayed. Unique to Muntin Block Display is the ability to link the Muntins Blocks from all Display Representations so that editing one of them edits them all. For this exercise, we will add muntins to the Model, Model High Detail and Elevation Display Representations. We will leave them out of Model Low Detail.

2. Select one of the Windows, right-click and choose **Edit Window Style**, and then Click the Display Properties tab.

3. Select the Model High Detail Display Representation (it is not bold currently) and then place a check mark in the Style Override box.

4. Click the Muntins tab.

5. Verify that "Automatically Apply to Other Display Representations and Object Overrides" is checked (see Figure 11–41).

Figure 11–41 *Overrides are applied to two Model Display Representations, and Elevation is active*

6. In the Muntins Block Display area, click the Add button, and then immediately click OK twice.

All we have done here is add the Muntin Block, but we will wait to configure it until we have added one to each Display Representation. This way, when we edit it in one Display Representation, the edit will apply to the others as well. This synchronized editing cannot occur if the Muntins Block does not exist in a particular Representation.

7. Select the Model Display Representation (it is not bold currently) and then place a check mark in the Style Override box.

8. Repeat the same steps to add a Muntins Block and then immediately click OK.

At this point there should now be an override applied to both Model and Model High Detail, and Elevation should be bold.

9. Verify that Elevation is active (bold) and selected in the Display Representation list and place a check mark in the Style Override box next to it.

10. Click the Muntins tab.

11. In the Muntins Block Display area, click the Add button.

This time we will configure the Muntins Block, so don't dismiss the dialog box. Because we had the "Automatically Apply to Other Display Representations and Object Overrides" box checked in the Model and Model High Detail Representations, any edits we make here in Elevation will automatically apply to those Representations. In the Window Pane area at the top left of the dialog box, you can designate which window pane should receive muntins. Choices include Top, All, or a specific pane designated by index number.

12. Choose the "Other" radio button and then choose the "All" radio button.

 NOTE If you need muntins in both Top and one other pane, you must add two Muntins Blocks. This is not an issue for this Window Style, since it has only one pane.

In the Lights area, you can set the number of lights you wish in each direction.

For Lights High use **3** and for Lights Wide use **2** (see Figure 11–42).

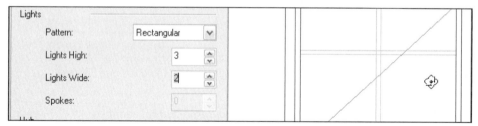

Figure 11–42 *Add a Muntins Block and configure the settings*

Notice that the horizontal and vertical muntins cross one another in the Viewer.

13. Put a check mark in the Clean Up Joints and the Convert to Body check boxes (see Figure 11–43).

Notice that all intersections between horizontal and vertical muntins cleanup when Clean Up Joints is turned on.

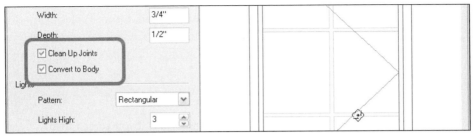

Figure 11–43 *Use Clean Up Joints to resolve intersections*

14. From the Pattern list, experiment with each of the four choices and then settle on one that you like.

15. Click OK three times to return to the drawing and view the effect of the changes (see Figure 11–44).

Figure 11–44 *Pick a pattern type to complete the muntins*

16. Save and Close the *Door and Window.dwg* file.

RE-IMPORT STYLES

You may recall from the last chapter that when you drag a style to a tool palette, it records the path to the file where the style is saved. In this case, the two tools that we just made point back to the *Door and Window.dwg* file. Since we have just made some additional edits to the Window style here in the *Door and Window.dwg* file, our *First Floor New* file is no longer up to date. This situation allows us to see another benefit of the tool palette system.

17. Return to the *First Floor New* file. If you closed it, reopen it from Project Navigator now.

18. On the View toolbar, choose **SW Isometric View** (or choose **3D Views > SW Isometric View** from the View menu).

19. Zoom in on the two Windows in the addition.

20. On the Project Tools palette (or Shared Project Tools if you copied them there), right-click the **New Addition Single Casement** tool and choose **Re-import 'New Addition Single Casement' Window Style**.

The two Windows should now inherit the change from the source file and display with muntins. As you can see in Figure 11–45, even though we performed the muntins edits with the Elevation Display Representation active, the "Automatically Apply to Other Display Representations and Object Overrides" feature automatically applied those edits to the Model Display Representation as well. Change the current Display Configuration to High Detail, and you will see them there as well. However, since we did not apply them to the Model Low Detail Representation, the muntins will disappear if you load Low Detail. Give it try. At any future time, we can return to the *Door and Window.dwg* file, make edits, save them and the use the Re-import feature to apply those changes across the project.

Figure 11–45 *The results of re-importing the style with muntins*

Here again, we are merely showcasing an alternative method to the Project Standards approach. Tools can be used to keep styles in synch as shown here. However, the tool system cannot update the styles without manual intervention. To update a style with a tool you must open the file that needs updating and then perform the steps listed here. This is fine for small projects with few custom styles that need only infrequent updates. If you have more styles, frequent updates or large project teams, the Project Standards are a much more efficient and powerful tool. Furthermore, while not covered in this text, Project Standards can be set to either semi-automatic or automatic updates in which the update process can become partially or completely automated. Consult the online help for more details.

21. Save and Close the *First Floor New* file.

 ## CREATE A CUSTOM MULTI-VIEW BLOCK

Plumbing fixtures have been added in the previous chapters. As we saw in those exercises, plumbing fixtures are ADT Multi-View Block (MVB) objects. ADT ships with a vast library of pre-built Multi-View Blocks. However, there will undoubtedly be those circumstances when the symbol required for a specific item is not available in the default library. In this case, you can build your own Multi-View Block content. We will look at the steps here to build a custom Multi-View Block for the new whirlpool bathtub on the second floor of the residential project. To make a Multi-View Block, you must first create an AutoCAD block for each unique view your Multi-View Block will require. Be certain that all of the insertion points line up in the same spot, and create elevation views "upright" in the correct plane. Plan and elevation views are typically drawn two-dimensionally (see Figure 11–46) and often contain a flat surface behind the linework. (An AEC Polygon object can be used for this.) This surface will conceal objects behind the Multi-View Block when it is used to generate 2D Section\Elevation objects (see chapter 16 for more on 2D Section/Elevation objects). In this exercise, the required Auto-CAD blocks have been provided in a file already. The blocks were adjusted from files provided by a plumbing fixture manufacturer. Each manufacturer builds their CAD files differently, so the amount of rework you will need to do and your results in using them will vary. We will therefore focus on the steps to create the Multi-View Block once the blocks are available and complete.

 NOTE Several manufacturers provide AutoCAD Blocks and/or ADT Styles and Content. You can see some of the providers listed in the Content Browser. Choose Content Browser from the Window menu (not the one on Project Navigator) and look for the *Architectural Desktop and VIZ Render Plugins* catalog.

1. On the Project Navigator in the *Elements* folder, double-click to open the file named *Whirlpool Tub.dwg*.

2. On the tool palettes, click the Project Tools tab to make it active.

Four blocks are inserted in this drawing. The red targets indicate the insertion points. Notice that the same relative corner is used consistently.

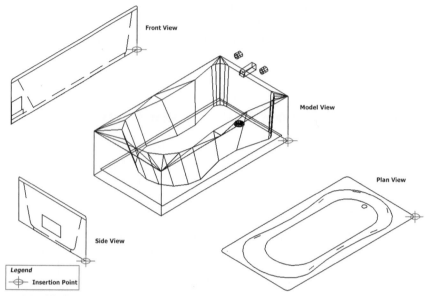

Figure 11–46 *Create the AutoCAD blocks with proper alignment and insertion points*

3. From the Format menu, choose **Style Manager**.

4. Expand *Multi-Purpose Objects*, and then select *Multi-View Block Definitions*.

5. On the right side, right-click and choose **New** and call it **Whirlpool Tub**.

6. Right-click on the *Whirlpool Tub* style, choose **Edit**

7. On the General tab, change the Description to **Mastering Autodesk Architectural Desktop 2006 - Whirlpool Tub – Residential Project** and then click the View Blocks tab.

In the View Blocks tab are listed the Display Representations of the Multi-View Block object type on the left, a column in the middle that will list any blocks that are loaded, and a column at the right listing the seven possible View Directions. Building a Multi-View Block is actually quite simple. Choose a Display Representation, load a block, and then check which View Directions should trigger the display of that block. In our case, we have a plan, front, side, and 3D model View Block. The General Display Representation is used to set up all orthographic drawings (plans, sections, and elevations). The Model Display Representation is exclusively for 3D. Plan High Detail and Plan Low Detail are for scale-dependent symbols (symbols that change size and/or level of detail in response to drawing scale). Finally, the Reflected Display Representation is used for reflected ceiling plans. In this case, we only need to add Blocks to the General and Model Representations.

8. Click the Add button.

9. In the list of blocks, choose **Whirlpool_Tub_P**, and then click OK.

The "_P" suffix indicates that this is the "Plan" block. Notice that all of the check marks under View direction are checked by default.

10. Leave Top and Bottom checked, and clear the rest.

11. Click Add again, choose **Whirlpool_Tub_F**, and then click OK.

12. Leave Front and Back checked, and clear the rest.

13. Click Add again, choose **Whirlpool_Tub_S**, and then click OK.

    Leave Left and Right checked, clear the rest (see Figure 11–47).

Figure 11–47 *Add the Plan and the Side View Blocks*

14. Click Add again, choose **Whirlpool_Tub_M**, and then click OK.

15. Leave Other checked, and clear the rest (see Figure 11–48).

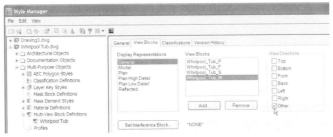

Figure 11–48 *View Blocks for the General Display Representation*

NOTE Even though the General Display Representation is typically used for plans and elevations, occasionally you will view the Multi-View Block in elevation, but from an oblique angle (that is, not 90°). These oblique views will trigger the display of the block loaded for in the "Other" view Direction. Loading the 3D model block (_M) will ensure that the Multi-View Block displays something in this instance.

16. On the left, click the Model Display Representation.

17. Click Add, choose **Whirlpool_Tub_M**, and then click OK.

18. Leave all of the View directions checked here (see Figure 11–49).

This Display Representation is used exclusively for 3D applications; therefore, regardless of View Direction; we want the 3D block to be displayed here.

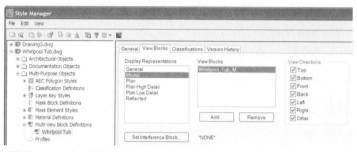

Figure 11–49 *The completed View Block configuration*

19. Drag the new **Whirlpool Tub** Multi-View Block Definition from the Style Manager and drop it on the Project Tools palette.

 NOTE If the 11 Residential [11 Residential Metric] palette is not in front, click OK to exit the Style Manager, and click the Project Tools tab on the tool palettes to make it active. Then return to Style Manager and the Multi-View Block Definitions category and perform the drag and drop.

A warning will appear indicating that the tool will fail unless you save the file.

20. Click OK to dismiss the Style Manager.

21. Save and Close the *Whirlpool Tub.dwg* file.

 IMPORTANT Don't forget to Save the Whirlpool Tub file, or the tool will not work

22. On the Project Tools palette, right-click the new **Whirlpool Tub** tool and choose **Properties**.

As before, this tool maintains a link back to the definition in the *Whirlpool Tub.dwg* file.

23. Click on the Layer Key worksheet icon, and choose **PFIXT** from the list.

24. Beneath Location, choose **Yes** from the Specify Rotation on Screen list.

If you are satisfied with the default icon generated for this tool, click OK to dismiss the Properties dialog box. Otherwise, use the following steps:

▶ Scroll to the bottom of the worksheet and right-click in the Viewer.

▶ Choose **Preset Views > NW Isometric**.

▶ Right-click again and choose **Shading Modes > Gouraud Shaded**.

▶ At the top of the worksheet, right-click directly on the existing icon (directly below the label "image") and choose **Refresh image** (see Figure 11–50).

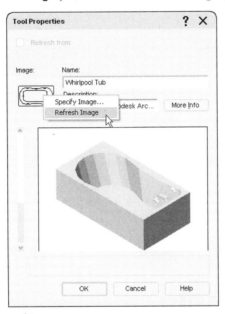

Figure 11–50 *Refresh a tool image*

25. Click OK to dismiss the Tool Properties worksheet and complete the tool configuration.

As before, if you wish, cut and paste this tool to the Shared Project Tools palette in the project based Content Browser Library. See the note above and the topics in Chapter 10 for more details.

26. In Project Navigator in the Constructs folder, open the *Second Floor New* file.

27. Click the new Whirlpool Tub tool and add the tub to upper corner of the bathroom on the left (see Figure 11–51).

    Notice that the tub is added on the correct layer and that it prompts you for rotation after you place it. This is the result of the settings that we added in the Tool Properties worksheet.

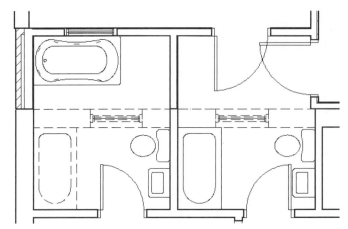

Figure 11–51 *Adding the new tub to the plan*

28. Using the icons on the Views toolbar or the commands on the **View > 3D Views** menu, change the drawing from **Top**, to **Left**, to **Front** and then any Isometric View.

Notice the way the new Multi-View Block changes to match each respective viewpoint. When you are satisfied that the MVB has been created properly, return to Top view.

29. Save and Close all project files.

ADDITIONAL EXERCISES

Additional exercises have been provided in Appendix A. In Appendix A you will find suggestions for adding a bay window to the dining room of the Residential Project and Column enclosures to the shell of the Commercial Project (see Figure 11–52). It is not necessary that you complete these exercises to begin the next chapter, they are provided to enhance your learning experience. Completed projects for each of the exercises have been provided in the *Chapter 11/Complete* folder.

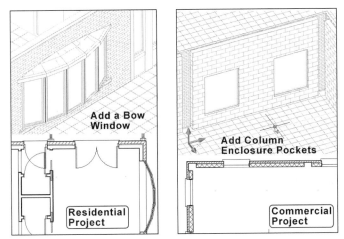

Figure 11–52 *Appendix A additions to the projects*

SUMMARY

Door and Window positions are controlled by their anchor settings.

Basic type and shape settings occur on the Dimensions and Design Rules tabs of the Door Style and Window Style dialog boxes.

Door and Window position can be manipulated in place with grips and right-click functionality, or in worksheets from the Properties palette.

The Edit in View tools provide a powerful way to isolate and edit a selected portion of the model in a convenient plan, section or elevation view.

Profiles can be used to customize the shape of Doors and Windows.

Profiles can include lites (rings set as voids) in Door styles.

Each default component of the Door and Window can have its own Layer, Color, Linetype, and Lineweight.

AutoCAD blocks can be used to customize component display in any view.

To establish custom scaling points within the custom block, include two scaling points on the Defpoints layer within the AutoCAD block definition to establish scaling.

Window muntins can be added parametrically without the need for a separate display block.

A Multi-View Block is composed of one or more AutoCAD blocks designed to represent the object from different viewing angles.

A Multi-View Block tool can be made with presets for Layer and rotation.

CHAPTER 12

Horizontal Surfaces

INTRODUCTION

In this chapter, we will add the various horizontal surfaces of the building, specifically, Roofs, Slabs and Spaces. Architectural Desktop offers two types of Roof objects: the standard (one-piece) Roof and the Roof Slab. With a variety of parameters including Rise, Run, and Edge conditions, these tools allow us to model complete roof structures and assemblies. Horizontal planes and slabs can be built with the Slab object. Slabs share many parameters in common with Roof Slabs—they can be flat or pitched, and they can vary in thickness and include holes. Slabs are beneficial for 3D modeling of complex horizontal planes and for generating sections. Floors and ceilings can be modeled using Slab or Space objects, depending on your project needs. Space objects are well suited to floor plans and provide an easy means of tracking room area, finishes and other Schedule properties.

OBJECTIVES

- Build Roofs.
- Learn to create and modify Slab objects.
- Add floor Slabs to the Project model.
- Understand Space objects.
- Add Spaces to existing files.
- Add Space labels.

CREATING ROOFS

There are two types of Roof object in Architectural Desktop: the standard Roof and the Roof Slab. The standard Roof object will be referred to here as the "one-piece" Roof. This is because it is comprised of a single monolithic piece of material, which includes all of the sloping faces. A Roof Slab is an individual sloping roof surface. The advantage of the Roof Slab over the one-piece Roof is that each edge of the Slab can be manipulated separately. As a general rule of thumb, it is useful to begin your roof design with the one-piece Roof and take it as far as its somewhat limited parameters will allow. This is easier because you will not need to be concerned with the mitering of edges and the intersection of roof planes. However, once the design progresses, you will likely need to convert it to Roof Slabs to gain more control and properly articulate the design. This basic prescription will be followed for the Residential Project in the sequence that follows.

INSTALL THE CD FILES AND LOAD THE CURRENT PROJECT

1. If you have not already done so, install the dataset files located on the Mastering Autodesk Architectural Desktop 2006 CD ROM.

   Refer to "Files Included on the CD ROM" in the Preface for information on installing the sample files included on the CD.

2. Launch Autodesk Architectural Desktop 2006 from the desktop icon created in Chapter 3.

If you did not create a custom icon, you might want to review "Create a New Profile" and "Create a Custom ADT Desktop Shortcut" in Chapter 3. Creating the custom desktop icon is not essential; however, it makes loading the custom profile easier.

3. From the File menu, choose **Project Browser**.

4. Click to open the folder list and choose your *C:* drive.

5. Double-click on the *MasterADT 2006* folder, then the *Chapter12* folder.

   One or two residential Projects will be listed: *12 Residential* and/or *12 Residential Metric*.

6. Double-click *12 Residential* if you wish to work in Imperial units. Double-click *12 Residential Metric* if you wish to work in Metric units. (You can also right-click on it and choose **Set Current Project**.) Then click Close in the Project Browser.

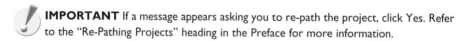 **IMPORTANT** If a message appears asking you to re-path the project, click Yes. Refer to the "Re-Pathing Projects" heading in the Preface for more information.

WORKING WITH THE ONE-PIECE ROOF

Roofs can be added manually by placing them point by point or by converting existing geometry into a Roof. In many circumstances, it is easier to convert existing geometry to a Roof. To do so, you would draw a polyline outline of the roof perimeter, and then convert it to a Roof. In this exercise, we will explore both techniques. We will first trace the footprint of the existing floor plan in the residential building manually with the Add Roof Command. Later we will convert some existing polylines for the roof over the new addition. As you work, you will indicate to ADT at each point whether you wish to create a sloped surface or a gable end.

CREATE A NEW CONSTRUCT

1. On the Project Navigator, right-click the *Constructs* folder and choose **New > Construct**.

2. Name the new Construct **Roof New**, with a Description of **Roof – New Construction**.

3. In the Assignments area, place check marks in both the "Roof – New" and "Roof – Existing" check boxes, and then click OK.

The roof of the existing house already exists in rough form in the *Roof Existing* file. However, since the roof will undergo major reconstruction, and the entire roof will be re-shingled in this project, we will recreate the entire Roof structure in this file. Later, if you wish to see Views of the project as existing before the new work, you will reference only the *Roof Existing* file. If you wish to see the house after new construction, you will reference the *Roof New* file instead. (See Chapter 5 for more information on Views.)

4. A new file named *Roof New* will appear. Double-click to open it.

5. From the Project Navigator, drag and drop the *Second Floor Existing* file into the drawing.

6. Repeat this process for the *Second Floor New* file as well. Zoom in on the XREFs.

We will construct the Roof of the Residential Project in this file. We have XREFed the second floor files into this drawing to assist us in placing the various portions of the Roof.

ADD ROOF

As we have already stated, we will begin with the one-piece Roof and lay out the existing portion of the Roof first. We will add this Roof point by point so that you can see the manual process. It is also possible to convert a polyline of the roof outline or the Walls of the Second Floor Existing file directly into a Roof. An example of this technique can be found in the "Create a Small Building" tutorial in Chapter 1.

7. On the Design tool palette, click the *Roof* tool.

If you do not see this palette or tool, right-click the tool palettes title bar and choose **Design** (to load the Design Tool Palette Group) and then click the Design tab.

8. On the Properties palette, in the Dimensions grouping set the Thickness to **10″ [250]** and the Edge cut to **Plumb**.

In the Next edge grouping, the Shape list includes two choices: Single Slope and Double Slope. Single Slope creates a hip or gable roof; Double Slope creates a Mansard or Gambrel roof.

9. Choose **Single Slope** for the Shape.

10. Set the Overhang to **6″ [150]**.

11. Set the Plate Height to **0″ [0]** and the Rise to **6″ [50]** (see Figure 12–1).

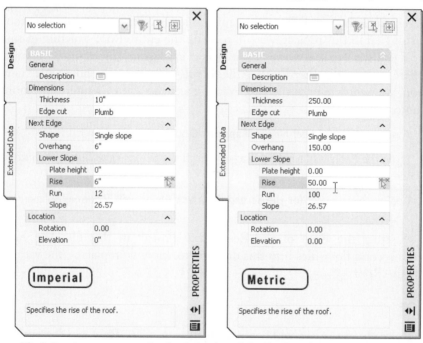

Figure 12–1 *Set the parameters for the first edge of the Roof on the Properties palette*

The Run is always 12 [100 in Metric], but we can assign the Rise to achieve the required roof pitch. The Overhang is how far the roof projects past the perimeter walls. And we are using a zero Plate Height because the *Roof* file will be referenced in at the correct height by the Project Management system.

12. Use an Endpoint Object Snap and set the first point of the Roof at the outside lower right corner of the existing house (see Figure 12–2).

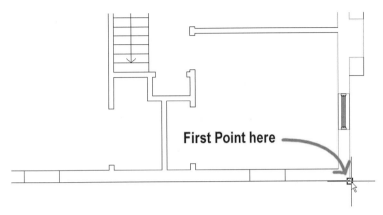

Figure 12–2 *Start at the lower right corner*

13. Snap the next point to the outside lower left corner.

14. Right-click in the drawing and choose **Gable**, and then at the "Is Gable" prompt, choose **Yes** to the dynamic prompt (if active) or right-click again and choose **Yes** (see Figure 12–3).

    This will make the next edge of the Roof a gable. To use this correctly, remember to turn it off once the edge is drawn.

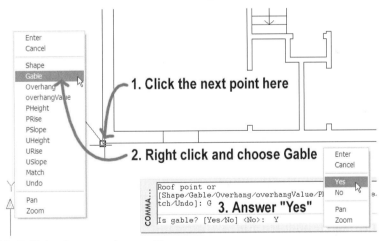

Figure 12–3 *Make the next edge a gable*

15. Click back to the drawing and set the next point at the intersection between existing and new construction on the left.

16. Right-click and choose **Gable** again, and this time, answer **No** (see Figure 12–4).

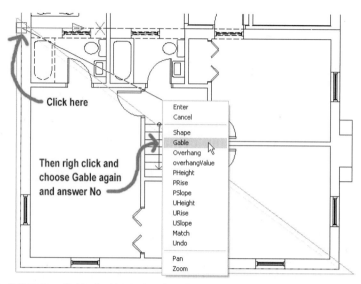

Figure 12–4 *Deselect Gable for the next edge*

17. Set the last point at the right between the existing and new construction (see Figure 12–5).

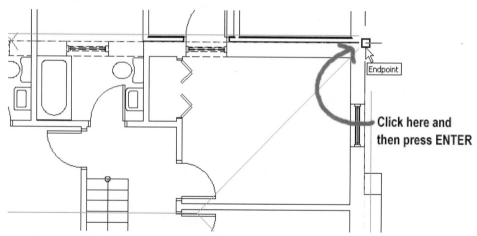

Figure 12–5 *Place the last point and then press* **ENTER**

18. Press ENTER to end the command.

Notice that the left side has drawn a gable end, but the right has a hip. This can be fixed easily with grips.

19. Click on the Roof, and highlight the grip at the ridge where the three hipped sides meet (see Figure 12–6).

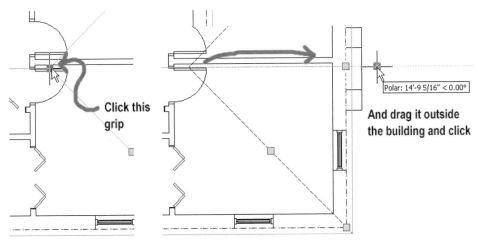

Figure 12–6 *Grip edit the right face to transform it to a gable*

20. Drag the grip all the way to the right past the edge of the Roof, click with your mouse to complete the Roof in-place edit.

   The right is now a gable end as well. Select the newly drawn Roof, right-click and choose **Object Viewer**. Have a look at it from all sides. Close the Viewer when finished.

CONVERT TO ROOF

For the new construction, the Roof is a bit more complex. Therefore, we will convert some polylines into Roofs.

21. On the Project Navigator, in the *Elements* folder, right-click the file named *Roof Outlines* and choose **Insert as Block**.

22. In the Insert dialog box, clear the check mark in the "Specify on Screen" check box for Insertion Point, (accept the default 0,0,0 insertion point) place a check mark in the "Explode" check box, and then click OK.

   Notice the two magenta rectangles that appear. By checking Explode, we have inserted the contents of the file as individual objects rather than grouped together as a block.

Any closed polyline can be converted to a Roof. We will convert these two closed polylines to Roofs.

23. On the Design tool palette, right-click the Roof tool and choose **Apply Tool Properties to > Linework and Walls**.

24. At the "Choose walls or polylines to create roof profile" prompt, click *one* of the rectangles and then press ENTER.

IMPORTANT Don't select both polylines at the same time or you will get a single Roof from their combined shape.

25. At the "Erase the layout geometry" prompt, choose **Yes** (or type **Y** and press ENTER).

26. On the Properties palette, verify that the parameters have matched those of the previous Roof. If not, change any settings as required to match the previous Roof.

27. Right-click and choose **Deselect All**.

28. Repeat the steps on the other polyline (see Figure 12–7).

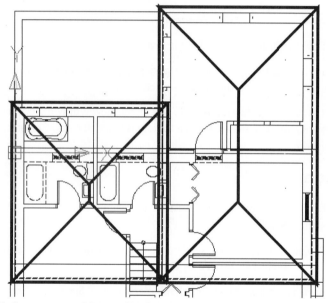

Figure 12–7 *Convert two polylines to Roofs*

The default shapes of the converted Roofs will be hip Roofs. By using the grip editing technique covered above, we can transform them into gable Roofs.

29. Use grips to edit the hip roof ends into gable roof ends (see Figure 12–8).

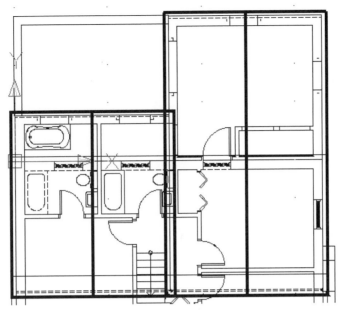

Figure 12–8 *Grip edit the ends to form gables*

> 30. On the Views toolbar, click the Back icon (or choose **3D Views > Back** from the View menu).
>
> Take a look at the overall relationship of each Roof to one another (see Figure 12–9).

This will give you a better sense of what we have so far.

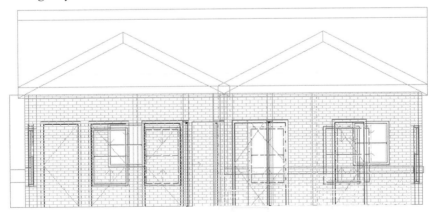

Figure 12–9 *Viewing the Roofs from the back*

> 31. Select all three Roof objects, right-click, and choose **Object Viewer**.
>
> 32. Orbit the Roofs around in 3D.

Study them from all sides (see Figure 12–10).

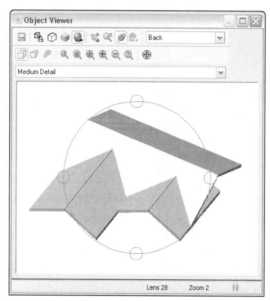

Figure 12–10 *Bring the Roof into the Object Viewer*

The one-piece Roof is a relatively simple object. It is not Style based and has few parameters. However, do not be fooled into thinking that the one-piece Roof is too simple to be valuable. It is precisely this simplicity that makes it valuable. Because it is a single continuous object, you can easily modify critical parameters like the Roof pitch and change the faces from hips to gables quickly. As you make these changes, all faces automatically join and miter correctly because it is a single continuous object. This is very valuable in the early stages of design where changes are frequent. Use the one-piece Roof in early design phases to "rough out" the roof configuration. Do not convert to Roof Slabs until the project needs require it. Before moving on to Roof Slabs, let's take a look at a few one-piece Roof properties.

33. Select any Roof, and modify some of its parameters on the Properties palette.

34. Click the Edges/Faces worksheet icon in the Dimensions grouping.

Individual edges can be edited in this dialog box. For instance, to make a saltbox style Roof, change the Plate Height of just one Edge. The Overhang parameter can be edited one edge at a time here as well. Some of the values are calculated by the software as a result of other parameters. For instance, you will not be able to edit the Eave value, Segments, or Radius. The Eave is a result of the pitch. Change the Rise, Run, or Angle in the Modify dialog box or in the lower portion of this tab, and the Eave will adjust accordingly. Segments and Radius would be available only if you converted a polyline with curved segments to a Roof. Even though it would convert the arc segment of the polyline into several straight segments, this parameter remains available for edit so that you can increase or decrease the quantity of segments used to describe the curve. The only problem with editing edges this way is that there is no way to tell which edge is which. For this reason, the Modify Edges command is a better choice.

35. Finish your experiments and click OK to close the worksheet.

36. Undo any changes made in either the Properties palette or the Edges/Faces worksheet.

37. On the Views toolbar, choose Top (or choose **3D Views > Top** from the View menu).

MODIFY ROOF EDGES

38. Select any Roof, right-click, and choose **Edit Edge/Faces**.

39. Click a single Edge of the Roof that you wish to edit and then press ENTER.

    Notice that there is now a single entry in the Edit Edges dialog box, making it impossible to edit the wrong edge (see Figure 12–11).

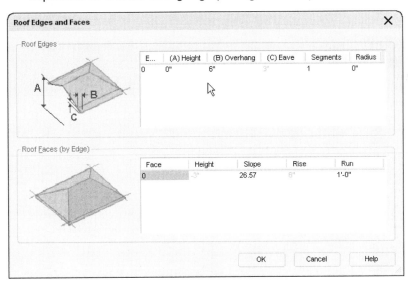

Figure 12–11 *Only a single edge is now available to edit*

40. Make any changes you wish and then click OK.

41. Undo the changes and Save the file.

WORKING WITH ROOF SLABS

As you have most likely noted, the parameters of the one-piece Roof are limited. To build a Roof achieving the complexity of the typical architectural project will often require several Roofs. However, separate one-piece Roofs are not able to interact with one another. The one-piece Roof gives very little control over eave, soffit, and fascia conditions. For these situations, we have the Roof Slab. Roof Slabs offer all of this additional functionality.

CONVERT TO ROOF SLABS

Our design has progressed to the point where the one-piece Roofs no longer suit our needs. We will convert them to Roof Slabs.

1. Select all three of the Roofs, right-click and choose **Convert to Roof Slabs**.

    A small Convert to Roof Slabs worksheet will appear.

2. Place a checkmark in the "Erase layout geometry" checkbox and then click OK.

Although it is possible to keep the original one-piece Roofs after conversion, this would only be advisable if you intended to spin off more than one scheme from the original Roof. If planning to do so, you can choose to keep the layout geometry after conversion.

    A flurry of grips will appear, as well as some additional linework (see Figure 12–12).

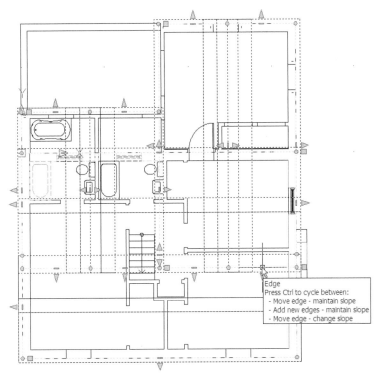

Figure 12–12 *The one-piece Roofs converted to several separate Roof Slabs*

As usual, the grips have a variety of shapes and functions. Hover your mouse over each grip to see a tool tip of its function. If you click any grips and make any edits, be sure to undo before continuing.

> All of this additional linework makes it is a bit difficult to see what is going on. Perhaps it is time that we turned off the display of the XREFs.

> 3. At the bottom right corner of the drawing on the Drawing Status Bar, right-click the Manage XREFs quick pick and choose **XREF Manager** (see Figure 12–13).

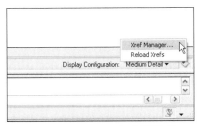

Figure 12–13 *Use the XREF Quick Pick to open the XREF Manager*

4. Select both the *Second Floor New* and the *Second Floor Existing* XREFs, click the Unload button, and then click OK.

The second floor plan will disappear. Now that we have the entire Roof in rough form, the XREF is no longer needed. However, just in case, we chose Unload so that we can quickly retrieve it should design needs require it again.

5. On the Shading toolbar, choose Flat Shaded, Edges on (or choose **Shade > Flat Shaded, Edges on** from the View menu).

TRIM THE ROOF SLAB

At the moment, the two Roofs for the new addition pass through the existing Roof. In this sequence, we will look at ways to trim the Roof Slabs to the intersection point instead.

6. On the Application Status Bar (at the bottom of the screen as shown in Figure 1–1 in Chapter 1), right-click the POLAR button and choose **Settings**.

7. Verify that there is a check mark in the "Polar Tracking On" check box.

8. Verify also that the Increment angle is set to **30°** and the additional angles are set as shown in Figure 12–14.

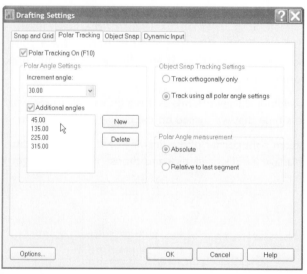

Figure 12–14 *Turn on Polar Tracking and verify angle settings*

9. Click the Object Snap tab.

10. Be sure that "Object Snap On," "Object Snap Tracking On," "Endpoint" and "Midpoint" are all checked, and then click OK (see Figure 12–15).

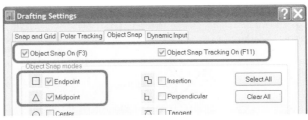

Figure 12–15 *Set up Object Snap settings*

11. On the Shapes toolbar, click the Polyline icon (or type **PL** at the command line and press ENTER).

12. At the "Start point" prompt, click the Endpoint at the left of the plan where the two Roof Slabs meet (see Figure 12–16).

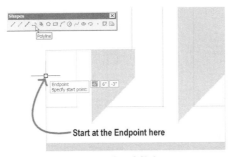

Figure 12–16 *Start the polyline where the Roof Slabs meet*

13. Using tracking, acquire the ridge point and track it down to the 45° valley (see Figure 12–17).

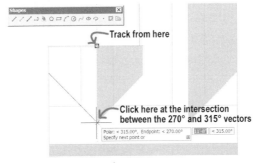

Figure 12–17 *Set the next point using tracking*

14. Complete the polyline using Figure 12–18 as a guide (see Figure 12–19 for the result).

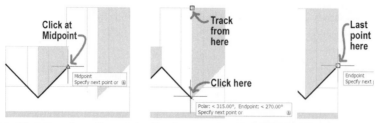

Figure 12–18 *Follow the valleys of the Roof to complete the polyline*

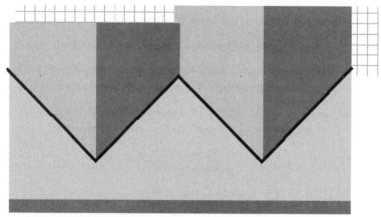

Figure 12–19 *The final polyline (polyline enhanced in the figure for clarity)*

15. On the Views toolbar, click the NW Isometric icon (or choose **3D Views > NW Isometric** from the View menu).

16. On the Shading toolbar, click the 2D Wireframe icon (or choose **Shade > 2D Wireframe** from the View menu).

17. Select the Slab shown in Figure 12–20, right-click, and choose **Trim**.

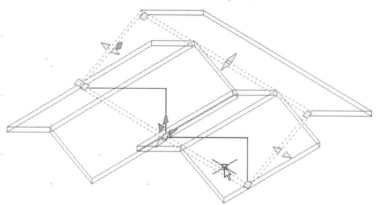

Figure 12–20 *Select the Slab to trim*

18. At the "Select trimming object" prompt, click the polyline.

19. At the "Specify side to be trimmed" prompt, click down and to the left of the Roof (see Figure 12–21).

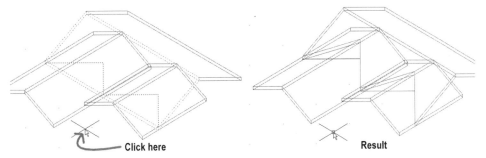

Click here   Result

Figure 12–21 *Exaggerate your pick when designating the side to trim*

 TIP Because we are working in a 3D view, click far enough away from the object so that ADT trims the correct side. If the wrong side is trimmed, Undo and try again clicking further away this time.

Notice that the Roof Slab has been trimmed to the shape defined by the polyline as projected along its sloped faces.

 NOTE Occasionally, you will need to Undo and start a particular sequence over. ADT objects do not always automatically regenerate properly. To force an ADT object to update, use the Regenerate Model icon on the Standard toolbar (or type **objrelupdate** at the command line and press ENTER twice).

20. Erase the polyline.

MITER THE ROOF SLABS

Now that the edge has been cut to the shape of the polyline, we can miter the intersecting Slabs with the trimmed one.

21. Select the Roof Slab we just trimmed, right-click, and choose **Miter**.

22. At the "Miter by" prompt, choose **Edges**.

23. At the "Select edge on first slab" prompt, click the edge along the trim line closest to you (see Figure 12–22).

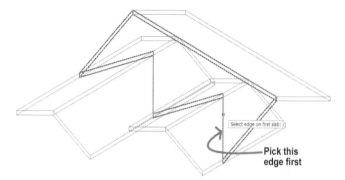

Figure 12–22 *Select the first edge of the miter*

24. At the "Select edge on second slab" prompt, click the concealed edge of the intersecting Slab (see Figure 12–23).

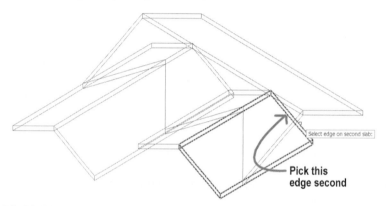

Figure 12–23 *Select the second edge of the miter*

25. Repeat the same steps for the remaining three miters (see Figure 12–24).

Be sure to use the Edges option and not the Intersection option. The Edges option gives you the control to select each edge you wish to miter. The Intersection option searches for the intersection without your input. For simple miters, Intersection will work well, but for complex situations like the ones we have here, use the Edges option.

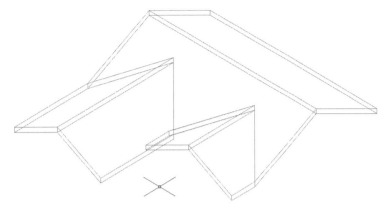

Figure 12–24 *The completed mitered Roof Slabs*

26. Save the file.

IMPORT ROOF SLAB STYLES

In addition to being able to trim and miter edges as seen in the previous sequence, we can apply articulated conditions at the edges of the Roof Slabs. These are called Roof Slab Edge styles. As with all other ADT objects, we can also apply materials to our Roof Slabs. The fastest way to get started with both of these tasks is to import some pre-made styles from the Content Library. We could use either of the two methods that we have covered so far in this book: Content Browser or Style Manager. In this case, importing a tool from Content Browser will be a little quicker. (With Style Manager, we would need to import from several categories, but with Content Browser there is just one in this case.) Furthermore, the Content Browser content is already saved as tools.

27. On the tool palettes, click the Project Tools palette tab to make it current.

   If you do not see this palette or tool, right-click the tool palettes title bar and choose **12 Residential [12 Residential Metric]** (to load the project tool palette group) and then click the Project Tools tab.

In Chapters 10 and 11 we discussed the two project-based palettes included in the 12 Residential [12 Residential Metric] tool palette group. You can refer back to those chapters for detailed discussions. You may recall that the Project Tools was typically used as our "scratch pad" palette and is saved locally on our own machine, while the Shared Project Tools is linked to a source catalog that lives in the project folder structure. If you work in a team environment, this palette keeps all team member's tools synchronized. The process requires that you build new tools on the Project Tools palette and then if you wish to share them with the team, that you copy them to the shared catalog where they can be refreshed to each user's workstation. For this exercise, we will work with the local version—Project Tools. Feel free to add them to the Shared Tools Palette at the end of the lesson.

It is also important to remember that there are actually two different Content Browser Libraries that we have explored in this text. (A Content Browser Library is simply a home page that has pointers to all of the various tool catalogs that you can access. Many of these catalogs live in common locations on your workstation that are accessible to anyone who logs in. Often these catalogs will live on a network server—check with your CAD Manager to be sure.) Architectural Desktop creates a Content Browser Library on each machine for each person that logs into a machine (typically named after your Windows login name). This is the "default" Content Browser Library and is accessed from the Navigation toolbar, the Window menu or by pressing CTRL + 4. The other type of Content Browser Library is a "project-based" Content Browser Library. It is accessed from the Content Browser icon on the Project Navigator. Once opened, both types of library behave exactly the same way. The difference is simply that one is generic and *not* specific to any particular project and the other *is* specific to a particular project.

 NOTE For the Content Browser icon on Project Navigator to call a project-based Content Browser Library, one must be setup in the project. This setting is usually included in the template project used to create a project.

28. Open the Content Browser (the "default" one for this exercise—click the icon on the Navigation toolbar or press CTRL + 4).

29. Click on *Design Tool Catalog – Imperial* [*Design Tool Catalog – Metric*].

30. Navigate to the *Roof Slabs and Slabs* category and then the *Roof Slabs* category.

A Roof Slab style controls the parameters of the Roof Slab itself. A Roof Slab Edge style controls the shape of the individual edges. The tools in the default Content Browser allow us to quickly import both. In this exercise, we will use styles from two different tools.

31. Using the eyedropper icon, drag the **04 – 1x8 Fascia** [**100 – 25x200 Fascia**] tool onto the Project Tools palette.

    This will create a new tool from this style.

32. Repeat these steps and drag the **04 – 1x8 Fascia + Soffit** [**100 – 25x200 Fascia + Soffit**] tool this time.

    You should now have two Roof Slab tools added to your Project Tools palette (see Figure 12–25).

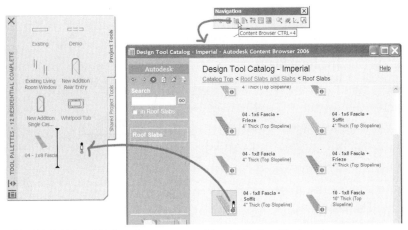

Figure 12–25 *Add Roof Slab styles to your residential Project Tools palette*

33. Close the Content Browser.

APPLY ROOF SLAB STYLES

Now let's apply the tools to various Roof Slabs.

34. Select the four Roof Slabs that comprise the roof of the new addition (see Figure 12–26). Do not select the existing house Roof.

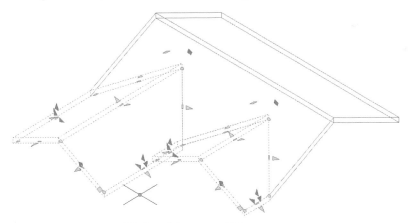

Figure 12–26 *Select the Roof Slabs of the new addition*

35. Right-click the **04 – 1x8 Fascia + Soffit** [**100 – 25x150 Fascia + Soffit**] tool and choose **Apply Tool Properties to > Roof Slab**.

Several things have occurred with this step (see Figure 12–27). The most obvious is the appearance of Materials on the Roof Slab surfaces. If you zoom in a bit, you will also note that the horizontal edges of the Roof Slabs now have edge profiles (Roof Edge styles) applied to them and that the Roof Slabs have shifted down.

Even less obvious is a change in the Roof's thickness. Some of these effects were desired. Let's address some of those that weren't.

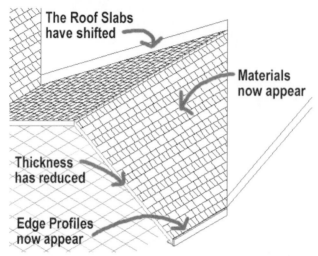

Figure 12–27 *The Roof Slabs have shifted down and then now have Materials and edge profiles*

36. With any of the four Roof Slabs selected, right-click and choose **Edit Roof Slab Style**.

37. On the General tab, rename the Style to **MADT New Construction**.

38. Click the Design Rules tab.

39. Change the Thickness to: **8″ [200]**, and the Thickness Offset to: **0″ [0]** and then click OK.

    Notice that Roof Slabs have now shifted back to flush condition with the other Roof Slabs and that the thickness has increased as well.

40. Select the remaining two Roof Slabs (the ones for the existing house) and on the Properties palette, change their Style to **MADT New Construction** (see Figure 12–28).

Notice that this time, there was no shift and that the edges did not receive Roof Edge styles. This is because we applied the style of the Roof Slab only on the Properties palette, rather than applying both the Roof Slab and Roof Slab Edge styles as the tool did. Remember, the command on the tool's right-click menu reads "apply tool properties to" meaning that it applies *all* of the tools properties, not just the style as we have just done manually here. Let's fine-tune the edge assignments.

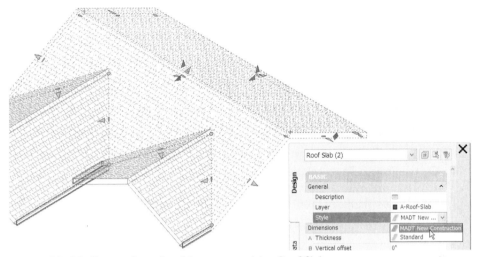

Figure 12–28 *Change the style of the two remaining Roof Slabs*

APPLY EDGE CONDITIONS

41. Zoom in on the valley between the two small Roofs (the new construction roof).

    Notice that an Edge style has been applied in a position that is concealed.

42. Select the Roof Slab that this edge belongs to, right-click, and choose **Edit Roof Slab Edges** (see the top of Figure 12–29).

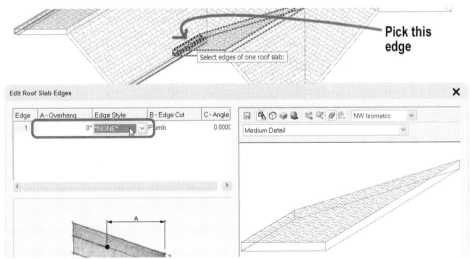

Figure 12–29 *Select a Roof Slab to edit edges*

43. At the "Select edges of one roof slab" prompt, click the edge with the fascia and press ENTER.

44. In the Edit Roof Slab Edges dialog box, change the A – Overhang value to **0″** **[0]**, click the Edge Style list, choose ***None***, and then click OK (see the bottom of Figure 12–29).

We can use this same technique (without changing the overhang) to apply Edge styles to the fascia boards of the gable ends as well. However, we will need a different Edge style for this.

45. On the Project Tools palette, right-click the **04 – 1x8 Fascia [100 – 25x200 Fascia]** tool and choose **Import '04 – 1x8 Fascia' Roof Slab Edge Style [Import '100 – 25x200 Fascia' Roof Slab Edge Style]** (see Figure 12–30).

 NOTE If yours says "Re-import," this likely indicates that you dragged the tool into the drawing before dragging it to the palette. You can choose Re-import, or just press the ESC key to cancel. Re-import means that the style is already present in your drawing and if you choose Re-import, it will retrieve the original Style definition from the remote file and overwrite the version currently resident in your drawing. In this case, there is no harm in doing so. However, consider the ramifications of this action carefully when doing this on "real" projects.

Figure 12–30 *Import a Roof Slab Edge style for use in the current drawing*

46. Select the same Roof Slab, right-click, and choose **Edit Roof Slab Edges**.

47. At the "Select edges of one roof slab" prompt, click the leading gable edge facing you and press ENTER.

48. In the Edit Roof Slab Edges dialog box, click the Edge Style list, choose **1x8 Fascia [25x200 Fascia]**, and click OK.

49. Repeat these steps, and move around the model and apply edges as appropriate (see Figure 12–31).

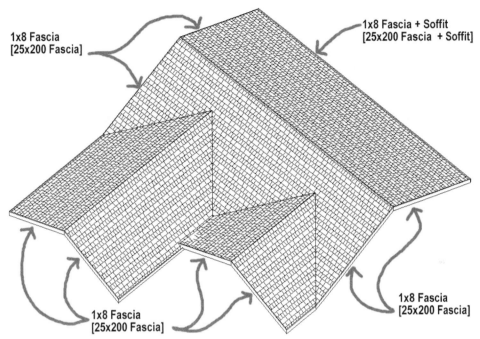

1x8 Fascia
[25x200 Fascia]

1x8 Fascia + Soffit
[25x200 Fascia + Soffit]

1x8 Fascia
[25x200 Fascia]

1x8 Fascia
[25x200 Fascia]

Figure 12–31 *Apply Edge styles as indicated*

ADD HOLES AND FINISHING TOUCHES

The file that we inserted from the *Elements* folder above contained a hidden layer with a few additional polylines. We will use these now to fine-tune the Roofs.

50. Turn on the Layer named A-Temp.

    Two magenta polylines appear.

51. Use the 3D Orbit icon on the Navigation toolbar to tilt the model down slightly.

    Click the 3D Orbit icon on the Navigation toolbar, then place the mouse in the small handle at the top of the green circle and drag down slightly. Press ESC to exit.

52. Select the leftmost Roof Slab, right-click, and choose **Hole > Add**.

53. When prompted, select the magenta rectangle (you may need to zoom in with your mouse wheel a bit), press ENTER and answer **Yes** to erase the layout geometry.

    A hole should appear in the Roof Slab.

54. Select the long gabled Roof Slab (opposite the one we just added the hole to), right-click, and choose **Trim**.

55. At the "Select trimming object" prompt, click the "L" shaped magenta polyline.

56. At the "Specify side to be trimmed" prompt, click up and to the right of the Roof Slab (see Figure 12–32).

 TIP For the "side to be trimmed" prompt, it is best to click far from the polyline to be certain that ADT trims the correct part of the slab.

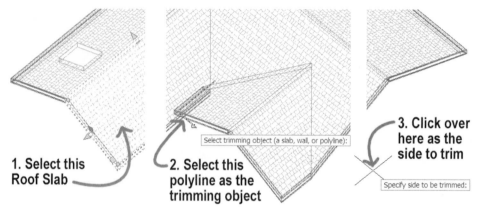

1. Select this Roof Slab

Select trimming object (a slab, wall, or polyline):

2. Select this polyline as the trimming object

3. Click over here as the side to trim

Specify side to be trimmed:

Figure 12–32 *Trim a slab using the provided polyline*

57. Erase the magenta polyline and then Save and Close the file.

PROJECT THE WALLS TO THE ROOF

Now that we have modeled the Roof, we can project the top edges of the walls in the second floor plan to follow the underside of the Roofs.

58. On the Project Navigator, Open the *Second Floor Existing* file.

59. On the Project Navigator, drag and drop the *Roof New* file into the current drawing.

    The hatching in the Roof Slab Materials may delay the regeneration a bit. To speed this up, click the small Surface Hatch Toggle icon on the Drawing Status Bar (shown in Figure 1–2 in Chapter 1).

60. From the Views toolbar, click the SW Isometric icon (or choose **3D Views > SW Isometric** from the View menu).

61. From the Shading toolbar, click the Hidden icon (or choose **Shading > Hidden** from the View menu).

Notice that the Walls do not follow the roof line. Notice also that there is another magenta polyline in this file (see Figure 12–33).

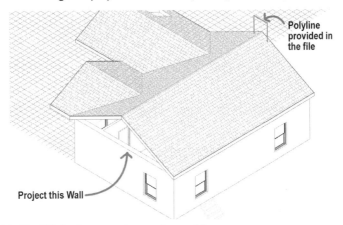

Figure 12–33 *The Walls do not project to the roof line*

62. Select the Wall under the gable end (as indicated in Figure 12–33), right-click and choose **Roof/Floor Line > Modify Roof Line**.

63. Right-click and choose the **Auto project** option.

 NOTE Auto project is used to project to any ADT object; recall Chapter 7 when we projected a Wall to the under side of the Stairs.

64. At the "Select Objects" prompt, select the *Roof New* XREF and then press ENTER.

65. Press ENTER again to complete the command.

   If necessary, right-click the Wall again, choose **Roof/Floor Line > Edit in Place** and manipulate the vertices to fine-tune the shape.

 TIP If you have trouble projecting the Wall to the *Roof New* file, try dragging in the *Roof Exist* file and project to it instead. Detach the *Roof Exist* file when finished.

66. Change to SE Isometric view.

67. Repeat the **Roof/Floor Line > Modify Roof Line** command.

   Use the Project to Polyline this time, and project the Wall with the chimney to the magenta polyline. You can erase the polyline when you're done (see Figure 12–34).

Figure 12–34 *The Walls now project to roof line.*

68. Detach the XREF, return to Top view and then Save and Close the *Second Floor Existing* file.

   If you like, you can return to the *Roof New* file and trim the Roof Slabs of the existing house around the chimney. If you have trouble with the Roof Slab Trim option that we used above, try the Cut option on the same right-click menu instead.

UPDATING THE RESIDENTIAL PROJECT MODEL

As you recall, in Chapter 5 we built several files for each of the projects. In both projects we created a Composite Model View file that gathered together all of the other files. Since we have added a new Construct here in this lesson, we should return to the Composite Model View and make some updates.

UPDATE THE COMPOSITE MODEL VIEW

1. On the Project Navigator, click the Views tab.

2. Right-click on the *A-CM00* View file and choose **Properties**.

3. On the left side of the Modify View dialog box, click the Content item.

4. Clear the check mark in the *Roof Existing* check box, place a check mark in the *Roof New* check box, and then click OK (see Figure 12–35).

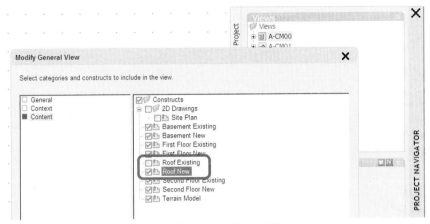

Figure 12–35 *Edit the Content of the Composite Model View*

5. Double-click *A-CM00* to open it.

6. From the Views toolbar, click the NW Isometric icon (or choose **3D Views > NW Isometric** from the View menu). See Figure 12–36.

As you can see, there is still some work to be done. There are no floors in the house, the back exterior walls do not project to the Roof, and the patio has no Roof at all. Feel free to project the walls to the Roof and add a Roof to the patio. There is an Additional Exercise for this in Appendix A. In the next sequence, we will shift our attention to Slab objects. We will switch to the Commercial Project for these explorations. At the end of the chapter, we will return to the Residential Project to have a look at Space objects.

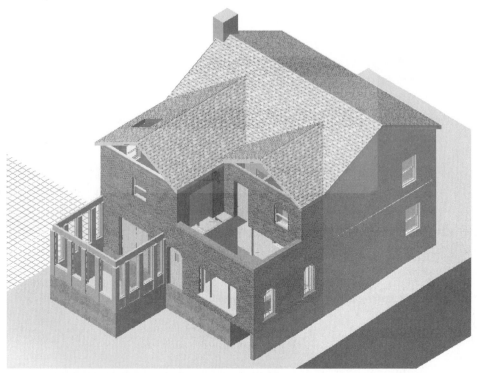

Figure 12–36 *The Residential Model is nearly complete*

7. Repeat these steps to update the *A-CM01* model and the *A-FP03* Roof Plan View as well.

8. Save and Close all Residential Project files.

 NOTE Remember to keep one empty file open, if you have no drawings open ("zero doc state") then click the Qnew icon to create a new empty file.

CREATING SLABS

Slabs can be used to represent any horizontal surface in your building model, whether that surface is flat or sloped. Slabs do not give much benefit in plan views. They are most useful for deck type structures like parking structures, for quick studies of stacking, flat roofs or for use in building sections. Therefore, we will create our Slabs in separate files and XREF them back to our composite model. In this way, you can isolate the Slab objects from the plan files and use them in the Views where they give the most benefit: the 3D views and sections. Let's have a look at Slabs in the Commercial Project.

LOAD THE COMMERCIAL PROJECT

Be sure that all files from the Residential Project have been closed and saved. However, you must make sure that a single empty file remains open. The Project Browser will not run in zero-doc state. The simplest way to do this is to close all files, and then use the Qnew icon to create a new empty file.

1. From the File menu, choose **Project Browser**.

2. Click to open the folder list and choose your *C:* drive.

3. Double-click on the *MasterADT 2006* folder, then the *Chapter12* folder.

   One or two commercial Projects are listed: *12 Commercial* and *12 Commercial Metric*.

4. Double-click *12 Commercial* if you wish to work in Imperial units. Double-click *12 Commercial Metric* if you wish to work in Metric units. (You can also right-click on it and choose **Set Current Project**.) Then click Close in the Project Browser.

 IMPORTANT If a message appears asking you to re-path the project, click Yes. Refer to the "Re-Pathing Projects" heading in the Preface for more information.

CONVERT TO SLABS

A few files have been added to the Commercial Project since the last chapter. Two Constructs for Slabs have been added in the Structural category. We also have a new Architectural Construct for the Roof of our building.

5. On the Project Navigator, click the Constructs tab and then expand the *Constructs\Structural* folder.

6. Double-click to open the *02 Slab* file.

   This file contains three polylines.

Slabs and Roof Slabs are nearly identical in interface and functionality. Like many ADT objects, Slabs can be added point by point or can be converted from existing geometry. In most cases, you will likely have some form of outline already in progress when you decide to create a Slab. This outline can be a polyline or a series of Walls. We will create some Slabs for our commercial model in the various new Constructs already mentioned. As a general rule of thumb, Slabs should be created in separate files from the basic Wall geometry Constructs. This is because the edges of the Slabs, which frequently coincide with the Walls, can prove distracting when working in plan view.

Be sure that you have Dynamic input on for this sequence. (be sure the DYN on the Application Status Bar is pushed in). Refer to Chapter 1 for more details.

7. On the Design tool palette, right-click the Slab tool and choose **Apply Tool Properties to > Linework and Walls**.

    If you do not see this palette or tool, right-click the tool palettes title bar and choose **Design** (to load the Design tool palette group) and then click the Design tab.

8. At the "Select walls or polylines" prompt, select the large outer polyline, and press ENTER (see Figure 12–37).

9. At the "Erase layout geometry" prompt, choose **Yes**.

10. At the "Creation mode" prompt, choose **Direct**.

The Direct creation mode converts the polyline in its current location. The Projected method prompts for a height above or below the existing polyline to create the Slab. Projected Slabs can also be created with a pitch. Direct conversions have no pitch applied, they match whatever geometry you are converting.

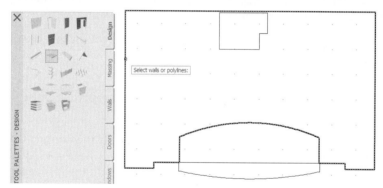

Figure 12–37 *Select the large outer polyline to convert.*

11. At the "Specify slab justification" prompt, choose **Top**.

A Slab has been created in place of the polyline and remains selected for editing. As with Roof Slabs earlier, styles can be used to specify varying edge conditions.

12. On the Properties palette, beneath the Dimensions grouping, verify that the Thickness is set to: **4″** [**100**].

13. In the Location grouping, change the Elevation to: **-4.5″** [**-112**].

    This setting will set the Slab directly on top of the Bar Joists that we added in Chapter 6, and allow 1/2″ [12] for the finish floor material (this will come from the Space object below).

14. Right-click and choose **Deselect All**.

15. Repeat this process selecting the small curved polyline at the bottom of the screen this time.

16. Change its Thickness to: **3″** [**75**] and its Elevation to: **-3.5″** [**-87**].

If you wish, experiment with the many grip shapes on Slabs. Be sure to undo all changes back to the original shape when you are through experimenting. If you want to see how things align with the first floor structure, drag the *01 Grid* Construct into the file, and then use the **Isolate Objects** > **Edit in Elevation** or **Edit in Section** routine to view the file from the side. Zoom in on the top of the Bar Joists and note that the Slab sits directly on top of them. When you have finished exploring, return to Top view (Exit Edit in View) and then Detach the XREF.

Let's cut a hole for the stairs and elevators to penetrate.

17. Select the large Slab, right-click, and choose **Hole > Add**.

18. When prompted, select the internal polyline and then press ENTER.

19. At the "Erase layout geometry" prompt, choose **Yes**.

20. Save and Close the file.

If you brought in the *01 Grid* file above, you probably noted that the joists will require adjustment. We will discuss this in the "Additional Exercises" heading below.

UPPER FLOORS

The second floor Slab shape accommodates a two-story atrium lobby space at the entrance to the building. The Slab for the third and fourth floors is similar but without this atrium space. Another Construct for the third floor has been provided with a polyline in this alternate shape.

21. On the Project Navigator, in the *Constructs\Structural* folder, double-click to open the *03 Slab* file.

    There are only two polylines in this file.

22. Repeat steps 7 through 14 on the large outer polyline.

23. Repeat steps 17 through 19 on the inner polyline to cut a hole.

24. Save and Close the file.

25. On the Project Navigator, in the *Constructs\Structural* folder, right-click the *03 Slab* file and choose **Copy Construct to Levels**.

26. In the Copy Construct to Levels dialog box, place a check mark in Level 4 and then click OK.

27. Right-click the new file named *03 Slab (2)* and choose **Properties**.

28. Rename it: **04 Slab**, change the Description to **Fourth Floor Slab**, and then click OK.

29. Click the Re-path icon at the bottom of the Project Navigator and then click Re-Path in the Reference File Re-path Queue dialog box.

 CAUTION Remember, this step is critical any time you move or rename a file in Project Navigator (see the "Create the Upper Floors" topic in Chapter 5 for more information).

BUILDING THE ROOF PLAN FILE

For simplicity at this stage, we will use a single flat Slab to represent the roof of the Commercial Project. In reality, there certainly would be sloped surfaces for drainage. At issue is whether the slight slope of a flat roof warrants the use of Roofs and Roof Slabs. As a rule, a Roof plan is a simple drawing. Your needs might be better served to use a single flat Slab for the 3D model and overall building section, and simply draft the slope of the Roof plan manually in the Roof Plan View file. This might sound like an odd recommendation in a book that has stressed the virtues of parametric design. However, the Roof and Slab objects really are a means to an end. Roof objects make good sense when working out a complex roof structure with many intersecting planes, hips, ridges and valleys. We also gain the benefit of creating linked sections and elevations directly from this geometry. (Refer to Chapter 16 for complete information on linked 2D Section/Elevation objects.) However, with a flat roof, the slope is so slight that a flat slab can do the job required by 1/8"=1'-0" [1:100] and 1/4"=1'-0" [1:50] drawings just as well. The true slope will then be expressed in Details. At small scales, 1/4" in 12 [1 in 50] slopes will not be perceived and the effort expended to generate correct mitering and sloping would be largely wasted.

By adding embellishments later directly in the Roof Plan View file, we achieve the requirements of a Roof Plan without adding unnecessary complexity to the model. Furthermore, by adding these embellishments only to the Roof Plan View file, we do not need to worry about freezing or thawing layers in any other files. This is because the extra linework will occur *only* in the Roof Plan file.

IMPORT SLAB STYLES

Another new Construct has been provided with the Commercial Project. The *Roof* file contains a few items to get us started.

1. On the Project Navigator, in the *Constructs\Architectural* folder, double-click *Roof* to open it.

This file includes two polylines that we will convert to Slabs and parapet walls, and it already contains some Walls for the Stair tower. In the preceding exercise, we used the Standard Slab style. In this sequence, we will visit the Content Browser again and drag over some Slab Style tools appropriate for our Roof file. We will add these tools to our 12 Commercial [12 Commercial Metric] tool palette via its linked catalog as we did previously in the "Refresh a Tool Palette" heading of Chapter 8.

2. Open the default Content Browser (not the one on Project Navigator, click the icon on the Navigation toolbar or press CTRL + 4).

3. Click on the *Design Tool Catalog – Imperial* [*Design Tool Catalog – Metric*] catalog.

4. Navigate to the *Roof Slabs and Slabs* category and then the *Slabs* category.

5. Right-click the **Cant** tool and choose **Copy**.

6. Return to ADT and on the Project Navigator, click the Project Browser icon.

   This is the Project-based Content Browser Library this time.

7. Click on the Project Catalog.

8. Click on the Shared Project Tools palette link to open it.

9. Right-click and choose **Paste**. This process copies a tool from the default Content Browser Library to the Project Library. Please note however that the tool's reference still points to a style in the default library. If you will not need to edit the style for the project, this approach is fine. If the Style needs to be modified for the project, then the style should be copied to the project Standards Drawing—*Commercial Styles* [*Commercial Styles – Metric*] and then a new tool created from the modified Style there.

10. Close the Content Browser.

11. In ADT, on the tool palettes, click the Shared Project Tools palette tab to make it current.

    If you do not see this palette or tool, right-click the tool palettes title bar and choose **12 Commercial** [**12 Commercial Metric**] (to load the 12 Commercial [12 Commercial Metric] tool palette group) and then click the Shared Project Tools tab.

12. Click the small refresh icon to refresh the palette and update it with the new tool.

A new *Cant* Slab tool will appear on the Shared Project Tools palette.

CONVERT POLYLINES TO SLABS

Using this new tool, let's convert the polyline in this file to a Slab.

13. On the Shared Project Tools palette, right-click the **Cant** tool and choose **Apply Tool Properties to > Linework and Walls**.

14. At the "Select walls or polylines" prompt, select the large outer polyline, and then press ENTER.

15. At the "Erase layout geometry" prompt, choose **No**.

 NOTE Normally we have answered Yes here. In this case, we are answering No because we will use the polyline again later.

16. At the "Creation mode" prompt, choose **Direct**.

17. At the "Specify slab justification" choose **Top**.

18. On the Properties palette, within the Location grouping, set the Elevation to **-6″ [-150]**.

 Metric Note: If you are performing these steps in Metric units, the **Cant** Slab style is missing its Material designation in the Content library. To fix this, open the Style Manager on the Format menu, and click the Open Drawing icon. Locate the *Material Definitions (Metric).dwg* file, expand Multi-Purpose Objects, and from there, import the **Concrete.Cast-in-Place.Exposed Aggregate.Medium** Material Definition. Finally, edit your **Cant** Slab style and assign this Material to Slab component.

The small curved polyline at the bottom of the screen needs to be converted a bit differently. We don't want the canted edges here. So we will first convert it to Standard style and then manually apply the *Cant* style via the Properties palette. This will have the effect of applying the Material parameters to the Slab without applying the **Cant Slab Edge** style.

19. On the tool palettes, click the Design tab to make it active.

20. Right-click the Slab tool and choose **Apply Tool Properties to > Linework and Walls**.

21. Convert the small curved polyline at the bottom of the screen this time. Answer **Yes** to the "Erase layout geometry" prompt, and use **Direct** for Creation mode, but **Bottom** for Justification this time.

NOTE We have been using Top for most of the Slabs because they are positioned relative to the floor lines of their respective levels. However, this particular Slab sits at the top of the Front Façade Curtain Wall, and by applying it to Bottom, it will be easier to move with the Curtain Wall should design changes dictate.

22. On the Properties palette, change the Style to **Cant** .

23. Take both Slabs into the Object Viewer to compare.

    Notice that both Slabs share the same Material, but the large Slab has canted edges while the small one does not.

CONVERT WALLS TO SLABS

Slabs can also be generated from Walls.

24. Returning to the Shared Project Tools palette, right-click the **Cant** tool again and choose **Apply Tool Properties to > Linework and Walls**.

25. At the "Select walls or polylines" prompt, select the four outer Walls of the Stair tower and then press ENTER (see Figure 12–38).

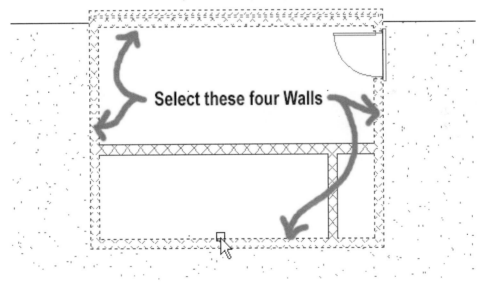

Figure 12–38 *Select a Wall to convert to Slab*

26. At the "Erase layout geometry" prompt, choose **No**.

 NOTE Here again we are answering No since we do not want to erase the Walls.

27. At the "Specify slab justification" prompt, choose **Top**.

28. At the "Specify wall justification for edge alignment" prompt, choose **Right**.

This means that the Slab will use the Right edge (in this case the inside edge) of the Walls in conversion.

29. At the "Specify slope direction prompt" press ENTER to accept the default.

30. On the Properties palette, beneath the Location grouping, set the Elevation to **11'-0"** [**3350**].

This moves the Slab to the correct vertical location (see Figure 12–39).

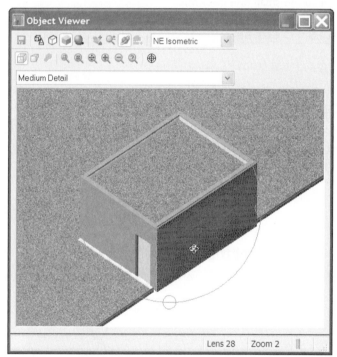

Figure 12–39 *View the results of the Stair tower Slab*

31. Save the file.

CREATE A PARAPET WALL

Now that we have all of our Slabs in place, let's generate a parapet Wall around the perimeter of our Roof. A Parapet Wall style has already been included in our *Commercial Styles* [*Commercial Styles – Metric*] file and a tool for it appeared on our Shared Project Tools palette above when we refreshed it.

32. Right-click in the drawing and choose **Basic Modify Tools > Offset**.

33. At the "Specify offset distance" prompt, "type" **10″** [**250**] and then press ENTER.

34. At the "Select object to offset" prompt, hold down the CTRL key and click the edge of the large Slab.

    The CTRL key activates the cycle mode and allows you to select the polyline that is underneath the Slab edge (the slab edge should highlight), click again near the blue slab edge—the polyline at the edge will highlight this time.

35. The polyline is magenta, and the Slab is blue. When you see the magenta edge on top, press ENTER to exit the cycle mode and select the polyline.

36. At the "Specify point on side to offset" prompt, click a point anywhere outside the Slab (see Figure 12–40).

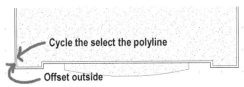

Figure 12–40 *Offset the edge of the polyline to form a polyline outside*

    A new polyline will be created. If you click the polyline and it is on the A-Slab [S-FloorSlab-G] layer, undo and try again. This indicates that you offset the Slab instead of the polyline.

37. Right-click and choose **Basic Modify Tools > Trim**.

38. Using Figure 12–41 as a guide, trim the three segments around the Stair tower.

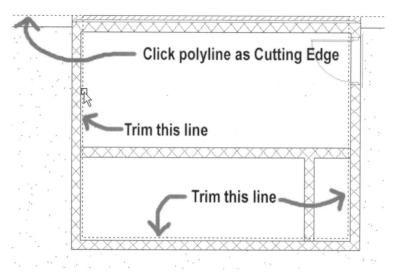

Figure 12–41 *Trim off unneeded polyline segments*

39. Stretch the two endpoint grips of the polyline to the corner endpoints of the Walls at the Stair tower (see Figure 12–42).

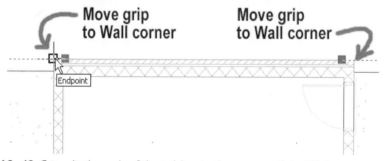

Figure 12–42 *Grip edit the ends of the polyline to the corners of the Walls*

40. With the polyline still selected, right-click the **Parapet** tool on the Shared Project Tools palette and choose **Apply Tool Properties to > Linework**.

41. At the "Erase layout geometry" prompt, choose **Yes**.

The polyline will be replaced with **Parapet** Walls.

On the Properties palette, verify that the Base height is set to **6′-0″** [**1800**] and that the Justify is set to **Baseline**. If you zoom and pan a bit you will notice that the Parapet Walls fit snuggly around the Slab. If they don't, you probably offset the Slab above rather than the polyline. If this is the case, please undo back and try again. Remember the CTRL key to cycle when offsetting the polyline. Also, remember that the polyline is magenta and the Slab blue. The best way to verify that you offset the correct one is to select the new polyline after you offset and verify that it is on Layer 0 and inserted at an Elevation of zero.

ADD A WALL SWEEP PROFILE

42. From the Views toolbar, click the SE Isometric icon (or choose **3D Views > SE Isometric** from the View menu).

    Notice that the Parapet Wall is composed of three components, one of which is a cap component at the top (see Figure 12–43).

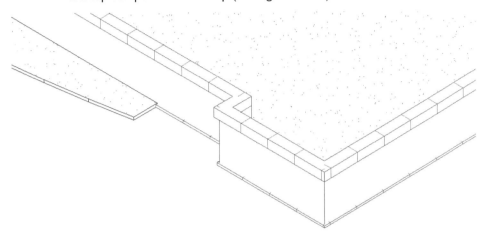

Figure 12–43 *The Parapet Wall Style includes three stacked components*

43. Turn on the Layer named A-Temp and then zoom in on the center Wall at the front of the building.

    There is a small magenta polyline at the position of the Parapet cap.

Wall components can have a "sweep" applied to them. A sweep is a custom cross section shape that is applied to the full length of the Wall component. Sweeps are applied to individual Wall segments, not to styles. Let's take a quick look at sweeps. In the following sequence, we will add the magenta polyline shape as a sweep for the cap of all of the Parapet Walls. If you wish, you can customize the shape to your own preferences rather than use the one provided.

44. Select the *Parapet* Wall visible on screen.

45. Right-click and choose **Select Similar**.

    All of the Parapet Walls should highlight.

46. Right-click and choose **Sweeps > Add**.

47. In the Add Wall Sweep dialog box, choose **Cornice** from the Wall Component list.

 CAUTION Choosing the Cornice component is important, if you forget to do this, you will be sweeping the wrong component.

48. Be sure that **Start from scratch** is chosen for Profile Definition and type **Parapet Cap** for the New Profile Name.

49. Verify that both Apply Roof/Floor Lines to Sweeps and Miter Selected Walls are chosen (see Figure 12–44).

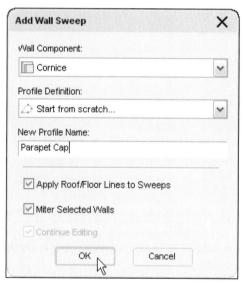

Figure 12–44 *Add a Wall sweep to the Cornice component of all selected Walls*

▶ **Apply Roof/Floor Lines to Sweeps**—Means that you will be able to modify swept components with the Roof/Floor Line edit commands.

▶ **Miter Selected Walls**—Will form nice clean intersections between all swept Wall segments. Without this setting, all sweeps would produce simple straight extrusions.

50. At the "Select a location on wall for editing" prompt, click a point on the Parapet Wall near the magenta polyline (see Figure 12–45).

It does not have to be exactly on the magenta polyline.

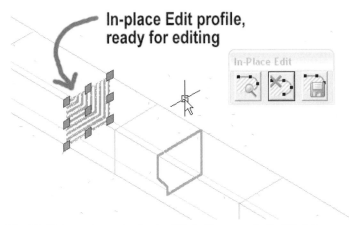

Figure 12–45 *Click a point to begin the Edit in Place mode for Wall Sweeps*

Feel free to experiment with the grips on the In-place Edit profile. Watch the grayed out Walls beyond respond in real time. You can also add or remove vertices with the CTRL key options as usual. If you are happy with the profile shape that you have devised in your experimentation, skip the next Step; otherwise proceed when ready.

51. Select the In-place edit profile, right-click and choose **Replace Ring**. When prompted, select the magenta polyline and then answer **Yes** to "erase the layout geometry."

52. On the In-Place Edit toolbar, click the Save all changes icon to complete the operation.

    Zoom out and take note of the nicely mitered corners.

Looking at the model, perhaps the parapet is a bit too tall. This is an easy change to make.

53. Select one of the Parapet Walls, right-click and choose Select Similar. On the Properties palette, change the Base Height to: **4'-0"** [**1200**].

Notice that the Cornice dropped with the Wall to the new height. Recalling the lessons of Chapter 10, this Wall Style has the vertical offsets of the Cornice component set relative to the top of the Wall rather than the Baseline. This give use the flexibility to adjust the height of the Parapet on the fly as design changes occur. Right-click the Wall, edit the Wall Style and study the parameters on the Components tab to see for yourself.

    That completes the Roof model.

54. Save and Close the file.

UPDATING THE COMMERCIAL PROJECT MODEL

As we did above for the Residential Project, it is time to update the Composite Model Views for the Commercial Project. In addition to the files that we have worked on here, some updates have been made to the *Ground* file. A Slab has been added and some Materials applied. Feel free to explore the objects in this file to understand their properties and usage better.

You may recall that in the Commercial Project, we actually have three Composite Model files. One is the generic 3D Model, while the other two are designed specifically to the needs of Elevations and Sections respectively. In this exercise, we will update the *A-SC01 – Architectural Building Sections Model*. Viewing a full building section is the best way to see the impact of our new Slab objects.

UPDATE THE COMPLETE BUILDING SECTION MODEL VIEW

1. On the Project Navigator, click the Views tab.

2. Right-click on the *A-SC01* View file and choose **Properties**.

3. On the left side of the Modify View dialog box, click the Content item (see Figure 12–46).

    Be sure that all Constructs are selected. If not, Place a check mark in the Constructs folder at the top of the list and then click OK

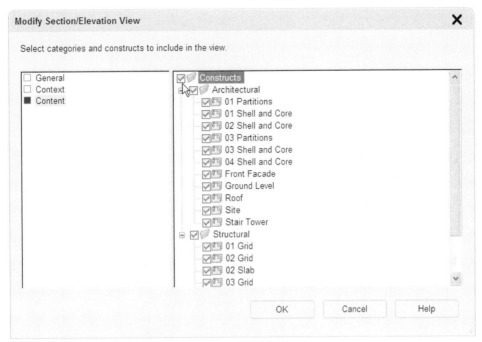

Figure 12–46 *Edit the Content of the Composite Model View*

4. Double-click *A-SC01* to open it.

5. From the Views toolbar, click the SE Isometric icon (or choose **3D Views > SE Isometric** from the View menu). (See Figure 12–47).

It may take some time for this Model to regenerate. Material Surface Hatching can slow things down a bit.

NOTE You can dramatically speed up the regen times of a Model like this by toggling off the Surface Hatch. The Surface Hatch toggle quick pick icon is located on the Drawing Status Bar at the bottom right corner of the screen. This is shown in Figure 1–2 in Chapter 1.

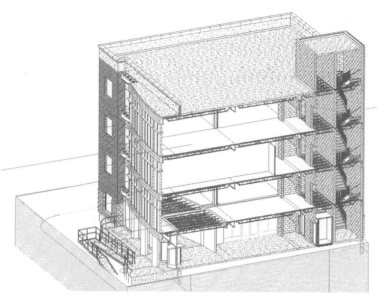

Figure 12–47 *Our Live Section through the Model has come a long way*

As you can see, we have come a long way on our Commercial Project. Feel free to zoom in to the various intersections between materials and see how the details are shaping up. Feel free to open any files, make adjustments and then reload the XREFs here to see the results. For instance, you may want to open the *01 Partitions* file and increase the height of the Walls around the atrium space. On the other hand, you might prefer a balcony overlook, in which case you might want to add some guard rails to a new Construct named *02 Partitions*. The decision is left to you. Naturally, one big problem that still remains is the joist framing. We will discuss this at the end of the chapter as an additional exercise.

6. Repeat the steps used here to update the *A-CM00 – Composite Building Model* as well. For the *A-EL01 – Architectural Building Elevations* Model, add only the *Roof* Construct, it does not need the Slabs.

7. Save and Close all Commercial Project files.

NOTE Remember to keep one empty file open. If you have no drawings open ("zero doc state") then click the Qnew icon to create a new empty file.

CREATING FLOORS AND CEILINGS

While Slabs are useful for modeling horizontal planes in a building model, they do not necessarily lend themselves to representing the individual floor and ceiling planes or the actual rooms of an interior model. They certainly can be used for this, but Space objects are better suited to this task. In addition, Space objects are very useful in tracking square footage and other "room specific" Schedule Data as you design a floor plan. Chapter 15 is devoted to Schedules and Schedule Data.

GENERATING SPACES FROM EXISTING WALLS

Space objects can be used before a floor plan is even laid out as a bubble-diagramming tool. In that capacity, the Space objects would ultimately assist in the creation of the Walls. However, it is often desirable to add Spaces after Walls and other objects have already been drawn. The Space Auto Generate Tool is designed specifically for this purpose.

IMPORT SPACE STYLES FROM A LIBRARY FILE

Space objects are used to track areas in plan and provide floor and ceiling surfaces in sections and 3D. To create spaces quickly, we will use the shape defined by the Walls to generate the Spaces. A Space object is typically used to represent a single room or contiguous space in a plan. We will begin by opening the Style Manager and importing some Space Style tools.

1. On the Project Navigator, click the Constructs tab and in the *Constructs\Architectural* folder, double-click *03 Partitions* to open it.

2. From the Format menu, choose **Style Manager**.

3. At the top of the Style Manager, click the Open Drawing icon (shown in Figure 10–2 in Chapter 10).

4. Click the Content icon in the Outlook bar at the left, and then double-click the *Styles* folder.

5. If you are working in Imperial units, double-click the *Imperial* folder, select the file named *Spaces - Commercial (Imperial).dwg* and then click Open. If you are working in Metric units, double-click the *Metric* folder, select the file named *Spaces - Commercial (Metric).dwg* and then click Open.

    The remote file will appear with a shortcut icon next to the drawing icon indicating that it is being access remotely (See Figure 10–4 in Chapter 10 for an example).

6. Click the plus sign (+) next to the remote file, expand the *Architectural Objects* folder and then highlight *Space Styles*.

Several Space Styles are included here that are suitable for commercial projects. You may have also noticed additional files for *Residential*, *Medical* and *Educational*. Each of those contains several additional Space Styles. Feel free to explore their contents later.

7. Hold down the CTRL key and click the **CLOSET,** CONFERENCE_MEDIUM, CORRIDOR, DINING_ROOM, ENTRY_ROOM, LIBRARY, MECH_ROOM, OFFICE_MEDIUM and WORKSTATION_SMALL style names in the right-hand pane.

   9 Styles total should be highlighted (See left side of Figure 12–48).

8. Right-click any one of the highlighted names and choose **Copy**.

9. On the left pane, locate the current drawing (*03 Partitions*), right-click on it and choose **Paste** (See right side of Figure 12–48).

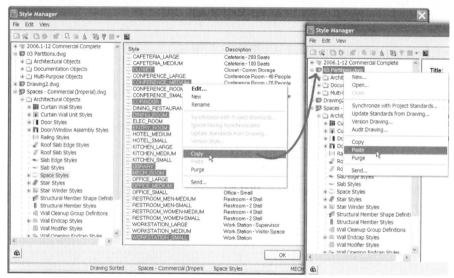

Figure 12–48 *Copy several Space Styles from the Library file and paste to 03 Partitions*

   It is not necessary to expand the folders of *03 Partitions* first. The pasted Space Styles will automatically paste to the correct location.

10. Click OK to dismiss the Style Manager.

GENERATING SPACES IN THE COMMERCIAL PROJECT

The Generate Spaces routine can be used to find the outline of existing Walls and other geometry and create a Space object to fill that shape.

11. On the Design tool palette, click the **Space Auto Generate** tool.

    If you do not see this palette or tool, right-click the tool palettes title bar and choose **Design** (to load the Design tool palette group) and then click the Design tab.

12. Move your mouse around the plan. Do *not* click yet.

Notice how the rooms highlight when the cursor is within them.

13. Move your mouse over the corridor in the center of the plan. Do *not* click yet.

    The corridor as well as the neighboring open plan spaces like the reception and conference room spaces should highlight.

14. In the Generate Spaces dialog, choose **OFFICE_MEDIUM** from the Style list and then click within the one of the offices to the left of the plan.

15. Repeat by clicking in each of the office spaces (7 total—5 along the left of the plan and 2 along the bottom) (See Figure 12–49).

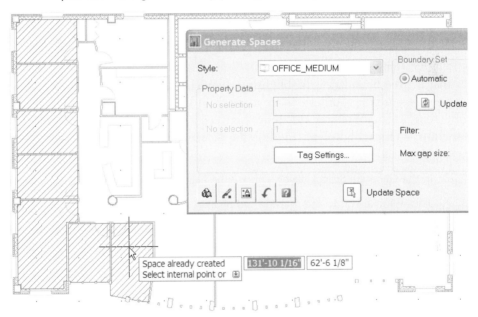

Figure 12–49 *Click in each office to create Spaces*

16. In the Generate Spaces dialog, choose **CLOSET** from the Style list and then click within the each of the two closets at the top of the plan.

There are plenty more Spaces to generate, but first we will modify the large open plan Space just created.

17. Make the room to the right of the closets using the **DINING_ROOM** Style.

18. Use the **MECH_ROOM** Style for the small square Space just above the reception area (See Figure 12–50).

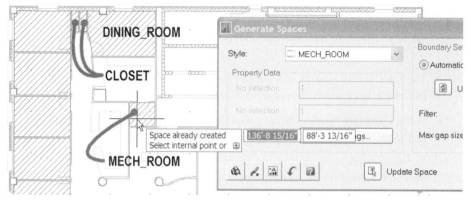

Figure 12–50 *Add the remaining "enclosed" non-open plan Spaces*

19. Move your mouse over the corridor in the center of the plan again. Do *not* click yet.

Notice that the red boundary is highlighting the inside edge of all of the counter-tops. This is because the countertops are made from Wall Styles. The Space Auto-Generate routine cannot tell the difference between a true Wall and a Wall used as a countertop. We can however, modify the selection set used by the routine manually.

20. In the Generate Spaces dialog, choose the Manual radio button.

    This will take you to the drawing and prompt for a selection.

21. Use a crossing window and select the entire plan. Then hold down the SHIFT key, and deselect each of the countertops.

22. Release the SHIFT and press ENTER when finished.

23. Move the mouse into the remaining spaces now.

    Notice that the desired shape will now highlight.

24. Change the Style to **CORRIDOR** and add a Space in the corridor/open plan area.

25. Use the **LIBRARY** Style for the last remaining Space (the one with the counter across the top Wall.

26. Click Close to exit the Generate Spaces dialog and then Save the file.

MODIFY A SPACE

We now have Spaces in all our rooms, but the large open plan Space should be cut up into smaller Spaces that better reflect the actual rooms.

27. On the Project Navigator, in the *Elements* folder, right-click the *Space Outlines* file and choose **Insert as Block**.

    Use 0,0,0 for the Insertion Point and check the "Explode" box before clicking OK. A magenta polyline should appear in the center of the plan.

28. Select the large Space, right-click and choose **AEC Modify Tools > Crop**.

29. At the "Select object(s) to form crop boundary" prompt, click the magenta polyline (at the corridor) and then press ENTER.

30. At the "Erase selected linework" prompt, right-click and right-click and choose **No** (or type **N** and press ENTER).

AEC Modify Tools work with profile-based objects such as Spaces or polylines. The Crop option uses the crop boundary object as you have seen here to cut away everything outside the boundary. We can now use the Walls and the magenta polyline to recreate the Spaces in the conference room, reception and work station areas.

31. On the Design tool palette, click the **Space Auto Generate** tool.

32. In the Generate Space dialog, choose the manual radio button again.

    Make the same selection that we did above, using the SHIFT key to deselect all countertops and then press ENTER.

33. Add the remaining three Spaces as indicated in Figure 12–51.

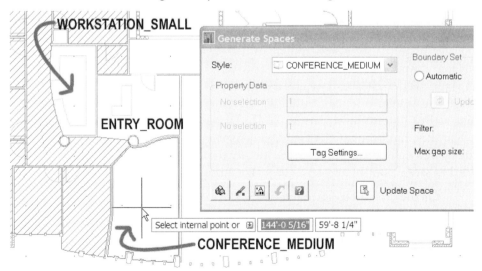

Figure 12–51 *Add Spaces to each of the remaining rooms using the Styles indicated*

34. Click Close to complete the routine.

35. Erase the magenta polyline.

36. Select the **WORKSTATION_SMALL** space in the center of the tenant space, right-click and choose **AEC Modify Tools > Divide**.

37. Using the Midpoint Osnap, draw a straight line across the middle of the Space.

    This will cut the Space into two.

Spaces have parameters for both the hatching we see in plan, and the physical dimensions of their floor and ceiling planes in 3D and section. If you select any Space, you will note that on the Properties palette, there is a Floor boundary thickness, a Ceiling boundary thickness and a Space height parameter. Spaces default to 9'-0" [2750] tall. We will change this next. Also, the default 6" [150] thick floor boundary is too thick based upon the Slabs we built above. We need the floor boundaries of the Spaces, which we will use as the finish floors, to be 1/2" [12] thick.

38. Select all Space objects and on the Properties palette, change the Space Height to **8'-0"** [**2400**] and change the Floor boundary thickness to **1/2"** [**12**].

39. Save and close the file.

    If you wish to see this change in the section, open the *A-SC01* View file now.

USING DISPLAY THEMES

In Chapter 2, we took a brief look at Display Themes. A Display Theme is a new object type in ADT 2006 that applies a display override to some or all objects in a drawing. The change is temporary and applies only as long as the Display Theme remains active. Like most objects in ADT, Display Themes are style-based objects. They key into Property Set Data (see Chapter 15) attached to the objects and use rules based upon those properties to modify the display. A few styles have been provided as sample out-of-the-box. Let's take a brief look at them now.

1. On the Project Navigator palette, click the Views tab and then double-click the *A-FP03* View file to open it.

    The third floor plan opens and all of the Spaces added to the *03 Partitions* file appear via XREF in this drawing.

2. On the tool palettes, right click the title bar and choose **Document** to load the Document tool palette group and then click the Scheduling tool palette to make it active.

3. On the Scheduling tool palette, click the ***Theme by Space Type – Commercial*** tool.

4. At the "Upper left corner of display theme" prompt, click a point off to the side of the plan and then press ENTER.

A Display Theme legend will appear at the location where you clicked. All of the Space objects in the *03 Partitions* XREF will now display with color coded solid hatch patterns. If you study the legend, you will see that all of the Space styles included in the *Spaces - Commercial (Imperial).dwg* [*Spaces - Commercial (Metric).dwg*] are included even though not all of them are used in this project. When you build the Display Theme Style, you decide which criteria it should include. In this case, this particular Style includes all of the commercial styles. We could, if we wanted, modify it to eliminate the ones that we are not using. There are two other Display Themes on the Scheduling tool palette. Let's look at one more.

5. On the Scheduling tool palette, click the ***Theme by Space Size*** tool.

6. At the "Upper left corner of display theme" prompt, click a point to side of the plan and then press ENTER.

This Theme uses different colors and codes by size rather than type. Notice that only one Theme can be active at a time. When you added the second one, the first was automatically disabled. You can right-click either one and chose to apply or disable them.

7. Close and Save all Commercial Project files.

CAD Manager Note: In this project, we have placed the Spaces directly to the *03 Partitions* Construct. In larger projects, it may be desirable to move the Spaces to their own Construct with a name such as *03 Spaces* or *03 Spaces and Grids*. This will allow for greater flexibility in staffing issues and work flow. If both the Spaces and the Walls are in the same Construct, only one individual can be editing that particular floor plate at a given time including both floor and ceiling plans. However, if you separate the Spaces out to their own Constructs, you can have two individuals working simultaneously on the same floor plate in separate Construct files. One would be responsible for Walls and Doors, while the other would maintain Spaces and Ceiling Grids (see Chapter 13). Usually the decision is made on a per-project basis and project size is the major determining factor.

LOAD THE RESIDENTIAL PROJECT

Be sure that all files from the Commercial Project have been closed and saved. However, you must make sure that a single empty file remains open. The Project Browser will not run in "zero-doc state." The simplest way to do this is to close all files, and then use the Qnew icon to create a new empty file.

1. From the File menu, choose **Project Browser**.

2. Click to open the folder list and choose your *C:* drive.

3. Double-click on the *MasterADT 2006* folder, then the *Chapter12* folder.

    One or two residential Projects are listed: *12 Residential* and/or *12 Residential Metric*.

4. Double-click *12 Residential* if you wish to work in Imperial units. Double-click *12 Residential Metric* if you wish to work in Metric units. (You can also right-click on it and choose **Set Current Project**.) Then click Close in the Project Browser.

 IMPORTANT If a message appears asking you to re-path the project, click Yes. Refer to the "Re-Pathing Projects" heading in the Preface for more information.

GENERATE SPACES

We will follow much the same process that we did above and add Spaces to the residential project.

5. On the Project Navigator, in the *Constructs* folder, double-click the *First Floor New* file to open it.

    Be sure that the drawing is displaying in plan view without shading. If it is not, choose Top from the Views toolbar and 2D Wireframe from the Shading toolbar.

Let's import some Space styles from the Residential Styles library into the current drawing.

6. On the Design tool palette, right-click the Space tool and choose **Space Styles**.

7. In the Style Manager, click the Open Drawing icon, (use the Content link to browse to your *Content* folder) and open the file named: *Spaces - Residential (Imperial).dwg* [*Spaces - Residential (Metric).dwg*].

8. Expand the *Architectural Objects* folder in the *Spaces - Residential (Imperial).dwg* [*Spaces - Residential (Metric).dwg*] file, right-click the *Space Styles* entry, and choose **Copy**.

9. Right-click on the *First Floor New.dwg* file in the tree view at left and choose **Paste**.

    If a "Duplicate Names Found" dialog box appears, simply click OK.

10. Click OK to dismiss the Style Manager. (Answer No if asked to save a file.)

This technique copied all Space Styles from the library rather than a selection of them like we did above. We now have some Space styles with which to work. We are ready to generate some Spaces.

11. On the Design palette, click the **Space Auto Generate** tool.

12. Verify in the list next to Selection Filter that **Walls, lines, arcs, polylines, and circles** is selected.

Walls, lines, arcs, polylines, and circles is the default choice for this command. If the boundaries that it finds are incorrect with this setting, try Walls only.

13. From the Style list, choose **FAMILY_ROOM** [**LIVING ROOM – SMALL**].

14. At the "Select internal point" command line prompt, click anywhere in the center of the large space at the right.

15. Change the Style in the Generate Spaces dialog box to **DINING_ROOM**, and click anywhere in the center of the large room at the left.

16. Continue adding Spaces choosing appropriate Styles from the list as you go.

    For the porch, choose any Space Style for now.

17. Click Close to complete the command.

18. Select the Space that you added to the porch, right-click and choose **Copy Space Style and Assign**. Name the new Style **Sun Room** and then click OK.

DESIGN CHANGES AND GENERATING SPACES

Space objects are not linked to the Wall objects. Therefore, as design changes occur, it is necessary to return to generate Spaces in order to update the perimeter and area of the Space objects.

MODIFY SPACES WITH GENERATE SPACES

1. In the center of the new addition, move the Wall between the two closets down **12″** [**300**].

   Notice that the Space object did not change.

2. On the Design palette, click the **Space Auto Generate** Tool.

When modifying existing Spaces, first set the correct style for the Space object you wish to change. For instance, in this case, we changed the closets; therefore, select *CLOSET* from the Style list before clicking the Update Space icon.

3. Change the Style to **CLOSET** [**CUPBOARD**].

4. Click the Update Space icon (see Figure 12–52).

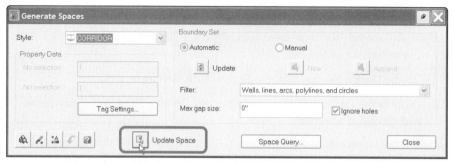

Figure 12–52 *To update a Space, click the Update Space icon*

5. At the "Select a space object to update" prompt, click on one of the existing closet Space objects.

6. At the "Select internal point" prompt, click anywhere within the area of the modified closet.

   You will see the Space object update immediately. It will conform to the new shape of the room.

7. Repeat the steps for the other modified Space and then click Close.

8. Undo any edits from this sequence.

Be sure to adjust the height of the Spaces added to the Residential files. Space Objects default to 9'-0" [2750]. For this particular residence, the ceiling height should be 8'-0" [2400].

9. Open Second Floor New and add Spaces there as well.

10. Save and Close all Residential Project files.

ADDITIONAL EXERCISES

Additional exercises have been provided in Appendix A. In Appendix A you will find exercises to add Slabs to the *Basement New* file and a Roof to the porch of the Residential Project. You will also find an exercise to add Spaces to the other project files. It is not necessary that you complete these exercises to begin the next chapter, they are provided to enhance your learning experience. Completed projects for each of the exercises have been provided in the *Chapter12/Complete* folder.

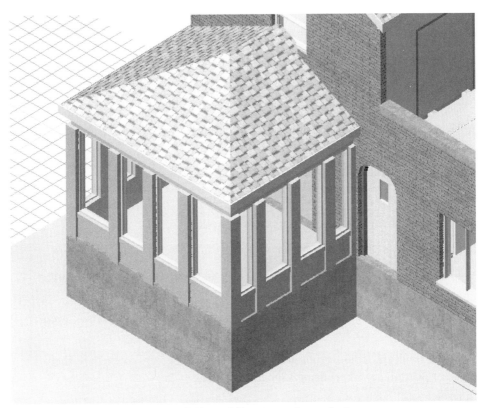

Figure 12–53 *Update projects in Additional Exercises – Appendix A*

SUMMARY

Always rough out a roof using the one-piece Roof first. When design needs warrant, convert it to Roof Slabs.

Eaves can be made square by choosing the plumb option in Properties.

Roof Slabs offer greater control and flexibility than the one-piece Roof but must be manually mitered.

Roof Slab edges can be customized with Roof Slab Edge styles.

Use Slab objects for virtually any horizontal surface.

Slab edges can be customized with Slab Edge styles.

Keep Slabs in separate XREFs to keep them from cluttering your plan files.

Use a flat Slab with 0 slope for flat roofs. Articulate the plan view with lines for drainage in a roof plan drawing.

Generate Spaces allows Space objects to be created from existing Walls; even those contained in XREFs.

Toilet partitions and countertops are Wall objects, so be sure to exclude them from the selection set when using Generate Spaces.

Display Themes offer a quick a convenient way to override Display parameters for presentation-type drawings.

As the design changes, use Update Space to keep the Space objects up to date.

Construction Documents

In this section, we take our building model into construction documentation. The next several chapters explore creating reflected ceiling plans, generating live schedules and dimensions, extracting sections and elevations from the building model and working with Details and the Detail Component Manager. All of this data is organized onto Sheets, ready for printing or export to electronic formats such as DWF.

Section III is organized as follows:

CHAPTER 13

Creating Reflected Ceiling Plans

INTRODUCTION

In this chapter, we will explore generation of a reflected ceiling plan (RCP) in Architectural Desktop. To facilitate the creation of reflected ceiling plans, ADT includes a special Display Representation specifically designed for reflected ceiling plans, called "Reflected." In addition to this Display Representation, the Ceiling Grid object is also available. Use this object to create suspended ceiling layouts. There are also many drag and drop content items, like lighting symbols and ceiling fixtures, at our disposal.

OBJECTIVES

A Display Configuration specially designed to display all objects with their "Reflected" Display Representation active is included in the standard ADT template files. Before work on the reflected ceiling plan can begin, this Display Configuration must be made active. (Review Chapter 2 for a detailed discussion of Display Configurations.) Working in the Commercial Project, we will add Ceiling Grid objects and consider ways to crop and center them within the shapes of the spaces in which they occur. The collection of content available for lighting fixtures will also be explored. The following topics will be explored in detail:

- Learn to switch to reflected ceiling plan Display Configuration.
- Create and modify Ceiling Grid objects.
- Clip Ceiling Grids to the shape of the room.
- Center Ceiling Grids in rooms.
- Add anchored light fixtures.
- Mask a Ceiling Grid with lights.

WORKING IN THE REFLECTED DISPLAY CONFIGURATION

Consistent with the other features of Architectural Desktop, Ceiling Grids have built-in intelligence and are displayed only when a Reflected Display Configuration is active. This makes it possible for the floor plan data and the reflected ceiling plan data to easily co-exist in the same model file. Ceiling information can be placed in the same Construct as the Walls and Doors or if desired, ceiling plan information can be placed in a separate Construct and XREFed back to the composite model, plan Views and Sheet files. In larger projects with multiple team members, this is often a good practice, as you can have one team member working on the floor plan data, while another completes ceiling plan information in a separate file. Should you decide to separate the ceiling information, you will also need to have the Spaces (created in the last chapter) resident in the ceiling information Construct instead of the partitions Construct. In smaller projects, it makes more sense in most cases to have both the Walls and the ceiling information together in the same Construct. In this small project, ceiling plan information will be added directly to the *03 Partitions* model file as were the Spaces in the last chapter.

INSTALL THE CD FILES AND LOAD THE CURRENT PROJECT

If you have already installed all of the files from the CD, simply skip down to step 3 below to make the project active. If you need to install the CD files, start at step 1.

1. If you have not already done so, install the dataset files located on the Mastering Autodesk Architectural Desktop 2006 CD ROM.

   Refer to "Files Included on the CD ROM" in the Preface for information on installing the sample files included on the CD.

2. Launch Autodesk Architectural Desktop 2006 from the desktop icon created in Chapter 3.

If you did not create a custom icon, you might want to review "Create a New Profile" and "Create a Custom ADT Desktop Shortcut" in Chapter 3. Creating the custom desktop icon is not essential; however, it makes loading the custom profile easier.

3. From the File menu, choose **Project Browser**.

4. Click to open the folder list and choose your *C:* drive.

5. Double-click on the *MasterADT 2006* folder, then the *Chapter 13* folder.

   One or two commercial Projects will be listed: *MADT Commercial* or *MADT Commercial Metric*.

6. Double-click *MADT Commercial* if you wish to work in Imperial units. Double-click *MADT Commercial Metric* if you wish to work in Metric units. (You can also right-click on it and choose **Set Current Project**.) Then click Close in the Project Browser.

IMPORTANT If a message appears asking you to re-path the project, click Yes. Refer to the "Re-Pathing Projects" section in the Preface for more information.

CHANGE THE DISPLAY CONFIGURATION

1. On the Project Navigator, in the *Constructs\Architectural* folder, double-click the *03 Partitions* file to open it.

2. On the Drawing Status Bar, open the Display Configuration pop-up menu and choose **Reflected** (see Figure 13–1).

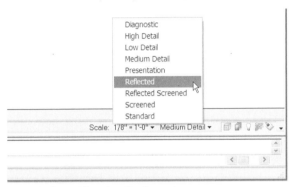

Figure 13–1 *Load the Reflected Display Configuration*

Notice that the graphical display of several of the objects in the drawing has changed (see Figure 13–2). (It may take a few seconds for the Display Configuration to switch.)

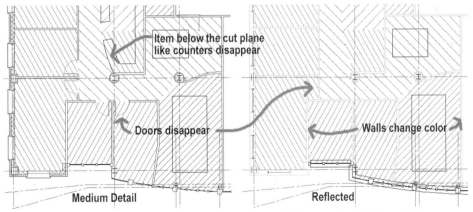

Figure 13–2 *Most objects change display properties with the switch to Reflected (Your screen appearance might vary slightly)*

NOTE See Chapter 2 for more information on the Display System.

ADD COLUMN GRID REFERENCE AND CONFIGURE BUBBLES TO DISPLAY IN RCP

It might be helpful while working on the ceiling to have the Column Grid displayed for reference. As you recall, the Column Grid is contained in another Construct (which in turn references a typical *Column Grid* Element file). We will add an Overlaid XREF to the *03 Grid* Construct.

3. On the Project Navigator palette, expand the *Constructs\Structural* folder.

4. Drag the *03 Grid Construct* and drop it into the drawing window.

 IMPORTANT Resist the urge to simply overlay the Column Grid Element file itself. In the current situation, this would yield the same result; however, if at some later point in the project, the third floor Grid were to become unique, the third floor ceiling would no longer be coordinated.

 NOTE When you drag a Construct into another Construct as we have done here, ADT will automatically use XREF Overlay, which is exactly what we need in this case. In this way, the *03 Grid* file will only be visible here in *03 Partitions*, and if it is needed in other files, must be XREFed separately by those files.

As you can see, there is an issue with the Column Grid display. The Column Grid lines are displayed, but the Column Bubbles are not. Typically it will be useful to have your Column Bubbles displayed in your ceiling plans. To fix these, we need to make a simple modification to the Column Grid Bubble multi-view block.

5. On the Project Navigator, in the *Elements\Structural* folder, double-click the *Column Grid* Element file to open it.

    This is the typical *Column Grid* Element file (created in Chapter 6) that is nested within the *03 Grid Construct*.

6. Select any one of the Column Grid Bubbles, right-click, and choose **Edit Multi-View Block Definition** (see Figure 13–3).

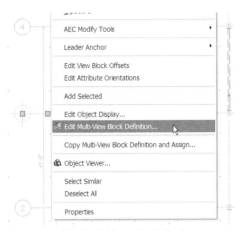

Figure 13–3 *Edit the Column Bubble Multi-View Block Definition*

7. In the Multi-View Block Definition Properties dialog box, click the View Blocks tab.

   Notice that there are several Display Representations listed on the left. (Review "Create a Custom Multi-View Block" in Chapter 11 for more information.) The Reflected representation is for reflected ceiling plans.

8. Select Reflected from the list of Display Representations.

   Notice that there are no View Blocks loaded on the right.

9. Click the Add button.

10. In the Select A Block dialog box, choose **BubbleDef**, and then click OK (see Figure 13–4).

    This is the same block that is loaded in the General Display Representation used for plans.

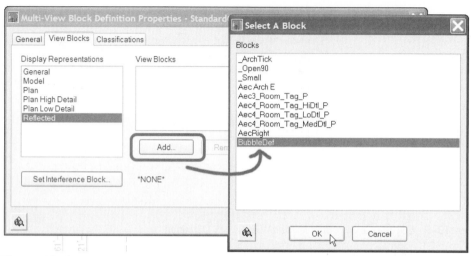

Figure 13–4 *Add the same View Block used for Plans to the Reflected Display Rep*

11. Click OK to return to the drawing.

12. Save and Close the *Column Grid* file.

   An XREF update alert will appear at the lower left corner of the screen (see Figure 13–5).

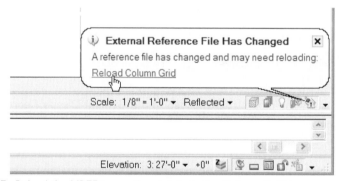

Figure 13–5 *Reload the XREF to see the update to the Column Bubbles*

13. Click the Reload Column Grid link (see Figure 13–5).

   The Column Bubbles should now appear in the Reflected Display Configuration.

CAD Manager Note: To make this change permanent for all drawings, edit your office standard template file(s). The Column Grid Bubble is named **StandardGridBubble**. This Multi-View Block is created automatically by ADT on the fly when a Column Grid is labeled. If you wish to customize it, you must create your own version, name it **StandardGridBubble** and save it in your office standard template file. Another useful edit to make to this MVB in your template is to add a View Block to each of the Plan, Plan High Detail and Plan Low Detail Display Representations. Make the scale of the Plan View Block for your Medium Detail Display (typically 1/8"=1'-0" [1:100]—making the text within it about 12" [300] tall), make the Low detail block twice as big (typically 1/16"=1'-0" [1:200]) and the High Detail block half as big (typically 1/4"=1'-0" [1:50]). When you do this, the Column Bubble will change scale with the drawings automatically in different Plan View files to which it is referenced.

WORKING WITH CEILING GRID OBJECTS

Ceiling Grid objects can be used for any type of grid pattern occurring on a reflected ceiling plan. Most typically, they are used for 2x2 [600x600] or 2x4 [600x1200] suspended ceilings. Ceiling Grid objects are added and modified in the same way as other ADT objects. Ceiling Grids are rectangular in shape; however, they can be clipped to the shape of the room in which they are placed. Both polylines and Space objects can be used as clipping boundaries. Most other parameters of Ceiling Grids are very similar to Column Grids, already covered in Chapter 6.

CAD Manager Note: Ceiling Grids are 2D objects designed for use in construction documents. Although they can be moved to the correct Z height and viewed from a 3D vantage point, being comprised of a grid of lines, they will vanish in renderings. If you plan to create hidden line renderings and walkthroughs, the Ceiling Grid placed at the correct Z height will prove sufficient. However, if you will be generating shaded or rendered output, use Space objects, covered in the last chapter, and apply a Material to the ceiling surface. The focus of this chapter and the entire Section III of this book is the creation of 2D Construction Documents.

ADD A CEILING GRID

We are ready to begin adding Ceiling Grid objects to our *03 Partitions* Construct.

> 1. Zoom in on the room in the top right corner of the tenant space (labeled Break Room in Figure 13–6).

Figure 13–6 *Use these designations to locate individual rooms in the suite*

 NOTE The labels shown in Figure 13–6 will be referenced throughout the remainder of this tutorial for convenience. Refer to this figure throughout the exercise. We will discuss adding Room Tags to our project in Chapters 14 and 15.

2. On the Design palette, click the **Ceiling Grid** tool.

3. On the Properties palette, within the Dimensions grouping, set both the X - Width and Y - Depth settings (which control the overall size of the Grid), to **20'-0"** [**6000**].

4. Verify that within the X Axis and Y Axis groupings, both Bay size settings (which control the size of each tile), are set to **2'-0"** [**600**] (see Figure 13–7).

 NOTE If the Bay size fields are not available, choose **Repeat** from the Layout type list. Like Column Grids, Ceiling Grids can use Space evenly to establish the modulation of bays. However, because most Ceiling Grids use a fixed pre-manufactured tile size, it would be rare to use this feature.

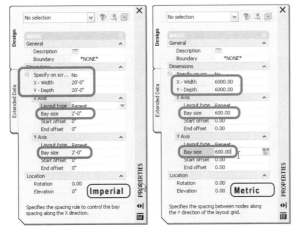

Figure 13–7 *Configure the parameters of the Ceiling Grid on the Properties palette*

5. Using an Endpoint Object Snap, set the Insertion point of the grid at the lower left corner of the Break Room, and then press ENTER to accept the default rotation of **0°** (see Figure 13–8).

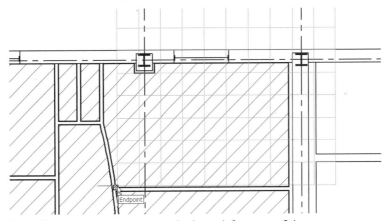

Figure 13–8 *Place the insertion point at the lower left corner of the room*

6. Press ENTER to complete the command.

There are several things to note about the Ceiling Grid just placed:

▶ It does not follow the shape of the room.

▶ It is bigger than the room.

▶ It is not centered.

7. Click on the Ceiling Grid to activate its grips.

8. Notice the grips at the four corners.

We can adjust the overall size of the grid with these grips. (You might have to zoom out to accommodate the entire grid on screen.)

9. Click the grip at the lower left corner of the Ceiling Grid.

10. Begin dragging the grip down and to the left, but don't click yet (see Figure 13–9).

11. Notice that the Ceiling Grid dynamically adds or removes bays on the right side and along the top as you drag. The bottom left corner moves.

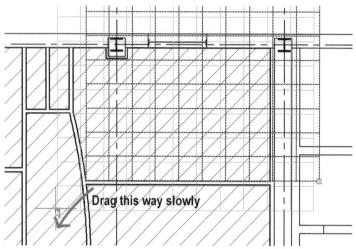

Figure 13–9 *Grip edit the size of the grid*

12. Drag down and to the left far enough to add a bay or two in each direction and then click to set the corner. (Do this without Object Snaps.)

The Ceiling Grid should now completely cover the space, including the small curved section at the bottom left corner of the room. Repeat the process on the upper right corner and notice the slightly different behavior. The lower left corner is the insertion point. When you dragged it, the entire Grid shifted and bays were added or removed from the opposite sides of the Grid. When you drag the upper right corner, it only adds or removes bays; the grid will stay anchored to the insertion point at the lower left. Furthermore, the bays are added or removed from the same side this time. In the next sequence, we will clip the Ceiling Grid to the shape of the room, so be sure to finish your grip editing with the grid extending past the outside the room in all directions. Having the grid overlap on all sides gives us "extra play" when clipping and centering.

CLIP THE GRID TO THE ROOM SHAPE

Now that the Ceiling Grid covers the entire room, we still have the larger issue of the Ceiling Grid's not following the shape of the room. We will use the Clip Ceiling Grid command to resolve this issue.

13. Select the Ceiling Grid, right-click, and choose **Clip**.

"Ceiling grid clip" will appear onscreen as a dynamic prompt or at the command line with three options: "Set boundary," "Add hole" and "Remove hole." Use Set boundary to clip the grid to the shape of the room or any other closed boundary (polyline) you wish. Add hole does the opposite: It creates a hole by cropping out a section of the grid within. Remove hole deletes a clipping boundary or hole.

NOTE You must keep the clipping boundaries. Erasing the boundary eliminates the Clip. If you do not wish to have the clipping boundaries displayed, put them on a separate layer that you freeze or designate as No Plot. Or use Masking Blocks instead. If you are clipping to Spaces, you can use Display Control to turn the Spaces off, or make them non-plotting.

14. From the dynamic prompt choose **Set boundary**.

15. At the "Select a closed polyline or space entity for boundary" prompt, click on the Space object beneath the Grid. (Click the 45° hatching to select it.)

16. Notice that the Ceiling Grid is now clipped to the shape of the room (see Figure 13–10).

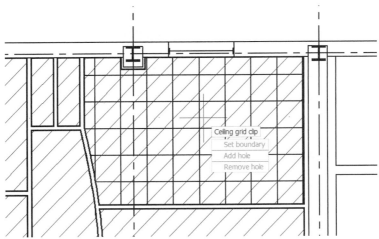

Figure 13–10 *The Ceiling Grid now conforms to the shape of the room*

17. Press ENTER to end the Ceiling Grid Clip command.

ADD CEILING GRID WITH GRID CLIPPING

The previous sequence illustrates the concept of clipped Ceiling Grids. In this sequence, we will clip a Ceiling Grid while it is being added to the drawing. This saves the step of our having to manually clip the grid later as previously demonstrated.

18. On the Design palette, click the **Ceiling Grid** tool.

    Use all of the same settings as before but don't place the grid yet.

19. On the Properties palette, within the General grouping, choose **Select object** from the Boundary option list (see Figure 13–11).

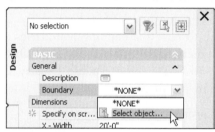

Figure 13–11 *Designate the clipping boundary as you add the Ceiling Grid with the Select object option*

20. At the "Select a space or closed polyline for boundary" prompt, click on the hatching in the Office in top left corner (to the left of the Break Room).

21. Move the cursor around and notice how the Ceiling Grid is dynamically cropped behind the shape of the Office space (see Figure 13–12).

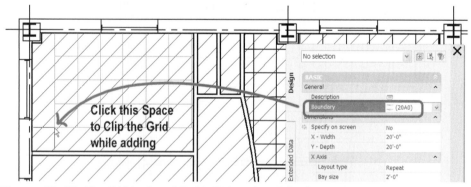

Figure 13–12 *The Grid is clipped "behind" the edge of the Space*

22. Move the mouse down and to the left past the lower left corner of the space.

23. Click outside the space to set the insertion point.

Clicking outside the Space, down and to the left achieves the same goal as above in ensuring that the clipped Grid covers the entire Space.

24. Press ENTER to accept the default rotation of **0°**.

The new Ceiling Grid is now clipped to the Office Space and the Add Ceiling Grid command is still active and ready to place another Grid. However, notice that it is still clipped in the same Space.

25. Pan the drawing down to the room beneath this one.

You can return to the Properties palette and change the Boundary setting to **Select object** again. However, you can also access this option on the right-click menu.

26. Right-click in the drawing and choose **Set Boundary** (or press the down arrow key to access the dynamic prompt options).

27. Click on the hatch in the next Office Space (see Figure 13–13).

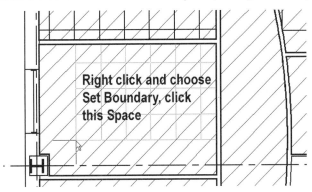

Figure 13–13 *Two clipped grids in the Office spaces*

The two Ceiling Grids placed so far were not centered within their rooms. A Ceiling Grid can be centered within its clipping boundary during placement in the drawing. This option will center the grid to the Space or Polyline bounding box. The bounding box is the smallest rectangle that would completely contain the entire object at its widest and tallest dimensions. For this reason, this trick works best with rectangular or nearly rectangular clipping boundaries. This applies to all of our office Spaces here. Remember that you can always move a grid after it has been placed to re-center it as necessary.

28. Right-click in the drawing again and choose **SNap to center** (or press the down arrow key for the same option).

29. Accept the default rotation by pressing ENTER.

NOTE When you use the SNap to center option, the rotation default will be entered from the clipping boundary as well. Simply press ENTER to accept this default. If you choose a different rotation, you will lose the centering.

30. Press ENTER to end the command.

EDIT THE HATCH PATTERN OF SPACES

At this point in the process, the drawing is feeling a bit cluttered. When you use Space objects as clipping boundaries as we are here, the hatching of the space can be distracting. We cannot turn the hatching off, because it would make it difficult to select the Spaces as boundaries as we would have to click at the edge of the room, which might interfere with the XREF Walls. This would be the same difficulty that we would have if we were using polylines as boundaries. However, we can edit the hatching to make it less prominent.

1. Select any Space object, right-click, and choose **Edit Object Display**.

2. Click the Display Properties tab (see Figure 13–14).

   Notice that the active Display Representation is **Reflected** and that the Drawing Default is the property source for this Display Representation. For the Spaces in a reflected ceiling plan, this will make our task easy. (See Chapter 2 for more information on the Display System.)

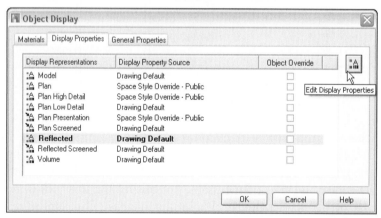

Figure 13–14 Edit the Drawing Default properties of the Reflected Display Representation

3. Click the Edit Display Properties icon.

4. On the Layer/Color/Linetype tab, click on the color of the Net Hatch component (currently ByBlock).

   This will open the Select Color dialog box.

5. Choose a gray color for the Hatch component (such as Color 9 or 254) and then click OK (see Figure 13–15).

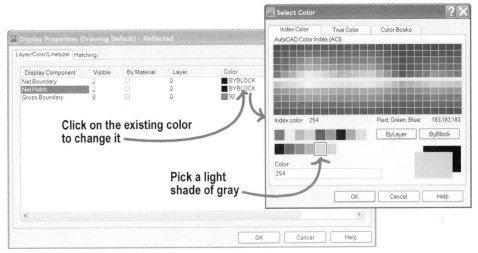

Figure 13–15 *Make the hatching display a gray color*

6. Click the Hatching tab.

7. Change the Scale of the hatching to **3′-0″ [900]** and the Angle to **60°** (see Figure 13–16).

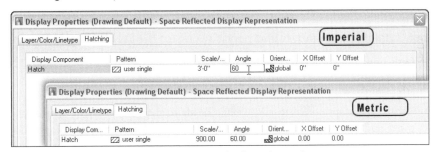

Figure 13–16 *Widen the hatching spacing and adjust the angle*

8. Click OK twice to return to the drawing (see Figure 13–17).

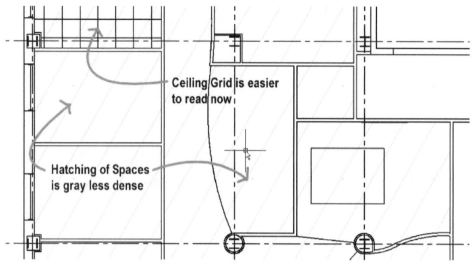

Figure 13–17 *The Space objects now display less obtrusive hatching as a result of the display settings*

 TIP If the drawing does not appear to change, choose Regen Model from the View menu.

Notice that this change affected all Spaces regardless of their Space Styles. This is because Reflected is active and as we saw, Reflected uses Drawing Default. This is just another example of the power and pervasiveness of the ADT Display System.

Let's now add several more Ceiling Grids.

9. On the Design palette, click the **Ceiling Grid** tool.

10. Repeat the steps above to add Ceiling Grids to the Reception, all Offices and the Copy Room. (Refer to the labels in Figure 13–6 earlier in the chapter for assistance locating each of these rooms).

Remember to right-click and choose **Set boundary** and **SNap to center** for all Grids. Notice that despite the use of the Snap feature, the non-rectangular spaces might still need some adjustment. We will perform these adjustments below. Remember: progressive refinement.

11. Press ENTER when finished.

12. Save the file.

CLIP A CEILING GRID TO A POLYLINE

In addition to Space objects, Ceiling Grids can be clipped to closed polylines. A closed polyline is already located in the Conference Room area.

13. Zoom in on the Conference Room (in the bottom right corner of the tenant space).

14. On the Design palette, click the **Ceiling Grid** tool.

15. Change the X - Width to **10′-0″** [**3000**], leave the Y - Depth at **20′-0″** [**6000**], and set both the X Axis > Bay size and the Y Axis > Bay size to **1′-0″** [**300**].

16. Choose **Select object** from the Boundary drop-down list.

17. At the "Select a space or closed polyline for boundary" prompt, click the edge of the rectangle in the middle of the Conference Room.

18. Right-click, choose **SNap to center**, and then press ENTER to accept the default rotation (see Figure 13–18).

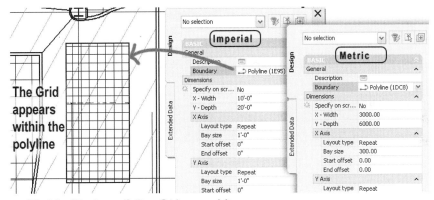

Figure 13–18 *Clipping a Ceiling Grid to a polyline*

19. Press ENTER to complete the command.

If the amount of clipped tile on any edge is too small, you can move the Grid half a tile and it will remain clipped to the boundary.

CAD Manager Note: While clipping Grids to polylines is effective, the technique exposes one of the weaknesses in the integration between ADT and AutoCAD technologies. Architectural Desktop uses the Display System to control the display of objects, AutoCAD uses Layers. This topic was discussed in detail in Chapter 2. When we clip a Ceiling Grid to a polyline, the Grid, which is an ADT object, will be controlled by the Display System, while the polyline will rely on layers. Unfortunately there is no way to make the Display System control the display of the polyline. Therefore you will need to manually keep track of its layer, turning it on in ceiling plans and off in plans. If you wish to avoid this, you have two options. First, you can consider the suggestion made above and separate the ceiling model components from the plan components into two different Constructs. In this way, while the polyline would still be managed by a layer, it would live in a ceiling-only Construct and would therefore only appear in View files that showed the ceiling. The other approach would be to convert the polyline to a Space object. You can use Copy and Assign to create a custom Space Style and on the Properties palette, you can turn off the floor boundary of this Space object (make it zero). Furthermore, you can edit the Plan Display Representations of this Space at the Style level and turn off all components so that it shows only in Reflected. If you employ this approach, be careful that you do not select this Space when scheduling as it will throw off your square footage values.

ADD A CLIPPING HOLE

Clipping "holes" can be added to Ceiling Grids to represent portions of the ceiling that don't have lay-in ceiling or areas where a suspended drywall ceiling may occur. A closed polyline is already located in the Reception Area.

20. Select the Ceiling Grid in the Reception Space, right-click and choose **Clip**.

21. Right-click again and choose **Add Hole.**

22. When prompted, select the edge of the rectangle in the Reception area (see Figure 13–19).

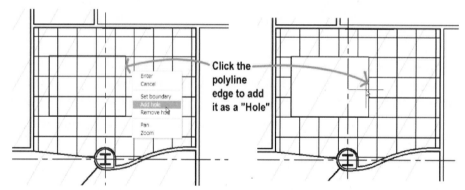

Figure 13–19 *Clip the Ceiling Grid with a hole*

23. Press ENTER to end the Clipping command.

There is now a void in the center of the Reception area Ceiling Grid.

24. Save the file.

MANUALLY CENTER A CEILING GRID IN A SPACE

In the previous exercise, you learned to add clipped Ceiling Grids and use the SNap to center feature to center those Grids in their boundaries. However, some of the Grids still require adjustment to their position. In this exercise, we will look at manual centering and make other adjustments as well. Apart from the Snap feature shown above, there is no special command or technique needed to center Ceiling Grids in a Space. Simply use the standard Move command and locate the grid where you want it. Grids will remain clipped even as you move them. Use standard AutoCAD Object Snaps and techniques to achieve desired placement.

1. Zoom in on the office in the upper left corner whose Grid is not centered.

2. Right-click in the drawing and choose **Basic Modify Tools > Move** (or simply type **M** and press ENTER at the command line).

3. At the "Select objects," prompt, select the Ceiling Grid in the Office and then press ENTER.

4. At the "Specify base point or displacement" prompt, use the Intersection Object Snap and click any Ceiling Grid intersection.

TIP You can hold the SHIFT key and right-click to choose an override OSNAP. Overriding the OSNAP setting this way helps guarantee that you won't inadvertently pick the wrong point.

5. At the "Specify second point of displacement" prompt, use the Center Object Snap and snap to the Center of the Space object.

TIP To snap to the Center of a Space object, move the mouse over any of the diagonal hatching lines of the Space object until the Center Snap marker appears.

The grid is now centered in the room. Notice also that the grid remains clipped to the space.

FINE-TUNE THE GRID POSITION

Perhaps the amount of cut tile is too much on the left and right. The grid can be fine-tuned further by moving it again one-half a tile in any direction.

6. Repeat the Move command.

7. Select the same grid and press ENTER.

8. Pick any point for the Base point prompt.

9. Make sure either POLAR or ORTHO is turned on.

10. Move the mouse directly to the right horizontally, type **12″ [300]**, and then press ENTER.

If you prefer the cut tile sizes on the left and right better after this move, keep the change; otherwise, Undo.

CENTER THE "OLD-FASHIONED" WAY

Sometimes using the Center Object Snap is not an option, or it does not give the desired results. In this case, draw a simple line diagonally across the room. You can also use the following technique if you want to indicate a specific "point of beginning." This is the best approach on the non-rectangular spaces, or when a Ceiling Grid spans several spaces with underpinned walls.

TIP To create a layout with underpinned walls, let the grid span over several spaces and clip it to a polyline instead of a Space object. Use the Add Hole option of the Ceiling Grid Clip command above and choose the overlapping walls as the "Hole." Try to keep it to a few Walls per grid to keep regen times in check.

11. Create a temporary non-plotting layer and make it Current.

Use the Layer Properties Manager icon on the Layer Properties toolbar. See the "Create a New Layer Based on a Layer Standard" section in Chapter 10 for more information on creating layers. However, your layer can be non-standard for this exercise.

12. Zoom in on any room.

13. From the Shapes toolbar, click the Line icon (or type **L** and press ENTER at the command line).

14. Snap the Endpoints of the line to two opposite corners of the room.

15. Move the Ceiling Grid in the room as instructed above, but this time, snap the grid to the line's Midpoint (see Figure 13–20).

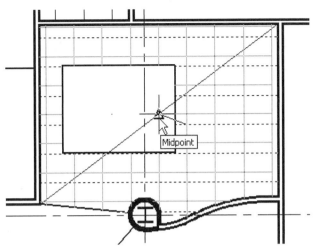

Figure 13–20 *Using a diagonal line to center a grid*

  TIP If you are working in the Reception Space or any Space with a "Hole" boundary, you can move the polyline boundary of the Hole as well.

GRIP EDIT A BOUNDARY

Grip clipping boundaries remain live objects. This means that you can edit the boundary, and it will affect the shape of the Ceiling Grid that is displayed within.

16. Zoom in on the Reception space (the one where we added the Hole above).

17. Click on the Space object in the Reception space.

18. Click the grip in the lower left corner of the space.

19. Drag the grip point straight down and click.

     Notice that the Ceiling Grid changes with the edit. The Ceiling Grid of the Reception space now encroaches on the Corridor.

20. Right-click and choose **Deselect All** or simply press the ESC key (see Figure 13–21).

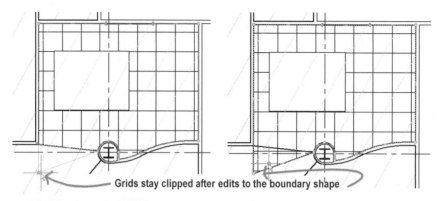

Figure 13–21 *Grip editing the space affects the grid*

21. Click the rectangle in the middle of the Reception space.

22. Grip edit it to a new shape (see Figure 13–22).

    Notice the change in the Ceiling Grid.

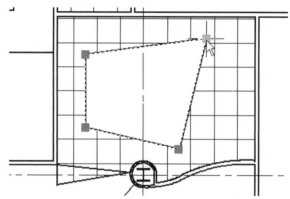

Figure 13–22 *Holes can also be adjusted with grips*

23. Right-click and choose **Deselect All** or simply press the ESC key.

CHANGE GRID BAY SIZES

2x2 [600x600] grids can easily be changed to 2x4 [600x1200] and vice versa.

24. Select the Ceiling Grid in any Office, right-click and choose **Properties**.

TIP Select several at once if you wish.

25. On the Properties palette, change the Bay size within the Dimensions >> X Axis grouping to **4'-0" [1200]** and then press ENTER (see Figure 13–23).

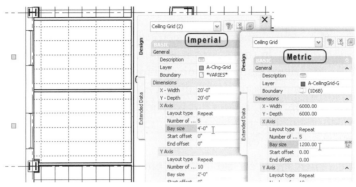

Figure 13–23 *Change the Ceiling Grip Bay size*

26. Save the file

Repeat any of the above techniques on other spaces to fine-tune the ceiling layout.

ADDING ANCHORED LIGHT FIXTURES

Now that we have our Ceiling Grids placed and centered, it is time to add lighting fixtures to our reflected ceiling plan. Lighting fixtures are typically made from Multi-View Blocks and/or Mask Blocks. We have already discussed Multi-View Blocks in several areas of this text; a Mask Block is an ADT object designed to "cover" up other ADT objects. A Mask Block is composed of two pieces: masks, which can be attached to any ADT object in order to cover it up or "white it out," and additional graphics, which can be any object or objects (AutoCAD or ADT) sitting on top of the mask. Mask Blocks can be used for almost any purpose. However, in the most common usage and in the context of this chapter, the Mask Blocks we will explore are lighting fixtures, specifically 2x4 [600x1200] lighting fixtures that when attached to a 2x2 [600x600] grid will cover up the line in the middle of the light. These particular Mask Blocks contain a mask in the shape of the overall 2x4 [600x1200] fixture, and a Multi-View Block as their additional graphics, sitting on top. These objects are included in the Content Library and are accessed through Content Browser or the AutoCAD DesignCenter. However, in the case of Mask Blocks, they function better when accessed from DesignCenter. Pay close attention to the command line prompts when dragging in lighting fixtures. In many cases, the light fixtures will anchor to the Ceiling Grid as they are inserted. In this case you will usually be prompted to select a node or something similar. Anchoring fixtures helps us to create a unified assembly of ceiling components.

ACCESS ELECTRICAL CONTENT

As with most ADT content, lighting and electrical content can be found in the Content Browser. However, in some cases it is useful to use the DesignCenter to access Content. As we have already mentioned, most of the lighting symbols use an Anchor to attach themselves to the Ceiling Grids. When you drag these fixtures from the Content Browser, it does not properly execute the Anchor functionality. Therefore, to take fullest advantage of this Content, we will use the DesignCenter to access ceiling content.

1. From the Insert menu, choose **DesignCenter** (or press CTRL + 2).

The DesignCenter will appear. DesignCenter behaves similarly to the other palettes: it can float, be docked and has the Auto-hide feature. Across the top, it has a toolbar and a string of tabs. The DesignCenter is part of the core AutoCAD functionality and as such has many more functions than we will go into here. To access the ADT Content library (the same library that we have been accessing with Content Browser up till now), you must click on the Custom tab.

2. Click on the AEC Content tab.

On the left side is a tree view, which can be toggled on and off with an icon on the toolbar at the top of the DesignCenter. There are also Preview and Description panes in the DesignCenter window. Both of these can be toggled on and off with icons on the toolbar as well (see Figure 13–24).

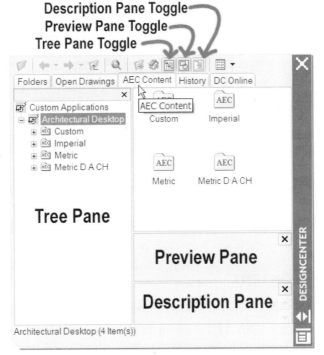

Figure 13–24 *Access Electric Content in the DesignCenter*

3. Using the tree pane, expand the *Architectural Desktop* root item to reveal the *Imperial* and *Metric* folders.

4. Continue to navigate to the *Imperial\Design\Electrical\Lighting\Fluorescent* [*Metric\Design\Electrical Services\Fluorescent\600x1200*] folder.

Note the various sub-folders within the Electric folder. Each contains a different category of Electrical symbols and content.

5. Highlight the first symbol by clicking *once* on its icon on the right (do not double-click).

Notice the larger Preview in the windowpane below. This is an interactive Viewer like the many other ADT viewers (see Figure 13–25).

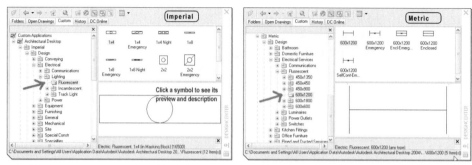

Figure 13–25 *Select an item to see its preview and description*

Right-click in the preview pane for the standard Viewer functions. The description field below includes the statement "in Masking Block." This helps you identify this symbol as a Mask Block before inserting it into the drawing.

6. Highlight other symbols and study their previews and descriptions as well.

DRAG AND DROP CONTENT

7. Turn off OSNAP ("pop up" the OSNAP button at the bottom of the screen).

OSNAPs are off when the OSNAP button at the bottom of the screen is popped up. The lighting symbol we are about to use automatically executes an Anchor as it is dropped into the drawing. Therefore, the OSNAPs are not necessary because the Anchor will afford us the requisite level of precision.

8. Click and hold down the mouse on the ***2x4 [600x1200 Enclosed]*** Fixture.

9. Drag the icon into the drawing window (see Figure 13–26).

Be sure to drag it from the icon and not the Preview pane (Viewer).

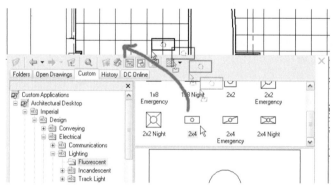

Figure 13–26 *Drag and drop the 2x4 light into the drawing*

10. At the "Select Layout Node" prompt, click on any grid intersection (see Figure 13–27).

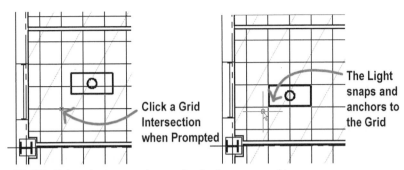

Figure 13–27 *Select the intersection as the Anchor parent object*

Notice that the lower left corner of the light fixture jumped to the grid intersection; this is the effect of the Anchor, in this case a Node Anchor.

 NOTE If the light jumps to the center of the room instead of a Grid intersection, undo and try again. This indicates that you anchored to the Space and not the Grid.

A Node Anchor establishes a point-to-point relationship between the parent object (in this case the Ceiling Grid) and the child object (in this case the light fixture). Specifically, the nature of the relationship means that the Ceiling Grid will affect the position of the light but the light has no effect on the position of the Ceiling Grid. Recall that columns also use a Node Anchor to anchor to Column Grids. (Refer to Chapter 6.)

11. Try moving the light using the standard Move command (see Figure 13–28).

Notice that, regardless of where you move it or whether or not you use Object Snaps, the light jumps to the nearest grid intersection. Try it a couple of times. Move it back to a good location when you are done.

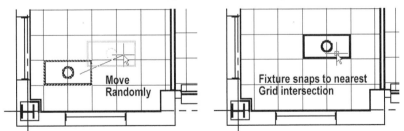

Figure 13–28 *Moving an anchored object simply shifts its Anchor to a different node*

Once the Anchor has been attached, it will control the position and orientation of the light regardless of which node it is anchored to.

CAUTION Lights can be anchored to the nodes on the portion of your Ceiling Grid that is concealed by the clipping boundary. In the drawing, this would appear as though the light had moved off the Grid. Move a light outside the drawing to see this.

12. Drag one light fixture into each room that has a Ceiling Grid and following the prompts, Anchor them to the Grids.

COPY LIGHT FIXTURES

Although an Anchor can be copied from node to node on the same parent grid, it cannot be copied from one grid to the next. Therefore, you must drag a light fixture from the DesignCenter for each room in order to establish the anchored relationship. Once anchored to the Ceiling Grid, additional lights within the same Grid can be simply copied or arrayed.

13. Zoom in to a single room.

14. Using either the AutoCAD **Copy** or **Array** command, copy the existing light fixture several times to create an appropriate lighting pattern (see Figure 13–29).

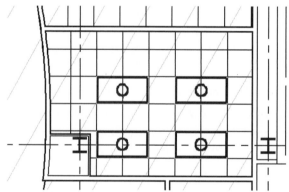

Figure 13–29 *Copy lights as required*

In the Copy Room, there is a counter running along the north Wall, so the off-center lighting layout shown in Figure 13–29 is appropriate. However, you will undoubtedly have some rooms where the Ceiling Grid layout does not suit the light pattern required. In this case, you can simply move the Grid one-half tile in either direction. This will shift all of the light while maintaining the centering. Give it a try.

ROTATE ANCHORED LIGHT FIXTURES

Anchored objects cannot be rotated in the normal way. There are two approaches to rotating lights: the parameters of the Anchor must be adjusted to affect the rotation, or the grid itself must be rotated. However, if you perform the latter, all lights attached to the grid will be rotated. If you are interesting in rotating only some of the lights, use the Anchor properties method rather than rotating the Ceiling Grid.

15. Zoom in on any room with a light fixture. Select the light, right-click and choose **Node Anchor > Set Rotation** (see Figure 13–30).

    You can select several lights at once if you wish.

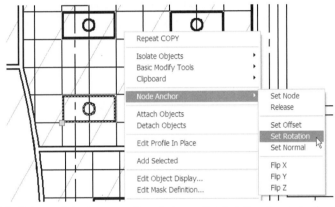

Figure 13–30 *Select a light (or several) and choose Set Rotation to rotate them*

Rotating the object along the Z axis will cause a plan rotation of the object relative to its node.

16. At the "Rotation angle about X axis" and the "Rotation angle about Y axis" prompts, press enter to accept the default of **0**°.

17. At the "Rotation angle about Z axis" prompt, type either **90**° or **-90**° depending on the direction that you need to rotate (see Figure 13–31).

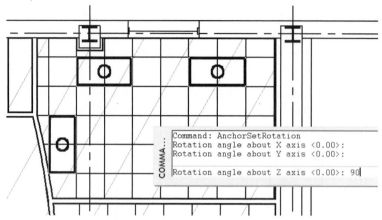

Figure 13–31 *The light has been rotated 90°*

18. Repeat any of the above steps as required to complete the remaining spaces.

ROTATE A GRID

If you would rather rotate all of the lights in a room, you can rotate the entire grid. This will take all of the anchored lights with it. However, be certain to pick your base point carefully, or you will have to re-center your grid in the Space when you are done.

 TIP Try using the Center of the Boundary Space, or one of the Grid Intersections as the Base Point of the Rotation.

19. Select any Ceiling Grid with anchored lights, right-click, and choose **Basic Modify Tools > Rotate**.

20. Follow the command prompts and rotate the Grid from its center **90°**.

    Notice that the lights have also rotated.

21. Repeat any of the steps outlined here to complete a lighting layout in each room.

22. Save the file when done.

ATTACHING MASK BLOCKS TO AEC OBJECTS

As was stated above, the light fixture we have inserted here is a Mask Block. This particular Mask Block consists of a 24"x48" [600x1200] rectangular mask with a light fixture multi-view block symbol for its "additional graphics." A multi-view block is used for the additional graphics in order to take advantage of display control. In this way, the lights will only appear when the Reflected Display Configuration is active. Let's complete our lighting layout by attaching the masks to the grid.

MASKING THE CEILING GRID

1. Zoom in on a room with a completed lighting layout.

2. Select all of the light fixtures in the room.

 CAUTION Be careful to select only lights, and not the Ceiling Grid or any other objects.

3. Right-click and choose **Attach Objects** (see Figure 13–32).

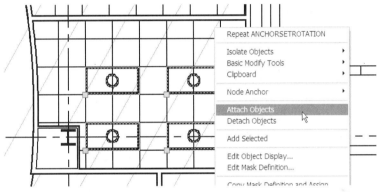

Figure 13–32 *Select the lights to attach*

4. At the "Select AEC entity to be masked" prompt, click on the Ceiling Grid within the room.

5. In the Select Display Representation dialog box that appears, confirm that you wish to mask the Reflected Display Representation (the only choice) by clicking OK.

If there were multiple Display Representations active, the Select Display Representation dialog box would list them all. In this case, the only Display Representation for the Ceiling Grid that is active is Reflected; therefore only one Display Representation is listed in the Select Display Representation dialog box.

Notice how the lights now conceal the grid lines (see Figure 13–33).

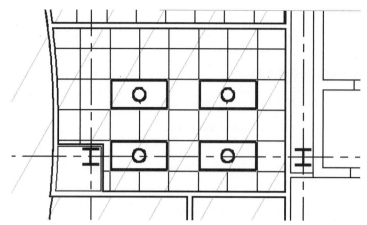

Figure 13–33 *The lights now conceal the Ceiling Grid*

Once you have attached the mask, you can move the object and it will stay masked.

6. Move one of the lights.

   Notice how the grid stays masked.

CAUTION Be careful not to move an object to a different Z coordinate. Masks must be coplanar to the object they are masking.

Although copying an anchored object also copies the Anchor, (as we saw above) copying a masked object does not attach the copied mask. Therefore, you must copy all of the lights first, and then attach the mask to them all at once.

7. Go room by room around the plan and repeat the steps just covered to attach the masks to all the Grids.

Once you have completed the steps covered here, you will have an assembly of ceiling components that behave as one. Moving and rotating the grid will also move and rotate the anchored lights, and any masked lights will remain attached. Test this out if you wish.

8. Move one of the Ceiling Grids with lights attached.

   Note that all lights move with the Grid and remain masked.

9. Undo the change.

If you ever need to detach a mask or Anchor, select the object, right-click, and choose **Detach Objects** to detach the mask, or **Node Anchor > Release** to release the Anchor. Note that objects can be anchored without being masked and masked without being anchored.

10. Save the file.

CONVERT A POLYLINE TO A MASK BLOCK

The only problem with the technique used above for the "hole" in the reception area ceiling is that it requires the polyline clipping hole object to remain in the drawing. If you delete the polyline, you will loose the hole. This means that you need to use Layers to manage the display of the polyline. Try switching temporarily back to the Medium Detail Display Configuration. Notice that when you do, all of the Grids and Lights disappear, but that the polyline remains. Return to Reflected before continuing.

There is another method to achieve the same result visually on the ceiling plan but have the hole disappear with the Grid when switching back to Medium Detail (or any other plan Display). We can create our own Masking Block—Masking Blocks can be used for more than just light fixtures. Let's convert the polyline "hole" in the reception Grid to a custom Masking Block instead.

11. Select the polyline in the reception area (the one we made into a hole above).

12. Right-click and choose **Convert To > Mask Block**.

13. In the Convert to Mask Block worksheet, type *Reception Area Ceiling Feature* for the New Name.

14. Choose the "Mask AEC Objects Automatically" option and place a checkmark in the "Erase layout geometry" checkbox (see Figure 13–34).

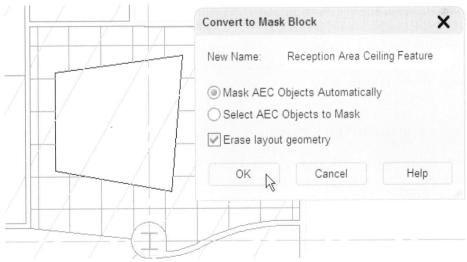

Figure 13–34 *Convert the polyline "hole" to a Mask Block*

15. Click OK to create the Mask Block and have it automatically attached to the Grid and Space.

You can move it and it will remain attached and continue to mask the Grid, just like the light fixtures. If you need to change the shape, select the custom Mask Block, right-click and choose **Edit Profile In Place**. You can then use the grips to modify the shape.

16. On the Drawing Status Bar, change the drawing to **Medium Detail**.

    Notice that this shape still displays. Let's correct this now.

17. Select the custom Mask Block, right-click and choose **Edit Mask Definition**.

18. On the Display Properties tab, place a checkmark in the Style Override checkbox next to the General Display Representation.

You have two options here. If you don't want to see this ceiling feature in plan at all, turn off the Boundary Profile component. If you would rather it appear in plans as dashed, edit the linetype, lineweight and plot style properties accordingly.

19. Turn off the Boundary Profile component (or make it dashed) and then click OK twice.

20. On the Drawing Status Bar, change the drawing back to **Reflected**.

We now have a custom ceiling feature that intelligently appears when we view Reflected, but disappears when we return to Medium Detail.

21. Save the file.

OTHER CEILING FIXTURES

There are many additional lighting and electrical fixtures that we could include on a reflected ceiling plan, such as incandescent lighting, track lighting, and surface mounted fixtures, to name a few. In this sequence, we will explore some of these items.

ADD SURFACE MOUNTED LIGHTS

The same fluorescent fixtures used above and anchored to the Grid can be inserted as surface mounted fixtures. To do so, simply skip the Anchor prompt when dragging in the fixture.

1. Zoom in to the workstations in the center of the plan.

   We have not placed a Ceiling Grid in this Space. It will have a drywall ceiling.

2. If you closed the DesignCenter, reopen it (CTRL + 2) now.

3. Navigate to the: *Imperial\Design\Electrical\Lighting\Fluorescent* [*Metric\Design\Electrical Services\Fluorescent\450x1350*] folder.

4. Drag the *1x4* [*450x1350 Enclosed*] light fixture icon into the drawing window.

5. Drop it into the lower workstation area (see Figure 13–35).

 NOTE This time drop it in roughly the spot where you want it to go. Because there will be no Anchor this time, it can be moved freely to the correct location.

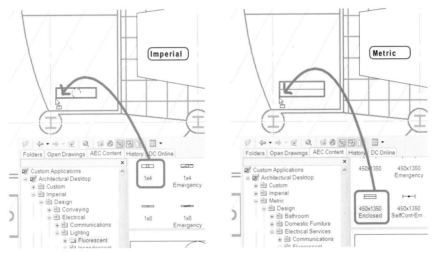

Figure 13–35 *Drag in a fixture for a surface mounted light*

6. When prompted to Select Layout Node, press ENTER.

Pressing ENTER will insert the light fixture without anchoring it at the location that you dropped it. Be certain to always choose a layout node when adding lights to a Ceiling Grid. Only press ENTER to skip the Anchor if there is no Ceiling Grid and you wish to represent a surface mounted light fixture. You will often need to perform an additional move to get it positioned just right when not anchored.

COPY SURFACE MOUNTED LIGHTS

Because the light just added is not anchored, it can be freely moved, rotated, and copied. Here is nice place for Array. We can use the standard AutoCAD Array command on the Basic Modify Tools > Array right-click menu, but for a simple linear array such as this, the AEC Modify Tools > Array (explored in Chapter 11) will be easier and provide more visual feedback. Feel free to use the AutoCAD Array instead if you wish.

7. Select the light, right-click and choose **AEC Modify Tools > Array**.

8. At the "Select an edge to array from or ENTER to pick two points" prompt, highlight one of the horizontal edges of the light fixture.

   In this case, it does not matter if you pick the top or bottom edge, only that you pick a horizontal edge—AEC Modify Array copies perpendicular to the edge you select.

9. At the "Drag out array elements" prompt, type **5′-0″ [1500]** into the dynamic dimension and then press ENTER.

 NOTE Remember that distances in Array are measured from center to center. The dimension we are using here includes the size of the light fixture

10. Drag the cursor up until there are four items total indicated on screen and then click (see Figure 13–36).

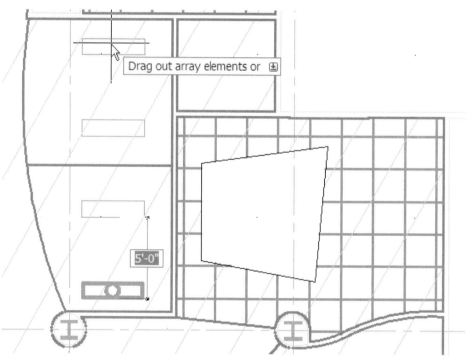

Figure 13–36 *Use the Array AEC Modify Tool to copy four equally spaced lights*

ADD TRACK LIGHTING

Many types of lighting fixtures are included in the Content library. Let's add some track lighting to the Conference Room.

11. Zoom in on the Conference Room (the Space in the lower right corner of the tenant space).

12. If you closed the DesignCenter, reopen it (CTRL + 2) now.

13. Navigate to the: *Imperial\Design\Electrical\Lighting\Track Light* folder.

 NOTE The Metric Content does not include track lighting. Please use the Imperial content for this exercise. The symbols, which were created in inches, will dynamically scale to an equivalent size in millimeters.

14. Drag the **4 Head – 10ft** track light symbol into the drawing.

15. Drop it into the conference room.

16. Press ENTER to accept the default rotation (see Figure 13–37).

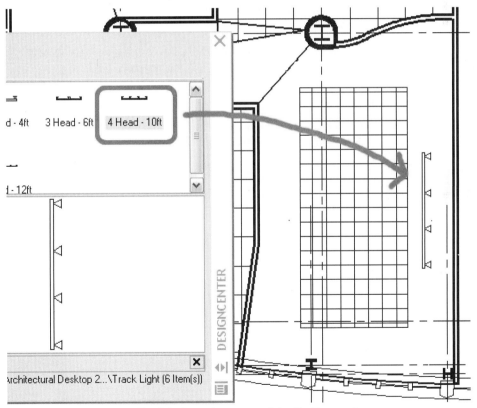

Figure 13–37 *Drop the Track Light in the center of the Conference Room*

17. Mirror it to the opposite side.

ADD A LAYOUT CURVE

A layout curve is similar to the Ceiling Grid, except that it is one-dimensional (linear) rather than two-dimensional. Any linear object, such as line, arc, polyline, or even Wall can be turned into a layout curve. First, you draw the base object; next, you convert it to a Layout Curve and establish the spacing rules for the nodes, and finally you Anchor items to the Layout. If the shape of the Layout changes, the anchored objects will follow.

TIP If you have been using the Divide and Measure commands in AutoCAD, you will want to begin using layout curves now instead. The layout curve functions like a parametric Divide and/or Measure tool.

18. Zoom in on the top end of the Corridor near the top left office.

19. Set the current layer to G-Anno-Nplt.

REMEMBER We are about to draw an AutoCAD entity, and therefore unlike ADT objects which auto-layer, we must designate the layer for the polyline manually.

20. On the Shapes toolbar, click the Polyline icon (or type **PL** and press ENTER).

21. Using an Endpoint OSNAP and Figure 13–38 as a guide, click the start point at the top left Endpoint of the Corridor.

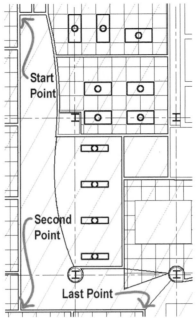

Figure 13–38 *Trace the Corridor with a polyline*

22. Pan down to the bottom end of the Corridor.

23. Click the second point at the bottom left corner Endpoint of the Corridor as indicated in Figure 10–38.

24. Click the last point just outside the Conference Room, as indicated by "Last Point" in Figure 10–38, and then press ENTER to end the command.

25. Right-click in the drawing and choose **Basic Modify Tools > Offset** (or type **O** and press ENTER).

26. At the "Specify offset distance" prompt, type **2'-0" [600]** and then press ENTER.

27. Click the polyline drawn in the last step and offset it to the middle of the Corridor.

28. Erase the original polyline and set the current layer back to Layer 0.

This polyline will now be used to create a layout curve, which will in turn control the spacing of lighting fixtures running down the Corridor.

29. Open the Content Browser (the default one, not the project based one— click the icon on the Navigation toolbar or press CTRL + 4).

30. Click on *Stock Tool Catalog*.

31. Navigate to the *Parametric Layout & Anchoring Tools* category.

32. Using the eyedropper icon, drag the **Layout Curve** tool into the drawing window.

33. At the "Select a curve" prompt, click on the polyline in the center of the Corridor (the one you just offset) (see item 1 in Figure 13–39).

    Nodes will be placed along the polyline equally, at a fixed spacing, or manually based on your input.

34. At the "Select node layout mode" prompt, choose **Space evenly** (see item 2 in Figure 13–39).

The spacing can begin directly at the start and end points of the polyline, or it can be set back from them by a specified increment.

35. Type **3'-0″ [900]** for the "Start offset" and press ENTER (see item 3 in Figure 13–39).

36. Type **3'-0″ [900]** again for the "End offset" and press ENTER (see item 4 in Figure 13–39).

Space evenly will divide the linear distance from the start offset to the end offset equally and place as many nodes in between as you specify.

37. For the Number of nodes type **10** and then press ENTER (see item 5 in Figure 13–39).

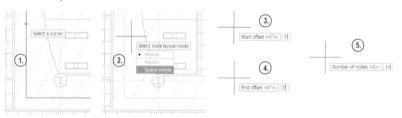

Figure 13–39 *Indicate the offsets and the number of nodes*

View the completed layout tool (see Figure 13–40).

Once created, a layout curve is very similar to a layout grid like the Ceiling Grid covered here, or the Column Grid covered in Chapter 6.

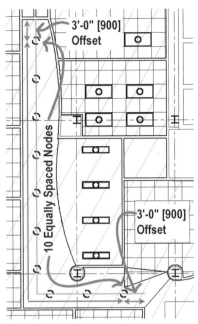

Figure 13–40 *The completed Layout Curve tool*

INSERT AN INCANDESCENT LIGHT FIXTURE

For this exercise, we will use the Imperial content for both Imperial and Metric projects. We will be assigning a unique scale factor to the symbol in appropriate units below.

38. If you closed the DesignCenter, reopen it (CTRL + 2) now.

39. Navigate to the: *Imperial\Design\Electrical\Lighting\Incandescent* folder.

 NOTE The Metric Content does not include incandescent lighting. Please use the Imperial content for this exercise. The symbols, which were created in inches, will dynamically scale to an equivalent size in millimeters.

40. Drag the **Ceiling** light fixture icon into the drawing window.

41. Drop it anywhere and press ENTER to accept the default rotation.

    Its insertion scale defaults a bit large. This can be easily adjusted.

42. Select the light fixture just inserted.

43. On the Properties palette, within the Scale grouping, set the X, Y and Z scale fields at **6″ [150]** (see Figure 13–41).

The value that we input for the scales of this Content translates into the diameter of the lighting symbol in this case.

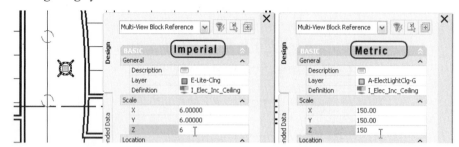

Figure 13–41 *Change the size of the light*

ANCHOR LIGHTS TO THE LAYOUT CURVE

We can now copy and anchor this light fixture to the layout curve created above to give us a series of light fixtures down the length of the corridor.

44. Switch back to the Content Browser (press CTRL + 4).

45. Your Content Browser should still be showing the *Parametric Layout & Anchoring* category of the Stock Tool Catalog. If it is not, navigate there now.

46. Using the eyedropper icon, drag the Node Anchor tool into the drawing window.

The Node Anchor has three options. The "Attach Object" option establishes a Node Anchor between a single object and a single node. The "Set Node" option is essentially a move command for Anchors. It allows you to move an existing anchored object to another node on the same layout. The "Copy to Each Node" option copies the object to every node on the layout. We will use this option here.

47. From the dynamic prompt, choose **Copy to Each Node**.

48. At the "Select object to be copied and anchored" prompt, select the Incandescent Light fixture inserted in the last sequence.

49. At the "Select layout tool" prompt, click on any one of the purple circles on the layout curve (see Figure 13–42).

Do not click the polyline; you must click directly on one of the magenta circles.

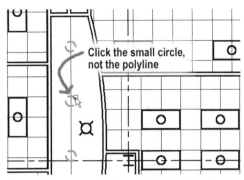

Figure 13–42 *Select the small magenta circle to finish the Anchor*

A light fixture is now anchored to each of the nodes along the polyline.

50. Press ENTER to end the command.

51. Grip stretch the polyline to see the benefit of this technique.

Notice that as you grip stretch the polyline, the nodes and their anchored lights move and re-space with it. This is very similar to the behavior of the Ceiling and Column Grids.

You can further manipulate it by selecting the layout curve itself (the purple circles), right-clicking, and choosing Properties. On the Properties palette, you can change the spacing mode to **Repeat** or **Manual,** or leave it to **Space evenly** and change the quantity of nodes. The trick is making sure you select the little magenta circles, which are the actual layout curve. The polyline is just the shape to which the layout is linked and whose shape it follows. Please note that increasing the number of Nodes will not automatically copy new light fixtures to those Nodes; however, removing Nodes will remove the anchored lights attached to them. Recall the similar "Modify Column Grids" exercise in Chapter 6.

Go Further

You can change the display properties of the Doors or any other object in the Reflected Display Configuration. Zoom in on one of the offices and notice that the Doors do not display. Open the Display Manager on the Format menu. Highlight the current Display Configuration and edit the Cut Height on the Cut Plane tab. By default it is above the Doors. This is why they don't show. You can also explore the Drawing Default display settings for Walls and Doors in the Representations by Object folder of the Display Manager. For example, you can select Wall on the left beneath the Representations by Object folder, double-click Reflected on the right side and edit the parameters on any of the tabs for Reflected. You can do the same for Doors (or any object). Just make sure that you are editing the Reflected Display Representation. For example, if you would rather not change the Cut Plane Height for Walls in Reflected, you could edit Doors in Reflected instead and deselect the Respect Cut Plane of Container Object when Anchored box on the Other tab. This will display Doors regardless of the cut height of Walls. Try it out.

RESTORE THE FLOOR PLAN

When you are finished working on the reflected ceiling plan, return to plan display.

52. On the Drawing Status Bar, open the Display Configuration pop-up menu and choose **Medium Detail**.

53. Save and Close the file.

Notice that all of the ceiling information disappears and the floor plan information returns. However, the polylines in the conference room and corridor are not ADT objects and therefore remain visible. If you want them invisible on screen, they must be turned off using layers. (Or you must convert them to appropriate ADT objects—see the CAD Manager Note in the "Clip a Ceiling Grid to a Polyline" heading above for more information on this). You could also assign the layer to No Plot, like the one in the corridor. So even though these polylines are showing on screen, they will be invisible when you print the drawing. If you find them distracting as you work, or you think they will be edited or deleted inadvertently, then turn the layer off.

ADD AN RCP VIEW AND SHEET TO THE PROJECT

Now that we have added reflected ceiling plan geometry to the Third Floor Partitions Construct, we need to create a View file for it so that it will be able to receive annotation unique to the reflected ceiling plan and then place it on its own Sheet. We will explore the philosophy behind this approach in much greater detail in the next chapter. For now, we'll simply create the required View and Sheet files so that the Project structure is up to date and ready to begin receiving annotations in the next chapter.

CREATE THE RCP VIEW FILE

1. On the Project Navigator, click the Views tab.

2. Right-click the Views folder and choose **New View Dwg > General**.

3. In the Add General View dialog, input **A-CP03** for the Name and type **Architectural Third Floor Reflected Ceiling Plan** for the Description and then click Next.

4. On the Context page, place a check mark in the Third Floor check box and then click Next.

5. On the Content page, be sure that all *Architectural* Constructs are selected, deselect the *Structural* category, and then select only the *03 Grid* Construct from the Structural category and then click Finish (see Figure 13–43).

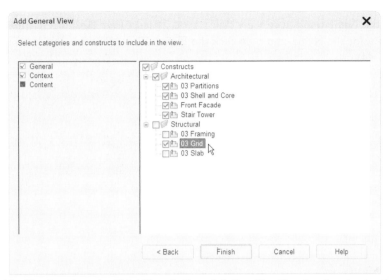

Figure 13–43 *Complete the Third Floor Reflected Ceiling Plan View by choosing its Content*

A new View file will appear on the Project Navigator. Click the Refresh icon at the bottom to re-sort the list and have the new *A-CP03* View appear in its correct alphabetical location. If you recall the exercises in Chapter 5 where we configured the Floor Plan Views, we still need to add a Model Space View and a Title Mark to this file, and configure it as an RCP. Let's first open the Third Floor Plan View file and borrow the Live Area Grid from it.

6. On the Project Navigator, double-click the *A-FP03* file to open it.

7. Select the Layout Grid (purple dashed rectangle) surrounding the Plan, right-click and choose **Clipboard > Copy**.

8. Close the *A-FP03* View file.

    It is not necessary to save the file.

9. On the Project Navigator, double-click the *A-CP03* file to open it.

10. Right-click and choose **Clipboard > Paste to Original Coordinates**.

As we did in the "First Floor Plan Model Space View" section of Chapter 5, we will create a Model Space View in this file based on the extents of the Layout Grid that we just pasted. We can assign properties to this Model Space View that will be used by the viewports when we drag this View to Sheet files.

11. On the Project Navigator, right-click the *A-CP03* View file and choose **New Model Space View**.

12. In the Add Model Space View dialog, type **Third Floor Reflected Ceiling Plan** for the Name, and then click the Define View Window icon (see Figure 13–44).

Figure 13–44 *Create a New Model Space View in the A-CP03 file*

13. Following the prompts in the drawing, snap to two opposite corners of the Layout Grid (the one just pasted) to define the View Window and then click OK to dismiss the Add Model Space View dialog box.

14. On the tool palettes, right-click the title bar and choose **Document** to load the Documentation Group.

15. Click the Callouts tab, and click the *Title Mark* (with name and number) tool.

16. At the "Specify location of symbol" prompt, click a point within the Layout Grid beneath the model and follow the remaining prompts to complete the routine.

The title that we assigned to the Model Space View should automatically appear in the Title Mark field.

17. On the Drawing Status Bar, choose **Reflected** from the Display Configuration pop-up list.

If you made modifications to the Reflected Display Configuration as noted in "Going Further" above, they will not automatically appear here when you choose Reflected. This is because those changes were applied to the *03 Partitions* Construct file and not the *A-CP03* View file. Two options are available to make those modifications available here. First, we can push our changes (using the "Update Standards from Drawing" command in the Display Manager) from the *03 Partitions* file to our *Commercial Displays* [*Commercial Displays – Metric*] Standards File (Created in Chapter 11) and then synchronize the project. This will apply those changes to the Reflected Display Configuration (and its nested Sets and Representations) to all drawings. The other approach would be to simply select the 03 Partitions XREF onscreen, right-click and choose **Edit Object Display** and then apply an override using the Reflected from the *03 Partitions* file. Please note, that if you choose the Standards approach, "Cut Plane" settings do not participate in Standards. Therefore, you will need to edit the Cut Plane manually in the *A-CP03* View file if required.

The Reflected Ceiling Plan View file is complete and ready to be dragged to a Sheet.

18. Save and Close the *A-CP03* file.

CREATE THE RCP SHEET FILE

Now that we have a reflected ceiling plan View, let's add a reflected ceiling plan Sheet file to our Project.

19. On the Project Navigator palette, click the Sheets tab.

20. Right-click on the *Plans* Sub Set beneath the *Architectural* Sub Set and choose **New > Sheet**.

21. For the Number, type **A-102** and for the Sheet Title, type **Reflected Ceiling Plans**. Click OK to accept the values and dismiss the New Sheet dialog box.

22. Double-click *A-102 Reflected Ceiling Plans* to open it.

23. On the Project Navigator palette, click the Views tab.

   Be sure that the "Third Floor Reflected Ceiling Plan" Model Space View is visible beneath the *A-CP03* View file. If it isn't, expand the plus (+) sign next to *A-CP03*.

24. Drag and drop the Third Floor Reflected Ceiling Plan Model Space View from the Project Navigator palette directly onto the Sheet Layout (see Figure 13–45).

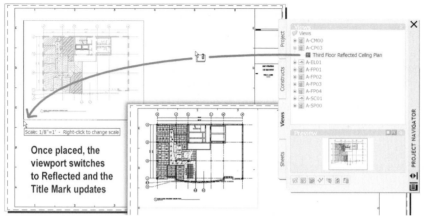

Figure 13–45 *Drag the Third Floor Reflected Ceiling Plan Model Space View onto the Sheet*

25. Click an insertion point on the Sheet to place the plan.

For now, we have only this single reflected ceiling plan. However, if we were to add additional RCP files later, we could easily rearrange and renumber the viewports on this Sheet. Notice that while you are still positioning the viewport, it shows the Medium Detail Configuration in the preview image. However, once you have placed the viewport it automatically switches to Reflected. The Sheet file is now complete.

26. Save and Close all project files.

 NOTE A more detailed explanation of the use of separate View files in the Project Navigator structure follows in Chapter 14. Please read on for more details on the benefits of this approach.

ADDITIONAL EXERCISES

Additional exercises have been provided in Appendix A. In Appendix A you will find an exercise to continue refining the Commercial Project Reflected Ceiling Plans and to change the way in which Doors display on the Reflected Ceiling Plans and add other ceiling items (see Figure 13–46). It is not necessary that you complete this exercise to begin the next chapter, it is provided to enhance your learning experience. Completed projects for each of the exercises have been provided in the *Chapter13/Complete* folder.

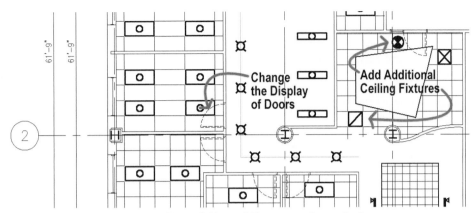

Figure 13–46 *Update project files in Additional Exercises – Appendix A*

SUMMARY

Ceiling Grids are rectangular Grid objects; to shape them to a room, use the clipping feature.

Use the Boundary option when placing a Ceiling Grid to add the grid and clip it in the same operation.

Add Grids first using rough positioning and move them later to center them, or use the SNap to Center feature.

Holes can be added to Grids to represent various design features.

Ceiling Grids can be easily transformed from 2x2 [600x600] to 2x4 [600x1200] and vice versa.

Ceiling Grids can be easily rotated.

Most Fluorescent fixtures in the Content Library are Mask Blocks capable of covering up part of the grid.

You can create your own custom Mask Blocks by drawing a closed polyline and then right-clicking it to convert.

Content Library light fixtures use Node Anchors to attach to the Ceiling Grids.

Anchored lights cannot be moved, copied, or rotated off their parent grid. Instead, use Anchor properties to rotate lights.

Add a new light from the DesignCenter for each unique grid.

Surface mounted lighting can be achieved by skipping the anchor prompt (press ENTER) when inserting the lights.

Layout curves are one-dimensional layout grids with regularly spaced nodes.

Even if Floor Plan and Ceiling Data are contained in the same Construct, they should be annotated in separate View files.

CHAPTER 14

Generating Annotation

INTRODUCTION

In this chapter, we will add annotation required for construction documents to our files. Information such as dimensions, targets, tags, labels, leaders, and notes are among the items we will explore here. In our continuing effort to work with and understand the ADT 2006 Project Management system, we will be adding all annotation to the View files that we have created in Project Navigator. The major benefit of this approach is a clean separation between modeling and annotation activities. Some firms prefer to add annotation directly to the Sheets. (Check with your CAD support personnel if you are in doubt as to which approach your firm employs.) However, it is recommended that, with the infrastructure already available in the Project Navigator, you seriously consider using the View files to place annotation. The entire Project Navigator infrastructure is designed around this philosophy.

OBJECTIVES

In this chapter, we will add annotation to project View files, add room tags to our commercial project and add dimensions to our residential plans using a variety of techniques. Our focus will be primarily AEC Tags and AEC Dimensions. We will also look at the Annotation tool palette for a sample of the documentation content provided. This chapter will explore the following topics:

- Add Room Tags.
- Use AEC Dimensions.
- Edit AEC Dimension styles
- Access Documentation Content.
- Add text leaders and notes.
- Add targets and tags.

ANNOTATION AND VIEW FILES

In Chapter 5, View files were defined as: *a "working" report of the building model akin to a particular type of drawing such as a "plan" or "section" which contains those project annotations like notes, dimensions and tags appropriate to the drawing type in question.* Now that we are about to begin adding annotations to our project files, it is time to elaborate further on this concept.

Let's consider some of the rationale behind the organizational philosophy of the Drawing Management system. It is intended that you make a separate View file for each unique type of drawing or document required in your construction document sets and other project deliverables. In this way, you can include completely unique annotation, dimensions and notes in each View and ultimately their associated Sheets. In so doing, you eliminate the traditional morass of layer configurations and other issues inherent in housing data required by several drawing types and disciplines within the same file. This approach solves many common problems inherent in the process that has evolved for most AutoCAD drawing files in production today. Consider the following common problems with construction document annotation in drawing files:

▶ Various drawing types often have similar but not identical annotation needs.

▶ Different professional disciplines have different presentation, documentation and annotation requirements.

▶ Annotation is often required at multiple scales.

▶ Project requirements often require multiple user access to files.

As a point of discussion, consider the simple matter of adding Room Tags to the Spaces. While it is true that this type of annotation would be required in nearly every type of plan (floor plan, reflected ceiling plan, MEP plans, etc.), each of these plans has its own unique and often incompatible characteristics.

For instance, in the floor plan, the best position for a Room Tag might be in the center of the room. However, when we switch to the reflected ceiling we notice that this position, while optimal for the floor plan, places the symbol directly on top of a light fixture or other notes, making the drawing difficult to read. If you move the tag for the reflected ceiling, it often ends up obscuring information in the floor plan. It is very difficult to find an optimal location for such "common" annotations that satisfy all drawing types, disciplines and scales.

Furthermore, you may wish to have one type of symbol for the floor plan (including both the Room Name and Room Number for instance), while in the reflected ceiling or the MEP plans you may wish to see the Room Number only. Traditionally, this would require a symbol with multiple embedded layers. While reliance on layer configurations is a workable solution, it often leads to improper settings in the Sheet files, which ultimately lead to wasted plots, wasted time and wasted money.

Another common annotation problem arises when a model file must be used at two different plotted scales, such as an overall plan and an enlarged detail plan. In this case, annotation must be scaled differently for each plotted scale. Traditionally, this would again require several scale-dependent layers, symbols and dimension styles—not to mention lots of time wasted coordinating all variations of the items in question.

With separate View files, the solution is simple. There is no conflict between drawing types because each drawing has its own unique annotation. With separate View files, each would have its own annotation symbols that appeared *only* in that View drawing. However, (and most importantly) both sets of annotation would reference the *same* Property Set Data whose source lives within the Construct. In other words, ADT tags merely reference the data that is attached to the objects. The data itself, whether graphical or non-graphical, belongs to the object itself, which means that it always *lives* in the Construct. (See Chapter 15 for more information on Property Sets.) The tag simply reads this data and displays it properly. In this way, the Room Names and Numbers remain synchronized throughout the set, even though each drawing has its own unique tag object. This setup frees us from all of the problems illustrated above. Each View drawing's tags can be in a different physical location relative to the associated ADT object (Space, Door, etc). Each of those tags can use a different graphical symbol, which displays the same or different properties, and each of those tags can be inserted at their own unique scale. Therefore, when the View is dragged to the Sheet, all annotations will appear the correct size for plotting without any further effort or layer changes required.

Finally, if Views are maintained for each type of drawing, many personnel issues are resolved as well. It becomes very easy to have two individuals working simultaneously on different annotation tasks at the same time. One person can work in a floor plan View, while another works in the reflected ceiling. Both are able to annotate their respective files independent of one another. However, since both of their respective View files reference the same Constructs, if one should open one of those Constructs and move a partition, delete a Door or change the underlying annotation property data (such as room name or number), the change would be updated in *both* Views.

For all these reasons and many more, the Drawing Management system in ADT is designed around the philosophy that each drawing type and scale requiring unique annotations should be configured within its own separate View file. One or more of these Views can be dragged to the same Sheet, so there will not necessarily be a one-to-one correspondence between the Views and Sheets. (Although there certainly could be a direct correspondence if project needs warranted it). We have already seen this throughout the Project Setup exercises in Chapter 5—for example, there are four separate floor plan View files for the Commercial Project, yet they all have been placed upon the same Sheet. When contemplating if a new View file is required, the three most important criteria are: *Discipline* (Architectural, Mechanical, Structural, etc), *Drawing Type* (Plan, Section, Elevation, etc) and *Scale*. Variation in any one of these warrants the creation of a new View file.

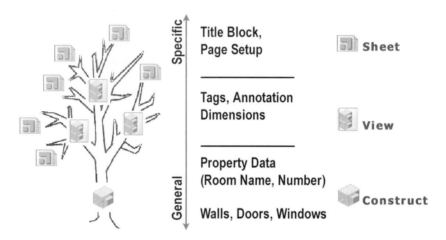

Figure 14–1 *The relationship of Constructs, Views and Sheets*

Think of the entire project structure like a tree. A tree has one trunk and many branches and leaves. Constructs are close to the trunk (they are unique), while the Views and Sheets branch out from them and are more numerous (see Figure 14–1). Another way to think of it is this: The more people who need a piece of data (the Walls, the Column Grid, the Room Names), the closer it should be to the trunk. The more specialized the item (Architectural First Floor Plan at 1/8″=1′-0″ scale), the further out on the branches it should be placed. Following this logic you can begin to make decisions about where certain project items ought to be placed. We have already discussed Walls, Doors, Windows and other "real" pieces of the model. They belong to Constructs and are required by everyone. Room designations are really two separate components: the label or tag itself, and the data to which the tag refers. The fact that a particular room is named "Conference Room East" is "real," and while we cannot physically "touch" the concept of the room's being named Conference Room East, no one can dispute that its designation as such is as "real" as the Walls of which it is composed. However, the label that indicates it as being "Conference Room East" is a component of a particular drawing, and is not "real" in the sense that tags are not constructed and installed with the building nor painted upon the conference room floor. Therefore, the tag is placed in a View file and references the data contained in the Construct file, which informs it that the Space in question is named "Conference Room East."

Views used in this way provide the means for us to capture a certain slice of the building model, annotate, embellish and configure it a certain way to document our intentions. Further out on the tree's branches lay the Sheets, which have the unique purpose of gathering our documentation intent in some deliverable format such as printed drawings.

ADDING ROOM TAGS

Let's explore some of the scenarios discussed in the previous section firsthand. In this exercise, we will add Room Tags to the Third Floor Plan and Reflected Ceiling Plan View files of the Commercial Project. We will then make a few adjustments to them and explore the flexibility inherent in having two separate View files.

INSTALL THE CD FILES AND LOAD THE CURRENT PROJECT

If you have already installed all of the files from the CD, simply skip down to step 3 below to make the project active. If you need to install the CD files, start at step 1.

1. If you have not already done so, install the dataset files located on the Mastering Autodesk Architectural Desktop 2006 CD ROM.

   Refer to "Files Included on the CD ROM" in the Preface for information on installing the sample files included on the CD.

2. Launch Autodesk Architectural Desktop 2006 from the desktop icon created in Chapter 3.

If you did not create a custom icon, you might want to review "Create a New Profile" and "Create a Custom ADT Desktop Shortcut" in Chapter 3. Creating the custom desktop icon is not essential; however, it makes loading the custom profile easier.

3. From the File menu, choose **Project Browser**.

4. Click to open the folder list and choose your *C:* drive.

5. Double-click on the *MasterADT 2006* folder, then the *Chapter 14* folder.

   One or two commercial Projects will be listed: *14 Commercial* or *14 Commercial Metric*.

6. Double-click *14 Commercial* if you wish to work in Imperial units. Double-click *14 Commercial Metric* if you wish to work in Metric units. (You can also right-click on it and choose **Set Current Project**.) Then click Close in the Project Browser.

IMPORTANT If a message appears asking you to re-path the project, click Yes. Refer to the "Re-Pathing Projects" section in the Preface for more information.

OPEN A VIEW FILE AND ADD ROOM TAGS

Adding tags is simple. Let's begin with the Third Floor Plan View file.

 IMPORTANT Be sure that the *03 Partitions* file is not open before continuing. The tagging routines that we will use here must have the ability to edit the referenced Construct files (in this case *03 Partitions*). If this or any other required Construct file has been opened in ADT by you or someone else, the tagging routine will fail.

1. On the Project Navigator, click the Views tab.

2. Double-click the *A-FP03* file to open it.

Before we begin adding annotation to this View file, we must verify the scale of the file. All annotation routines in Architectural Desktop take advantage of automatic scaling. Therefore, you need only assign the scale setting to the View file once and from that point on, all annotation will be properly scaled based upon that setting. This is the Third Floor Plan and we have already designated in Chapter 5 that it will be plotted at 1/8"=1'-0" [1=100mm]. Therefore, we simply need to verify that this scale is properly assigned.

3. On the Drawing Status Bar, verify that **1/8"=1'-0" [1:100]** is chosen. If it is not, choose it now (see Figure 14–2).

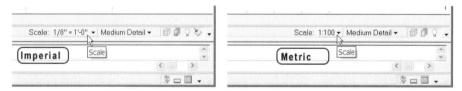

Figure 14–2 *Verify the drawing scale for Annotation items*

In Chapter 12, we applied a Display Theme to this file. Let's disable it now.

4. Select the Space Size Legend at the bottom right corner of the drawing, right-click and choose **Disable Display Theme**.

The colored solid fill patterns will disappear.

5. On the tool palettes, right-click the title bar and choose **Document** to load the Documentation Group.

6. Click the Scheduling tab, and then click the *Room Tag – Project Based* tool.

7. At the "Select object to tag" prompt, click on the hatch pattern of the Reception Space (on the right of the tenant space just below the elevator lobby).

8. At the "Specify location of tag" prompt, click in the center of the Reception Space (see Figure 14–3).

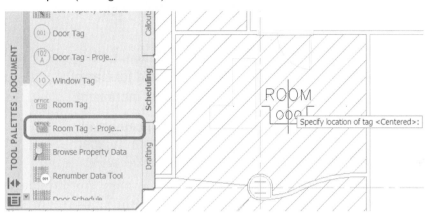

Figure 14–3 *Click the Reception Space to add the first Room Tag*

  TIP You can also right-click and choose Centered to automatically place the tag at the geometric center of the Space.

9. In the Edit Property Set Data worksheet that appears, collapse the RoomFinishObjects grouping.

10. In the RoomObjects grouping, type **Reception** for the Name (see Figure 14–4).

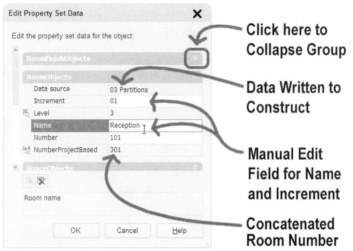

Figure 14–4 *Edit the name of the Space in the Edit Property Set Data worksheet*

When you add the Tag to the drawing, it attaches one or more Property Sets to the selected object (the Reception Space object in this case). A Property Set is a collection of data that is attached to an object for tagging or scheduling purposes. In this case, we are only concerned with the *RoomObjects* Property Set and specifically the "Name" field for the tag that we have chosen. Notice that the Data Source for this Property Set is *03 Partitions*. As was mentioned in the previous section, the actual data associated with a tag will be written to the Construct even though the tag is inserted within the View.

Take note of the Increment and Level fields. These two fields are concatenated together to form the value in the NumberProjectBased field (currently 301). This number will also appear in the tag when we click OK in this worksheet. The Increment field is editable, but it will automatically increment to the next number with each Space we tag. Do not be concerned that the Number field (also editable) does not match the NumberProjectBased field. The Number field is used by the non-project-based tags. There is another Room Tag on the Scheduling tool palette that is not project-based that uses this field instead. We will not be using it here; therefore, you can simply ignore the Number field. (See Chapter 15 for more information on Schedules and Property Set data.)

11. Click OK in the Edit Property Set Data worksheet.

   Notice the Room Tag that appears, as does the room number value: 301. The Name, however, still reads "Room." This will be the case until you complete the tagging routine. As soon as we are finished tagging and press ENTER, all of the room names will update to the values we type in (see Figure 14–5).

Figure 14–5 *While the Room Number does update, the Room Name will not update until the routine is complete*

12. When the "Select object to tag" prompt repeats, click the Conference Room Space next. Type **Conference Room** for the Name and then click OK.

 Notice that the auto-increment Number for this room is 302.

13. When the "Select object to tag" prompt repeats, right-click in the drawing and choose **Multiple** (or press the down arrow to access this option from the dynamic prompt).

14. At the "Select objects to tag" prompt, click the office to the left of the Conference Room.

 The "Select objects to tag" prompt will repeat.

15. Continue clicking the offices around the perimeter of the plan. Click each, one at time in a clockwise order. After clicking the last office, press ENTER.

You can use any selection method when you choose the Multiple option of the tagging routine, including windows and crossing windows. However, since this Property Set uses the auto-increment feature, clicking the Spaces one at a time will ensure that they are numbered in the order in which we clicked them.

 In the Edit Property Set Data dialog, the Increment field will now read "Varies." This is because several rooms have been selected and they all have a unique Number.

16. In the Name field, type **Office** and then click OK.

 Continue tagging until all rooms are tagged. Refer to Figure 13–6 in Chapter 13 for reference if necessary. Also add tags to the two closets at the top and the one above reception and the corridor.

17. When you are finished tagging, press ENTER to complete the routine (see Figure 14–6).

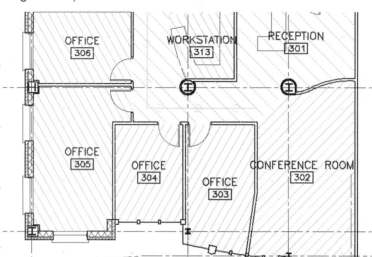

Figure 14–6 *All Property Set values appear when the tagging routine is complete*

Notice that once you have completed the routine, all of the Room Names fill in properly.

> If an alert balloon indicating that the *03 Partitions* XREF has changed appears, click on it to reload the XREF.

ADJUST ROOM TAG POSITION

Now that all of the Room Tags have been placed, you may not be happy with the precise position of some of them. You can easily move them around on screen to suit the needs of this Floor Plan View. Tags are anchored, but unlike other anchored objects that we have seen, the location of a tag can be modified with the standard AutoCAD MOVE command.

1. Zoom in on the top of the plan.

2. Using the standard AutoCAD Move command, move the two Room Tags for the two small closets outside of the plan and place them above their respective closet Spaces (see Figure 14–7).

TIP you can also use the grips on the tags to move them if you prefer.

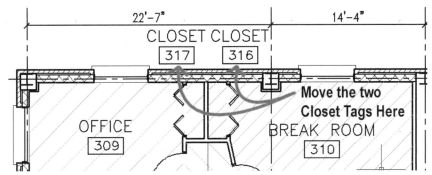

Figure 14–7 *Move the two Closet Room Tags outside of the Plan*

3. On the tool palettes, click the Annotation tab.

4. Click the Text (Straight Leader) tool.

   With Leaders, the first point you click is the location of the arrowhead. Therefore, we must click our first point within the Closet Space and end at the Tag. Object Snap Tracking will be helpful to keep the points aligned.

5. At the "Specify first point of leader line" prompt, use Object Snap Tracking, acquire the corner of the Room Tag, track down from it into the Closet Space and then click (see left panel of Figure 14–8).

6. At the "Specify next point of leader line" prompt, snap to the same endpoint acquired in the last step (see middle panel of Figure 14–8).

7. When the "Specify next point of leader line" prompt repeats, press ENTER to stop adding points.

8. At the "Select text width" prompt, simply press ENTER again to accept the default.

9. At the "Enter first line of text" prompt, press ENTER again to accept the default (see right panel of Figure 14–8).

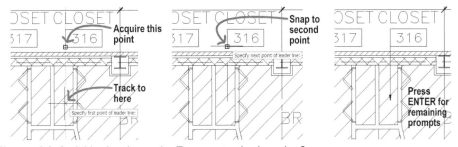

Figure 14–8 *Add a Leader to the Tag pointing back to the Space*

10. In the Mtext edit window, click the OK button to dismiss the editor.

This is a text leader tool. Normally you would specify with your mouse how wide you wanted the note on screen to be. Then you would type your text into the Mtext editor window that appears. However, in this case, we only want the Leader itself. To get just the leader, however, we must press ENTER at all of the default prompts. Do not press ESC. This will terminate the entire command including the Leader.

11. On the Standard toolbar, click the Match Properties icon.

12. At the "Select source object" prompt, click one of the Room Tags.

13. At the "Select destination object(s)" prompt, click the Leader and then press ENTER.

14. Select the Leader. On the Properties palette, beneath the Lines & Arrows grouping, choose **Dot** from the Arrow list (see Figure 14–9).

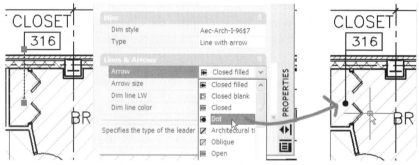

Figure 14–9 *Change the Leader to a Dot Arrowhead*

If you prefer a different arrowhead, feel free to choose it or to retain the original if you wish.

15. Copy the Leader to the other Closet Room Tag.

16. Repeat this entire process on any other ill-positioned Room Tags.

17. Save the file when finished.

EDIT ROOM TAG PROPERTY DATA

As you are moving tags around, perhaps you notice a room name that is not correct. It is easy to edit the Property Data of an object. The trick is in understanding that it is not the Tag that requires editing. Remember, the Tag merely *refers* to the Property Data within the object (Space object in the XREF file in this case). To make the edit, you must therefore edit the Space.

1. Locate a Space whose Name you wish to edit. It is not important which one.

2. Click on the Space of the room you wish to edit (click on the hatching).

Notice that the entire XREF highlights. This is expected since the Spaces are contained in the *03 Partitions* XREF file.

3. Right-click and choose **Edit Referenced Property Set Data**.

4. At the "Select objects" prompt, click on the Space you wish to edit again.

This time, the individual object (in this case a Space) within the XREF is highlighted (see Figure 14–10).

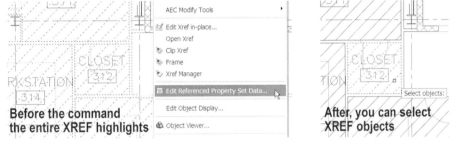

Figure 14–10 *Edit Referenced Property Set Data allows you to select the objects within the XREF*

5. Press ENTER to complete your selection and then in the Edit Referenced Property Set Data dialog, edit the Room Name.

6. Click OK to accept the change and return to the drawing.

If an alert balloon indicating that the *03 Partitions* XREF has changed appears, click on it to reload the XREF.

7. Make any additional changes you wish and then Save the file.

ADD ROOM TAGS TO THE REFLECTED CEILING PLAN

Adding Tags to the Floor Plan was a fairly simple process. Adding them to any subsequent View file is even easier. You only need to perform the process of adding tags and inputting Property Set values only the first time you add Tags. Since the data to which the Tag refers lives on the Construct and is passed to the Tag, you can add Tags in the other View files in a matter of seconds, and the data values will already be assigned.

1. On the Project Navigator, on the Views tab double-click the *A-CP03* file to open it.

As before, verify that the scale is set to **1/8"=1'-0"** [**1:100**].

2. On the Scheduling tool palette, click the ***Room Tag – Project Based*** tool.

3. At the "Select object to tag" prompt, click the Conference Room Space.

4. At the "Specify location of tag" prompt, press ENTER to accept the default of Centered.

   In the Edit Property Set Data dialog, scroll down and note that all of the fields are already filled in with the same values input while working in the Floor Plan.

5. Click OK to return to the drawing.

6. At the return of the "Select object to tag" prompt, right-click and choose **Multiple**.

7. Using a Crossing Window, select all of the Spaces and then press ENTER.

A message reading: "1 object was already tagged with the same tag. Do you want to tag it again?" will appear. This is referring to the Tag that you just added to the Conference Room. Since we do not want two tags in the Conference Room, we will click "No" here.

8. In the dialog that appears, click No.

9. In the Edit Property Data dialog, click OK and then press ENTER to complete the tagging routine (see Figure 14–11).

Notice that all of the Tags have appeared with the same values that they had in the First Floor Plan. This is again because the data to which these tags are attached lives in a common Construct file: *03 Partitions*.

 NOTE In order to use the "Multiple" Tag option, you must place the first tag manually. That is why we tagged one Space (the Conference Room) first.

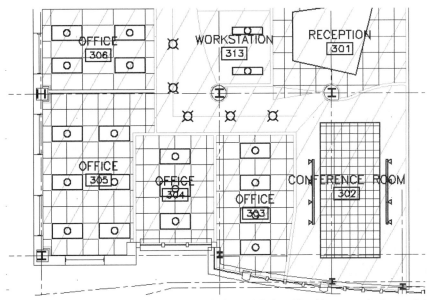

Figure 14–11 *All Property Set values in the Reflected Ceiling Plan View match those of the Floor Plan*

10. Using the process above, edit the Referenced Property Set Data of one of the rooms and change the Name.

If you closed the *A-FP03* View file, reopen it from Project Navigator now.

If you left it open, use the Window menu to switch to it. An XREF alert should appear. Reload the XREFs.

11. Zoom in on the same Space that you just renamed in the *A-CP03* file.

Notice that the name has changed here as well.

12. Close *A-FP03* and return to the *A-CP03* View file.

New in ADT
2006

CREATE AN ALTERNATE TAG FOR THE RCP

In the "Annotation and View Files" section above, we also mentioned the possibility of using a different symbol in one View file than is used in another yet keeping them linked to the same data. Let's explore that now.

1. On the File menu, choose Open.

Browse to: *Chapter14\Standards\Content* and open *Commercial Styles [Commercial Styles – Metric]*.

We have visited this file on a number of occasions in previous chapters. We will continue in this way to keep all of our project-based content in a consistent location, however, you could perform the following steps in any project file.

If you prefer, you may use the text and shape creating in the following steps have been provided in a separate file in the *Elements* folder of Project Navigator. To use them instead, click the Constructs tab on the Project Navigator palette, and in the *Elements* folder, right-click the file named *Room Tag* and choose **Insert as Block**. Choose "Specify on screen" for the Insertion point and check the "Explode" checkbox. When prompted, click a point on screen to place the objects and then Zoom in on them and then skip down to Step 7.

2. On the Annotation palette, click the **Text Tool**.

3. At the "Select MText insertion point" prompt, click a point on screen and then press ENTER to accept the default width of 0 (zero).

4. At the "Enter first line of text" prompt, type: **Number** and press ENTER twice.

5. Select the piece of text and on the Properties palette, choose Middle center for the Justify. Set the Height to **1″ [100]**.

6. Zoom in on the text and draw whatever shape (rectangle, oval, pillbox) you wish to surround the text.

    Keep the size of the shape a little larger than the text all the way around.

USING THE DEFINE SCHEDULE TAG WIZARD

The Define Schedule Tag wizard is new to ADT 2006. With it, you simply draw the tag using normal geometry and text (as we have done here) and then it converts those items to a Schedule Tag with your input in a simple worksheet.

Be sure you are zoomed in on your text and graphics.

7. From the Format menu, choose **Define Schedule Tag**.

8. At the "Select object(s) to create tag from" prompt, select the text and surrounding geometry and then press ENTER.

9. In the Define Schedule Tag worksheet, input **Room Tag Number Only** for the Name.

Each piece of text in the selection will be listed in the table at the bottom of the worksheet. In this case, we have a single entry labeled: Number. Each piece of text in the tag can be left as is and remain text, or you can choose **Property** from the Type column. If you make it a Property, you then choose the Property Set and Property that you wish to reference.

10. From the Type column, choose **Property**. Set the Property Set to **RoomObjects** and finally choose **NumberProjectBased** from the Property list (see Figure 14–12).

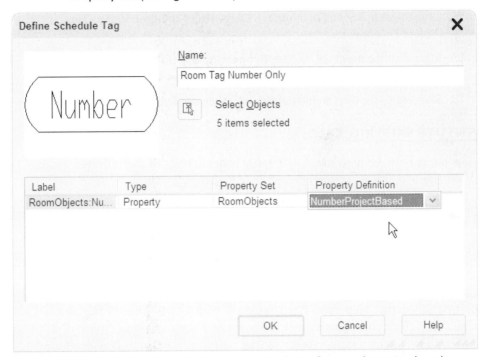

Figure 14–12 *Assign the piece of text as a Property that references the project based room number*

11. Click OK to complete the worksheet. At the "Specify insertion point" prompt, click a useful insertion point relative to the geometry (such as the top midpoint or center of the tag).

Click on the geometry and you will see it is no longer separate objects but rather a new Multi-View Block named *Room Tag Number Only*. The only problem with the way that the wizard creates the tag is that it does not show any graphics in the Reflected Display Representation by default. Let's address that before we add it to our Project Tools palette.

12. Select the new tag, right-click and choose **Edit Multi-View Block Definition**.

13. On the View Blocks tab, select Reflected on the left, click the Add button and then choose *Room Tag Number Only* from the list and then click OK.

14. Save the file.

Now let's add the tag to our Project Tools palette.

15. Right-click on the tool palettes title bar, load the **14 Commercial [14 Commercial Metric]** tool palette group and then click the Project Tools palette to make it active.

16. Drag the new **Room Tag Number Only** object and drop it onto the Project Tools palette.

17. Right click the new tool, choose **Properties** and change the Layer key to **AREANO** and then click OK.

18. Save and Close the *Commercial Styles [Commercial Styles – Metric]* file.

SUBSTITUTE EXISTING TAGS

The new tag is now ready to use. Let's apply it to the tags in the reflected ceiling plan View file.

19. Return to the *A-CP03* View file, right click new tag tool and then choose **Import 'Room Tag Number Only' Multi-View Block Definition**.

20. Select any Room Tag, right-click and choose **Select Similar**.

21. On the Properties palette, choose **Room Tag Number Only** from the Definition drop down list.

Notice that all of the Tags have been replaced with the Number Only symbol. However, they all retained the correct reference to their individual Space object's Room Number. This is because the new tag points to the same *NumberProject-Based* property that the original did. If you wish to make a change like this to your office standard, you can create any custom Tag and include it on the tool palettes of all users rather than just the project.

22. Repeat any of the fine-tuning steps above to move Tags and/or add Leaders as required to complete the drawing and make it legible (see Figure 14–13).

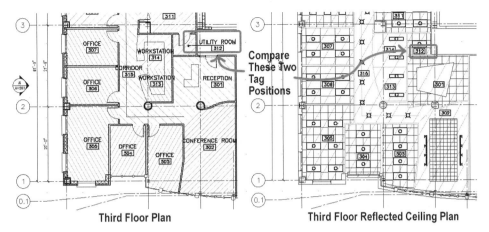

Figure 14–13 *All the Tags have been substituted with the new Number Only Tag*

As you can see, tagging objects in View files gives us the flexibility to use unique styles and positioning of annotation while maintaining 100 percent confidence that all data in multiple Views is coordinated back to its original source.

23. Save and Close all Commercial Project files.

DIMENSIONING ADT OBJECTS

Autodesk Architectural Desktop offers several tools for dimensioning. Since Autodesk Architectural Desktop is built on and around an AutoCAD foundation, you have all of the AutoCAD dimensioning functionality at your disposal. However, the ADT AEC Dimension object is also worthy of serious consideration for regular Construction Document dimensioning use. We will explore this tool in the tutorials that follow. However, for the sake of completeness, here is a brief overview of the various dimensioning tools available in ADT 2006.

▶ **AutoCAD Dimensions**—Dimensioning tools have been a part of AutoCAD since its earliest releases. This powerful and robust tool set can be used to dimension any geometry including ADT objects. However, you must add dimension lines point by point. If the **DIMASSOC** system variable is set to a value of **2**, then AutoCAD dimensions that are added to AutoCAD drafted entities (such as lines, arcs, circles and polylines) will remain associative to those entities. AutoCAD dimensions can be used to dimension ADT objects, but they will not remain associative. To get dimensions that remain associative to ADT objects like Walls, Doors and Windows, use AEC Dimensions (see below). AutoCAD dimensions are drawn on the current layer.

▶ **AEC Content Dimension tools**—Accessible via the Content Browser or the DesignCenter, a collection of AEC Content tools has been provided that are essentially AutoCAD dimensions that "auto-layer." As with all AEC Content, when you execute an AEC Content Dimension tool in your drawing, it will automatically place the dimensions created by the tool on the correct layer regardless of the current layer.

▶ **Wall Dimensions**—Available only at the command line (type **WALLDIM**), this tool applies a set of AutoCAD dimensions to a selection of Walls in the drawing. This tool provides a quick way to add a string of dimensions to a Wall or Walls. These dimensions come in on the correct layer. Since they are AutoCAD Dimensions, once placed, they can be edited using all standard AutoCAD Dimension techniques.

▶ **AEC Dimensions**—Are live Dimension objects that are updated automatically as the dimensioned geometry changes. AEC Dimensions sport many useful features including: the ability to edit the text and lines of AEC Dimension objects, the ability to dimension basic geometry as well as ADT objects, and the ability to attach AEC Dimensions across XREFs. In addition, AEC Dimensions are scale-dependent and use display control. This means that the same string of AEC Dimensions can be displayed at different drawing scales. For all of your linear dimensioning needs, it is quite practical to consider AEC Dimensions over all of the other options. AEC Dimensions do not currently support Radial, Diameter or Angular dimensions. Use AutoCAD dimensions for these tasks.

LIVE DIMENSIONING WITH AEC DIMENSIONS

AEC Dimensions are live Dimension objects that are updated automatically as the dimensioned geometry changes. No manual grip editing or stretching is required. In addition, AEC Dimensions use display control. Therefore, they can be toggled on and off by loading an appropriate Display Configuration, and can be configured to display at different scales. They automatically layer like all other ADT objects, and they work across XREFs, making them the perfect complement to our Project Management system.

CHANGE THE CURRENT PROJECT

To explore the features of AEC Dimensions, we will return to the Residential Project.

1. From the File menu, choose **Project Browser**.

2. Click to open the folder list and choose your *C:* drive.

3. Double-click on the *MasterADT 2006* folder, then the *Chapter14* folder.

   One or two residential Projects will be listed: *14 Residential* or *14 Residential Metric*.

4. Double-click *14 Residential* if you wish to work in Imperial units. Double-click *14 Residential Metric* if you wish to work in Metric units. (You can also right-click on it and choose **Set Current Project**.) Then click Close in the Project Browser.

 IMPORTANT If a message appears asking you to re-path the project, click Yes. Refer to the "Re-Pathing Projects" section in the Preface for more information.

ADD SPACE TAGS TO A VIEW

Following the procedure used above, let's add Tags to the First Floor Plan of the Residential Project. This time we will use a different Tag

1. On the Project Navigator, click the Views tab and then double-click *A-FP01* to open it.

    We have not opened this file since the end of Chapter 5. As you can see, it has progressed nicely.

2. Zoom in on the plan a bit.

    Verify that the scale of the drawing is: **1/4"=1'-0" [1:50]**.

Instead of the Room Tag that we used above, here we will go to Content Browser and retrieve the Space Tag. This Tag will show the name of the Space Style and its dimensions.

3. Open the Content Browser (click the icon on the Navigation toolbar or press CTRL + 4).

4. Click on Documentation Tool Catalog – Imperial and navigate to the *Schedule Tags* category and then the *Room & Finish Tags* category.

5. Using the eyedropper icon, drag the Space Tag tool into the drawing window.

6. Follow the prompts to tag one Space and then right-click and choose **Multiple**.

7. Select and tag the remaining Spaces using the Multiple option and then press ENTER to complete the routine.

8. Make any moves or other necessary adjustments.

This Tag is tied directly to the Space Style Name. Therefore we did not have to edit the Room Name. It also shows the dimensions of the Space. If you edit the size of the Space, the Tag will adjust to show the new values. Give it a try if you like. Remember though that the Spaces are in the Construct, so you will have to open it to make a change and then reload the XREFs to see it. Be sure to restore everything to the original sizes before continuing.

ADD AEC DIMENSIONS

Using the same litmus test that was applied above for Room Tags, Dimensions are Annotations and often have very unique needs per drawing. View files therefore are the ideal location for dimensions. Part of what makes the View file an attractive location for annotation and dimensions is that tools like AEC Dimensions and Schedule Tags (as we saw above) work across XREFs.

Remember from Chapter 5 that the purple dashed line surrounding our plan represents the edge of the viewport on the Sheet. Therefore, when we begin adding Dimensions, we will want to be certain that they do not encroach past this line.

9. On the tool palettes, click the Annotation tab to make the Annotation tool palette active.

10. Click the **AEC Dimension (2)** tool on the Annotation tool palette.

    If you do not see this palette or tool, right-click the Tool Palettes title bar and choose **Document** (to load the Documentation Tool Palette Group) and then click the Annotation tab.

11. At the "Select geometry to dimension" prompt, click the horizontal masonry Wall at the top of the plan (between the addition and the Sun Porch), and then press ENTER (see Figure 14–14).

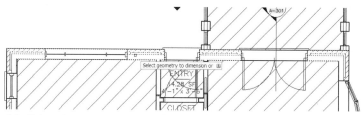

Figure 14–14 *Select the exterior masonry Wall between the addition and the Sun Porch*

    Notice that you are able to select a single Wall even though the Wall is part of an XREF file.

12. At the "Specify insert point" prompt, click a point just above the Porch (see Figure 14–15).

 NOTE When prompted to "Specify insert point," the point you pick is the point where the first chain of dimensions will be placed (the one closest to the object being dimensioned).

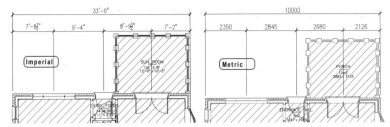

Figure 14–15 *Generate a string of AEC Dimensions*

13. Click anywhere on the dimensions just created.

Notice that this is a single continuous object, unlike an AutoCAD dimension.

14. Repeat these steps to add a string of AEC Dimensions on the left and right sides of the plan.

On the left, select just the vertical masonry Wall, and on the right, try selecting both the vertical masonry Wall and the vertical curtain Wall of the porch (see Figure 14–16).

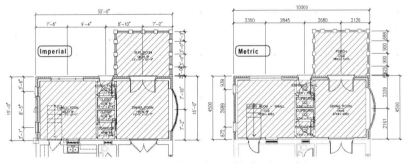

Figure 14–16 *Add AEC Dimensions to the right and left sides of the plan*

TIP When selecting the Curtain Wall, be careful to select the Curtain Wall itself and not the nested infill Windows.

EDIT THE AEC DIMENSION STYLE

Like most ADT objects, AEC Dimensions are style-based. In this case, using the *AEC Dimension (2)* tool has automatically assigned the *2 Chain* style to all of the AEC Dimensions we have added so far. AEC Dimension styles determine the number of strings that AEC Dimension objects have, between a minimum of one and a maximum of ten. They also control the specific points that will be dimensioned and the AutoCAD Dimension style that will be used to display all of the dimension components.

15. Select any AEC dimension string, right-click, and choose **Edit AEC Dimension Style**.

16. Click the Chains tab.

    The number of chains can be adjusted on the Chains tab.

17. Change the number of Chains to **3** and then click OK.

Notice the addition of a third chain to all dimension strings. If you were going to keep this chain, you would likely want to configure it or the second one a little differently than it defaulted. However, we are going to undo it for now.

18. Undo the last command.

19. Select any AEC Dimension, right-click and choose **Select Similar**, and then right-click again and choose **Copy AEC Dimension Style and Assign**.

20. On the General tab, name the new Style *MADT Residential* and then click the Display Properties tab.

By now this dialog box should be looking familiar. As you can see, there are several Display Representations for AEC Dimensions, and most of them have a Style-based override applied. For now we will focus on Plan, which is the active Display Rep.

21. With Plan selected, click the Edit Display Properties icon.

22. Click the Layer/Color/Linetype tab.

There are four components to an AEC Dimension string.

- **AEC Dimension Group**—Contains the actual Dimension objects. This is what we see in the drawing and recognize as the Dimension objects. There are no overrides assigned to the AEC Dimension group, so they will default to the Layer, Color, Linetype, Lineweight and Plot Style of the overall AEC Dimension object.

- **AEC Dimension Group Marker**—An icon used to help identify each separate group of AEC Dimensions when a drawing contains several. It sits to the right of the AEC Dimension group. This component is off and assigned to a non-plotting layer by default.

- **Removed Points Marker**—Shows dimension points that have been manually removed. This component is off and assigned to a non-plotting layer by default, making it appear as though the points have been removed. Turn it on to restore removed points.

- **Override Text & Lines Marker**—When you override the text and/or lines of an AEC Dimension object, a small marker will appear on a non-plotting layer to indicate that the dimension string has been manually modified. You should *NOT* turn this component off (see below for more information).

You may have noticed that all of the Windows and Doors are currently dimensioned to the center. Since this is a masonry Wall, it may be better to dimension the rough opening instead.

23. Click the Contents tab.

24. Select Chain 1 at the bottom and then select an object type in the "Apply to" list.

 NOTE Remember, Chain 1 is the one closest to the dimensioned geometry.

25. Explore several of the Object types and their respective settings to get a better sense of how the AEC Dimensions operate.

26. When you are through exploring, click "Opening in Wall" from the "Apply to" list.

27. Clear the check mark from the Center of Opening check box, and place a check mark in the Opening Max. Width check box instead (see Figure 14–17).

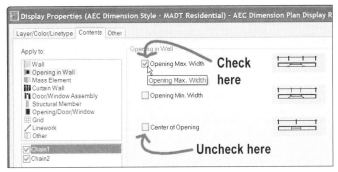

Figure 14–17 *Change AEC Dimension Contents parameters for Opening in Wall*

28. Select Wall from the "Apply to" list and keep Chain 1 selected beneath it. On the right side, remove the check mark from the Wall Intersections check box.

29. Click OK twice to return to the drawing and see the result (see Figure 14–18).

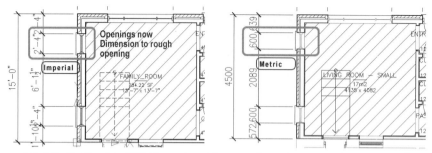

Figure 14–18 *Configuring AEC Dimension objects to dimension to rough opening rather than center*

30. Save the file.

EDIT INDIVIDUAL AEC DIMENSIONS

AEC Dimensions remain linked to the objects they are dimensioning. Therefore editing AEC Dimensions simply requires editing the drawing. AEC Dimension objects adjust automatically.

31. Select *First Floor New* XREF, right-click, and choose **Edit XREF in-place**.

32. In the Reference Edit dialog box, confirm that *First Floor New* is highlighted and then click OK.

33. Select one of the Doors or Windows and move or resize it.

    Notice that no change to the AEC Dimensions has occurred yet. They will update as soon as we save the changes back to the XREF.

34. On the Refedit toolbar, click the Save back changes to reference icon and then click OK in the confirmation dialog box.

    Note how the drawing and the AEC Dimensions update (see Figure 14–19).

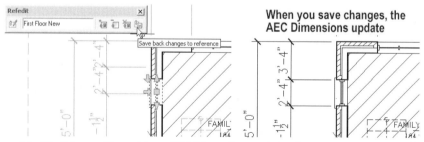

Figure 14–19 *Edit an object with an AEC Dimension, and the dimension is automatically updated when the XREF is saved*

35. Keep the changes if you wish, or repeat the process and reverse the change.

CAUTION You can also use Undo, but be sure to only undo the change to the geometry (not the entire Refedit session), and then click the Save back changes to reference icon on the Refedit toolbar to execute the change. If you don't, the AEC Dimensions will not update properly.

You can also grip edit AEC Dimensions to fine-tune placement.

36. Click the AEC Dimension string at the top of the drawing.

Two grips appear on the right: The triangular shaped one allows you to move all Chains of the dimension string closer or farther away from the dimensioned geometry. The round grip activates the In-place Edit mode, which we will see below.

37. Click the triangular shaped grip and drag all Chains up enough to allow room to dimension the porch (see Figure 14–20).

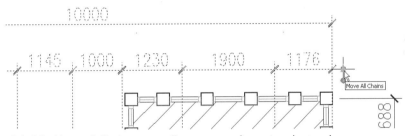

Figure 14–20 *Move all Chains up to allow room to dimension the porch*

AEC Dimensions are very flexible. You can attach new objects to the string at any time. You can also remove them.

38. Right-click the same AEC Dimension object and choose **Attach Objects**.

39. At the "Select Building Elements" prompt, click the horizontal Curtain Wall of the porch (be certain to select the Curtain Wall and not the nested Window infills) and then press ENTER.

The AEC Dimension object now includes points to dimension both the exterior masonry Wall and the porch. However, this may be a bit too cluttered to read well.

40. Right-click the same AEC Dimension object and choose **Detach Objects**, and then click the Curtain Wall again when prompted (or simply undo).

41. Using the same grip edit method from above, move all Chains down closer to the masonry Wall.

42. Click the **AEC Dimension (2)** tool on the Annotation tool palette.

43. At the "Select geometry to dimension" prompt, click the horizontal Curtain Wall of the porch (be certain to select the Curtain Wall and not the nested Window infills) and then press ENTER.

44. At the "Specify insert point" prompt, click a point above the porch (see Figure 14–21).

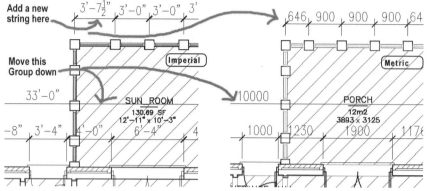

Figure 14–21 *Detach the Curtain Wall from the main string and add a new AEC Dimension object instead*

45. Select the new AEC Dimension string, right-click, and choose **Properties**. Change the Style to **MADT Residential**.

OVERRIDE TEXT AND LINES AND OTHER FINE-TUNING

The new AEC Dimension group added to the Curtain Wall only has one Chain. It would be nice to see the overall dimension of the porch as well as the dimensions shown. There are some other tweaks that we can make to this group as well.

46. Select the new AEC Dimension string, right-click, and choose **Edit AEC Dimension Style**.

47. On the Display Properties tab, be sure Plan is selected and then click the Edit Display Properties icon.

48. On the Contents tab, select the "Curtain Wall" entry in the "Apply to" list and then click on Chain 2.

    Notice that all check boxes for Chain 2 are cleared. This is why there is no second chain displayed for Curtain Walls.

49. Place a check mark in the Overall check box (see Figure 14–22).

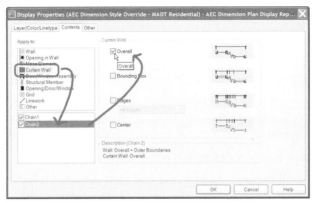

Figure 14–22 *Configure AEC Dimensions to add an overall dimension to Curtain Walls in Chain 2*

50. Click OK twice to return to the drawing and see the result (see Figure 14–23).

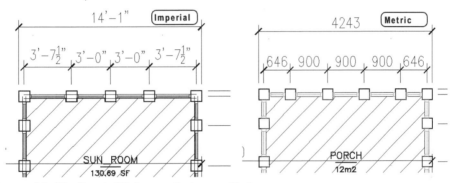

Figure 14–23 *Curtain Walls now display two Chains*

This is better, but it still needs work.

51. Select the same dimension, right-click, and choose **Add Dimension Points**.

52. At the "Pick dimension points on screen to add" prompt, snap to the endpoint of the outside corner on each side of the Curtain Wall and then press ENTER (see Figure 14–24).

53. At the "Select Dimension chain" prompt, click the lower (running) Chain.

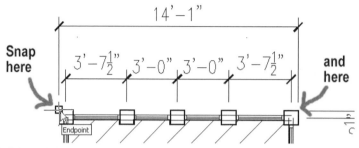

Figure 14–24 *Add dimension points to the two outside corners*

We now have two small dimensions at either end that dimension half of the post. This might be a bit busy. Let's remove the point at the center of the end posts so that the two dimensions at the end include the full width of the corner posts.

54. Select the same dimension, right-click, and choose **Remove Dimension Points**.

55. At the "Select extension lines to remove dimension points" prompt, click the two extension lines passing through the center of the corner posts and then press ENTER.

That is better still, but one more edit is in order.

56. Select the same dimension, right-click, and choose **Override Text & Lines**.

57. At the "Select dimension text to change" prompt, click on the left dimension of Chain 1 (see Figure 14–25).

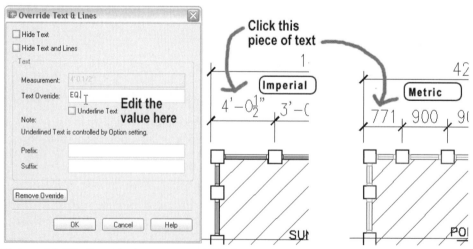

Figure 14–25 *Override the text to read "EQ."*

58. In the Override Text & Lines dialog box, type **EQ.** in the Text Override field and then click OK.

Notice the small horizontal bar that appears above the text in the drawing when you apply the override. This is the Override Text & Lines Marker defined above. This marker is on a non-plotting layer and is very useful to flag dimension text that has been manually edited on screen. Do *NOT* turn this component off.

59. Repeat the process on the other side (see Figure 14–26).

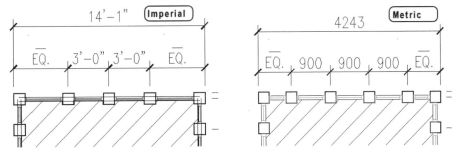

Figure 14–26 *The completed AEC Dimension group for the porch, including text overrides*

60. Perform similar edits on the Curtain Wall dimension string on the porch's right side.

   Return to the Override Text & Lines dialog box if you wish and experiment with other settings.

61. Save the file.

AEC DIMENSIONS IN-PLACE EDIT

Sometimes you need to fine-tune the location of the individual dimension Chains, the text or the extension lines. This can all be done easily with in-place grip editing.

62. Select the main horizontal dimension string (the one dimensioning the horizontal Wall).

63. Click the small round Edit in-Place grip (see Figure 14–27).

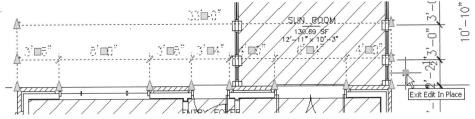

Figure 14–27 *Click the edit in-place grip to reveal grips for all dimension components*

Notice the array of grip points that appears. A triangular shaped grip appears at each extension line and dimension Chain. A square grip appears at each piece of text.

64. Click the Move Chain (triangular shaped) grip for the overall dimension (33′ 0″ [10,000])

65. Move it up past the dimensions of the porch and click again (see Figure 14–28).

Experiment further with the various grip points. Unfortunately, if you move text with the grips to make the dimension more legible, ADT does not create a leader pointing the text back to its dimension. You will need to manually draft the leader using the AutoCAD leader command.

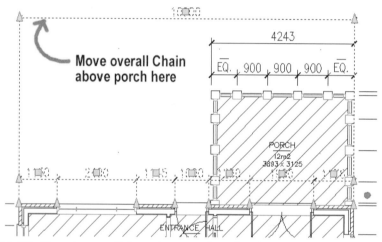

Figure 14–28 *Move the overall dimension away from the plan*

66. Save the file.

EXPLORING DISPLAY SETTINGS

In addition to remaining linked to the objects they are dimensioning, AEC Dimensions interact with the display system. A major benefit of this is the ability to use the same dimensions at two different scales. To understand this functionality, let's explore the display settings of AEC Dimensions.

WORK WITH AEC DIMENSION DISPLAY

1. On the Drawing Status Bar, open the Display Configuration pop-up menu and choose **High Detail**.

2. Zoom and pan around the drawing and study the results.

3. On the Drawing Status Bar, open the Display Configuration pop-up menu and choose **Low Detail**.

4. Zoom and pan around the drawing again and study the results.

As you zoom and pan around the drawing, note several important points. First, the scale of the dimensions has halved at High Detail and doubled at Low Detail. Another thing you will notice is that none of the in-place edits (including text overrides) that we just performed is visible here. This is because these types of edits are applied to the Display Representation. Therefore, you will have to repeat any edit that you wish to see at the other scales. However on the positive side, since it is likely that you will require more detailed dimensions at high detail, and very schematic overall dimensions at low detail, this is probably not as big an issue as it might at first seem.

CAD Manager Note: The settings in AEC Dimension Styles can be set to match the most common scales used for dimensioning in your office. The three built-in Display Representations are configured as follows: Plan Low Detail is 1/16″=1′-0″ [1:200], Plan is 1/8″=1′-0″ [1:100] and Plan High Detail is 1/4″=1′-0″ [1:50]. You can change this if you wish, but be forewarned, it is no small task. AEC Dimensions are fairly straightforward because the scaling is controlled at the style level in the Display Properties. However, if you deviate from the default scales, you will need to reconfigure the Display Properties of each style you use in your templates and tool palettes. If you require more scales than three, you can add your own custom Display Representations to the AEC Dimension object type in Display Manager. The real work in choosing alternate scales comes in when you begin modifying the existing Scale-Dependent Multi-View Blocks. In that case you will need to edit each of the View Blocks for each of your Multi-View Content items. This can be potentially dozens of symbols or more.

5. Select an AEC Dimension string, right-click, and choose **Edit Object Display**.

6. Click the Display Properties tab.

   Notice that Plan Low Detail is currently active (bold).

7. Click the Edit Display Properties icon.

8. Click the Other tab (see Figure 14–29).

   In the AutoCAD Dimension Settings area, notice that ***AEC-Arch-1-192*** [***AEC-Arch-M-200***] is assigned as the Dimension Style. Notice also that the Distance between Chains is set to 6′-0″ [1800].

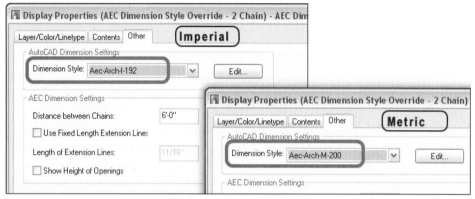

Figure 14–29 *Dimension Styles are assigned to AEC Dimension Styles on the Other tab*

AEC Dimensions refer to an AutoCAD Dimension style for settings such as arrowhead size and type, extension and dimension line settings. (For more information about AutoCAD Dimension styles refer to the online help.) The most important setting in the Dimension style with respect to their use in AEC Dimensions is the **DIMSCALE** setting.

9. Next to the Dimension Style dropdown list, click the Edit button.

   The Dimension Style Manager will appear, listing all dimension styles in the current drawing.

10. Click on **AEC-Arch-I-192 [AEC-Arch-M-200]** in the list and note the preview.

11. Click on **AEC-Arch-I-96 [AEC-Arch-M-100]** and note the preview, and then do the same for **AEC-Arch-I-48 [AEC-Arch-M-50]**.

As you can see from the previews, *AEC-Arch-I-192 [AEC-Arch-M-200]* is twice the size of *AEC-Arch-I-96 [AEC-Arch-M-100]*, which is twice the size of *AEC-Arch-I-48 [AEC-Arch-M-50]*. Keep this in mind as we complete the next several steps. Feel free to explore any of the dimension settings in the Dimension Style Manager. Simply select a dimension style and then click the Modify button to view and/or edit its settings. The **DIMSCALE** value is configured on the Fit tab in the Scale for Dimension Features area.

12. Click Close to dismiss the Dimension Style Manager.

13. Click Cancel to dismiss the Display Properties dialog box.

14. From the Display Representation list, select **Plan High Detail**, and then click the Edit Display Properties icon.

Notice that the Dimension Style used in this Display Representation is **AEC-Arch-I-48** [**AEC-Arch-M-50**]. Notice also that the Distance between Chains is set to 1'-6" [450] (see Figure 14–30).

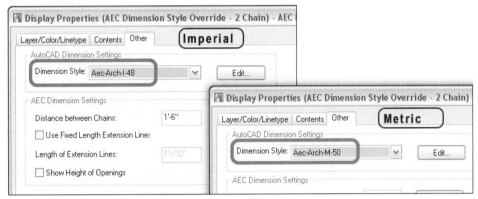

Figure 14–30 *The trick to AEC Dimension's scale-dependent behavior lies in the dimension style assigned to the Display Rep*

15. Repeat the steps again on the Plan Display Rep.

    Note the differences.

16. Click Cancel to dismiss the Display Properties dialog box.

CONFIGURE THE LOW DETAIL DISPLAY REP OF THE AEC DIMENSION STYLE

For Low Detail display to truly be "low," the dimensions should be simplified. At 1/16"=1'-0" [1=200mm] scale it is likely that we would only wish to see some overall dimension strings.

17. Select any AEC Dimension, right-click, and choose **Edit AEC Dimension Style**.

18. On the Display Properties tab, select Plan Low Detail and then click the Edit Display Properties icon.

19. Click the Contents tab.

20. In the "Apply to" list, choose "Wall" and Chain 1. Remove the check boxes from all options on the right.

21. Select Chain 2, deselect Length of Wall and check Overall instead (see Figure 14–31).

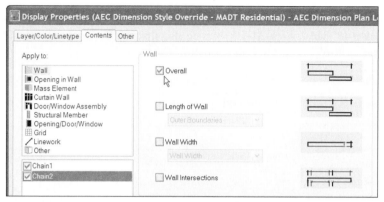

Figure 14–31 *Configure Plan Low Detail to display overall dimensions only*

22. Repeat the same process for the Curtain Wall and the Opening in Wall entries.

The overall Chains (Chain 2) are a bit far away from the Walls in the default configuration, particularly since we are now turning off Chain 1.

23. Click the Other tab.

24. Change the Distance between Chains in the AEC Dimension Settings area to **0**.

This will place the single overall string that we have configured here in the same spot as Chain 1 in the Medium Detail configuration.

25. Click OK twice when finished to view the results.

26. Use Edit in-Place to fine-tune the positions of individual strings (see Figure 14–32).

 CAUTION Don't use the Move All Chains grip; if you do, it will move the dimensions in all Display Reps.

 NOTE If moving individual Chains with grips gives unexpected results, undo and try attaching an Object Level Display Override to the Dimension instead. On the Other tab, you can type in a number for the distance between Chains.

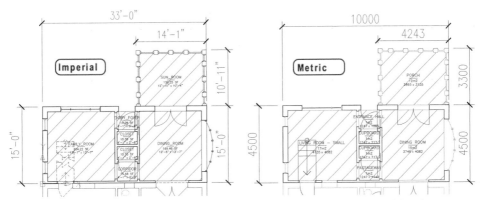

Figure 14–32 *Fine-tune the placement of Chains to complete the Low Detail Display*

27. Change the current Display Configuration back to **Medium Detail**.

    Notice that all of the overrides made in this display are still applied.

28. Save the file.

USING AEC CONTENT DIMENSIONS

For non-linear dimensions, like angular, radial, diameter and so on, you can use the tools in the Content Library. There is no functional difference between these tools and the AutoCAD equivalent (on the Dimension toolbar), except that the ADT variety will automatically layer. However, this is reason enough to use these tools instead of the AutoCAD counterparts on the Dimension menu.

CAD Manager Note: Adjust the layer setting used by all of the ADT Dimension tools by editing the *DIMLINE* layer key. All of the ADT Dimension tools use the current Dimension style active in the drawing. Therefore, make sure that the current DimStyle in the template is the company standard.

ACCESS AEC CONTENT DIMENSIONS (LAYER-KEYED AUTOCAD DIMENSIONS)

1. On the Annotation tool palette, click the Angular tool (see Figure 14–33).

2. Follow the command line prompts and dimension something in the drawing like the angle of the Windows in the Bow Window.

Since this is a standard AutoCAD dimension routine, you will not be able to use the select objects options; AutoCAD dimensions do not work through XREFs as AEC Dimensions do. Therefore, you must press ENTER to get the Select vertex option.

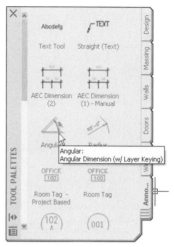

Figure 14–33 *The AEC Content Dimension tools*

3. The layer will not change until the command is complete, usually after you have pressed ENTER.

Going Further

Repeat any of the above techniques on the second floor – A-FP02 View file. You can also switch to the Commercial Project and try them out in A-FP03 as well.

DOCUMENTATION CONTENT

Within the ADT Content library, there are scores of symbols, routines, and macros designed to add the myriad annotation symbology required by architectural construction documents. Many of these routines are simple macros that add an AutoCAD block or polyline boundary, layered and scaled properly for the drawing. The Content Browser and the DesignCenter can be used to interface with this documentation content. Because most of this content is composed of AutoCAD entities that do not use display control, it is again beneficial to add these items in the View files. This will keep display management simple and layer management to a minimum. In the following sequence, we will continue to articulate the First Floor Plan View file *A-FP01*. See the introduction at the start of this chapter for more information on this topic.

VERIFY THE DRAWING SCALE

Most of the AEC Content that we will explore is not scale-dependent but rather scales physically based on the settings in the current drawing. There are two types of automatic scaling used by AEC Content and configured in Drawing Setup:

▶ **Drawing Scale**—Uses the scale factor equivalent to the intended plotting scale, such as 96 for 1/8"=1'-0" or 100 for 1:100. The quickest way to set this is on the Drawing Status Bar as you saw at the start of this chapter.

▶ **Annotation Plot Size**—Sets a desired fixed height for text within symbols and scales all of the symbols accordingly to match this desired text height for final plots.

1. From the Drawing Status Bar, click the Open Drawing menu and choose **Drawing Setup** (see the left side of Figure 14–34).

2. Click the Scale tab and verify that the Drawing Scale is set 1/8"=1'-0" [1:100] and the Annotation Plot Size is set to 3/32" [2.5] (see the right side of Figure 14–34).

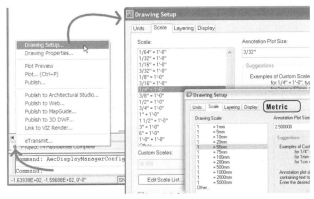

Figure 14–34 *Verify Drawing Scale and Annotation Plot Size settings*

 TIP If you are not interested in changing the Annotation Plot Size, it is faster to change the current Scale with the pop-up menu on the Drawing Status Bar. (See above).

3. Click OK.

ADD TEXT AND LEADERS

A sampling of frequently used annotation content is provided on the Annotation tool palette. To see the complete collection of all that is available, open the Content Browser and browse the various categories of the *Design Tool Catalog – Imperial [Design Tool Catalog – Metric]*.

1. On the Annotation tool palette, click the **Text** tool.

2. Click two points within the existing house to set the width of the text block.

3. Press ENTER to open the Mtext editor. On the Text Formatting toolbar that appears, choose **Title-Normal** from the Style list.

4. From the Font list, choose **Arial Black** and set the Text Height to **18″** [**450**].

5. Type **Existing House** in the text edit worksheet, and then click OK (see Figure 14–35).

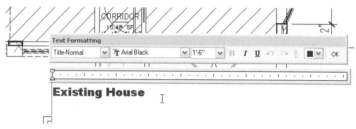

Figure 14–35 *Editing text in the text edit worksheet*

6. On the Annotation tool palette, click the **Straight (Text)** tool.

7. At the "Specify first point of leader line" prompt, click a point near the existing kitchen window.

8. Drag out the leader and click again, press ENTER, click a point to indicate the width of the text block and then type the following note: **Existing Window to be removed. Install New Hardwood Pass-through**.

9. Adjust the position of the note to a suitable location (see Figure 14–36).

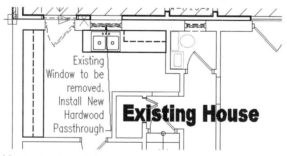

Figure 14–36 *Add a text note with leader*

ADD WINDOW TAGS

We have already added some Room and Space tags above. Let's add some Window tags here.

10. On the Scheduling tool palette, click the Window Tag tool.

11. At the "Select Object to Tag" prompt, click on the lower Window to the left in the new addition.

    In the Edit Property Set Data dialog, edit any values you wish and then click OK.

12. At the "Specify location of tag" prompt click a point just outside the Window.

13. Repeat the steps for the remaining Windows. (It is not necessary to tag the Windows of the Porch.)

    The Tags may not have the exact numbering that you desire. We can renumber them sequentially with the Renumber Data Tool on the Documentation palette.

14. On the Documentation palette, click the **Renumber Data** Tool.

15. In the Renumber Data dialog box, choose **WindowObjects** from the Property Set list, **Number** from the Property list, leave both Start Number and Increment set to **1**, and then click OK (see Figure 14–37).

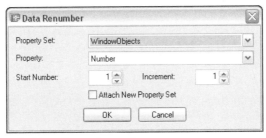

Figure 14–37 *Using the Data Renumber tool*

16. Following the command prompts, click the objects in the order that you would like them numbered and then press ENTER.

 NOTE This tool can be used to renumber any auto-increment Property like the Room Tags used in the Commercial Project at the start of the chapter as well.

In the next chapter, we will explore Schedules and Tags in further detail.

17. Save and Close the file.

Going Further

There are other tools on the Annotation tool palette, and dozens more in the Content Browser library. Feel free to try some others in this file and in the other View files in this project. We will explore the Keynote functions in Chapter 17. If you wish you can load the Commercial Project and annotate it as well. See Appendix A for suggested exercises. In addition to dimensions, targets and tags, we can also annotate our drawing sets with schedules. This topic is covered in detail in the next chapter.

ELEVATION LABELS

1. On the Project Navigator, double-click the *A-CM01* file to open it.

This is the Section and Elevation Composite Model built in Chapter 5. Since that time, we have swapped out the *Existing Roof* file, for the New Roof file in Chapter 12. Otherwise, there have been no changes to this file.

2. Zoom in on East Elevation.

3. On the Annotation tool palette, click on the **Plan Elevation Label (1)** tool.

 NOTE This symbol is meant to be used on plans, but it works equally well on elevations and sections. There are other choices of symbol in the Content library. Feel free to choose a different one if you wish. They all work the same way.

4. At the "Specify insertion point" prompt, use Object Snap Tracking and line it up with the first floor line as shown in Figure 14–38.

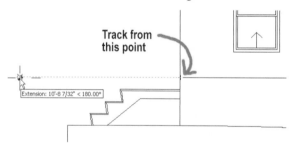

Track from this point

Extension: 10'-8 7/32" < 180.00°

Figure 14–38 *Line up the Elevation Label with the First Floor Level*

5. Click a point to place the symbol.

6. In the Add Elevation Label dialog box, type **First Floor** in the Prefix field, and **AFF** in the Suffix field.

7. In the bottom left corner of the Add Elevation Label dialog box, click the Define UCS icon.

8. When prompted, click the endpoint of the ground line in the Elevation as the Base point of the UCS.

9. Pull the cursor straight up and click to set the Z Direction (see Figure 14–39).

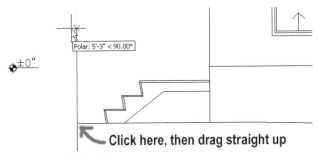

Polar: 5'-3" < 90.00°

±0"

Click here, then drag straight up

Figure 14–39 *Define a UCS for Elevations*

10. At the "Enter name for UCS" prompt, type **East Elevation** as the name of the UCS and then press ENTER.

11. Click OK in the Add Elevation Label dialog box to complete the routine (see Figure 14–40).

Figure 14–40 *Verify all values and then click OK to add the Label*

Repeat the steps to add additional labels. This time the UCS will already read "East Elevation," so you will not need to define an additional UCS. Simply type in prefixes such as "Second Floor" and "Eave" and place the labels (see Figure 14–41). The nice thing about these routines is that although they reference a UCS, they do not actually make it active. Therefore, your current UCS in the drawing remains the World Coordinate System for all other drawing operations. Try moving one of the labels. You will notice that the elevation changes. This is because it is linked to that UCS. Add an Elevation Label to the basement level; notice that it automatically registers with a negative number.

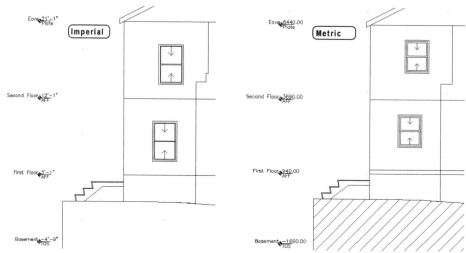

Figure 14–41 *Labels added to all key points on the Elevation*

If you later decide to reference another benchmark rather than the ground line we picked here, simply redefine the UCS, and all of the labels that reference it will update automatically.

12. Save and Close all project files.

ADDITIONAL EXERCISES

Additional exercises have been provided in Appendix A. In Appendix A you will find exercises to add dimensions and annotation to each of the remaining View files in both the Residential and Commercial Projects. It is not necessary that you complete these exercises to begin the next chapter, they are provided to enhance your learning experience. Completed projects for each of the exercises have been provided in the *Chapter14/Complete* folder.

SUMMARY

Add all annotation content to View files to take full advantage of the ADT Project Management features.

Create a new View file for each different type of drawing, in each discipline at each unique scale.

Room Tags are added with a simple Content routine.

Multiple Tags can be added at the same time after the first one is placed.

The Same Tag added to different View files will always remain synchronized to the Property Set Data stored in the Construct file.

Creating new tags is easy with the Define Schedule Tag wizard.

AEC Dimensions are updated automatically with the objects as they change.

AEC Dimensions work with ADT and AutoCAD objects.

AutoCAD associative dimensions work only on AutoCAD entities and not on ADT objects.

AutoCAD dimensions do not use display control; AEC Dimensions do.

AEC Dimensions can be displayed at different scales dynamically.

AEC Dimensions can dimension objects through XREFs.

Break marks, north arrows, bar scales, revision clouds, and scores of additional content are available as drag and drop content from the Content Browser, tool palettes or DesignCenter.

CHAPTER 15

Generating Schedules

INTRODUCTION

Schedules are an important part of any architectural document set. Creating a Schedule presents many unique challenges, among them deciding what to schedule, how to format it, and most importantly, how to keep the information up to date and accurate. ADT Schedule objects are ideally suited for creating any type of architectural schedule such as Door Schedules, Equipment Schedules, Wall Schedules, and Room Finish Schedules. Schedules can also play an important role in tracking key quantity information within the project at any phase regardless of whether the Schedule is intended for publication or printing. In this chapter, we will explore the many facets of the Schedule Table tool set that Architectural Desktop provides.

OBJECTIVES

In this chapter, we will be working on our Commercial Project. We will begin by first adding a simple Space Inventory Schedule using the default styles. With this Schedule style as our lab, we will explore the many tools and functions available in the complete Schedule Table tool set. At the completion of the exercise, we will have fully customized the Schedule both in format and in the data it tracks. At the end of the chapter is an advanced tutorial on creating a Project-Based Door Schedule and associated tags for the complete building. We will also revisit the new Display Theme objects herein as well. The following topics will be explored in this chapter:

- Learn to use the built-in Schedule Tables and Styles.
- Understand Property Sets.
- Work with Schedule Table Styles.
- Learn to adjust formatting and presentation of Schedule Data.
- Explore the pros and cons of automatically updating Schedules.
- Work with Display Themes.
- Explore Project-Based Schedule tools.

OVERVIEW AND KEY FEATURES

Schedule Table objects offer many features and benefits:

▶ **Schedules can be linked to ALL objects**—When you decide what to schedule, both standard AutoCAD objects (such as lines, polylines, and circles) and AEC objects (such as Walls, Doors, and Windows) can be included in a Schedule. This allows for much greater flexibility than AEC objects alone would.

▶ **Fully customizable**—The formatting of the Schedule Table object and the object properties that it tracks can be customized to suit specialized needs, from simple to complex.

▶ **Dynamic live link**—Reporting Property Set Data to a Schedule Table object is maintained with a "direct link" to the drawing data being scheduled. This guarantees that Schedules are kept up to date.

THE SCHEDULE TABLE TOOL SET

The Schedule Table tool set in Autodesk Architectural Desktop consists of a collection of interconnected components. Before you begin to work with Schedules, it is important to understand the function of each of these components.

▶ **Object**—Any object in the drawing whose properties you wish to track in a Schedule.

▶ **Property Set Definition**—Establishes a set of object data specific to a particular type of object or objects. This data will determine the available columns of an associated Schedule Table. Property Set Definitions are user defined and can be applied at the style or object level.

▶ **Property Data**—The data from one or more Property Set Definitions attached to a specific object. This data can feed one or more Schedules.

▶ **Property Data Format**—Transforms raw Property Set Data from the objects into the desired presentation format prior to its appearing in the Schedule Table or other Schedule components. For example, expressed as raw data, the width of a typical door is 36 units in Imperial measure. Using the appropriate Property Set Data Format, the expression on the Schedule would read 3′–0″.

▶ **Schedule Table Style**—Like other ADT Styles, the Schedule Table style controls the configuration and overall appearance of the Schedule Table object itself. The style also determines the type of objects to which the Schedule Table can link.

▶ **Schedule Table**—The actual Table object itself. This includes the borders, titles, headers, row and column cell divisions, and cell data (text). Each row item within the Schedule Table is linked to an object within the drawing.

▶ **Object Tag**—As required by construction documentation needs, tags *can* be included in drawings and function as an additional report of the attached Property Data. When used, Property Data from the object is fed to the tag. Multi-View Blocks defined specifically to receive data from a Property Set Definition can be used as tags.

PUTTING IT ALL TOGETHER

Objects (AutoCAD or ADT) populate the drawing. One or more Property Set Definitions establish data for each column we wish to see in the Schedule (see Figure 15–1). Property Data is attached to each of the individual drawing objects or Object styles (ADT objects only). These objects are selected to appear in the Schedule Table. As appropriate, each piece of data is formatted by a Property Data Format. Finally, a report on object Property Data is generated in the form of a Schedule Table or an Object Tag (or even exported to Excel). A Schedule Table Style governs the Schedule Table's overall format and appearance.

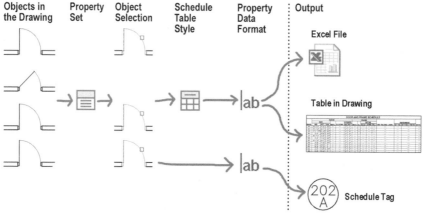

Figure 15–1 *The inter-relationship of components in the Schedule Table tool set*

GETTING READY TO USE SCHEDULE TABLES

From a thousand-foot level, getting started with Schedules involves a few basic steps:

Choose—First you must decide what information you want to schedule.

Format/Link—The styles and components from which the Schedule will be generated must be imported or built next.

Report—Generate the Schedule and/or insert the tags, which will remain linked and up to date.

 Manager Note: Many Schedule Table Styles have been included with ADT. Start with these styles to become acquainted with the tool. Custom solutions can be created from scratch or by modifying the existing offerings. For this reason, it is beneficial to become familiar with the configuration and intent of the samples provided with the software. You can build tool palettes containing the Schedule Table Styles that your firm uses frequently.

ADDING SCHEDULE COMPONENTS

In the following tutorial, we will begin with a Schedule Table of Spaces showing department, geographical information, areas, and quantities in the third floor of the Commercial Project. We will use this to explore the many features of the Schedule Table tool set. Following the precedent established in the previous chapter, we will add Schedule Tables and Tags in View files.

INSTALL THE CD FILES AND LOAD THE CURRENT PROJECT

If you have already installed all of the files from the CD, simply skip down to step 3 below to make the project active. If you need to install the CD files, start at step 1.

1. If you have not already done so, install the dataset files located on the Mastering Autodesk Architectural Desktop 2006 CD ROM.

   Refer to "Files Included on the CD ROM" in the Preface for information on installing the sample files included on the CD.

2. Launch Autodesk Architectural Desktop 2006 from the desktop icon created in Chapter 3.

If you did not create a custom icon, you might want to review "Create a New Profile" and "Create a Custom ADT Desktop Shortcut" in Chapter 3. Creating the custom desktop icon is not essential; however, it makes loading the custom profile easier.

3. From the File menu, choose **Project Browser.**

4. Click to open the folder list and choose your *C:* drive.

5. Double-click on the *MasterADT 2006* folder, then the *Chapter 15* folder.

   One or two commercial Projects will be listed: *15 Commercial* and/or *15 Commercial Metric.*

6. Double-click *15 Commercial* if you wish to work in Imperial units. Double-click *15 Commercial Metric* if you wish to work in Metric units. (You can also right-click on it and choose **Set Current Project**.) Then click Close in the Project Browser.

 Important If a message appears asking you to re-path the project, click **Yes**. Refer to the "Re-Pathing Projects" heading in the Preface for more information.

CREATE A SCHEDULE VIEW FILE

Let's begin by creating a View in which to generate our first Schedule Table. We are going to create a View file of the third floor that will be very similar to the other third floor Views we have already created. The major difference will simply be the way it is used in our project and the type of data it contains.

1. On the Project Navigator, click the Views tab.

2. Right-click the *Views* folder and choose **New View Dwg > General**.

3. Name the View: **A-SH03**, and give it a Description of: **Architectural Third Floor Schedules** and then click Next.

4. Choose the third floor and then click Next.

5. Clear the check boxes for all Constructs except *03 Partitions* and then click Finish.

This gives us a new View file of the third floor that contains only the *03 Partitions* Construct. We have deliberately omitted the others for the time being. We will load them in later.

ADD A SCHEDULE TABLE

6. On the Views tab, double-click the new *A-SH03* file to open it.

7. Zoom the drawing out a bit to allow room to place the Schedule Table to the left.

8. On the Scheduling tool palette, click the *Space Inventory* tool (see Figure 15–2).

   If you do not see this palette or tool, right-click the Tool Palettes title bar and choose **Document** (to load the Documentation Tool Palette Group) and then click the Scheduling tab.

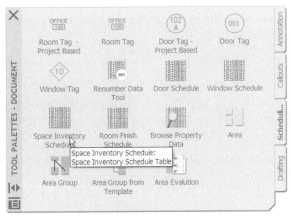

Figure 15–2 *Add Schedule Table from the Scheduling tool palette*

As with any ADT object, when adding a Schedule Table, the Properties palette includes several settings. Many are preset by the tool that we picked. For instance, since we clicked a Space Inventory Schedule tool, the Style will naturally be pre-set to *Space Inventory*. The Scale is also pre-set to the current scale factor of the drawing (see the "Verify the Drawing Scale" topic in the previous chapter for more information on drawing scale). There are some additional settings available when adding Schedule Tables that warrant discussion here:

- ▶ **Update automatically**—When **Yes** is chosen here, the Schedule Table will dynamically update as changes to the drawing take place. Use this feature sparingly because it can have a negative impact on the performance of your system.

- ▶ **Add new objects automatically**—Automatically adds newly created drawing objects directly to the table if they meet the original table selection criteria. For example, in a Door Schedule, if a new Door were added to the drawing, it would also appear on the Schedule immediately.

- ▶ **Scan xrefs** and **Scan block references**—Allow objects nested within blocks or XREFs to also be included in the Schedule.

- ▶ **Layer wildcard**—Allows the selection of objects for the Schedule to include only those on specified layers. Any standard Windows wildcard can be used, such as (*) for any character, question mark (?) for a single character and so on.

You can change any of these settings after placing the Schedule. When scheduling objects in XREFs, make sure the Scan xrefs is set to **Yes**. If you wish to select only certain objects within the XREF, you can use the pipe (|) character in your Layer wildcard. For example, **01 Partitions|*** would find objects on any layer within the first floor partitions XREF only.

9. Change "Update automatically" and "Add new objects automatically" to **Yes** (see item 1 in Figure 15–3).

   Verify that Scan xrefs is also set to **Yes**. Leave the remaining pre-sets as they are.

10. At the "Select objects or ENTER to schedule external drawing" prompt, click on the *03 Partitions* XREF in the drawing and then press ENTER to end the selection (see item 2 in Figure 15–3).

11. At the "Upper left corner of table" prompt, click a point off to the left in the drawing to place the corner of the Schedule (see item 3 in Figure 15–3).

Figure 15–3 *Set the Parameters, select the XREF and place the Schedule at the default scale*

12. Press ENTER at the "Lower right corner (or RETURN)" prompt.

    This makes the Schedule Table the default size without scaling it. Schedule Tables will be scaled automatically to the Drawing Scale (on the Drawing Status Bar).

A Schedule Table object will appear in the drawing. However, most of the fields in the Schedule are showing a question mark (?) (see Figure 15–4). The reason is that the required Property Set Data has not been attached to the individual Space objects yet (see the following sequence). Since there is no data yet attached to the Spaces for the missing fields, the Schedule shows a question mark instead. Property Set Data can be attached to objects within the drawing in a few different ways.

SPACE INVENTORY								
SPACE	LOCATION						AREA	QTY
	SITE	BUILDING	FLOOR	ZONE	DEPARTMENT	OWNER		
CLOSET	?	?	?	?	?	?	8.66 SF	2
CONFERENCE_MEDIUM	?	?	?	?	?	?	330.18 SF	1
CORRIDOR	?	?	?	?	?	?	293.26 SF	1
DINING_ROOM	?	?	?	?	?	?	181.16 SF	1
ENTRY_ROOM	?	?	?	?	?	?	212.55 SF	1
LIBRARY	?	?	?	?	?	?	155.44 SF	1
MECH_ROOM	?	?	?	?	?	?	35.43 SF	1
OFFICE_MEDIUM	?	?	?	?	?	?	242.26 SF	1
OFFICE_MEDIUM	?	?	?	?	?	?	165.92 SF	1
OFFICE_MEDIUM	?	?	?	?	?	?	132.12 SF	1
OFFICE_MEDIUM	?	?	?	?	?	?	139.45 SF	1
OFFICE_MEDIUM	?	?	?	?	?	?	141.62 SF	1
OFFICE_MEDIUM	?	?	?	?	?	?	140.16 SF	1
OFFICE_MEDIUM	?	?	?	?	?	?	143.19 SF	1
WORKSTATION_SMALL	?	?	?	?	?	?	84.97 SF	1
WORKSTATION_SMALL	?	?	?	?	?	?	74.44 SF	1
								17

Figure 15–4 *Several fields in the Schedule Table are missing data*

13. Save the file.

PROPERTY SET DATA

Property Set Definitions (as defined in "The Schedule Table Tool Set" section above) determine which Property Set Data will be available to the objects and Schedules in the drawing file. Property Set Definitions can be referenced by styles or directly by individual objects. In the current example, question marks occur in our Schedule where object-based Property Sets are referenced. The question marks indicate that the required data has not yet been attached. Attaching object-based Property Sets can be accomplished using a few techniques.

▶ Attach Schedule Tags to Objects (which will also attach the Property Sets)

▶ Attach All Property Sets at once via the Schedule Table right-click menu

▶ Attach Property Sets Manually, Object by Object

Generally the first option requires the most effort and should typically be avoided. The tagging option occurred automatically in the last chapter while we were tagging objects. We will see additional examples below. The first option was performed in the last chapter in the "Adding Room Tags" heading. (We did not elaborate on the attachment of Property Data there as we will below.) The second option will be explored next.

ATTACH PROPERTY SET DATA VIA THE SCHEDULE TABLE

The fastest way to attach all of the required Property Sets is via the Schedule Table itself.

1. Click on the Schedule Table, right-click and choose **Add All Property Sets** (see Figure 15–5).

The change to the Schedule Table will be subtle. In place of the questions marks will now be a double dash (–) within each field.

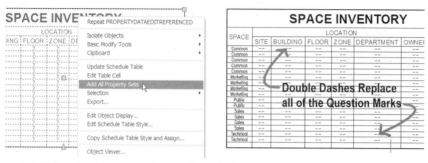

Figure 15–5 *Attach All Property Sets via the Schedule - All the Question Marks are replaced by a place-holder value (–)*

A double dash is simply a null value (see right side of Figure 15–5). The Property Set is attached, but a value for the field has not been input yet, or is not applicable, so the default value (–) is displayed in the field. All of these fields are now ready to receive typed in values. Unlike the columns on the left and right (which are automatic fields – see below), these fields are all manual type-ins.

We can edit the values assigned to each field in a few ways. Let's take a look.

2. Select the *03 Partitions* XREF, right-click and choose **Edit Referenced Property Set Data** (see the left side of Figure 15–6).

3. At the Select Objects Prompt, click on one of the Office Space objects and then press ENTER (see the right side of Figure 15–6).

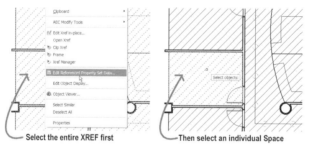

Figure 15–6 *Edit Referenced Property Set Data to edit the values of one or more Spaces*

The Edit Referenced Property Data worksheet will appear.

Each Property Set attached to the selected Space is indicated with a gray title bar bearing the name of the Property Set Definition. All of the Properties contained within each Property Set are grouped beneath this gray bar. You can use the small double arrow icon at the right side of each of these groupings to collapse and expand the group (see Figure 15–7). You will recognize several of the names of the

Properties within the *GeoObjects* grouping as several of them appear as columns in our Schedule Table. These are the fields that we will edit. However, notice that some of the fields available in *GeoObjects* do not appear on the Schedule. This illustrates the point that defining a property within a Property Set does not require it to be used by the Schedule Table.

4. In the Edit Referenced Property Data worksheet, fill in any of the fields within the **GeoObjects** grouping that you wish and then click OK (see Figure 15–7).

 TIP If the Schedule Table does not update on its own, select it, right-click and choose Update Schedule Table.

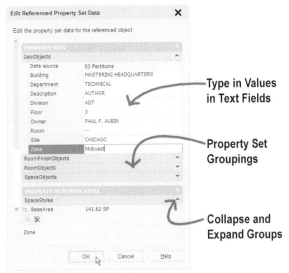

Figure 15–7 *Edit Referenced Property Set Data worksheet editing GeoObjects fields*

The Edit Referenced Property Set Data command can be used on one object at a time or a selection of several. However, if you choose more than one object, remember that any field you edit will actually apply to *all* objects in the selection. This can be very handy and also potentially problematic. Pay close attention to your object selections for the best results.

5. Repeat the process selecting more than one Space to edit this time (see Figure 15–8).

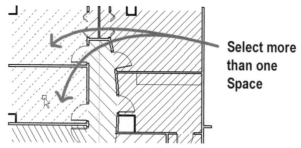

Figure 15–8 *Repeat Edit Referenced Property Set Data and select more than one Space*

6. Edit the **GeoObjects** field for the group selection and then click OK.

Notice that any change you made has been applied to all objects in the selection. Remember, the Property Sets are actually attached directly to the objects within the Construct. Therefore, in addition to this technique, you can also open the XREF and edit the Properties by selecting the objects directly.

7. Select the XREF, right-click and choose **Open XREF**.

The *03 Partitions* Construct file will open on screen.

8. Select one of the Office Spaces, right-click and choose **Properties** and then click the Extended Data tab (see Figure 15–9).

Notice that the same information that was included in the Edit Referenced Property Set Data worksheet is contained here, because it *is* the same data. You can edit it the same way. This is simply another method to edit it. Go ahead and edit some of the values if you like; you will see any changes reflected in the steps to follow.

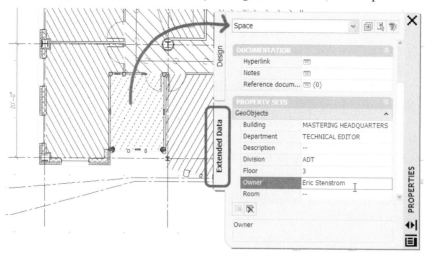

Figure 15–9 *Edit Property Data on the Properties Palette for selected objects*

As before, you can collapse any Property Sets that you are not currently editing to make the palette easier to manage. As we have noted above, all of the fields of the *GeoObjects* Property Set are editable text fields and are the ones shown in the Schedule in *A-SH03*. There are other Property Sets attached to the Spaces in the *03 Partitions* Construct however; namely *RoomFinishObjects*, *RoomObjects* and *SpaceObjects*. If you scroll down to and view some of these, you will notice that *RoomFinishObjects* and *RoomObjects* both have editable text fields. In the *SpaceObjects* Property Set however, you will notice that all of the properties there have a small lightning bolt icon next to them and *cannot* be edited. These are Automatic Properties that come directly from the drawing (see Figure 15–10). New to ADT 2006, Style-based property sets now appear on the Properties palette for easy access. You can see this on the right side of the figure. Automatic and Style-based Properties will be discussed more below.

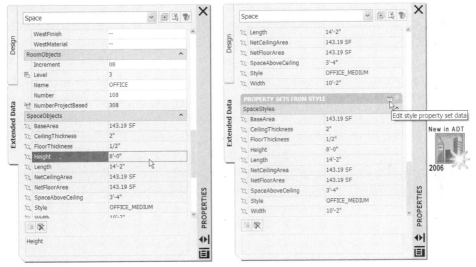

Figure 15–10 *A Lightning Bolt icon indicates an Automatic Property read directly from the object geometry*

The *RoomFinishObjects*, *RoomObjects* and *SpaceObjects* Property Sets were actually attached to the Spaces in this file during tag insertion in the previous chapter (see the "Adding Room Tags" heading in Chapter 14 for more information). Now that we have a clearer understanding of what Property Sets are, we can see that we were adding much more than tags alone in that exercise.

9. Save and Close the *03 Partitions* file.

    The *A-SH03* View file should still open on screen. If it is not, double click it on the Project Navigator palette to re-open it now. If necessary, right-click the

Schedule Table and choose **Update Schedule Table** to see the edits (see Figure 15–11).

SPACE	LOCATION						AREA	QTY
	SITE	BUILDING	FLOOR	ZONE	DEPARTMENT	OWNER		
CLOSET	--	--	---	--	---	---	8.66 SF	2
CONFERENCE_MEDIUM	--	--	---	--	---	---	330.18 SF	1
CORRIDOR	--	--	---	--	---	---	293.26 SF	1
DINING_ROOM	--	--	---	--	---	---	181.16 SF	1
ENTRY_ROOM	--	--	---	--	---	---	212.55 SF	1
LIBRARY	--	--	---	--	---	---	155.44 SF	1
MECH_ROOM	--	--	---	--	---	---	35.43 SF	1
OFFICE_MEDIUM	--	--	---	--	---	---	242.26 SF	1
OFFICE_MEDIUM	--	--	---	--	---	---	165.92 SF	1
OFFICE_MEDIUM	--	--	---	--	---	---	132.12 SF	1
OFFICE_MEDIUM	CHICAGO	MASTERING HEADQUARTERS	3	MIDWEST	TECHNICAL	ERIC STENSTROM	139.45 SF	1
OFFICE_MEDIUM	CHICAGO	MASTERING HEADQUARTERS	3	MIDWEST	TECHNICAL	PAUL F. AUBIN	141.62 SF	1
OFFICE_MEDIUM	CHICAGO	MASTERING HEADQUARTERS	3	MIDWEST	MARKETING	---	140.16 SF	1
OFFICE_MEDIUM	CHICAGO	MASTERING HEADQUARTERS	3	MIDWEST	MARKETING	---	143.19 SF	1
WORKSTATION_SMALL	--	--	---	--	---	---	84.97 SF	1
WORKSTATION_SMALL	--	--	---	--	---	---	74.44 SF	1
								17

Figure 15–11 *View the edits to the Schedule Table*

ADD SCHEDULE TAGS

As stated at the beginning of the "Property Set Data" heading, there are three basic methods used to attach object-based Property Sets: object-by-object, via tagging and via the Schedule Table. We explored the Schedule method above. The tagging method was used in the last chapter without elaboration. Therefore, let's take another look at attaching Property Sets via tagging now.

All of the referenced Spaces that we have on screen already have Property Sets attached. Let's edit the Properties of the *A-SH03* View file and call in the *Shell and Core* Construct to explore tagging as a method of attaching Property Sets.

1. On the Project Navigator palette, right-click the *A-SH03* file and choose **Properties**.

2. On the left side, click the Content link, place a check mark in the *03 Shell and Core* check box and then click OK (see Figure 15–12).

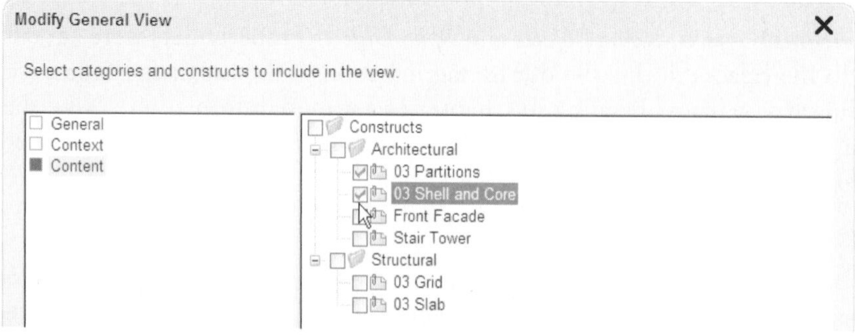

Figure 15–12 *Add a Construct to the Current View file*

The addition of this Construct has made the Schedule update. At the top of the Schedule there is now a new line that shows a quantity of 6 in the QTY column, but question marks for all other values. Notice also that the total in the Quantity column has updated to 23 (see Figure 15–13). What these two changes indicate is that the Spaces within the newly added XREF have been automatically added to the Schedule (a setting we enabled in the "Add a Schedule Table" heading above), but the Spaces within the XREF do not have Property Sets attached. Let's use the tagging method to attach the Property Sets.

SPACE	LOCATION						AREA	QTY
	SITE	BUILDING	FLOOR	ZONE	DEPARTMENT	OWNER		
?	?	?	?	?	?	?	? SF	6
CLOSET	--	--	--	--	--	--	8.06 SF	2
CONFERENCE_MEDIUM	--	--	--	--	--	--	330.18 SF	1
CORRIDOR	--	--	--	--	--	--	293.26 SF	1
DINING_ROOM	--	--	--	--	--	--	181.16 SF	1
ENTRY_ROOM	--	--	--	--	--	--	212.55 SF	1
LIBRARY	--	--	--	--	--	--	155.44 SF	1
MECH_ROOM	--	--	--	--	--	--	35.43 SF	1
OFFICE_MEDIUM	CHICAGO	MASTERING HEADQUARTERS	3	MIDWEST	MARKETING	--	140.16 SF	1
OFFICE_MEDIUM	--	--	--	--	--	--	165.92 SF	1
OFFICE_MEDIUM	CHICAGO	MASTERING HEADQUARTERS	3	MIDWEST	TECHNICAL	ERIC STENSTROM	139.45 SF	1
OFFICE_MEDIUM	CHICAGO	MASTERING HEADQUARTERS	3	MIDWEST	TECHNICAL	PAUL F. AUBIN	141.62 SF	1
OFFICE_MEDIUM	--	--	--	--	--	--	132.12 SF	1
OFFICE_MEDIUM	CHICAGO	MASTERING HEADQUARTERS	3	MIDWEST	MARKETING	--	143.19 SF	1
OFFICE_MEDIUM	--	--	--	--	--	--	242.26 SF	1
WORKSTATION_SMALL	--	--	--	--	--	--	84.97 SF	1
WORKSTATION_SMALL	--	--	--	--	--	--	74.44 SF	1
								23

Figure 15–13 *A new line of 6 "unknown" objects appears at the top of the Schedule and updates the total*

3. On the Drawing Status Bar, verify that the Scale setting reads **1/8"=1'-0"** **[1:100]**.

    Always double check this before adding tags, as they will scale to this setting.

4. Open the Content Browser (click the icon on the Navigation toolbar or press CTRL + 4).

We will use the Space Tag that we used in the Residential Project in Chapter 14.

5. Click on *Documentation Tool Catalog – Imperial* [*Documentation Tool Catalog – Metric*].

6. Navigate to the *Schedule Tags* category and then the *Room & Finish Tags* category.

7. Using the eyedropper icon, drag the Space Tag tool into the drawing.

8. At the "Select object to tag" prompt, select one of the Spaces in the *Core* such as the elevator lobby.

9. At the "Specify location of tag" prompt, click a location in the middle of the Space.

10. If the Edit Property Set Data dialog box appears, note the appearance of both the **GeoObjects** and **SpaceObjects** Property Sets and then simply click OK.

TIP The automatic appearance of the Edit Property Set Data dialog box can be turned off in Options (Tools menu) on the AEC Content tab (see below).

11. Press ENTER to end the command.

Notice the change in the Schedule. (Reload XREFs, and Update Schedule Table as necessary.) A new "ENTRY_ROOM" row has appeared with a double dash (–) replacing each of the question marks (see Figure 15–14).

SPACE	LOCATION						AREA	QTY
	SITE	BUILDING	FLOOR	ZONE	DEPARTMENT	OWNER		
?	?	?	?	?	?	?	? SF	5
CLOSET	--	--	--	--	--	--	8.66 SF	2
CONFERENCE_MEDIUM	--	--	--	--	--	--	330.18 SF	1
CORRIDOR	--	--	--	--	--	--	293.26 SF	1
DINING_ROOM							184.15 SF	1
ENTRY_ROOM	--	--	--	--	BUILDING COMMON	--	82.69 SF	1
ENTRY_ROOM	--	--	--	--	--	--	212.55 SF	1
LIBRARY							155.44 SF	1

Figure 15–14 *Adding a tag to the Space also attaches the required Property Sets*

The Schedule fields were updated just as they were when the Property Sets were added in the earlier sequence. Adding Schedule Tags actually runs a series of commands that include the attachment of any required Property Sets.

ADD A TAG TO OBJECTS WITH EXISTING PROPERTY DATA

Tags can be added to objects even after the Property Set Data has been added to them. In this case, the tag simply refers to the Property Set Data that are already attached.

12. Return to the Content Browser again.

13. Drag the Space Tag tool into the drawing again.

TIP you may want to add the Space Tag tool to your Project Tools tool palette as we did in the previous chapters. In this way, you can avoid having to re-open the Content Browser each time.

14. At the "Select object to tag" prompt, select one of the Space objects within the tenant space whose Property Sets you previously edited.

15. At the "Specify location of tag" prompt, place the tag in the space and then press ENTER. (If the Edit Property Set Data dialog box appears, click OK.)

Notice that nothing has changed in the Schedule, but the tag correctly displays the area of the Space as reported by the Schedule. In addition, if the Edit Property Set Data dialog box appeared, it correctly showed the information that you previously added to the Schedule.

COPYING TAGS MANUALLY (SHOULD BE AVOIDED)

At this point, it might be nice to add tags to the rest of the Spaces. Copying tags from existing tags in the drawing is NOT recommended. Simply copying a tag from an existing object to another does not associate the data within the tag with the new object. Proximity to an object does not determine its relationship to the Property Set Data within the object. A Tag Anchor is used to create this link. Let's try an experiment.

16. Copy the tag just added to another room using the standard AutoCAD Copy command. (Use the **Copy** in Basic Modify Tools, do *not* use Copy and Paste.)

17. Change the current Display Configuration to **Diagnostic** (use the pop-up menu on the Drawing Status Bar, see Chapter 2 for more information).

Notice the change in the drawing. As its name implies, the Diagnostic Display Configuration is used to help you troubleshoot problems in the drawing. One of its most useful features is the display of the Anchor Tag to Entity (or Tag Anchors) as seen here by the curved lines that connect the tags to a corner of Space to which they are anchored. Notice that both the original tag and the copied tag are anchored to the same Space object. The copy operation copied only the tag; it did *not* attach the Anchor to the new Space.

If you are having difficulty seeing to which Space the Tag Anchor is pointing, try temporarily moving the Space. (You will need to open the *03 Partitions* XREF or use **Edit XREF in-place** to do this.) When you return to the *A-SH03* View file, you will see the curved Tag Anchors follow. Be sure to undo this change when you are satisfied (see Figure 15–15).

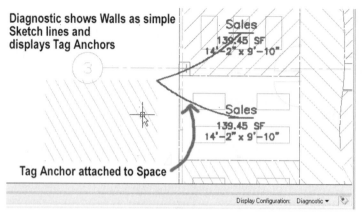

Figure 15–15 *Try moving a Space in the 03 Partitions Construct to see the Tag anchor more clearly*

18. Reverse any changes that you made, and delete the copied tag.

Be sure to restore the **Medium Detail** Display Configuration before continuing.

When you add a tag from the tool palette or Content Browser, it is not merely inserting a symbol; it actually runs Schedule Tag routine— **AecScheduleTag**. This command imports and attaches the required Property Sets and correctly anchors the tag to the object it is tagging. This process also allows the correct values to be input within the attributes of the tags.

Tags are a new tool type in ADT 2006. A Tag is a Multi-View Block object as we saw in the previous chapter. In previous versions of ADT, MVBs used as tags did not "know" that they were tags. To use them as tags required detailed and complex configuration of DesignCenter AEC Content items and tool palette tools. As we saw in Chapter 14, it is now very easy to create a custom tag using the wizard on the Format menu. Once a tag is created, it will "know" that it is a tag in the drawing. Select any tag, right click and note the appearance of the "Tag Anchor" right-click menu. If the tag is already attached to an object, this menu will show options to attach and release the Anchor. If the tag is not attached to any object, this menu will give only one option—**Set Object**. You can experiment with the tags on screen. Try releasing the Anchor on one and then right-clicking the tag to see this menu. When you drag a tag to a tool palette, the resultant tool automatically contains tag insertion properties. Again this was explored in Chapter 14.

CAD **Manager Note:**   In the "Using the Define Schedule Tag Wizard" heading in Chapter 14, we created a Tag Tool on our Project Tools palette. Tools can also be made that only attach the Property Set and don't insert a tag. This can be valuable in instances where you want an easy way to apply Property Set Data to objects without needing to add a Schedule or Tags.

ADD THE REST OF THE SPACE TAGS

Adding tags is a very effective and simple way to add Property Set Data to objects that require it. However, the process can be slow and tedious if you place them one at a time, and more so if the Edit Property Set Data dialog box pops up in between each insertion. In most cases, you will want to quickly add all of your tags to the drawing, and then progressively refine the Property Set Data over time as more design detail becomes known. We have two tricks to facilitate speedy tag insertion. First, we can turn off the setting that makes the Edit Property Set Data dialog box appear (if it is not off on your system already), and we can use the Multiple feature of the Tag add routine (as we did last chapter) to add a collection of tags all at once.

19. From the Format menu, choose **Options** (or right-click in the command line and choose **Options** from the shortcut menu).

20. On the AEC Content tab, clear the Display Edit Property Set Data Dialog During Tag Insertion check box and then click OK (see Figure 15–16).

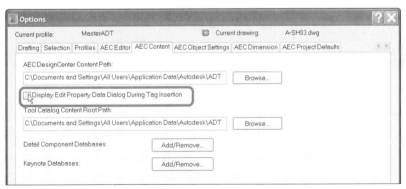

Figure 15–16 *The AEC Content tab of Options*

21. Bring the Content Browser to the front again (CTRL + 4), and drag the Space Tag tool into the drawing again.

 NOTE If you followed the tip above and made a tool on your Project Tools palette for the Space Tag tool, click it instead.

22. Following the prompts as above, click a Space without a tag and place the tag in that room. Press ENTER automatically place the tag at the center of the Space object.

   Notice that the Edit Property Set Data dialog box did not appear this time.

As we saw in the last chapter, after you have placed the first tag with the Schedule Tag Add command, you will be able to use the Multiple option. This option is not available on the first click, so you always have to place the first one manually. After the first tag, you can choose the Multiple option and then select several objects to tag all at once. Let's use this option for the remaining tags.

23. Right-click in the drawing and choose **Multiple**.

24. At the "Select objects to tag" prompt, use a crossing window and surround the entire plan and then press ENTER.

   A dialog box like the one shown in Figure 15–17 will appear alerting you that some Spaces already have tags.

Figure 15–17 *Multi-tag recognizes that some Spaces are already tagged*

25. In the AutoCAD alert dialog box, click the No button.

26. Press ENTER to end the Schedule Tag command.

Notice that all of the tags appear in the center as well. The Multiple option will match the insertion point of the first tag placed. Inevitably some of the tags will need to have their placement fine-tuned. Use any of the techniques covered in the last chapter to relocate a tag.

27. Move any tags as required for legibility.

28. The Schedule Table should update automatically. However, if it does not, right-click it and choose **Update Schedule Table**.

FINE-TUNE THE SCHEDULE'S SELECTION

Sometimes you will wish to exclude some objects from the Schedule, even though they are of the same type as those you are scheduling. For instance, you might want to exclude existing doors from a door Schedule and show only new ones, or as in this case, you might want only occupied areas, not including shafts or other "unused" spaces.

 NOTE This chapter will not cover basic BOMA or BOCA requirements for definitions on tagging or auditing space. Terminology used to identify space in this tutorial is used for instructional purposes only. If you need to provide your clients with industry or regulatory space information, please check with those guidelines for proper space tracking requirements.

29. Select the Schedule Table, right-click, and choose **Selection > Remove**.

30. At the "Select objects" prompt, click any Space within the building *Core*.

 NOTE Notice how the entire 03 Shell and Core XREF highlights. This command is not able to select the individual Spaces within the XREF. However, depending on what goal you have for the Schedule in this View file, this may be fine. In other words, if you only intend to report the Spaces for the tenant that occupies the suite on the third floor, then removing the entire *Core* from the Schedule Table's selection is appropriate. However, if you wish instead to remove only certain Spaces from the *Core*, then you will need to use a different technique. Layer filtersor Classifications can be used in that case. In the "Building a Project-Based Door Schedule (Advanced)" heading below, we will see an example of layer filters.

31. Press ENTER to complete the sequence.

Notice that the several Spaces have been removed from the Schedule. However, these Space objects remain in the drawing. This process allows you to fine-tune the selection of objects used to generate the Schedule. It does *not* change the actual objects in any way.

Within the Schedule Tables > Selection sub-menu is other useful tools: Add, Select Again and Show.

▶ **Add**—Allows you to add objects to the Schedule Table selection and is the opposite of the Remove option shown here.

▶ **Select Again**—Allows you to replace the Schedule selection with a completely new set of objects.

▶ **Show**—Allows you to click a line item in the Schedule and have the corresponding item in the drawing highlighted.

If you were to Add the Core back to the selection, you would notice that the original data was restored. As we have already stated, these commands affect only the selection of objects that appear in the Schedule, *not* the objects or their associated Property Set Data.

32. Practice each of these tools to get a sense of their function.

Be sure to restore the original selection when you are finished exploring.

ADD AND DELETE SCHEDULE ITEMS

Adding newly created items to a Schedule is simple. If Add New Objects Automatically is set to **Yes** in the Schedule object's settings, then new objects added to the drawing will be added automatically to the Schedule if they meet the original selection criteria of the Schedule Table. Otherwise, the new objects will need to be manually added to the Schedule selection using the tools covered above.

33. Select the *03 Partitions* XREF, right-click and choose **Open XREF**.

34. On the Design tool palette, click the Space tool.

    If you do not see this palette or tool, right-click the Tool Palettes title bar and choose **Design** (to load the Design Tool Palette Group) and then click the Design tab.

35. Add a new Space next to the drawing using any parameters.

36. Save the *03 Partitions* file and then double click *A-SH03* on the Project Navigator palette to switch to this file.

37. Reload the *03 Partitions* XREF, and Update the Schedule Table.

Notice the addition of the new item in the Schedule. (It should be the first line of the Schedule.) Several question marks appear for this new object, because like the objects in the previous sequences, it is not yet linked to the object-based Property Data. In the same fashion, items deleted from the drawing will be automatically removed from the Schedule.

38. Return to *03 Partitions* and delete the new Space.

39. Save and Close *03 Partitions*. In *A-SH03*, reload the XREF, and Update the Schedule.

Notice the removal of the line of question marks from the Schedule. The object has been removed from both the drawing and the Schedule. Adding and deleting Spaces from the drawing, as we have done in this sequence, and the impact this has on the Schedule is very different from manipulating the Schedule's selection as we did in the last topic.

 NOTE Please do not confuse the two operations, because they are quite different. When manipulating the Schedule Table selection, we are only affecting the content of the Schedule itself; when adding and deleting objects from the drawing, we are affecting both the drawing and the Schedule.

AUTOMATIC AND MANUAL PROPERTY SETS

Editing the Property Set Data attached to an object is simple. Edits will dynamically apply to all pieces in the data chain (object, Schedule, and tag.) Editing Property Set Data can be done in a few ways; we have already seen some. However, you will not be able to use the methods covered so far on all Properties. The method used to edit depends on what types of Property Sets are attached to the object in question.

PROPERTY SETS CAN BE EITHER AUTOMATIC OR MANUAL AND CAN BE ATTACHED TO EITHER OBJECTS OR STYLES

Every AutoCAD or ADT object has a collection of unique defining properties inherent to its specific object type. This is true of basic drafted entities like lines and circles and architectural objects like Walls and Doors. These *automatic* properties include characteristics such as length, width, style, color, and layer, to name a few. A Property Set Definition is capable of tracking any of the object's inherent properties in an *automatic* property definition. For example, if an object is resized in the drawing, an automatic property defined to track the object's length, width, or height would immediately report this change. This is very powerful indeed.

However, as useful as automatic properties are, it is often desirable to include data in the Schedule that cannot be acquired automatically. This type of data can be included in the Property Set using a non-automatic *manual* property. Examples include cost, fire rating, and material finish.

Automatic and Manual properties are articulated in the dialog boxes and worksheets with the following icons:

Automatic property—Automatic properties may not be edited on the Properties palette or the Schedule. This data is only available on the object itself in the main drawing window through normal AutoCAD or Architectural Desktop editing such as grip editing or object properties (on the Design tab of the Properties palette).

Manual property—These properties *are* available for edit on the Properties palette, or on the Schedule Table itself. Editing in either place affects the entire data chain. Since they are not tied to any inherit or geometric properties, edits to the object such as grip editing or object properties, have no impact on Manual Properties. There are actually five additional types of Automatic Property. Many of these offer powerful advanced functionality which will fall outside the scope of this book. We will see some of these in the "Building a Project-Based Door Schedule (Advanced)" topic below.

Formula property—An automatic property that processes the data in some way before reporting it. Examples include concatenation of two or more other properties or even running complex programmatic calculations.

Formula properties have received a completely new interface in Architectural Desktop 2006. It is now much easier to add complex formulas containing VB Script code. In addition, the new dialog even allows you to input sample values and test the validity of a formula directly in the Property Set. This will save time and reduce error trapping. For more information on this feature, consult the online help.

 Location property—An automatic property that reads Property Set Data from a particular (often adjacent) Space, Area or AEC Polygon.

 Classification property—An automatic property that reads the Classification or a particular piece of Classification Property Set Data of an object. Classifications are user-defined categories that can be assigned to ADT object styles.

 Material property—An automatic property that reads the Material Definition or a particular piece of Material Definition Property Set Data of an object.

 Project property—An automatic property that reads data from the current project database (such as floor number, project name, etc).

 NOTE Although we will not include extensive coverage of these five automatic Property Set types we will see examples of Formula, Location and Project Properties in the "Building a Project-Based Door Schedule (Advanced)" topic below. You are encouraged to experiment with these Properties on your own if you wish.

Both automatic and manual properties can be defined within "style-based" or "object-based" Property Set Definitions. A style-based Property Set is attached to an ADT object style and therefore applies automatically to all objects belonging to that style. For instance, if you assign a "Wall Type" property to a Wall Style Property Set, then you can change the Wall Type of all the Walls that belong to that style by simply editing the Wall Type Property at the style level. AutoCAD objects cannot use style-based Property Sets. Object-based properties, on the other hand, are attached individually to each object and must be edited individually by selecting each associated object. The *GeoObjects* Property Set used above is an example of an object-based Property Set.

IDENTIFY AND EDIT OBJECT-BASED MANUAL PROPERTIES

All of the properties under the Location heading in the Schedule we have on screen (which belong to the *GeoObjects* Property Set) are manual (non-automatic) properties.

> Continue in the *A-SH03* View file. If you closed it, double click it on Project Navigator to re-open it.

1. Click on the *03 Partitions* XREF, right-click and choose **Edit Referenced Property Set Data**.

2. At the "Select Objects" prompt, click on any Office Space and then press ENTER.

This is the technique that we used above. All of the manual Properties will appear as editable text fields. All Properties in the *GeoObjects* Property Set are manual.

3. Type **Jane Doe** in the Owner field and then click OK.

> If the Schedule does not update automatically, right-click it and choose **Update Schedule Table**.

Multiple spaces can be selected before editing as well. This will allow you to edit the same property for several spaces at once. This technique was covered above as well.

 NOTE If the fields in your Edit Property Set Data dialog box are unavailable for edit, you have at least one object in your selection that does not have Property Set Data attached. Use any of the methods covered above to attach the Property Sets to the objects.

EDIT A TABLE CELL

Manual properties can also be edited directly within the Schedule Table object's cells.

4. Click the Schedule Table object, right-click, and choose **Edit Table Cell**.

5. Click the double dash (–) in the Site column (directly on the Schedule in the same row as Jane Doe).

6. In the Edit Referenced Property Set Data dialog box, change the value to **Chicago** and then click OK.

> Notice the change to the Schedule. Also notice that regardless of how the data is entered, it appears in the Schedule as uppercase, because of the Property Data Format applied to these fields. (Refer to the "Schedule Table Tool Set" section above for a definition of Property Data Format.)

With the Edit Table Cell command still active, take note of the command line, which reads "Select Schedule Table item (or the border for all items), hover for information, or CTRL-select to zoom." If you click on any borderline of the Table, it will call the Edit Property Set Data dialog box and allow the global edit of all objects in the Schedule (similar to the process completed in the last sequence). Exercise caution here, because in some cases, you will not want to perform a global edit.

7. Click on any Table border.

8. In the Building field, type **Mastering Headquarters** and then click OK (see Figure 15–18).

Figure 15–18 *Select any border of the Schedule Table using Edit Table Cell to edit all items*

Notice the change occurs to all items in the Schedule.

With the Edit Table Cell command still active, if you hover over an item, you will get information pertinent to that item.

9. Hover your cursor over the Zone column for any space.

Notice the small tool tip that appears.

10. Hover your cursor over the Area column for any space.

Notice that the message is different this time. Here you are being told that you cannot edit this cell directly because it has an automatic source.

11. Press ENTER to end the command.

12. Using either method, change some more values before proceeding to the next step.

EDIT AUTOMATIC PROPERTIES

Automatic properties can only be edited by directly editing the object. This is because an automatic property is a direct link to some inherent property of the object. If you wish to see this, open the *03 Partitions* Construct.

13. Select the *03 Partitions* XREF, right-click and choose **Open XREF**.

 NOTE If you prefer, you can also use Edit XREF in-place functionality to complete this sequence.

14. Click to select any Space and then perform a grip edit on the Space to change its size and shape.

15. Save *03 Partitions*, switch to *A-SH03*, reload the XREF and then update the Schedule Table.

The Area column of the Schedule noted the change and updated accordingly. The tags also now reflect the new dimensions.

16. Return to the *03 Partitions* file and undo the change to return the Space to normal.

17. Save and Close *03 Partitions* and then reload the XREF and update the Schedule Table.

EDIT STYLE-BASED PROPERTY DATA

As was noted previously, Property Sets can also be assigned to an object style. This allows properties to be assigned globally to all objects referencing a particular style. The likelihood of accidentally missing an object is greatly diminished with style-based properties. The real power of style-based properties comes from the ease of editing them. Should the design change, the edit needs to be made only once at the style level to update the entire drawing and Schedule at once. Like object-based Property Sets, style-based Property Sets can be both automatic and manual.

1. Open the *03 Partitions* XREF, Select one of the office Spaces (in the upper left corner), right-click, and choose **Edit Space Style**.

2. Click the General tab.

3. Click the Property Sets button.

There is a single Property Set named *SpaceStyles* attached to this Style. Notice that all its Properties come from an automatic source (they all have the small lightning bolt icon and none can be edited in this dialog box). They must be edited on the drawing objects as we did in the sequence above.

4. Click Cancel twice to return to the drawing.

We also mentioned above that you could now access Style-based properties on the Extended Data tab of the Properties palette.

5. Select the same Space.

6. On the Properties palette, click the Extended Data tab, and scroll to the bottom of the Property Set list.

7. Click the small Edit Style Property Set Data worksheet icon.

A worksheet containing the same automatic properties will appear. Click Cancel to dismiss it when finished.

Before closing *03 Partitions*, let's take a look at another style-based Property Set that contains manual properties.

8. Within the tenant space, click one of the interior partitions.

9. On the Properties palette, on the Extended Data tab notice the "Property Sets From Style" heading and worksheet icon.

10. Click the "Edit style property set data" worksheet icon (see Figure 15–19).

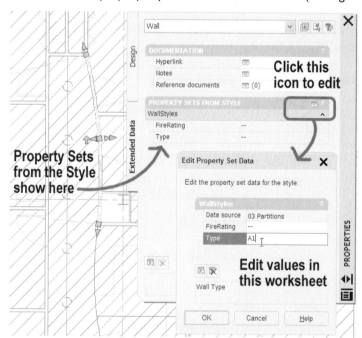

Figure 15–19 *Editing a style-based manual properties on the Properties palette*

There is also a single Property Set attached here. However, both of its Properties are manual (they have editable type in text fields). If we were to edit the Type field for instance, that change would apply across the Style to all Walls belonging to **Wall Style Stud-X.**

11. Edit the Type and then click OK to return to the drawing.

12. Select a different Wall, and repeat the process.

Note that the value you edited for the first Wall appears here as well. This is a Style-based Manual property. This means that you can input any manual value you wish, but it will apply to all objects belonging to that Style.

13. Save the *03 Partitions* Construct file (do not Close it yet).

 NOTE There is a Display Theme that applies to the FireRating property seen here. We first used this Display Theme back in Chapter 2. Display Themes were discussed in Chapter 12 and will be explored further below.

 ## PROPERTY SET DEFINITIONS

Property Set Data attached to objects is comprised of one or more object properties defined by one or more Property Set Definitions. Property Sets establish the link between objects and the Schedules that report them. A Property Set Definition determines how a Property Set will be applied: object-based or style-based, what properties it contains and how they are configured. Analyzing the Property Set Definitions of the sample Content provided in ADT is a good way to begin understanding how they work.

EXPLORE A PROPERTY SET DEFINITION

1. Return to the *A-SH03* drawing (choose it from the Window menu, press CTRL + TAB or double click it on Project Navigator).

2. From the Format menu, choose **Style Manager**, expand the *Documentation Objects* category, and then select *Property Set Definitions*.

3. In the Style Manager, right-click **GeoObjects** from the list on the right and choose **Edit**.

 TIP you can also double-click on the Style name in Style Manager to edit it.

4. Click the General tab.

The text in the Description field reads: "Object-based geographic schedule properties for all objects." This Property Set is defined for use in tracking the geographic location of objects within the project. This is a good example of how Property Sets and Schedules can be used as an aid in facilities management.

5. Click the Applies To tab.

A Property Set can apply to either individual objects or object styles and definitions, but not to both at the same time. At the top of the Applies To tab, note that this particular Property Set applies to objects (see Figure 15–20).

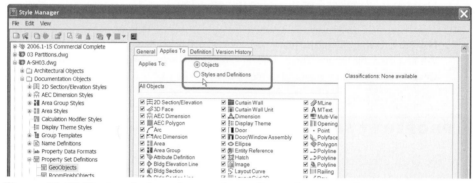

Figure 15–20 *Viewing whether a Property Set applies to objects or styles*

6. Click the radio button next to Styles and Definitions and notice the shift in the list below.

When "Objects" is chosen, the list shows all of the object types available in ADT (including all of the standard AutoCAD entities). When "Styles and Definitions" is chosen, the list shows all of the styles and definitions available in ADT (this list does *not* include AutoCAD styles). The buttons below each list, Select All and Clear All, are provided to allow easier selection within the lists. Because geographic location is not a property unique to any particular object or style, the *GeoObjects* Property Set is object-based and applies to all objects.

7. Click the Objects radio button to select it.

8. Click the Select All button to be sure all entities are checked.

9. Click the Definition tab (see Figure 15–21).

Each individual property of the Property Set is configured on this tab.

A Property Set can have one or many properties. The only limit is practicality. All properties in this list are user defined. You can tell this by reading the entry in the Automatic column (third column from the left). An entry of No means the property is not automatic and therefore a manual property. An entry of Yes means the property is an automatic property. There are also icons in the left column next to each entry like those pictured above in the "Identifying Automatic and Manual Properties" topic. When a property is manual, you will be able to set its "name," "description," "type," "default," and "format." When it is automatic, you will be able to set the "name," "description," and "format," as well as establish to which object property it should link. Refer to Figure 15–21, and review the following terms:

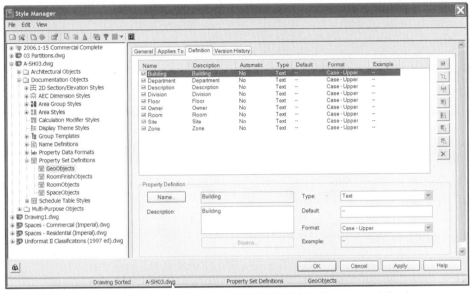

Figure 15–21 *The Definition tab of the Property Set Definition Properties dialog box*

▶ **Name**—Should be descriptive, short, and *not* include spaces.

▶ **Description**—A longer more descriptive version of the name used to convey the intention of the property. Spaces can be used in descriptions.

▶ **Format**—References the list of available Property Set Data Formats to convert raw data to the desired presentation format (such as Feet and Inches, Case - Upper).

▶ **Type**—(Manual properties only) sets the kind of data the property represents. Types include auto increment, real numbers, integers, text, and true/false.

▶ **Default**—(Manual properties only) an initial value input automatically in the Property Data. Default is typically two dashes (–), which is recommended to facilitate Table Cell editing. It can be left blank, but if no value is assigned, the user will be unable to use Edit Table Cell because there will be no way to select the "existing" value to edit.

For Manual properties, there are several types, as mentioned above. Auto Increment will begin at whatever value you designate and step sequentially—there are both numeric and character options. Real Numbers can be any numeric value with any quantity of decimal places. Integers allow whole numbers only (no decimals). Text is used for any character or numeric values that do not fit into one of the other categories. Any value can be input in a text field. True/False is a binary "on/off" type value. Use this for any property that has only two possibilities. A Property Set Data Format can be designed that includes any two values desired, such as Yes/No or In/Out.

CAD Manager Note: Property names should never include spaces if you want to create a custom tag to accompany your Schedule. This is because AutoCAD attributes are used to define the text fields within the tags, and AutoCAD attribute names do not support spaces. The property name is used for internal configuration purposes, so not using spaces shouldn't be a problem. The description however can contain spaces and should be written clearly in language that makes each property's intention clear to the everyday user.

The *GeoObjects* Property Set does not offer much variety. All of the properties are manual (non-automatic) text values.

10. Click Cancel to exit the Property Set Definition Properties – GeoObjects dialog box.

11. On the left side in the tree, expand the *03 Partitions* file, then the *Documentation Object* category and then the *Property Set Definitions* category. Finally, click on **FrameStyles.**

 NOTE if you closed *03 Partitions* above, click OK to close Style Manager, re-open *03 Partitions* on the Project Navigator, and then return to Style Manager to continue.

12. Click the Applies To tab.

This Property Set applies to Styles and Definitions, and it is much more focused than the previous one. It applies only to Door, Door\Window Assembly and Window Styles.

13. Click the Definition tab.

Study the list of properties. Notice that there are some automatic and some manual properties.

14. Select the FrameWidth property in the list.

This is an automatic property, as you can see from the lightning bolt icon and the Yes in the Automatic column. Notice that Type, Default, and Example all become unavailable. These settings are controlled automatically as well.

15. Click the Edit Source button at the bottom (see Figure 15–22).

In the Automatic Property Source dialog box, notice the check mark in the box next to Frame Width for Door and Window. Door\Window Assembly has nothing checked. Take a moment to review the complete list of properties.

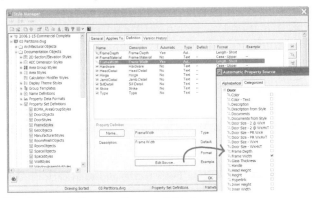

Figure 15–22 *The Automatic Property Source dialog box*

16. Click OK to dismiss the Automatic Property Source dialog box.

17. Continue to explore the various properties and their settings. When you are satisfied, click Cancel to return to the drawing.

SCHEDULE TABLE STYLES

Like all style-based ADT objects, the format and configuration of the Schedule object is controlled by a Schedule Table Style. The function of Schedule Table Styles can be broken down into two major functions: configuring the data content of the Schedule, and graphically formatting the look of the Schedule Table itself. Let's begin our exploration of the Schedule Table Style with its graphic formatting.

SCHEDULE TABLE STYLE GRAPHIC FORMATTING

Of the many tabs in the Schedule Table Style Properties dialog box, Default Format, Layout, and Display Properties control the format and visual characteristics of the Schedule. The Default Format tab is used to establish the text defaults for the entire Schedule Table style. On the Layout tab, you can assign text override formatting for the title and headers of the Schedule. Finally, Display Properties for Schedules is the same as it is for other ADT objects—here you control the Layer, Color, Linetype, Lineweight and Plot Style parameters of all Schedule components.

EXAMINE SCHEDULE TABLE TEXT FORMATTING

The Schedule Table style allows for a different Text style for each major Schedule component, including the main title, headings, and the data entries. Typical Auto-CAD Text Styles are used within the Schedule Table Style to determine text formatting (font, width factor, etc.) of a Schedule's components. Because they have not included a means to edit Text Style settings from within the Schedule Table Style dialog box itself, it is useful to remember to set up any Text Styles needed prior to editing the Schedule Table Style.

1. On the Project Navigator palette, on the Views tab, double click *A-SH03* to open it.

 NOTE If you left this file open above, then this action will simply make that file active.

2. From the Format menu, choose **Text Style** and then click on the Style Name list.

Notice that there are three styles for Schedules included in the list: *Schedule-Data*, *Schedule-Headers* and *Schedule-Title*. These Text styles have been set up for use in the sample Schedule Table Styles provided with the software.

3. Choose the **Schedule-Data** Text Style from the list.

   Study the settings and look at the preview.

4. Repeat for the other two styles as well.

As you can see, these styles are well suited to their tasks. For instance the Schedule-Title Style uses a bold font, while the Schedule-Data Style does not. If you wish, you can change the settings; otherwise, click Cancel to return to the drawing.

EXPLORE THE SCHEDULE TABLE STYLE DEFAULT FORMAT

5. Select the Schedule object in the drawing, right-click, and choose **Edit Schedule Table Style**.

6. Click the Default Format tab (see Figure 15–23).

This tab establishes the basic text parameters and other formatting considerations of the Schedule. Use this tab to set the way you wish the rows of data to be displayed. Titles and headers are formatted on the Layout tab (see below).

Figure 15–23 *The Default Format tab of the Schedule Table Properties dialog box*

▶ **Style**—Allows the choice of any previously defined Text style. This Text style is used for the entire Schedule unless overrides are attached on the Layout tab.

▶ **Alignment**—Choose from all the standard text alignments, such as left or center.

▶ **Height**—Sets the height of the text throughout the Schedule.

 NOTE Do not use scale factors; the Schedule Table object is already scaled using the value from Drawing Setup dialog box (also on the Drawing Status Bar). However, it is important to note that the Annotation Plot Size value (from Drawing Setup dialog box) does not have any effect on Schedules.

▶ **Gap**—Sets the space around the text within each cell. This is applied to all sides. For example, if the text Height is 1/8" [3] and the Gap is 1/16" [1.5], then the total height of the cell will be 1/4" [6].

▶ **Rotation**—Choose from either horizontal or vertical. Vertical is sometimes used for column headers, but rarely on the Default Format tab.

▶ **Matrix Symbol**—When using Matrix columns, (such as in a residential finish Schedule), a choice of symbols is available (see below).

▶ **Use for True/False**—Some Property Set Data return a true/false value. If this box is checked, a symbol from the matrix symbol list will be used for true values; the cell will be blank when the value is false.

▶ **Fixed Width**—This value forces the width of cells to a set number of units. Text will wrap to this width if the value is too long to fit. This will enlarge the height of the cell. If the width value remains 0, then cells will widen as the value within them grows. The final width of a cell will be determined by the widest entry in a particular column.

 TIP Be careful when assigning a fixed width (other than 0). Although the data within the cell will wrap, long words might run over the cell borders and bleed into the next cell. This is because the word wrapping does not have the ability to hyphenate words. There is no way to prevent this other than to use a variable width (width=0).

OVERRIDE FORMATTING

7. Click the Layout tab.

   The title that appears at the top of the Schedule is entered here.

8. Change the table Title to **SPACE ALLOCATION SCHEDULE** (see Figure 15–24).

 NOTE The title will appear as you type it, uppercase or lowercase. A Property Data Format is not used for titles or headers.

Figure 15–24 *A specific title can be assigned to the Schedule Style on the Layout tab*

The three buttons below the title field on the Layout tab can be used to assign text overrides to the title and the headers of the Schedule.

9. Click the Override Cell Format button next to Title.

   The fields in this dialog box match the ones those on the Default Format tab. Override values will appear in red to help distinguish them from defaults.

   Notice that the Style, Height, and Gap already have overrides applied. The effect is to make the title larger and bolder (see Figure 15–25).

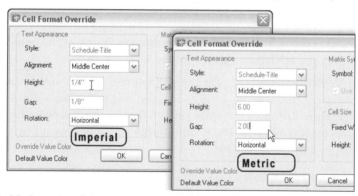

Figure 15–25 *Overriding Schedule title cell format properties*

10. Click OK to return to the Layout tab.

CONFIGURE BORDERS, COLORS, AND LINEWEIGHTS

Every component of the Schedule Table can be modified to achieve the exact graphic display and printed output desired. The Display Properties tab gives access to settings such as Layer, Color, and Lineweight for each component. The settings established with the default out of the box content are well conceived. Regardless, let's have a look to see if there is something you wish to change.

11. Click the Display Properties tab.

> This reveals the standard ADT Display Properties tab. Notice that Schedule Tables have just one Display Representation: General. This is appropriate considering that Schedules are typically drawn the same regardless of the drawing on which they are displayed.

CAD Manager Note: Typically, you will wish to enforce a single office-wide standard regarding the graphical display of Schedules. In this case, the goal is best accomplished by assigning all Schedule Table Display settings at the Drawing Default level. Editing at the style or object level would work against uniform consistency but may be appropriate in specialized scenarios, or on a per-project basis.

12. Click the Edit Display Properties button.

Several components are shown here. As you can see, each piece of the Schedule Table can be configured individually.

13. Select the Title Row Line component.

14. Change its Color, Lineweight, and Plot Style to match the Outer Frame component (see Figure 15–26).

Display Component	Visible	By Ma...	Layer	Color	Linetype	Linewei...	Lt Scale	Plot Style
Outer Frame			0	■ 234	ByBlock	0.70 mm	1.0000	Extra Wide
Title Row Line			0	■ 234	ByBlock	0.70 mm	1.0000	Extra Wide
Header Row Line			0	■ 222	ByBlock	0.25 mm	1.0000	Medium

Figure 15–26 *Manipulating the Display Props of Schedule objects*

15. Review the remaining settings and make any changes as desired.

16. Click OK twice to return to the drawing. If the drawing is not updated on its own, choose **Regen All** from the View menu.

> Note the changes to the title and the row line under the title.

17. To see the lineweights displayed, click the Lineweight (LWT) toggle button at the bottom of the screen on the Application Status Bar (see Figure 15–27).

Lineweight display in model space is relative to screen size and not truly indicative of final plotted appearance. Paper space layouts are much better suited to accurate on-screen display of lineweights. (Refer to Chapter 18 for information on using layouts.)

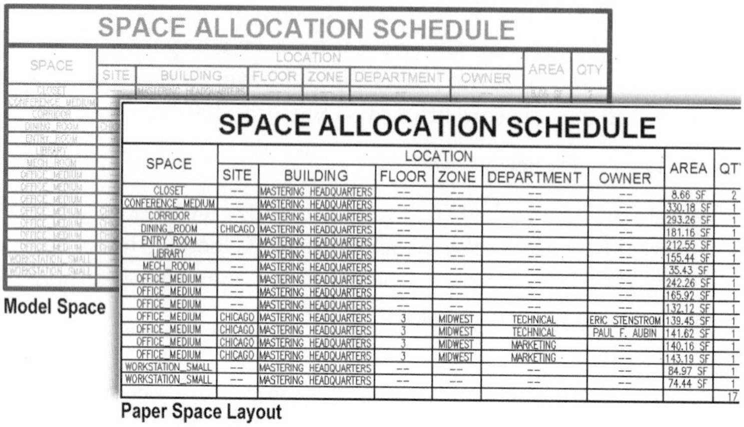

Figure 15–27 *Comparing lineweight display in model space and paper space*

CAD Manager Note: The Color, Lineweight and Plot Style were all assigned in the preceding exercise. Some firms use color; others use the Lineweight property, and still others use Plot Style to assign Lineweights for plotting. Use whichever setting is appropriate for your firm's standards in real practice.

SCHEDULE TABLE STYLE DATA CONTENT

The tabs covered so far in the Schedule Table Style relate to a Schedule's graphical display. The remaining tabs: Applies To, Columns, and Sorting, establish what data the Schedule contains and how it is presented. Let's begin with the Applies To tab.

DETERMINE WHICH OBJECT TYPES APPEAR IN THE SCHEDULE TABLE

1. Select and right-click the Schedule object, and choose **Edit Schedule Table Style**.

2. Click on the Applies To tab.

The Applies To tab establishes the link to a particular type of object or objects. This works in the same way that it does in the Property Set Definition. By clicking the same object type in both the Property Set Definition and the Schedule Table style, you bind the two together. You can check more than one item in Applies To. In fact, when creating a new Property Set Definition or Schedule style, the Applies To defaults to *All Objects*. In order for a Property Set to be available to the Schedule Table style, the Property Set Definition must apply to at least all of the items that the Schedule style applies to. The Property Set Definition can apply to more than the Schedule, but not vice versa.

Notice that this Schedule has a check mark in the Space entry only. Therefore, any Property Set Definition that applies to at least Spaces will be available on the Columns tab. Again, Property Sets can apply to more object types than the Schedule style, the way *GeoObjects* does for instance, but they must *at least* apply to Spaces to be included in this particular Schedule style.

DETERMINE WHICH PROPERTIES WILL APPEAR IN THE SCHEDULE TABLE

The Columns tab establishes the columnar structure of the Schedule Table, and any headers. Each column contains the data of a single Property from a single Property Set. Each column can reference Properties from a different Property Set or all columns can reference Properties from the same Property Set. Any combination is also possible.

3. Click the Columns tab.

4. Scroll horizontally and make note of the various columns.

If you can, position this dialog box so that the Schedule Table in the drawing is visible beyond. Then compare the columns listed in the Columns tab with those actually in the drawing. You can also resize the Schedule Table Style Properties dialog box. Modifying any of these columns will have a direct effect on the structure of the Schedule in the drawing (see Figure 15–28).

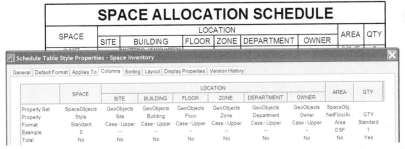

Figure 15–28 *Position the Schedule Table Style Properties dialog box so that the Schedule in the drawing is also visible*

5. Click the ZONE column (it will be highlighted), and then click the Modify button at the bottom of the dialog box (see Figure 15–29).

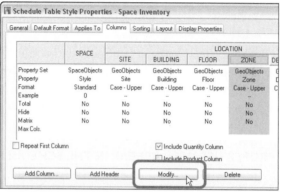

Figure 15–29 *Selecting columns to modify in the Schedule Table Style Properties dialog box*

Each column of the Schedule refers to a single property of a single Property Set. The first two items in the Modify Column dialog box show exactly which Property Set and specific Property this column accesses. The Schedule Table style can change the way a Property is formatted in the Schedule. The Heading box allows you to enter any column heading desired. This is the actual text that will appear at the top of the column on the Schedule itself. This does not change the Property Set Definition in any way.

6. Change the heading from ZONE to **REGION** and then click OK.

The Data Format dropdown list shows all of the Property Data Formats currently available in the drawing. Its value defaults to whatever was set in the Property Set Definition. Notice that this column is assigned to *Case – Upper*. This means that regardless of how the data is input by the user, it will be formatted in upper case in the Schedule Table. Please note however, that this formatting does not apply to the titles and headers. These must be typed in uppercase if you wish them displayed as such.

OVERRIDE CELL FORMATTING

7. Select the Area column and then click the Modify button.

8. Click the Override Cell Format button.

This button gives access to the same settings outlined above in the Default Format tab. Any of the values discussed previously can be edited here. However, the overrides apply to this column only.

9. Change the Alignment to **Bottom Right**.

10. In the Fixed Width field, type **1 1/2″ [30]** and then click OK.

11. Back in the Modify Column dialog box, place a check mark in the Total check box (see Figure 15–30).

Figure 15–30 *Configuring the Area column for a total at the bottom*

12. Click OK twice more to return to the drawing.

Notice the change to the Zone and Area columns of the Schedule; Zone is now REGION and all of the values in the AREA column now line up properly at the right side. There is a total area at the bottom of the AREA column and there is more room to the left of the column. If your Schedule now overlaps the plan, move it over to the left.

MOVE, DELETE, AND ADD COLUMNS

13. Right-click the Schedule object, choose **Edit Schedule Table Style** again, and return to the Columns tab.

14. Click the REGION heading and drag it on top of the FLOOR heading (see Figure 15–31).

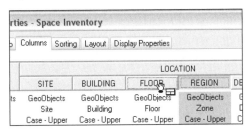

Figure 15–31 *Drag columns to change their order*

15. Click OK to view the change.

Notice the new order of the REGION and FLOOR columns.

16. Right-click the Schedule object, choose **Edit Schedule Table Style** again, and return to the Columns tab.

17. Select the BUILDING column and then click the Delete button at the bottom of the window.

18. Click OK when asked to confirm and then click OK again to return to the drawing.

Notice the removal of the BUILDING column.

19. Right-click the Schedule object, choose **Edit Schedule Table Style** again, and return to the Columns tab.

20. Click the Add Column button at the bottom of the dialog box.

In the Add Columns dialog box, available Properties are listed at the left. Unavailable items (grayed out) have already been added to the Schedule. Column properties for the selected item can be edited on the right before clicking OK. Use the area at the bottom to decide where in the existing column order to place the new column (or if you prefer, you can drag it after you place it).

21. Select the Categorized tab and scroll through the list of properties.

22. Locate the **RoomObjects** Property Set and select **Name** (Room Name).

23. In the Heading box type **ROOM NAME** (remember uppercase).

24. At the bottom of the dialog box, choose **Insert Before** and then choose **SpaceObjects:Style** from the Column list (see Figure 15–32).

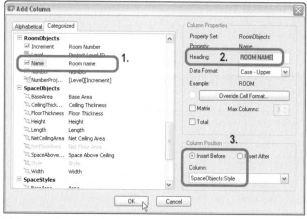

Figure 15–32 *Adding a column to the Schedule style*

25. Click OK and note the new column at the left.

26. Click OK again to view the change in the drawing.

There is now a Room Name column at the beginning of the Schedule. The Room Names have already been filled in based on the Property Sets we added in Chapter 14 (see the "Adding Room Tags" heading in Chapter 14 for more information).

27. Modify the alignment of some of the text columns to make them left justified (review the "Override Cell Formatting" topic above for assistance if required).

WORK WITH A MATRIX COLUMN

In this sequence, we will change the format of the DEPARTMENT column to a "Matrix" column, which offers a more graphical display of the information in that column. Before we can gain much value form this change, we need to finish editing the values in this column.

28. Using the procedures outlined above, fill in the values in the DEPARTMENT column.

You can edit the Spaces directly in *03 Partitions* on the Properties palette, or close *03 Partitions* and use Edit Referenced Property Set Data command or Edit Table Cell from the *A-SH03* file. Remember, you cannot use these tow methods if *03 Partitions* is open. For the offices along the bottom and left side, and the two Workstation Spaces, use a mix of Department names like: **Marketing, Sales** and **Technical**. Make the Reception, Conference Room and Corridor use the name: **Public**. Make the Department name for the Break Room, Work Room and Utility Room: **Common**. Make the Closets the same as the Spaces to which they are attached (see Figure 15–33).

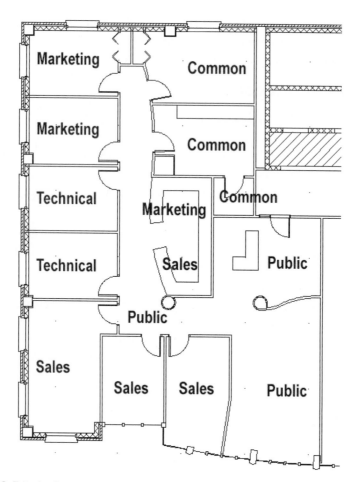

Figure 15–33 *Edit the Department names of all Spaces*

29. Right-click the Schedule object, choose **Edit Schedule Table Style** again, and return to the Columns tab.

30. Select the DEPARTMENT column and then click the Modify button.

31. Put a check mark in the Matrix check box and change the "Max Columns" to **6** and then click OK (see left side of Figure 15–34).

32. Click the Default Format tab and in the Matrix Symbol area, open the list of symbols and choose **Dot** (see right side of Figure 15–34).

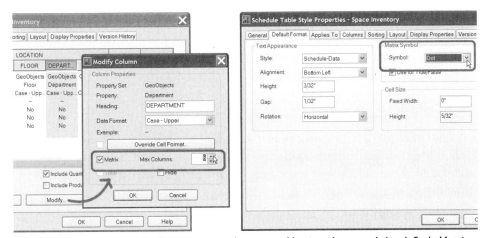

Figure 15–34 *Changing the Department column to a Matrix column and the default Matrix symbol to a Dot*

33. Click the Layout tab.

34. Next to Matrix Headers, click the button labeled Override Cell Format (don't click the check box).

35. From the Style list, choose **Schedule-Header**.

36. From the Rotation list, choose **Vertical** and then click OK (see Figure 15–35).

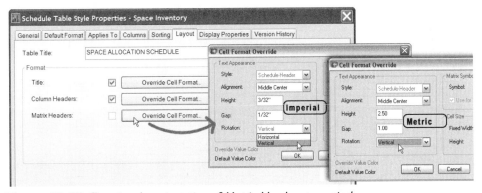

Figure 15–35 *Changing the orientation of Matrix Headers to vertical*

37. Click OK again to view the results.

Several changes have occurred in the Schedule Table. The individual department names are running vertically as sub-headings beneath the DEPARTMENT column heading. A dot appears in the Schedule as each department occurs. Matrix Headers only occur for Properties actually in use. To test this out, select the XREF, right-click and choose **Open XREF,** (this opens *03 Partitions*) then select any Space object in the drawing, and choose Standard as the Style on the Design tab of Properties palette. Save and Close the *03 Partitions* file. Update the Schedule Table if necessary. Notice the addition of a new Matrix column named Standard. Repeat the steps to reverse the change.

SORT A SCHEDULE TABLE

There is one more tab left to consider in the Schedule Style. This is the Sorting tab. Here we can designate the column(s) from which to sort the data in the Schedule Table. You can choose ascending or descending, and you can sort by more than one column if you wish.

38. In the *A-SH03* View file, select the Schedule, right-click, and choose **Edit Schedule Table Style**.

39. Click the Sorting tab.

40. Select SpaceObjects:Style and click Remove.

41. Click the Add button.

42. Choose RoomObjects:Name and then click OK twice to return to the drawing and view the change.

    The Schedule now sorts alphabetically (ascending) by Room Name (see Figure 15–36).

SPACE ALLOCATION SCHEDULE

ROOM NAME	SPACE		LOCATION			DEPARTMENT						OWNER	AREA	QTY
			SITE	REGION	FLOOR	COMMON	MARKETING	PUBLIC	SALES	TECHNICAL				
BREAK_ROOM	DINING_ROOM		CHICAGO	--	--	●						--	181.16 SF	1
CLOSET	CLOSET		--	--	--		●					--	8.66 SF	1
CLOSET	CLOSET		--	--	--	●						--	8.66 SF	1
CONFERENCE_ROOM	CONFERENCE_MEDIUM		--	--	--			●				--	330.18 SF	1
CORRIDOR	CORRIDOR		--	--	--			●				--	293.26 SF	1
OFFICE	OFFICE_MEDIUM		CHICAGO	MIDWEST	3		●					--	140.16 SF	1
OFFICE	OFFICE_MEDIUM		CHICAGO	MIDWEST	3		●					--	143.19 SF	1
OFFICE	OFFICE_MEDIUM		--	--	--				●			--	242.26 SF	1
OFFICE	OFFICE_MEDIUM		CHICAGO	MIDWEST	3						●	ERIC STENSTROM	139.45 SF	1
OFFICE	OFFICE_MEDIUM		CHICAGO	MIDWEST	3						●	PAUL F. AUBIN	141.62 SF	1
OFFICE	OFFICE_MEDIUM		--	--	--				●			--	165.92 SF	1
OFFICE	OFFICE_MEDIUM		--	--	--				●			--	132.12 SF	1
RECEPTION	ENTRY_ROOM		--	--	--			●				--	212.55 SF	1
UTILITY_ROOM	MECH_ROOM		--	--	--	●						--	35.43 SF	1
WORK_ROOM	LIBRARY		--	--	--	●						--	155.44 SF	1
WORKSTATION	WORKSTATION_SMALL		--	--	--		●					--	84.97 SF	1
WORKSTATION	WORKSTATION_SMALL		--	--	--					●		--	74.44 SF	1
													2489.47 SF	17

Figure 15–36 *The completed Space Allocation Schedule*

The Schedule customization is now complete. Edit the Property Set Data of the remaining Spaces to complete the Schedule. Tweak any other settings you wish to fine-tune the Schedule style to meet your own standards.

43. Save the file.

SPECIAL COLUMN TYPES AND FEATURES

There are some additional items on the Columns tab of the Schedule Table Style Properties dialog box worthy of mention.

- **Quantity**—Identical objects can have their line items in the Schedule grouped together and expressed as a quantity rather than being listed separately. The Schedule used in this tutorial uses a quantity column. In order to be quantified, duplicate line items must be *identical* in all columns. Include a quantity column by putting a check mark in the Include Quantity Column check box.

- **Product**—When a quantity column is used, a product column can also be added by clicking the Include Product Column check box. This is useful for costing columns and other similar columns. A product column will multiply the quantity Column by one other column such as a cost column and display the product of the two. If you want more robust calculations, explore the Formula Property type. With a Formula Property you can include any valid VBScript expression and have the values calculated and included in a Schedule Table.

CAD Manager Note: If you wish to use this Schedule Table style in other drawings, edit it one more time and rename it to something unique and descriptive. Save the file. Create a new tool palette or use an existing one and make it active. Drag this Schedule Table from the drawing onto a tool palette. It will make a new tool from it. However, remember that this tool will reference the current drawing, so save before you build the tool and consider copying the style to another more central file first. Use the Style Manager to copy and paste this style to that drawing, then drag from that drawing to your tool palette.

ADD A MODEL SPACE VIEW

In the next topic, we will do an advanced tutorial adding a complete project-based Door Schedule. In that exercise, we will create a new Sheet file for our Project in which to insert the Door Schedule. We will then add the Schedule created here to that Sheet. In preparation for that, let's make a Model Space View that defines only the area occupied by our Space Allocation Schedule to be included on the Sheet.

44. On the Project Navigator, right-click on *A-SH03* and choose **New Model Space View**.

45. In the Add Model Space View dialog, type: **Third Floor Space Allocation Schedule** for the Name and accept the remaining defaults.

46. Click the Define View Window icon on the right and in the drawing, click two points defining a rectangular region just a bit larger than the Schedule Table.

47. Click OK to complete the Model Space View.

We will drag this View onto the *A-601 Schedules* Sheet below. For now, that completes our work in this file.

48. Save and Close the file.

BUILDING A PROJECT-BASED DOOR SCHEDULE (ADVANCED)

The remainder of this chapter gets more advanced. If you are not interested in linking Door numbers to Room numbers, this topic can be safely skipped. In this topic, we will explore a project-based Door Schedule of the entire building. There are many exciting features of the Schedule Table tool set that we will explore. For example, in the exercise that follows, we will add Door Tags to the First Floor Plan and the Third Floor Plan View files, build a new Sheet file for Schedule Tables and add a complete Door Schedule of the entire Commercial Project. This process is involved but very beneficial.

Following the logic used above, we could create our new Door Schedule within its own View file. In order to schedule the entire project, this would need to be a View much like our Composite Model and include all of the Constructs that contained Doors. There is nothing wrong with that approach, and you are free to follow it if you wish. However, there is an even better approach that takes advantage of an exciting feature of Schedule Tables: the ability to schedule a remote drawing. Let's explore this feature next.

CREATE A NEW SHEET AND SCHEDULE A REMOTE FILE

A Schedule Table can be set to schedule a remote drawing file. In this case, the Schedule becomes a one-way report with a live link to the data contained in the drawing that it schedules. When you use this feature, it is unnecessary to create a View file for Schedules and then reference that View file to a Sheet. Instead, you simply add the Schedule Table directly to a Sheet file in paper space, and link it to your composite building model (or any other file that you wish to schedule).

While this approach might seem contradictory to the definition of Sheets presented in Chapter 5, actually it is not. The fact that this Schedule, once added and linked to a remote file, is a one-way report, means that it will require no further intervention from a user. Therefore, it is fine (and preferable) to have it directly on the Sheet. Every time that Sheet is opened, it will automatically gather the latest

Schedule Data, and be "ready to print" – exactly what we would expect from a Sheet file.

1. On the Project Navigator, click the Sheets tab.

2. Right-click on *Architectural – Schedules and Diagrams* Sub Set and choose **New > Sheet**.

3. Type: **A-601** for the Number and type: **Schedules** for the Sheet title, and then click OK (see Figure 15–37).

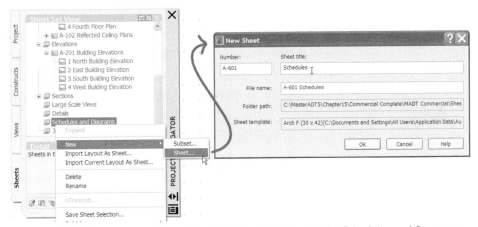

Figure 15–37 *Create a new Sheet file for Project Schedules in the Schedules and Diagrams Subset*

4. Double-click the new *A-601 Schedules* to open it.

   New Sheet files open into a single paper space layout named the same as the file, which contains a title block.

5. On the Scheduling tool palette, click the **Door Schedule Project Based** tool.

   If you do not see this palette or tool, right-click the Tool Palettes title bar and choose **Document** (to load the Documentation Tool Palette Group) and then click the Scheduling tab.

6. At the "Select objects or ENTER to schedule external drawing" prompt, press ENTER.

   A rectangle will appear at your cursor and the command prompt will request the "Upper left corner of table."

7. Click a point near the top left corner of the sheet and then press ENTER to accept the default size.

   An empty Door Schedule will appear showing only the title and headers.

8. Select this Schedule and on the Properties palette, click the Design tab, and scroll down to the Advanced grouping.

9. In the External Source grouping, change Schedule external drawing to **Yes**.

An additional field labeled "External drawing" will appear.

10. Open the menu in External drawing (click on *None*).

A list of all of the View files in the Views folder will appear. If the file that you wish to schedule does not appear on the list, you can also choose Browse at the bottom of the list and choose any drawing.

We are going to choose the Composite Building Model to schedule. This is because it contains all of the Constructs for the entire project. Therefore, if we schedule it, we will be sure to include all of the Doors.

11. From the list, choose the Composite Model file: ..\Views\A-CM00.dwg (see Figure 15–38).

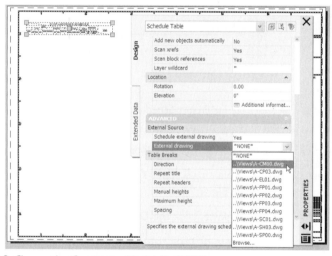

Figure 15–38 *Choose the Composite Model (A-CM00)*

12. With the Schedule still selected, right-click and choose **Update Schedule Table**.

When the Table completes the update, there will now be several rows of data. The Schedule Table has extracted information about every Door in the project! If you select this Schedule Table again and check its properties, you will notice that they are all grayed out. This means that we will not need to do anything more to this Schedule Table. Whenever the Doors in the project change, those changes will be read directly by this Schedule Table each time the *A-601 Schedules* Sheet file is opened—usually for printing.

MANIPULATE SCHEDULE TABLE SIZE AND SCALE

13. Select the Schedule to reveal its grips.

14. Hover over any of the four grips at the corners.

Notice the dynamic dimensions that appear. The overall size of the Schedule Table is revealed as well as the scale. As you can see, since we added this Schedule Table in paper space, the scale is set to **1**.

15. Click the small triangular grip at the bottom edge of the Schedule.

16. Drag it up and click again.

Notice that this grip changes the Maximum page height. This particular Schedule is not very tall, but on very large projects where there are hundreds of Doors, the Schedule can get quite long. Use this grip to "wrap" the Schedule on the Sheet. Additional parameters for this feature appear on the Properties palette.

That is all that we need to do to this Schedule Table for now. However, if you zoom in on it and look at the data, you will see that we still have quite a bit of work in the project files. Our Doors do not have any object-based data.

17. Save and Close the file.

ADD TAGS TO XREFED DOORS

We are now ready to add Door Tags. As we saw above in the Space Allocation Schedule tutorial, adding tags does more than just insert a Multi-View Block. This is how we will also import all of the object-based Property Set Data that is required for our Door Schedule, attach it to the Doors, and anchor the tags to those Doors. For this sequence, we will use the tags on the Annotation tool palette.

18. On the Project Navigator, click the Views tab.

19. In the *Views* folder, double-click the *A-FP01* file to open it.

You will note that Space objects have been added to this file. They will be important to the process of tagging Doors. You will also note that the *Core* file XREF has been unloaded. We will reload it later in the tutorial.

There are two Door Tags on the Annotation tool palette. The Door Tag is a simple auto-incrementing Door Tag much like the Window Tag that we used in the last chapter. The Door Tag – Project Based is a more robust Tag that references the Room number of a neighboring Space object. In this way, we are able to link the Door numbers to the Room numbers.

20. Zoom in on the central lobby space (just below the elevators and above the curved Wall).

21. On the Scheduling tool palette, click the **Door Tag – Project Based** tool.

22. At the "Select object to tag" prompt, click one of the two Doors facing into this Space.

23. At the "Specify location of tag" prompt, click a point near the Door.

 TIP It will be easier to place the tag is you turn off OSNAP

Don't exit the command yet.

Notice that number appears as "?A." As you recall, the question mark indicates that the Property Set Data has not yet been added, yet we have just tagged the Door, which is supposed to attach the required Property Set Data. The problem is that this particular tag uses a more advanced Property Set. The Door Number property is a Formula property (see the definition above in the "Automatic and Manual Property Sets" topic). The formula in use here is a simple concatenation of the adjacent room number and a letter suffix. The question mark in this case is indicating that the Property Set is missing from the Space, not the Door.

The command line again reads "Select object to tag."

24. Tag the Door on the opposite side of the lobby, the three revolving Doors at the building entrance and the exit door near the Stair and then press ENTER to complete the command (see Figure 15–39).

Right-click and use the Multiple option if you wish.

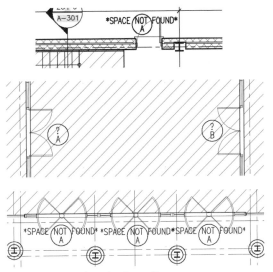

Figure 15–39 *Various errors appear in the Door Tags*

 NOTE Most likely an XREF alert will appear stating that your reference files have changed. For now, it is OK to ignore these. When you tag XREFs, it is actually adding the Property Set Data to the XREF files, and they have therefore changed. However, graphically there will be no change to the drawing, so an update at this time is not necessary.

The problem seems a bit worse on the three revolving Doors. Let's deal with each issue separately starting with the retail lobby Doors. We need to make two adjustments in order to correct the Door Tag numbers. The Door Tags use a Location Property (see the definition in the "Automatic and Manual Property Sets" heading above) to query the Room Number of the Space in which they swing. So first, we must attach the required Property Sets to the Spaces. The easiest way to accomplish this is to tag the Spaces.

25. On the Scheduling tool palette, click the **Room Tag – Project Based** tool.

26. Follow the prompts as before and tag each of the Spaces and then press ENTER to exit the command.

    Start with the main entrance lobby at the bottom of the plan (the one with the three revolving Doors), tag the retail lobby next, and then the two retail tenant spaces. Tagging them in this order will ensure that they are numbered as such.

27. Using any of the methods covered in this chapter; edit the Room Names of each of the four Spaces.

    The room tags should update with the new names.

EDIT THE NUMBER SUFFIX

When you finish with the room tags, the door tags in the center of the plan (in the retail lobby Space) should update immediately). If they do not, on the Standard toolbar, click the Regenerate Model icon and then press ENTER (or choose Regen Model from the View menu). The three revolving Doors will remain incorrect.

Even though the two retail lobby Doors did update, they will both have the same number (or specifically the same "Number Suffix").

28. Click on the retail lobby Space to select the *01 Partitions* XREF, right-click, and choose **Edit Referenced Property Set Data**.

29. At the "Select Objects" prompt, click the retail lobby Door on the right side of the plan and then press ENTER.

30. In the Edit Referenced Property Set Data dialog, change the Number Suffix to: **B** and then click OK (see Figure 15–40).

Figure 15–40 *Edit the Number Suffix of the second Door in the retail lobby*

ADJUST THE LOCATION GRIP

The three revolving Doors in the main entrance lobby still do not show a Room Number. They are also all assigned to Number Suffix "A" like the Doors above.

31. Click on any of the three revolving Doors, right-click and choose **Open XREF**.

    The *01 Shell and Core* Construct will open.

32. Click any one of the Revolving Doors.

In addition to the normal grips, Note the star-shaped grip with the curved tail attached to it. This is a Location Grip. It indicates into which space the Door belongs. ADT will search your model for a Space, Area or AEC Polygon beneath this grip. The first one that it finds will be reported as this Door's location. In this case there is no Space outside of the building. This is why we received the error message in the *A-FP01* file.

33. Click the location grip and drag it into the building (see Figure 15–41).

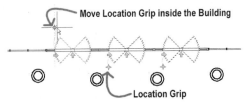

Figure 15–41 *Move the Location grip into the building*

Perform the same action on the exit door by the Stair. It will not update right away, but later when we reload the *Core* it will.

34. Repeat for all three revolving Doors and then Save and Close the *01 Shell and Core* file.

35. Back in the *A-FP01* file, reload the *01 Shell and Core* XREF by clicking the link in the balloon that appears.

The three revolving Door tags should update to the correct room number, but they still all contain the "A" suffix.

36. Repeat the process outlined in the "Edit the Number Suffix" heading above to change the Number Suffix of two of the revolving Doors to " **B**" and " **C**" respectively (see Figure 15–42).

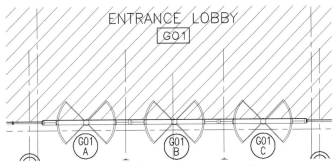

Figure 15–42 *Edit the Number Suffix of each of the revolving Doors*

PROPERTY SETS AND ELEMENT FILES

The way that we have structured our Commercial Project, the *Core* file is a typical configuration that repeats on four floors. There are five Doors in the Core Element file. However in the project, this is really 20 Doors since the Element repeats four times. If you returned to the *A-601 Schedules* Sheet file and made a count of the total number of Doors, you would see that the actual quantity of 20 Doors is being properly represented within the Schedule. However, in order to correctly label each of these Doors, we really need each of the five Doors in the Element to realize that they really represent four Doors each. To accomplish this, we must first add the Property Sets to the objects within the Element file, and then perform the normal steps to attach the actual project-based Properties in the various Constructs and Views. The first values in the Element file itself will become overridden by the ones we apply to the Constructs via the View file.

If you recall our discussion in the last chapter regarding View files, it was noted even though we tag in the View file, the data to which the Schedule tags refers actually lived within the Construct. The same is true in the case of the Elements. Even though Property Sets can and must be attached to the Doors within the Element file itself, ultimately it is not this data that is read into the tags and Schedules, but rather the data that lives within the Construct. Therefore, even though there are really only five Door objects, they are able to correctly represent a total of 20 actual project Doors. This occurs because Property Set Data overrides are attached to each of the Doors from the *Core* Element file within each floor's respective Construct files. These overrides are the data that is then passed to the schedule.

In order to make this work properly, when typical Element files are involved in objects you wish to schedule, you must follow a fairly strict process. First open the Element and attach the Property Sets to the object within. It is not important what values you assign to these Properties as each Construct will ultimately carry its own values; it only matters that the Property Sets be attached. Once this is done, the normal procedures covered throughout this chapter can then be employed.

37. On the Project Navigator palette, click the Constructs tab.

38. In the *Elements\Architectural* folder, double click *Core* to open it.

New in ADT All we need to do here is attach the required Property Sets. We have covered a few techniques to do this in this chapter so far. However, unless you 2006 know exactly which Property Sets are needed, and are certain that they are already resident in the file, or you know from where to import them, the easiest way to add the Property Sets using the tag tools. Naturally we don't want to actually add the tags in this file, but we can use the tag tools to apply the Property Sets only.

39. On the Scheduling tool palette, right-click the **Room Tag – Project Based** tool and choose **Apply Tool Property Set Data to Objects**.

40. At the "Select objects to apply property set data to" prompt, click each of the Spaces and then press ENTER to exit the command.

Be sure to click the Spaces in the order you wish them to be numbered.

Notice that the first part of the Room Number currently reads "NA." This first part of the Room Number is a Project Property (see the definition in the "Automatic and Manual Property Sets" heading above). It is configured to read the level from the Project Database. Since we are currently working directly in the Element file, this value is not applicable (NA). Elements do not have a level designation. This value will be replaced with the correct level indication within each Construct.

Even though we stated above that the actual values of the Properties assigned to the Element file were not important, the values assigned here will be used as a default within each Construct. Therefore, if you do not want to type "Elevator Lobby" for the Room Name in each of the four Floor Plan Constructs, you can input that value here and it will be used as the default in each Construct when the Construct's Property Set Data override is applied.

41. Select the lobby Space next to the elevators and on the Properties palette, click the Extended Data tab. Type: **Elevator Lobby** for the Room Name.

42. Repeat the process for each Space giving each one a Room Name.

43. On the Scheduling tool palette, right-click the ***Door Tag – Project Based*** tool.

44. Follow the prompts as before and click each of the Doors and then press ENTER to exit the command.

45. Edit the Number Suffix of the three Doors of the left Corridor Wall making them **NA02A**, **NA02B** and **NA02C**.

    Since we are working directly in the file with the Doors, simply select the Door and edit the Property Sets on the Extended Data tab of the Properties palette.

46. Move the Location Grip for the Stair Door into the Corridor Space (see Figure 15–43).

 NOTE Figure 15–43 shows tags inserted for clarity and for suggested Room Names. If you prefer, you can add tags as well and then delete them after you are finished editing all of the Property Sets. Deleting the tags does not delete the attached Property Sets.

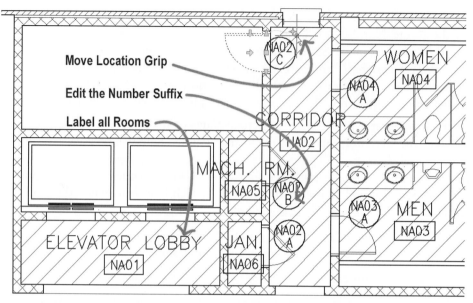

Figure 15–43 *Add tags to the Element file as a means of attaching all required Property Sets*

We now have all of the Property Sets needed attached to the various objects within the *Core* Element file.

47. Save and Close the *Core* Element file.

TAG THE FIRST FLOOR CORE

The First Floor Plan file (*A-FP01*) should still be open on screen. If it is not, open it now from the Views tab of the Project Navigator palette. Now that Property Sets have been attached to the Core Element file, let's attach the required over-rides to those Properties for the First Floor Plan.

48. On the Drawing Status Bar, right-click the XREF Quick Pick icon and choose **Reload Xrefs**.

The previously unloaded *Core* file should re-appear.

49. On the Scheduling tool palette, click the **Room Tag – Project Based** tool.

50. Follow the prompts as before and tag each of the Spaces in the Core ending with the Space in the Stair Tower and then press ENTER to exit the command.

Use the Multiple option if you wish and be sure to click the Spaces in the order you wish them to be numbered.

Notice that the Names and Numbers assigned within the Element file have been assigned here as well, however, all of the Numbers correctly begin with the prefix "G" (for Ground Level) rather than "NA" as was the case in the Element file. If you are not sure where the "G" comes from, review the "Establish the Project Framework" topic in Chapter 5. It is the Level ID, and can be viewed or edited on the Project tab of the Project Navigator palette. Click the Edit Levels icon.

51. Using the Edit Referenced Property Set Data command, change the Increment value of the Stair Tower to: **7**. Change the Name to: **Stair**.

52. On the Scheduling tool palette, click the Door Tag – Project Based tool.

53. Follow the prompts as before and tag each of the Doors and then press ENTER to exit the command.

54. Fine tune the position of any tags as desired and edit the suffix of the exit door to **D** (see Figure 15–44).

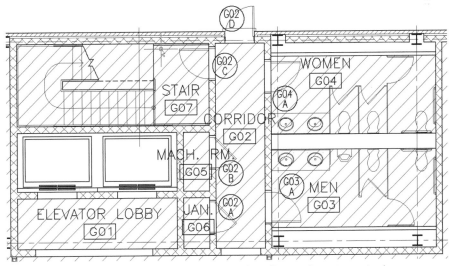

Figure 15–44 *Add tags to the First Floor Core – Virtually no editing is required*

Notice how it was not necessary to edit the door tags this time. This is because they read their default values from those in the Element file. If we needed to edit them further on a particular floor, we could do so, but the advantage of adding good default values to the Element file first is that we often will not need to change them later.

55. Save and Close the *A-FP01* file.

TAG THE UPPER FLOORS

Compared to this floor, the upper floors will be easy.

56. On the Project Navigator, on the Views tab, double-click *A-FP02* to open it.

57. Repeat the process just followed in the "Tag the First Floor Core" topic.

58. Save and Close the *A-FP02* file.

 NOTE The Stair Tower is a Spanning Construct. It contains a single Space object in-serted at the lowest Stair with a height equal to four stories plus the roof—a very tall Space. As we have said, Property Set Data lives on the Construct. Therefore, you will notice that this particular Space maintains the prefix of the First Floor: "G."

59. On the Project Navigator, on the Views tab, double-click *A-FP03* to open it.

For the *Core*, use exactly the same process. For the Tenant Space, the room tags were added already in Chapter 14. Simply add the door tags.

60. Repeat the above process for the *Core*.

61. Tag the Doors starting with the entry Door to the Suite.

62. Right-click and choose Multiple, and tag the rest of the Doors in the tenant suite.

Notice how all of these Doors automatically read the room number of the adjacent Spaces. Also notice that the room numbers reference the Level from the Project database—all of the numbers on this floor begin with a "3." If you are unhappy with the default Space referenced by a particular tag, open the *03 Partitions* file on the Project Navigator, and edit the Location grip.

63. Save and Close the file.

64. Repeat once more for *A-FP04* and *A-FP05* (the Roof Plan).

65. Save and Close all files.

OPEN THE SCHEDULE SHEET AND VIEW THE RESULTS

We have finished numbering and editing all of the Doors in the Project. Let's check our progress in the Schedule Table and re-open$ A-601$ and see how things are shaping up.

66. On the Project navigator, click the Sheets tab and then double-click *A-601 Schedules* to open it.

Notice that all you have to do is open the file and the Schedule automatically up-dates. Much of the data has filled in nicely. There are still several rows of question marks, however. It turns out that the original criteria that we used to make the Schedule Table selection were too broad. This Schedule is reporting *all* Doors in our entire model. This currently includes all of the toilet stalls in the rest rooms as well as the actual Doors. This situation is easily resolved.

67. Select the Schedule and on the Properties palette (on the Design tab), within the Selection grouping, type ***Door** [***Door-G**] for the Layer wildcard and press ENTER (see Figure 15–45).

DOOR AND FRAME SCHEDULE

MARK	DOOR					LOUVER		FRAME						FIRE RATING LABEL	HARDWARE		NOTES
	SIZE			MATL	GLAZING			MATL	EL	DETAIL				SET NO	KEYSIDE RM NO	
	WD	HGT	THK			WD	HGT			HEAD	JAMB	SILL				
?	?	?	?	—	?	0"	0"	—	—	—	—	—	?	?	?	?
G01A	5'-8"	8'-8"	2"	—	—	0"	0"	—	—	—	—	—	—	—	—	—
G01B	5'-8"	8'-8"	2"	—	—	0"	0"	—	—	—	—	—	—	—	—	—
G01C	5'-8"	6'-8"	2"	—	—	0"	0"	—	—	—	—	—	—	—	—	—
G02A	5'-7"	7'-9 1/2"	1 3/4"	—	—	0"	0"	—	—	—	—	—				
G02B	5'-7"	7'-9 1/2"	1 3/4"	—	—	0"	0"	—	—							
G04A	3'-0"	7'-0"	1 3/4"	—	—	0"	0"	—	—							
G04B	3'-0"	7'-0"	1 3/4"	—	—	0"	0"	—	—							
G04C	3'-0"	8'-0"	1 3/4"	—	—	0"	0"	—	—							
G05A	3'-0"	8'-0"	1 3/4"	—	—	0"	0"	—	—							
G08A	3'-0"	7'-0"	1 3/4"	—	—	0"	0"	—	—							
202A	3'-0"	8'-0"	1 3/4"	—	—	0"	0"	—	—							
202A	3'-0"	7'-0"	1 3/4"	—	—	0"	0"	—	—							
202A	3'-0"	7'-0"	1 3/4"	—	—	0"	0"	—	—							
203A	3'-0"	8'-0"	1 3/4"	—	—	0"	0"	—	—							
204A	3'-0"	8'-0"	1 3/4"	—	—	0"	0"	—	—							
301A	3'-0"	6'-10"	1 3/4"	—	—	0"	0"	—	—							
302A	3'-0"	7'-0"	1 3/4"	—	—	0"	0"	—	—							
302B	3'-0"	7'-0"	1 3/4"	—	—	0"	0"	—	—							
302C	3'-0"	8'-0"	1 3/4"	—	—	0"	0"	—	—							
303A	3'-0"	7'-0"	1 3/4"	—	—	0"	0"	—	—							
304A	3'-0"	7'-0"	1 3/4"	—	—	0"	0"	—	—							
304A	3'-0"	8'-0"	1 3/4"	—	—	0"	0"	—	—							
305A	3'-0"	7'-0"	1 3/4"	—	—	0"	0"	—	—							
306A	3'-0"	7'-0"	1 3/4"	—	—	0"	0"	—	—							
307A	3'-0"	7'-0"	1 3/4"	—	—	0"	0"	—	—							
308A	3'-0"	7'-0"	1 3/4"	—	—	0"	0"	—	—							
309A	3'-0"	7'-0"	1 3/4"	—	—	0"	0"	—	—							
310A	4'-6"	7'-0"	1 3/4"	—	—	0"	0"	—	—							
310B	4'-6"	7'-0"	1 3/4"	—	—	0"	0"	—	—							
313A	3'-0"	7'-0"	1 3/4"	—	—	0"	0"	—	—							
313B	4'-6"	7'-0"	1 3/4"	—	—	0"	0"	—	—							
314A	3'-0"	7'-0"	1 3/4"	—	—	0"	0"	—	—							
314B	3'-0"	7'-0"	1 3/4"	—	—	0"	0"	—	—							
402A	3'-0"	7'-0"	1 3/4"	—	—	0"										

Schedule Table

BASIC

General

Description	
Layer	A-Anno-Schd
Style	Door Schedule Proj...
Scale	1.00000
Update automa...	No

Selection

Add new object...	No
Scan xrefs	Yes
Scan block refer...	Yes
Layer wildcard	*Door

Location

Figure 15–45 *Add a more restrictive Layer Wildcard*

The Schedule Table should update and remove all extraneous toilet stall Doors. All that remains to do is add our Space Allocation Schedule from earlier in the chapter to this Sheet.

68. On the Project Navigator palette, click the Views tab.

69. Click the small plus sign next to *A-SH03* to expand its Model Space Views.

70. Drag the Third Floor Space Allocation Schedule Model Space View onto the Sheet.

71. Save and Close all project files.

New in ADT 2006

UNDERSTANDING DISPLAY THEMES

In the "Using Display Themes" heading of Chapter 12, we used some of the out-of-the-box Display Theme Styles. As we saw in that section, a Display Theme applies an override to the Display Properties of one or more ADT objects (as defined by the Display Theme Style) and presents the Theme in a legend inserted on the drawing.

1. On the Sheets tab of Project Navigator, in the *Schedules and Diagrams* subset, create a new Sheet named *A-602 Display Themes*.

2. Open the new Sheet, drag the *A-FP03* Floor Plan View file and place it on this new Sheet.

3. Copy the Viewport twice (for a total of three) and position them on the Sheet.

   If necessary, choose **Regenall** from the View menu to refresh the display.

4. On the Scheduling tool palette, click the **Theme by Space Size** tool and when prompted, select the first Viewport.

5. At the "Upper left corner of display theme" prompt, place the legend in proximity to the Viewport and then press ENTER to accept the default size and scale.

6. Repeat this process using the other two Display Theme tools on the Scheduling tool palette (**Theme by Fire Rating** and **Theme by Space Type – Commercial**) and apply them to the other two viewports.

7. Save the file.

The first thing that you will notice is that different Display Themes can be applied to different Viewports yielding a completely different graphical look, even when the contents of the Viewports are the same. Now let's edit one of the Display Theme styles to get an understanding of how it works. The *Theme by Fire Rating* Display Theme is perhaps the simplest one on screen. Let's start with that one.

8. Select the **Theme by Fire Rating** legend, right-click and choose **Edit Display Theme Style**.

9. Click the Design Rules tab (see Figure 15–46).

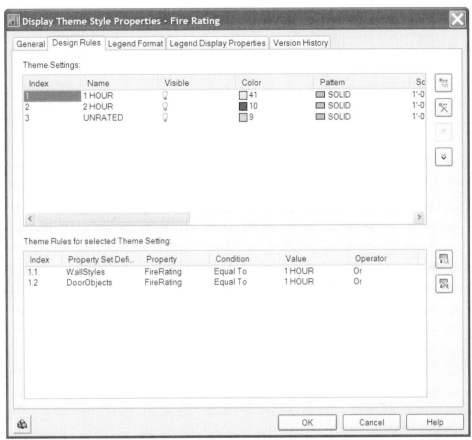

Figure 15–46 *The Display Theme Design Rules theme objects based on Property Set values*

A Display Theme contains one or more components at the top, that each have one or more Rules at the bottom. In this case, there are three components: "1 Hour," "2 Hour" and "Unrated." Selecting any one of these components at the top reveals two rules (4 for unrated) for each at the bottom; one for Wall objects, and another for Door objects. Each Rule contains a criterion composed of a Property from a Property Set and a Condition against which it is tested. If the object type and property in question meets the condition, then the object is themed. If it does not meet the criterion, then it is not affected. In this case, the first component at the top, looks at Wall and Door objects. If either object type has their FireRating Property set to "1 Hour" then the objects are colored color number 41. Color 10 is used if the FireRating equals "2 Hour" and if it is blank, or set to a double dash (–) they are colored gray (color 9). Currently there are no one or two hours Walls in our project.

> The Display Theme looks at the style-based **WallStyles** Property Set and the object-based **DoorObjects** Property Set.

10. Using the techniques covered above, open the *Core* Element file and edit the **WallStyles** Property Set of core Walls to be "1 Hour" and/or "2 Hour" rated.

For the complete effect, edit the Doors Property Sets as well.

11. When the XREF balloon appears, reload the *Core* file.

The Theme will update immediately.

Let's look at the *Theme by Space Type – Commercial* Style next. In Chapter 13, we imported several out-of-the-box Space Styles and applied them to the Spaces in this project. The legend however, shows all Space Types, not just those we are using. We can edit the Legend and remove the Styles that we aren't using.

12. Select the **Space Type – Commercial** Style legend, right-click and choose **Copy Display Theme Style and Assign**.

13. On the General tab, rename the Style to **MADT Space Style**.

14. On the Design Rules tab, in the top pane, highlight Index 1 – Cafeteria Large and then click the Remove Component icon on the right.

15. Repeat for each Style that is not in use in our project and then click OK.

CREATE A CUSTOM DISPLAY THEME

Like most Styles, it is easy enough to create your own Display Theme Style—simply begin with an existing one and modify it. Choose an existing one that is either close in terms of the Design Rules so that you will have minimal editing, or pick one that already has a color scheme that you like. Ideally you can find one that has a little of both. There are some additional Display Theme tools in the *Documentation Tool Catalog - Imperial* [*Documentation Tool Catalog – Metric*] in the Content Browser.

16. Copy one of the Viewports to create a fourth Viewport and then add a Display Theme to it.

For example, try **Theme by Space Type – Medical** in the Content Browser.

We are going to build our Theme based on the *GeoObjects* Property Set. However, it is not currently part of this drawing. The easiest way to add it is to use the *Space Inventory* Schedule tool that we used above.

17. On the Document tool palette, click the **Space Inventory** tool and then press ESC without adding it to the drawing.

18. Right-click the Theme and choose **Copy Display Theme Style and Assign**.

19. Name it **MADT Theme by Department** and then click the Design Rules tab.

20. Rename the first component to Marketing; at the bottom, next to Index 1.1, choose **GeoObjects** from the Property Set list, **Department** from the Property list and type **Marketing** in the Value field.

21. Repeat this for each of the Departments that we used above.

22. Delete all unused components (see Figure 15–47).

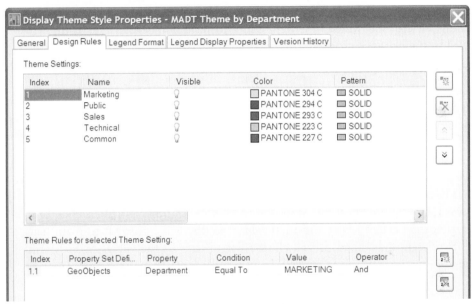

Figure 15–47 *Build a custom Display Theme to show Departments*

23. Click OK to see the results.

If the Theme does not update immediately, select its legend, right-click and choose Apply Display Theme and then select the Viewport when prompted.

Display Themes offer a vast array of possibilities of which we have only scratched the surface of here. Please feel free to experiment further. All of the out-of-the-box themes utilize solid fill hatching and pantone colors, however, if you scroll to the right you will note that other properties are also available like hatch patterns, linetype, lineweight and plot style.

24. Close and Save all commercial project files.

ADDITIONAL SCHEDULE TOOLS

Several other tools are available to manipulate Schedules once they have been added to the drawing.

RENUMBER DATA

This command allows you to renumber any Property Set that uses automatic increment (such as Room Number). Simply click the Renumber Data tool on the Documentation tool palette and choose the property to renumber, the number to start at, and the amount to increment by. Follow the command prompts to select the objects in the order in which you want them to be renumbered. This command was used in the previous chapter.

EXPORTING SCHEDULES

Schedule Tables can be exported to popular file formats like Microsoft Excel. To do so, select a Schedule Table in the drawing, right-click, and choose Export. In the dialog box that appears, choose a file format such as Microsoft Excel 97 (*.xls) from the Save As Type list. Type a name for the export file or click Browse to set a new location and name in separate dialog box. Click OK to complete the export. If you are working in Imperial units, a Format dialog box will often appear, indicating that Excel does not support the Architectural Units format. Choose Convert to Formatted Text, put a check mark in Apply to All Columns, and then click OK. Excel cannot understand Feet and Inch values, so this will convert it to plain text, which Excel can understand. If you choose not to convert the values to text, they will be in raw numeric format instead. In other words, 6'-0" will become 72.

 Caution: This is an "export" command. The Excel spreadsheet is NOT linked back to the drawing objects in any way.

OUT OF DATE MARKER

Automatic update is a valuable feature of Schedules. The advantage of automatic update is clear; however, it can cause a significant drain on drawing performance. Manual update imposes no such performance detriment, but it does require user intervention to keep the data in the Schedule current. When you choose manual update, it is imperative that the Display Configuration settings of the Schedule Table objects be configured properly to alert the user when the Schedule needs updating. Schedule Tables contain a Display Component called Out of Date Marker. Edit the Display Properties of Schedule Tables to turn this on or off. When this is turned on, and Automatic Update is turned off, a line will appear through the Schedule when the data in it no longer matches the drawing. To update, you then right-click the Schedule and choose Update Schedule Table.

CAD **Manager Note:** There is no harm in having the Out of Date Marker turned on even when Schedules are set to Automatic Update. However, in contrast, it could be very damaging to productivity to have the component turned off in situations where it is needed. Making this modification to your template files will serve as a valuable preventive measure. Regardless, get people in the habit of performing a manual update before critical submissions.

ADDITIONAL EXERCISES

Additional exercises have been provided in Appendix A. In Appendix A you will find exercises to add additional Schedule Tables in both the Residential and Commercial Projects. It is not necessary that you complete this exercise to begin the next chapter, it is provided to enhance your learning experience. Completed projects for each of the exercises have been provided in the *Chapter15/Complete* folder.

SUMMARY

Schedule Tables link directly to drawing objects.

In order for the correct data to appear in the cells of the Schedule, an associated Property Set must be attached to the objects (or their styles) listed in the Schedule.

Objects can be added and deleted from Schedules by adding or removing them from the Schedule selection.

Adding or deleting the objects in the drawing changes both the drawing and the Schedule.

Style-based properties are edited at the object-style level and affect all objects of the same type in the Schedule and the drawing.

Property Set Definitions can apply to objects or styles.

Individual properties within the Property Set Definition can be Automatic or Manual.

Automatic properties can use formulas and track physical location relative to Spaces and Areas.

The text formatting of Schedule objects can be customized with AutoCAD Text styles and Format settings in the Schedule Table style.

Layers, Colors, Linetypes, and Lineweights are all fully accessible for Schedule Table styles with the Display Properties settings.

Schedule columns and headers can be moved and edited to suit specific needs.

Schedule Tables can be exported to Microsoft Excel; however, the data is not linked back to ADT.

When using manual update, make sure that the Out of Date component is active in the Display Properties.

Doors tags can reference the Room numbers of adjacent Spaces.

Room numbers can automatically reference the level from the Project database.

When you add a Schedule Table to a Sheet, you can reference a remote drawing that will update automatically each time the file is opened

Display Themes key into Property Sets attached to objects. Edit the Display Theme Style to modify this behavior.

CHAPTER 16

Generating Sections and Elevations

INTRODUCTION

There are two basic approaches to generating sections and elevations in Autodesk Architectural Desktop 2006: sections and elevations can be generated from the ADT model as a linked graphical "report" of the data contained within it, or the ADT model can be viewed "live" in an Elevation or Section Display Configuration. When generating sections (the "report" approach), ADT offers both a three-dimensional Section/Elevation object suitable for presentation drawings and "cut away" perspectives, and a two-dimensional Section/Elevation object useful for inclusion in design development and construction documents. Live sections are also very useful for design and presentation purposes. The 2D Section/Elevation object is style based and robust. It will be the main focus of this chapter.

OBJECTIVES

In this chapter, we will look at the 2D Section/Elevation object in detail. We will cover ways to make this tool produce top-quality sections and elevations from your ADT model. We will work in the Residential Project, as we work through the process of creating Sections and Elevations. At the end of the chapter, we will also look briefly at Interior Elevations and Live Sections. In this chapter, we will explore the following topics:

- Learn to Add Section/Elevation lines.
- Working with Callouts.
- Generate a 2D Section/Elevation object.
- Work with 2D Section/Elevation Styles.
- Update 2D Section/Elevation objects.
- Understand Edit Linework and Merge Linework commands.

WORKING WITH 2D SECTION/ELEVATION OBJECTS

2D Section/Elevation objects are useful for generating section and elevation drawings from an ADT model suitable for design development and construction documents. 2D Section/Elevation objects have a broad scope; use them for full building sections and elevations, interior elevations, as an underlay for wall sections and even to get started with details. To create a 2D Section/Elevation, you must first add a Section/Elevation Line object to indicate where you wish the section cut to occur and in which direction it ought to look. Section/Elevation lines are added when you run the tools on the Design tool palette and through the various Callout tools on the Callouts tool palette. Callouts are robust routines that add required annotation and cross-link it throughout the project. The tools on the Design palette add only the Section/Elevation Line object with no cross-referenced annotation. You configure and fine-tune the appearance of the 2D Section/Elevation object in much the same way as other ADT objects. You can edit its style, change its Display Properties and/or edit the actual component linework within the object.

THE BLDG SECTION LINE AND THE BLDG ELEVATION LINE

In this exercise, we will revisit the Section and Elevation Composite Model View file that we created back in Chapter 5 for the Residential Project. At that time, we built a composite model of the entire project and cut four elevations and two sections using the Callout routines. One of those elevations, the East Elevation, has been removed from this file. We will begin our exploration of 2D Section/Elevation by recreating this elevation.

The Bldg Section Line and Bldg Elevation Line objects are actually three-dimensional "boxes." The purpose of this three-dimensional box is to determine what portion of the Building Model will be included in the Section or Elevation.

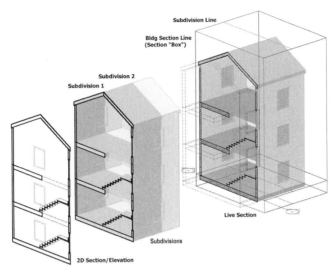

Figure 16–1 *Bldg Section Line at right with a Live Section, middle showing subdivisions and left showing 2D Section/Elevation*

In the illustration shown in Figure 16–1, the Bldg Section Line is sized to give a full building cross-section. If the size of the box is adjusted, in both plan and Z heights, we can effectively create an entirely different type of drawing like the interior elevation shown in Figure 16–2.

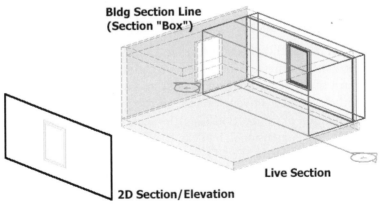

Figure 16–2 *Resize the Bldg Section Line to give an Interior Elevation*

The Display Representations of the Bldg Section Line and Bldg Elevation Line objects contain three sub-components. Two of the three subcomponents: the boundary and the subdivisions are used merely for purposes of configuring the 2D Section/Elevation object (as shown in the above two figures). The third component, the defining line, is the only component that would potentially have value when printed; at least for Sections (you will likely not want it printed for Elevations). The Defining Line can be used in conjunction with a Section Bubble in your drawings, or you can simply use the bubbles by themselves. By default, all of these components are on a non-potting layer. The defining line is the "cut line," drawn through the building in plan view. It forms a plane three-dimensionally that determines where the bold cut line in the Section will be. The boundary *is* the "box" (referred to above). Nothing outside the boundary is included in the Section or Elevation. The subdivisions are drawn as lines parallel to the back edge of the Section box and determine where the Lineweight zones occur (see Figure 16–3). We will look more carefully at subdivisions below.

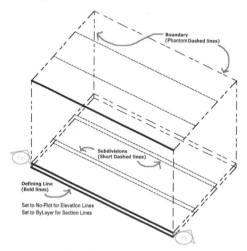

Figure 16–3 *The components of the Bldg Section Line and Bldg Elevation Line objects*

Another way to think of the Bldg Section Line and Bldg Elevation Line objects is like the field of view for a camera. When you look through your camera lens, you can only see so much of the scene both side to side and moving back. If you imagined mapping the field of view of your camera on the ground, you would have a pretty good approximation of the Section/Elevation Line object. Even though this boundary is actually three-dimensional, by default ADT will automatically include the entire height of the model in the 2D Section/Elevation object. This default can be changed if required. Ironically, when this option is active ("Use model extents for height" is set to **Yes** on the Properties palette), no height will be shown graphically in 3D views since it includes *all* of the height.

INSTALL THE CD FILES AND LOAD THE CURRENT PROJECT

If you have already installed all of the files from the CD, simply skip down to step 3 below to make the project active. If you need to install the CD files, start at step 1.

1. If you have not already done so, install the dataset files located on the Mastering Autodesk Architectural Desktop 2006 CD ROM.

   Refer to "Files Included on the CD ROM" in the Preface for information on installing the sample files included on the CD.

2. Launch Autodesk Architectural Desktop 2006 from the desktop icon created in Chapter 3.

If you did not create a custom icon, you might want to review "Create a New Profile" and "Create a Custom ADT Desktop Shortcut" in Chapter 3. Creating the custom desktop icon is not essential; however, it makes loading the custom profile easier.

3. From the File menu, choose **Project Browser**.

4. Click to open the folder list and choose your *C:* drive.

5. Double-click on the *MasterADT 2006* folder, then the *Chapter16* folder.

   One or two residential Projects will be listed: *16 Residential* and/or *16 Residential Metric*.

6. Double-click *16 Residential* if you wish to work in Imperial units. Double-click *16 Residential Metric* if you wish to work in Metric units. (You can also right-click on it and choose **Set Current Project**.) Then click Close in the Project Browser.

 IMPORTANT If a message appears asking you to re-path the project, click **Yes**. Refer to the "Re-Pathing Projects" heading in the Preface for more information.

ADJUST THE BLDG ELEVATION LINE

1. On the Project Navigator, click the Views tab, and then double-click *A-CM01* to open it.

This is the Section and Elevation Composite Model View file that was created in Chapter 5. As was mentioned above, the East Elevation has been removed from this file and we will re-create it below. For the time being, let's focus our attention on the Bldg Elevation and Bldg Section Lines that already appear in this file.

2. Click to select any one of the three elevations or the two sections.

   Notice a dashed red line will also highlight in the plan. This is the Bldg Section or Bldg Elevation Line associated with the selected section or elevation.

3. Right-click and choose **Deselect All**.

Let's take our camera analogy from above a bit further. You are able to adjust what the camera sees by adjusting the focal length of your lens and your f-stop for the depth. With grips, you can make the similar adjustments to what the Bldg Section/Elevation Line boundary "sees." The three grips on the back edge (left side in this case) change the "depth of field." However, to keep the back edge parallel to the defining line, you should always use the middle triangular shaped grip. The two ends of the defining line (right side in this case), can be used to widen or narrow the field of view. The middle grip of the defining line can be used as the edge of the Section Elevation Line boundary. A small gray grip appears on the defining line to toggle on and off the "Use Model Extents for Height" feature.

4. Zoom in on the model.

5. Click to select the Bldg Elevation Line running horizontally along the top of the plan.

Several grips of varying shape will appear. Hover your mouse over each grip to see its function (see Figure 16–4).

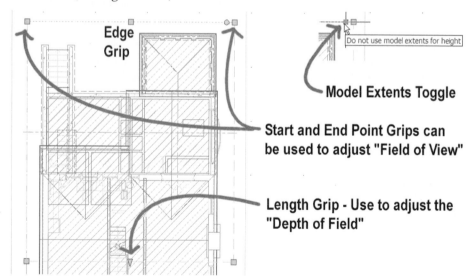

Figure 16–4 *Examine the Bldg Elevation Line Object Grips*

6. Experiment with the Grip points if you wish.

When you are finished experimenting, undo any grip edits you made.

MAKE THE SECTION/ELEVATION LINE EASIER TO READ

The display properties of the Section/Elevation line are assigned to a non-plotting layer by default. If you wish, you can edit this in the same way as other ADT object. Bldg Section Line and Bldg Elevation Line objects are not style-based.

Therefore, you can either edit the Drawing Default display properties of these objects or apply object level overrides.

7. Select any Bldg Elevation Line object in the drawing, right-click and choose **Edit Object Display**.

Like most ADT objects, there are several Display Representations. However, unlike more complex objects like Doors and Walls, the Bldg Section Line and Bldg Elevation Line objects show the same three components in all Display Reps.

8. Click the Edit Display Properties icon.

    There are three components:

    ▶ **Defining Line**—This is the cut line akin to the cut plane for plans. Geometry cut by this line will typically be rendered in a heavy lineweight automatically.

    ▶ **Subdivision Lines**—As you recede back from the defining line, you can define "zones" or subdivisions that have different (often receding in thickness) lineweights as they move further from the defining line (see below).

    ▶ **Boundary**—This is the outer edge of the Section/Elevation Line object.

The Boundary Line is off by default. This is fine for our purposes here. However, later in the chapter when we add sub-divisions to some of our elevations, it might be easier to read if the Subdivision Lines are a different color.

9. Assign a different color such as magenta to the subdivisions and then click OK twice (see Figure 16–5).

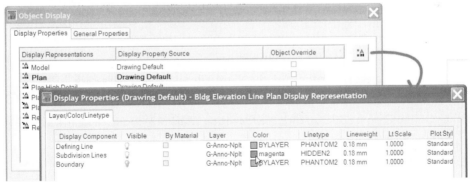

Figure 16–5 *Assigning colors to the Display Components of the Section/Elevation Line object*

We don't have any subdivisions yet, so this change will not become evident until later.

10. Save the file.

GENERATING A 2D SECTION/ELEVATION OBJECT USING A CALLOUT

Now that we have an understanding of how the Bldg Section Line and Bldg Elevation Line objects are used to generate sections and elevations, let's re-create the East Elevation in the Section and Elevation Composite Model file. Although this can be done manually, it is much easier to use a Callout routine as we did in Chapter 5. We will review that process next.

Section and Elevation objects in ADT can be thought of as graphical "reports" of the data in the drawing model. The zone defined by the Bldg Section Line object and an object selection made while generating the 2D Section/Elevation determines the specific objects that will be included in this report. Using our camera analogy, it is the "snapshot."

We will begin in the First Floor Plan where we will add an elevation Callout. The Callout routines are powerful tools that are capable of performing several steps in one operation. The typical Callout routine will add one or more Callouts to the current drawing, create a Bldg Elevation Line (in the same or another View file), add a Model Space View with association to the 2D Section/Elevation, add a Title Mark and then Generate a 2D Section/Elevation object.

ADD A CALLOUT

You typically execute the Callout routine within the drawing where you wish to have the Callout appear. Since indicators for elevations and section are typically placed in plans, in this case we will begin in the First Floor Plan View file. However, we want the 2D Section/Elevation object to be created in the *A-CM01* View file (which was created for that purpose in Chapter 5). The Callout routine will prompt us for this location at the appropriate time and then generate the 2D Section/Elevation object within that file automatically.

1. On the Project Navigator, click the Views tab, and then double-click A-FP01 to open it.

2. On the Callouts palette, click the **Elevation Mark A2** tool (see item 1 in Figure 16–6).

    If you do not see this palette or tool, right-click the Tool Palettes title bar and choose **Document** (to load the Documentation Tool Palette Group) and then click the Callouts tab.

    Several notes have been added to this file to aid you in this task. These labels are on the G-Anno-Nplt layer so they won't print, but you should erase them when you are finished.

3. At the "Specify location of elevation tag" prompt, click the Midpoint of the cyan rectangle, as indicated by the "SNAP to MIDPOINT HERE" leader, (see item 2 in Figure 16–6).

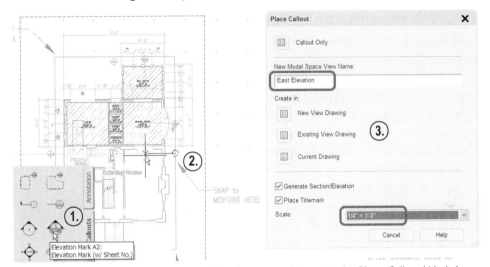

Figure 16–6 *Add a Callout and then fill in the required items in the Place Callout Worksheet*

4. At the "Specify direction of elevation" prompt, move the mouse directly to the left (using POLAR or ORTHO) and then click.

   This determines which way you want the elevation to "look."

5. In the Place Callout worksheet, type: **East Elevation** for the New Model Space View Name (see item 3 in Figure 16–6).

 NOTE This Model Space View Name must be unique in your project. Project Navigator will not allow you to create a New Model Space View using a name that already exists in the Project, even if it is within a different drawing.

6. From the Scale list at the bottom, choose **1/4″=1′-0″ [1:50]** (see item 3 in Figure 16–6).

   Be sure that both "Generate Section/Elevation" and "Place Titlemark" are checked.

When you instruct the Callout routine to "Generate Section/Elevation" it will create the 2D Section/Elevation object for you. Place Titlemark will add a Title Mark Callout to the drawing you specify in the Create in area above and that Title Mark will be scaled to whatever you indicate in the Scale list. This scale will also be assigned to the New Model Space View named at the top of the worksheet. For more information and other examples of Callouts, see the "Create the Building Elevation View" heading in Chapter 5.

7. Click the Existing View Drawing icon in the center of the worksheet (see Figure 16–7).

   An Add Model Space View worksheet will appear listing all of the View files within the current project.

8. Select *A-CM01* and then click OK (see Figure 16–7).

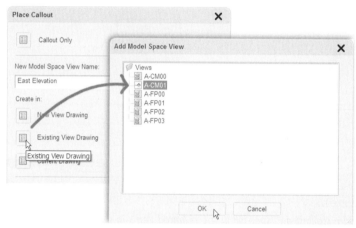

Figure 16–7 *Choose the Section and Elevation Composite Model – A-CM01*

9. At the "Specify first corner of elevation region" prompt, snap to the upper left corner of the cyan rectangle as indicated by the "FIRST CORNER OF ELEVATION REGION" text leader.

10. At the "Specify opposite corner of elevation region" prompt, snap to the lower right corner of the cyan rectangle as indicated by the "OPPOSITE CORNER OF ELEVATION REGION" text leader. *Do not press* ENTER *or hit the* ESC *key. You are not done with the Callout routine yet!*

Look at the command line and notice the message that has appeared. It will read:

** You are being prompted for a point in a different view drawing **

When you create elevations using the Callout routine to an existing or new View file, you still must indicate within the current drawing where you would like the elevations to be created in that file. This prompt serves to inform you of that. It is usually best to pick a point off to the side of the plan. Since we already have several elevations and sections in the *A-CM01* View file. ADT will highlight these temporarily until you click the insertion point. Each elevation and section will show the name of its associated Model Space View. In this case, we want to click a point between the North and South Elevations that already exist in that file. The text leader at the right side of the plan will help you to do this.

11. At the "Specify insertion point for the 2D elevation result" prompt, click a point in the drawing to the right of the floor plan where indicated by the "PLACE INSERTION POINT OF ELEVATION NEAR HERE" text leader (see Figure 16–8).

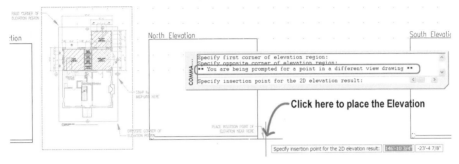

Figure 16–8 *Pick an insertion point for the 2D Section/Elevation within the A-CM01 drawing*

Here in the First Floor Plan, a Callout has appeared where we indicated. Nothing else has appeared to have changed. The Bldg Elevation Line and the 2D Section/Elevation object were both created in the *A-CM01* View file. We will need to open that file to see them. Before we leave the First Floor Plan however, let's cleanup a bit.

12. Erase the four text leaders and their associated notes. Erase also the cyan rectangular Layout Grid that we used to set the boundaries of the elevation.

    Erase only the inner (and smaller cyan colored) Layout Grid. Do not erase the outer purple dashed one.

If you wish to reference the East Elevation from the other Floor Plans, copy the Callout that we just added to the clipboard and then on Project Navigator, open each Floor Plan View: *A-FP00, A-FP02* and *A-FP03*, right-click and choose **Clipboard > Paste to Original Coordinates** in each of those files.

13. Save and Close all Floor Plan View files.

14. On the Project Navigator palette, open the *A-CM01* file if it is not already open.

15. Zoom in on the newly created East Elevation (third elevation from the right).

As you can see, we have come a long way since Chapter 5! This Elevation is nearly OK as is, but we can certainly find a few things to enhance the drawing. Re-Position the Elevation for the Sheet

Notice that there is a horizontal line with a note pointing to it. This line matches up with the first floor of the other Elevations in this file. In Chapter 14 we added Elevation Labels to this file. In order for those heights to be correct relative to this new elevation, we must move it to line up precisely with the others in this file. The line has been provided to simplify the process. We will now move the Elevation to line up with this line.

16. Click on the East Elevation 2D Section/Elevation object.

Notice the dashed gray box that surrounds the elevation. This is the Model Space View that is associated with this elevation. It has a grip on each edge that can be used to edit the extents of this Model Space View.

17. Click the square Location grip at the left side of the elevation.

    Use this grip to move the elevation within its Model Space View.

18. Move the mouse down and snap using Intersection or Perpendicular to the provided guide line (see Figure 16–9).

    Zoom as required.

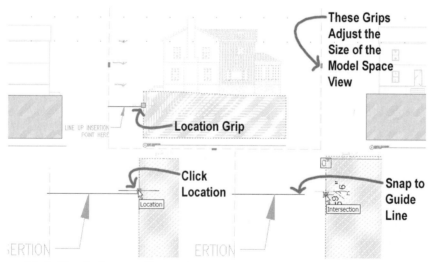

Figure 16–9 *Use the Location Grip to align the new elevation to the others*

If you prefer, the AutoCAD Move command can be used. If you use Move, set the Base Point at the Insertion point of the elevation and then snap perpendicular to the guide line.

 NOTE The Location Grip is useful for moving the elevation object within its Model Space View boundaries. If you wish to move the elevation *and* its associated Model Space View together, then use the AutoCAD Move command instead.

19. Erase the line and the note.

20. Save the file.

UPDATING AND MODIFYING SECTION/ELEVATIONS

2D Section/Elevation objects remain linked to the building model. When changes occur in the building model, you can perform an update on the 2D Section/Elevation objects. This greatly reduces the amount of rework required to keep Sections and Elevations up to date.

 NOTE 2D Section/Elevation objects are not live data; they can be edited for their own purposes as we will see below, but manual update is required to keep them current with the state of the model. There is also a batch process routine available that will update all elevations and sections within an entire project. From the View menu, choose Refresh Sections/Elevations to access it.

MODIFY THE MODEL

Suppose a design change caused the position of one of the windows to move. Let's open the Second Floor New file, make this change, and then return to the Elevation and update it to see the change.

Be sure to keep the *A-CM01* file open.

1. On the Project Navigator, click the Constructs tab and then double-click the *Second Floor New* file to open it.

2. Zoom in to the Bedroom in the top right corner.

3. Select the Window in the right vertical Wall of the Bedroom.

4. Using the Location grip (square one in the middle), move the Window up **2′ 0″** [**600**] (see Figure 16–10).

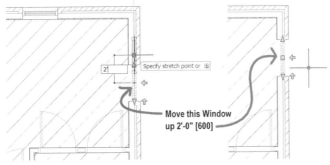

Move this Window up 2'-0" [600]

Figure 16–10 *Move a Window on the Second Floor*

5. Save and Close the file.

REFRESH THE ELEVATION

Back in the *A-CM01* file, you should receive an alert that the *Second Floor New* XREF has changed at the lower right corner of your screen (see Figure 16–11).

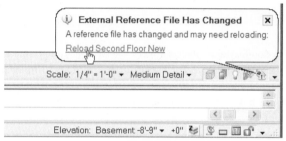

Figure 16–11 *An XREF Alert will appear*

6. Click the link on the balloon to reload the XREF.

 TIP You can also right-click the XREF Quick pick icon for a menu to open the XREF Manager. Use this technique if the balloon does not appear.

7. Select the East Elevation (the one we added above), right-click, and choose **Refresh**.

Keep your eye on the second floor Window as the elevation refreshes.

Notice the update to the Elevation. The Window on the second floor has moved 2'-0" [600] to right (see Figure 16–12).

Figure 16–12 *Reload the XREFs and then refresh the Elevation to see the change*

 IMPORTANT It is very important that you first update the XREFs, and then refresh the Elevations. If you refresh first, you will see no change in the Elevation since the change took place in a remote file. We have to load those remote changes before the Elevation can "see" them.

MODIFY THE CHIMNEY

One of the advantages of studying Elevations is that you can discover flaws in the design or the drawing components used to portray the design. For this reason it is good practice to set up the Elevations and Section early in the project and refresh them often throughout the design process. (This was certainly the rationale behind the setting up of the current file back in Chapter 5.) Here we have such an example. The Chimney was created with Plan Modifiers which are effective for the plan representation and when there are no sloped surfaces in elevation. However, if we want to have the chimney taper properly as it goes up, we will need to model it differently. Let's add a simple Body Modifier to what we already have thus far.

8. On the Project Navigator, click the Constructs tab and then double-click the *Second Floor Exist* file to open it.

9. Click the Wall on the right (the one with the chimney) right-click and choose **Isolate Objects > Edit in Elevation**.

10. At the "Select linework or face under the cursor or specify reference point for view direction" prompt, move the cursor close to the outside edge of the chimney. When the blue construction line appears, click the mouse (see Figure 16–13).

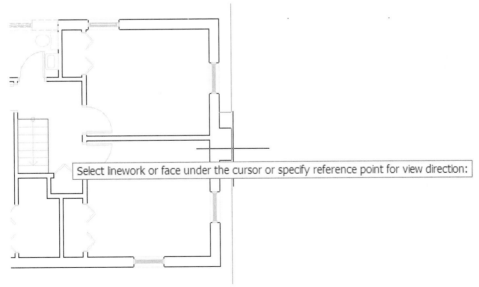

Figure 16–13 *Select the outside edge of the chimney for the reference view location*

11. At the "Specify elevation extents" prompt, drag to the left just beyond the opposite side of the Wall and then click again. (You just need to include the full thickness of the selected Wall).

All objects except the selected Wall will be hidden and the drawing will switch to an elevation view looking at the chimney.

12. On the Shapes toolbar, click the Polyline icon (or type **PL** and then press ENTER).

13. Using the illustration in Figure 16–15 as a guide, draw a polyline describing the shape of the desired taper of the chimney.

Use Object Snaps and Object Snap Tracking to draw with precision.

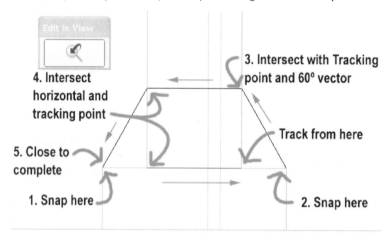

Figure 16–14 *Draw a polyline to describe the shape of the desired chimney taper*

14. When the polyline is complete, select it, right-click and choose **Convert To > Mass Element**.

15. At the "Erase selected linework" prompt, choose **Yes**.

16. At the "Specify extrusion height" prompt, type 1'-3" [380] and then press ENTER.

You should now have an extruded Mass Element in place of the polyline.

17. Select the Wall, right-click and choose **Body Modifier > Add**.

A Body Modifier is a piece of 3D geometry that modifies the shape of a Wall. You can apply it to the mass of the Wall using a choice of operations such as "Additive," "Subtractive" and "Replace." If you anticipate needing to place a Door, Window or other opening into the space occupied by the Body Modifier, choose "Additive Cut Openings."

18. At the "Select objects to apply as body modifiers" prompt, click the Mass Element and then press ENTER (see the left side of Figure 16–14).

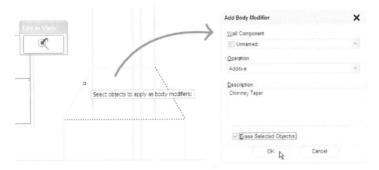

Figure 16–15 *Apply the Mass Element to the Wall as a Body Modifier*

19. In the Add Body Modifier worksheet, choose Additive from the Operation list and type Chimney Taper in the Description field.

20. Place a check mark in the Erase Selected Object(s) checkbox and then click OK (see the right side of Figure 16–14).

 NOTE You may want to view the wall in the Object Viewer at this point to verify that everything is correct. For instance, if the profile snapped to the wrong face of the chimney, select the Wall, right-click and choose **Body Modifier > Edit in Place** to correct.

21. Click the Exit Edit in View icon on the Edit in View toolbar to exit the Edit in Elevation mode and return to the previous plan view.

22. Save and Close the file.

23. Back in the *A-CM01* file, repeat the steps above in the "Refresh the Elevation" topic to reload the XREF and then refresh the elevation.

ADD A ROOF WINDOW

In Chapter 12 we added a "hole" in the Roof for a Roof Window. However, the hole has yet to be filled with a Window. An Element file named Roof Window has been provided here to use for this purpose. It does not actually contain a Window at all. The Roof Window in that file is made from a custom Roof Slab Style. There is a very thin Slab with a glass material applied and then a Roof Slab Edge Style surrounding it to represent the window frame with a painted metal material. The reason that an actual Window was not used for this is because Windows do not display well at an angle other than vertical. Also with this solution, it is very easy to match the slope of the Roof, simply use the Rise and Run parameters on the properties palette. The one provided is already set to the correct slope.

24. On the Project Navigator, on the Constructs tab, double-click the *Roof New* file to open it.

On the Project Navigator, on the Constructs tab, locate the *Roof Window* file in the *Elements* folder.

25. Right-click *Roof Window*, and choose **Insert as Block**.

26. Make sure that Specify On-screen is not checked for any of the values, and verify that the X, Y and Z Insertion point is set to **0**, the Scale to I and the Rotation is: **0**.

27. Place a check mark in the Explode check box and then click OK (see Figure 16–16).

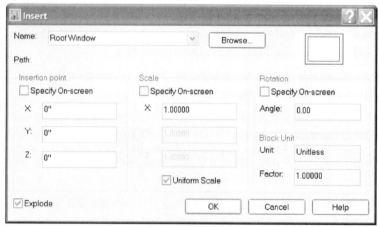

Figure 16–16 *Insert the Roof Window Element file as a block*

Notice the Roof Window appears at the location of the hole. If you wish, experiment with this object. As mentioned above, it is a Roof Slab, and is therefore fully editable in all the ways that we covered in Chapter 12.

28. When you are ready, Save and Close the *Roof New* file.

29. Back in the *A-CM01* file, repeat the steps above in the "Refresh the Elevation" topic to reload the XREF and refresh the elevation (see Figure 16–17).

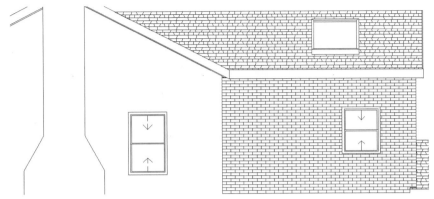

Figure 16–17 *The updated Elevation with the chimney modifications and the Roof Window in place*

30. Save the *A-CM01* file.

APPLY A MATERIAL BOUNDARY

The hatching on the Elevation was generated automatically based on the Material Definitions applied to the various pieces of the model (see below for more information on Material Definitions). Sometimes showing hatching all the way across the surface of the Elevation can be a bit too busy. We can limit the extent of the hatching on each 2D Section/Elevation object very simply.

31. Draw a closed polyline like the one shown in Figure 16–18.

Make sure that you draw only three sides and then right-click and choose **Close** for the last side. The exact size or shape is not important.

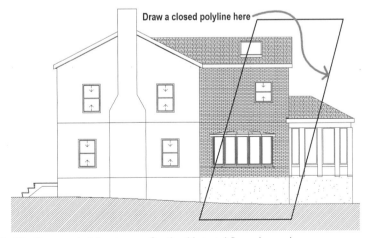

Figure 16–18 *Draw a polyline to use for the Material Boundary edge*

32. Select the Elevation, right-click, and choose **Material Boundary > Add**.

33. At the "Select a closed polyline for boundary" prompt, click the polyline that you just drew.

34. At the "Erase selected linework" Prompt, choose **Yes**.

The 2d Section/Elevation Material Boundary dialog box will appear.

35. Accept all defaults for now and click OK (see Figure 16–19).

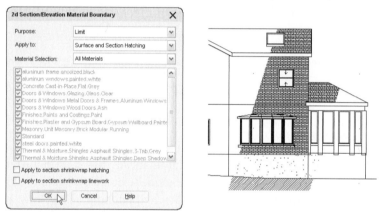

Figure 16–19 *Applying a Material Boundary with the default settings*

Material Boundaries are quite exciting, but the default settings do not give us such a good affect in this case.

36. Select the Elevation, right-click, and choose **Material Boundary > Edit in Place**.

You can grip edit the shape of the Material Boundary in real time. Give it a try (see Figure 16–20).

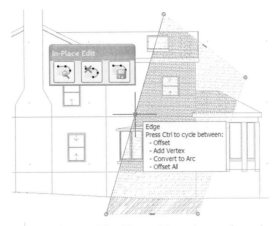

Figure 16–20 *Grip edit the shape of the Material Boundary and see the results in real time*

Hover over each grip to see its CTRL key options. Try some of them out. You can undo any of these edits.

37. With the In-Place Edit boundary selected, right-click and choose **Edit Material Boundary**.

This will call the 2d Section/Elevation Material Boundary dialog box again.

38. Change the Purpose to **Erase**.

This will show the hatching everywhere except the region within the polyline. It is like reversing the polyline.

39. Choose **Surface Hatching Only** from the Apply to list.

This will exclude the hatching on the terrain that we are actually sectioning through and apply the boundary only to hatches seen in elevation.

40. Leave Material Selection set to **All Materials** and then click OK (see Figure 16–21).

With this chosen, all materials will be affected by the boundary. If you wish you can choose **Selected Materials** instead and choose from the list of materials only those that you wish to affect.

Figure 16–21 *Reverse the effect of the boundary and apply it only to hatches in Elevation*

41. On the In-place Edit toolbar, click the Save All Changes icon to complete the operation.

Continue to perform In-place Edits and experiment with the boundary until you are fully satisfied. For instance, you can draw another closed polyline, repeat the **Material Boundary > Add** process above and then edit each one in the In-place Edit mode. Try also changing the Material Selection to just a few selected materials as well. Also, by applying a Material Boundary and choosing one of the Linework options, you can even use a Material Boundary to erase or crop the linework of an elevation.

42. When you are satisfied with the Material Boundary, Save the file.

2D SECTION/ELEVATION STYLES

2D Section/Elevation objects are controlled by styles in much the same way as other ADT objects. By means of the style, we will be able to adjust the graphical display of virtually all of the linework in the 2D Section/Elevation object. By adding layers and lineweights to the linework of the section, we will help the key features to "punch" and achieve an overall better read from the Section/Elevation object.

EXPLORE THE 2D SECTION/ELEVATION STYLE

A large part of the configuration of the 2D Section/Elevation Style is the display control. As we have seen in previous chapters, display control allows us (within specific Display Representations) to assign Layer, Color, Linetypes, Lineweights and Plot Styles to the various components of an ADT object. Many objects offer additional display parameters as well, such as Cut Plane, Custom Block Display and Hatching. 2D Section/Elevation objects have only a single Display Representation called "General." Currently all of the Sections and Elevations in this drawing use the 2D Section/Elevation style named "2D Section Style 96 [2D Section Style 100]." This is the default used by all Callout routines. Let's take a quick look at the Display settings of the Standard style.

1. Select the East Elevation, right-click and choose **Edit 2D Section/Elevation Style**.

2. Click the Display Properties tab.

   Notice that General is the only Display Representation, and that a Style Override has been applied.

3. Click the Edit Display Properties icon (see Figure 16–22).

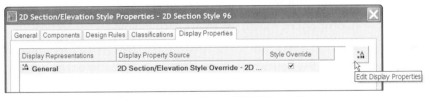

Figure 16–22 *Attach the Style override and then Edit Display Props*

4. Click the Layer/Color/Linetype tab and scroll through the list of components.

When a 2D Section/Elevation is cut, all of the geometry will be rendered to the components listed here based on how they fall relative to the defining (cut) line and other criteria covered below. Here is a brief description of each component:

▶ **Defining Line**—Used for objects cut through by the Bldg Section/Elevation line. Usually assigned a bold lineweight.

▶ **Outer Shrinkwrap**—Similar to the shrinkwrap component of Walls. This is the outermost edge of the all objects on the defining line.

▶ **Inner Shrinkwrap**—Similar to the shrinkwrap component of Walls. This is the innermost edge of the all objects on the defining line. It can be thought of as "holes" in the defining line.

▶ **Shrinkwrap Hatch**—A hatch pattern that fills in the space between the outer and inner shrinkwrap.

▶ **Surface Hatch Linework**—All of the hatching on surfaces shown in elevation.

▶ **Section Hatch Linework**—All of the hatching of materials cut through at the defining line. Section Hatch Linework differs from the Shrinkwrap Hatch in that it applies individually to each model component, rather than the entire space between the shrinkwrap components.

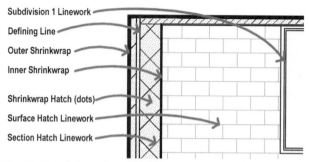

Subdivision 1 Linework

Defining Line

Outer Shrinkwrap

Inner Shrinkwrap

Shrinkwrap Hatch (dots)

Surface Hatch Linework

Section Hatch Linework

Figure 16–23 *The 2D Section/Elevation Style Components in context*

- ▶ **Hidden**—Used for objects concealed from view.

- ▶ **Erased**—Used to store linework that is manually edited and erased.

- ▶ **Unknown Component**—Used for errors encountered by the Section objects.

- ▶ **Subdivision 1 – 10**—Zones beyond the defining line used to display receding lineweights.

5. Click the Hatching tab.

    There is a single entry here, the Shrinkwrap Hatch. If you turn on the Shrinkwrap Hatch, configure its pattern and dimensions here. It is currently set to Solid Fill. We'll skip the Other tab for now.

EDIT A 2D SECTION/ELEVATION STYLE

Now that we have seen some of the settings available for 2D Section/Elevation styles, let's make a few minor changes.

6. Click back on the Layer/Color/Linetype tab and turn off the Shrinkwrap Hatch component (dim the light bulb icon).

Notice that the Defining Line, Outer Shrinkwrap and Inner Shrinkwrap components have their layer set to 0 and most of the other properties set to ByBlock. ByBlock is an old AutoCAD term. This setting means that the particular property in question will be inherited from the parent object. (It is called ByBlock simply because Blocks were the first objects to be able to do this.) In other words, the Defining Line does not have its own explicit Color, Linetype, Lineweight or Plot Style designation; rather it inherits these properties from the 2D Section/Elevation object itself. The Shrinkwrap components do likewise (with the exception of Color which is set explicitly to Red). Therefore, the properties of the 2D Section/Elevation object will inherited by these components. If the linetype, lineweight or Plot Style of the 2D Section/Elevation object itself changes, then the linetype, lineweight and/or Plot Style of all Defining Line and Shrinkwrap components will also change. (This is also true for Color in the case of the Defining Line.)

It is rare that properties like Color, Linetype, Lineweight and Plot Style are assigned directly to objects in the drawing. Typically, individual objects use the setting: ByLayer. ByLayer means that the object will inherit the properties of the layer upon which it resides. This is standard industry practice. Therefore, when the internal components of an object are set to ByBlock, and the object itself is set to ByLayer, the net result is that the components also behave as if they were set to ByLayer. Typically you will find the Color, Linetype, Lineweight and Plot Style settings assigned either directly in the Display Properties dialog box or via the Layer. You should never apply these properties directly to the objects themselves. This is generally considered bad practice and will not be looked on favorably by most CAD Managers and co-workers. In other words, there is little good reason to select the 2D Section/Elevation object and makes its color Green. Rather, you would move the 2D Section/Elevation object to a Green colored layer.

It is however, fairly common for colors, linetypes, lineweights, and plot styles to be assigned to object components with a particular Display Rep. For instance, as you can see here, the two Shrinkwrap components have been assigned to Color Red, while the Shrinkwrap hatch is assigned Color 9. For instance, to better understand which components are being rendered to Inner and Outer Shrinkwrap, it might be helpful to assign them different properties.

7. Select Outer Shrinkwrap, and set its Color to **154**, its Lineweight to **0.70 mm** and its Plot Style to **Extra Wide**.

8. Select Inner Shrinkwrap, and set its Color to **45**, its Lineweight to **0.70 mm** and its Plot Style to **Extra Wide** (see Figure 16–24).

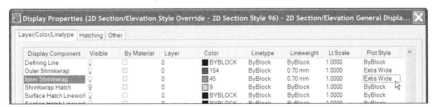

Figure 16–24 *Change the Lineweight and Color for the Shrinkwrap components*

9. Click OK twice when finished to return to the drawing.

The most obvious change was the elimination of the solid fill hatch from the terrain object beneath the elevation.

CAD Manager Note: There are various approaches to the process of manipulating the graphical display of a 2D Section/Elevation object. Conventional AutoCAD wisdom uses a series of layers; each assigned a unique color, which in turn is assigned a Lineweight while plotting. We can also have the layer assign the Lineweight or Plot Style properties or assign all three. Finally, all of these properties could be assigned directly within the 2D Section/Elevation dialog box without need for further layers. It is a matter of CAD management philosophy. Regardless of the CAD management philosophy implemented at your firm, ensure that your standards and templates are documented so the staff is clear on the method. In this exercise, we will explore the process by using a little of each technique. If you decide to use layers, make sure that all of the layers that you need are created in the file before you begin editing the 2D Section/Elevation style. You will not be able to create a new layer or edit the properties of an existing layer while in the Display Properties dialog box.

10. Save the file.

UPDATE THE SECTIONS

To better see the effects of the edits we made to the Shrinkwrap components, let's shift our attention to the Sections for a while.

11. Pan over to the left side of the drawing.

12. Select the both of the Section objects (to the left of the model), right-click and choose **Refresh** (see Figure 16–25).

Figure 16–25 *Refresh the two Sections*

You will see an immediate change in the color that outlines the cut objects.

13. Zoom in on each Section and study them.

These need a bit more work than did our Elevations. The most obvious problem is that the Roof is missing. This is because at the time that these Sections were cut, we were using the *Roof Existing* file. You may recall that in Chapter 12 we built the *Roof New* file and updated this Composite Model to reference it instead of the original file. We did not however, update the Sections and Elevations at that time. Therefore, rather than use Refresh, we must regenerate these Sections and the other three Elevations. Unfortunately, regenerating 2D Section/Elevation objects must be done one at a time.

14. Select the Longitudinal Building Section, right-click and choose **Regenerate**.

All of the XREFs we need are included in the selection for this Section except the *Roof New* file. Rather than replace the entire selection, we will simply add to it.

15. Click the Select Additional Objects icon (see Figure 16–26).

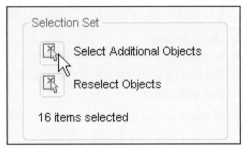

Figure 16–26 *You can add to the existing selection by clicking Select Additional Objects*

16. At the "Select objects" prompt, click the *Roof New* file in the model at the right.

    You may need to zoom in a bit. It should be easy to select, it has the roof shingle hatching. If you have the surface hatching toggled off, try clicking on one of the ridge or valley lines to select it.

17. Press ENTER to return to the Generate Section/Elevation dialog box and then click OK to regenerate the Section.

18. Repeat this process on the other Building Section.

ADJUST THE BLDG SECTION LINE POSITION

When we added the Bldg Section Line, we did not have the porch yet. If you zoom in on the Transverse Building Section, you will notice that the porch has been cropped out of the section. Zoom in on this area in the model and you will discover the reason (see Figure 16–27). The Section line passes directly through the middle of a Wall (and even though we cannot see it here, the Door as well). This problem is very easy to correct.

> Simply widen the Bldg Section Line with the grips and then Refresh the Section.

19. Click the square End grip at the top of the Section Line and drag it up using POLAR or ORTHO.

20. Click just outside the porch.

21. Select the Section, right-click and choose **Refresh**.

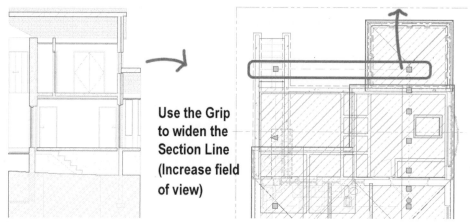

Figure 16–27 *The Bldg Section Line needs to be widened to include the porch*

If the section line now cuts through one of the porch piers, you can simply move the entire Bldg Section Line slightly and then refresh again.

22. Click the Section object and use the small magenta edge grips (attached to the gray dashed lines) to widen the extents of the Model Space View.

FINE-TUNE THE MODEL

There is one other very obvious issue with the model illuminated by these Sections: The Terrain Model has not been cut to receive the house. You may recall that we used the Drape command in Chapter 5 to build the *Terrain Model* file. The Drape command simply creates free-form Mass Elements. Therefore, it is a simple matter to subtract geometry equivalent to the shape of our model to form the cut in the Terrain. To do this, we will use the actual Basement files and some of the techniques we learned earlier. To properly cut away what we don't want from the *Terrain Model* file, we need a "solid" version of our Basement. (Imagine filling the Basement with concrete, and then using this to subtract away the terrain.) To get that from the files we have, we will enable the Volume Display Rep of Space objects as we did in Chapter 2.

Leave the *A-CM01* file open while you perform these steps.

23. On the Project Navigator, double-click the *Terrain Model* Construct to open it.

24. From Project Navigator, in the *Constructs* folder, drag the *Basement Existing* file and drop it into the drawing window of the *Terrain Model* file. Repeat for *Basement New*.

They should insert in exactly the correct location.

In Chapter 2, in the "Add a Set" heading, we learned how to make Space objects display a 3D solid representing their volume. By repeating that process now, and enabling something similar for the Stair leading up from the Basement, we will have a solid object to subtract from our terrain model.

25. From the Format menu, choose **Display Manager**.

26. Expand *Sets*, select Model (it will be bold) and then on the right side, on the Display Representation Control tab, select Space.

27. Clear the Model check box and instead place a check mark in the Volume check box and then click OK.

28. Select the *Basement New* file, right-click and choose **Open Xref.**

29. Select the Stair object leading out of the Basement, right-click and choose **Edit Object Display**.

30. On the Display Properties tab, be sure Model is selected and click the Edit Display Properties icon.

31. Turn on the Clearance component and then click OK twice to return to the drawing.

    The Stair treads will appear to "grow" up as the Stair clearance is represented three-dimensionally with a volume.

32. Save and Close the file and then back in the *Terrain Model* file, reload the XREF to *Basement New.*

We are now ready to subtract the volume of these two XREFs from the mass of the *Terrain Model.*

33. Select the large terrain Mass Element.

    A blizzard of grips will appear. Time and space do not permit us to discuss them here, but feel free to explore on your own.

34. Right-click and choose **Boolean > Subtract**.

35. At the "Select objects to subtract" prompt, click on both the *Basement Existing* and *Basement New* files and then press ENTER.

36. At the "Erase layout geometry" prompt, choose **No**.

We don't want to erase the XREFs. Erasing them is not the same as detaching and unloading. If you no longer want a particular XREF, Detach it. If you just want it to be invisible for a while, you can maintain the XREF by choosing Unload instead.

You should now have a hole in the terrain for the excavation. If you find it difficult to see the results, select just the Terrain Mass Element, right-click and choose **Object Viewer** (see Figure 16–28).

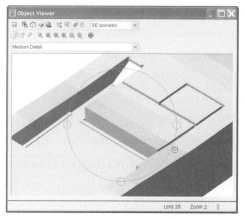

Figure 16–28 *Producing the excavation in the Terrain Model with a Boolean Subtraction*

Re-open *Basement New* and turn off the Clearance Component of the Stair object, Save and Close the file. You can also restore the display of Spaces in the *Terrain Model* back to Model instead of Volume as well.

The Boolean functions use the active display graphics. This is why we changed both Spaces and Stairs in their respective files to a volumetric display. Had we not done this, the basement would have appeared to be filled with dirt, since the Walls and slabs would be the only objects to subtract.

37. On the Drawing status bar, click the XREF quick pick.

    This will open the XREF Manager.

38. In the XREF Manager, select the *Basement New* XREF and then click the Detach button. Repeat for *Basement Existing*.

39. Click OK to close the XREF Manager.

40. Save and Close the file.

41. Back in the *A-CM01* file, repeat the process from above to first reload the XREFs and then refresh the two Sections (see Figure 16–29).

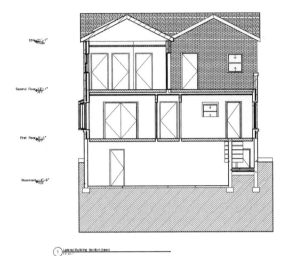

Figure 16–29 *After Reloading the XREFs and Refreshing the Sections, the Terrain now displays with the proper excavation*

42. Save the *A-CM01* file.

FREEZE DEMOLITION

Look carefully at the lower right portion of the Longitudinal Section and you will see the existing exterior Stair seeming to float in the Basement. We assigned this to a Demolition layer back in Chapter 10. It is, however, not showing as dashed here, and it is hiding a portion of the Section beyond it. Let's remove these items from the section.

43. On the Layer Properties toolbar, click the Layer Manager icon.

44. Locate all "demo" layers and freeze them.

> The layer names will have XREF names as a prefix, like First Floor Existing A-Wall-D [First Floor Existing A-Wall-GR] for example. Demo layers might end in "-Demo" or simply "-D" ["-Demo" or "-GR"]

You can use a Layer Properties Filter to make this task easier. On the toolbar across the top of the Layer Properties Manager, click the New Property Filter icon (you may need to pause over each icon to see a tool tip). In the Layer Filter Properties dialog that appears, give the filter a name like: **Demo Layers**. Click in the Name field, and type: ***D** [***R**]. You can add more than one criterion, so click in the Name field beneath that one and type: ***Demo** (see Figure 16–30). Notice that this filter will also capture the threshold layers like A-Door-Thld. This is because this layer also ends in "d." This is fine as our goal here is to shorten the list of layers so we can easily locate the demolition layers.

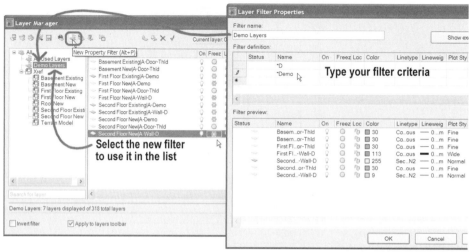

Figure 16–30 *Add a New Layer Property Filter*

45. After you have frozen all "demo" layers, click All in the tree on the left of the Layer Properties Manager to return to an unfiltered list and then click OK to apply the changes and dismiss the Layer Properties Manager.

 NOTE Be sure to use Freeze, turning them Off will not work.

46. Refresh the two Sections.

The Basement Stairs should no longer appear in the Sections.

SUBDIVISIONS

Subdivisions are physical zones within the Bldg Section/Elevation Line boundary that designate different Display properties when the Section/Elevation is generated. Typically, each zone is assigned a lighter lineweight as it moves farther from the defining line. In this way, you can begin to introduce depth to your Sections and Elevations.

REGENERATE THE REMAINING ELEVATIONS

The North and West Elevations are good places to explore subdivisions because the plan steps back in a few places. Before we begin, let's add the *Roof New* file to these Elevations as we did with the Sections above.

1. Select the North Elevation, right-click and choose **Regenerate**.

2. From the Style to Generate list, be sure that: *2D Section Style 96* [*2D Section Style 100*] is selected.

3. As we did above, click the Select Additional Objects icon, pick the *Roof New* file in the drawing (remember, click the eaves or valleys for easy selection) and then click OK to finish and regenerate the elevation.

4. Repeat this on each of the other two elevations.

ADD SUBDIVISIONS

5. Zoom the drawing so that you can see both the North Elevation and the house model comfortably on screen.

6. Click to select the North Elevation.

Notice how one of the Bldg Elevation Lines in the model has highlighted in red. This is the Bldg Elevation Line used to generate this elevation.

7. Select the highlighted Bldg Elevation Line, right-click and choose **Properties**.

At the top of the Properties palette, "All (2)" will appear in the selection list.

8. From the selection list at the top of the Properties palette, choose **Bldg Elevation Line (1)**.

9. In the Dimensions grouping, click the Subdivisions worksheet icon (see Figure 16–31).

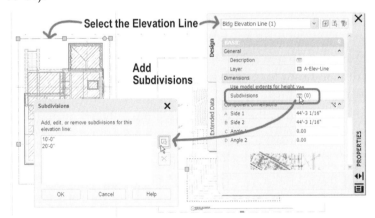

Figure 16–31 *Add a subdivision*

10. In the Subdivisions worksheet, click the Add icon.

A Subdivision will appear automatically set to 10'-0" [3000]. It may be edited here, but it is usually easier to grip edit it in the drawing instead.

11. Click Add again.

This one will come in at 20'-0" [6000].

12. Click OK to dismiss the worksheet and add the subdivisions to the drawing.

Since we changed the Display Properties of the Bldg Section/Elevation Line object earlier in this chapter, the two Subdivision lines come in magenta. This makes them easier to see and edit. Notice that each subdivision has a triangular shaped grip. If you hover over this grip, the dimension of the subdivision will appear. The Subdivision lines are located pretty well where the default dimensions placed them. The one further away at 20'-0" [6000] sits in the space between the back Wall of the addition and the Wall of the patio on the second floor. This will make the linework of Wall behind the patio Wall lighter than the back Wall of the addition. The closer subdivision may need some adjustment. The transition between one subdivision and another is abrupt. If the subdivision cuts across the hip portion of the porch Roof you will get a sharp line on that Roof surface indicating the break between subdivisions. In this case, you would want to move the subdivision using the grip back a bit to cut through the ridge of the porch roof.

> Using the grips on the subdivisions, make any necessary adjustments to their position.

13. Refresh the North Elevation.

You should see a shift in color with the porch reading in one color, and the rest of the elevation using a different color as it recedes from us.

14. Repeat the process on the West Elevation. Be mindful of where the Roof ridges and valleys occur as you are placing the subdivision lines.

The difference in color is very subtle in the default 2D Section/Elevation Style. Furthermore, the color used for Subdivision 3 is the same as Subdivision 2, so it is not as obvious where this break occurs.

Let's edit those colors so that the transition from one subdivision to the next becomes more recognizable.

15. On the Layer Properties toolbar, click the Layer Manager icon.

16. Change the color of A-Sect-Thin to **20** and the color of A-Sect-Fine to **31** and then click OK.

These colors will have no impact on plotting. The Lineweight and Plot Style settings will determine how these layers plot. We are choosing brighter colors for the items that are closer to use and softer colors for those that recede back. This is being done simply to make the drawing easier to read on screen.

 NOTE The default out of the box for ADT is Named Plot Styles, if you use Color-Dependent Plot Styles in normal production, then a change like this might affect plotting.

17. Select either elevation, right-click, and choose **Edit 2D Section/Elevation Style**.

18. Click the Display Properties tab and then the Edit Display Properties icon.

19. Select Subdivision 1, and set its Layer to **A-Sect-Medm**, select Subdivision 2, and set its Layer to **A-Sect-Thin**. Subdivision 3 can remain **A-Sect-Fine** (see Figure 16–32).

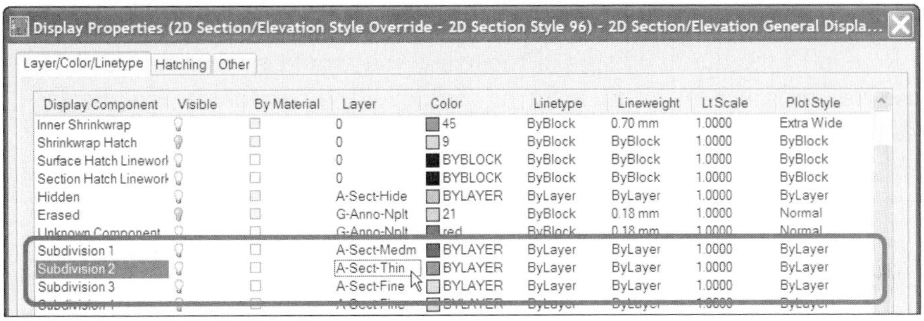

Figure 16–32 *Making the Surface Hatching use Subdivision properties*

20. Click OK to return to the drawing.

You will see some of the linework change to the new color. Notice how all of the Hatching remains unchanged. If you wish, you can make the Surface Hatching use the same properties as its Subdivision. To do this, edit the Style again, choose Edit Display Properties, click the Other tab and then place a check mark in the "Use Subdivision properties for surface hatching" check box. Although this might at first seem a logical option to enable, if you try it you will see that there will no longer be any contrast between the Surface Hatching and the Subdivision linework. Also, the hatching in the foreground (Subdivision 1) will display much too bold. It is not recommended that you use this setting.

21. Save the file.

2D SECTION/ELEVATION STYLES DESIGN RULES

Subdivisions allow us to address the overall needs of the Section/Elevation object, but they don't offer enough flexibility. If a piece of geometry falls within a particular subdivision, it will use those display properties assigned to that subdivision regardless of any other display settings the object itself may have. Design rules can be used to give more control over precise graphical display.

DOOR AND WINDOW SWING DESIGN RULES

Design rules are linked to object colors in the model. A design rule will search the drawing model for a particular color and render the objects of that color to a particular component when generating the 2D Section/Elevation. To take fullest advantage of this functionality, you need to be familiar with which colors have been assigned to objects in the template files.

If you look at the Doors shown in the North Elevation, you will notice that their swings display in dashed lines despite the particular subdivision in which they occur. As we mentioned above, the Section_Elev Display Set is used to generate Elevations and Sections. If you were to edit this Display Set in the Display Manager (Chapter 2), you would learn that it uses the Door object's Elevation Display Representation. If you were to then open any Construct, and edit any Door's Elevation Display Rep and examine its Display Properties, you would further find that the Doors use color number 54 for their Swings by default. Therefore, to build a Design Rule for Door Swings, it must make reference to Color Number 54. This has been done already in the default 2D Section Style 96 [2D Section Style 100] Style that we are using here. This is why Door swings are dashed.

1. Select any Section or Elevation on screen, right-click and choose **Edit 2D Section/Elevation Style**.

2. Click the Design Rules tab.

   Take notice of Rule 1.

This rule states that if an item is found on color 54, and it is visible, then it should be rendered by the 2D Section/Elevation Style to the "Swing Lines" component (see Figure 16–33). Remember, the 2D Section/Elevation object performs a hidden line removal when it generates the section or elevation. Therefore, this rule (because the objects must be visible) ignores all items that are not visible regardless of color. The "Swing Lines" component is a custom component added to this style specifically for Door and Window Swings. You can see this on the Components and Display Properties tabs.

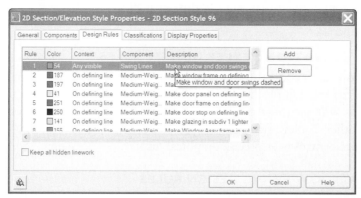

Figure 16–33 *Many Design Rules have been included in the default 2D Section/ Elevation Style*

Feel free to further examine the other rules on this screen. The descriptions are helpful in explaining the purpose of the rule. In order to understand and ultimately to build your own rules, the most critical requirement is a color that you can assign to a unique purpose. In this case, color 54 has been reserved in the default templates to Door and Window Swings only. If you were to assign some other object to color 54, and then generate an elevation from it using this style, those objects would render dashed in the elevation object. Using this logic, let's make a simple Design Rule.

While making Door and Window Swings dashed for swinging Doors and Windows is useful, perhaps the small arrow swing indicator on the double hung Windows is not desirable. We can create a Design Rule to erase all of these automatically provided that we can assign a unique color to this component in the model. We will use color 56 for this purpose. However, since we are already editing the 2D Section/Elevation Style, let's add the rule first.

3. Click the Add button.

   Rule 14 will appear.

4. Next to Rule 14, in the Color column, click color Red.

5. In the Select Color dialog box, choose color **56** (you can just type the number), and then click OK.

6. In the Context column, choose **Any visible** from the dropdown list.

7. In the Component column, choose **Erased** from the dropdown list.

8. In the Description field, type **Erase Double Hung Window Swings** (see Figure 16–34).

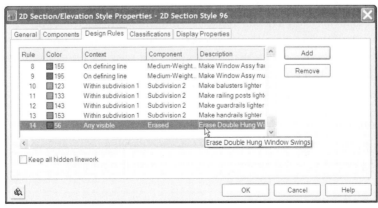

Figure 16–34 *The Door Swings rule completed*

9. Click OK to return to the drawing.

At the moment, there will be no change evident.

10. On the Project Navigator palette, click the Constructs tab, and then double click *First Floor Existing* to open it.

11. Select any Double Hung Window in the drawing, right-click and choose **Edit Window Style**.

12. On the Display Properties tab, place a check mark in the Style Override check box next to Elevation (it will not be bold).

    In the Display Properties dialog, on the Layer/Color/Linetype tab, notice that the Swing color is currently 54, as we noted above.

13. Change the color of the Swing component to **56** and then click OK twice to return to the drawing.

14. Save and Close the file and then repeat the process in the *Second Floor Existing* Construct.

15. Back in the *A-CM01* View file, Reload the XREFs.

16. From the View menu, choose **Refresh Sections/Elevations**.

17. In the Batch Refresh 2D Section/Elevations, be sure that Current Project is chosen and then click the Start button.

    In the background behind the dialog, you will see each elevation and section highlight as it is being refreshed.

This is a very powerful command that can be used to refresh every 2D Section/Elevation object in the entire project. It will scan every Project drawing and then refresh the sections and elevations. In this case, the current file is the only one that contains 2D Section/Elevation objects. Therefore, we could also have simply selected them all and manually chosen Refresh from the right-click menu. Use caution with this command as it will refresh every section and elevation. If for some reason there are elevations or sections that you do not wish to refresh, do not use this command, or use the Folder option instead.

18. In the Batch Refresh 2D Section/Elevations, click the Close button.

19. Zoom and Pan the drawing and examine the results.

    Notice that Door and Window Swings are still dashed except the Double Hung Windows in the existing house which have disappeared.

 NOTE Other Double Hung Windows are unaffected. If you wish however, you can edit those Styles and change the Swing color to 56 and then refresh the elevations to apply the rule to them as well.

CAD Manager Note: You will want to reserve a pool of colors for this type of editing. By setting some colors aside, you will ensure that someone in the office will not inadvertently choose a color already being used by other objects. This will help you avoid future hassles in trying to track down the reason that the section design rules do not work correctly. You can also pre-assign as many colors as you wish in the appropriate Display Representations to facilitate 2D Section/Elevation object generation. Don't forget to map those same colors to the Design Rules in your 2D Section/Elevation styles. Materials make configuring Design Rules a bit more challenging. When a particular component is set to By Material, it will look to the Material Definition for its color, rather than the component color. It may not always be possible to reserve a color for a Design Rule and for its purposes in a Material Definition. In this case, you may need to build a Custom Display Set for generating 2D Section/Elevations.

 ## EDITING MATERIALS

We have already seen quite a bit of materials throughout the chapters of this book. Perhaps it is time we took a slightly more formal look at them. This topic is brief, but should prove sufficient to give you some idea of what the materials functionality in Architectural Desktop is all about. A Material Definition is a complete collection of display settings (plan, section, elevation and 3D rendering), designed to portray a single real-life material. Several dozen pre-made materials have been defined and included with ADT out of the box. We have seen brick, concrete, rood shingles, drywall, paint, wood and more so far in the exercises in this book. Materials are assigned to objects via their object style. For example, each component of a Wall style is assigned to a particular Material Definition. When you assign a brick or concrete material to a component in a Wall style, the material controls the hatching and linework used to represent that brick in plan, section and elevation. In addition, materials also have the ability to reference high-quality photo-realistic textures that appear within ADT in shaded viewports and remain attached when you link your ADT Model to VIZ Render. VIZ Render is a stand-alone application included in the box with ADT 2006. It is used to generate photo-realistic renderings and animation (see Chapter 20 for more information). Perhaps the best feature of materials is that they allow you to think of your ADT Models in real-life terms. If your project will use three types of brick and two types of CMU, you define these five materials in ADT and use them in your project wherever they occur. One set of settings will govern your brick in all plans, sections, elevations and 3D renderings. If brick number two changes mid design from red to blond brick, you simply need to edit the Material Definition, and all of the objects using it will update.

EDIT THE AIR GAP MATERIAL DEFINITION

If you zoom in on the Sections again, and take a close look at the exterior Walls, you will notice that the Air Gap for the Wall Style is rendering "hollow" and that in some cases you can see linework for other Walls beyond. In addition, since the Air Gap is being represented as a void, this space is also being shrink-wrapped by the Section. To see the shrinkwrap best, toggle on the Lineweight Display temporarily—click the LWT icon on the Application Status Bar. Let's modify the Material Definition used for Air Gap to make this void less prominent in the Sections.

Leave the *A-CM01* file open while you perform these steps.

1. On the Project Navigator, open the *First Floor New* Construct.

2. Select one of the exterior masonry Walls, right-click and choose **Edit Wall Style**.

3. Click the Materials tab.

Each material of the Wall style is listed here with its material assignment.

4. Click on the Air Gap component and, at the top right corner of the dialog box, click the Edit Material icon (see Figure 16–35).

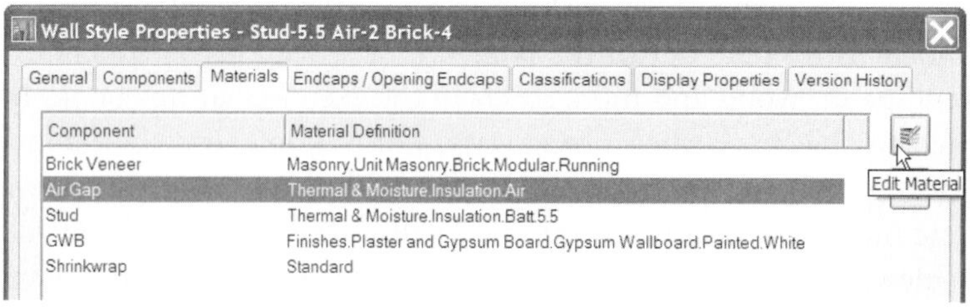

Figure 16–35 *You can edit Material Definitions directly from the Materials tab*

Material Definitions contain only a General and a Display Properties tab.

Even though there are eight Display Representations for Material Definitions, the one we want to edit is the one used by the Section_Elev Display Set. This is General Medium Detail (which also happens to be the currently active Display Rep). To see this for yourself, choose Display Manager from the Format menu, expand Sets and select the Section_Elev Set. On the Display Representation Control tab, notice that General Medium Detail is checked for Material Definition.

5. On the Display Properties tab, with General Medium Detail selected, click the Edit Display Properties icon.

6. Click the Layer/Color/Linetype tab.

Material Definitions contain eight Display Components:

▶ **Plan Linework**—This is the linework used to draw the shape of the component in plan. For instance, for Walls it would be the two parallel lines and the Endcap shape of the component.

▶ **2D Section/Elevation Linework**—This is the linework of components that you are looking at "beyond" in Sections and Elevations, in other words, any linework contained in the subdivisions.

▶ **3D Body**—This is the used to determine the linework for objects cut by the defining line in 2D Section/Elevation objects. It is also used to hide the object behind the material. In Live Sections and Models, this is the actual 3D object displayed for each material.

▶ **Plan Hatch**—Any plan hatching within the Plan Linework components, such as the hatching for brick in plan.

▶ **Surface Hatch**—Used for all surfaces that you are looking at "beyond" in Sections and Elevations, in other words, any hatching contained in the subdivisions.

▶ **Section Hatch**—This is the hatching for objects cut by the defining line.

▶ **Sectioned Boundary**—In Live Sections, this is the outline of the defining line.

▶ **Sectioned Body**—In Live Sections, this is the portion of the 3D object that is sectioned away. Render materials can be applied to this to make it display.

As you can see, in this case nearly every component is turned off. In order for other objects to not appear beyond the Air Gap void and to prevent it from shrinkwrapping, we must turn on the 3D Body component. As was mentioned above, one of the functions of the 3D Body is to hide other components behind it.

7. Click the light bulb icon next to "3D Body" to turn it on.

    By turning on only the 3D Body and not the hatching, the Air Gap linework will still display as a gap, but it will no longer show objects behind nor will it shrinkwrap.

8. Click OK three times to return to the drawing.

EDIT THE GLAZING MATERIAL

While we are in the *First Floor New* file, it might also be a good idea to make a slight modification to the glazing material used by the Windows. When we section a Window, you probably don't want the shrinkwrap to outline the glass. This makes the glass too bold. Let's edit the glass material to fix this.

9. Select one of the Windows, right-click and choose **Edit Window Style**.

10. Click the Materials tab.

11. Click on the Glass component and at the top right corner of the dialog box, click the Edit Material icon.

12. On the Display Properties tab, with General Medium Detail selected, click the Edit Display Properties icon.

13. Click the Other tab.

14. Below the 2D Section Rules item, place a check mark in the "Exclude from 2D Section Shrinkwrap" check box (see Figure 16–36).

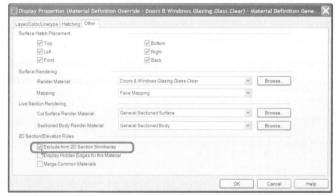

Figure 16–36 *Excluding the glass material from the 2D Section/Elevation Shrinkwrap*

As you can see, this is also the tab where you choose Render Materials and set up Live Sectioning Rules. See Chapter 19 for more information on these settings.

15. Click OK three times to return to the drawing.

16. Save the file.

EXPORT CHANGES TO THE SECOND FLOOR

Both of these changes are needed on the *Second Floor New* file as well.

17. On the Project Navigator, open the *Second Floor New* Construct.

18. From the Format menu, choose **Material Definitions**.

The Style Manager will open filtered to the *Material Definitions* item beneath the *Multi-Purpose Objects* folder of the *Second Floor New* file. We can use Style Manager to copy the modified definitions from the *First Floor New* file and update those contained in the *Second Floor New* file.

19. On the tree at left, below the *First Floor New.dwg* entry, expand the *Multi-Purpose Objects* folder and then highlight *Material Definitions*.

20. Hold down the CTRL key and select ***Doors & Windows.Glazing.Glass.Clear*** and ***Thermal & Moisture.Insulation.Air***.

    They should both be highlighted.

21. Right-click and choose **Copy** (see left side of Figure 16–37).

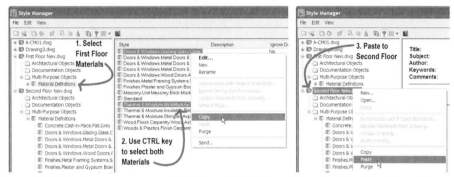

Figure 16–37 *Copy the two modified Material Definitions from the First Floor New file*

22. On the left side in the tree view, right-click on the *Second Floor New.dwg* entry and choose **Paste** (see right side of Figure 16–37).

An alert dialog box will appear warning you that the two Definitions already exist in the destination drawing. By choosing the Overwrite Existing option, we will be updating them to match the new versions that edited in the First Floor.

23. In the Import/Export – Duplicate Names Found dialog box, choose **Overwrite Existing** and then click OK (see Figure 16–38).

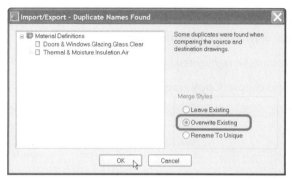

Figure 16–38 *Choose the Overwrite existing option to update the Second Floor with the Definitions from the First*

24. Click OK to close the Style Manager and accept the changes.

If you would like to verify that the changes have been copied to the *Second Floor New* file, fee free to repeat the steps used above to edit the Materials in the *First Floor New* file.

CAD Manager Note: The process outlined here is the manual way to update a Style from one project file to a newer version from another project file. The same result could be achieved by enabling Project Standards for the residential project as we have done in the commercial project. If you were to do so, you would need to designate a drawing as the Standards Drawing for the residential project such as the *Residential Styles [Residential Styles – Metric]* file in the *Standards\Content* folder at the root of the project folder. Open that file and add the modified Material Definitions to it. Next, you would need to configure Project Standards as we did for the commercial project in Chapter 8. In the Configure Project Standards dialog we could assign the *Residential Styles [Residential Styles – Metric]* file as our Standards Drawing. Be sure to check 'Material Definitions as one of the Style Types to synchronize. Finally, perform synchronization.

25. Save and Close both the *First Floor New* and the *Second Floor New* files.

UPDATE THE SECTIONS

The *A-CM01* file should still be open, if you closed it, use the Project Browser to re-open it.

Back in the *A-CM01* file, you should receive an alert that the *First Floor New* and the *Second Floor New* XREFs have changed at the lower right corner of your screen (similar to Figure 16–11).

26. Right-click the XREF quick pick (small binder clip icon), and choose **Reload XREFs**.

27. Select both Sections, right-click, and choose **Refresh** (see Figure 16–39).

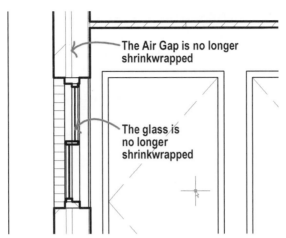

Figure 16–39 *Windows and Air Gaps no longer have shrinkwrap applied (toggle LWT to see clearly)*

If necessary, you may need to move the Bldg Section Line objects slightly to get them to cut through a Window to see the results.

EDIT BRICK COURSING

Zoom in on the East Elevation at the point where the First and Second Floors meet. Notice that the brick coursing does not match. This is because the height of the Walls on the First Floor is 9'-0" [2750] and not an even brick dimension.

28. On the Project Navigator, open the *Second Floor New* Construct.

29. Select the right vertical exterior Wall, right-click and choose **Isolate Objects > Edit In Elevation**. Follow the prompts like we did for the chimney above.

30. Select the brick Wall, right-click and choose **Edit Wall Style**.

31. On the Materials tab, select the Brick Veneer component and then click the Edit Material icon.

32. With General Medium Detail highlighted, click the Edit Display Properties icon and then click the Hatching tab.

33. In the Y Offset field of the Surface Hatch component, type: **-1.33" [-34]** (see Figure 16–40).

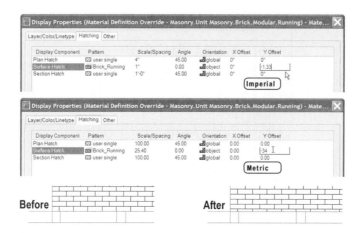

Figure 16–40 *Edit the Y Offset of the Material Definition Surface Hatching to shift the Brick Coursing*

34. Click OK three times to return to the drawing.

You will see the results of the Brick Hatch Offset right away in the model file (see bottom of Figure 16–40). Let's return to the Section and Elevation Composite Model file and refresh the elevation to see the overall effect.

35. Click the Exit Edit in View icon to return to the previous view and restore the hidden objects.

36. Save and Close the *Second Floor New* file.

    Back in the *A-CM01* file, you should receive an alert that the *Second Floor New* XREF has changed.

37. Reload the XREF and then Refresh the East Elevation.

ELIMINATE THE LINES BETWEEN FLOORS

The brick coursing looks a lot better, but you may have noticed that there is a line between floors in the elevations. This line is the top and bottom edge of the individual Walls for the First and Second Floors. We can however, configure that Material to remove this line for us. We do this by instructing the Material to merge with similar Materials. We can make this change directly in the Section and Elevation Composite Model file.

38. On the Format menu, choose **Material Definitions**.

All Materials used in the sections and elevations of this file are automatically imported into this file. To make the common Materials from different XREFs merge together, we edit the imported Material Definition here.

39. Double click on the Material Definition named: ***Masonry.Unit Masonry.Brick.Modular.Running*** to edit it.

40. On the Display Properties tab, with General Medium Detail selected, click the Edit Display Properties icon.

41. Click the Other tab and place a checkmark in the "Merge Common Materials" check box (see Figure 16–41).

Figure 16–41 *Merge Common Materials to eliminate the "Lines between floors"*

42. Click OK three times to return to the drawing.

43. Select the East Elevation and choose **Refresh**.

Notice that the line between the brick of the First Floor and the brick of the Second Floor has disappeared (see Figure 16–42).

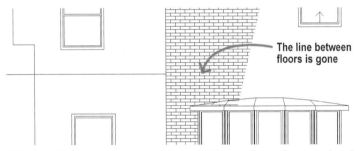

Figure 16–42 *The Brick now appears continuous with the Materials merged and the line between floors removed*

 NOTE While this technique produces the desired merged brick effect, we have now created a slightly different version of the Material in the elevation file than occurs in the Constructs. It should however, not cause any problems.

SHOWING HIDDEN MATERIALS BEYOND

In some cases, you wish to have a particular Material show even though it is hidden from view. For instance, it is often desirable on elevations to have foundation Walls and footings appear dashed below grade. There is a setting in the Material Definition to control this.

44. On the Format menu, choose **Material Definitions**.

45. Double click on the Material Definition named: **Concrete. Cast-in-Place.Flat.Grey** to edit it.

46. On the Display Properties tab, with General Medium Detail selected, click the Edit Display Properties icon.

47. Click the Other tab and place a checkmark in the "Display Hidden Edges for this Material" check box.

48. Also place a checkmark in the "Merge Common Materials" check box as we did with Brick.

This Material will now show dashed when ever it is hidden rather than being removed as other hidden linework is. Also, this material will merge as the Brick did above.

49. Click OK three times to return to the drawing.

50. Select the North Elevation and choose **Refresh**.

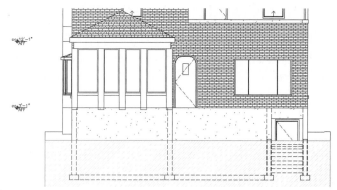

Figure 16–43 *The Concrete Material now shows dashed automatically when hidden*

Notice that the foundation Walls and Footings now display as dashed lines (see Figure 16–43).

51. Refresh the remaining elevations and Save the file.

CAD Manager Note: If you wish, you can include your most commonly used Material Definitions with the merge and/or display hidden edges options enabled in your Section/Elevation View Template file. On the Project Navigator palette, on the Project tab, click the Edit Project Details icon to access the settings for the project. There, you can assign a different template file to General, Section/Elevation and Detail View files. You may include certain Materials in this file that have the settings you wish enabled already configured. This will not guarantee that users will not need to edit these settings for some Materials. Rather this approach would simply eliminate a few steps for the most commonly used Materials.

New in ADT
2006

CREATE A NEW MATERIAL DEFINITION AND TOOL

It might be nice to have a more appropriate hatch pattern for the Terrain Model in our Sections. To do this, let's open the *Terrain Model* file and create a new Material Definition.

52. From the File menu, choose **Open**. Browse to the *Standards\Content* folder at the root of the project folder.

53. Open the *Residential Styles [Residential Styles – Metric]* file.

54. Right-click the tool palettes title bar and choose **16 Commercial [16 Commercial Metric]** (to load the 16 Commercial [16 Commercial Metric] tool palette group) and then click the Project Tools tab.

55. From the Format menu, choose **Material Definitions**.

56. Right-click on the **Standard** Material and choose **Copy**, right-click again and choose **Paste**.

57. Double click on **Standard(2)** to edit it.

58. On the General tab rename it to: **MADT.Site.Earth**.

59. On the Display Properties tab, make sure General Medium Detail is selected and then click the Edit Display Properties icon.

60. On the Layer/Color/Linetype tab, set the Color of the Section Hatch to **44**, its Lineweight to **0.18 mm** and the Plot Style to **Fine** (see Figure 16–44).

Display Component	Visible	Layer	Color	Linetype	Linewei...	Lt Scale	Plot Style
Plan Linework		0	■ BYBLOCK	ByBlock	ByBlock	1.0000	ByBlock
2D Section/Elevation Linework		0	■ BYBLOCK	ByBlock	ByBlock	1.0000	ByBlock
3D Body		0	■ BYBLOCK	ByBlock	ByBlock	1.0000	ByBlock
Plan Hatch		0	■ BYBLOCK	ByBlock	ByBlock	1.0000	ByBlock
Section Hatch		0	□ 44	ByBlock	0.18 mm	1.0000	Fine
Sectioned Boundary		0	■ red	ByBlock	ByBlock	1.0000	ByBlock
Sectioned Body		0	□ 9	ByBlock	ByBlock	1.0000	ByBlock

Display Properties (Material Definition Override - MADT.Site.Earth) - Material Definition General Medium Detail Display ...

Layer/Color/Linetype   Hatching   Other

Figure 16–44 *Set the properties of the Section Hatch*

61. Click the Hatching tab.

62. Click on the entry in the Pattern column next to Section Hatch (currently User Single).

    This will call a Hatch Pattern dialog box.

63. In the Hatch Pattern dialog box, choose **Predefined** from the Type list and **Earth** from the Pattern list, and then click OK.

64. In the Scale/Spacing column next to Earth, type **40 [40]** (see Figure 16–45).

Display Com...	Pattern	Scale/...	Angle	Orient...	X Offset	Y Offset
Plan Hatch	user single	300.00	45.00	global	0.00	0.00
Surface Hatch	user single	300.00	0.00	object	0.00	0.00
Section Hatch	EARTH	40	45.00	object	0.00	0.00

Display Properties (Material Definition Override - MADT.Site.Earth) - Material Definition General Med

Layer/Color/Linetype Hatching Other

Figure 16–45 *Assign a scale to the Earth hatch*

65. Click OK to return to the Style Manager.

66. Drag the **MADT.Site.Earth** Material from Style Manager to the Project Tools palette.

67. In the warning that appears, click OK. Click OK to dismiss the Style Manager and then Save the file.

You have created a new Material Definition and a tool referencing it. We can now use this tool to quickly apply this new material to other objects like the Mass Element in our *Terrain Model* Construct.

68. Close the *Residential Styles* [*Residential Styles – Metric*] file.

New in ADT
2006

APPLY A MATERIAL TOOL TO AN OBJECT

69. On the Project Navigator, open the *Terrain Model* Construct.

As we have already seen, the Terrain object is a Mass Element. Mass Elements have materials assigned to them via their styles like all other ADT objects. Our primary concern in this topic is the Section Hatch for the Mass Element. We will apply our new Material Definition to this Mass Element.

70. Select the Terrain Mass Element, right-click, and choose **Copy Mass Element Style and Assign**.

71. On the General tab, name the Style: *Terrain* and then click OK to return to the drawing.

72. On the Project Tools palette, click the new **MADT.Site.Earth** Material tool.

73. When prompted, click the Mass Element. Choose **Style** from the Apply to drop down list and then click OK.

If you want to see the result right away, try **Isolate Objects > Edit in Section**. The process is very similar to Edit in Elevation except that you pick two or more points to define the section line rather than highlighting existing geometry. To try it, select the Mass Element, right click and choose **Isolate Objects > Edit in Section**.

Following the prompts, click two points that cut across the Mass Element and then press ENTER. You will be zoomed to the section cut and will see the new hatch pattern applied on the cut surface. Click the Exit Edit in View icon to restore the previous view and exit the Edit in View mode.

74. Save and Close the file.

75. Back in the *A-CM01* file, reload the XREF and refresh the Sections.

76. Following the steps in the "Apply a Material Boundary" topic above, Draw some polylines and add a Material Boundary to each side of the Section (see Figure 16–46).

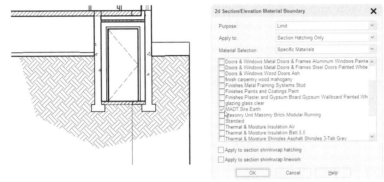

Figure 16–46 *You can add Material Boundaries to crop out part of the Earth hatching.*

To make a Material Boundary like the one shown here, draw the polyline boundary, right-click the Section and choose **Material Boundary > Add**. Select the polyline and answer **Yes** to erase it. Choose to apply it to Section Hatching only, and then pick **Selected Materials** for the Material Selection choice. Finally check only the *MADT.Site.Earth* material and then click OK. Repeat on the other side with a new polyline. You can add as many Material Boundaries as you want to a single Section or Elevation. Add some more for the surface hatching on the Roof. For instance, you would not want to see shingles on the inside of the house.

77. Save the file.

EDITING AND MERGING LINEWORK

Subdivisions give a broad-brush level of control over 2D Section/Elevation Display Properties. Design rules and materials allow for more focused refinement of the 2D Section/Elevation. The final level of control over the display of Sections and Elevations comes from the Edit Linework and Merge Linework commands. With these commands, you can "reach into" the 2D Section/Elevation object and edit the actual linework from which the Elevation is comprised. The editing capabilities are limited to changing selected linework from one component index to another, deleting unwanted linework while editing, and adding new linework while using Merge. In most cases these will prove sufficient to fine-tune the Section/Elevation in acceptable fashion.

ERASE 2D SECTION/ELEVATION LINEWORK

It is nice to have a bold line at the ground line, but we probably don't want it to continue all the way around, particularly if you have added Material Boundaries to crop the Earth hatching. Let's use Edit Linework to remove it.

1. Select the Section and then click the small round gray Edit Linework grip.

    When you enter the Edit Linework mode, the hatching will temporarily disappear. Don't worry—it will return when you are finished.

2. Select the bottom horizontal line, and the two vertical lines that make up the bottom of the terrain object, as well as two stray vertical lines within the terrain boundary in the Section (see Figure 16–47).

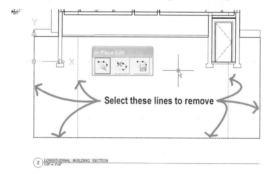

Figure 16–47 *Select bottom, sides and stray lines of the Terrain to erase*

3. Press the DELETE key to erase the three lines.

4. On the floating In-Place Edit toolbar, click the Save All Changes icon (see Figure 16–48).

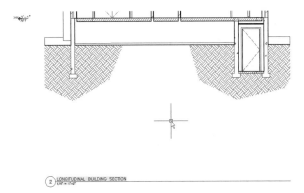

Figure 16–48 *The Section with terrain edges removed*

> 5. Repeat these steps on the other section and all elevations.

Zoom around the drawing and look for other lines to erase.

MODIFY 2D SECTION/ELEVATION LINEWORK

Sometimes you will want to edit the way a particular line appears. The lines of the gables facing us on the North Elevation could be a little bolder.

> 6. Zoom over to the North Elevation, select it, and then click the Edit Linework grip.

> 7. Select the lines of the top edges of the left-hand gable as shown in Figure 16–49.

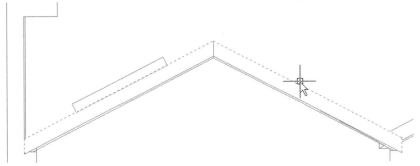

Figure 16–49 *Select the lines of the gable*

> 8. Right-click and choose **Modify Component**.

In the Select Linework Component dialog box, the Linework Component of the selected lines is currently set to Subdivision 2. This makes sense considering how we configured the subdivisions above. To make these lines display a bit bolder, we can move them up one subdivision. In the Linework Component list are all of the display components of the 2D Section/Elevation object. Simply choose Subdivision 1 to make these lines that physically fall within Subdivision 2 display as though they were in Subdivision 1.

9. Choose **Subdivision 1** from the Linework Component list and then click OK (see Figure 16–50).

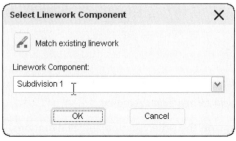

Figure 16–50 *Choose the new component for display*

10. Repeat the process on the gable at the right, choosing **Subdivision 2** this time.

11. On the In-Place Edit toolbar, click the Save All Changes icon.

MERGE LINEWORK

Sometimes to edit linework, it is easier to re-draw it. You can draw any linework with the standard tools on the Shapes toolbar, and then merge them into the Section. When you do, you will be prompted to choose a component, just like the Modify Component commands above.

12. Draw any shape from the shapes toolbar, and then right-click an Elevation and choose **Linework > Merge**.

13. When prompted, select the linework and then press ENTER.

14. In the Select Linework Component dialog box, choose a component from the Linework Component list and then click OK.

The selected shapes will now be part of the Elevation. Sometimes it is easier to use this technique to edit the linework of the 2D Section/Elevation objects. For instance, zoom in on the porch on the East Elevation. Notice how the Wall of the foundation is overlapping the pilasters of the porch. You can try erasing the unwanted lines with Edit Linework, but the line of the foundation Wall is one continuous line. Instead try this:

15. Draw some small line segments with the Intersection Osnap as shown in Figure 16–51.

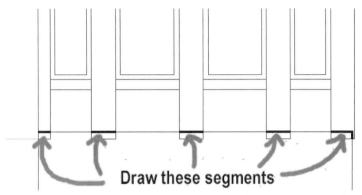

Figure 16–51 *Draw several small segments at the points that you want erased*

16. Select the Elevation, right-click, and choose **Linework > Merge**.

17. At the "Select objects to merge" prompt, select all of the lines drawn in the previous step and then press ENTER.

18. In the Select Linework Component dialog box, choose **Erased Vectors** from the Linework Component list and then click OK (see Figure 16–52).

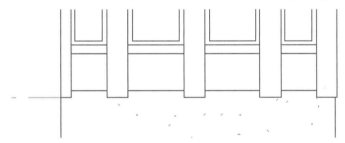

Figure 16–52 *By merging lines to the Erased Vectors Component, you can erase the underlying linework*

UPDATE 2D SECTION/ELEVATION OBJECTS WITH USER EDITS

Inevitably the design will change and the section will need to be updated. When the update occurs, the 2D Section/Elevation object will automatically re-apply all user edits. However, sometimes the nature of the change is such that some of the edits will no longer be in the same physical location as the newly regenerated model geometry. When this occurs, user edits can be saved to a second 2D Section/Elevation object and merged back into the section after the update. To do this effectively, you need to build another 2D Section/Elevation style. The basic process is as follows:

1. Create a New 2D Section/Elevation style (**Format > Style Manager**) named *User Edits*.

2. In this new style, On the Display Properties tab, turn on *all* components, especially Erased. Set the Erased component to an alert color such as Magenta.

3. Regenerate (not Refresh) the Section and choose the **User Edits** style from the "Style for User Linework Edits if Unable to Reapply" list (see Figure 16–53).

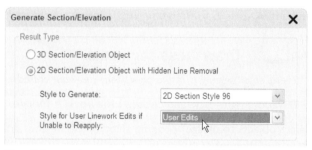

Figure 16–53 *User Edits can be saved to another style if ADT is unable to re-apply them automatically*

4. After the update, explode the User Edits Section, move the linework into place on the Section, and then reapply it with the **Linework > Merge** command (see Figure 16–54).

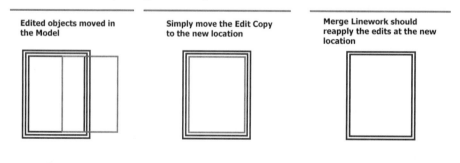

Figure 16–54 *Move saved copy of user edits into position and then Merge Linework*

CAD Manager Note: Although "off" is the proper setting for Erased Vectors in styles used in production, it is recommended that one style be maintained for saving user edits that has the Erased component turned on and set to color 6-Magenta. You can name the style "User Edits." This will make it possible to easily see and merge the second section after an update. You can even do this as the Drawing Default, because all of your production styles will have style-level overrides. The choice is up to you. All 2D Section/Elevation styles, including this one, should be saved as part of the office standard template files.

It is important to have the colors used for Design Rules established and well documented. Properly configuring and documenting this powerful tool will enhance the leverage of creating and managing Sections and Elevations from the ADT model.

5. Save and Close the *A-CM01* Section and Elevation Composite Model file.

DRAG AN ELEVATION TO A SHEET

We began this chapter by re-creating the East Elevation. As a result, it must be placed on its Sheet over again. This will also complete the cross-referencing of the field codes within the Elevation Callouts on the plans.

6. On the Project Navigator palette, click the Sheets tab and then double click *A-202 Elevations* to open it.

7. On the Project Navigator palette, click the Views tab and then expand the plus (+) sign beneath *A-CM01* to reveal the Model Space Views contained within.

8. Drag the East Elevation Model Space View and drop it on the Sheet.

9. At the "Insertion point" prompt, click a point to place the Elevation on the Sheet.

10. Save and Close the file.

    If you wish, open any of the Floor Plan View files to see the updated Callouts referencing the East Elevation as drawing 1 on Sheet A-202.

 ## CREATING INTERIOR ELEVATIONS USING CALLOUTS

Interior Elevations are achieved simply by manipulating the size and shape of the Bldg Section and Bldg Elevation lines. By default ADT uses the full extents of the model for the height of the Elevation or Section. By changing this setting to a fixed height that matches your ceiling height and cropping the sides of the Section/Elevation Line to the size of the room, your result will be an interior Elevation. This can be achieved automatically using the Callouts designed for this purpose.

USE AN INTERIOR ELEVATION CALLOUT

1. On the Project Navigator, open the *A-FP01* View file.

As you recall, this is our First Floor Plan View file. We will create the Callouts here and through the routine create a new View file scaled for Interior Elevations.

2. On the Callouts palette, click the ***Interior Elevation Mark B1*** tool.

   If you do not see this palette or tool, right-click the Tool Palettes title bar and choose Document (to load the Documentation Tool Palette Group) and then click the Callouts tab.

3. At the "Specify location of elevation tag" prompt, click a point within the Dining Room Space.

4. At the "Specify direction for first elevation number" prompt, move the mouse straight up with POLAR or ORTHO and then click.

This will be the direction of Elevation 1. There will be four total going clockwise from the first. The Place Callout worksheet will appear. We have seen this worksheet on several previous occasions. In the New Model Space View Name field there is only one name suggested this time. You can still enter the name for all four Model Space Views. Simply separate the names with a semicolon. It is recommended that you input four unique names, other wise you will get names that simply have a numeric suffix appended to the end, such as Elevation, Elevation (2), etc.

5. In the New Model Space View Name field, type: **North Dining Room Elevation;East Dining Room Elevation;South Dining Room Elevation;West Dining Room Elevation**.

 Caution: *Be certain to include the semicolon separators between each name and do not put a space after the semicolons.*

6. Accept the remaining defaults and then click the New View Drawing icon (see Figure 16–55).

Figure 16–55 *Setting up interior Elevations in the Callout Worksheet*

Again, by now in this text, we have created several View files, so this process should be review. We are creating a new View file because the interior elevations are a different scale than any of the other Views we have created so far.

   7. Name the new View: **A-EL01** with a Description: **Architectural Interior Elevations**.

   8. Click Next, verify that only the First Floor (both Existing and New Divisions) is selected, click Next again and verify that *First Floor New* and *First Floor Existing* Constructs are selected and then click Finish.

   9. At the "Select space(s) and area(s)" prompt, select the Dining Room Space (click the hatching) and then press ENTER.

The full prompt here includes an option to "ENTER to pick region to specify elevation line width." If you have Space objects as we do here, then simply select the Space or Spaces. The Bldg Elevation Lines will be created to match the width and height of the Space object. If you press ENTER instead, you will be prompted to select a rectangular region, which will be used to determine the widths of the elevations and later you will be prompted for the height. Regardless of the option you select, you will be prompted next for the Depth of the elevations.

   10. At the "Specify elevation line depth" prompt, drag into the room slightly and then click.

If you wait a second or two, you will see some temporary graphics being drawn on screen to represent the extent of the elevations. Just get them approximately correct with your mouse click. You can always adjust each bldg Elevation Line later as required. At this point the now familiar "** You are being prompted for a point in a different view drawing **" prompt will appear.

11. At the "Specify insertion point for the 2D elevation result" prompt, click a point to the right side of the plan, drag slightly to the right (using POLAR or ORTHO) and then click again.

As with other Callout routines, you will see a progress bar as the elevations are generated.

12. On the Project Navigator palette, double click the new *A-EL01* View file to open it (see Figure 16–56).

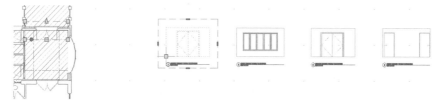

Figure 16–56 *Interior Elevations generated automatically in their own View file*

Using the same Callout Routine, create additional interior elevations of the other major Spaces in the First Floor Plan. Generate any additional elevations in the same *A-EL01* View file (choose the Existing View Drawing icon this time). Create interior elevations of the Second Floor Plan in a new View file named *A-EL02*. You can drag them to the same Sheet.

13. Refer to the process in the "Create the Elevation Sheet File" heading in Chapter 5 and create a new Sheet in the Enlarged Views Sub Set. Name it *A-401 Interior Elevations* and drag the *A-EL02* View file onto it. Follow the prompts to place each Model Space View in sequence.

Note the update of all Callout numbers and references throughout the Project files.

14. Save and Close all open Project files.

LIVE SECTIONS

The A-SC00 View file in the Commercial Project was set up to use a Live Section. A Live Section operates on the actual building model data and can be edited directly, unlike the 2D Section/Elevation object. A Live Section *is* the model. Live Sections make great design tools. Adding one is simple—just add a Bldg Section Line where you would like the Section to be cut, right-click it, and choose Enable Live Section. (There is a generic Bldg Section Line tool on the Design tool palette that adds a Bldg Section Line without a Callout.) Switch to 3D View and have a look. It is that simple! Even though the Live Section that we added was of the entire Commercial Building model and has proved very useful for viewing our progress as we have updated our model, to fully see the benefit of a Live Section you should add one to one of your Construct files. They only display when you are in 3D. In plan (Top View) the entire model will show.

Live Sections are applied to the active Display Configuration. So if you wish to have it applied to a Configuration other than the currently active one, be sure to choose the appropriate Display Configuration prior to enabling the Live Section. Load the Commercial Project and view the Live Section in *A-SC00*. Create one in a Construct in either project if you wish and then play around.

SHEET FILES

The Sheet files for plotting Elevations and Sections were created and configured in Chapter 5. Open them now to view the fruits of your labors in this chapter. A-201, A-202 and A-203 display the Elevations, while A-301 is the Sections. All files were saved with the paper space layout active and Plot Styles displayed. Toggle on the Lineweight display to get a good preview of how these drawings will plot.

ADDITIONAL EXERCISES

There is plenty more work to be done on all of the Sections and Elevations that have been worked on here. Additional exercises have been provided in Appendix A. In Appendix A you will find exercises to add additional Sections and Elevations in the Commercial Project (see Figure 16–57), as well as further refine those created here in the Residential Project. It is not necessary that you complete these exercises to begin the next chapter, they are provided to enhance your learning experience. Completed projects for each of the exercises have been provided in the *Chapter16/Complete* folder.

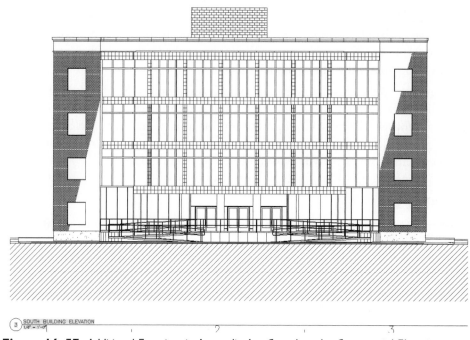

SOUTH BUILDING ELEVATION
1/8" = 1'-0"
3

Figure 16–57 *Additional Exercises in Appendix A—Complete the Commercial Elevations*

SUMMARY

The Section Line boundary determines how much of the model will be considered by the Section.

The Section must be updated to reflect changes to the floor plan.

Callouts completely automate the process of adding a Bldg Section Line and then generating the Section or Elevation including placement of a linked Callout.

2D Section/Elevation styles can be assigned to the Section and Elevation to give them more graphical definition.

Subdivisions are used to assign lineweight and other properties to objects as they recede from the defining line.

Hatching generates automatically from the Material Definitions in the Model.

When cutting a Section or Elevation from XREF models, remember to first reload the XREFs and then update the Section/Elevation.

2D Section/Elevation styles offer a great way to manage Layer, Color, and Linetype settings on 2D Section/Elevation objects.

Layers used in the 2D Section/Elevation style must exist in the drawing prior to being assigned in the Style Display Props.

Design rules offer a way to assign custom display settings to objects within a 2D Section/Elevation regardless of the subdivision in which they occur.

Edit Linework allows the 2D Section/Elevation object to be directly edited.

Merge Linework allows custom drawn linework and previously edited linework to be incorporated into the 2D Section/Elevation object after it has been updated.

Interior Elevations use the same tools with different parameters assigned to the Bldg Elevation Line. Callout with automated routines are provided.

CHAPTER 17

Generating Details and Keynotes

INTRODUCTION

We could not complete our discussion of Construction Documents in ADT without a look at Details and Keynotes. All of the ADT objects covered so far serve overall drawing and modeling needs. Even a very complex and detailed Wall Style can really only accommodate the needs of general plans, sections and elevations at low and medium levels of detail. When you are ready to begin drawing construction details, ADT offers a robust detailing module and a complete keynoting system. In this chapter we will explore both of these tools to help round out the CDs for our two projects.

OBJECTIVES

In this chapter, we will look at the Detail Component Manager and its associated tools and tool palettes. The Keynoting system, which is tied to a Microsoft Access database, will allow us to quickly annotate our details and View files. We will also complete our exploration of the Callout tools begun in the earlier chapters of this book with a look at the Callout routines designed to assist in detailing. In this chapter, we will explore the following topics:

- Explore Keynote Assignments.
- Add Reference Keynotes.
- Add Sheet Keynotes.
- Learn to use Detail Callouts.
- Use the Detail Component Manager.
- Explore the Detailing Tool System.

DETAILS

The concept behind detailing in ADT is simple. Standard AutoCAD drafted entities are created on the fly from exact specifications contained in a central database. These components are not connected to the model in any way, but models built with ADT objects such as Walls, Doors and Windows provide an excellent framework for creating such details. Using the tools on the Callouts tool palette, we can crop out areas of the model that we wish to detail. A 2D Section/Elevation object of the designated area is generated using a special non-plotting 2D Section/Elevation Style (see the previous chapter for more information). We then use the robust and fully extensible Detail Component Manager to draft our details directly on top of this sketch 2D Section/Elevation framework. The basis of the Detail Component Manager is a collection of industry standard building components (bricks, steel shapes, bolts, fasteners, insulation, etc) in a variety of standard sizes.

CREATE A DETAIL VIEW FILE

We will begin in the Residential Project where we will create a Typical Wall Section. The first step in the process is to decide from what point in the model you wish to reference the detail. Then use a Callout routine to create a new View file for details from this region. Once we have this rough linework as a guide, we add Detail Components directly in the View file.

INSTALL THE CD FILES AND LOAD THE CURRENT PROJECT

If you have already installed all of the files from the CD, simply skip down to step 3 below to make the project active. If you need to install the CD files, start at step 1.

1. If you have not already done so, install the dataset files located on the Mastering Autodesk Architectural Desktop 2006 CD ROM.

    Refer to "Files Included on the CD ROM" in the Preface for information on installing the sample files included on the CD.

2. Launch Autodesk Architectural Desktop 2006 from the desktop icon created in Chapter 3.

If you did not create a custom icon, you might want to review "Create a New Profile" and "Create a Custom ADT Desktop Shortcut" in Chapter 3. Creating the custom desktop icon is not essential; however, it makes loading the custom profile easier.

3. From the File menu, choose **Project Browser**.

4. Click to open the folder list and choose your *C:* drive.

5. Double-click on the *MasterADT 2006* folder, then the *Chapter17* folder.

    One or two residential Projects will be listed: *17 Residential* and/or *17 Residential Metric*.

6. Double-click *17 Residential* if you wish to work in Imperial units. Double-click *17 Residential Metric* if you wish to work in Metric units. (You can also right-

click on it and choose **Set Current Project**.) Then click Close in the Project Browser.

 IMPORTANT If a message appears asking you to re-path the project, click **Yes**. Refer to the "Re-Pathing Projects" heading in the Preface for more information.

 NOTE The Metric Complete version of the project (provided with the dataset files from the Mastering Autodesk Architectural Desktop 2006 CD ROM) varies slightly from the results indicated here and those included in the Imperial Complete version.

USING DETAIL CALLOUTS

There are several Callout tools specifically designed for details. We can use any of these as a starting point. Please note that you can start a detail without first using a Callout. However, the Callout approach gives you a nice framework upon which to construct your detail.

7. On the Project Navigator, click the Views tab, and then double-click *A-CM01* to open it.

This is the Section and Elevation Composite Model View file that was created in Chapter 5 which was extensively modified in the last chapter. You can generate callouts from the model directly, or from an existing elevation or section. We will do the latter here.

8. Zoom in on the Longitudinal Building Section.

9. On the Callouts tool palette, click the *Detail Boundary B* tool.

If you do not see this palette or tool, right-click the Tool Palettes title bar and choose **Document** (to load the Documentation Tool Palette Group) and then click the Callouts tab.

10. At the "Specify one corner of detail box" click just above the eave to the left (see Figure 17–1).

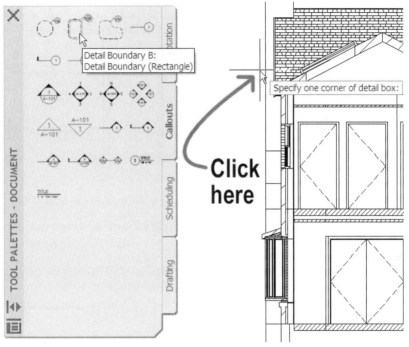

Figure 17–1 *Click a point at the Roof Eave*

11. At the "Specify opposite corner of detail box" prompt, click a point below and to the right of the footing (see Figure 17–2).

Try not to include the Roof Window in the region.

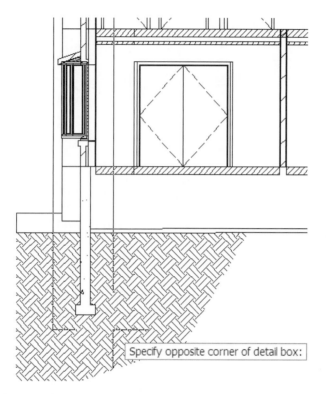

Specify opposite corner of detail box:

Figure 17–2 *Click the other point below the footing*

12. At the "Specify first point of leader line on boundary" prompt, click a point to the left of the boundary and then press ENTER (see Figure 17–3).

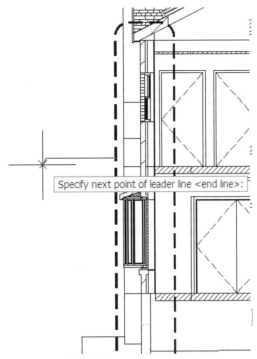

Figure 17–3 *Indicate the location of the Callout*

13. In the Place Callout worksheet, type **Typical Wall Section** for the New Model Space View Name.

14. At the bottom of the worksheet, change the Scale to: **1"=1'-0"** [**1:10**].

15. In the Create in area, click the New View Drawing icon (see Figure 17–4).

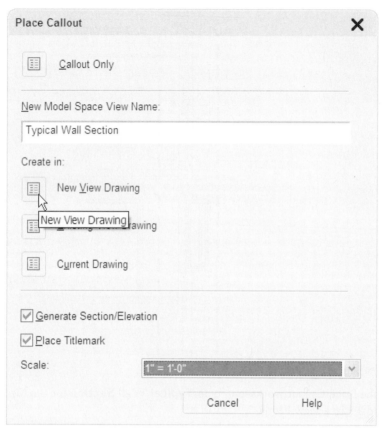

Figure 17–4 *Configure settings for a New View Drawing in the Place Callout worksheet*

The Add Detail View Wizard will appear. This is identical to the other View Wizards that we have seen; however, it uses different default settings. (You can also optionally assign a different template to Detail Views in the Project's Properties.)

16. In the Add Detail View wizard, type **A-DT01** for the Name and **Architectural Details** for the Description, and then click Next.

On the Context page, the Detail View wizard has selected all Levels and Divisions. However, in our case, this is a new construction detail and it is therefore not necessary to include the Existing Division. It would not actually cause any detriment to our detail if we were to leave the Existing Division selected; it will just help to simplify the View file we are generating a bit by excluding unnecessary XREFs.

17. On the Context page, right-click on the Existing Division column and choose **Clear Model Division** (see Figure 17–5). Click Next.

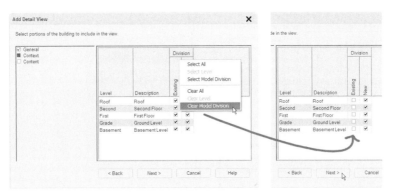

Figure 17–5 *Clear the Existing Division to exclude all of its Constructs from the Detail View file*

> 18. On the Content page, clear the check box next to the *2D Drawings* folder and then click Finish.

The routine is almost complete and a polyline boundary surrounding the designated area has appeared on the section. A detail callout has also appeared. As with other Callout routines used in previous chapters, this one shows temporary placeholder values. At the command line will be the now familiar notice:

You are being prompted for a point in a different view drawing message.

> 19. At the "Specify insertion point for the 2D section result" prompt, click a point to the right of the Longitudinal Section (see Figure 17–6).
>
> The Detail Callout will now appear with a "?" in both the Sheet and Detail Numbers. This is correct for now. Later, when you add this detail to a Sheet, those question marks will update to the correct values.

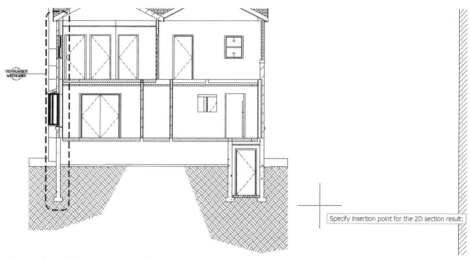

Figure 17–6 *Insert the new 2D Section/Elevation result to the left of the drawing*

20. Save and Close the *A-CM01* file.

NOTE These same Callout routines can be used to create Enlarged Plan drawings and Plan Details. To create an Enlarged Plan drawing, clear the checkmark for Generate Section/Elevation and Create in a New View Drawing. The new View drawing will use an XCLIP to crop all but the detail area. If you leave Generate Section/Elevation checked, you will get a plan view 2D Section/Elevation object. This is a useful starting point for Plan Details. Please see the "Challenge Exercise" at the end of this chapter for more information.

SIMPLIFY THE DETAIL BACKGROUND

Let's now open the *A-DT01* View file and see the results. We want to create a complete Wall Section detail from the footing to the roof. However, since the full height of the Wall is too tall to be inserted on the Sheet at the 1″=1′-0″ [1:10] scale we designated, we will need break lines along the height of the section. We can use the new Cut Line tools in ADT 2006 to assist us in this.

21. On the Project Navigator, click the Views tab, and then double-click *A-DT01* to open it.

Locate the new wall section (it will be drawn in blue linework).

Notice that all of the linework in these 2D Section/Elevation objects is a cyan color. This indicates that all of the linework of the section is on a non-plotting layer. This is done deliberately since the section is intended to be an underlay upon which you can build a detail using ADT Detail Components. If you wish to see how this is accomplished, select one of the 2D Section/Elevation objects, right-click and choose **Edit 2D Section/Elevation Style.** Click the Display Properties tab and click the Edit the Display Properties icon. On the Layer/Color/Linetype tab, notice that all components are assigned to a non-plotting layer. Now let's locate the Bldg Section Line from which this 2D Section/Elevation object is generated.

22. Toggle off Surface Hatch Display with the toggle icon on the Drawing Status Bar (see Figure 17–7).

Figure 17–7 *Toggle off the surface hatching*

23. Click on the 2D Section/Elevation object

Look carefully at model to the right and locate the Bldg Section Line object. It will be highlighted in red while the associated section object is selected. We also witnessed this behavior in Chapter 16. (Notice that the location of this Bldg Section Line matches the location of the Longitudinal Building Section from which it was generated in *A-CM01*).

24. Click on the Bldg Section Line object on the right (it should still be highlighted in red).

Notice how deep the Bldg Elevation Line is. This does not pose a major problem, but the sketch linework of our section will appear cleaner if we reduce the depth of the Bldg Elevation line. We can do this interactively on screen with the Length grip (see the left side of Figure 17–8) or on the Properties palette (see the right side of Figure 17–8). If you still have both the 2D Section/Elevation object and the Bldg Elevation Line object select, the Properties palette, will read All (2) at the top.

25. On the Properties palette, choose **Bldg Elevation Line (1)** from the object list at the top.

When you have more than one type of object selected at the same time, the Properties palette reveals only generic parameters. Use this technique to edit the object specific parameters of a particular class of object in a group selection.

26. In the Component Dimensions grouping, change both Side 1 and Side 2 to: **1'-0" [300]** (see Figure 17–8).

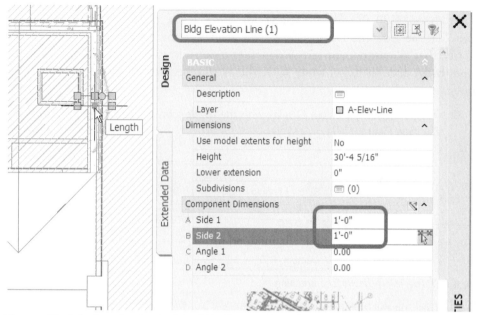

Figure 17–8 *Reduce the depth of the Bldg Elevation Line Object*

When we Refresh the section, some of the extraneous lines beyond will be removed, making the section easier to read.

27. Right-click and choose **Deselect All**.

28. Select the 2D Section/Elevation object, right-click and choose **Refresh**.

ADDING BREAK MARKS

Now that we have simplified our sketch section background, we can add some break lines to crop unnecessary portions of the detail. This will allow us to exclude items like Windows so that our section is more generic in nature, as is appropriate for a "typical" detail. In this View file, the overall floor-to-floor height will remain unaltered. We will use the Cut Line tool to crop out the portions that we don't wish to include on the final printed detail. We will then rely on Model Space Views and Paper Space Viewports to crop out typical portions of the Wall so that the detail will fit vertically on the Sheet. Before we can begin, we need to create a custom Cut Line tool. The tool provided on the Annotation tool palette includes a break line on only one side. We need one that draws a break on both sides.

1. On the Annotation tool palette, right-click the *Cut Line (1)* tool and choose **Copy**.

If you do not see this palette or tool, right-click the Tool Palettes title bar and choose **Document** (to load the Documentation Tool Palette Group) and then click the Annotation tab.

2. Right-click the tool palettes title bar and choose **17 Residential [17 Residential Metric]** and then click the Project Tools palette.

3. Right-click and choose **Paste** (see Figure 17–9).

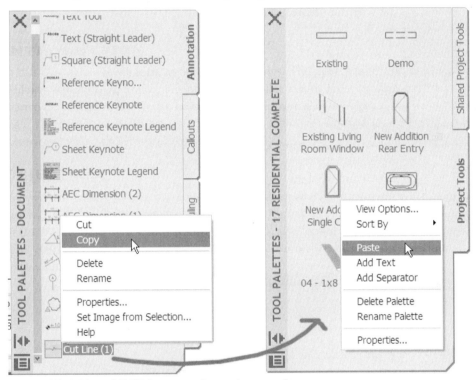

Figure 17–9 *Using OSNAPS to move the sections together*

4. Right-click the new tool and choose **Properties**.

5. At the top of the worksheet, rename the tool to *Cut Line (2)*.

6. At the bottom of the worksheet, choose **Dual Break** from the Type list and then click OK (see Figure 17–10).

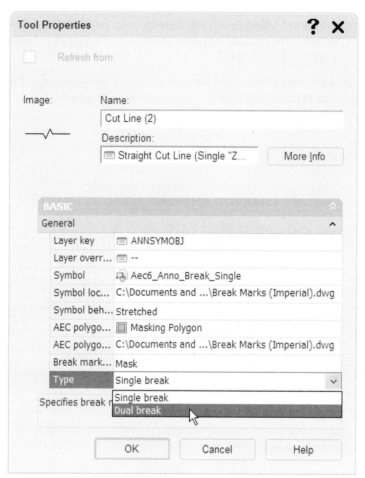

Figure 17–10 *Change the new Cut Line (2) tool to a Dual Break*

7. Click the new **Cut Line (2)** tool.

8. At the "Specify first point of break line" prompt, click a point below and to the left of the eave.

9. At the "Specify second point of break line" prompt, using POLAR or ORTHO, drag straight across the detail to the right and then click again.

10. At the "Specify break line extents" prompt, drag down past the bottom of the window and above the second floor and then click (see Figure 17–11).

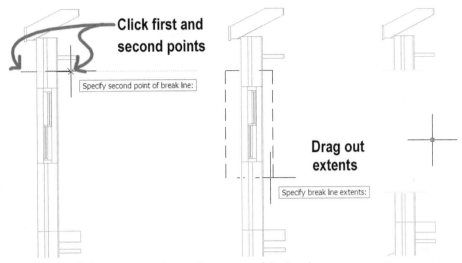

Figure 17–11 *Click points to designate the extents of the break*

11. Repeat twice more to create two more breaks as indicated in Figure 17–12.

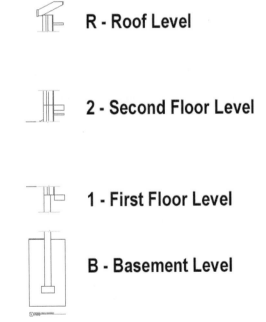

Figure 17–12 *Add additional break marks to create four zones*

We are now ready to begin adding Detail Components on top of our 2D Section/ Elevation sketch.

NOTE For the remainder of this tutorial, the labels on the right of Figure 17–12 will be used for reference. Therefore, if the tutorial calls for work to be done in the area of the second floor, it will simply direct you to work in region 2.

12. Click near any of the breaks.

Notice that a highlighted shape masks the underlying geometry. This is an AEC Polygon object. AEC Polygons are ADT objects that are very similar to polylines except they also have surfaces. This surface can be colored, hatched or made into a mask as we have here. Once the Cut Lines are drawn, you can use the grips on the masking polygon to adjust its shape as required.

13. Save the file.

CREATE A DETAIL

We are now ready to begin adding Detail Components to our drawing. We will use the 2D Section object as a guideline to help us place our Detail components. This will make the creation of the detail go quickly, and ensure greater accuracy and fidelity of the model.

DETAILING TOOL PALETTES

You will create Detail Components from tools on tool palettes (in the same way as other ADT objects) or with the Detail Component Manager. Since many of the most common components used in details are already available on tool palettes, we will begin our detail using them.

1. Right-click the Tool Palettes title bar and choose **Detailing** and then click the Exterior tab.

   This will load the Detailing Tool Palette Group and the Exterior tool palette.

2. Zoom in on region 1 of the detail (as shown in Figure 17–12).

3. On the Exterior tool palette, click the ***Standard Brick - 3/8" Jt [Standard 65mm Brick - 10mm Jt]*** tool.

A single brick in section view will be attached to your cursor. You can start placing the brick right away, or you can make adjustments to it on the Properties palette first.

NOTE Please note, however, that unlike other ADT tools, if you begin placing a Detail Component, and then change parameters on the Properties palette before completion of their placement, the routine will start again and cancel the placement of the Detail Components in progress.

4. In the Specifications grouping, verify that Show mortar reads: **Yes**.

5. In the Mortar grouping, from the Right joint type list, choose **Flush [Struck Flush]** (see bottom right side of Figure 17–13).

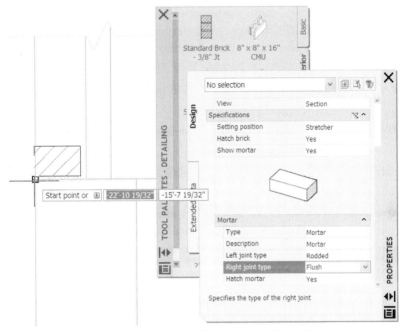

Figure 17–13 *Change the mortar joint type and begin placing the brick*

6. At the "Start point" prompt, click the bottom outside edge of the brick in region 1 of the section (see left side of Figure 17–13).

7. Drag the mouse straight up to the top of the detail and click at the Endpoint of the brick just beneath the roof eave and then press ENTER to complete the routine (see Figure 17–14).

Zoom and Pan as necessary.

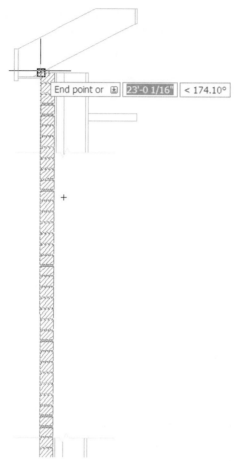

End point or ⊡ 23'-0 1/16" < 174.10°

Figure 17–14 *Drag up to draw continuous courses of brick; click a point near the eave*

Notice that as you move the mouse, bricks will be drawn filling in from the point where you started to wherever you click the "End point." When you are finished, your bricks might be showing on top of the breaks that we built above. This is easy to address.

8. Click near the edge of each of the masking polygons.

   The edge lines up with the endpoints of the Cut Lines.

9. Right-click and choose **Basic Modify Tools > Display Order > Bring to Front** (see Figure 17–15).

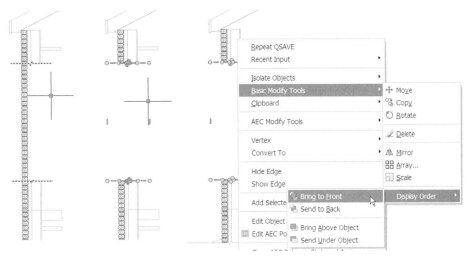

Figure 17–15 *Bring the masking polygon to the front*

If you wish, you can wait till the end to perform this step since we are likely to need to repeat it when we add more components. Also, feel free to delete the hidden bricks if you wish.

ADDING WOOD FRAMING

Sometimes the component you need is not configured on the palette exactly as you need it. This is no problem as you can easily modify the parameters of any tool directly on the Properties palette before you place the component. For instance, the framing of the exterior Wall uses 2x6 [50x150] framing. The tool on the tool palette is for 2x4 [50x100] framing. We can simply use this tool and then change the size.

10. On the Interiors tool palette, click the **2x4 [50 x 100mm Nominal]** tool.

11. On the Properties palette, choose **2x6 [50 x 150mm Nominal]** from the Description list.

12. From the View list, choose **Elevation**.

13. Right-click in the drawing and choose **X Flip**. (You can also press the down arrow and choose **X Flip** from the dynamic prompt).

14. At the "Start point" prompt, click the bottom of the stud within the air gap in region 1 (see Figure 17–16).

Figure 17–16 *Draw the Stud component from the air gap side*

15. At the "End point" prompt, drag straight up using POLAR or ORTHO and click within the sloped rafter at the roof (see Figure 17–17).

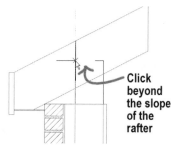

Click beyond the slope of the rafter

Figure 17–17 *Draw the stud all the way to the roof rafter*

16. On the Properties palette, change the View to **Section**.

17. Right-click in the drawing and choose **Rotate** (or use the down arrow and the dynamic prompt).

18. At the "Rotation" prompt, type **90** and then press ENTER.

19. In region 1, snap the lumber to bottom of the stud drawn above to make a bottom plate for the Wall.

20. Using the underlying section object guidelines, snap another bottom plate at the base of the second floor Wall (see Figure 17–18).

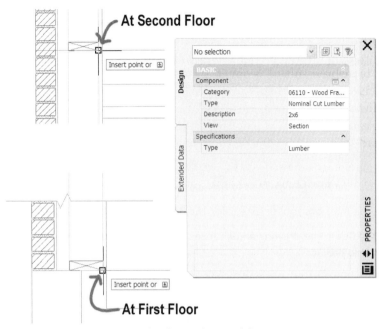

Figure 17–18 *Add bottom plates at the first and second floors*

21. Right-click in the drawing and choose **X Flip**.

22. Using the Intersection OSNAP, add a double top plate at the point where the roof rafter intersects the vertical stud (drawn above) and then press ENTER (see Figure 17–19).

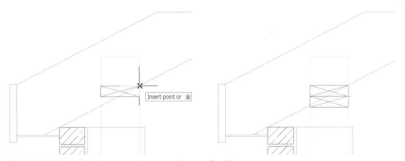

Figure 17–19 *Add a double top plate at the roof rafter*

23. Select the vertical stud drawn above, right-click and choose **AEC Modify Tools > Trim**.

24. At the "Select the first point of the trim line" prompt, click the same intersection used to place the top plate.

25. At the "Select the second point of the trim line" prompt, use POLAR or ORTHO to create a horizontal trim line.

26. At the "Select the side to trim" prompt, click anywhere above the trim line (see Figure 17–20).

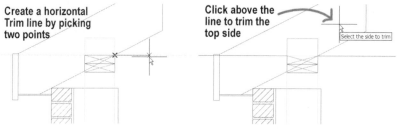

Figure 17–20 *Use the AEC Modify Tools to trim off the excess stud at the roof*

27. On the Interiors tool palette, click the **5/8" Plywood [15mm Plywood]** tool.

28. Right-click in the drawing and choose **X Flip** (or use the arrow keys and the dynamic prompts).

29. At the "Start point" prompt, snap to the bottom left Endpoint of the first floor bottom plate.

30. At the "End point" prompt, snap to the Endpoint of the underlying section object to the right (see Figure 17–21).

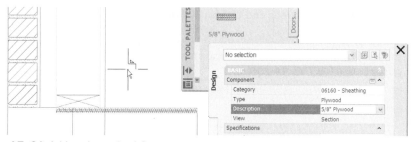

Figure 17–21 *Add a plywood subfloor*

31. Repeat this process on the second floor.

32. Select the vertical stud component, right-click and choose **Add Selected**.

33. On the Properties palette, choose **2x10 [50 x 250mm Nominal]** from the Description list.

34. Draw a floor joist beneath each of the two plywood subfloors just drawn.

TIP Don't forget to right-click and change the reference point to Left.

35. Change the Description back to **2x6 [50 x 150mm Nominal]** and draw a ceiling joist starting as the double top plate in region R, and then press ENTER to complete the routine (see Figure 17–22).

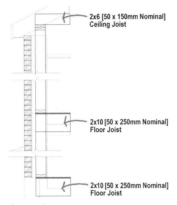

Figure 17–22 *Add Floor and Ceiling Joists*

36. Save the file.

ADD ROOF RAFTERS

The Roof Rafters present a bit of a challenge since we don't have all of the reference points that we need. This is easily solved by adding some construction lines.

37. On the Shapes toolbar, click the Construction Line icon.

38. At the "Select linework under cursor" prompt, mouse over one of the sloped rafter lines and click.

39. At the "Specify offset" prompt, click the point where the top plate intersects the rafter and then press the ESC key (see Figure 17–23).

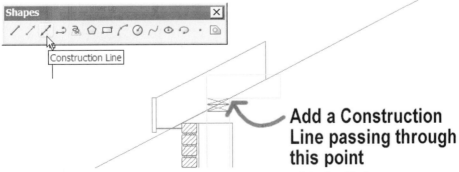

Figure 17–23 *Add a Construction Line following the slope of the Roof Rafter*

40. Select the Ceiling joist, right-click and choose **Add Selected**.

41. On the Properties palette, change the Description to **2x8 [50 x 200mm Nominal]**.

42. At the "Start point" prompt, using the Nearest OSNAP, click a point on the construction line to the left of the roof eave.

43. At the "End point" prompt, using the Nearest OSNAP, click a point on the construction line to the right side of the detail and then press ENTER.

   Make sure the rafter is too long in both directions (see Figure 17–24).

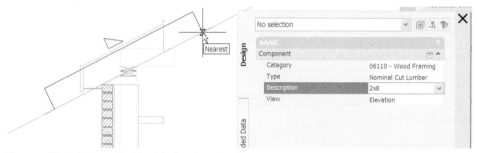

Figure 17–24 *Draw the Roof Rafter "too long"*

44. Follow the process in the "Adding Wood Framing" section above, and use the **AEC Modify Tools > Trim** command to trim off the unwanted parts of the roof rafter (see Figure 17–25).

 IMPORTANT Do not use the normal AutoCAD TRIM command for this operation. AEC Modify Tools preserve the integrity of the Detail Component object, but this is not necessarily the case with normal AutoCAD commands.

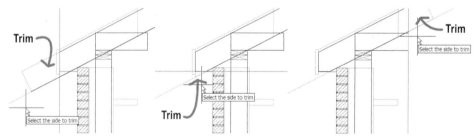

Figure 17–25 *Use AEC Modify Tools to trim the roof rafter to the correct shape*

45. Offset the construction line up **5″ [125]** and using it and **AEC Modify Tools > Trim**, chamfer the corner of the ceiling joist.

 TIP You can use the AEC Modify Tools, such as Trim on standard AutoCAD entities as well as AEC Objects.

46. Delete the construction lines.

47. Click on the newly trimmed roof rafter and on the Properties palette, click the Extended Data tab.

Notice that all of the Detail Component information has been retained on this object even though we have given it a custom shape. This is true for all Detail Components. You can always see the Detail Component Extended Data by selecting the component and viewing its properties on the Extended Data tab of the Properties palette this way. Try this on some other Detail Components. Let's make a few finishing touches to the roof framing members.

48. Return to the Design tab of the Properties palette.

49. Select the sloping roof rafter, right-click and choose **AEC Modify Tools > Subtract**.

50. When prompted, select both of the top plate members and press ENTER. Answer **No** to erasing them.

We can show the ceiling joist as if it were behind the roof rafter.

51. Select the horizontal ceiling joist, right-click and choose **AEC Modify Tools > Obscure**.

52. At the "Select obscuring object(s)" prompt, click the sloping roof rafter and then press ENTER (see Figure 17–26).

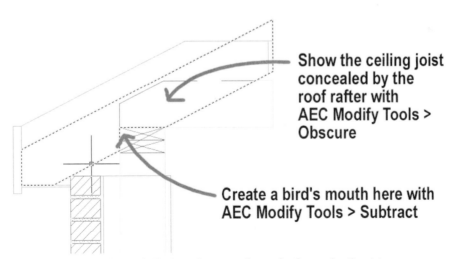

Show the ceiling joist concealed by the roof rafter with AEC Modify Tools > Obscure

Create a bird's mouth here with AEC Modify Tools > Subtract

Figure 17–26 *Use AEC Modify Tools to fine tune the roof rafter and ceiling joist*

The end of the rafter that overlaps the sloping rafter will now be dashed and on a different layer.

53. Save the file.

ADD REMAINING FRAMING COMPONENTS

There are still some framing components required. We can use the Add Selected functionality to add these.

54. Using Add Selected, add a double top plate for the first floor snapping to the bottom of the second floor joist.

55. Using Add Selected, add a sill plate beneath the first floor joist.

56. Using Add Selected, add a joist in section at each floor.

57. Using Add Selected, add plywood sheathing from the sill plate to the double top plate at the ceiling joists and along the top edge of the roof rafter (see Figure 17–27).

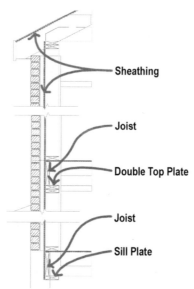

Figure 17–27 *Add remaining framing components*

58. Select the vertical stud, right-click and choose **AEC Modify Tools > Divide**.

59. Follow the prompts and use Figure 17–28 as a guide to divide the vertical stud into two pieces at the double top plate.

60. Using **AEC Modify Tools > Trim**, remove the piece of the upper vertical stud that overlaps the floor joist and floor sheathing (see Figure 17–28).

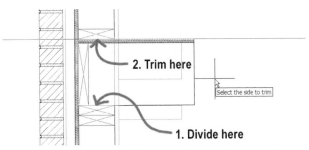

Figure 17–28 *Divide the vertical stud and then trim the portion at the floor joist*

61. Select any lumber component, right-click and choose **Add Selected**.

62. On the Properties palette, change the Description to **1x8 [25 x 200mm Nominal]**, change the View to **Section** and then beneath the Specifications Grouping, change the Type to **Plank**.

63. Add this component to the soffit of the roof.

64. Save the file.

DETAIL COMPONENT MANAGER

Although all of the Detail Components that we have added so far have come from tools, the Detail Component Manager (see Figure 17-29) is the primary interface to the ADT Detailing System. Tools are simply "shortcuts" to specific items in the Detail Component Manager. If an item that you wish to add cannot be found on an existing tool palette, use the Detail Component Manager. You may also add new sizes to existing components in the database with Detail Component Manager, and you can drag items from it to tool palettes for quick future access.

1. From the Insert menu, choose **Detail Component Manager**.

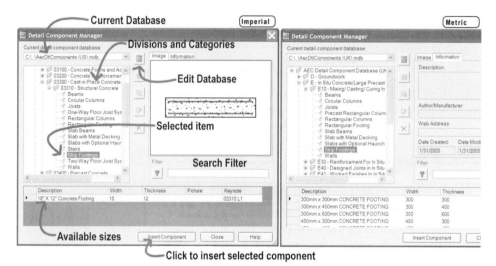

Figure 17–29 *The Detail Component Manager*

2. Browse to the correct category and item:

   In Imperial: *Division 3 – Concrete*, then *Cast-in Place*, then *Structural Concrete* and finally *Strip Footings*.

   In Metric: [*E - In Situ Concrete/Large Precast Concrete*, then *E10 - Mixing/ Casting/ Curing In Situ Concrete*, then *Strip Footings*]

In imperial, there is only one size listed for Strip Footings and it is not the size that we need. However, it turns out that we can easily add sizes to the database. Once we add an item, it will be available from this point on.

CAD Manager Note: All Detail Components are saved in a Microsoft Access database file. New sizes that are added become new fields in the database. You will likely want to locate the standard database for the office on your server. You can do this in the Options dialog. You may also want to consider access rights to this central database. However, if you limit user write access to the database, you must provide a process by which users can add the additional components that their designs require. New to ADT 2006, you can now edit the database directly from the Detail Component Manager dialog. Use this functionality to add sizes to existing components (see below) and to add custom groups and components. You can also edit components directly in the database using the edit icon. This will call a Component Properties wizard which will give access to all of the component's custom parameters.

New in ADT
2006

3. At the top of the dialog, click the Edit Database icon.

   A blank line—with a star (*) next to it—will appear at the bottom of the list of sizes.

4. Click in the last line of the size list (the blank one that just appeared) in the Description column.

5. Type **20″ x 10″ Concrete Footing [500mm x 250mm CONCRETE FOOTING]** and then press the TAB key.

6. Type **20 [500]** in the Width column, and press TAB again, type **10 [250]** in the Thickness column and press TAB again.

7. Right-click in the Keynote column and choose **Edit**. Pick the same keynote listed for the other sizes and then click OK.

8. Click the Edit Database icon again, to exit edit mode. When prompted to save changes, click Yes.

9. With this new item selected, click the Insert Component button.

10. On the Properties palette, in the Rebar – Longitudinal grouping, change the Bars value to **2**. In the Rebar – Lateral grouping, change the Edges to **Center**.

11. Place the Footing component relative to the upper left corner of the section underlay and then press ENTER.

 NOTE The Footing that you are placing will be larger to the right and project lower than the outline provided by our underlain 2D Section/Elevation object. This is OK. The detailing process has revealed a flaw in the design of the foundation. There is a suggested exercise to edit this in the model in Appendix A.

ADD THE FOUNDATION WALL

Let's return to the tool palette for the Concrete Foundation Wall. There is not a specific tool for this, however, on the Basic tool palette; there is a general tool for each major category of Detail Component.

12. On the tool palettes, click the Basic tab and then click the *03 Concrete [E - In Situ Concrete/Large Precast Concrete]* tool.

13. On the Properties palette, choose **Walls** from the Type list.

14. From the Description list, choose **12″ Concrete Wall [300mm CONCRETE WALL]**.

15. Verify that View is set to Section, and from the Show Reinforcing list, choose **No**.

16. At the "Start point" prompt, click the Midpoint of the top of the footing.

17. At the "End point" prompt, snap Perpendicular to the bottom edge of the sill plate.

BRICK HAUNCH

Let's cover the floor construction at the first floor with some bricks.

18. Copy seven bricks down to fill in the space between the foundation Wall and the brick veneer.

 NOTE Even though the bricks are covered by the masking polygon, you can still select and copy them.

Approximately three bricks will overlap the foundation wall. Let's create a haunch for them.

19. Draw a rectangle starting at the point where the sheathing intersects the foundation wall and with its opposite corner aligned with the point where the overlapping brick intersects the foundation wall (see Figure 17–30).

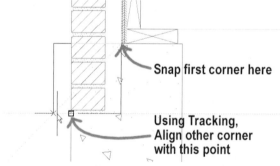

Figure 17–30 *Draw a rectangle relative to the bricks and the sheathing*

20. Select the foundation wall, right-click and choose **AEC Modify Tools > Subtract**.

21. At the "Select object(s) to subtract" prompt, click the rectangle and then press ENTER.

22. At the "Erase selected linework" prompt, choose **Yes**.

FOOTING KEY

23. Draw another rectangle **4″ [100]** square and then move it so that its geometric center is snapped to the Midpoint of the top edge of the footing (see the left side of Figure 17–31).

24. Select the footing, right-click and choose **AEC Modify Tools > Subtract**. Select the square to subtract, but do not erase it this time (see the middle of Figure 17–31).

25. Select the foundation wall, right-click and choose **AEC Modify Tools > Merge**. Select the square to merge and do erase it this time (see the right side of Figure 17–31).

 NOTE Be careful not to mistake the underlying 2D Section/Elevation linework as part of the detail at this point. The 2D Section/Elevation linework is on a non-plotting layer.

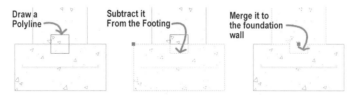

Figure 17–31 *Add a key to the footing using AEC Modify Tools*

26. Save the file.

ADD INSULATION

27. On the Interiors tool palette, click the **3-1/2″ Acoustical Batt Insulation [89mm Acoustical Batt Insulation]** tool.

28. On the Properties palette, choose **5-1/2″ Acoustical Batt Insulation [140mm Acoustical Batt Insulation]** from the Description list.

29. At the "Start point" prompt, snap to the top Midpoint of the sole plate.

30. At the "Next point" prompt, snap to the bottom Midpoint of the top plate. Press ENTER twice to complete the routine.

31. Repeat the steps on the next floor.

 TIP Use Add Selected to avoid having to reset parameters on the Properties palette.

Zoom in on the floor structure of the first floor in region 1.

32. Using Add Selected, select and add batt insulation. Right-click and choose **Right**.

33. At the "Start point" prompt, snap to the bottom Endpoint of the first floor plywood subfloor on the right side of the detail.

34. At the "Next point" prompt, snap to the top right Endpoint of the sectioned joist.

35. At the "Next point" prompt, move the mouse down and snap to the bottom right corner of the same joist. Press ENTER twice to complete the routine (see Figure 17–32).

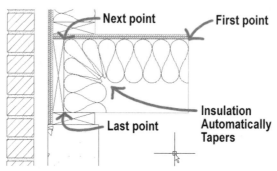

Figure 17–32 *Add batt insulation around a corner – the shape automatically conforms to the turn*

We will use the Add Selected command and create and taper the Batt Insulation along the ceiling joist.

36. Zoom into the roof line in Region R. Select the batt insulation, right-click and choose **Add Selected**.

37. Right-click and choose **Right**.

38. Pick the first Endpoint in the lower left hand corner of the ceiling joist and the second point in the lower right hand corner of the ceiling joist and then press ENTER.

39. At the "Select first point of first taper boundary" prompt, snap to one lower left Endpoint of the ceiling joist chamfer.

40. At the "Select second point of first taper boundary" prompt, snap to the other Endpoint of the ceiling joist chamfer and then press ENTER (see Figure 17–33).

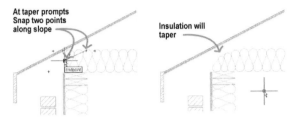

Figure 17–33 *Taper Batt Insulation with two points*

41. Save the file.

FLASHING

To draw items like flashing, or any type of membrane, draw a polyline of the desired shape first. Then insert the Detail Component as normal. You will be prompted to select objects, at which point you can select the polyline to convert it. This will automatically change the polyline to the predefined layer and add the proper keynote detail component associations.

1. At the base of the brick cavity, draw a polyline to represent flashing.

2. From the Insert menu, choose **Detail Component Manager**.

3. In the Filter field, type **flashing** and then press ENTER.

   The tree list on the left side will shorten to show only flashing items.

4. Select a flashing component and then click Insert Component.

5. At the "Select Objects" prompt, click the polyline you just drew and then press ENTER twice to complete the command.

   To see the changes applied, select the new flashing component, and check the properties of this converted polyline. Choose the Extended Data tab and observe the Component information and keynote information.

COMPLETE THE DETAIL

The detail still needs drywall on the inside, building paper on the outside of the sheathing, compacted gravel at the footing, soil, rigid insulation on the inside of the foundation wall, furring strips and drywall at the ceiling, and building paper and shingles on the roof. At this point, you should be familiar enough with the Detailing tool palettes and the Detail Component Manager to locate and add all of these remaining components. ADT 2006 adds Masonry Anchors, Control Joints, Sealant and Backer Rods, Caulking, Metal Soffit Vents and much more to the

Detail Component database. Add any of these items as well. Do this now before moving on.

If you would like to have a bold outline around the entire detail, you can use the Shrinkwrap tool on the Basic tool palette of the Detailing tool palette group. This tool will create a polyline outline around any object(s) that you select. To add a bold shrinkwrap around the outer edge of the entire detail, simply select all detail components, being careful not to select the Cut Lines, the underlain 2D Section or the masking polygons. You can then edit the resultant polyline to suit your needs and desired graphical appearance of the detail.

6. On the Annotation tool palette (in the Document tool palette group) click the **Cut Line (1)** tool.

7. Following the prompts, add three cut lines to the right of the detail at the first floor, the second floor and the roof.

8. Repeating the steps covered above, use **Basic Modify Tools > Display Order > Bring to Front** to move the masking polygons back on top as required (see Figure 17–34).

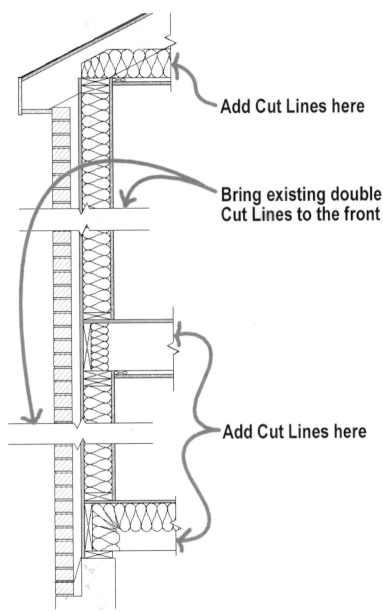

Add Cut Lines here

Bring existing double Cut Lines to the front

Add Cut Lines here

Figure 17–34 *Add single Cut Lines and bring the existing double Cut Lines to the front*

9. Save the file.

MODIFY A DETAIL

You can modify an existing detail using all of the same tools that were used to create it. One additional tool that is very handy when modifying Detail Components is "Replace Selected." Since Detail Components are not "true" objects, but rather very smart drafted entities, they cannot take advantage of many of the normal progressive refinement techniques promoted throughout this book. However, with Replace Selected, they come very close. We saw above that any Detail Component will remember its parameters (visible on the Properties palette) when you select them later. You can view them on the Extended Data tab of the Properties palette and when you choose Add Selected, a new Component routine is started using the same parameters. Replace Selected is the same basic concept except that instead of adding a new Component and keeping the original, this command will erase the original selected Detail Component, and populate the Properties palette with all of its original settings so that you can add a new Component in its place. You use this when you need to make a change to a Component size, or some other parameter. Replace Selected is often the quickest way to progressively refine details and Detail Components.

Be sure to also rely heavily on the AEC Modify Tools as we have above. While some components can be edited with grips in the normal fashion, others cannot. The AEC Modify Tools, however, work on *all* detail components. You can use the Trim function to cut a detail component, and crop and subtract can be used (as we saw above) to apply the shape of another object to the selected detail component as a means of editing its shape. Obscure (also shown above) is also a very handy tool that will make a selected detail component appear to be behind another one.

KEYNOTES

Once you have generated elevations, sections (in Chapter 16) and details (this chapter), your next task will be to add notes and annotation to them. Some techniques for this have already been covered in Chapter 14. There we added Elevation labels to the elevations and sections to indicate their floor heights and we added some basic text leaders. You can also use the dimensioning techniques covered in that chapter on your elevations, sections and details as well. In this topic, we will explore the ADT Keynoting feature.

KEYNOTES ASSIGNED TO STYLES

Keynotes are standard notes that are pre-assigned to both ADT object styles and Detail Components. Keynotes are stored in a Microsoft Access database that can be modified and added to as the needs of your firm dictate. They can be added with or without text leaders as reference keynotes, or in keynote symbols tied to individual Sheets called Sheet keynotes. Let's begin by taking a look at the Keynote references within ADT Styles.

I. On the Project Navigator, open the *First Floor New* Construct.

2. Select one of the exterior masonry Walls, right-click and choose **Edit Wall Style**.

3. Click the General tab.

4. On the right side of the tab is a field that lists the currently assigned Keynote. There is also a Select Keynote button that you can click to change the Keynote designation (see Figure 17–35).

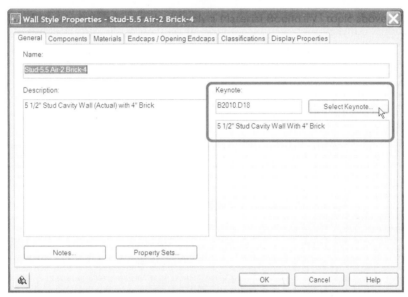

Figure 17–35 *You can assign a Keynote to an ADT object Style on the General tab*

CAD Manager Note: Keynotes reference a Microsoft Access database file (*.mdb*). ADT ships with a few databases and you can also create your own. Database files can be saved and accessed from a network file server so that all members of your firm access the same keynotes. The path to the Keynote databases is configured in the Options dialog on the AEC Content tab. There you can add multiple Keynote Database paths. You can also assign custom Keynote databases to the current Project. This can be done on the Project Navigator palette on the Project tab. Click the Edit Project icon and then the Add/Remove button next to the "Project Details Component Databases" and "Project Keynote Databases" entries. If you wish to create your own custom Keynote database, template files have been provided in the ADT *Template* folder. By default this is located in the following path:

C:\Documents and Settings\All Users\Application Data\Autodesk\ADT 2006\enu\Template\Details and Keynotes

Use caution when creating Databases based on these templates. You should only attempt this if you are already familiar with Microsoft Access.

New in ADT
2006 Architectural Desktop 2006 makes editing Keynote databases (and migrating custom ones from ADT 2005) much simpler. If you already have a custom Keynote database that was in use with ADT 2005, use the utility provided in the Autodesk group of the Windows Start menu to migrate it to 2006. Choose **Start > All Programs > Autodesk > Autodesk Architectural Desktop 2006 > Detail Component – Keynote Database Migration Utility**. Use this tool to merge your existing database(s) into the 2006 one. Also in the same location on the Start menu, is the **Keynote Editor**. Use this tool to open existing keynote databases and add or edit keynote entries. You can also create a custom database using this tool.

> 5. Click the Select Keynote button.

You can choose any Keynote from the list that appears. Notice that at the top of this dialog, you have the option to choose a database, and then all of the Keynotes within that database will be listed. At the bottom, you can type in a key word and then click the filter icon to shorten the list as we did above in the Detail Component Manager. On the right side are icons to edit the database directly in this dialog (see Figure 17–36).

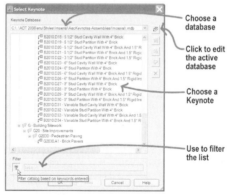

Figure 17–36 *Filter the Keynote list by typing in a key word*

> 6. Click Cancel to return to the Wall Style Properties dialog.
>
> 7. In the Wall Style Properties dialog, click the Materials tab.
>
> 8. Select the Brick Veneer component and then click the Edit Material icon.
>
> 9. In the Material Definition Properties dialog, click the General tab.
>
> 10. Notice that the Material Definition also has a Keynote designation.
>
> 11. Click Cancel three times to return to the drawing.
>
> 12. Close the *First Floor New* file. It is not necessary to save it.

ADD KEYNOTES TO SECTIONS AND ELEVATIONS

The Keynotes referenced by the Wall Styles and Material Definitions that we just explored can be accessed by the tools on the Annotation tool palette.

1. On the Project Navigator palette, on the Views tab, double-click the *A-CM01* file to open it.

2. Zoom in on the East Elevation.

3. On the Annotation tool palette, click the **Reference Keynote (Straight Leader) [Keynote (Straight Leader)]** tool.

   If you do not see this palette or tool, right-click the tool palettes title bar and choose **Document** (to load the Documentation tool palette group) and then click the Annotation tab.

4. At the "Select object to keynote" prompt, click on the Brick Hatching on the elevation.

5. The Select Keynote dialog will appear (see Figure 17–37).

Figure 17–37 *The Select Keynote dialog jumps directly to the item referenced by the Material Definition*

The Keynote choice will jump directly to the Keynote category indicated by the Material Definition, in this case "Clay Masonry Units [F10/110 - CLAY FACING BRICKWORK]"

6. Indented beneath the selected category, select **Standard Brick – 3/8″ [F10/110 - CLAY FACING BRICKWORK]** Joint and then click OK.

7. At the "Select first point of leader" prompt, click a point for the leader and then follow the remaining prompts to place the note.

8. Continue this process to add Keynotes to all the sections and elevations (see Figure 17–38).

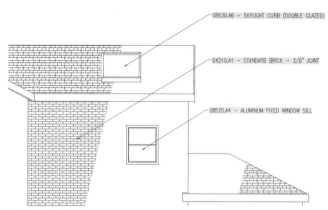

Figure 17–38 *Add Keynotes to the elevations and sections*

CHANGE KEYNOTE DISPLAY

You can change the way that Keynotes display in your drawings. They can show the Keynote number, the Keynote text or both.

9. From the View menu, choose **Keynote Display > Reference Keynotes**.

The three choices are: "Reference Keynote - Key only," "Reference Keynote - Note only" and "Reference Keynote - Key and Note." Figure 17–38 shows "Reference Keynote - Key and Note" display. In Figure 17–39, "Reference Keynote - Key only" is shown on the left and "Reference Keynote - Note only" is shown on the right. In addition, you may also choose your desired format in lower or upper case.

10. Choose your desired Keynote Display option and then click OK.

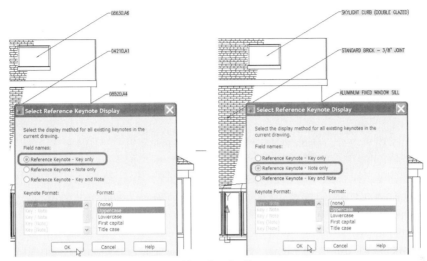

Figure 17–39 *Changing the method of Display for Keynotes*

11. Save and Close the file.

KEYNOTING PLANS

In addition to keynoting elevations and sections, you can also keynote plans.

12. On the Project Navigator palette, on the Views tab, double-click the *A-FP01* file to open it.

This is the First Floor Plan View file. We added some simple notes and dimensions to it in Chapter 14. Let's add a few keynotes to it now. You may have noticed that when we added keynotes to the elevations, they automatically used the keynote assigned to the Material Definition. When you keynote in plan, you get the choice to use either the keynote assigned to the Wall Style or the ones assigned to the individual components (via Materials).

13. On the left side of the plan, add a Reference Keynote pointing to the exterior Wall.

14. In the Select Element to Keynote dialog, place a checkmark in the Style check box (see Figure 17–40).

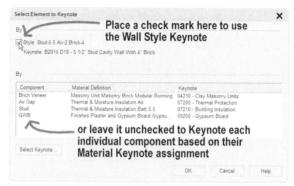

Figure 17–40 *Keynoting in Plan gives the option of the Wall Style or Component Material Keynotes*

 TIP Before placing any keynotes, check the Drawing Scale of the file (on the Drawing Status Bar). Keynotes, like other ADT annotation, use the drawing scale to automatically scale the text and the leader.

If you want to wrap the keynote to more than one line, you can select the text, and then change the width in the Properties palette. Make the width some value other than zero. When the width of text is set to zero, it does not word wrap and continues on in one long line. After you change it to a value other than zero you can then fine tune the width with the grips.

15. Save and Close the file.

KEYNOTING DETAILS

Keynoting Details uses the same tools and procedures. Since we used Reference Keynotes in the examples above, we will use Sheet Keynotes here. Either type of Keynote, Reference or Sheet, can be used in either situation; we are switching here merely to show both types in practice. A Sheet Keynote will be assigned a numeric designation that will be used for each instance of a particular note. Therefore if you note the same brick five times on a Sheet, they will all use the same numeric designation that refers back to a Keynote Legend (see below). These designations will be applied when the drawings are added to a Sheet.

Return to *A-DT01*. If you closed it, re-open it from Project Navigator now.

1. On the Annotation tool palette, click the **Sheet Keynote** tool.

2. At the "Select object to keynote" prompt, click on a Detail Component to select and Keynote it.

3. At the "Select first point of leader" prompt, click the point where you wish the arrowhead to be located.

4. At the "Specify next point of leader line" prompt, click your next leader point. Continue to add leader points if you wish and press ENTER when finished.

A Keynote symbol will appear. It will currently display a question mark (?) within its field code because this detail has not yet been added to a Sheet. These fields will update once this detail has been dragged to a Sheet (see below).

5. Continue to add as many Sheet Keynotes as you wish.

Be sure to keynote the same type of Component in more than one location on the detail. For instance, keynote the brick at the top (region R) and the bottom (Region B) and perhaps even the middle (region 1 or 2) of the detail (see Figure 17–41).

 TIP Create a Construction Line on each side of the detail using the icon on the Shapes toolbar. Snap to this construction line when placing Keynotes to keep them all lined up neatly.

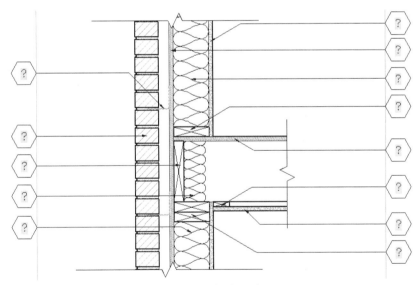

Figure 17–41 *Add several Sheet Keynotes to the Detail*

6. Locate the underlying 2D Section and click on it. Verify that all of the notes fit within the dashed gray boundary that appears. If they don't use the grips to adjust the width of the boundary.

These grips adjust the size of the Model Space View that we will drag to the Sheet next.

7. When you have finished Keynoting, Save and Close the file.

ADD DETAILS TO SHEETS

Let's add our detail to a Sheet and then create some Keynote Legends.

8. On the Project Navigator palette, click the Sheets tab.

9. In the *Architectural > Details* Sub Set, double-click the *A-501 Details* Sheet to open it.

   The Sheet was provided for convenience and it is currently empty.

10. On the Project Navigator palette, click the Views tab.

11. Click the plus (+) sign beneath the *A-DT01* View file to reveal the Model Space Views contained in the file.

12. Drag Typical Wall Section Model Space View onto the Sheet and snap its lower left corner to the lower left corner of the Sheet to place it.

   The Viewport created will be too tall for the Sheet.

13. Copy (**Basic Modify Tools > Copy**) the Viewport three times next to itself (you will have four total).

14. Use the grips on the viewports to reduce the height only and crop each Viewport to one portion of the detail (B, 1, 2 and R) (see Figure 17–42).

   Be careful not to change the width of any of the Viewports, maintaining the width will allow us to stack them up and keep everything properly aligned.

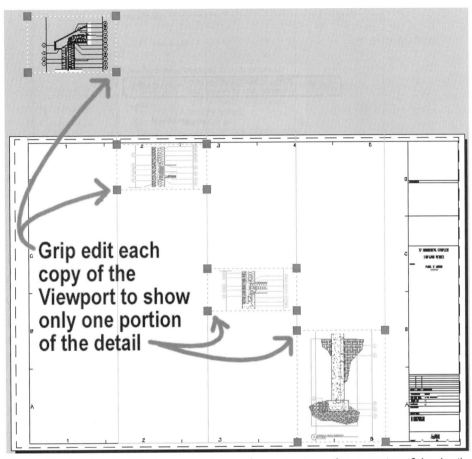

Figure 17–42 *Copy the Viewport and adjust each copy to view only one section of the detail each*

15. Move all of the Viewports to the left edge of the Sheet and use object snaps to stack them up on top of each other (see Figure 17–43).

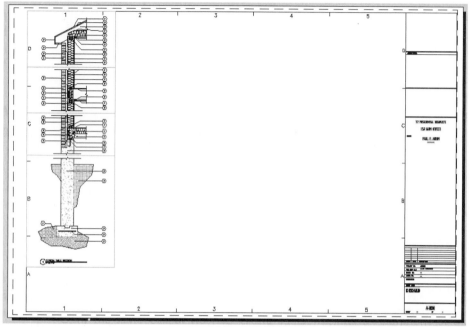

Figure 17–43 *Move all of the Viewports together and snap them to one another*

16. Save the file.

NOTE The Metric Complete version of the project (provided with the dataset files from the Mastering Autodesk Architectural Desktop 2006 CD ROM) varies slightly from the results indicated here and those included in the Imperial Complete version.

KEYNOTE LEGENDS

When you use Keynotes, you can place a legend on your Sheet files that references all of the Keynotes showing on a particular Sheet. If they are Reference Keynotes, the legend is organized logically by category. If they are Sheet Keynotes, then the Legend will enumerate the Keys and update all of the field codes throughout the set.

SHEET KEYNOTE LEGEND

1. On the Annotation tool palette, click the **Sheet Keynote Legend** tool.

2. At the "Select keynotes to include in the keynote legend" prompt, make a window around all Sheet Keynotes (across all four Viewports) and then press ENTER.

This can be done from Paper Space *without* making the Viewport active.

3. At the "Insertion point of table" prompt, click a point on the Sheet to place the Table (see Figure 17–44).

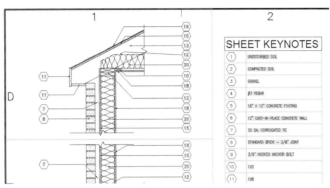

Figure 17–44 *Place a Sheet Keynote Legend to enumerate and describe the Sheet Keynotes*

4. Save and Close the file.

REFERENCE KEYNOTE LEGEND

5. On the Project Navigator palette, click the Sheets tab.

6. In the *Architectural > Elevations* Sub Set, double-click the *A-202 Elevations* Sheet to open it.

7. On the Annotation tool palette, click the **Reference Keynote Legend** tool.

8. At the "Select keynotes to include in the keynote legend" prompt, make a window around all Reference Keynotes and then press ENTER.

This can be done from Paper Space *without* making the Viewport active.

9. At the "Insertion point of table" prompt, click a point on the Sheet to place the Table (see Figure 17–45).

TIP After selecting the Keynotes to include in the legend, right-click and choose Sheets. A Worksheet will appear in which you can select several or all of your Sheets to include in the Legend.

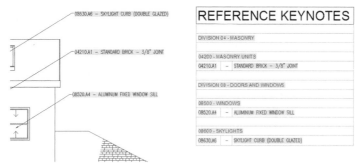

Figure 17–45 *Place a Reference Keynotes Legend*

   10. Repeat for any of the other Sheets that have Keynotes.

   11. Save and Close all files.

INSERT A SHEET LIST

We have one more Table to insert (or rather update). At this point, we are nearing the end of the book and have added many Views and Sheets since starting our two projects back in Chapter 5. From the Sheet Set, we can insert a Sheet List. In Chapter 5, we did this. However, it is time to update that list to reflect the new Sheets that have been added to the project.

   12. On the Project Navigator palette, click the Sheets tab.

   13. In the General Sub Set, double-click the *G-100 Cover Sheet* to open it.

   14. Right-click on the Sheet List in the lower right corner of the Sheet and choose **Update Sheet List Table**.

      The Table will update to reflect the added Sheets.

The object that is placed here is an AutoCAD Table object. It is a style-based object similar to the type of table that you might add in Microsoft Word. The style used here is generated automatically for the Sheet Index and Keynote Legends. While this object at first appears similar to an ADT Schedule Table as covered in Chapter 15, it is not the same type of object, nor does it share the same features. Tables do not interface with Property Sets, nor do they use ADT Display Control. They can have Field codes inserted within their cells. ADT Schedule Tables cannot. So while these two objects are similar in use and function, they both have unique uses.

   15. If you wish, follow the steps in Chapter 5 and plot the set.

   16. Save and Close all residential project files.

CHALLENGE EXERCISE

Section III has introduced us to Tags and Dimensions in Chapter 4, Schedules, Property Sets, Tags and Display Themes in Chapter 15, Callouts, Sections and Elevations in Chapter 16 and finally Details, Keynotes and Legends here in Chapter 17. While presented individually across four chapters, all of these tools work together to give us a complete Construction Documents package. In this exercise, we will use several of the tools from each of these chapters to create an enlarged plan detail of the commercial project.

INSTALL THE CD FILES AND LOAD THE CURRENT PROJECT

If you have already installed all of the files from the CD, simply skip down to step 3 below to make the project active. If you need to install the CD files, start at step 1.

1. From the File menu, choose **Project Browser.**

2. Click to open the folder list and choose your *C:* drive.

3. Double-click on the *MasterADT 2006* folder, then the *Chapter17* folder.

    One or two commercial Projects will be listed: *17 Commercial* and/or *17 Commercial Metric.*

4. Double-click *17 Commercial* if you wish to work in Imperial units. Double-click *17 Commercial Metric* if you wish to work in Metric units. (You can also right-click on it and choose **Set Current Project**.) Then click Close in the Project Browser.

     IMPORTANT If a message appears asking you to re-path the project, click **Yes**. Refer to the "Re-Pathing Projects" heading in the Preface for more information.

CREATE A PLAN DETAIL USING A CALLOUT

The first step in our process is to create a plan view detail. We can use the same Callout routine that we used at the start of this chapter for this purpose.

5. On the Project Navigator, click the Views tab, and then double-click *A-FP01* to open it.

This is the First Floor Plan View file. We will create a Callout around the Core area in this file.

6. Zoom in on the building core.

7. On the Callouts tool palette, click the **Detail Boundary B** tool.

If you do not see this palette or tool, right-click the Tool Palettes title bar and choose **Document** (to load the Documentation Tool Palette Group) and then click the Callouts tab.

8. At the "Specify one corner of detail box" click just above and to the left of the Stair tower.

9. At the "Specify opposite corner of detail box" prompt, click a point below and to the right of the Men's Restroom.

Click both points close to but outside of the masonry Walls.

10. At the "Specify first point of leader line on boundary" prompt, click a point to the left of the boundary and then press ENTER (see Figure 17–46).

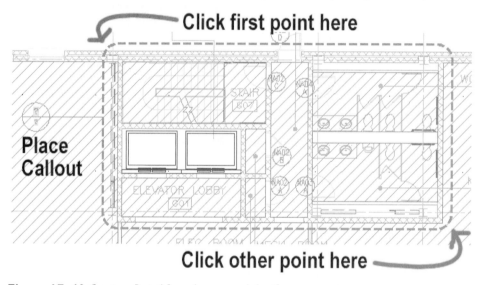

Figure 17–46 *Create a Detail Boundary around the Core area*

11. In the Place Callout worksheet, type **Enlarged Core Plan** for the New Model Space View Name.

12. At the bottom of the worksheet, remove the check mark from the Generate Section/Elevation checkbox.

13. Change the Scale to: **1/4″=1′-0″** [**1:50**].

14. In the Create in area, click the New View Drawing icon (see Figure 17–47).

Figure 17–47 *Deselect "Generate Section/Elevation to create an Xclipped Plan View file"*

The Add Detail View Wizard will appear. This is identical to the other View Wizards that we have seen; however, it uses different default settings. (You can also optionally assign a different template to Detail Views in the Project's Properties.)

15. In the Add Detail View wizard, type **A-EP01** for the Name and **Architectural Enlarged Floor Plans** for the Description, and then click Next.

16. On the Context page, the Detail View wizard has selected the First Floor to match the current file. Click Next.

17. On the Content page, be sure that all Constructs are selected and then click Finish.

The routine is almost complete and a polyline boundary surrounding the designated area has appeared on the plan. The last prompt requests the corners of the Model Space View. When you clear the "Generate Section/Elevation" checkbox, you get a plan detail view in which all XREFs are clipped to the shape of the Callout boundary. However, you will likely want an area larger than this as a Model Space View so that the Viewport of this enlarged plan has enough room for notes, dimensions and a title mark.

18. At the "Specify first corner for model space view" prompt, click a point above and to the left of the Callout Boundary allowing plenty of room on both sides.

19. At the "Specify opposite corner for model space view" prompt, click a point down and to the right of the Callout Boundary again allowing plenty of room on both sides (see Figure 17–48).

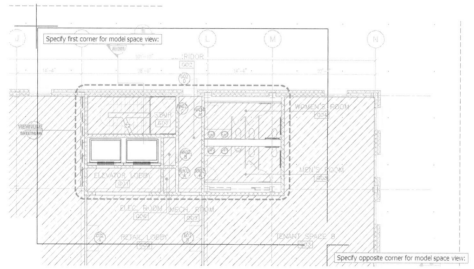

Figure 17–48 *Create a Detail Boundary around the Core area*

20. On the Views tab of Project Navigator, double-click to open *A-EP01*.

21. Manually drag in the *Elevators* file from the *Elements* folder on the Constructs tab.

Notice that the XREFs here have been clipped to the shape indicated by the Callout Boundary. Also note that like the other Callouts, this one created a title mark beneath the plan area. The drawing is ready to receive embellishment. Use the techniques covered in this and the previous three chapters to add Tags, Dimensions, Elevation and Section Callouts and Keynotes. You can also add Schedules or Display Themes if you wish. You might want to select the Grid XREF, right-click it and choose Xclip and adjust the Clip so that some of the column bubbles show. When you are finished, drag the View to a new Sheet. You can find additional

Tags and Annotation routines in the Content Browser in the *Documentation Tool Catalog – Imperial* [*Documentation Tool Catalog – Metric*] catalog.

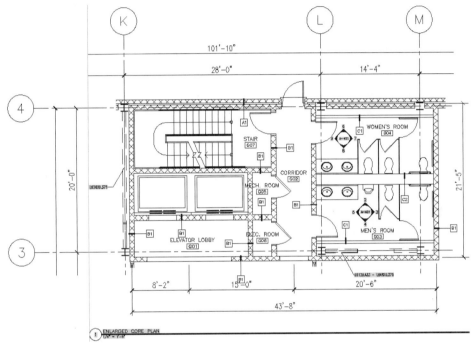

Figure 17–49 *Add annotation to the enlarged Core Plan*

22. Save and Close all commercial project files. Plot the set if you wish.

ADDITIONAL EXERCISES

There are certainly plenty more details to be drawn in both of our projects. Feel free to create additional details, keynotes and legends. Additional exercises have been provided in Appendix A. In Appendix A, you will find exercises to add additional Sections and Elevations in the Commercial Project, as well as further refine those created here in the Residential Project (see Figure 17–50). It is not necessary that you complete these exercises to begin the next chapter, they are provided to enhance your learning experience. Completed projects for each of the exercises have been provided in the *Chapter17/Complete* folder.

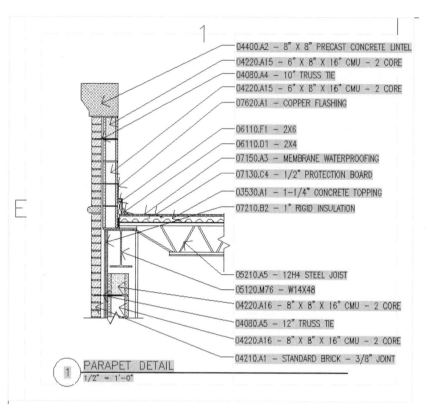

04400.A2 — 8" X 8" PRECAST CONCRETE LINTEL
04220.A15 — 6" X 8" X 16" CMU — 2 CORE
04080.A4 — 10" TRUSS TIE
04220.A15 — 6" X 8" X 16" CMU — 2 CORE
07620.A1 — COPPER FLASHING

06110.F1 — 2X6
06110.D1 — 2X4
07150.A3 — MEMBRANE WATERPROOFING
07130.C4 — 1/2" PROTECTION BOARD
03530.A1 — 1-1/4" CONCRETE TOPPING
07210.B2 — 1" RIGID INSULATION

05210.A5 — 12H4 STEEL JOIST
05120.M76 — W14X48
04220.A16 — 8" X 8" X 16" CMU — 2 CORE
04080.A5 — 12" TRUSS TIE
04220.A16 — 8" X 8" X 16" CMU — 2 CORE
04210.A1 — STANDARD BRICK — 3/8" JOINT

1 PARAPET DETAIL
   1/2" = 1'-0"

Figure 17–50 *Additional Exercises in Appendix A—Add Details*

SUMMARY

Details in ADT are created from a series of predefined components saved in a central database file.

Use Callouts to designate a region of the Model that you wish to Detail.

The Callout tools will create a guideline "sketch" 2D Section/Elevation object on top of which you can draft your detail.

Details with break lines can be created from several smaller 2D Section/Elevation objects lined up one on top of the other.

The Detailing tool palettes contain a wide variety of preconfigured Detail Components.

Use the Detail Component Manager to access the complete database of Detail Components.

New sizes can easily be added to the database directly within the Detail Component Manager.

Use Add Selected and Replace Selected to progressively refine details.

AEC Modify Tools offer the ability to customize the shape of any standard Detail Component.

Keynotes can be assigned to object Styles and Material Definitions.

The Keynote Assignments made within the model carry through to the linework and hatching on the 2D Section/Elevation objects.

Reference Keynotes can include a Key, a Note or both.

Sheet Keynotes are enumerated at the time that they are added to a Sheet and a Sheet Keynote Legend is added.

Keynote Legends and Drawing List Tables can be added to automatically display an inventory of Keynotes and/or Sheets in the Sheet Set.

CHAPTER 18

Area Takeoffs

INTRODUCTION

Most building projects require calculation of critical areas in some form. The extent of such need will vary depending on whether you seek to calculate the building Floor Area Ratio (FAR), determine leasing rates based on Building Owners and Managers Association (BOMA) standards, or generate a Proof of Areas report for building department officials. There are almost as many ways to extract this information from an ADT drawing as there are reasons to do it. In this section, we will look specifically at the AEC Area object and Group tools included with ADT.

There is often confusion among ADT users regarding when to use Space objects and when to use Area objects with regard to Area Takeoffs. Both objects can report their included area in a schedule. The Space object has the ability to use a SPACEQUERY command and export directly to Microsoft Access. New to ADT 2006, the Space object even has the ability to show three separate boundaries: Net, Useable and Gross. The Area object can export information directly to a spreadsheet in Excel format. So which one is best for area calculations of your drawings? If you need only a simple report of areas for each individual room in a building, either tool could be used. However, the Area object is more attractive for complex area takeoffs such as those required by building codes and leasing agents because it is purpose-built for these applications. This is true despite the new useable and gross boundaries of Space objects in 2006.

OBJECTIVES

The primary goal of this chapter is to explore what is required to incorporate Areas and Area Groups into your firm's workflow—specifically the Project Navigator workflow. Herein we will explore Areas, Area Groups and the procedures required to use them successfully. In this chapter, we will explore the following topics:

- Understand the relationships between Areas and Area Groups.
- Learn how to create and manipulate Areas and Groups.
- Understand and use Area Group Templates.
- Explore the possibilities of Display Control settings for Areas.

- Utilize Area and Area Group Property Sets and Schedule Tables.
- Generate reports of Area information.

BASIC AREA TECHNIQUES

Area objects are two-dimensional shapes designed specifically to report building area information. They graphically represent positive and negative areas and, with a special "Decomposed" Display Rep, can even show an area broken down into a series of simplified triangular regions with dimensions for generating area proof (this is not commonly needed in the United States, but is required by building codes in Germany and other countries). Area objects can also be grouped together using Area Groups. An Area Group is a simple "wrapper" object, used to report the total area of several Area objects and/or other Area Groups. You can create Areas from existing geometry in your Building Model, but in many cases, to get the exact Area required, you will often need to manipulate the Area manually. Grip editing works best for this sort of manipulation. Area Objects are typically used for documentation purposes and as such, they should typically be added to a Project's View files.

ADT users are often confused as to which object should be used in projects: the Space object or the Area object. It is the position of this text that both Space objects and Area objects can (and often should be) used in the same project. However, it is possible to choose to use only one or the other. You will need some mechanism to represent the rooms in your buildings. Attached to these rooms at a minimum will be the properties of name and number. You can of course add considerably more properties as we saw in Chapter 15. Spaces are the most commonly used object for this purpose. They are well suited to the task of representing the physical characteristics of a "room," particularly when we consider their three-dimensional components of floor, ceiling and height. The Space object basically represents the interior volume of space enclosed by a room's walls, floor and ceiling. In addition to these physical and graphical qualities, the floor area, ceiling area and volume of a Space object can be readily extracted. Also, as mentioned in the introduction, useable and gross boundaries can optionally be enabled in an ADT 2006 drawing file (see the online help for more information).

If you are not interested in the physical three-dimensional qualities of Space objects, and prefer to specify the boundaries of the "room" based on leasing formula calculations or other specific criteria that do not necessarily follow the finished surfaces of the room, you can consider using Area Objects in place of Spaces to represent your rooms. If you choose this approach, your Area objects will transcend their documentation qualities to become the "actual" rooms (albeit 2D versions of such). In this case, you should change the approach outlined here and add the Areas to a Construct rather than a View. Area Tags and Area Schedule Tables will still be placed in a View, but the Areas themselves would be placed in the Constructs, as the Spaces were in previous chapters.

Even though either object can be used successfully as a "room" object, or close approximation to it, it is the position of this text that the complete needs of the Building Information Model are best served by using both objects to their own advantages in the same project. Use Spaces to represent rooms in a physical and three-dimensional sense, and measure and quantify space as required by building codes or other criteria using Areas and Area Groups. If you adopt this approach, then place your Areas directly in View files designed exclusively for that purpose. This is the approach instructed throughout this chapter. Ultimately it is a matter of office standards and procedures. You may even use different approaches on different types of projects.

INSTALL THE CD FILES AND LOAD THE CURRENT PROJECT

If you have already installed all of the files from the CD, simply skip down to step 3 below to make the project active. If you need to install the CD files, start at step 1.

1. If you have not already done so, install the dataset files located on the Mastering Autodesk Architectural Desktop 2006 CD ROM.

    Refer to "Files Included on the CD ROM" in the Preface for information on installing the sample files included on the CD.

2. Launch Autodesk Architectural Desktop 2006 from the desktop icon created in Chapter 3.

If you did not create a custom icon, you might want to review "Create a New Profile" and "Create a Custom ADT Desktop Shortcut" in Chapter 3. Creating the custom desktop icon is not essential; however, it makes loading the custom profile easier.

 NOTE Unlike the other chapters in this book, this chapter does not include separate Imperial and Metric datasets. Regardless of your preference in units, please load the Imperial project listed below.

3. From the File menu, choose **Project Browser**.

4. Click to open the folder list and choose your *C:* drive.

5. Double-click on the *MasterADT 2006* folder, then the *Chapter18* folder.

    A single Project will be listed: *18 Commercial.*

6. Double-click *18 Commercial* to make it current. (You can also right-click on it and choose **Set Current Project**.) Then click Close in the Project Browser.

 IMPORTANT If a message appears asking you to re-path the project, click **Yes**. Refer to the "Re-Pathing Projects" heading in the Preface for more information.

AREA CREATION

Areas can be drawn with the Area tool on the Scheduling tool palette, or created from existing objects like polylines or Walls.

1. On the Project Navigator, click the Constructs tab, and in the *Elements* folder, double-click *Area Objects* to open it.

2. On the Scheduling tool palette, click the Area tool.

   If you do not see this palette or tool, right-click the Tool Palettes title bar and choose **Document** (to load the Documentation Tool Palette Group) and then click the Scheduling tab.

3. Click on the left side of the screen to add the first point.

4. Click to right of this point to draw a horizontal segment, and then click again above to form an "L" shape.

5. Right-click and choose Ortho to complete the Area.

6. At the "Point in direction of close" prompt, click to the left.

   The Ortho option works the same as Ortho Close for Walls (see Chapter 4); the next point you click will determine the direction of the next segment and then a final segment will be drawn from there perpendicular to the first.

While drawing Areas, you can right-click and draw Arc segments. Add an arc segment to the next shape.

7. Draw another closed Area completely within the first one.

8. Leave some space and draw another closed Area next to the first two. Press ENTER to complete the command (see Figure 18–1).

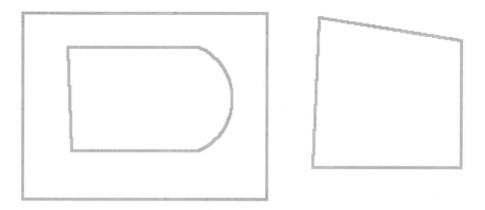

Figure 18–1 *Creating Area Objects*

Areas can contain both positive and negative rings, and can include more than one shape.

> 9. Select the first Area that you drew (the ortho-closed one), right-click and choose **AEC Modify Tools > Subtract**.
>
> 10. At the "Select object(s) to subtract" prompt, click the inner shape (the one we added an arc segment to above) and press ENTER.
>
> 11. At the "Erase selected linework" prompt, choose **Yes**.

Notice that the inner ring turns magenta when the command completes. This indicates that it is a void Area. When creating such Areas, be careful not to let positive and negative areas overlap. Such overlap will cause an error marker to appear as shown in Figure 18–2.

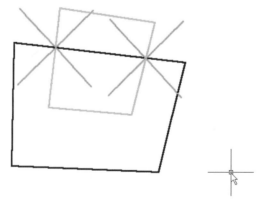

Figure 18–2 *Error markers when positive and negative rings intersect*

NOTE The magenta color assigned to the Subtractive boundary represents a slight variation on the out-of-the-box Display Configuration. By default, both additive and subtractive boundaries are the same color. Here, the dataset was modified to make subtractive boundaries easier to read. To make this change, the Object Display properties of Area objects was modified (right-click any Area and choose **Edit Object Display**) and in the Plan Display Representation's properties, the Subtractive boundary component was changed to Magenta.

12. Repeat the steps using **AEC Modify Tools > Merge** this time. Select the shape drawn last to the immediate right.

13. At the "Erase selected linework" prompt, right-click and choose **Yes**.

14. Click on the final shape when finished.

Notice that all three rings are now part of the same Area object. This is referred to as a compound Area object.

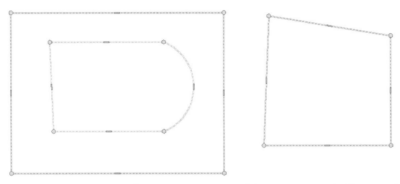

Figure 18–3 *Using AEC Modify Tools to convert several Areas into a single Compound Area object*

MODIFY AREA OBJECTS

The shape of Area Objects can be modified easily with grips and the AEC Modify Tools. To the right side of the current file are several additional Area objects each with a number designation next to them. Shape 1 is a compound Area including three rings: two positive and one negative very much like the one we just created.

15. Select Area 1, right-click and choose **AEC Modify Tools > Trim** (see panel "A" in Figure 18–4).

16. When prompted, click the midpoint of the square ring at right as the first point of the trim (see panel "B" in Figure 18–4).

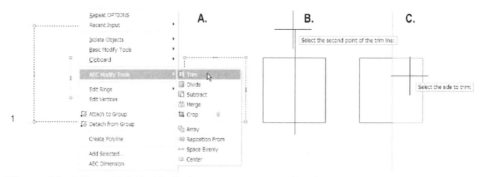

Figure 18–4 *Using AEC Modify Tools to trim a portion of an Area*

17. Click a second point straight up from the first. It does not need to snap.

18. At the "Select the side to trim" prompt, click a point to the right side of the Area object to trim it away (see panel "C" in Figure 18–4).

A similar operation is Divide, which will break the Area into two pieces but keep both as separate Area objects rather than remove one.

19. Select the same object, right-click and choose **AEC Modify Tools > Divide**.

20. Following the prompts, pick two points across the top of the Area as shown in Figure 18–5.

The top portion will now be a separate Area object.

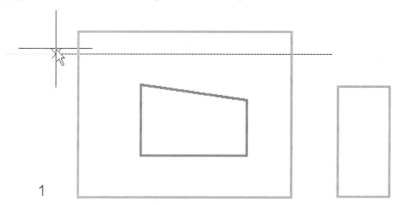

Figure 18–5 *Divide an Area into two separate Area objects*

21. Repeat the same steps, but this time cut across the negative portion of the Area object.

Notice that since the negative ring is no longer completely contained by the positive one, the shape of the new Areas simply incorporates the void and the edge is no longer magenta. Move the resultant Area objects a bit to see this more clearly (see Figure 18–6).

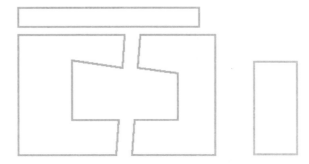

Figure 18–6 *Result of dividing across the void Area*

There are several other AEC Modify Tool operations that we can perform. Shape 2 includes four separate Area objects. The two at the top are compound Areas with negative "holes" in the center. The two at the bottom are separate overlapping shapes.

22. Select the large Area at the bottom in region 2 (labeled in the drawing and shown in Figure 18–8), right-click and choose **AEC Modify Tools > Subtract**.

23. When prompted, select the small Area in the middle and press ENTER.

24. At the "Erase selected linework" prompt, right-click and choose **Yes**.

Notice that the Area in the middle turns red, just like the first demonstration. This indicates that it is now a negative portion of the larger area. Another way to say this is that the larger Area is now a donut. To see this more clearly, we can manipulate the Display Properties of the Area object.

25. Select the same Area, right-click and choose **Edit Object Display**.

26. Click the Edit Display Properties icon on the right (see Figure 18–7).

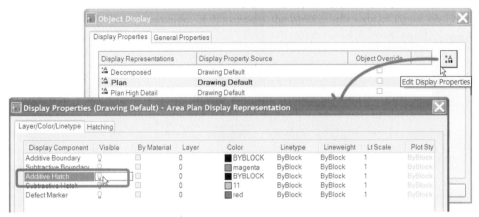

Figure 18–7 *Edit the Display Properties of Area Objects*

27. Turn on the Additive Hatch component (click the dim light bulb) and then click OK twice to return to the drawing (see Figure 18–7).

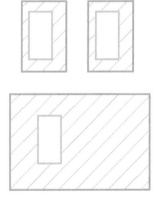

2

Figure 18–8 *Areas displayed with Additive hatching turned on*

We turned on the Additive hatching so that it is much easier to see the "donut" effect of these Area shapes. You can also turn on Subtractive hatching if you wish. If you do, you should pick a hatch pattern that is not used for anything else and make the color the same magenta as the rings. This will help the Subtractive Hatch stand out against the additive portions of the Area objects.

 NOTE Some of these types of modifications have already been performed to the Area objects in the files used in this chapter. The out-of-the-box settings for Areas use different colors than seen here.

28. Move the Area at the top right (in region 2) down into 20524-18-002.tif the larger Area beneath it.

Make sure that it is completely contained in the larger Area as shown in Figure 18–9.

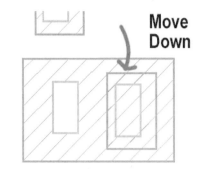

Figure 18–9 *Move the upper Area down into the larger one*

29. Select the larger compound Area, right-click and choose **AEC Modify Tools > Merge**.

30. When prompted, select the small overlapping Area (the one we just moved), and erase the layout geometry.

The result is not very exciting. Since you joined a "donut" with a solid, it simply became solid.

31. Undo the Merge command.

32. Repeat the process, this time using **AEC Modify Tools > Subtract** this time (see Figure 18–10).

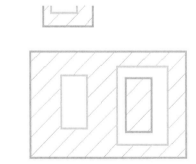

Figure 18–10 *The result of subtracting a "donut" from a solid*

This time the result is much more interesting (see Figure 18–10). Anything that was solid (additive) in the small Area was cut away from the solid portion in the large Area. However, since there was a hole in the small area, it left some of the material behind when carved away from the large one, forming a type of "island."

You can also manipulate the additive and subtractive rings after they are created. Therefore, if you would like to turn a void into a solid, you can simply "reverse" the ring.

33. Select the Area, right-click and choose **Edit Rings > Reverse Ring**.

34. At the "Pick a ring to reverse" prompt, click on any internal subtractive (red) ring.

The void will hatch to indicate that it is now solid. Do a few more experiments if you wish. If you choose **Reverse Profile**, this will reverse the entire shape; anything that was positive will become negative, and vice versa. Use the **Remove Ring** option to delete one of the rings, particularly if you now have two overlapping positive rings. To join the large bottom shape with the remaining small shape above it, use the **AEC Modify Tools > Merge** command. This will yield a compound shape like Shape 1.

Shape 3 contains two overlapping Areas. Use this shape to experiment with the Crop command.

35. Select one of the two Areas in region 3.

36. Right-click and choose **AEC Modify Tools > Crop**.

37. At the "Select object(s) to form crop boundary" prompt, select the overlapping Area and then press ENTER.

38. At the "Erase selected linework" prompt, right-click and choose **Yes**.

You will be left with a small Area the size of just the overlapping portion that they shared. You can use any AutoCAD or ADT object in conjunction with the AEC Modify Tools. They do not have to be other Areas. Recall that they were used extensively in the previous chapter as well.

CREATE AN AREA OBJECT FROM OTHER OBJECTS

Areas can be generated from existing objects. This is useful if you wish to measure the area of a particular object or objects at a specified height.

1. Pan the drawing to the right and locate a small room with four Walls and a Roof.

2. On the Scheduling toolbar, right-click the Area tool and choose **Apply Tool Properties to > Linework and AEC Objects** (see Figure 18–11).

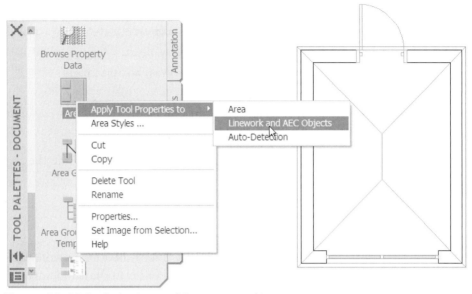

Figure 18–11 *Apply the Area tool Properties to objects*

3. At the "Select objects to convert" prompt, select the Roof object and then press ENTER.

4. In the Convert to Area worksheet that appears, type **10'-6"** in the Cut Plane Height field and then click OK (see Figure 18–12).

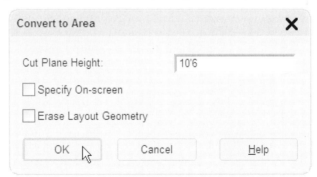

Figure 18–12 *Moving the mouse interactively sets the height and follows the slope of the Roof*

Do not check "Specify On-screen" or "Erase Layout Geometry" for this conversion.

You can make Areas from virtually any ADT object this way. You will always get to assign the Cutplane where the Area is extracted. This can prove a valuable way to generate takeoffs of your Model. You can also generate areas with this command from the Walls that bound the rooms in a plan.

5. Delete the Area just created and then right-click the Area tool again and choose **Apply Tool Properties to > Auto Detection**.

6. At the "Select bounding objects that will form the boundary of the areas" prompt, make a window selection around all four Walls and then press ENTER.

7. At the "Pick internal point" prompt, click a point inside the room.

8. Press ENTER to complete the routine.

Notice that an Area object has been created along the inside perimeter of the Walls.

GRIP EDIT AREA SHAPE

You can use grips to fine-tune the exact shape. For instance, you could grip edit the Area to the centerline or outside edge of the Walls. The Edit Vertices command is also very useful. Using this command, you can add or delete Vertex points from the Area's shape. This is a very helpful way to manipulate an Area after creation.

9. Select the Area that was just created, right-click and choose **Edit Vertices**.

10. At the "Pick vertex to add or shift-select to remove" prompt, click the Endpoint at the Window and then repeat on the other side of the Window (see Figure 18–13).

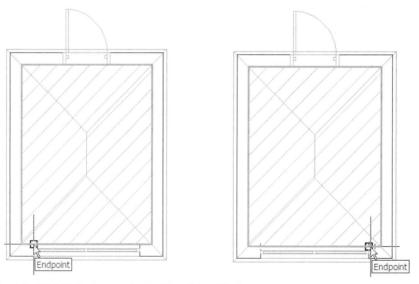

Figure 18–13 *Add Vertices at both sides of the Window*

11. Press ENTER to end the routine.

12. Click on the Area. Select the edge grip at the middle of the Window.

13. Drag it down to the opposite side of the Window and then click (see Figure 18–14).

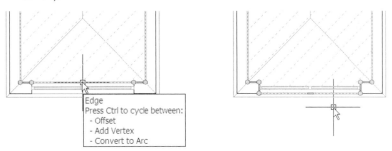

Figure 18–14 *Drag the Edge Grip to move and edge and create new connected segments*

14. Click the same edge grip, and then press the CTRL key twice.

    This will convert this edge to an arc.

15. Drag the edge down a bit and then click (see Figure 18–15).

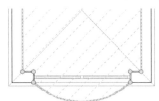

Figure 18–15 *Convert an Edge to an Arc Segment*

DECOMPOSED DISPLAY REPRESENTATION

In Germany and some other German-speaking countries, building departments require a plan showing "proof of areas." This plan usually shows all area calculations triangulated with dimensions on each leg of the various triangles. This can be accomplished with Area objects by enabling a special Decomposed Display Representation.

16. On the Drawing Status Bar, load the **Diagnostic** Display Configuration (see Figure 18–16).

17. Review the information provided in this Display Configuration. When done, reset the Display back to **Medium Detail**.

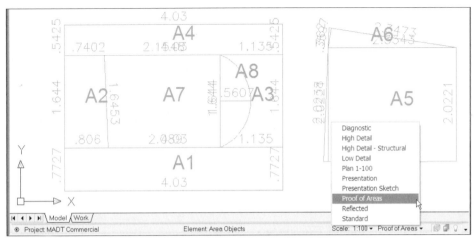

Figure 18–16 *The Decomposed Display Representation shows "Proof of Areas" (Note: the figure shows the "Proof of Areas" Display Configuration available in the Metric D A CH template file)*

The Diagnostic Display Configuration shows the Decomposed Display Representation for Area objects. There is a "Proof of Areas" Display Configuration that is only available in the *Aec Model (Metric D A CH Ctb).dwt* template file. If you wish to use Proof of Areas in your work, you can copy it from Metric D A CH to other template files using the Display Manager (see Chapter 2).

UNDERSTANDING THE RELATIONSHIP OF AREAS TO GROUPS

Area Groups are used to group Areas. You can use an Area Group to measure the total area of several contained Area objects. Let's add a few Groups to this file to see the way that they behave.

1. Click on the Area in region 2 and look at the values in the Dimensions grouping of the Properties palette.

   Notice that the Base Area equals 59 SF.

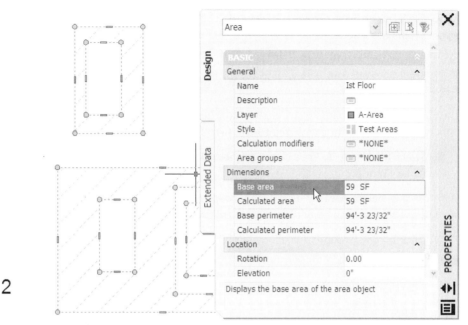

Figure 18–17 *You can see the Base Area of an Area object on the Properties palette*

2. On the Scheduling tool palette, click the Area Group tool.

3. At the "Specify insertion point" prompt, click a point anywhere on screen to place the Area Group Marker and then press ENTER.

Before Area objects have been attached to an Area Group, only a small rectangular marker and a text label will appear. You can name each Group anything you wish. Once you attach Area objects to the Area Group, they will be linked to the Area Group Marker.

4. Select the new Group Marker, and on the Properties palette, change the Name to: **Group A**.

Notice that the Base Area of the Group is currently zero.

5. With Group A still selected, right-click and choose **Attach Areas/Area Groups**.

6. At the "Select Areas and/or Area Groups to Attach" prompt, select the Area object in region 2 and then press ENTER.

The entire Area object will fill in with a solid hatch.

This behavior again represents a modification to the out-of-the-box Display settings. The Object Display Properties of Area Group objects in this file have been modified and the Entity and Hatch components have been toggled on.

7. Click on the Area Group and view the Dimensions values on the Properties palette.

   Notice that the Base Area is also equal to 59 SF (however, it shows a more precise value with a few decimal places this time—the value above was rounded off).

The Base Area of the Area Group is the same as the Base Area of its contained Area object. Let's attach another Area to the Group and then check the Base Area.

8. Select the Group A Area Group, right-click and choose **Attach Areas/Area Groups**.

9. At the "Select Areas and/or Area Groups to Attach" prompt, select the small square Area object in region 3 and then press ENTER.

10. Click on the Group A Area Group object and check the Base Area.

    It is now 68.2 SF.

11. Select Group A, right-click and choose **Add Selected**.

12. Add two more Area Groups and then press ENTER. Rename them **Group B** and **Group C**.

13. Select the Group B Area Group, right-click and choose **Attach Areas/Area Groups**.

14. At the "Select Areas and/or Area Groups to Attach" prompt, select all of the Area objects in region 1 and then press ENTER.

    The Base Area of the Group B Area Group should be 127 SF.

CHANGING AREA GROUP DISPLAY PROPERTIES

Area Groups are currently configured to display solid fill hatching in the same color as the layer upon which they are inserted. We can assign a unique color to each Group by assigning them to Styles and editing the Style Display Properties.

15. Select one of the Group objects, right-click and then choose **Edit Object Display**.

16. On the Display Properties tab, click the Edit Display Properties icon and then turn on the Area Connection Line component on the Layer/Color/Linetype tab.

17. Change the color of the Area Connection Line to Green and then click OK twice to return to the drawing (see Figure 18–18).

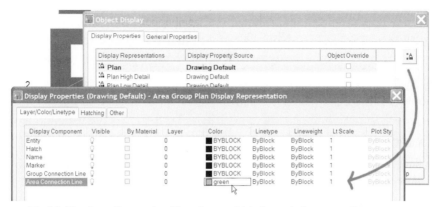

Figure 18–18 *The Area Connection Line shows which Areas belong to a Group*

18. Select Group B, right-click and choose **Copy Area Group Style and Assign**.

19. In the Edit Area Group Style Properties dialog, rename the Style **Group B** and then click the Display Properties tab.

20. Place a checkmark in the Style Override check box next to the Plan Display Representation.

21. In the Display Properties dialog, choose a different color for the Hatch component and then click OK twice to return to the drawing.

22. Repeat these steps for each of the other Groups.

It should now be much easier to tell each Group from one another and with the Area Connection Lines turned on, we can see which Areas are attached. If you wish, you can experiment with other Display Parameters as well, for instance, if you wish to use a hatch pattern other than solid fill.

You can also attach Area Groups to other Area Groups. Let's take a look at this now.

23. Select Group C, right-click and choose **Attach Areas/Area Groups**.

24. At the "Select Areas and/or Area Groups to Attach" prompt, select Group A and Group B and then press ENTER.

TIP When selecting Group A and B to attach, click on the Group Marker. If you select at the Areas, it will select the nested Areas instead.

The Base Area of the Group C Area Group should be 195.2 SF—the total of Groups A and B.

Notice that a Group Connection Line appears that connects the Group C Marker to the Groups A and B Markers. Notice also that the hatching of Group C has applied to the other two Groups. It might make sense to use a pattern other than solid fill for the Group C style.

25. Save and Close the *Area Objects* file.

AREA TOOLSET

Once you have mastered the mechanics of working with Area and Area Group objects, you must then establish how best to use them in your organization. There are many building code requirements that Area objects and Groups can help to satisfy. In addition, they have potential as a tool for generating takeoffs of all sorts of critical building quantity information.

To get started, let's look at the available tools (some of which we have already seen in the preceding tutorial). The Area object tool set consists of five distinct Style types:

- **Area Styles**—Control the name and graphical properties of an Area object (see above).

- **Area Group Styles**—Control the name and graphic display properties of an Area Group object (see above).

- **Group Templates**—Predefined collections of prenamed and styled Area Groups organized hierarchically.

- **Name Definitions**—Lists of names that can be associated with Areas and Area Groups to help standardize the naming of individual Area objects.

- **Calculation Modifier Styles**—Formulas that are applied to the actual area and perimeter values derived from Area objects.

The Area object is the basic component of the tool set. As we saw above, it is a simple 2D shape that behaves much like a polyline and is therefore simple to use and understand. However, the other tools, and more importantly, the relationships between each of the tools, are a bit more complicated. In addition to each of these tools, you will work with Schedules, Property Sets and Area Evaluations to export the data from ADT and into a program such as Microsoft Excel for analysis and reporting.

EXPLORING THE GROUP TEMPLATE

Although there are many ways that you can incorporate Areas into your workflow, if you do the same type of Area calculations frequently, the best approach is to build a template file that includes all of the basic Styles you need most often and then use that template in your Area View files. Group Templates can be a great help for this. A Group Template contains a pre-built collection of Area Groups and their hierarchical structure. They are useful for pre-assigning the Name Definitions to each Group as well (see Figure 18–19).

A subcategory, the *Areas* folder, and three Area View files have been added to the Commercial project for this chapter. The file prefix of "AA" has been used for these files. (AA was chosen since it is the AutoCAD shortcut for the AREA command and is likely to not be used elsewhere in file names.) Areas are documentation objects. They annotate the project in a certain way and typically you would not want them to be seen in any file other than files devoted to the needs of area takeoffs. Therefore, the proper location for Area and Area Groups objects in the ADT Project structure is within their own dedicated View file. You would create such a View file for each floor of the project. This has been done already to save steps. The process of creating Floor Plan Views has been covered extensively throughout this text. Refer to the lessons in Chapter 5 and elsewhere for more information.

1. On the Project Navigator, click the Views tab. In the *Views\Areas* folder, double-click *A-AA02* to open it.

This is a View file of the second floor of our Commercial Project. This file contains an Area Group structure based on a pre-built Group Template and some Schedule Tables designed specifically to report on the Areas and Groups in this file. As you can see in the Group structure at the top, there are several levels of nested Groups. We saw in the exercises above that Area Groups can contain Area objects or other Area Groups. When interpreting the requirements of BOMA, you end up with the potential for several nested Area Groups. This hierarchical structure has been built into a Group Template and is included in the Area View files used in this chapter. To see this template and its settings, use the Style Manager.

2. On the Format menu, choose **Style Manager**.

3. Expand *Documentation Objects* and within the *Group Templates* node, double-click **Leasing** to edit it and then click the Content tab (see Figure 18–19).

4. Expand all items on the left side and select an item such as "Store Areas."

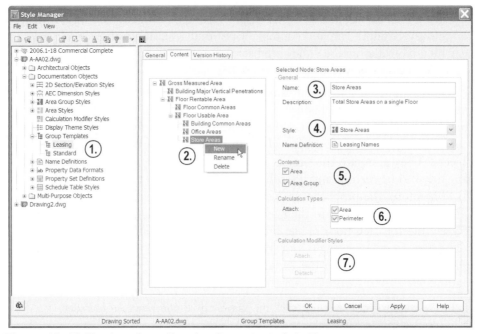

Figure 18–19 *The Area Group Template Properties in Style Manager*

Notes on Figure 18–19:

1. Navigate to **Leasing** using standard Style Manager techniques.

2. All Area Groups are listed hierarchically. Right-click a node on the left in the tree to create a New Area Group, or to Delete or Rename an existing sub-group.

3. Add a Name and Description for the selected Node. (This will be the name and description of a single Area Group when used in the drawing. We edited this above on the Properties palette).

4. Choose an Area Group Style and an Area Name Definition for the selected node. With an Area Name Definition assigned, individual Area objects that are later attached to this Group will be able to reference names from that list. This will help to maintain consistency in naming.

5. By default, Area Groups can include both Areas and other Area Groups. If you wish, you can clear one of the check boxes and make the current Group able to contain only Areas or only Area Groups.

6. By default, Area Groups calculate both Area and Perimeter of the objects they contain. You can clear either one of these check boxes if you wish to limit what it calculates.

7. Calculation Modifier Styles can be attached to either Areas or Area Groups. They are attached directly to the objects themselves, not through the Styles. An alternative to Calculation Modifiers is Formula Properties in Property Set Definitions (see Chapter 15 for more information).

8. Click OK to dismiss the Style Manager.

CAD Manager Note: If you wish to create your own Group Template, you must first build the Group Styles and Name Definitions that you wish to use. Both of these Style types can be created within the Style Manager. You should consider giving each of your custom Area Group Styles a unique color and/or hatch pattern. Name Definitions are useful if you wish to standardize the labeling of Area objects. When an Area is attached to an Area Group, it will gain access to the Name Definition list assigned to that Group. In this way, users will be able to simply choose a Name from a list rather than typing it in. Once you have these resources built, create a Group Template. The names appearing in the tree at the left of the Group Template are the labels that will appear next to each Area Group Marker in the drawing (in the preceding lesson, this was Group A, Group B and Group C). Often these names will be the same as the Area Group Style, but this is not required. At each node, select the Area Group Style, Name Definition and other settings on the right as shown in Figure 18–19. To create a nested Group, simply right-click on a Group in the tree and choose **New**—a new Area Group node will appear beneath the selected group, ready to be configured. Save the Group Template and include it in your office standard drawing template files (*.dwt*) or create a Group Template tool on the tool palettes for it that links to a library file (see the "Using the Styles in Other Drawings" section in Chapter 11 for more information on creating a linked tool palette tool).

EXPLORING THE SCHEDULE TABLE

Some Schedule Table objects have been included in this file. The top one "applies to" Area objects that appear on the *A-Area-Grss* layer. This is used specifically to report the Gross Building Area. A special Area Style has been created and the Gross Building Area displays on screen with a blue outline. There is only one Area using this Style and it has been placed on the *A-Area-Grss* layer in order to isolate it on the top schedule. The remaining Area objects are on the default *A-Area* layer. The middle schedule is set to filter to this layer and therefore includes all Areas *except* the Gross Building Area object. The last schedule lists the Area Groups.

CAD Manager Note: If you prefer, you can use Classification Definitions to filter your Schedule Tables. Classifications offer an alternative means of filtering ADT objects for Display and Scheduling purposes. Create and modify Classification Definitions in the Style Manager. Classifications are assigned to the Style of an object on the Classifications tab. For more information, consult the online help.

9. Select the top Schedule Table object.

10. On the Properties palette, in the Selection grouping, notice that the Layer wildcard is set to **A-Area-Grss** (see Figure 18–20).

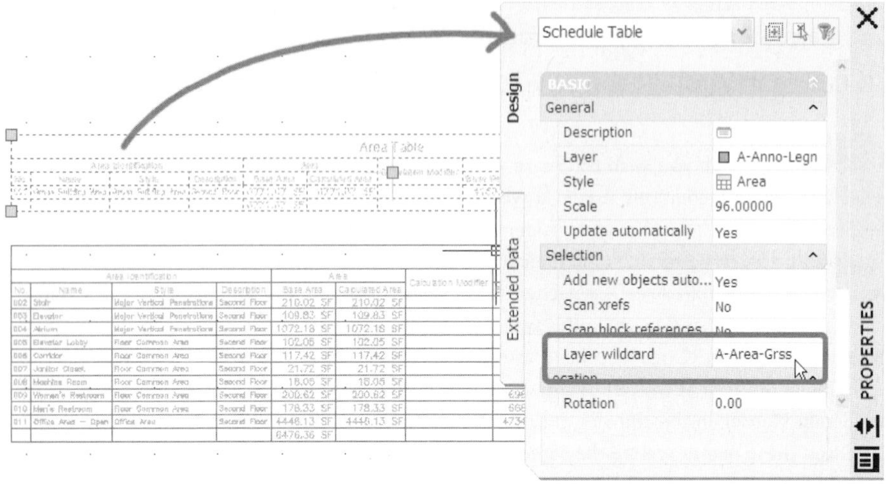

Figure 18–20 *Using a layer wildcard allows only Areas on that layer to be included in the Schedule*

11. Right-click and choose **Deselect All**.

12. Select the Schedule in the middle and note its Layer wildcard.

The middle Schedule is set to layer: **A-Area**; this is the default layer on which Areas are created.

NOTE The default layer varies with the Layer Standard being used. If you use a Layer Standard other than AIA 256 Color, your default Area layer will be different. Make adjustments to your Schedule filtering accordingly.

13. Right-click and choose **Deselect All**.

14. Select the Schedule at the bottom and note its Layer wildcard.

The bottom schedule uses the default Layer wildcard: (*) which means "all". This means that it will include objects on any layer. A Layer wildcard is not required on the bottom Schedule Table because its style "applies to" Area Groups only and not Area objects. Both of the top two Schedules "apply to" Areas. Therefore, the Layer wildcard was used to help those two schedules discern between different types of Areas, namely, those that occur on certain layers. Refer back to Chapter 15 for more information on the settings such as "Applies To" in Schedule Table Styles.

EXPLORING AREA STYLES

Graphically a small rectangular Marker with a text label is used to represent a Group within the drawing. These icons are organized hierarchically in a tree-style organization based on the Group Template explored above. Each one also includes a line to show how it is attached to other groups. Those lines are colored green in these sample files (by default they are turned off). Another line is used to link Area Groups to the Areas they contain. Those lines are colored cyan with a dashed line-type in these files (by default these are also turned off).

Another important tool to help us control, understand and convey our Area and Group structure is the Display Properties. In these files, the Display Properties of each Area Style and Area Group Style have been configured to make it easy to understand at a glance the Style to which a particular object belongs (see Figure 18–21). Table 18–1 summarizes these settings. For more information on how to configure these Display Settings, see the "Changing Area Group Display Properties" section above.

Table 18–1 *Commercial Project Area Styles Display Settings*

No	Area Style	Color	Pattern	Angle
2	Gross Building Area	5 - Blue	Turned OFF	
4	Major Vertical Penetrations	222	User Single	90
7	Office Area	183	User Single	60
8	Store Area	132	User Single	30
9	Building Common Area	83	User Single	120
11	Floor Common Area	23	User Single	150

Notes on Table 18–1:

1. The colors and hatch settings shown in Table 18–1 are used for the Additive Boundaries and Additive Hatch for Area objects in the Plan and Plan High Detail Display Representations.

2. Similar colors are used for Area Group Styles with like names in the Plan, Plan High Detail and Plan Low Detail Display Representations. Hatching for all Area Group Styles is set to 90°.

3. In Plan Low Detail, additive Hatch has been turned off for Area objects and set to Solid Fill for Area Groups.

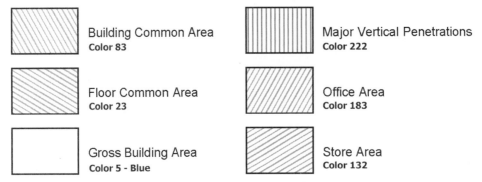

Figure 18–21 *Varying the angles and colors of the hatch patterns helps identify each of the Styles easily on screen*

Each Display Configuration in this file shows Areas and Groups a bit differently. Medium Detail is the currently active Display Configuration. Let's take a look at the others.

15. On the Drawing Status Bar, load the **High Detail** Display Configuration.

The display of the Areas has adjusted to reflect the scale, but more importantly, the Groups now show only their icons. Use this Configuration when you wish to view the Areas without their Connection Lines or the superimposed Area Group hatching.

16. Load the **Low Detail** Display Configuration.

Hatching is now solid fill. This Configuration could be used to plot a color presentation key plan of the critical Area divisions in a plan.

17. Load the **Medium Detail** Display Configuration again.

As we have already seen, both Areas and Groups are shown here. The hatching has been configured in these datasets to make it easier to work with both object types displayed and hatched. The Area objects are each coded by color and hatch, as indicated above. The Groups are shown here with a simple hatch pattern with wide spacing at 90 degrees. This enables you to easily select either the Area or the Group without the need to use complex selection techniques. You might need to zoom in while selected in some cases.

CAD Manager Note: Please remember that the Display Properties showcased in these datasets are not the out-of-the-box settings. If you create Areas and Area Groups using the out-of-the-box template files, they will display differently. If you wish, you can save one of the files here as an "Area Creation Template" or you can use Project Standards to copy the settings for Areas and Area Groups from these files to your own templates. Feel free to modify any of the settings as appropriate for the specific uses in your firm. Also note that the dataset used here has been synchronized using

Project Standards to ensure that all of the Area and Area Group settings are consistent within each project file.

CALCULATE RENTABLE AREA

Area calculation needs and requirements will vary from one jurisdiction to the next. They sometimes vary from one client to the next. The following tutorial explores the use of the Area and Area Group objects in calculating the rentable areas of the Commercial project.

In the remainder of the tutorial, we will look at an example of using the Area Objects and Groups to meet the calculation requirements of the Building Owners and Managers Association (BOMA), which are prevalent in the United States and Canada for determining rentable areas in buildings. Included in the Project files for this chapter are some drawings with prebuilt Area Group Templates and Area Styles. This will make the process flow much more smoothly. As was mentioned previously, this portion of the lesson is provided in Imperial units only. *Tools developed in this chapter are designed to be an illustration of technique in performing area calculations based on the guidelines outlined in the publication "Standard Method for Measuring Floor Area in Office Buildings," © 1996, Building Owners and Managers Association International. The following is solely the interpretation of the author. To avoid misinterpretation, information in this chapter should not be used without acquiring the complete ANSI/BOMA Z65.1-1996 publication.*

CREATE A GROUP TEMPLATE TOOL

As with any style-based ADT object, you can create a tool from a Group Template. The advantage of doing so is that it can then be easily used in any drawing in the project while ensuring consistency in style usage throughout all drawing files.

1. On the Scheduling tool palette, right-click the Area Group from Template tool and choose **Copy**.

2. Right-click on the Tool Palettes title bar and choose New Palette. Name the new palette **Chapter 18**.

3. Right-click on the tool palettes title bar and make the 18 Commercial tool palette group active.

4. Right-click on the Project Tools palette and choose **Paste**.

    The Area Group from Template tool will appear on the palette.

NOTE An additional shared project tools palette named: "Leasing Area Tools" has been added to the project. This palette contains tools for each of the basic Area content items mentioned so far in this chapter.

5. Right-click the new Area Group from Template tool and choose **Properties**.

6. In the Properties worksheet, change the name to **MADT Leasing Group Template**.

7. From the Template Location list, choose **Browse** (see left side of Figure 18–22).

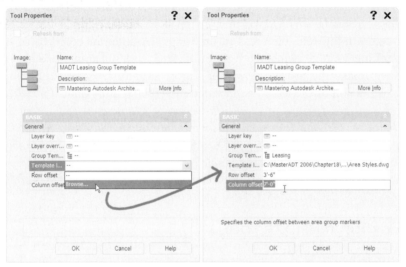

Figure 18–22 *Choose Browse to set the tool's Style location to a remote library file*

8. Browse to the *C:\MasterADT 2006\Chapter18\MADT Commercial\Elements* folder, choose the *Area Styles.dwg* file and then click Open.

9. From the Group Template list, choose **Leasing**.

10. For the Row offset, type: **3'-6"** and for the Column offset: **7'-0"** and then click OK (see right side of Figure 18–22).

The tool is ready to use.

If you wish, you can add this tool to the shared "Leasing Area Tools" tool palette following a process similar to the one we have used in previous chapters. Right-click the new *MADT Leasing Group Template* tool and choose **Copy**. Click the Content Browser icon on the Project Navigator, and navigate to the *Project Catalog > Leasing Area Tools* category. Right-click and choose **Paste**. Back in ADT, refresh this palette.

USE AN AREA GROUP TEMPLATE

As we have seen above, a Group Template is like any other Style in ADT. It is accessed through the Style Manager and can be imported from or exported to other files. It can be referenced remotely by tools on tool palettes. For convenience, the files in this dataset already include a sample Group Template. Using the Group Template is easy. Simply insert it into the drawing and then attach Areas to its Groups. Let's do this now in the First Floor.

11. On the Project Navigator, click the Views tab. In the *Views\Areas* folder, double-click *A-AA01* to open it.

The first floor has an atrium entrance, a main lobby and a core in the rear. To either side are open retail spaces. Most of the spaces have already been marked by Area objects. You will notice that there are no Area Groups in this file and that the retail space to the right has not been drawn yet.

12. On the Project Tools palette, click the **MADT Leasing Group Template** tool.

13. At the "Specify insert point" prompt, click a point above the plan. (The exact location is not important.) Press ENTER to complete the command (see Figure 18–23).

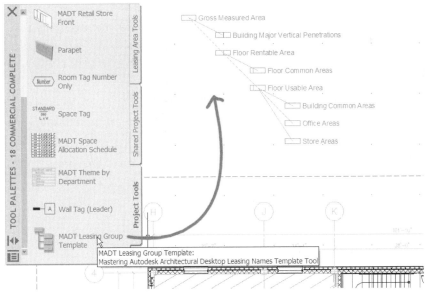

Figure 18–23 *Place a collection of nested Area Groups using a Group Template*

You will now have eight Area Groups inserted into the drawing and linked to one another. All we need to do now is attach the Area objects (already in the drawing) to the appropriate Groups. If you frequently use the same grouping structure, the productivity boost from using a Group Template cannot be overstated.

14. Zoom in on the bottom Schedule Table.

Notice that all of the Area Groups have already populated the Schedule. This is because the schedule references Style-based Property Set Data and all of the pre-built Area Group Styles in this file already have the appropriate Property Set attached. For more information on Property Sets and Schedule Tables, refer to Chapter 15. All of the area values are currently zero (see Figure 18–24). This is because (as in the exercise above) there are no Area objects attached to any of the Groups yet.

Area Group Table		
No.	Name	Calculated Area
3	Gross Measured Area	0.00 SF
4	Building Major Vertical Penetrations	0.00 SF
5	Floor Rentable Area	0.00 SF
7	Office Areas	0.00 SF
8	Store Areas	0.00 SF
9	Building Common Areas	0.00 SF
10	Floor Usable Area	0.00 SF
11	Floor Common Areas	0.00 SF

Figure 18–24 *The Group structure populates the existing Schedule, but the values read zero until Areas are attached to the Groups*

ATTACHING AREAS TO GROUPS

15. Select the "Building Major Vertical Penetrations" Group icon at the top of the drawing (from the collection of Groups just added), right-click and choose **Attach Areas/Area Groups**.

16. Select the two Area objects at the Stair and Elevators and then press ENTER (see Figure 18–25).

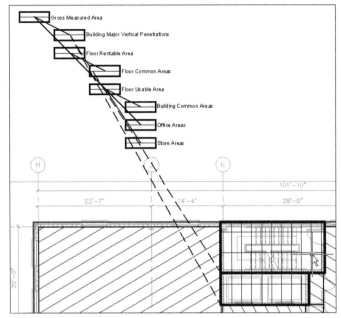

Figure 18–25 *Attach the two vertical circulation spaces to the appropriate Area Group*

17. Pan over to the Schedules and check the numbers (see Figure 18–26).

Notice in the Area Table that the Stair is 210.00 SF and the Elevator is 109.31 SF. The sum of these two values is 319.31 SF, which is in fact the value that shows in the Area Group Table for Building Major Vertical Penetrations.

| Area Identification | | | | Area | | C |
No.	Name	Style	Description	Base Area	Calculated Area	
002	Stair	Major Vertical Penetrations	First Floor	210.00 SF	210.00 SF	
003	Elevator	Major Vertical Penetrations	First Floor	109.31 SF	109.31 SF	
004	Atrium	Building Common Area	First Floor	834.89 SF	834.89 SF	
005	Main Lobby	Building Common Area	First Floor	660.75 SF	660.75 SF	
006	Elevator Lobby	Building Common Area	First Floor	101.04 SF	101.04 SF	
007	Store Area 1	Store Area	First Floor	2089.18 SF	2089.18 SF	
009	Corridor	Floor Common Area	First Floor	116.51 SF	116.51 SF	
010	Janitor Closet	Floor Common Area	First Floor	22.23 SF	22.23 SF	
011	Machine Room	Floor Common Area	First Floor	18.56 SF	18.56 SF	
012	Women's Restroom	Floor Common Area	First Floor	202.50 SF	202.50 SF	
013	Men's Restroom	Floor Common Area	First Floor	177.00 SF	177.00 SF	
				4541.96 SF		

These two total item 4

The total does not match item 3

Area Group Table

No.	Name	Calculated Area
3	Gross Measured Area	319.31 SF
4	Building Major Vertical Penetrations	319.31 SF

Figure 18–26 *Comparing the values in each Schedule Table*

This value also appears in the Gross Measured Area field as well. This is because the Building Major Vertical Penetrations Group is a member of the Gross Measured Area Group and is currently the only member. Compare the total at the bottom of the Area Schedule with the value for the Gross Measured Area Group in the Group Schedule. The current value for Gross Measured Area in the Area Group Table is 319.31 SF, while the total at the bottom of the Area Table is 4541.96 SF. This is because all of the Areas in the file are included in the Area Table while only the two that we have currently grouped appear in the Area Group Table. This is a good way to check that you haven't missed Areas, since the Gross Measured Area of the building should include all of the Areas in the drawing.

18. Using Figure 18–27 as a guide, attach the remaining Areas to the appropriate Groups.

TIP You can also select the Area object(s) first, right-click and then choose **Attach to Group**.

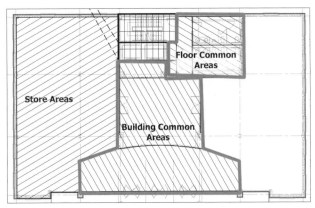

Figure 18–27 *Attach the remaining Areas to Groups*

Compare the total at the bottom of the Area Schedule with the value for the Gross Measured Area Group in the Group Schedule. Both values should now read 4541.96 SF (see Figure 18–28). It appears that we have successfully grouped all Areas.

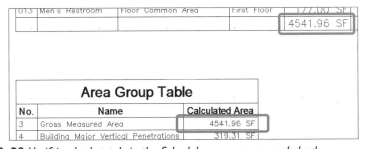

Figure 18–28 *Verifying both totals in the Schedule serves as a good check*

 NOTE If these two totals do not match, double check your attachments with Figure 18–27 to be sure that you didn't miss any.

ADD A NEW AREA

Even though the totals in the Schedule confirm that all Areas in the drawing have been attached to Groups, there is one Area missing for the Retail Space at the right. (The Schedule cannot report this situation, unfortunately.) Let's add it now.

19. On the Leasing Area Tools tool palette, click the **Store Area** tool.

On the Properties palette, in the General grouping, click to open the Name list. Notice that all of the Names in the list are residential in nature and inconsistent with the current Project. This is because the new Area that we are adding is not yet attached to a Group in this file. When we attach it, the associated Name Definition list from this file will become available. We can actually attach this new Area to a Group *before* we begin drawing it.

20. On the Properties palette, in the General grouping, click the worksheet icon next to the "Area groups" item (see the right side of Figure 18–29).

21. In the Area Groups worksheet, expand the list completely and place a checkmark in the box next to the "Store Areas" Group (see the left side of Figure 18–29).

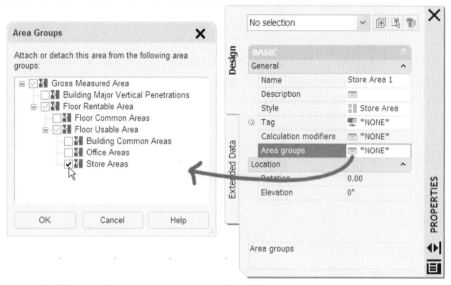

Figure 18–29 *Expand the Group list and check the Store Areas Group*

22. Click OK in the Area Groups worksheet.

23. On the Properties palette, in the General grouping, open the Name list again.

Notice that the list now comes from the Name Definition that is attached to the Area Group object. This is why we attached the Area to the Group first. Names are not critical to the use of Areas. They are really an extra bit of information attached to the Area objects. However, if you choose to use Names, using the associated Name Definition list is a good way to ensure consistency in naming.

24. Choose **Store Area** from the Name list (see Figure 18–30).

Figure 18–30 *Choose a predefined Name from the list*

25. Begin clicking points around the open retail space on the right using Figure 18–31 as a guide.

The Area should measure to the inside face of the right and rear exterior Walls, and also the Core Walls. It should measure to the center of the common Walls between the retail space and the lobby. Finally, the Area should project all the way to the outside face of the street frontage. Be sure to right-click and switch to Arc on the segment along the atrium.

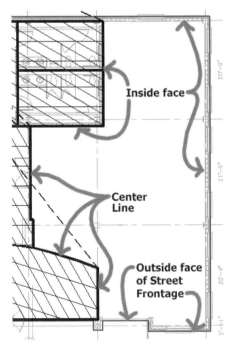

Figure 18–31 *Create an Area for the right side Retail Space*

Look at the Schedule again.

Notice that since we added the Area directly to the Group, both Schedules have updated and the totals still match (6229.60 SF). However, notice that there is a question mark in the number column of the new Area. This is because the number property is an object-based property. For more information on object-based Property Sets, see the "Property Set Data" section in Chapter 15.

26. Select the Schedule Table, right-click and choose **Add All Property Sets**.

If you wish, you can change the Name of each of the Groups that pertain to the "Floor." In other words, we currently have a Name: "Floor Common Areas." You can select the Group Marker for this Group and on the Properties palette, change the Name to "First Floor Common Areas." Do this for all "Floor" names (see Figure 18–32).

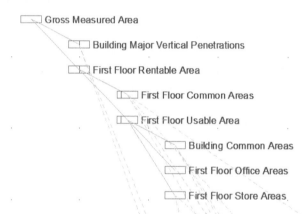

Figure 18–32 *Rename the Area Groups to pertain to the First Floor*

27. Save and Close the *A-AA01* View file.

The Schedule updates to reflect this change. If you wish, you can use the Renumber Data Tool on the Scheduling tool palette to renumber all of the Area objects in whatever order you wish.

EDIT THE SECOND FLOOR AREA PLAN

The Elevator Lobby space on the second floor is incorrectly included as part of the Building Common Areas Group. It ought to be included in the Floor Common Areas instead.

28. On the Project Navigator, click the Views tab. In the *Views\Areas* folder, double-click *A-AA02* to open it.

NOTE If you left the Second Floor Area View open above, this action will simply make that file active.

29. Select the Area just beneath the Elevators (Elevator Lobby).

30. On the Properties palette (Design Tab), click the worksheet icon next to Area Groups.

This opens the worksheet that we saw above showing the Area Group hierarchy.

31. Expand the list, clear the "Building Common Areas" check box and place a checkmark in the "Second Floor Common Areas" check box instead (see Figure 18–33).

Figure 18–33 *Re-assign the Group for this Area*

 TIP This is also a good way to see to which Group(s) an Area is attached.

32. Click OK to close the Area Groups worksheet.

Note the changes to the Schedule Tables.

33. Save and Close the *A-AA02* View file.

CREATE THE FOURTH FLOOR AREA PLAN

The Area View file for the fourth floor has been created and included in Project Navigator in the *Areas* folder. It does not yet contain any Areas or Groups.

34. On the Project Navigator, click the Views tab. In the *Views\Areas* folder, double-click *A-AA04* to open it.

You may have noticed that in the previous files, the floor plan background appeared in a dark gray color. The same is true here for the *Column Grid* and *Front Fa\c¢cade* XREFs. This was done deliberately to make the Area and Area Group objects stand out better. However, the *Shell and Core* file is not displaying this way in the current floor. Let's fix this now. The technique used to create this effect is to apply an Object Display Override to the XREFs.

35. Select the *Shell and Core* XREF on screen, right-click and choose **Edit Object Display**.

36. In the Object Display dialog, click the XREF Display tab and place a checkmark in the "Override the display configuration set in the host drawing" checkbox and then pick **Screened** from the list that appears. Click OK to close the dialog.

Overriding the XREF Display allows you to select a Display Configuration from the XREF file to use for the selected XREF rather than have it display using the default applied to the current host drawing. This was done on the stair tower in Chapter 7. If the Screened display makes the background too dark for your taste, you can edit the color of the A-Anno-Scrn layer. This layer is assigned to color 250 by default. On a black background, it can appear almost invisible. Feel free to change this color to a lighter shade of gray such as 252 or 253. The Screened Display Configuration assigns all ADT objects to this layer. For more information on Display Configurations, refer to Chapter 2. Let's add Areas and Groups to the fourth floor now.

37. Select the Core Areas in one of the other files, Copy them to the clipboard and then choose **Clipboard > Paste to Original Coordinates** in the *A-AA04* file.

38. Draw the Area for the open Office Space.

The fourth floor does not have an atrium space. If you wish, you can copy both the Open Office Space and the Atrium Space from the *Second Floor Area Plan (A-AA02)*, paste them to original coordinates, and then use the **AEC Modify Tools > Merge** function to join them together. Be sure that the Style of the final Area is *Office Area*. Use Grips to edit the shape as necessary.

39. Use the ***MADT Leasing Group Template*** tool created above to add Area Groups to the file.

40. Rename all "Floor" Groups to "Fourth Floor."

41. Attach all Areas to Groups.

42. Copy and Paste to Original Coordinates the Gross Building Area object (the one with the blue outline surrounding the building) from *A-AA03* to the fourth floor.

The plan, the Groups and the Schedules should all update to reflect your changes as you work.

43. Save and Close the file.

CREATING AREAS FROM WALLS

The following process is a bit involved, but the individual steps are simple to execute. The goal is to quickly generate a large collection of Area objects from an existing floor plan. Two factors make this process complex: First, the Areas that we need for proper leasing calculations must go to the centerline of the Walls and not the inside face as the Area object's Auto Detection routine does. Second, some of the bounding Walls are in XREF files that are not supported by Auto Detection. These two issues will be common in many medium-scale to large-scale projects. Therefore, the following process has been developed as a way to minimize the total amount of work required to achieve results. As has been mentioned, minimizing the workload does not necessarily mean minimizing the steps.

This tutorial is broken down into three major parts. We must first open the file containing the Walls for the third floor and copy them out to a temporary file. Next we will generate Area objects from these Walls using a modified Display Configuration and finally, we will move the Areas back to the appropriate file.

One final note is worth mentioning. This process is intended for larger projects. We are using the third floor of our commercial project simply because of all the dataset files at our disposal, it has the most Walls—which is important to make the technique understandable. However, in reality, you would only need one large Area for this entire tenant space. Unless of course, the occupant of each office had to pay their own rent. So if this were a real project, you would likely take the Area from the second floor as a starting point, copy and paste it to the third floor, use the AEC Modify tools to cut it in half at the demising wall and then use grips to fine-tune the shape. Feel free to do this first for practice if you wish. At some point however, if you find yourself with a project that has a large number of departments, or zones that need to be calculated and charged separately and the Walls are already drawn, then the following technique will prove helpful.

BUILD A TEMPORARY WORKING FILE

1. On the Project Navigator, click the Constructs tab. In the *Constructs\Architectural* folder, double-click *03 Partitions* to open it.

2. Select any interior partition in the tenant suite (the Wall by the elevators is easy to select), right-click and choose **Select Similar**.

3. Right-click again and choose **Clipboard > Copy**.

4. Close the file; you do *not* need to Save it.

5. On the Project Navigator, click the Constructs tab. In the *Elements* folder, and double-click *Third Floor Outline* to open it.

This file already contains a previously drawn polyline of the Building perimeter line. This line represents the rentable area of the entire third floor. (On the masonry walls, it goes to the inside face, and on the curtain wall it goes to the face of the glass). It also contains a small arc for the division between the workstations and the corridor. These have been provided here to save time (see Figure 18–34).

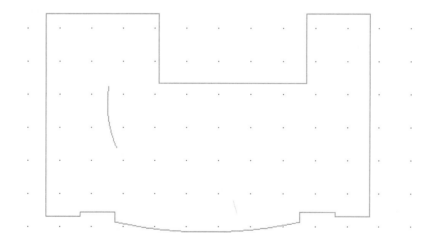

Figure 18–34 *The file contains a polyline of the floor rentable area and an arc for the workstations*

6. Right-click in the drawing and choose **Clipboard > Paste to Original Coordinates**.

We will need to cleanup the plan a bit. There are Walls that are not required to generate proper Areas, such as the column enclosures.

7. Delete all column enclosures, including the round ones (see the left side of Figure 18–35).

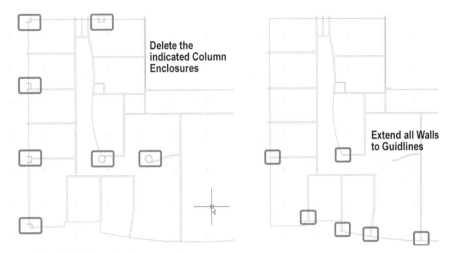

Figure 18–35 *Delete all column enclosures*

8. Extend all Walls as required to the perimeter polyline (see the right side of Figure 18–35).

TIP A very useful addition to the Trim and Extend commands in recent releases of AutoCAD, gives you the ability to use a crossing window to trim or extend several objects at once. At the "Select object to Trim/Extend" prompt, simply click from right to left. Try it out. Also remember that you can hold down the SHIFT key to switch from Trim to Extend and vice versa.

LOAD THE DIAGNOSTIC DISPLAY CONFIGURATION

As we mentioned above, we need to have most of the Areas in the tenant suite measured to the centerline of Walls. If we use the Auto Detection routine that we used at the beginning of the chapter, it detects the inside face of the Walls. Therefore, we will use the AutoCAD BOUNDARY command instead. To use this command, we need geometry that represents the centerline of the Walls. In this case the easiest way to do this is to load the Diagnostic Display Configuration.

9. On the Drawing Status Bar, load the **Diagnostic** Display Configuration.

Diagnostic shows the Sketch and Graph Display Representations for Walls. The Sketch Display includes a Marker (small yellow wedge at the midpoint of each Wall). This will interfere with the BOUNDARY command, so let's turn it off.

10. Select one of the Walls, right-click and choose **Edit Object Display**.

11. Select the Sketch Display Representation at the bottom of the list and then click the Edit Display Properties icon.

12. On the Layer/Color/Linetype tab, turn off the Marker and Baseline components and turn on the Centerline component (see Figure 18–36).

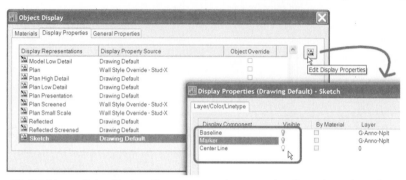

Figure 18–36 *Turn off the Marker Component and turn on the Centerline of the Sketch Display Representation*

In our case, all of our Walls use Center justification, so turning on the Centerline and turning off the Baseline was not entirely necessary, but it is a good habit nonetheless.

13. Click OK twice to return to the drawing.

The small yellow triangle markers should no longer be visible.

14. Perform any final cleanup necessary.

The goal is to have all "gaps" closed. The command that we will use to generate the Area objects is very similar to the HATCH command. Therefore, if the spaces are not completely closed, it will fail. You can use Extend, Trim and grip editing as necessary. You can also draw lines as needed to close small gaps. Also, remember that you can easily manipulate the Area objects later using their grips. These Walls are a means to an end, not an end in themselves, so don't get carried away making them perfect. Sometimes it is easiest to just add a small Wall or line segment to close a gap. Use this technique to "close" the Conference Room and Reception areas off from the corridor. Remember that in this Display Configuration, Walls will look just like lines. Therefore, this task should be very simple (see Figure 18–37).

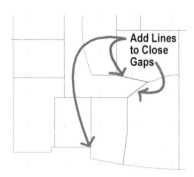

Figure 18–37 *Add Lines to fill in gaps*

GENERATE THE AREA OBJECTS

Before we can generate the Areas, we must create a polyline in each room. This step is unfortunately necessary since Auto Detection routine uses the true width of the Walls regardless of the Display Set currently active. However, the AutoCAD BOUNDARY command "looks" only at the graphics on screen. Therefore, we will first create polylines from our single-line Walls and then convert the polylines to Area objects.

15. At the command line type BO and then press ENTER.

16. In the Boundary Creation dialog, click the Pick Points icon.

17. At the "Select Internal Point" prompt, click once within each closed space of the tenant space on the left. Do not click on the tenant space on the right or the core.

 NOTE The current layer is 0. However, this will prove unimportant in the coming steps.

Each space should highlight as you click. If you get an error, go back and check to see that the space is closed and that all gaps are filled.

18. Select one of the polylines created, right-click and choose **Select Similar**.

19. Hold down the SHIFT key and click to deselect the outer magenta polyline shape.

20. On the Leasing Area Tools tool palette, right-click the **Office Area** tool and choose **Apply Tool Properties to > Linework and AEC Objects**.

21. In the Convert to Area worksheet, set the Cutplane Height to **0**, place a checkmark in the "Erase Layout Geometry" checkbox and then click OK.

Several Areas will appear. They will be displaying in the Decomposed Display Representation explored above. This is because we still have the Diagnostic Display Configuration active. It is not important to change anything, however, as we are now ready to move these Areas to the Third Floor Area Plan View file. However, if you wish, you can reload the Medium Detail Display Configuration before we do.

MOVE THE AREAS TO THE CORRECT FILE

The final step in creating the Areas is to move them to the correct file.

22. Select any one of the Area objects, right-click and choose **Select Similar**.

23. Right-click and choose **Clipboard > Cut**.

24. On the Project Navigator, click the Views tab. In the *Views\Areas* folder, double-click *A-AA03* to open it.

25. Right-click and choose **Clipboard > Paste to Original Coordinates**.

26. Select one of the new Areas (click near the corridor to easily select one), right-click and choose **Select Similar**.

27. Hold down the SHIFT key, and deselect the tenant space on the right.

28. Right-click and choose **Attach to Group** and attach them to the *Tenant A Office Areas* Area Group (see Figure 18–38).

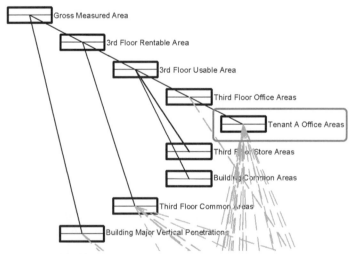

Figure 18–38 *Attach all new Areas to the Tenant A Office Areas Group*

29. Using grips, tweak any of the Area Boundaries as required.

30. Select the Area Table Schedule, right-click and choose **Add all Property Sets** to update the object-based Property Sets.

   If you wish, use the Renumber Data Tool to renumber the Areas.

31. Check the totals in the Schedules, apply Names to the various Areas and then Save and Close the file.

It is not necessary to save the *Third Floor Outline.dwg* file. Again, it is stressed that although there were several steps involved in the previous process, the result was less labor intensive and therefore faster than the process of manually tracing each room's centerlines to generate the Area object.

AREA TAGS

Several sample Tags have been provided out-of-the-box with ADT. You can find some in the Content Browser in the Documentation Catalogs (both Imperial and Metric). Use them like other ADT tags to add labels to each Area if you wish to print your Area Plans or simply for on-screen reference. Most of the tags that apply to Spaces will also work with Areas as well. (This is because their underlying Property Sets apply to both Areas and Spaces.)

There are also several examples contained in the *Metric D A CH catalog*. If you installed the Metric D A CH content with ADT, use your Content Browser to browse to the Sample Palette Catalog - Metric D A C H and then browse to the *Document < Scheduling* category to locate a variety of samples. You will also find a few Display Themes in this catalog that override the display of Areas. The items in this catalog are functionally the same as all other Tags in ADT. They do however have a few interesting features to note:

▶ These Tags show different View Blocks in Low Detail and High Detail Display Configurations and do not display at all in Medium Detail.

▶ The Property Data Formats used by the Property Sets in these Tags are based on Metric units.

▶ Default values used in the Property Sets and in the icons reference German terminology for Area measurement. (See the "DIN 277 Standard Terminology" sidebar.)

DIN 277 STANDARD TERMINOLOGY

The default value shown in D A CH Area Tags (HNF 1) is an abbreviation used in the DIN 277 (Deutsche Industrie Norm) German Standard.

HNF 1 - Hauptnutzfläche, which is similar to the dominant portion of a usable area in a building. There are a variety of such classifications: HNF 1 includes living rooms, HNF 2 - office areas, HNF 3 - manufacturing areas and so on.

Other designations include:

NNF – Nebennutzfläche, which includes Floor common areas like sanitary rooms and wardrobes.

HNF + NNF = usable area

VF – Verkehrsfläche, which is an area for "connection" like corridors, stairs, elevators.

The DIN standard is very detailed. Following it closely is very important to acquiring a building permit in Germany.

Tag the Areas in your Area Plans if you wish. Refer to the "Adding Room Tags" section in Chapter 14 detailed steps to tagging.

PROCESSING AREA DATA

Calculation Modifiers are simple mathematical formulas that are attached to either the Area objects or the Area Groups. You can assign these formulas to work on the Area, the Perimeter or both. The values generated are referred to as the "Calculated Area" and the "Calculated Perimeter." You might use Calculation Modifiers to assist with meeting the requirements of a particular building code or with performing quantity takeoffs.

More complex formulas can be created using Property Set Definitions containing Formula Properties. This was touched upon briefly in Chapter 15. Both of these techniques open the doors to attaining some very specific and useful information extracted directly from the Building Model. Although time and space do not permit us to explore the topic further in this text, the reader is encouraged to explore these powerful features further.

GENERATING REPORTS

Now that we have seen nearly the complete Area tool set in action, it is time to learn how we can work with all of this data. If you are able to solve all of your calculation and reporting needs directly using these tools, you will need only to set up a sheet file that organizes and presents the data in a way suitable for printing. However, in most cases you will want to export the data from ADT for further manipulation and presentation.

There are two basic ways to export the data from Areas and Area Groups in ADT: by using the Area Evaluation feature and by directly exporting a Schedule Table. Area Evaluations provide a rather interesting report, complete with bitmapped graphic representations of each Area and Group. The problem with them is that the format is a bit inflexible. An Excel template file (*.XLT) is used to designate the format of the columns in the report, and an ADT interface allows you to select which data to include.

AREA EVALUATION REPORTS

1. On the Project Navigator, click the Views tab. In the *Views\Areas* folder, double-click *A-AA03* to open it.

2. On the Scheduling tool palette, click the Area Evaluation tool.

Notice the tree structure at the left. This includes all of the Groups and their attached Areas. If you had something selected when you opened this dialog box, it is already selected here in the tree.

3. Close the dialog box, select the Third Floor Usable Area Group marker in the drawing and then click the Area Evaluation tool again.

Notice that now the entire Third Floor Usable Area Group is already selected.

4. In the Evaluation area, click the Evaluation Options button.

In the dialog box that appears, you can decide which information to include in the exported report, whether to include images and which Excel template to use for column formatting (see Figure 18–39).

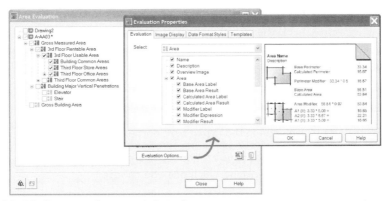

Figure 18–39 *Choosing what to include in the report*

5. On the Evaluation tab, choose **Area** from the Select list, and check all boxes except the Decomposed Image box.

6. Choose **Area Group** from the Select list, and check all boxes.

Most of the defaults on the remaining tabs will give satisfactory results. The Image Display tab allows you to set the size of the bitmap image and set the Display Configuration from which it is generated. If you wish to format any of the data differently in Excel than the way it is in the drawing, click the Data Format Styles tab and make your choices. If you would like to include the "raw" unformatted value of the area or perimeter, you can check the "Additional Exact Value" box (see Figure 18–40).

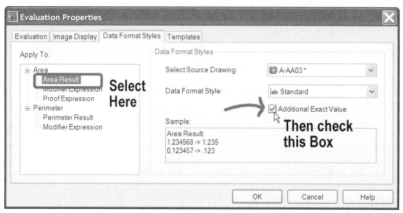

Figure 18–40 *Add the Additional Exact Value to your report*

Finally, you can choose a different Excel template to determine the organization of the columns in the report. There are several samples provided with ADT. The default location for these templates is: *C:\Documents and Settings\All Users\Application Data\Autodesk\ADT 2006\enu\Template\Evaluation Templates*

If you have installed your templates and content on your server, they will be located on the server as well. Check the online help for information on how to customize these templates. It is not necessary to choose a template in the Evaluation Properties dialog. ADT will prompt you to select a template when you generate the report if none is selected here.

7. Click OK to dismiss the Evaluation Properties dialog.

8. Click the small Excel icon in the lower right corner of the dialog box.

9. If the Open template dialog appears, choose one of the templates listed. (For this exercise, it is not important which one you choose.)

10. If an alert appears asking if you wish to make this your default template, choose **No**.

11. In the Save Excel Evaluation File dialog, type a name and choose a location (such as your desktop) to save the file and then click the Save button.

12. When the export is complete, Close the Area Evaluation dialog and then Open the Excel file.

 NOTE If you do not have Microsoft Excel installed on your system, you can still export the report. However, to view it, you will need either Excel or an application that reads Excel format. There is also an icon for exporting to a text file (instead of Excel) next to the Excel icon.

The report will look something like Figure 18–41.

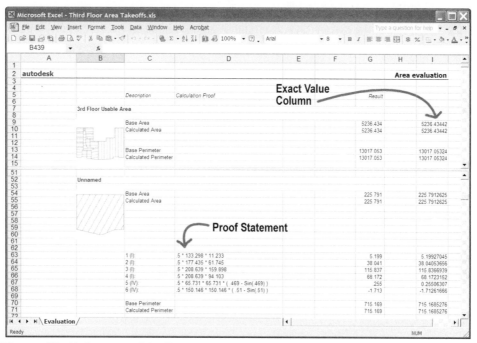

Figure 18–41 *A portion of the Evaluation Report as it appears in Microsoft Excel*

Repeat the steps, and try clearing several of the items in the Select list. Try using different templates and see how you can influence the results.

Although you can build your own Excel template and rearrange the columns, you do not have any control over the order in which the rows appear. The order will be Area - Proof - Perimeter. However, you can choose not to include some of these rows. You need to do any post-processing manually in Excel. This tool serves specific area reporting needs in certain localities. If the report does not give you the data or formatting options you require, consider exporting your Schedule Table instead.

EXPORTING A SCHEDULE TABLE

In addition to Area Evaluations, we can export Schedule Tables directly to Excel. Here we get a bit more flexibility in organizing the columns, including the ability to add headers.

13. Select the Area Table (the Schedule in the middle) right-click and choose **Export**.
14. Click the Browse button and choose a name and location (such as your desktop) to save the Excel file.
15. Click OK in the Export Schedule Table dialog to complete the export process.
16. Open the Excel file and explore the results (see Figure 18–42).

Architectural or Engineering Units formats are not natively supported in Excel. If you are using either of these in any column of the Schedule, you will receive a dialog box asking how you wish to format the exported data. If you intend to perform calculations on the data, choose "Use Unformatted Decimal Value." If you will simply print the report from Excel, you can choose "Convert to Formatted Text."

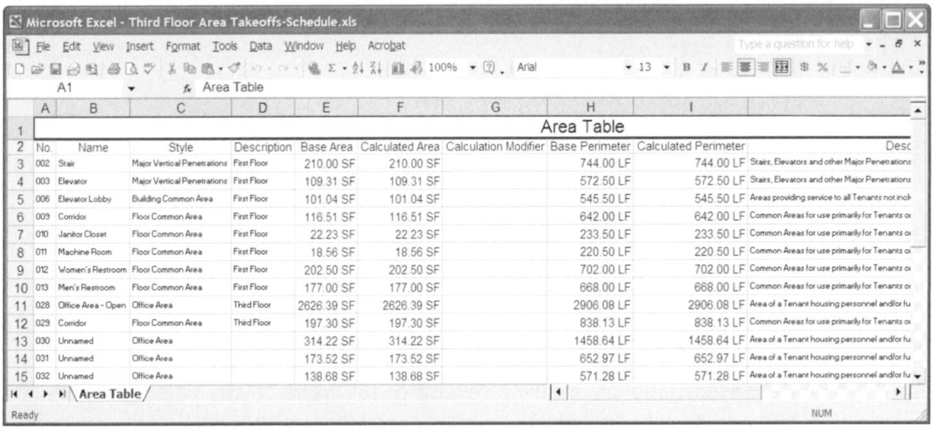

Microsoft Excel - Third Floor Area Takeoffs-Schedule.xls

Area Table

No.	Name	Style	Description	Base Area	Calculated Area	Calculation Modifier	Base Perimeter	Calculated Perimeter	Desc
002	Stair	Major Vertical Penetrations	First Floor	210.00 SF	210.00 SF		744.00 LF	744.00 LF	Stairs, Elevators and other Major Penetrations
003	Elevator	Major Vertical Penetrations	First Floor	109.31 SF	109.31 SF		572.50 LF	572.50 LF	Stairs, Elevators and other Major Penetrations
006	Elevator Lobby	Building Common Area	First Floor	101.04 SF	101.04 SF		545.50 LF	545.50 LF	Areas providing service to all Tenants not incl
009	Corridor	Floor Common Area	First Floor	116.51 SF	116.51 SF		642.00 LF	642.00 LF	Common Areas for use primarily for Tenants o
010	Janitor Closet	Floor Common Area	First Floor	22.23 SF	22.23 SF		233.50 LF	233.50 LF	Common Areas for use primarily for Tenants o
011	Machine Room	Floor Common Area	First Floor	18.56 SF	18.56 SF		220.50 LF	220.50 LF	Common Areas for use primarily for Tenants o
012	Women's Restroom	Floor Common Area	First Floor	202.50 SF	202.50 SF		702.00 LF	702.00 LF	Common Areas for use primarily for Tenants o
013	Men's Restroom	Floor Common Area	First Floor	177.00 SF	177.00 SF		668.00 LF	668.00 LF	Common Areas for use primarily for Tenants o
028	Office Area - Open	Office Area	Third Floor	2626.39 SF	2626.39 SF		2906.08 LF	2906.08 LF	Area of a Tenant housing personnel and/or fu
029	Corridor	Floor Common Area	Third Floor	197.30 SF	197.30 SF		838.13 LF	838.13 LF	Common Areas for use primarily for Tenants o
030	Unnamed	Office Area		314.22 SF	314.22 SF		1458.64 LF	1458.64 LF	Area of a Tenant housing personnel and/or fu
031	Unnamed	Office Area		173.52 SF	173.52 SF		652.97 LF	652.97 LF	Area of a Tenant housing personnel and/or fu
032	Unnamed	Office Area		138.68 SF	138.68 SF		571.28 LF	571.28 LF	Area of a Tenant housing personnel and/or fu

Figure 18–42 *Open the exported Schedule Table in Excel*

17. Save and Close all files when you are done.

If you like, add Model Space Views around the various schedules in *A-AA01* through *A-AA04* and then drag them to a new Area Takeoffs Sheet. You can even insert the Excel files back into this sheet as OLE objects.

SUMMARY

Creating and manipulating Area objects is similar to working with polylines and other profile-based ADT objects.

AEC Modify Tools can be used to manipulate an Area object's shape in many useful ways.

Areas can show both positive and negative rings in a single compound Area object.

Group Templates provide a complete set of Area Groups, predefined and attached to one another and ready to use.

Areas can be created from existing objects such as bounding Walls and other 3D objects like Roofs and Slabs.

Each Display Configuration can show Areas and Groups differently to meet different display needs.

Area Tags can be added to Area objects if desired following the same process as other Tags.

Reports can be generated directly to Excel from the Areas and Groups themselves using an Area Evaluation or through the Export of Schedule Tables.

SECTION IV

Generating Output

In this final section, we will wrap up our exploration of Autodesk Architectural Desktop 2006 with an overview of plotting, a look at Autodesk VIZ Render, and a brief conclusion.

Section IV is organized as follows:

CHAPTER 19

Plotting and Publishing

INTRODUCTION

Conceptually, plotting in Autodesk Architectural Desktop is no more difficult than printing a document from any other computer program. The major difference lies in the complexity of the data being sent to the printer. The data flow for plotting AutoCAD objects is as follows: Objects are drawn and assigned properties (Layer, Color, Linetype, Lineweight and Plot Style). This object data is then sent to the plotter. The major contrast between AutoCAD plotting and printing in other software applications is the use of "Plot Style Tables." AutoCAD object properties interact with a Plot Style Table; the result of this interaction often modifies the look of the objects in the final plot. This is because the Plot Style Table can optionally change an object's properties at the time of printing—so that your colored lines on screen become pure black only when printing for example. There is no equivalent to this in other applications.

You can preview the result of this interaction of parts with great accuracy before committing to final plot in a "paper space layout." This gives you much better control over the entire process. If you don't take advantage of this, what you see on screen may *not* be what you see in print.

With Architectural Desktop objects, there are a few more interactions happening before the data hits the plotter. As we have seen throughout this text, all ADT objects have one or more Display Representations. It is the Display Representation that actually determines what graphical entities are drawn to the screen and sent to the plotter. Display Representations also are responsible for the assignment of all AutoCAD properties like Color, Layer and Plot Style. If you want to print a plan, the Display Representation sends 2D plan graphics; if you want a model, it sends 3D. Therefore, with ADT objects, the flow is as follows: the Display Representation determines the Object components that will be drawn; it then assigns properties to each of these components; the assigned properties interact with the Plot Style Table; and the result is sent to the plotter.

Understanding how this data flow works can help achieve more successful plots. It will certainly aid you in troubleshooting plotting problems. Plotting troubles with

AutoCAD entities can therefore be addressed most often in layers or plot style tables. For ADT objects, look to the Display Representation first.

OBJECTIVES

This chapter gives an overview of the plotting process. We will look at Paper Space Layouts, Page Setup, Plot Style Tables, Sheet Sets, eTransmit, Transmittal Setups, Plotting, Publishing and 3D DWF Export. The following topics will be explored:

- Understand data flow in Plotting
- Understand Layouts
- Work with Page Setup Manager
- Transmit Sheet Sets
- Publish construction documents
- Sheet Sets
- 3D DWF Output

 ## SHEET FILES

A "Sheet File" is an ADT drawing set up specifically for plotting purposes. The goal is to create a "ready-to-plot" drawing file that can be opened and printed without need for detailed configuration or checking. We have created several Sheets throughout this book. XREFs and Paper Space Layouts are used to create sheet files. Several sheet files can be gathered together and organized into "Sheet Sets." Sheet Sets are fully incorporated into the ADT Project Navigator. Several sheet files were built to generate our cartoon set at the beginning of this book. (Review Chapter 5 for complete information on Sheet files). We have continued to add additional sheets to both projects with each successive chapter.

INSTALL THE CD FILES AND LOAD THE CURRENT PROJECT

If you have already installed all of the files from the CD, simply skip down to step 3 below to make the project active. If you need to install the CD files, start at step 1.

1. If you have not already done so, install the dataset files located on the Mastering Autodesk Architectural Desktop 2006 CD ROM.

    Refer to "Files Included on the CD ROM" in the Preface for information on installing the sample files included on the CD.

2. Launch Autodesk Architectural Desktop 2006 from the desktop icon created in Chapter 3.

If you did not create a custom icon, you might want to review "Create a New Profile" and "Create a Custom ADT Desktop Shortcut" in Chapter 3. Creating the custom desktop icon is not essential; however, it makes loading the custom profile easier.

3. From the File menu, choose **Project Browser**.

4. Click to open the folder list and choose your *C:* drive.

5. Double-click on the *MasterADT 2006* folder, then the *Chapter19* folder.

One or two commercial Projects will be listed: *19 Commercial* and/or *19 Commercial Metric*.

6. Double-click *19 Commercial* if you wish to work in Imperial units. Double-click *19 Commercial Metric* if you wish to work in Metric units. (You can also right-click on it and choose **Set Current Project**.) Then click Close in the Project Browser.

 IMPORTANT If a message appears asking you to re-path the project, click **Yes**. Refer to the "Re-Pathing Projects" heading in the Preface for more information.

LAYOUTS

Layouts provide the means to emulate a sheet of paper and the composition of all its components organized to scale. The sheet is typically set up to facilitate plotting. The ADT drawing environment consists of model space and paper space. Model space emulates "real space." It is a full-scale, full-size environment without physical limit. The model space environment is 3D, but the actual model needn't be. Any drawing created in this environment is referred to as a "model," regardless of whether it is two-dimensional or three-dimensional. In contrast, paper space is an environment made up of one or more "paper space layouts," or simply layouts. A layout is a "2D only" drawing environment. Typically, a layout is used to organize the various components that comprise the printed sheet. These often include a title block, general notes and one or more Viewport objects. The viewports are used to show the model from various vantage points and at specific architectural scales. In this capacity, a layout is a page layout/plotting tool. The main goal is to provide very accurate preview capabilities, which eliminate the need to generate endless "test" plots. This helps save paper, time, and money.

▶ Access layouts by clicking the tabs at the bottom of the screen.

▶ Create a new layout from the right-click menu on one of the existing layout tabs and choose **New layout** or **From template**.

WORKING LAYOUTS

It is also possible to set up "working" layouts, where the model is viewed in a particular Display/Layer Configuration conducive to the task at hand. One of the advantages of this approach is that you can work in the drawing in a format that is very similar to the way it will appear when plotted. The template files used to generate Elements, Constructs and Views in ADT 2006 all contain a "Work" layout tab. To be most effective, you should set a scale to the viewports in this layout and lock them while you work. You can even edit the Page Setup to "Display Plot Styles" in the layout (which usually makes the drawing appear in black and white). This is most effective when your AutoCAD background color is set to White, so that lines appear black. There is also the "Maximize Viewport" icon that allows you to temporarily fill the screen with the contents of a particular viewport. The focus of this chapter is plotting, so we will not look further at this technique, but you are encouraged to experiment with it on your own.

 ## PAGE SETUP MANAGER

Like all Windows applications, ADT uses Page Setup to configure the printer, page size, scale, and other output settings. Page Setup in ADT has some unique features that other Windows programs do not share. For instance, a Page Setup configuration in ADT can be saved and recalled later. Page Setup configurations are created, edited and deleted from the Page Setup Manager dialog. Let's look at some of the key features of Page Setup.

EXPLORE PAGE SETUP

1. On the Project Navigator, open the *A-101 Floor Plans* Sheet file.

   This is the Sheet file generated in Chapter 5 containing all of the floor plans for the Commercial Project.

As you may recall, this Sheet file XREFs floor plan View files in model space and has four viewports in the Plot Layout. Each viewport is configured to "look" at just one of the plans. This occurred automatically when we dragged and dropped the various floor plan View files into the Sheet. (Unique layers for each XREF were added automatically by ADT to accommodate the process.)

2. From the File menu, choose **Page Setup Manager** (see Figure 19–1).

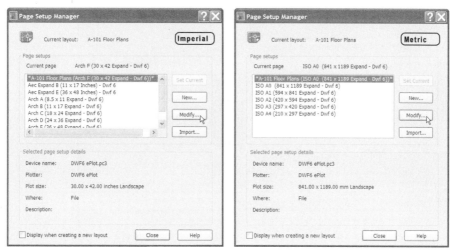

Figure 19–1 *The Page Setup Manager dialog*

In this dialog are listed all of the Page Setups saved in the current drawing. The template file used to create this Sheet is responsible for adding all of the Page Setups we see here.

You might want to consider creating your own Named Page Setup configurations. You can add them to your office standard Sheet template file. They will then appear in addition to or in place of the Page Setups shown in Figure 19–1. Making a Page Setup active will set all of the plotting settings automatically. This can be a great way to standardize plotting settings.

> The current Page Setup: **Arch F (30 x 42 Expand - Dwf 6)** [*ISO A0 (841 x 1189 Expand - Dwf 6)*] is listed at the top and selected in the list.

> 3. Be sure that **Arch F (30 x 42 Expand - Dwf 6)** [*ISO A0 (841 x 1189 Expand - Dwf 6)*] is selected and then click the Modify button (see Figure 19–1).

Outlined here is a brief explanation of each setting in the Page Setup dialog box. You can change any of the settings you wish to meet your specific needs. As a general rule of thumb, move through the dialog on the left first, moving top to bottom, then make any adjustments on the right. Refer to Figure 19–2 as you work through each description.

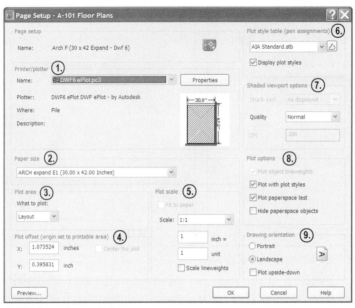

Figure 19–2 *The Page Setup dialog*

1. Printer/Plotter

▶ **Name**—All plotters available on your system will be listed here. This list often includes all of the printers in the Windows Printers folder as well. Windows Printers have a small printer icon next to them, while AutoCAD
plotters have a small plotter icon.

▶ **Properties**—This button gives access to hardware settings in the printer/
plotter driver. You can optionally save your changes to a new plotter configuration file (*.pc3*).

2. Paper size

▶ Choose from a list of standard sheet sizes. This list is unique to each plot device. Therefore, it is critical that you choose your plotter/printer first.

3. Plot area (What to plot list)

▶ **Layout**—99 percent of the time you should choose Layout as the Plot area.
This setting prints the entire Layout area as defined by the paper size. This
guarantees an exact fit to the paper. This setting is recommended.

▶ **Extents**—Sets the print area to the outer edge of the actual Drawing
objects.

▶ **Display**—Sets the print area to the view currently on screen.

▶ **View**—Available if Named Views have been created in the drawing. Drawing
Management creates a Named View of each drawing dragged to the Sheet.

▶ **Window**—Use the mouse and pick a rectangular print area in the drawing.

4. Plot offset (origin set to printable area)

▶ **X and Y**—Shift the plot relative to the lower left corner of the paper.

5. Plot scale

▶ **Scale**—Long list of predefined plotting scales.

 NOTE Most often, you will choose 1:1 (shown as 1'-0" = 1'-0" in Imperial) as the plot scale when working from a layout. This is because drawings have already been scaled in the viewports. If you are plotting from model space, choose the appropriate scale here.

▶ **Custom**—Type a scale factor ratio if the scale you wish to use is not in the list.

▶ **Scale lineweights**—Adjust the thickness of the lineweights relative to the scale chosen. This is very useful for half-size plots.

6. Plot style table (pen assignments)

▶ Drop down list includes all of the Plot Style Tables available on your system. There are two types of Plot Style Tables: Named and Color Dependent. Drawings can use only one type at a time. Check with your CAD support personnel for the type your firm uses. Check the online help for complete information on the differences between the types. ADT 2006 ships with Named Plot Styles set as the default.

▶ **Edit icon**—This icon loads the Plot Style Table Editor with the current Plot Style loaded for editing. When you save your changes in the Plot Style Editor, you will be returned to Page Setup.

▶ **Display plot styles**—With this function turned on (check mark in the box), the layout becomes a live preview of the drawing as it will actually appear when plotted. This setting used in conjunction with Lineweight display and properly scaled viewports is highly recommended.

7. Shaded viewport options

▶ **Shade plot**—With this option, you can plot the image shaded as it appears in the drawing, using Hidden, Shaded, or Gouraud Shaded.

▶ **Quality**—Several preset qualities from Draft to High quality are available for shaded plotting. The higher the quality, the longer it will take to plot.

▶ **DPI**—If you choose "Custom" quality, you can set any DPI (Dots Per Inch) that you wish for plotting shaded viewports.

8. Plot options

- ▶ **Plot with lineweights**—Use this option to toggle on and off lineweights where available.

- ▶ **Plot with plot styles**—Check this to use the settings in the Plot Style Table. Recommended.

- ▶ **Plot paperspace last**—Check this to make sure paper space objects are not concealed by model space objects. Recommended.

- ▶ **Hide objects**—Use if the drawing is 3D to create a hidden line rendering. (If you are plotting 3D from a layout, use the Hideplot feature of the viewport instead.)

9. Drawing orientation

- ▶ **Portrait**—Sheet oriented short side horizontal.

- ▶ **Landscape**—Sheet oriented long side horizontal.

- ▶ **Plot upside-down**—Image rotated 180° on the sheet.

4. Click the Preview button to get an onscreen plot preview of the current drawing.

5. Right-click and choose **Exit** to leave the Plot Preview and return to the Page Setup dialog.

6. Click OK to accept any changes and return to the Page Setup Manager. 1.

If you are finished in the Page Setup Manager, click the Close button, otherwise, select a different Page Setup to Modify, click New to create a New Page Setup, or click Import to browse to another drawing and import its Page Setups.

VIEWPORTS

Viewports are used to crop out sections of the building model and present them on the layout sheet. Use the Viewports submenu on the View menu to manually create viewports of varying sizes and shapes. Entities such as closed polylines and circles can be converted to viewports. Use the Properties palette to assign a scale to the viewport. Each viewport can have its own Scale, Layer, and Display Configuration settings. If you are using the Project Management features of ADT 2006 as we have throughout this book, Viewports are created for you and automatically scaled properly when you drag a View file or a Model Space View saved within a View file (on the Views tab of the Project Navigator) onto a Sheet file. This is the recommended way to set up Sheet files and Viewports (see Chapter 5 for more information).

PLOT STYLE TABLES

Plot Style Tables can assign a variety of plotting attributes, such as color, half tone, lineweight, and join and end styles to the final plotted linework in your drawings. Join and end style, and in some cases halftone, must be set in the Plot Style Table. Settings such as Lineweight and Linetype can (and often should) be set in the drawing instead of the Plot Style. The specific approach varies from office to office. There are two types of Plot Style Table: Named Plot Style Tables (*.stb*) and Color Dependent Plot Style Tables (*.ctb*). Named Plot Style Tables are user defined and can be assigned ByLayer, within Display Representations, or directly to individual objects. Named Plot Styles are the ADT 2006 default and have been used throughout this book. This varies from traditional AutoCAD usage where Color Dependent Plot Style Tables are more commonly used. A Color Dependent Plot Style Table assigns plotting attributes to objects based on their color as they are printing. In other words, a permanent mapping exists between each of the 255 AutoCAD colors and a Plot Style within the Color Dependent Plot Style Table.

EDIT A PLOT STYLE TABLE

1. Return to Page Setup Manager (File menu), Modify the current Page Setup, and then click the Edit icon next to the Plot style table list.

2. Click the Form View tab (see Figure 19–3).

Plot styles are listed on the left. They are named for the function they serve. For instance, Bold uses a bold 1.40 mm Lineweight. In the Properties area, you can configure the properties listed below for the selected Plot Style on the left. (Select and configure multiple entries with the SHIFT and CTRL keys.)

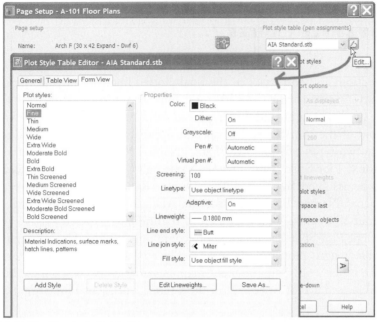

Figure 19–3 *The Plot Style Table Editor*

The Plot Style Table can be used to affect the following printing attributes:

▶ **Color**—Color of the ink used to plot items of this style.

▶ **Dither**—Enables the process of simulating colors unavailable on your plot device by mixing colors that are available.

▶ **Grayscale**—Translates the value of the color to an equivalent shade of gray.

▶ **Pen #**—For pen plotters, assigns the pen to use.

▶ **Virtual pen #**—For plotters with pen plotter emulation, assigns the pen to use.

▶ **Screening**—Uses a lower intensity of ink expressed in a percentage from 1 to 100. 100 is full ink, 1 is almost none.

▶ **Linetype**—Assign hardware linetypes if your plotter supports them.

▶ **Adaptive**—Works with the hardware linetypes to make the linetype wrap around corners.

▶ **Lineweight**—Overrides the Lineweight property in the drawing.

▶ **Line end style**—Choose from a list of end shapes.

▶ **Line join style**—Choose from a list of corner conditions.

▶ **Fill style**—Choose from a list of patterns.

If you use Color-based plotting instead, you will have all of the same settings, but instead, there will be 255 colors listed on the left. You cannot add to, delete from, or rename the list. It is fixed.

TIP If you want to convert a Named Plot Style drawing to Color Dependent or vice versa, use the CONVERTPSTYLES command at the command line.

CAD Manager Note: If you use an outside service bureau to manage your plotting, please consult with your vendor to determine the best setup parameters for plotting.

Most of the settings in the Plot Style Table have the option to "Use object" setting. For instance, if you wanted to use AutoCAD linetypes instead of hardware linetypes, you would choose **Use object linetype** and not turn on Adaptive. For construction documents on a typical modern ink jet plotter, you will use Plot Style Tables to force all colors to use Black ink when plotting. Some firms set lineweights in the drawing as a property of the layers; others use the lineweight option in the Plot Style Table to override the setting in the drawing. Whichever method your firm uses, be sure you realize that the Plot Style Table gets final "say." If a lineweight is assigned both in the drawing and in the table, the Plot Style Table's lineweight will win.

AIA STANDARD PLOT STYLE TABLE (ADT 2006 DEFAULT)

The AIA Standard Plot Style Table – *AIA Standard.stb*, is the Autodesk Architectural Desktop 2006 default. There are two basic groups of style: The Black ink styles and the Screened styles. Within each group is a collection of Lineweights configured according to industry standard recommendations like the U.S. National CAD Standard (see Figure 19–4 and Table 19–1). All of the End and Join styles are the same, Square and Miter respectively. Every Named Plot Style Table has the Normal Plot Style, which cannot be edited. It simply plots the linework as it appears in the drawing with no change. Therefore all of its settings are set to Use object default. The Standard Plot Style is used for certain non-plotting objects in the default ADT templates. These objects will appear in blue even when Display Plot Styles is turned on in Layouts. Finally, the Invisible Ink Plot Style is another way to achieve a "No-Plot" setting. This is useful when it is not desirable to use a layer, such as for certain display control settings.

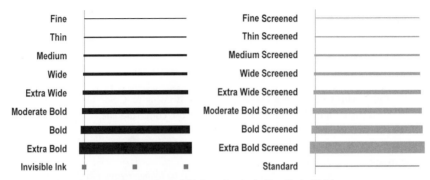

Figure 19–4 *Plot Styles in the Default AIA Standard.stb Plot Style Table*

Table 17–1 *Plot Style Settings for AIA Standard.stb*

Normal	Color By Object	Lineweight By Object	Screening By Object
Fine	Black	0.18	100
Thin	Black	0.25	100
Medium	Black	0.35	100
Wide	Black	0.50	100
Extra Wide	Black	0.70	100
Moderate Bold	Black	1.00	100
Bold	Black	1.40	100
Extra Bold	Black	2.00	100
Thin Screened	Black	0.25	50
Medium Screened	Black	0.35	50
Wide Screened	Black	0.50	50
Extra Wide Screened	Black	0.70	50
Moderate Bold Screened	Black	1.00	50
Bold Screened	Black	1.40	50
Extra Bold Screened	Black	2.00	50
Standard	Blue	0.18	100
Invisible Ink	Black	By Object	0
Fine Screened	Black	0.18	50

3. Click Save and Close to save changes to the current table. (Choose **Save As** to create a new table with the changes.)

4. Click OK to close the Page Setup dialog and then click Close to dismiss the Page Setup Manager.

PLOTTING

If you wish to plot a single drawing file, you can use the Plot command on the File menu. The dialog that appears is nearly identical to the Page Setup dialog pictured in Figure 19–2. The only difference is that the right side is hidden from view. You can expand the right side with the small icon at the bottom right corner of the plot dialog. To see this, choose **Plot** From the File menu (see Figure 19–5).

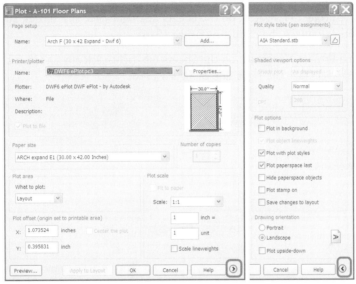

Figure 19–5 *The Plot dialog is nearly identical to the Page Setup dialog*

Simply choose a Page Setup from the list at the top, verify and modify any settings as required and then click OK to plot. Although drawings can be printed one at a time this way, it is much more effective to "Publish" drawings from your Sheet Set contained in Project Navigator. You will get much more control and reliability.

PUBLISH A SHEET SET

Our drawing set has come a long way since we first plotted it in Chapter 5. Let's review that process now and explore some of the many features available on the Sheet tab for publishing drawing sets.

5. On the Project Navigator, click the Sheets tab.

As we have seen already, the Sheets tab incorporates the AutoCAD Sheet Set functionality. A Sheet Set gathers a collection of drawing Layouts for plotting. The Sheet Set can be organized into Subsets. The order of the drawings as they appear in the Sheet Set will be the order in which they will list when adding a Sheet List (see Chapter 5) and the order in which they will plot. Using the functions inherent to the Sheet Set, you can print the entire set of project drawings, or any subset you wish without the need to open and plot each one individually. Let's take a look at what functions are available at each level of the Sheet Set.

6. On the Project Navigator, on the Sheets tab, right-click the *MADT Commercial* Sheet Set node at the top of the list (see the left side of Figure 19–6).

7. On the Project Navigator, on the Sheets tab, right-click any Subset, such as Architectural (see the middle of Figure 19–6).

8. On the Project Navigator, on the Sheets tab, right-click any Sheet, such as *A-100 Site Plan* (see the right side of Figure 19–6).

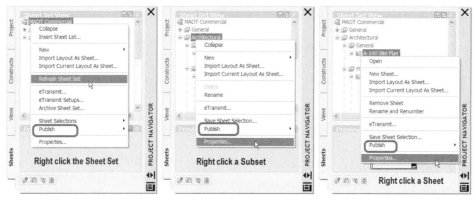

Figure 19–6 *Right-click menus vary at each node of the Sheet Set*

Notice that at each level of the Sheet Set, certain options such as Publish remain available while others are not. From whatever level you choose Publish, all of the items contained within it will be chosen. For instance, to Publish the entire project, right-click at the top on *MADT Commercial*; to Publish only the *Architectural* Subset, right-click it and choose **Publish** and so on. You can also make custom selections with the SHIFT and CTRL keys. For instance, let's say that for a Design Development submission, you only need the Plans and Elevations, not the Site, nor any Details.

PUBLISH A SHEET SET TO DWF

1. Select the *Architectural – Plans* subset, hold down the CTRL key and then click the *Architectural – Elevations* subset.

Once you have a custom Selection, you can right-click to access the Publish and other options, or you can save the selection for future retrieval. Right-click the *MADT Commercial* Sheet Set node to restore saved selections later.

2. With both items selected, right-click and choose **Save Selection**. In the dialog, name it **Design Development**.

Since plotters vary widely, in this exercise, we will plot to a DWF file. If you prefer, feel free to create a hard copy plot of the set. When we explored the settings in the Page Setup we saw that the *A101 Floor Plans* file is set up to print to the DWF6 ePlot.pc3 plotter. (This is the default of all the files in our project.) *DWF6 ePlot.pc3* is a digital plotter designed to create a DWF file. A DWF (Design Web Format) is a highly compressed, vector-based file format designed for viewing and distributing drawing files over the Internet and by email. What makes the DWF file so powerful is that it is a vector-based, high-quality drawing file that is read only. This means it can be distributed to consultants and clients without fear of unauthorized editing. DWF preserves access to Layers and can even include the Property Set Data attached to the objects within the drawing. DWF files can also be embedded in Web pages for viewing in a browser with the plug-in provided free from Autodesk. Anyone with a copy of the Autodesk DWF Viewer software, available as a free download on the Autodesk Web site (*http://www.autodesk.com/*), can view, zoom, pan, turn on and off layers, and print the DWF file. If the recipient has a copy of Autodesk DWF Composer, they can add redline comments to the DWF file, which can then be loaded back into ADT. Creating a DWF is simple, because it is the same as printing to a hard copy device. Before creating our DWF, let's make a modification to the Publish Options.

3. Right-click the *MADT Commercial* Sheet Set node at the top and choose **Publish > Sheet Set Publish Options** (see Figure 19-7).

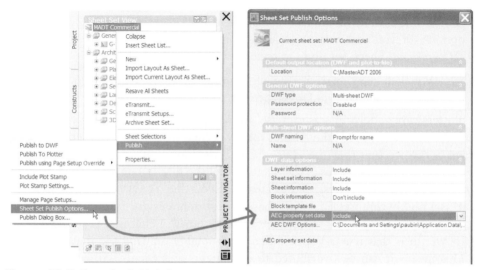

Figure 19–7 *Sheet Set Publish Options*

4. In the Default output directory, choose Publish locally, click the browse icon (…) and then choose your *C:\MasterADT 2006* (or another location of your choosing).

5. Choose Multi-sheet DWF in the DWF type area.

6. In the DWF data area, choose **Include** from the "Layer information," "Sheet set information," "Sheet information" and the "AEC property set data" options.

7. Click in the AEC DWF Options field, click the small the Browse icon (…) and then place a check mark in the "Publish Property Set Data" check box. Click OK twice.

Configuring those options will make a more robust DWF. All of the Property Set Data and Sheet information will now be included with the DWF file. When this DWF is opened in Autodesk DWF Viewer, the recipient will be able to select any object that has Property Set Data and view it directly.

8. Right-click the *MADT Commercial* Sheet Set node at the top and choose **Publish > Publish to DWF**.

As we saw in Chapter 5, this is all we need to do to publish the entire set. Use the same process to publish a sub set or custom selection. When the Publish is complete, you can view a report or the DWF file itself by right clicking on the Plot and Publish icon on the Application Status Bar (see Figure 19–8).

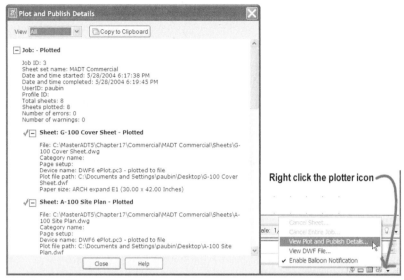

Figure 19–8 *View Plot and Publish Details*

If you wish to view the DWF, right-click the Publish icon and choose View DWF File (you can also double-click it from wherever you saved it). In the Autodesk DWF Viewer, select a Sheet that you wish to view on the left. Click the Selector icon on the toolbar and then click on an object in the drawing. They will highlight red under you cursor as you mouse over. When an object is selected, you can view its Property Set Data on the left. It you are using DWF Composer, you can add redline markups (see Figure 19–9).

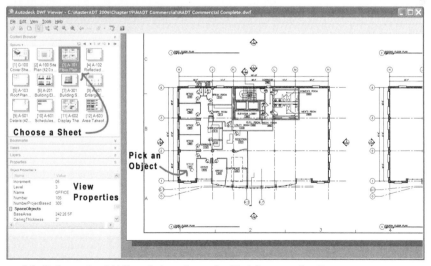

Figure 19–9 *Open the DWF and view the embedded Property Set Data*

3D DWF

New to Architectural Desktop 2006 is the ability to publish 3D Models to DWF files. Use this feature to output a 3D Model containing Property Set data in a compact DWF file. Just like the 2D DWF set published above, all your recipients need is a copy of the free Autodesk DWF Viewer application to view, orbit, query and print the 3D model.

9. On the Project Navigator, on the Views tab, double-click the *A-CM00* file to open it.

10. On the Drawing Status Bar, click the Open Drawing Menu icon and choose **Publish to 3D DWF**.

11. In the AEC 3D DWF Publishing Options dialog, click the browse icon and choose a location to save the file in the DWF File Name field.

12. In the 3D DWF Organization area, choose "Object Type and Style."

Two options exist here. When you publish to 3D DWF, you can select the objects in the published file. The choice made here determines how those published objects will be grouped and therefore how they will be selected in the DWF. You should try both options to determine your own preference.

13. Place a check mark in the "Include Properties from AEC DWF Options" checkbox.

14. Click the Edit AEC DWF Options button, and as before, place a check mark in the "Publish Property Set Data" check box and then click OK twice (see Figure 19–10).

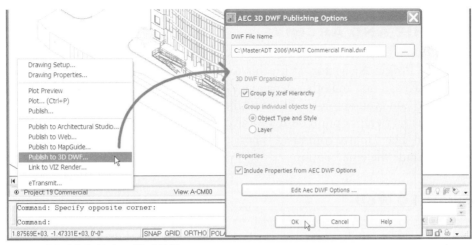

Figure 19–10 *Publish a 3D DWF*

15. At the " " prompt, select all objects and then press ENTER.

 NOTE Depending on your hardware specifications, it may take a little while for the 3D DWF to generate. Please be patient.

16. Open the new DWF file as before.

You will have slightly different icons this time. You can select objects in the DWF and view their properties, you can also orbit the model, zoom, pan and isolate objects. A tree hierarchy of all objects appears on the left. Use it to select objects, and right-click to access isolate options (see Figure 19–11).

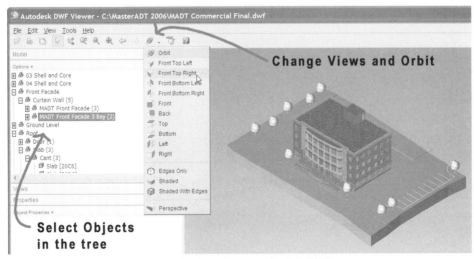

Figure 19–11 *View the model and select objects in the 3D DWF*

ETRANSMIT AND ARCHIVE

If you wish to send the actual drawing files to a recipient, you can use the eTransmit from the Project Navigator. With this tool, you can gather all of the drawings, their XREFs, fonts, and other dependent files and package them all up into a single ZIP file and even automatically send the ZIP file as an email attachment. Use the eTransmit Setups option on the right-click menu to determine which settings you want. The Archive command is nearly identical to eTransmit except that it does not include the email option.

PUBLISH TO WEB

If you generate lots of DWF files and wish to include them in a project Web site for online viewing on the Internet, use the Publish to Web Wizard on the Format menu to generate the DWFs, rather than plotting them manually. This wizard will create the DWFs and accompanying HTML Web pages ready to post to the Web. This wizard has many options. Be sure to check it out.

ADDITIONAL EXERCISES

Additional exercises have been provided in Appendix A. In Appendix A you will find exercises to plot the Residential Project.

SUMMARY

The properties of objects are assigned through a variety of hierarchical settings, starting with the object Display Representation and ending with the Plot Style Table.

Sheet files are valuable tools to enhance plotting productivity.

Use Layouts and Page Setup to set up the "sheet of paper" exactly as you want it plotted.

Page Setups can be saved and named for easy retrieval.

Use viewports to crop out individual views of your building model.

Each viewport can be any shape, and it can have its own layer and unique Display Configuration.

The Plot Style Table gets the "final word" in plotting attributes like color and lineweight.

Assign Plot Style Tables to your layouts to make the drawing "black and white"
for plotting.

Use the Publish utility to create groups of files that can be printed together unattended.

Create a multi-sheet DWF file that includes embedded Property Set Data.

Publish any ADT file to a 3D DWF and view it in Autodesk DWF Viewer

eTransmit and Archive options are available for the Sheet Set.

CHAPTER 20

Working with
Autodesk VIZ Render

INTRODUCTION

Included in the box with Autodesk Architectural Desktop 2006 is the VIZ Render application. This application is a stand-alone rendering and animation application fully consistent with ADT 2006 in look, feel, and function. Use ADT to build your geometry, and use VIZ Render to present it in still renderings or walkthrough animation. VIZ Render boasts realistic photometric lighting, radiosity, raytracing, and a robust material editor. Models built in ADT are linked to VIZ Render and remain linked (much like an XREF), so that changes to the ADT Model can be reloaded to the VIZ scene at anytime.

OBJECTIVES

In this chapter, we will take a brief look at the process of linking models to VIZ Render. This chapter is intended to assist you in generating a quick high-quality rendering. It is *not* a comprehensive exploration of all of VIZ Render's functionality, but it should serve to get you started on the right foot. If you would like to learn more about VIZ Render, pick up a copy of Mastering VIZ Render – A Resource for Autodesk Architectural Desktop Users, by Paul F. Aubin and James D. Smell. The following summarizes topics covered in this chapter:

- Understand VIZ file linking.
- Work with materials.
- Update the ADT Model.
- Generate rendered output.
- Create a quick animation study.

UNDERSTANDING VIZ RENDER

VIZ Render offers a powerful "linking" feature that allows your ADT Model to be referenced directly into VIZ Render and manipulated directly. This feature is similar to XREFs, but more flexible. Like XREFs, file linking combines the ability to remain linked to the original file with the ability to update and receive the latest changes to the geometry. In contrast to XREFing, you can actually manipulate the individual components of the linked model, to add or edit materials.

The linking process involves the following basic steps:

1. Link an ADT Model into a VIZ Render scene.

2. Add and adjust Material Mappings.

3. Add lights and a camera.

4. Generate a rendering.

 NOTE VIZ Render is a separate software package that comes with Architectural Desktop. If you performed the standard ADT 2006 install, it was installed on your system. If you don't have it installed, locate your original ADT 2006 CDs or DVD to install it.

VIZ RENDER AND AUTODESK VIZ

The feature set of VIZ Render (VIZr) is quite extensive. This short exercise is intended to help you quickly bring a Model into VIZr, add some basic materials, lights and cameras, and then generate a rendering. VIZ Render has much in common with its more robust sibling: Autodesk VIZ. However, it is a separate application with its own features and functionality. In addition, VIZ Render creates a file with a *.drf* file extension. This file format is compatible with current versions of Autodesk VIZ or 3ds Max (3d studio max) but is not supported in older versions. Despite these differences, you may still find some useful information in Autodesk VIZ and 3d studio max publications such as *Inside 3D Studio VIZ* by Ted Boardman and Jeremy Hubbell or *Mastering 3D Studio VIZ* by George Omura.

FILE LINKING

We will begin our discussion with file linking. File linking can create very large files; particularly as the number of XREFs in ADT grows—this can sometimes make linking unmanageable. Do your best to solve this problem before it starts. If you have developed a carefully planned project file strategy as recommended in Chapter 5, you should be able to isolate just those portions of your project that you are interested in rendering. For instance, when creating a rendering of the building exterior, having good separation in your XREFs (such as we have in the Commercial Project) will allow you to link in only the exterior portions of your project and leave out all of the interior floor plates and partition walls. You might even be able to get away with linking in only one or two of the exterior walls if you are doing a static view (no animation). If you're planning to render an entryway, you can use a Live Section to isolate just the portion you need.

INSTALL THE CD FILES AND LOAD THE CURRENT PROJECT

If you have already installed all of the files from the CD, simply skip down to step 3 below to make the project active. If you need to install the CD files, start at step 1.

1. If you have not already done so, install the dataset files located on the Mastering Autodesk Architectural Desktop 2006 CD ROM.

   Refer to "Files Included on the CD ROM" in the Preface for information on installing the sample files included on the CD.

2. Launch Autodesk Architectural Desktop 2006 from the desktop icon created in Chapter 3.

If you did not create a custom icon, you might want to review "Create a New Profile" and "Create a Custom ADT Desktop Shortcut" in Chapter 3. Creating the custom desktop icon is not essential; however it makes loading the custom profile easier.

3. From the File menu, choose **Project Browser**.

4. Click to open the folder list and choose your *C:* drive.

5. Double-click on the *MasterADT 2006* folder, then the *Chapter20* folder.

   One or two residential Projects will be listed: *20 Residential* and/or *20 Residential Metric*.

6. Double-click *20 Residential* if you wish to work in Imperial units. Double-click *20 Residential Metric* if you wish to work in Metric units. (You can also right-click on it and choose **Set Current Project**.) Then click Close in the Project Browser.

 IMPORTANT If a message appears asking you to re-path the project, click **Yes**. Refer to the "Re-Pathing Projects" heading in the Preface for more information.

 NOTE Only the Residential Project is included with this Chapter. There is also not a "Complete" version of the entire project, rather a final VIZ Render (DRF) file and rendering files (JPG and AVI) have been provided in the *Chapter20* folder. If you wish to render the Commercial Project, use the version from Chapter 19.

LINK TO AUTODESK VIZ RENDER

1. On the Project Navigator, click the Views tab and open the *VIZ Render* View file.

The *VIZ Render* View file has been added to the project for this exercise. An additional Construct has also been added named *Neighbors*. This file contains some simple block models of the two neighboring houses. This will help give the scene a bit more context. The *VIZ Render* View file is a copy of the complete *A-CM00* Composite Model File including the *Neighbors* Construct. The original *A-CM00* does not include *Neighbors*.

2. On the Drawing status bar, click the Open Drawing Menu quick pick and choose **Link to Autodesk VIZ Render** (see Figure 20–1).

VIZ Render will launch, and the current ADT drawing (*VIZ Render.dwg*) will be linked into the current scene. VIZr uses default settings to link in the geometry of the ADT file.

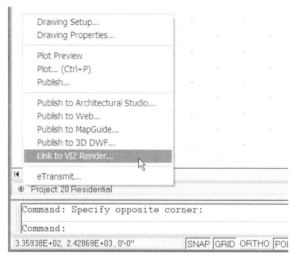

Figure 20–1 *Link to Autodesk VIZ Render on the Drawing Status Bar*

THE VIZ RENDER USER INTERFACE (UI)

The Autodesk VIZ Render user interface shares many features in common with ADT 2006. However, there are also plenty of similarities to Autodesk VIZ. (See the VIZ Render and Autodesk VIZ sidebar above for information on Autodesk VIZ.) Similarities to ADT include the standard Windows title bar and minimize/maximize controls, tool palettes, pulldown menus and toolbars. Similarities to Autodesk VIZ include the Command Panel, the Animation Time Slider, status bar and animation controls. The Command Panel has been preserved for those familiar with Autodesk VIZ, but most of its functions can be found in pulldown menus and toolbar icons (see Figure 20–2).

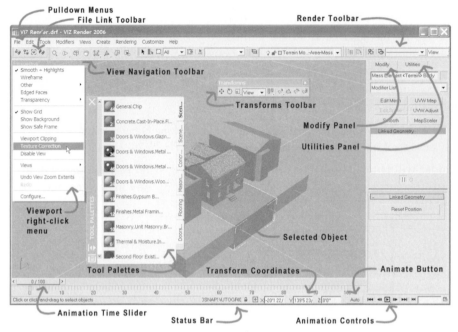

Figure 20–2 *The Autodesk VIZ Render user interface*

Tool palettes in VIZ Render function similarly to the way that they do in ADT. They are used in VIZr to store materials (see below). The small number next to an icon on the tool palette indicates how many times that material is applied to an object in the scene. You can add palettes and manipulate the contents of existing ones. Right-click on tools and palettes to get options similar to those found in ADT. The Content Browser is also used with VIZr to store libraries of materials for use in scenes and addition to palettes (see below).

A few additional Autodesk VIZ Render interface options are also worthy of mention. By default, VIZr opens to a single Perspective viewport; press ALT + W to toggle to a four viewport screen view. Press ALT + W again to toggle back. If you are in a four viewport mode, use your right mouse button to change active viewports. This will change the active viewport without changing the object selection. If you left-click, you will also change viewports, but you are likely to deselect your current selection and select some other object. Using right-click to change views prevents this. Just like in ADT, you can use your wheel mouse in VIZr to zoom and pan. Roll the wheel to zoom, and hold it down and drag to pan. Unlike in ADT, these functions also work while in perspective viewports!

RELOAD LINKED FILES

The defaults used when a file is linked work best for most models, but you can change some of these settings when you reload a linked ADT Model if you like.

1. On the File menu, choose **File Link Manager**.

In the File Link Manager, notice the ADT drawing listed in the Linked Drawings field. If a small red flag appears on the linked file's icon, it indicates that the file has changed and needs to be reloaded in VIZ Render to see the change.

2. In the File Link Manager, click one of the two Reload buttons:

▶ **Reload**—Click to reload the entire scene from the ADT model.

▶ **Use scene material definitions**—Toggle this on to use the materials as defined in VIZr rather than the version linked in from ADT.

▶ **Use scene material assignments (on Reload)** —Toggle this on to preserve material assignments made in VIZr that are different from the ones in the linked ADT model.

 NOTE These functions can also be found on the File Link toolbar.

The File Link Manager dialog box is used to manage existing links. If the Show Reload Options check box is selected, the File Link Settings dialog box will appear when you click either Reload button.

3. Click the Reload button (see Figure 20–3).

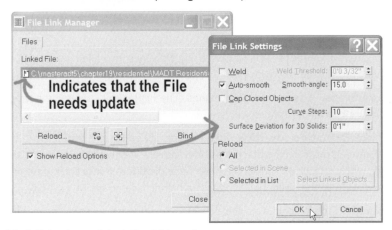

Figure 20–3 *Reloading a linked file yields options*

Options in the File Link Settings dialog box include the following:

▶ **Weld**—Determines whether to join nearby vertices of converted objects into a single object according to the Weld Threshold setting. Notice that Weld is turned off by default. Many VIZ experts advocate the use of Welding when linking AutoCAD files. This is good advice for AutoCAD files that do not have many ADT objects, but importing ADT objects with Welding turned on can cause unusual anomalies to appear on the surfaces. This setting should stay off unless you absolutely need it.

▶ **Auto-smooth**—Helps VIZ decide which surface intersections should render smooth (curved) and which ought to be corners. When you turn smoothing on, the angle designates the threshold of smoothing. When the angle between the face normals of two surfaces is smaller than the smoothing angle, the surfaces will smooth (see Figure 20–4). A face normal is an imaginary line (vector) pointing away from a surface at a 90° angle. The positive direction of this vector is considered the "front" of the surface, while the negative is the back.

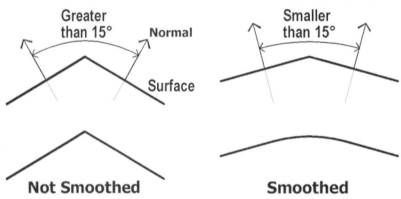

Figure 20–4 *Smoothing occurs when the angle between face normals is smaller than the Smooth-angle setting*

▶ **Cap Closed Objects**—Turns closed 2D Polylines into filled surfaces. You will not often need this with ADT Models.

▶ **Curve Steps**—Used to convert splines. The higher this number the more curvature converted splines will have.

▶ **Surface Deviation for 3D Solids**—When 3D solids are contained in the ADT model, this setting determines how closely VIZr will approximate those solids with surfaces. A smaller number here produces a more accurate approximation of the solid, but produces a higher face count.

▶ **Reload**—In the Reload area, the default is to reload the entire scene. However, you can choose to reload only the objects currently selected in the VIZr scene, or you can choose to select objects from a list. (Selecting from a list is a bit challenging because the names of the ADT objects are a bit generic.)

If you want more information on any of these settings, click the small help icon on the title bar, or press the F1 key to call the VIZ Render Help.

4. Click Cancel to dismiss the File Link Settings dialog box.

5. Close the File Link Manager.

6. From the File menu, choose **Save** (or press CTRL + S).

Here we opened the File Link Manager even though the ADT Model had not changed. File Link is very similar to XREFs. If the ADT model should change, those changes can be reloaded into VIZ Render using the File Link Manager or the File Link Toolbar.

VIZ Render will automatically use the same file name as the ADT file (except with a *.drf* extension) and save it to the same location (in this case *C:\MasterADT 2006\Chapter20\MADT Residential\Views [C:\MasterADT 2006\Chapter20\ MADT Residential Metric\Views]*). If you wish to change the name or location, choose Saveas from the File menu instead. If the File Link later has trouble reloading, you can redirect the path to the ADT file using the File Link Manager.

MATERIALS

Materials in VIZr are a set of parameters that, when applied to an object, make it appear like a particular real-life material. These parameters include color, shading, texture, shadow effects, transparency, and much more. Material Definitions in VIZr are hierarchical and can be very simple, designating only basic color and texture, or extremely complex and multi-layered constructs that subtly blend the effects of bitmap images and complex mathematical algorithms.

ADT objects can also be complex. As we have seen, all ADT objects contain one or more components. Often each of those components will require a different material designation. For instance, a Door may have a wood panel and a metal frame. Material assignments made to objects in ADT using Material Definitions, are automatically transferred over to Autodesk VIZ Render. This makes the application of materials to your scenes an easy task. Often you will only need to make some minor modifications to existing material assignments or add a material to an object that does not have an assignment in the ADT Model.

ASSIGN MATERIALS

Assigning materials to objects in VIZr is a simple drag and drop process. A fairly extensive library is provided in the default installation. You access it like most other ADT libraries via the Content Browser.

1. On the Tools menu, choose **Content Browser** (or click the icon on the Tool Palettes toolbar).

2. Click on the Render Material Catalog.

3. Navigate to the *Sitework* category and then the *Planting* category (see Figure 20–5).

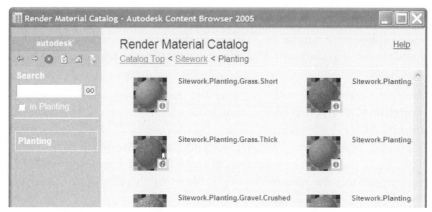

Figure 20–5 *Planting category of the Autodesk® Sample Render Material Catalog*

4. Using the eyedropper icon, drag the **Sitework.Planting.Grass.Thick** material into the VIZ Render scene, keep the mouse button depressed.

5. Hover your mouse (still an eyedropper icon) over the Terrain object.

   You will see a small cursor tip appear: Mass Element <Site> Body (see Figure 20–6).

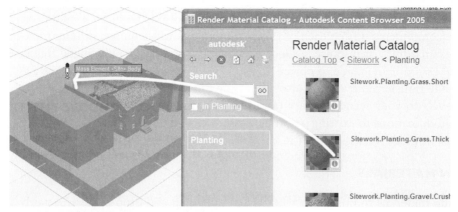

Figure 20–6 *Drop the material onto the Terrain object*

6. When you see the "Mass Element <Site> Body" tip, release the mouse to assign the material to the object.

The material applies to all three Terrain objects because they all share the same Mass Element Style in the original ADT file. You can also use the Materials tool palette to apply materials to objects.

7. Click on each tab of the Materials palette to see the available materials.

To apply a material from the palette, select an object in the scene, and then right-click the tool for the material that you wish to apply.

8. Select one of the two gray neighbors' houses.

    Simply click on it to select. The neighbors' houses are simply Mass Elements—like a single block of wood in a traditional physical model.

9. On the Materials tool palette, on any tab, right-click any tool and choose **Apply to Selected** (see Figure 20–7).

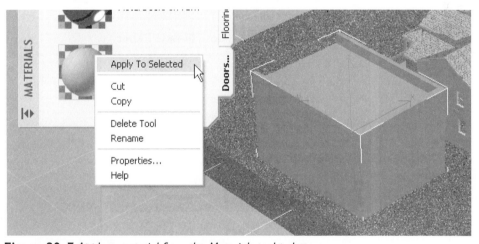

Figure 20–7 *Apply a material from the Materials tool palette*

Depending on the material you choose, you may get an Assigning Material dialog box warning stating that: "A material with the same name already exists in the scene. Do you want to…" In this case, you can simply choose the "Replace Existing" option and then click OK. However, in some cases you will prefer to choose the other option and rename the material to a new name. This alert occurs because some of the tool palettes import their materials from external library files (the way many of the tools in ADT do) and therefore, they might be calling in a different version of the material from the one you are using in the scene. In our case, we have not modified any materials, so it not a concern to us.

    Notice that the material applies to both neighbors' houses. Again they share the same style in ADT.

If you like the new material, you can leave it. Otherwise, click the Scene - In Use tab, and apply the *General.Chip* material to them to return them to the way they were.

CREATE A NEW MATERIAL

Creating new material can be very simple, or very complex. In this exercise, we will create a very simple material. We will create a "White Foam Core" material for the two neighboring houses.

10. Click on the Scene – Unused palette.

NOTE When you use keyboard shortcuts in VIZ Render, unlike in AutoCAD/ADT, you do not press enter to execute them. Simply press the appropriate letter or key combination on the keyboard.

11. Right-click the **Create New** tool and choose **New > Paper**.

TIP In some instances, the material will not create initially. If this occurs, please quit and restart VIZ Render and it should work.

A new tool named *Paper* will appear beneath the **Create New** tool.

12. Right-click the new Paper tool and choose **Properties**.

13. At the top right of the Material Editor, (just beneath a button labeled "Architectural"), change the name from "Paper" to **White Foam Core** (see Figure 20–8).

Create a New Paper Material   **Edit the Material Properties**   **Rename the Material**

Figure 20–8 *Rename Paper to White Foam Core*

All materials in VIZ Render are of the type Architectural. Templates are used to set the most common settings for a particular class of material. For instance, glass is transparent; Stone Polished is opaque and shiny.

14. Click the small gray color swatch next to Diffuse Color (see Figure 20–9).

Diffuse Color is the "primary" color of the object that we see. When you say "that object is red" it is the Diffuse Color that you see as red.

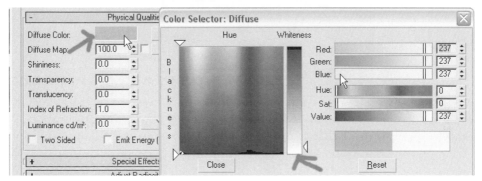

Figure 20–9 *Change the Diffuse Color*

15. In the Color Selector, pick a color much closer to white (drag the Whiteness slider down).

16. Click Close to close the Color Selector.

Foam core has a bit of sheen to it. Let's increase the shininess a bit to simulate this.

17. Use the spinner (up and down arrow) or simply type to set the Shininess to about **20** (see Figure 20–10).

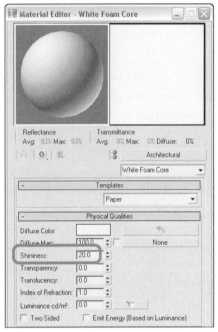

Figure 20–10 *Make the material a bit shiny*

There are plenty of other settings in the Material Editor. Materials can be made transparent, translucent, and even include their own luminance. In this simple material, we have assigned only a color, but materials can also use bitmaps like bricks or stone patterns for more realism. Many of the sample materials use maps. The None button next to the Diffuse Map item is where you choose a map.

18. Click the Close (X) box on the Material Editor to complete the material.

The new material remains on the Scene – Unused palette until you use it (apply it to an object) in the scene.

ASSIGN THE NEW MATERIAL

One of the neighbors' houses should still be selected. If it isn't, select one now.

19. On the Scene - Unused palette, right-click on the new *White Foam Core* tool and choose **Apply to Selected**.

Again, it will apply to both neighboring houses. Notice that the material has also moved from the Scene - Unused tab to the Scene - In Use tab. It will also have a small numeral 2 in the corner indicating that it is now assigned to two objects in the scene.

20. Click the Scene - In Use tab (see Figure 20–11).

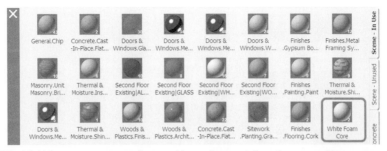

Figure 20–11 *When you assign a Scene - Unused material to the scene, it moves to the Scene - In Use palette*

    21. Right-click any of the other materials on the Scene - In Use tab and choose **Properties**.

When you edit the properties of a material in the current scene, you can explore the settings of the existing materials in the Material Editor. If you make any changes, they will apply immediately to the scene.

COPY A MATERIAL TO THE PROJECT CONTENT BROWSER LIBRARY

If you wish to use a material that you make or edit in another scene or back in ADT, then you will need to drag it to the Content Browser. You can drag to the default Content Browser Library that we opened above, or you can open the Project Catalog and add it there instead.

    Architectural Desktop should still be running and available on the Windows task bar. If you closed it, re-launch it now.

    22. Switch to ADT and on the Project Navigator, click the Content Browser icon at the bottom of the palette.

    23. Click on the Project Catalog icon and then click on *Shared Project Tools* palette.

    24. Switch back to VIZ Render.

    25. On the Scene - In Use palette, drag the **White Foam Core** tool from the palette to the Content Browser icon on the Windows task bar (see Figure 20–12).

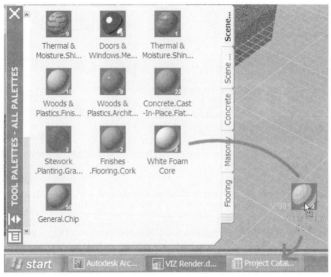

Figure 20–12 *Drag the White Foam Core tool to the Project Catalog on the Windows task bar*

When you drag it down to the task bar, the Content Browser will pop forward.

 NOTE If you receive the following error: "You cannot drag an item onto a button on the taskbar" be patient. It may take a second or two for the Content Browser to "pop forward"

26. Continuing to hold down the mouse button, move into the space of the Shared Project Tools window and then drop the Tool (see Figure 20–13).

 TIP If you have trouble dragging to the taskbar and then to the *My Tool Catalog*, right click the Content Browser title bar and choose Always on Top. In this way, you can drag directly to the Content Browser. Simply turn off Always on Top when finished.

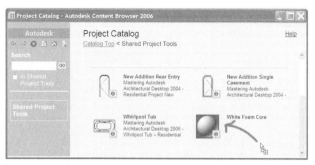

Figure 20–13 *Continue to drag and then drop it inside the Shared Project Tools palette in the Project Catalog*

If you prefer to create the tool in a non-project Catalog, you can open the default Content Browser from VIZr (on the Tools menu or the toolbar icon). There you can create a custom Catalog, or use the default *My Tool Catalog*. This catalog is created in the default installation of ADT. It is located in your *My Documents* folder in an *Autodesk\My Content Browser Library* sub-folder. If you do not have this, right-click and choose **Add Catalog** to create one. With this (or any Catalog of your choosing) open you can repeat the process above to drag a material tool from VIZr to a Content Browser Catalog. Material Tools in Catalogs may be dragged back into any VIZr scene or the ADT drawing window. When you wish to build custom materials in ADT, you must build the render material in VIZr first and then drag it over to ADT directly or (as shown here) drag it first to a Catalog and then into ADT.

Suppose you now wanted to apply this new *White Foam Core* Material to the objects in the ADT Model. You can execute the new tool in ADT and it will prompt you to create an ADT Material Definition on the fly.

27. Return to ADT and on the Shared Project Tools palette, click the small refresh icon in the bottom right corner.

    This will add the new *White Foam Core* tool to our Shared Project Tools palette.

28. On the Construct tab of the Project Navigator double-click to open the *Neighbors* Construct in the *Constructs\Architectural* folder.

29. On the Shared Project Tools palette, click the new *White Foam Core* tool.

30. In the "Create AEC Material" worksheet, choose Standard from the "AEC Material to use as a Template" drop down list

31. Verify that the name remains "White Foam Core" and that the "Convert to AEC Material Tool" checkbox is selected and then click OK (see Figure 20–14).

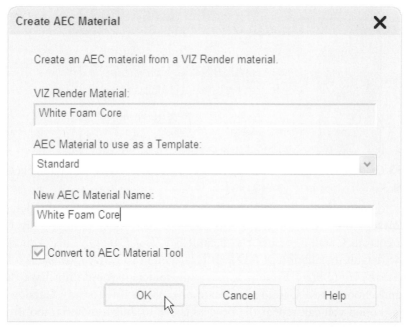

Figure 20–14 *Click a VIZr tool in ADT and it will be converted automatically to an ADT Material Definition*

It is very important to create such materials using a template material as the basis (choose an existing Material from the "AEC Material to use as a Template" list). This will save you a great deal of time and trial and error in configuring the myriad settings (such as lineweights, colors and hatching) available in an ADT Material Definition.

32. In the warning dialog that appears, click OK and then Save the file.

If you wish to make further edits to the material such as hatching, color or lineweight settings, choose **Material Definitions** from the Format menu and then edit the *White Foam Core* Material Definition.

33. Click the ***White Foam Core*** tool again and then click on one of the neighbors' houses when prompted.

Since we instructed ADT to convert this tool to an ADT Material Tool, we are now prompted with the "Apply Material to Components" worksheet that we first saw in Chapter 11.

34. In the "Apply Material to Components" worksheet, choose Style from the Apply to list and then click OK.

35. Save and Close the file. Reload the XREF in the *A-CM00* View file (in ADT), and then back in VIZr, reload the file link.

VIEWPORT NAVIGATION AND CAMERAS

Let's find a good vantage point for a rendering. Since all of the new work on this house is around the back, we should look at the scene that way. You can use the tools on the Viewport Navigation toolbar to orbit, zoom, and pan the scene.

USE ARC ROTATE

Arc Rotate is like Orbit in ADT. You use it to dynamically rotate and "spin" the model in 3D.

1. On the View Navigation Tools toolbar, click the Arc Rotate icon.

Depending on where you click and drag relative to the yellow circle, you will get a different result. If you click and drag in the middle of the circle, you move both up and down (arc) and spin round the model (rotate) at the same time. To constrain movement to just up and down, click and drag either the top or bottom handle. To constrain to rotation only while maintaining the height, click and drag one of the left or right handles (see Figure 20–15).

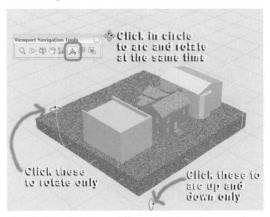

Figure 20–15 *Using Arc Rotate functions differently depending on where you click and drag*

2. Click one of the left or right handles and drag until your view is of the back of the house.

3. Click either the top or bottom handle and drag the model up and down a bit.

4. Use the wheel on your mouse to zoom in and out while arc rotate is active. You can also pan with the wheel (see Figure 20–16).

Figure 20–16 *Using Arc Rotate to view the back of the house*

5. Right-click anywhere in the Viewport to cancel the Arc Rotate command.

 NOTE Unlike ADT, you use the right-click here to cancel, not the ESC key.

ADD A CAMERA

6. Press ALT + W at the keyboard.

This will toggle the screen view to show four viewports.

 TIP To change whether the viewport shows in wireframe or shaded (smooth and high-light), right-click on the viewport label (top left corner of each viewport).

7. From the Create menu, choose **Cameras > Target Camera**.

8. In the Top viewport, click (and hold the left mouse button down) a point above and to the left of the house (where you want the viewer to "stand"), and drag toward the center of the Model (see Figure 20–17).

A cone will begin to expand out from your first pick point toward the target location. Be sure to use one fluid motion—do not release the mouse until it is where you want the target point to be (the point you are looking at). When you release the mouse button, the camera will begin moving vertically.

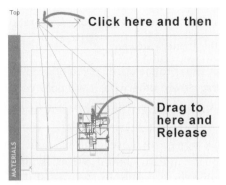

Figure 20–17 *Add a Camera in the Top viewport*

Look at the Left viewport and gently move the mouse.

9. When the Camera is at a good height (about half way up the height of the house) click to finish creating the Camera (see Figure 20–18).

It is not necessary to activate the Left viewport, simply watch it to guide your click. In summary, the motion for creating the Camera is click—drag—release—click.

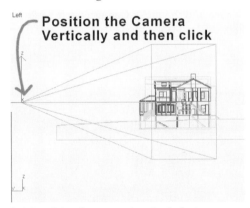

Figure 20–18 *Set the height of the Camera using the Left view as a guide*

10. Right-click in the Perspective viewport to activate it.

Be sure to use the right mouse button for this—the right button switches the active viewport without "dropping" the selection (the Camera in this case). The left mouse button will also switch the viewport, but you will end up deselecting the Camera. If you do this, click on the Camera to re-select it.

11. At the keyboard, press C to activate the Camera view in the Perspective viewport.

The Perspective viewport is now looking "through" the camera lens. Take a look at the Viewport Navigation icons. When you switch to Camera view, the icons change shape and function slightly.

12. View Navigation Tools toolbar, click the Dolly Camera icon.

13. Drag up slowly in the Camera viewport.

   The Camera will get closer to the target. If you drag down, it will move further away. Be sure to drag slowly; otherwise, it will dolly very fast.

Experiment with both Camera and Target until the view in the Camera viewport has a pleasing angle. When an object is selected in VIZ, you can drag one of the Axis markers attached to the object to move it along only that axis. To do so, right-click and in the Transform menu, choose Move. Then click and drag any axis in any viewport to fine-tune the position. In the case of the Camera, moving it in other views you will see the Camera viewport update interactively. Finally, you have additional controls for the Camera (and any VIZr selection) on the Modify Command Panel.

14. With the Camera still selected, click the Modify Command Panel.

Notice that you can change things like the stock lenses that are used. Experiment further with these settings if you wish.

15. Use the Orbit Camera and the Dolly Camera tools to fine-tune the view and make it match the image in Figure 20–19.

Camera02

Figure 20–19 *Adjust the Camera view to a satisfactory angle*

16. Save the file.

Any time you save the file, you can increment a version if you like. To do this, you chose Saveas from the File menu and then click the small plus (+) sign icon next to the file name. VIZr will save the file with a numeral after the name, so *VIZ Render.drf* would become *VIZ Render01.drf* automatically.

DAYLIGHT

Our file is a bit dark. Since this is an exterior scene, we will add a Daylight system and configure a few parameters to give the scene a bit more life.

ADD A DAYLIGHT SYSTEM

1. Right-click in the Top viewport to activate it and roll your wheel to zoom out a bit.

2. On the Create toolbar, choose **Daylight System**.

3. A Daylight Object Creation message will appear. Accept the recommendation by clicking Yes (see Figure 20–20).

Figure 20–20 *Accept recommended changes for Daylight System*

4. Click in the Top viewport just above the Model hold the mouse button down and drag up slightly.

A Compass rose object will appear next to the Model. This is used to indicate to VIZr, which direction is north. Drag slowly, as it resizes quickly. The exact size of the Compass Rose is not important. Just make it big enough to be legible and easily selected.

5. When the Compass Rose is a decent size, release the mouse (see Figure 20–21).

The command will still be active as now you need to indicate the "Orbital Scale" of the Daylight System. This is basically how far away the Daylight object emitter icon will be.

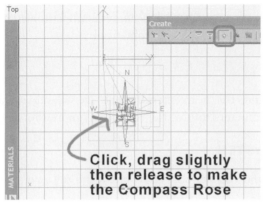

Figure 20–21 *Create the Compass Rose*

6. Move the mouse to position the Daylight Object away from the Model as far as you can (it is easier to see if you use the Left viewport as a guide, and drag to the top of this view) and then click (see Figure 20–22).

Figure 20–22 *Click and drag to position the Daylight object*

The Daylight System can be a bit tricky to place; in summary it is click—drag—release—move—click. As with all things in VIZr, the Daylight System will remain active after placement, and we can edit the parameters. All of these parameters are available in the Modify Command Panel at the right of the screen. The Orbital Scale value in the Command Panel adjusts how far from the Model the Daylight emitter object is placed. We can also set the time of day, the date, the geographical location and the direction of North.

ADJUST THE DAYLIGHT SYSTEM

7. On the Modify Command Panel, in the Site grouping, type **180** in the North Direction field, and then press ENTER.

This will spin the daylight system around 180° and give the back of the house a southern exposure. Here we can also adjust the time of day and date.

8. Set the time to **9:30 am** (see Figure 20–23).

We can also set the location of our project on the globe. The default location is San Francisco. Although warmer climates are certainly tempting, this book is authored in "sunny" Chicago.

9. Click the Get Location button and place a check mark in the Nearest Big City box.

10. Click near **Chicago, IL** on the map, and then click OK (see Figure 20–23).

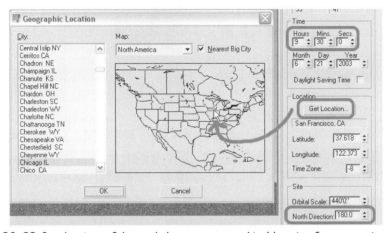

Figure 20–23 *Set the time of day and choose a geographical location for your project*

The Daylight should move slightly. You can choose a different city if you prefer, and adjust the time of day and date. All of these will have an effect on the final position of the Daylight object. Be sure the time is set to daylight hours.

CREATE A QUICK RENDERING

11. Right-click in the Camera viewport to activate it.

12. On the Main toolbar, click the Quick Render icon.

Depending on a variety of factors (amount of RAM, CPU speed, etc.), it can take a few moments to several minutes to render. Grab a cup of coffee while you wait. When you return, your rendering should look something like Figure 20–24.

NOTE Occasionally a warning about missing maps will appear. To correct this, click the browse button and search for the location of your maps on your system. You can also just click Continue to bypass the message.

Figure 20–24 *The Quick Render output*

It's OK, but definitely could use some more attention. First of all, the background is black. This not only makes the background quite dull, it means that the glass in the Windows has nothing to reflect.

ADD A BACKGROUND IMAGE

You can load any bitmap image in almost any common file format as a background. A simple sky image has been included in the Residential Project folder.

13. From the Rendering menu, choose **Environment**.

14. In the Background area, click the None button for Environment Map.

15. In the Browse dialog box, navigate to *C:\MasterADT 2006\Chapter20* folder.

16. Choose the *PuffyClouds1.jpg* file and then click Open.

17. At the bottom of the dialog box, click the Render button.

Notice that the map not only displays as a background, but it also reflects in the windows. Now that we have added a partly cloudy sky, let's set the Daylight System accordingly.

18. Select the Daylight System.

TIP If you have trouble selecting an object, press the H key to call the Select by Name dialog box.

19. On the Modify Command Panel in the Coverage grouping, choose **Partly Cloudy**.

20. Render the scene again.

Notice that this softens all of the shadows in the scene. Typically, the Partly Cloudy setting will give more realistic shadows.

21. Save the file.

RADIOSITY TIPS

Here are some tips on using radiosity. A critical radiosity setting is in the Environment Dialog (Choose Environment from the Render menu) under Logarithmic Exposure Control Parameters. When using a daylight system make sure the Exterior daylight check box is chosen.

On the Radiosity tab, the Reset All button is preferred over the Reset button because sometimes the radiosity process doesn't see all of the changes to a scene. Reset All dumps everything. The only exception to this is when you have a very complex subdivision mesh, which is going to take forever to re-process. Using the Reset button in that situation won't dump the existing mesh, it just updates the solution stored in it. You must click Start and rerun the solution after resetting.

Enabling Global Subdivision under the Radiosity Meshing Parameters should be left off while test rendering and then turned on in the final stages, it greatly improves the quality of the solution but the solution processing times go way, way up. Play with the size to add more detail to the solution.

You will want to turn off the radiosity display in the Interactive Tools grouping if you need to work in a wireframe viewport after you've enabled this.

Filtering allows you to anti-alias the light and shadows. It can be useful—typical settings in single digits are usually around 4 or 5. Turning it on will smooth the shadow edges out—you can watch the effect in a shaded viewport if you have Display Radiosity checked.

The other issue that is important is the reflectance values for materials—these occur in the Adjust Radiosity rollout in the Material Editor. This number is an indication of how bright a material appears and, more importantly, how much light it's pushing back into the scene. White paint should have a reflectance value of between 85 and 90. The reflectance is controlled by the Value of the diffuse color. It's quite common for materials to appear washed out when using radiosity—one workaround is to pick the color you want and then knock the Value down by about 40%.

Thanks to Michan Walker for the tips in this sidebar.

ADD RADIOSITY

Radiosity is a rendering processing algorithm that takes into account the reflection of light and light bounced off other surfaces in the scene. The results can be very realistic, but render times can be greatly increased. Let's process the radiosity in this scene to help remove some of the harshness from the shadows.

22. On the Rendering menu, choose **Radiosity**.

23. In the Radiosity dialog box, click the Start button at the top to process the radiosity solution.

This is another process that can be time consuming. When the processing is all done, you will need to Render again to see the result.

24. Click the Render button to see the effect of the radiosity (see Figure 20–25).

Figure 20–25 *After processing Radiosity, perform another Render*

Continue to tweak and fine-tune your results. When you get a result that satisfies you, click the Render tab of the Render dialog box, choose an Output size in the Output area, and click the Files button under Render Output to save the rendering to a file. A sample file is included in the *MasterADT 2006\Chapter20* folder named *Final Rendering.jpg*.

PERFORM A "SUN STUDY"

In addition to the robust rendering features, VIZ Render is also a full-featured animation package. Let's do a quick sun/shadow study of our residential addition.

25. Press H at the keyboard, and in the Select Objects dialog box, highlight **[Daylight01]** and click the Select button.

It should be at the very top of the list.

26. Click the Auto button at the bottom of the screen (see Figure 20–26).

Figure 20–26 *Turn on the Auto button to begin recording key frames for animation*

This will be a very simple animation. Across the bottom of the screen is a time slider starting at frame 0 and ending at frame 100.

 TIP You can change the total number of frames and other time settings in the Time Configuration dialog box. Click the small Time Configuration icon to access this dialog box (see Figure 20–27).

Figure 20–27 *Edit the number of frames and the frames per second by clicking the Time Configuration icon*

27. Drag the Time Slider all the way to **100**, (or type **100** in the Frame field next to the Time Configuration icon).

28. On the Modify Panel, in the Time area of the Daylight01 object, set the time to **18:00** and press ENTER.

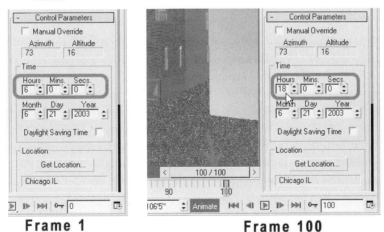

Figure 20–28 *Small red brackets appear to indicate an animated parameter*

When you press ENTER, the time parameters should become highlighted by small red brackets (see Figure 20–28), and two small Key Frame boxes should appear along the time slider.

29. Turn off the Auto button (click it again).

30. Slide the time slider back and forth to preview the animation in the viewport.

If you would like to render this animation to an AVI file, you will need to open the Render Scene dialog box, rather than use the Quick Render button.

31. Click the Render the Current Scene icon (next to the Quick Render icon) (see Figure 20–29).

Figure 18–29 *Click the Render Scene icon to call the Render dialog box and configure its parameters*

Be sure to use the settings as listed below and shown in Figure 20–30:

▶ In the Time Output area, choose Range **0** to **100**.

▶ In the Output Size area, click the 320x240 button.

▶ In the Render Output area, click the Files button, choose AVI as the file format and give the file a name and location. Click OK to accept all defaults in the Render Output dialog box and return to the Render Scene dialog box.

From the Viewport list, make sure a **Camera** viewport is chosen.

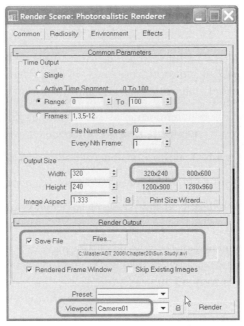

Figure 20–30 *Configure the settings to render a small sized animation file*

32. Click the Render button.

You will definitely need a coffee break this time. Even though the image size is only 320x240, VIZr must still generate 100 images and then compress them all into the AVI file. Depending on your hardware configuration, this could take some time. When the file is finished, open it up and play it. A sample file is included in the *MasterADT 2006\Chapter20* folder named *SunStudy.avi*.

Try additional renderings using the Presets on the Preset list. A Preset is a saved collection of render settings. You can get vastly different results from the same model, camera, materials and daylight when rendered to a different Preset.

FINAL WORD

We have only scratched the surface of what Autodesk VIZ Render can do. With its complex materials, radiosity lighting, and high-end rendering capabilities, there is no limit to what you can produce. We also didn't cover many of the newest tools included in ADT 2006 such as renderable splines, walkthrough assistant and scene states. The primary focus of this chapter has been to assist you in making the most of the model produced in ADT and linked into VIZr. The inclusion of Autodesk VIZ Render in the box with ADT 2006 is very exciting indeed. Have fun exploring! Be sure to check out the plug-ins included with VIZr in the Content Browser. For more information on VIZ Render, pick up a copy of Mastering VIZ Render for Autodesk Architectural Desktop Users co-authored by Paul F. Aubin and James D. Smell.

ADDITIONAL EXERCISES

Additional exercises have been provided in Appendix A. In Appendix A you will find an exercise for completing a rendering of the Commercial Project dataset. This exercise is provided for your information and practice.

SUMMARY

Linking an ADT Model to VIZr is a simple one-click process.

All of your ADT materials will come across to the linked file in VIZ Render.

You can edit the materials or add your own in the Material Editor.

Drag custom materials to the Content Browser for use in ADT or other VIZr scenes.

A large collection of sample VIZr materials has been included in the Autodesk Sample Render Material Catalog.

Add a simple Daylight System and process a radiosity solution for high-quality rendered results.

You can use the VIZr Animation functionality to perform Sun Studies or other animations with ease.

Appendices

The Appendices in this section support the text of the book. Appendix includes the additional exercises referenced throughout the text. Appendix B summarizes the rules of Wall Cleanup in Chapter 9. Appendix E gives the solution to the additional Wall Cleanup exercise from Chapter 9. You can find many websites and other online resources listed in Appendix C and finally Appendix D gives some guidelines for sharing files with external consultants who may or may not be using Architectural Desktop.

Section V is organized as follows:

Appendix A   Additional Exercises
Appendix B   Wall Cleanup Checklist
Appendix C   Online Resources
Appendix D   Sharing Files with Consultants
Appendix E   Wall Cleanup Solution

Additional Exercises

INTRODUCTION

This Appendix includes several practice exercises furthering the topics covered in many of the chapters. It is intended that you will visit this appendix at the completion of each chapter. Once you have finished the lessons in a particular chapter, perform the exercises herein for that chapter. Each of the projects has been provided in completed form in a folder named *Complete* within each respective chapter's folder. You are encouraged to experiment in each of these exercises. The notes given here are merely guidelines for your further explorations. Feel free to perform other tasks not listed and experiment with other tools. You can use Saveas to explore alternatives.

 NOTE In all of these exercises, work in the non-complete versions of the files for the chapter in question. In some cases, these exercises have already been completed in the 'Complete' version from the CD. Not all exercises have been completed in the Complete CD versions.

INSTALL THE CD FILES AND LOAD THE CURRENT PROJECT

If you have already installed all of the files from the CD, simply skip down to step 3 below to make the project active as appropriate per the exercises that follow. If you need to install the CD files, start at step 1.

1. If you have not already done so, install the dataset files located on the Mastering Autodesk Architectural Desktop 2006 CD ROM.

   Refer to "Files Included on the CD ROM" in the Preface for information on installing the sample files included on the CD.

2. Launch Autodesk Architectural Desktop 2006 from the desktop icon created in Chapter 3.

If you did not create a custom icon, you might want to review "Create a New Profile" and "Create a Custom ADT Desktop Shortcut" in Chapter 3. Creating the custom desktop icon is not essential; however, it makes loading the custom profile easier. In some of the exercises that follow, you will work in stand-alone files provided in the *C:\MasterADT 2006\Appendix-A* folder. In other cases, you will work in the project files for the corresponding chapter. To work in a project for a particular chapter, you must first load the project files for that chapter.

3. From the File menu, choose **Project Browser**.

4. Click to open the folder list and choose your *C:* drive.

5. Double-click on the *MasterADT 2006* folder, then the appropriate chapter folder (depending on the particular exercise).

   You will find either the Metric or Imperial version of *Commercial* and *Residential*.

6. Double-click *XX Commercial* or *XX Residential* (where XX equals the Chapter number) if you wish to work in Imperial units. Double-click *XX Commercial Metric* or *XX Residential Metric* if you wish to work in Metric units. (You can also right-click on it and choose **Set Current Project**.) Then click Close in the Project Browser.

 IMPORTANT If a message appears asking you to re-path the project, click Yes. Refer to the 'Re-Pathing Projects' heading in the Preface for more information.

CHAPTER 4

RESIDENTIAL PROJECT

Here are some additional exercises to practice topics covered in Chapter 4 – Residential Project.

Basement Existing Conditions

The basement Existing is a simple file to create.

1. Start with the *First Floor Existing* file. Select all four exterior Walls, right-click, and choose **Clipboard > Copy**. Create a new file using the same template that you used in Chapter 4 for the First Floor file. Right-click in the new file, and choose **Clipboard > Paste to Original Coordinates**. Select the right vertical Wall, right-click and choose **Plan Modifiers > Remove**. Remove the firebox and the hearth (see Figure A–1).

2. Save the file as *Basement Existing.dwg*. Close the file.

Figure A–1 *Basement Existing Conditions Plan – Residential Project*

Second Floor Existing Conditions

The Second Floor Existing Conditions can be built in much the same way as the First Floor.

1. Start with the *First Floor Existing* file. Select all four exterior Walls, right-click, and choose **Clipboard > Copy**. Create a new file using the same template that you used in Chapter 4 for the First Floor file. Right-click in the new file, and choose **Clipboard > Paste to Original Coordinates**. Select the right vertical Wall, right-click, and choose **Plan Modifiers > Remove**. Remove the firebox and the hearth.

2. Save the file as *Second Existing.dwg*.

3. Using the process followed in the text of Chapter 4 for the First Floor, offset, trim, extend and fillet Walls as required to lay out the Second Floor following the dimensions in Figure A–2.

4. Use the Existing Conditions Wall style from the *First Floor Existing* file. Set the Width of Exterior Walls to **12″ [311]** and the interior Walls to **5″ [125]**. Set the height of all Walls to **9′-0″ [2750]**.

5. Use the Automatic Offset/Center option for Doors and Windows and an offset of **4″ [100]** unless noted otherwise. For the hinged Doors, use the Single – Hinged tool and choose the **2′-6″ x 6′-8″ [750x2200]** Standard Size. For the Bi-fold Doors, pick sizes appropriate to the various closets. The Windows are all Double Hung, **3′-0″ x 4′-8″ [900 x 1200]** with a Head Height of **6′-8″ [2200]**. The two bathroom Windows are **3′-0″ x 3′-0″ [900 x 900]**.

6. Save and close the file.

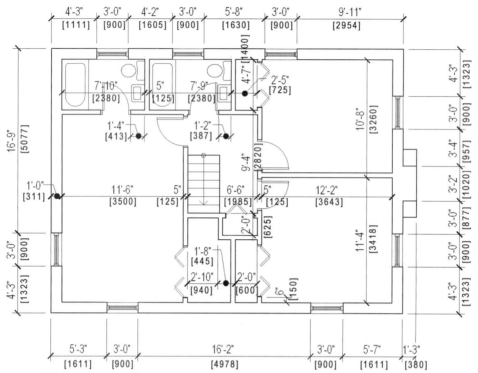

Figure A–2 *Second Floor Existing Conditions Plan – Residential Project*

COMMERCIAL PROJECT

Third Floor Build Out

1. In the *C:\MasterADT 2006\Appendix-A* folder (from the CD) open the file named *Chapter04 Commercial.dwg* [*Chapter04 Commercial - Metric.dwg*] and add the Walls shown in Figure A–3. Use the same techniques as above. For the curved Walls, simply draw them with the Wall tool, using a curved Wall segment, and use grips to edit the curves to the approximate shapes and dimensions shown.

2. Add Doors as appropriate using the standard tools and sizes.

3. Save the file as *Third Floor Build Out* and then close the file.

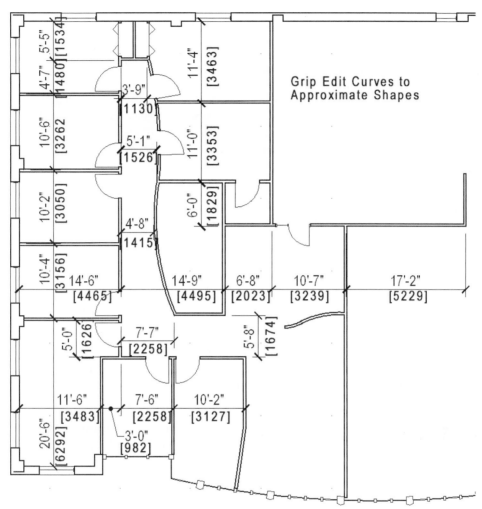

Figure A–3 *Third Floor Build out – Commercial Project*

CHAPTER 5

COMMERCIAL PROJECT

The two projects that we are building in this book do not take fullest advantage of the use of Elements in a Project Structure. With an Element, you can create a portion of the building model that is re-used in more than one location in the complete building model. For instance, a typical Stair Tower, or Typical Restroom layout are excellent examples. You can also use an Element to assist with Furniture Layout. In the *C:\MasterADT 2006\Appendix-A* folder (from the CD) are two files, *Office Furniture.dwg* [*Office Furniture – Metric.dwg*]. The files contain some sample furniture and a rectangle representing the typical office size of the Commercial Project.

Create a Furniture Layout Element file

1. From the *C:\MasterADT 2006\Appendix-A* folder, open *Office Furniture.dwg* [*Office Furniture – Metric.dwg*]. Using the Rectangle as a guide, create a typical furniture layout using the symbols provided in the file. (If you wish to add others, there are others in the library. The Content Library is covered in detail in later chapters).

 NOTE The lower left corner of the rectangle is 0,0. Name the file *Typical Furniture*.

2. On the Project Navigator, right-click the Elements folder and choose **Save current drawing as Element**.

3. Open the *Third Floor Build Out* file created in the previous exercise, and on the Project Navigator, right-click the Constructs folder and choose **Save current file as Construct**. Check Third Floor for the assignments.

4. Drag the *Typical Furniture* Element from the Project Navigator and drop it into the *Third Floor Build Out* file. Move it into position in one of the offices. The single grip at the insertion point of the file is helpful for this. Copy and mirror it into several offices. Save the *Third Floor Build Out* file.

5. Return to the *Typical Furniture* layout Element file, and make a change to the layout. Move a piece of furniture, add something, delete something. The exact change is unimportant. Save the *Typical Furniture* Element.

6. Return to the *Third Floor Build Out* file, open the XREF Manager (use the quick pick on the lower left corner of the Drawing Status Bar) and Reload the *Typical Furniture* XREF. Note the change. Continue to experiment.

7. Save and Close all files.

CHAPTER 6

RESIDENTIAL PROJECT

Using the Structural Member Catalog, we can import some shapes for the beams and columns needed in the Basement of the Residential Project.

Add Columns and Beams to the Basement

1. Open the Structural Member Catalog as done for the Column Grid in Chapter 6. Import (double-click to make styles) Shapes appropriate for a

column and beam in the Residential Basement. Add two Beams and two Columns as shown in Figure A–4 in the center of the *Basement Existing* file.

2. Use the Elevation setting on the Properties palette to get the height correct.

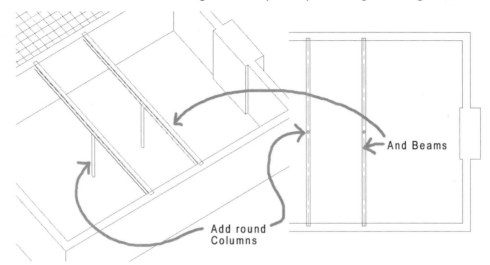

And Beams

Add round Columns

Figure A–4 *Third Floor Build out – Commercial Project*

Add a Beam Haunch

1. Draw a Mass Element (Design tool palette group > Massing tab) big enough for the overlap of the Beam on the Wall. Position it at the top of the Wall.

2. Right-click the Wall and choose **Body Modifier > Add**. When prompted, select the Mass Element. In the Add Body Modifier dialog box, choose Subtractive for the Operation. Put a check mark in the "Erase Selected Objects" check box to delete the Mass Element after the operation (see Figure A–5). The result will be that the Mass Element will carve away a haunch for the Beam.

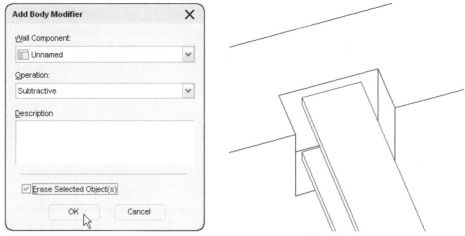

Figure A–5 *Add a Subtractive Mass Element as a Beam Haunch*

CHAPTER 7

RESIDENTIAL PROJECT

Add some Stairs to the new Residential Addition.

Add Basement Stairs

1. The small space between the two vertical Walls on the left of the *Basement New* file for the Residential Project needs some Stairs to the Basement. Use the Content Browser, *Design Tool Catalog – Imperial [Design Tool Catalog – Metric]*, Stairs and Railings category to import the Concrete Stair Style. Set the Width to **4'-4″ [1300]** and the Height to **4'-4″ [1254]**. Use Right justification and trace the edge of the Wall. Move the Stair down about **12″ [300]** from the top edge of the Walls. Make sure that the Elevation of the Stairs is set to **0**. Set the Tread size to **12″ [300]** (see Figure A–6).

2. When you are finished, save the file, open the Composite Model file *A-CM00*, and view the model from the NW Isometric view to see the results. The Stair should come up to the level of the terrain. In a later chapter, we will carve out the terrain where the Stairs occur. For now simply verify the heights in wireframe.

3. Save and Close all files.

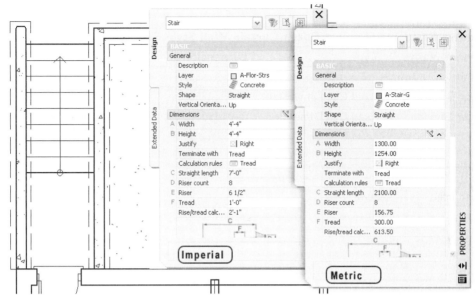

Figure A–6 *Settings the parameters for a Basement Stair in the New Construction*

COMMERCIAL PROJECT

Add some Railings to the *Ground* Construct file.

Convert Linework to Railings

1. Choose a Railing Style from the library and import it into the *Ground Level* Construct file. Right-click the Railing tool on the Design tool palette, and choose **Apply Tool Properties to > Linework**. Choose the Blue lines in this file. Erase the layout geometry. Select the resultant Railings and change their style to the style that you imported.

2. Verify that the heights are correct and if necessary adjust the Elevation of the Railing objects.

3. Save and Close all files.

CHAPTER 8

RESIDENTIAL PROJECT

Add a Porch to the new Residential Addition.

Create a Curtain Wall for a Screen Porch

Create a Curtain Wall style to use for a Screen Porch on the Residential Project. Work in the First Floor New file of the Residential Project. Define the following elements:

Table A–1 Residential Project Porch Curtain Wall Style Elements

Element Name	Type	Dimensions	Other
Divisions			
Screen Division	Fixed Cell Dimension	9'-0" [2750]	Shrink Top
Pilaster Division	Fixed Cell Dimension	3'-0" [900]	Shrink Left & Right Maintain half Cell Offset Start & End 8" [180]
Infills			
Screen Infill	Style	N/A	Porch Window Style Create a Window Style using 2" x 4" [50 x 100] Frame and a zero Sash
Frames			
Corner Pier Frame	Basic (No profile)	10" [250] wide x 10" [250] deep	X Offset = 5" [125] End Offset = 12" [300]
Half Pilaster Frame	Basic (No profile)	5" [125] wide x 10" [250] deep	End Offset = 12" [300]
Cornice Frame	Basic (No profile)	4" [100] wide x 12" [300] deep	
Knee Wall Frame	Use Profile	8" [200] wide x 24" [600] deep	Y Offset = 1" [25]
Mullions			
Pilaster Mullion	Basic (No profile)	10" [250] wide x 10" [250] deep	

1. Define all Element Definitions. Be sure to make the Window style and the Profile Definition required first. The easiest way to create the Profile Definition is to first create the Curtain Wall with a basic rectangular Frame definition, and then right-click it in the drawing and choose **Frame/Mullion > Add Profile**. Create a new Profile from scratch, and then add a few vertices to shape it as shown in Figure A–7.

Figure A–7 *Draw a Profile to use for the Knee Wall Frame*

2. Create a new style, name it: **Screen Porch**. For the Primary Grid, assign the **Pilaster Division.** Assign **Cornice Frame** to the Default Frame Assignment for **Top**, **Left** and **Right** (not **Bottom**). Set Mullion Assignment to **Pilaster Mullion**. For the Secondary Grid, use the **Screen Division**; the Cell Assignment should be the **Screen Infill** and the Frame Assignment is **Knee Wall Frame**. The Mullion Assignment does not matter since there are no Mullions in this Grid (see Figure A–8).

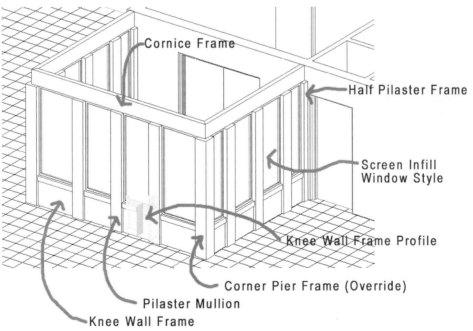

Figure A–8 *Draw a Profile to use for the Knee Wall Frame*

3. Use the right-click **Frame/Mullion > Override Assignment** to change the Corner Frames to the Corner Pier Frame and the Frames against the Wall of the house to the Half Pilaster Frame.

4. Draw a three-sided porch using the new style on the top right side of the plan of the *First Floor New* file.

5. Save and Close the file.

CHAPTER 9

COMMERCIAL PROJECT

Additional Wall Cleanup practice:

Fix Cleanup Problems on the Third Floor

1. Load the Commercial Project for Chapter 9. Open the *03 Partitions* Construct in the Project Navigator. Follow the "Rules" in Chapter 9 and see if you can solve all of the Cleanup Problems.

2. If you get stuck, see Appendix E for the solution.

CHAPTER 10

RESIDENTIAL AND COMMERCIAL PROJECTS

Add additional components.

Add Casework

Work in the First Floor Existing file for the Residential project; the upper left space is the kitchen. Work in the 03 Partitions file for Commercial. The Workstations and Copy Room in the middle of the plan need counters.

1. Open the Style Manager or Content Browser and load the *Wall Styles - Casework (Imperial).dwg* [*Wall Styles - Casework (Metric).dwg*] content file.

   All of the styles in this content file have their Baseline assigned to the back edge of the counter.

2. Add them the way you would any other Wall using the Baseline justification. However, before you begin to add countertops, choose **Options** from the Format menu, and on the AEC Object Defaults tab, turn off the Autosnap New Wall Baselines option. (If you don't, the ends of your countertops will snap to the centerline of your Walls.) Explore the Styles, the Endcaps, and the Components. Look at the Display Properties, and the Cut Plane height. Make changes as you see fit. (For instance, remove the bullnosed Endcap Style and use **Standard** instead.) Get a complete sense of how these styles work.

Also try the Casework in the Content Browser's Design catalogs. There are dozens of "drag and drop" content items within. Try them out in your drawings.

Add Equipment and Furniture

Equipment and Furniture can be found in Content Browser as well.

3. Drag in some appliances in the Residential Project and some office equipment in the Commercial Project.

CHAPTER 11

RESIDENTIAL PROJECT

Add additional components.

Add Bow Window

Load the Residential Project and open the First Floor New Construct. Add a Bow Window to the vertical Wall on the right side of the Dining Room. You will find a pre-made Bow Window style in the Content library.

1. Open the Content Browser and navigate to *Design Tool Catalog – Imperial.*

 NOTE Please note that there is no Bow or Bay Window style in the Metric catalog, but you can use the one in the Imperial catalog and simply change the dimensions to suit.

2. Open the *Doors and Windows* category, and then the *Windows* category.

3. Locate the Bow Window style (it is on page 2). Drag it into the drawing (or to a palette first if you like) to add it to the drawing.

4. Add it to the vertical Wall on the right side of the Dining Room as noted above. On the Properties palette, set the Width to **10'-0"** [**3000**], the Height to **5'-0"** [**1800**] and the Head Height to **6'-8"** [**2200**] (see Figure A–9).

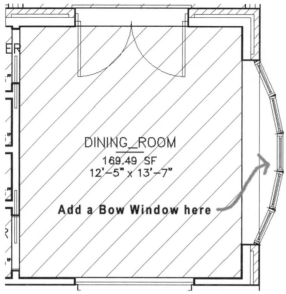

Figure A–9 *Add a Bow Window*

The Bow Window style makes use of a Custom Display Block named: **Aec_Window_Bow** . *This block contains several Window objects of the style* **Bow – Unit** *organized in a Bow configuration. You can insert this block and refedit it to change the configuration (quantity of Windows, type of Window, etc.). You can also open the Style Manager and edit the* **Bow – Unit** *Window style if you wish. For instance, you could follow the steps in Chapter 11 to add Muntins to the Window units. Experiment with both the Display Block and the Window style to see what you can achieve. For more information on Custom Display Blocks, refer to the "Adding Custom Display Blocks in Plan" topic in Chapter 11.*

 5. Save and Close the file when finished.

COMMERCIAL PROJECT

Adding Column Enclosures and using Wall Interference:

Build a Multi-View Block for Column Cutouts – Interference with Walls on Shell File

This exercise covers two concepts: Multi-View Block Interference blocks and Wall interference. An Interference condition enables one object to interact with the mass of another object. In the case of Wall objects, interference is used to make a separate object interact with the shrinkwrap of the Wall(s) to which it is attached. There are three possible shrinkwrap effects: Additive, Subtractive, and Ignore. Use Additive to make the shrinkwrap include (wrap around) the object, use Subtractive to make the object appear to carve away from the object, and use Ignore to have the shrinkwrap unaffected by the interfering object. In this example, we will carve out a pocket for each of the Columns in the shell Wall. Interference will be used to achieve this. You can create interfer-

ence between a Wall and any other AEC object. Mass Elements are often used for this. In this case, we will take advantage of the ability that multi-view blocks have to contain an Interference block.

1. Load the Commercial Project and open the *01 Shell and Core* Element file.

2. Create a Mass Element Box (from the Massing palette in the Design tool palette group) with the dimensions: **1'-6" x 1'-6" x 12'-0"** [**450 x450 x 3650**]. Using the Midpoint of one of the sides as a base point, create a block from this Mass Element named **Column Enclosure_M** .

3. Open the Style Manager, in the Multi-Purpose Objects folder, create a new Multi-View Block named **Column Enclosure**. Edit the Multi-View Block and add **Column Enclosure_M** to the General Display Rep in **Top** and **Bottom** views only. Also click the Set Interference Block and add the same block there as well. Drag the Multi-View Block to your Project Tools palette. Close the Style Manager, save the drawing and then use the new tool to add the Multi-View Block to the drawing.

4. In the Project Navigator, right-click on the *Column Grid* Element file in the *Structural* sub-folder and choose **XREF Overlay**. Move the Multi-View Block that you added in the last step into position relative to one of the columns. Copy it to the other columns as well.

5. Select one of the Walls, right-click, and choose **Interference Condition > Add**. Select all of the Multi-View Block Column Enclosures that overlap the selected Wall. For the Shrinkwrap effect, choose **Subtractive**.

6. At the corners, you need to perform the Interference for each of the Walls. However, choose **Subtractive** for the first Wall, and **Ignore** for the second (see Figure A–10).

The reason for this is that if you use Subtractive for both, the Subtractive on the second one will override the Subtractive of the first one. When you choose Ignore, it prevents this from happening.

7. Select the Multi-View Block, right-click and choose **Edit Multi-View Block Definition**. On the General Display Rep, deselect the Top box. This will make the multi-view invisible in this view. If you ever need to edit the definition, either use the Style Manager, or switch to Bottom view to show them and then select.

8. Save and Close the file. Repeat on the other floors.

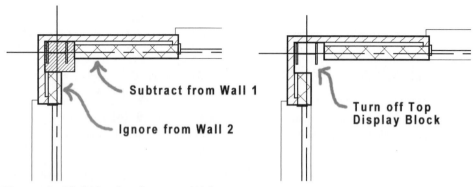

Figure A–10 *Adding Interference to Walls*

CHAPTER 12

RESIDENTIAL PROJECT

Add Floor Slabs and a Roof.

Add Basement Floor Slabs

1. Load the Residential Project and open the *Basement New* file. Add Floor Slabs to the Stair galley and the main Basement space of the new addition. Use the techniques covered in Chapter 12 to import an appropriate Slab style. You can simply trace the spaces with the Slab command, or draw polylines first and convert them.

2. Save and Close the file.

Add a Roof to the Porch

1. Open the *Roof New* file and add a Roof for the Porch.

2. On the Project Navigator, right-click the *First Floor New* file and choose **XREF Overlay**. This will add it at the correct height relative to the Roof. Using the techniques covered in Chapter 12, draw a Roof over the screen porch as shown in Figure A–11.

3. Detach the XREF, and Save and Close the file.

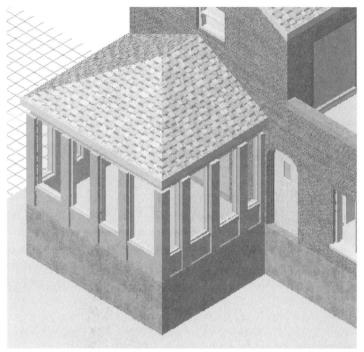

Figure A–11

Add Other Slabs (Optional)

Later in Chapter 16, when Sections are cut through the Residential Building Model, you will find that the Space objects added in Chapter 12 do not form a continuous material all the way across the floor plate. To achieve this effect in section, you will need to add floor Slabs to the Residential Model. This is not required, as you can also edit the linework of the Section (see Chapter 16 for more information).

1. Create a Construct for the First Floor named *First Floor Slab*. XREF the *First Floor New* file into it as a guide and then add Slabs for the floor structure. Be sure to place them at a negative Z height. Repeat this process for the Second Floor.

2. Add these new Constructs to the *A-CM00* and *A-CM01* Composite Model View files.

3. When you cut sections in Chapter 16, freeze the Space layers in the XREFs so that you will cut only through the Slabs.

4. Save and Close all files.

CHAPTER 13

COMMERCIAL PROJECT

Continue to refine the Third Floor Reflected Ceiling Plan. Add additional elements and configure Door Display in Reflected.

Edit Door Display in Reflected

There are a few ways that you can configure Doors to display in Reflected Ceiling Plans. Each firm has its own standards. Here are a few techniques to try:

1. Load the Commercial Project and open the *03 Partitions* file. On the Drawing Status Bar, load the **Reflected** Display Configuration.

2. Select one of the interior Walls, right click and choose **Edit Object Display**. On the Display Properties tab, highlight Reflected and then click the Edit Display Properties icon. On the Layer/Color/Linetype tab, turn on the Below Cut Plane component. Configure the Above Cut Plane component to your desired Layer, Color, Linetype and Plot Style settings (this is the "header" component). Click the Cut Plane tab, place a check mark in the "Override Display Configuration Cut Plane" check box and change the value to **6'-6"** [**1800**] and then click OK back to the drawing.

    The location of the openings within the Walls should now appear. Doors will also appear and be fully closed. You can change the way they display based on your preferences.

3. Select one of the Doors, right click and choose Edit Object Display. On the layer/Color/linetype tab, turn off components that you don't wish to see. If you do not want Doors to display at all in the Ceiling Plan, turn off all components and then click OK back to the drawing. If you wish some components to display, you can leave them on and change their settings. For instance, you may wish to leave the Frame, Swing and Panel on, but change them to a gray color and a dashed linetype and a lighter lineweight. Check the Other tab and review the "Override Open Percent". If this is selected and set to 0, then the Door swing will appear closed. Experiment with this feature and see that the Door swing can be configured differently between Display Configurations. Set to your office standard or desired effect. OK back to the drawing when finished (see Figure A–12).

4. Save and Close all files.

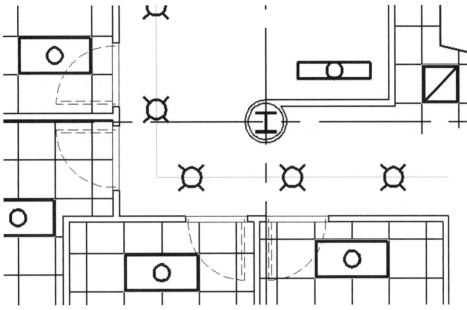

Figure A–12

Exercise

1. Finish adding all of your Ceiling Grids and lights.

2. Add some other ceiling fixtures (using the Content Browser or the DesignCenter). There are exit signs, diffusers, and other types of lights. The process for everything in the Content Browser and DesignCenter is "drag and drop" and then read the command line for options.

3. Try the other electrical items such as switches, outlets, bells, and fire alarms.

4. Save and Close all files.

You may also repeat any of the techniques covered in Chapter 13 on the Residential Project, as well, if you like. Files have been provided in the *Chapter13\Residential* folder for practice.

CHAPTER 14

RESIDENTIAL AND COMMERCIAL PROJECTS

Annotate the remaining View files.

Add AEC Dimensions and Annotation

1. Load either the Residential or the Commercial Project.

Open any of the View files for either the Commercial or Residential project. Use the techniques covered in Chapter 14 and perform the following:

2. Add AEC dimensions to each *Floor Plan* View file

3. Add detail clouds, notes and revision clouds using the tools on the palettes and the Content Browser.

4. Open the *Elevation and Section Composite Model* View files and add Elevation Labels to each Elevation and Section. Remember to define a new UCS for each Elevation and Section as required.

   There are also elevation tags, north arrows, bar scales, and many other symbols on tool palettes, Content Browser and in the DesignCenter. Try several of them to get comfortable with what is available. Be sure to work in the View files of each project.

5. Add Room Tags to other View files to practice techniques covered in the chapter.

6. Repeat on the other Levels.

7. Save and Close all files when done.

CHAPTER 15

RESIDENTIAL AND COMMERCIAL PROJECTS
Add additional Schedules.

Add Other Schedules

Included on the tool palettes are two other Schedule Table styles: Window Schedule and Room Finish Schedule. In addition, there are several more examples in the Content Library.

1. Load either project. Create additional Schedule Table View and Sheet files. If you wish to be able to use the Edit Table Cell functionality, create a View file for the Schedule that you wish to create; otherwise, simply make a new Sheet file in the "600" series (A-601, A-602, etc), and add a Schedule there using the procedure for adding the Door Schedule covered in Chapter 15.

2. Open the Content Browser (CTRL + 4) and Click on the *Documentation Tool Catalog – Imperial* [*Documentation Tool Catalog – Metric*]

3. Navigate to the *Schedule Tables* category (see Figure A–13).

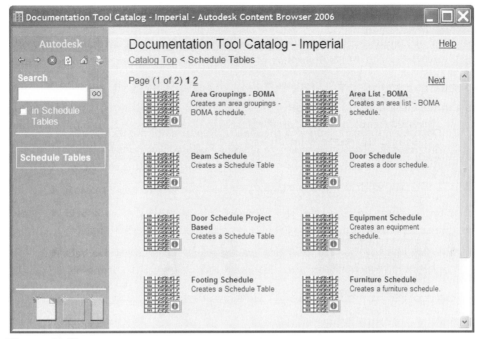

Figure A–13

4. Add some more Schedules. Follow the same basic procedures we followed in this chapter. If you are unhappy with the format of any Schedule, see if you can edit the style to change it.

5. Save and Close all files.

CHAPTER 16

RESIDENTIAL PROJECT

Refine Sections and Elevations.

Floors and Ceilings in Section

The Floor and Ceiling planes in our Sections come from the Spaces. You will notice that there are gaps. If you want the floor and ceiling planes to go all the way across, you could try adding slabs to your model. Add them directly to the Composite Model View so that they do not interfere with your plans. You can also simply use Edit Linework and edit the Section directly. Use both methods and explore the benefits of either.

Material Boundaries

Add more Material Boundaries to all Sections and Elevations as appropriate.

Other Refinements

Add notes and dimensions to the elevations and sections. Do general cleanup of any extraneous linework.

COMMERCIAL PROJECT

Update and add Elevations and Sections.

1. Update all of the Elevations in the *A-EL01*. Repeat all of the techniques discussed in Chapter 16 to create pleasing Building Elevations (see Figure A–14).

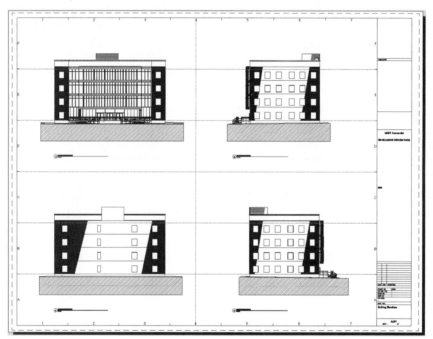

Figure A–14

2. Add Material Boundaries.

3. Add labels, notes and dimensions as appropriate.

4. Open *A-201* Sheet file to see the results. To make the *A-201* Sheet file load more quickly, unload all XREFs except the *A-EL00* XREF.

A Live Section was built in the *A-SC00* View file in the early stages of this project. You can continue to keep this section live and up to date as the project progresses, and even print it as part of the final set. However, it will probably be more useful in the later stages of Design Development and Construction Documents to cut some 2D Section/Elevation objects for the Building Sections instead.

1. Open the View file named *A-SC01*.

2. Switch to the **Medium Detail** Display Configuration.

3. In Top view, follow steps similar to those covered in Chapter 16 and update the Section.

4. If you wish to add some more Sections, create them from the Plan View files using the appropriate Callout tools on the tool palette. Be sure to choose the "Existing View Drawing" option to create the new sections in the *A-SC01* View file.

5. Add Subdivisions, Material Boundaries, Notes, Labels, Dimensions, and Edit Linework as appropriate on the existing and any new sections you create.

6. Save and Close all files.

CHAPTER 17

RESIDENTIAL PROJECT

The Typical Wall Section Detail created in the Residential project exposes an issue with the dimensions of the foundation Wall used. It is too narrow for the construction above it. This is exactly the kind of thing that early creation and blocking out of details is meant to reveal. Try correcting these problems in the *Basement New* Construct and then updating the underlying section beneath our detail to verify.

1. Open the *Basement New* Construct.

2. Select all Foundation Walls and using the **Copy Wall Style and Assign** right click functionality, create a new style for these Walls. Edit the components to match the sizes used for the detail components (**12″ [300]** Wall width, **20″x 10″ [500 x 250]**. Be sure to keep the reference points relative to the Baseline intact, in other words, change the Width of the Wall component to: **12″ [300]** and the Edge offset to: **-12″ [-300]**. (Refer to Chapter 10 for assistance with Wall Style editing.)

3. Open the *A-DT01* View file and Refresh the background 2D Section/Elevation object to coordinate it with the detail.

Commercial Project Details

Using the techniques covered in the chapter, create details in the Commercial project.

CHAPTER 18

RESIDENTIAL PROJECT

Add Area objects to the Residential Project to get total floor areas.

CHAPTER 19

RESIDENTIAL PROJECT

Plot the Residential Project.

Plot the Residential Project

1. Repeat the steps covered in Chapter 19 to plot the Residential Project.

   Save different Sheet Selection lists for different purposes; for example, one list might be for plans only, while another is the complete set. To do this, use the shift and ctrl keys to select one or more Sheets on the Sheets tab, right-click and choose **Save Sheet Selection**. Give it a name.

2. Create a 3D DWF of the Residential project.

CHAPTER 20

COMMERCIAL PROJECT

Render the Commercial Project.

Link to VIZ Render and Create Rendering

1. Load the Commercial Project (use the version from Chapter 19).

2. Open the *A-CM00* Composite Model file.

3. Choose Link to VIZ Render as shown in Chapter 20.

4. Add, edit and adjust materials as required. Add a Camera. Add a Daylight System.

5. Set a Viewport to the Camera (type C) and generate a test render (see Figure A–15). Add a background photograph. (Search on the web for a suitable *.jpg* file.) Run the Radiosity solution. Generate the final rendering.

6. Try animating the Daylight system as was done for the Residential Project in Chapter 20. Try animating the Camera this time to generate a walkthrough or a fly by. VIZ Render 2006 has a new Walkthrough Assistant—try it out.

7. Save and Close all files.

Figure A–15

Wall Cleanup Checklist

AUTOMATIC WALL CLEANUP

CLEANUP WILL OCCUR WHEN:

- ▶ The Graph Line of one wall intersects the Graph Line of another wall, or
- ▶ The Graph Line of one wall intersects the cleanup radius of another wall, or
- ▶ The cleanup radius of one wall intersects the Graph Line of another wall.

RULE 1 – USE WALL GRAPH DISPLAY

- ▶ Right-click on any Wall, and choose **Cleanups > Toggle Wall Graph Display**.

RULE 2 – SET THE DEFAULT CLEANUP RADIUS TO 0

- ▶ Properties palette and Wall tool defaults.

RULE 3 – PRACTICE GOOD CLEAN DRAFTING

- ▶ Avoid doubles.
- ▶ Use Offset, Fillet, Trim, and Extend.
- ▶ Use 'L' and 'T' cleanup tools (right-click a Wall, choose **Cleanups > 'L' Cleanup, 'T' Cleanup**).
- ▶ Use Object Snaps (F3).
- ▶ Autosnap (**Format > Options**: AEC Object Settings tab).

RULE 4 – TWEAK THE CLEANUP RADIUS

- ▶ Use the Cleanup Radius grip (visible when the graph is displayed) or the Properties palette.
- ▶ For Center and Baseline justification – set between ½ × ¾ × Wall Width.
- ▶ For Left and Right justification – set between ½ amd 1 × Wall Width.

OTHER CONSIDERATIONS

▌ Justification can impact Cleanup. Experiment with Left, Right, Center or Base-line to yield variations in Cleanup behavior.

▌ A Cleanup Group limits cleanup to Walls that are part of the same group.

▌ To Make XREFs cleanup with the host drawing, edit the Cleanup Group to ig-nore XREF Barrier.

▌ Complex Wall styles use component priorities to determine how their indi-vidual sub-components cleanup.

▌ Wall Cleanup will not occur between walls that are not in the same X,Y plane.

▌ In **Options > AEC Object Settings**, check to see if Autosnap Grip Edited Wall Baselines check box is deselected. This will assist in grip editing walls manually.

▌ If all attempts at Automatic Cleanup fail, use manual cleanup. (Right-click a Wall, choose **Cleanups > Add Wall Merge Condition**.)

MANUAL WALL CLEANUP

CONSIDERATIONS

▌ Try to trigger automatic cleanup before resorting to Wall Merge.

▌ Apply a Wall Merge by right-click (**Cleanups > Add Wall Merge Condi-tion**).

▌ Graph Display reveals merged Walls.

WALL COMPONENT PRIORITIES

CONSIDERATIONS

▌ Components with the same priority number will cleanup.

▌ Lower-numbered components will interrupt (pass through) higher-numbered components.

▌ Wall priorities will take precedence over drawing order. (The order in which Walls were drawn has no bearing.)

Online Resources

In this appendix are listed several online Web sites and other resources that you can visit for information on ADT and related topics. All of these URLs are also available in an HTML file in the Resources folder on the CD.

WEB SITES RELATED TO THE CONTENT OF THIS BOOK

http://www.paulaubin.com

Web site of the author. Includes information on this book and Aubin's other books like the *Mastering VIZ Render – for Autodesk Architectural Desktop Users* (co-authored with James D. Smell), as well as Aubin's other books on previous versions of ADT. Check there for ordering information and addenda.

http://www.autodeskpress.com

Web site for Autodesk Press. Visit for information on other CAD titles, online resources, student software, and more.

http://www.autodesk.com

Autodesk main Web site. Visit often for the latest information on Autodesk products.

WEB SITES OF RELATED INTEREST

http://discussion.autodesk.com

Autodesk Discussion Groups main page. Online community of Autodesk users sharing comments, questions, and solutions about all Autodesk products.

http://adt_blog.typepad.com/

Chris Yanchar, a product designer on the Architectural Desktop team shares and aggregates information on Architectural Desktop and architecture. At the Blog site you will also find links to How To's and articles about ADT features.

http://autodesk.blogs.com/between_the_lines/

Shaan Hurley, a Technical Marketing Manager for the AutoCAD group and manager of most of the beta programs for Autodesk shares his views on Auto-CAD, technology and life. Not everything is official Autodesk opinion, endorsement, or recommendation; it is a Blog from Shaan himself. This approach provides direct contact between customers and Autodesk personnel most familiar with the products.

http://modocrmadt.blogspot.com/

Matt Dillon based in San Antonio, Texas, Matt specializes in consulting with architectural and engineering firms to make the most of their ADT and other BIM software. His Blog includes many useful ADT resources, tutorails and musings.

CADalyst Magazine web forum

http://cadence.advanstar.com//ubbthreads/ubbthreads.php?Cat=

Online user forum hosted by CADalyst Magazine and moderated by Paul F. Aubin.

http://www.cadalyst.com/cadalyst/

Main home page for CADalyst Magazine. View magazines online or subscribe to print edition.

http://www.nibs.org

Web site of the National Institute of Building Sciences.

http://www.nationalcadstandard.org

Web site for information and purchase of the United States National CAD Standard.

http://www.aia.org

Web site of the American Institute of Architects.

ONLINE RESOURCES FOR ADT PLUG-INS, AND TRAINING

www.paulaubin.com

Website of the author. Paul F. Aubin Consulting Services provides Autodesk Architectural Desktop Training and Implementation services. Visit the site for details on services available. Contact Paul F. Aubin to discuss your ADT training and implementation needs: http://www.paulaubin.com/contact.php

www.cadmin.com

Cadmin is an Architectural Service company providing expertise in the Administration, Customization, and Standardization of an Architectural Cad Environment. Specializing in the management of AutoCAD Architectural Desktop,

Cadmin has enabled its customers, from the smaller residential designers to the large multi-national corporations, to successfully implement CAD technology.

www.stardsign.com

Stardsign CAD solutions provide pre-packaged (free) and custom applications to facilitate tasks in Architectural Desktop. Visit the AEC information Center for valuable must-have information for working with Architectural Desktop.

www.archidigm.com

The world's largest independent online source of architectural desktop information featuring tips, tricks, news, downloads, and a complete subscriber-only area with several eGuides on ADT and AutoCAD.

www.GreenBuildingStudio.com

Green Building Studio is a free web-service that provides architects and engineers using Autodesk Revit, Autodesk Architectural Desktop, or Autodesk Building Systems with early design stage whole building energy analysis and product information appropriate for their building design

www.e-specs.com

Built around our e-SPECS technology which links the project drawings to the specification documents, InterSpec has a variety of products and services to help you manage your construction specifications in the most accurate, efficient and cost effective manner possible.

http://tsc.wes.army.mil/products/standards/aec/aecstd.asp

The CADD/GIS Technology Center of the Department of Defense in their A/E/C CADD Standard release 2.0. This standard is compliant with the National CAD Standard 2.0 and available for free public download.

https://tsc.wes.army.mil

Technology Center Homepage.

ONLINE RESOURCES FOR VIZ RENDER

http://www.vizdepot.com/

Your resource for all of the latest in tutorials, news and support for all releases of Viz including ADT 2004/Viz Render, ADT 2005/Viz Render 2005, Max 6 and now Viz 2005. Please register with the site. Registration enables posting in the forums, adding images in the galleries as well as access to the many new tutorials and contents of the site. With this update brings many changes to the Vizdepot, including textures, downloads, scripts and more.

http://www.hermanmiller.com

http://www.3dcafe.com
http://www.turbosquid.com
http://www.erco.com
http://cgarchitect.com\

APPENDIX d

Sharing Files with Consultants

INTRODUCTION

In nearly every project, you will be faced at some point with the need to share digital design data with other individuals or software packages. For instance, preparing drawings for distribution to MEP and other consultants offers some challenges in ADT. The objects that make ADT so powerful do not exist in the base version of AutoCAD or in other CAD software. We must therefore translate ADT drawings to an AutoCAD-friendly format beforehand. This appendix is focused on the various issues related to distributing ADT drawings to consultants who may be using a variety of alternative ADT and AutoCAD versions.

OBJECTIVES

In this appendix, we will discover what is required to translate drawings into formats readable by consultants not using ADT. Due to the potentially destructive nature of the commands being covered, be sure to practice on files unrelated to any 'real' projects until you are familiar all the issues covered in this appendix. The following summarizes topics covered in this appendix:

- Understand file formats.
- Understand Object Enablers.
- Understand Proxy Graphics.
- Using Export to AutoCAD.

DETERMINING THE REQUIRED TRANSLATION

A recipient using Architectural Desktop 2006, 2005 or 2004 can open, edit and re-save any file that you send to them with nearly complete "round trip" ability. The new Display Theme object and Spaces with Net and Gross Boundaries enabled will appear in these versions as a "proxy" (see below) and Field Codes will appear like normal text in ADT 2004.

AutoCAD 2006 has built-in awareness of ADT objects—all required Object Enablers (see below) are included in the software. Therefore, if your recipients are using AutoCAD 2006 (or 2005), there is no translation required. However, it is important to note that even though they will be able to open ADT files they still do not have all of the ADT tools, and can perform only limited edits. It is also important that your AutoCAD recipients are made aware that exploding ADT objects will destroy them irrevocably.

Versions of AutoCAD older than 2005, versions of Autodesk Architectural Desktop older than 2004, and other AutoCAD based 'Desktop' products are *not* capable of properly displaying ADT 2006 objects natively. There are three ways to share ADT 2006 drawings with these non-ADT users: the consultant can install an Object Enabler; you can save proxy graphics into the ADT 2006 file; or you can use the **Export to AutoCAD** command prior to sending the consultant the file.

Use the following chart to determine the relevant issues.

Table D–1 *Determining Translation Requirements per Software Version*

Consultant Software and Version	Translation Required	ADT Objects Preserved
Autodesk Architectural Desktop		
ADT 2006	No Translation Required	Yes
ADT 2005	No Translation Required [1]	Yes
ADT 2004	No Translation Required [1&2]	Yes
ADT (3.3, 3.0, 2i 2.0)		Export to AutoCAD
ADT R1	Export to AutoCAD > R12 DXF Format	No
AutoCAD		
AutoCAD 2006	No Translation Required	Yes
AutoCAD 2005	No Translation Required	Yes
AutoCAD 2004	Object Enabler 2004	Yes
AutoCAD 2000, 2000i or 2002	Object Enabler 2004 or Export to AutoCAD	Yes No
AutoCAD R14	Export to AutoCAD > R12 DXF Format	No
Other CAD Packages		
Non-AutoCAD CAD software that reads DWG format	Export to AutoCAD (DWG 2000 File Format)	No
Non-AutoCAD CAD software that does not read DWG format	Export to AutoCAD (DXF 2000 or DXF R12 File Format)	No

1. Display Themes and Spaces with Net and Gross Boundaries enabled appear as Proxies

2. Field Codes appear as normal text with no background shading.

SHARING ADT FILES WITH OTHER ADT USERS

If you wish to share an ADT 2006 file with users of a previous version of ADT, you can freely share your files with users of ADT 2005 and 2004. They all share the same file format: 2004 DWG. The only caution one should have when sharing files with such users is that Display Themes and Spaces with Net and Gross Boundaries enabled appear as proxies in both 2004 and 2005. All field codes will appear as normal text to users of ADT 2004. If they should edit these "text" entities, they will destroy the underlying field code hidden beneath. AutoCAD Table objects (introduced in 2005) will appear as proxy objects to ADT 2004 users. In all other ways, sharing files between ADT 2006 and ADT 2005 or 2004 is seamless. Unfortunately, if your recipients are using a version of ADT prior to 2004, you will not be able to share your files with them without first translating them to plain AutoCAD files. The process of doing so—**File > Export to AutoCAD**, will destroy all ADT objects and their underlying intelligence and replace them with representative AutoCAD geometry.

UNDERSTANDING OBJECT ENABLERS

To share drawings with consultants that are using AutoCAD 2004, some other version of AutoCAD or an AutoCAD-based vertical product, Autodesk provides a free software plug-in called an *Object Enabler*. An Object Enabler, as its name implies, "enables" the otherwise unknown ADT objects directly within the vanilla version of AutoCAD (or other "Desktops") without the need for a copy of ADT. As mentioned above, an Object Enabler is not required for AutoCAD 2005 or 2006—it is already built into the software for those versions. Users of Object Enablers will be able to correctly view and print ADT objects and will be able to perform limited grip editing and Properties palette functions upon the ADT objects contained within a file that they receive. They will not be able to add new ADT objects to the drawing or perform most ADT specific commands. Without an Object Enabler, sharing drawings with consultants using other versions of AutoCAD will require the use of Proxy Graphics (simplified geometric approximations of the actual ADT object) or the need to Export to AutoCAD.

At the time of this writing, you can find the latest Autodesk¢\textregistered Architectural Desktop Object Enabler, online on the Autodesk web site at:

http://usa.autodesk.com/adsk/servlet/index?siteID=123112&
id=2753223&linkID=2475897

If this URL is not functioning, visit www.autodesk.com, click the search link, type "object enabler" and then click Search. This process should locate the latest download location of the Object Enabler.

UNDERSTANDING PROXY GRAPHICS

ADT objects generate the graphics used to represent themselves in different views automatically and based on the parameters of the specific object. The generic AutoCAD package is incapable of generating these graphics and would simply ignore the 'unknown' objects under typical circumstances. However, within ADT, a feature called Proxy Graphics can be turned on. When enabled, ADT creates two sets of graphics for the ADT object when the file is saved: the one used by ADT and the one saved with the file for generic AutoCAD. As the name *proxy* implies, proxy graphics are 'stand-in' objects that appear on screen only when the host application that created them is not present on the machine. In this case, that host application is Autodesk Architectural Desktop. A proxy graphic is not editable in any way. Therefore, if the recipient of the file with proxies needs to edit the data, this would not be a good choice of formats. To share Architectural Desktop drawings using proxy graphics successfully, it is important to understand the multiple Display Representations of Architectural Desktop objects and how they are stored as proxy graphics. Each object can store a single proxy representation. Therefore, it is important to configure your work session to be sure the file you are saving displays correctly for your AutoCAD recipients.

For persons needing only to view the data in the file for review purposes, proxies can provide a workable solution. This is because no special requirements are placed on recipients. They simply open files as they normally would. One small caveat: a message will typically appear alerting the recipient of the presence of proxy graphics and requesting their input on how to deal with them. The choices are to display the proxies, show the bounding box only, or ignore them. Please instruct your recipients to choose to **display proxy graphics**. If they choose not to display them, they will see nothing and will most likely call you to ask why you sent an empty file. If they choose bounding box, each object in the file will appear as a box that matches the overall size of the total object. This option will not reveal very detailed information and is not very useful.

Avoid using proxy graphics if you can. Again, these are stand-in graphics, not the real thing. They cannot be edited and they provide only limited visual information. Finally, proxies nearly double the size of ADT files. Depending on the file size without proxies, this can be significant.

TURNING PROXY GRAPHICS ON

To keep file sizes small, Proxy Graphics is turned off by default in ADT. For your recipients to see proxy graphics, you must turn it on and set the drawing view to the view you wish them to see (Plan, Model, etc.) prior to saving and sending the file.

1. Type PROXYGRAPHICS at the command line and then press ENTER.

2. Type I and press ENTER again.

3. Save the file and then send this file to your recipient.

EXPORT TO AUTOCAD

Sometimes the best approach in sharing data with consultants is simply sending along a standard AutoCAD drawing file. To do this, you must *remove* all ADT objects from the drawing. You could try to do this manually with Explode, but you would not achieve much success. The proper way to explode and remove all ADT objects from a drawing is to use the Export to AutoCAD commands on the File menu. This routine absolutely eliminates the AEC objects from a drawing file and replaces them with lines, arcs and circles. It saves the resultant file in one of several versions of DWG or DXF. Use this only if the file will not be returned to you for editing.

There is NO way to retrieve the ADT objects once this export has been performed.

USING EXPORT TO AUTOCAD

1. Open the file you wish to convert. (You can open within the Project Navigator or from the standard **File > Open** command.)

2. Be sure that the view direction and Display Configuration are set to the view that you wish your recipient to see.

   For example, if you want them to receive a Floor Plan, be sure that you set the drawing to Top view and Medium Detail.

If the recipient has AutoCAD 2006, 2005, 2004 or an AutoCAD 2006-based, 2005-based, or 2004-based vertical product:

3a. From the File menu choose **Export to AutoCAD > 2004 Format.**

4a. Confirm the file name and location and then click Save.

If the recipient has AutoCAD 2002, 2000i or 2000 or a vertical product based on these versions such as ADT 3.3, 3.0, 2i or 2.0:

3b. From the File menu choose **Export to AutoCAD > 2000 Format**

4b. Confirm the file name and location and then click Save.

If the recipient has AutoCAD R14 or a vertical product based on R14 such as ADT R1:

3c. From the File menu choose **Export to AutoCAD > R12 DXF Format**.

4c. Confirm the file name and location and then click Save.

If the recipient is using a non-AutoCAD-based product that does not read DWG, use one of the DXF options.

CAD Manager Note: CAD collaboration with vendors, clients, joint venture partners, and contractors is a big area of concern in the AEC industry. Before a new project is begun, schedule a Project CAD Standards Kick-Off meeting to discuss standards and setup. In this way, a clear guideline of the rules, proper workflow, and procedure can be established with every team member, inside and outside your office.

USING eTRANSMIT AND ARCHIVE PROJECT

If you wish to send an entire project (or any portion of it) in AutoCAD format, you can use the eTransmit or the Archive function on the Project Navigator. Simply load the project, decide which files you wish to send, and then use the right-click functions to eTransmit or Archive the project. Both functions are very similar in process, so once you have learned the use of one, the other will be very simple. eTransmit has a few more options. You can eTransmit from any tab of the Project Navigator. It also has the option to send the files directly as an email attachment. Let's look at an example from the Sheets tab, using eTransmit to send the entire project set.

1. On the Project Navigator palette, click the Sheets tab.

2. Right-click the Sheet Set node at the top (it has the same name as the currently active project; *MADT Commercial* for example) and choose **eTransmit Setups**.

3. Click the New button and name the New Transmittal Setup: **CD Submission** and then click Continue (see Figure D–1).

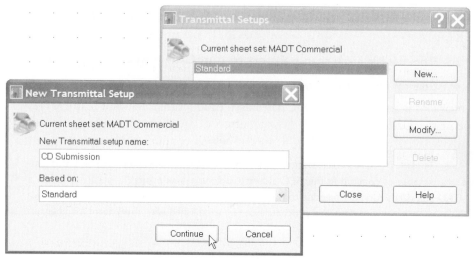

Figure D–1 *Create a New Transmittal Setup*

Several options will appear in the Modify Transmittal Setup dialog. At the top, in the Transmittal package type list is three options (see Figure D–2).

- ▶ **Folder (set of files)**—This option will copy all of the required files to a new folder.

- ▶ **Self-extracting executable** (*.exe)—This option will create a single compressed self-extracting archive file containing all of the selected files. Recipients simply double click the .exe file to extract all of the files.

- ▶ **Zip (*.zip)**—This option will create a single compressed archive file containing all of the selected files. To open the .zip file, the recipient must have compatible extraction software, such as WinZip or the native Windows XP Zip file capabilities.

In the File Format list is four options including two which match the Export to AutoCAD functions discussed above:

- ▶ **AutoCAD 2004 Drawing Format**—This option will simply save all drawing files in 2004 format without translation.

- ▶ **AutoCAD 2000 Drawing Format**—This option will perform a Saveas to AutoCAD 2000 file format. However, the results with ADT objects could be unexpected. This option is not recommended when files contain ADT objects.

- ▶ **AutoCAD 2004 Drawing Format with Exploded AEC Objects**—This option is the same as Export to AutoCAD discussed above. The result will be AutoCAD drawing files containing only AutoCAD entities with no ADT objects in AutoCAD 2004 format.

▶ **AutoCAD 2000 Drawing Format with Exploded AEC Objects**—This option is the same as Export to AutoCAD discussed above. The result will be AutoCAD drawing files containing only AutoCAD entities with no ADT objects in AutoCAD 2000 format. (This is the recommended setting for most external consultants.)

Next is the Transmittal file folder list. It defaults to the same location as the current project, or you can click the Browse button to save it to a different location. Beneath this is the Transmittal file name list, which contains three options:

▶ **Prompt for a filename**—This option simple prompts you with a dialog to input the file name as the eTransmit routine is executed.

▶ **Overwrite if necessary**—This option will overwrite any existing files with the same name(s) as those being created by the transmittal.

▶ Increment file name if necessary—This option will create a new transmittal file or files with an incremental suffix added to the file name(s).

The next set of options determines the organization used in the folder structure of the eTransmitted files or archive.

▶ **Use organized folder structure**—This option will match the eTransmitted folder structure to the folder structure chosen from the list. Use Browse to choose a different root folder.

▶ **Place all files in one folder**—This option is useful when the recipient has different server names than you and will be viewing the files in vanilla Auto-CAD. Using this option, you are able to keep all XREFs attached (not Bound) and not worry that their paths will be broken. This is because AutoCAD always searches the current folder for XREFs.

▶ **Keep files and folders as is**—This option uses the same folder structure as the original project without modification. This option is best for recipients that are using ADT 2006, 2005 or 2004.

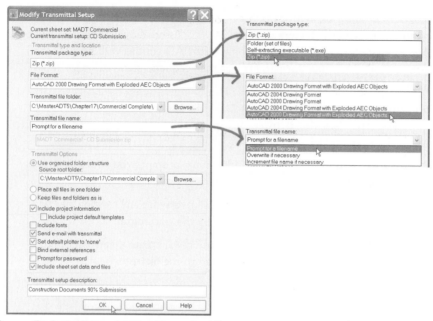

Figure D–2 *Configure the New Transmittal Setup*

▶ **Include project information**—Use this option to include the Project APJ and XML files with the eTransmittal. This is useful if your recipient is an ADT 2006, ADT 2005 or ADT 2004 user.

▶ **Include project default templates**—This option will include all of the Drawing Template File (DWT) with the transmittal.

▶ **Include fonts**—This option includes the required font files. If you have only used standard AutoCAD fonts in your drawings, this option is not necessary.

▶ **Send e-mail with transmittal**—This option will launch your default email application and send the file(s) created by the transmittal as an email attachment.

▶ **Set default plotter to 'none'**—This option is useful if your recipient does not have the same default plotter as you. They will need to choose a plotter in the files you send before printing.

▶ **Bind external references**—Use this option if you do not want External References included in the drawing files. All XREFs will be bound to their hosts, and then those files saved using whatever settings you designated above. This setting is useful with the two "Exploded AEC Objects" options above for recipients that need only a simple background file.

▶ **Prompt for password**—Use this option with the .*zip* and .*exe* options to require a password to open the archive. Don't forget to share the password with the recipient.

▶ **Include sheet set data and files**—This option will include the Sheet Set, all Sub Sets and all Sheet Set settings. This option includes the DST file with the transmittal.

4. Make your desired changes (use Figure D–2 as a guide) and then click OK to create the Transmittal Setup.

   Create alternative ones if you wish and then click Close to finish.

5. Make your desired changes (use Figure D–2 as a guide) and then click OK to create the Transmittal Setup.

6. Right-click the Sheet Set node at the top again and choose **eTransmit**.

   A progress bar will appear as the list for required files is complied.

7. Choose CD Submission on the right side.

Click the various tabs on the left to see all of the files that will be included in the transmittal. You can uncheck any files that you do not wish to include before clicking OK. You can also click the View Report button to see a detailed report of what will be included. There is also a notes field that you can type in any message you wish. This will be included as a text file with the transmittal and in the email message as well.

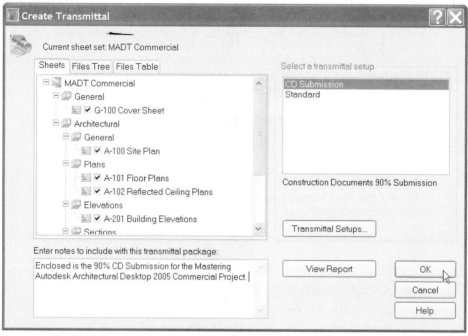

Figure D–3 *eTransmit the entire Project*

8. Click OK to generate the Transmittal. And then click Save in the dialog that appears.

When the *.zip* file creation is complete, you will be prompted to send the email. You can continue with the email and send a test email to yourself, or simply cancel. The Archive command works in nearly the same way except that it will not offer the email option. You can also eTransmit from any node on the Project Navigator, Constructs, Views, Sheets, and Sheet Sub Sets. Feel free to experiment.

PUBLISH TO DWF

In some cases, your recipient will not need a DWG at all—they only need to review a drawing or print it for example . In this case, consider sending them a DWF file instead. The DWF format is much more compact than DWG making it easier to email. You can plot the entire set of drawings to a single multi-sheet DWF file (see Chapter 5 and 19). Also, as we saw in Chapter 19, we can include Property Sets, Layers and Sheet Set data in our DWF files. Recipients need only install the free Autodesk DWF Viewer software to view and print the files. If they wish to add redmarks and comments, they can purchase the Autodesk DWF Composer software for a nominal fee. Redmarks created in DWF Composer are able to be loaded back into ADT for review. To create a DWF, refer to the tutorial in Chapter 19.

APPENDIX e

Wall Cleanup Solutions

COMMERCIAL PROJECT

Additional Wall Cleanup practice:

 NOTE Solutions to the Chapter 9 Wall Cleanup Exercise in Appendix A.

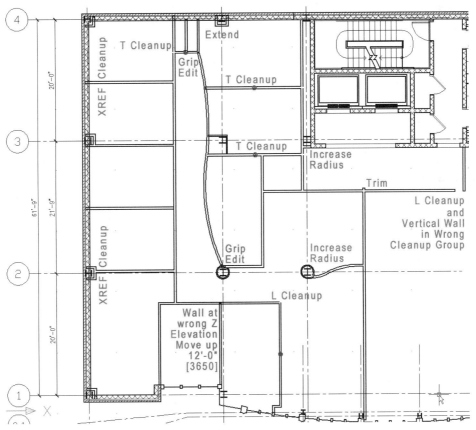

Figure E–1

INDEX

Numeric

2D Section/Elevation objects *See* Section/
Elevation objects, 518
3D DWF files, 34, 1036–1037, 1114
3D Views, 544–546

A

absolute coordinate system, 168
additive boundaries, 971, 989
ADT (Architectural Desktop) overview,
xxix–xxxi, 3–20
AEC Dimensions *See* dimensioning, 352
AEC Modify Array tool, 589–590
AEC Modify Tools, 48
AEC template files, 151–152
AIA CAD Layer Guidelines , 242
Air Gap, 519–522, 888–889
anchors
best practices, 345
Column bubbles, 347–352
Column Grids, 331–337, 343–347
lighting fixtures, 709–720
overview, 118–129
Railings, 396
tags/tagging, 797–798
Window, 584–588
animation software *See* VIZ Render, 1040
annotations *See also* Callouts; dimensioning;
tags/tagging
Column Grids, 345–353
Content libraries, 774–777
Elevation labels, 778–780
Keynotes, 946–958
scaling of, 144
View files, 737–741
APJ (Autodesk Project Information file), 240
Application Data folder, 510
application status bar, 39, 55
Arc Rotate feature, Viz Render, 1057–1058

Architectural Desktop (ADT) overview,
xxix–xxxi, 3–20
Architectural Subsets, 291
Archive command, 1038, 1109–1114
Area Evaluation reports, 1011–1013
Area Groups, 967–968, 980–987
Area objects
creating, 969–971, 976–978, 1003–1009
Decomposed Display Representation,
979–980
grip editing, 978–979
modifying, 971–976
overview, 967–968
rentable area calculation, 991–1003, 1010
Schedule Tables, 1014
tagging, 1009–1010
toolset, 984–991
arrays, 589–590, 714, 722
AutoCAD
best practices, 67–71
Drawing Recovery Manager, 137
exporting ADT to, 1108–1109
file sharing compatibility with ADT,
1104–1106
prerequisite skills, 67–71
vs. ADT, xxix
Autodesk Project Information file (APJ), 240
auto-hide palettes feature, 45
Auto-Import feature, Layer Key Styles,
148–150, 550
automatic offsets, 185–188
automatic property sets, 803–804, 807,
810–813
automatic Wall cleanup rules, 479–492
AutoSave, 137
Autosnap feature, 70, 486–489

B

background images in VIZ Render,
1064–1065

LICENSE AGREEMENT FOR AUTODESK

A Thomson Leaning Company

IMPORTANT-READ CAREFULLY: This End User License Agreement ("Agreement") sets forth the conditions by which Delmar Learning, a division of Thomson Learning Inc. ("Thomson") will make electronic access to the Thomson Delmar Learning-owned licensed content and associated media, software, documentation, printed materials and electronic documentation contained in this package and/or made available to you via this product (the "Licensed Content"), available to you (the "End User"). BY CLICKING THE "I ACCEPT" BUTTON AND/OR OPENING THIS PACKAGE, YOU ACKNOWLEDGE THAT YOU HAVE READ ALL OF THE TERMS AND CONDITIONS, AND THAT YOU AGREE TO BE BOUND BY ITS TERMS CONDITIONS AND ALL APPLICABLE LAWS AND REGULATIONS GOVERNING THE USE OF THE LICENSED CONTENT.

1.0 SCOPE OF LICENSE

1.1 Licensed Content. The Licensed Content may contain portions of modifiable content ("Modifiable Content") and content which may not be modified or otherwise altered by the End User ("Non-Modifiable Content"). For purposes of this Agreement, Modifiable Content and Non-Modifiable Content may be collectively referred to herein as the "Licensed Content." All Licensed Content shall be considered Non-Modifiable Content, unless such Licensed Content is presented to the End User in a modifiable format and it is clearly indicated that modification of the Licensed Content is permitted.

1.2 Subject to the End User's compliance with the terms and conditions of this Agreement, Thomson Delmar Learning hereby grants the End User, a nontransferable, non-exclusive, limited right to access and view a single copy of the Licensed Content on a single personal computer system for noncommercial, internal, personal use only. The End User shall not (i) reproduce, copy, modify (except in the case of Modifiable Content), distribute, display, transfer, sublicense, prepare derivative work(s) based on, sell, exchange, barter or transfer, rent, lease, loan, resell, or in any other manner exploit the Licensed Content; (ii) remove, obscure or alter any notice of Thomson Delmar Learning's intellectual property rights present on or in the License Content, including, but not limited to, copyright, trademark and/or patent notices; or (iii) disassemble, decompile, translate, reverse engineer or otherwise reduce the Licensed Content.

2.0 TERMINATION

2.1 Thomson Delmar Learning may at any time (without prejudice to its other rights or remedies) immediately terminate this Agreement and/or suspend access to some or all of the Licensed Content, in the event that the End User does not comply with any of the terms and conditions of this Agreement. In the event of such termination by Thomson Delmar Learning, the End User shall immediately return any and all copies of the Licensed Content to Thomson Delmar Learning.

3.0 PROPRIETARY RIGHTS

3.1 The End User acknowledges that Thomson Delmar Learning owns all right, title and interest, including, but not limited to all copyright rights therein, in and to the Licensed Content, and that the End User shall not take any action inconsistent with such ownership. The Licensed Content is protected by U.S., Canadian and other applicable copyright laws and by international treaties, including the Berne Convention and the Universal Copyright Convention. Nothing contained in this Agreement shall be construed as granting the End User any ownership rights in or to the Licensed Content.

3.2 Thomson Delmar Learning reserves the right at any time to withdraw from the Licensed Content any item or part of an item for which it no longer retains the right to publish, or which it has reasonable grounds to believe infringes copyright or is defamatory, unlawful or otherwise objectionable.

4.0 PROTECTION AND SECURITY

4.1 The End User shall use its best efforts and take all reasonable steps to safeguard its copy of the Licensed Content to ensure that no unauthorized reproduction, publication, disclosure, modification or distribution of the Licensed Content, in whole or in part, is made. To the extent that the End User becomes aware of any such unauthorized use of the Licensed Content, the End User shall immediately notify Delmar Learning. Notification of such violations may be made by sending an Email to delmarhelp@thomson.com.

5.0 MISUSE OF THE LICENSED PRODUCT

5.1 In the event that the End User uses the Licensed Content in violation of this Agreement, Thomson Delmar Learning shall have the option of electing liquidated damages, which shall include all profits generated by the End User's use of the Licensed Content plus interest computed at the maximum rate permitted by law and all legal fees and other expenses incurred by Thomson Delmar Learning in enforcing its rights, plus penalties.

6.0   FEDERAL GOVERNMENT CLIENTS

6.1   Except as expressly authorized by Delmar Learning, Federal Government clients obtain only the rights specified in this Agreement and no other rights. The Government acknowledges that (i) all software and related documentation incorporated in the Licensed Content is existing commercial computer software within the meaning of FAR 27.405(b)(2); and (2) all other data delivered in whatever form, is limited rights data within the meaning of FAR 27.401. The restrictions in this section are acceptable as consistent with the Government's need for software and other data under this Agreement.

7.0   DISCLAIMER OF WARRANTIES AND LIABILITIES

7.1   Although Thomson Delmar Learning believes the Licensed Content to be reliable, Thomson Delmar Learning does not guarantee or warrant (i) any information or materials contained in or produced by the Licensed Content, (ii) the accuracy, completeness or reliability of the Licensed Content, or (iii) that the Licensed Content is free from errors or other material defects. THE LICENSED PRODUCT IS PROVIDED "AS IS," WITHOUT ANY WARRANTY OF ANY KIND AND THOMSON DELMAR LEARNING DISCLAIMS ANY AND ALL WARRANTIES, EXPRESSED OR IMPLIED, INCLUDING, WITHOUT LIMITATION, WARRANTIES OF MERCHANTABILITY OR FITNESS OR A PARTICULAR PURPOSE. IN NO EVENT SHALL THOMSON DELMAR LEARNING BE LIABLE FOR: INDIRECT, SPECIAL, PUNITIVE OR CONSEQUENTIAL DAMAGES INCLUDING FOR LOST PROFITS, LOST DATA, OR OTHERWISE. IN NO EVENT SHALL DELMAR LEARNING'S AGGREGATE LIABILITY HEREUNDER, WHETHER ARISING IN CONTRACT, TORT, STRICT LIABILITY OR OTHERWISE, EXCEED THE AMOUNT OF FEES PAID BY THE END USER HEREUNDER FOR THE LICENSE OF THE LICENSED CONTENT.

8.0   GENERAL

8.1   Entire Agreement. This Agreement shall constitute the entire Agreement between the Parties and supercedes all prior Agreements and understandings oral or written relating to the subject matter hereof.

8.2   Enhancements/Modifications of Licensed Content. From time to time, and in Delmar Learning's sole discretion, Thomson Thomson Delmar Learning may advise the End User of updates, upgrades, enhancements and/or improvements to the Licensed Content, and may permit the End User to access and use, subject to the terms and conditions of this Agreement, such modifications, upon payment of prices as may be established by Delmar Learning.

8.3   No Export. The End User shall use the Licensed Content solely in the United States and shall not transfer or export, directly or indirectly, the Licensed Content outside the United States.

8.4   Severability.If any provision of this Agreement is invalid, illegal, or unenforceable under any applicable statute or rule of law, the provision shall be deemed omitted to the extent that it is invalid, illegal, or unenforceable. In such a case, the remainder of the Agreement shall be construed in a manner as to give greatest effect to the original intention of the parties hereto.

8.5   Waiver. The waiver of any right or failure of either party to exercise in any respect any right provided in this Agreement in any instance shall not be deemed to be a waiver of such right in the future or a waiver of any other right under this Agreement.

8.6   Choice of Law/Venue. This Agreement shall be interpreted, construed, and governed by and in accordance with the laws of the State of New York, applicable to contracts executed and to be wholly preformed therein, without regard to its principles governing conflicts of law. Each party agrees that any proceeding arising out of or relating to this Agreement or the breach or threatened breach of this Agreement may be commenced and prosecuted in a court in the State and County of New York. Each party consents and submits to the non-exclusive personal jurisdiction of any court in the State and County of New York in respect of any such proceeding.

8.7   Acknowledgment. By opening this package and/or by accessing the Licensed Content on this Website, THE END USER ACKNOWLEDGES THAT IT HAS READ THIS AGREEMENT, UNDERSTANDS IT, AND AGREES TO BE BOUND BY ITS TERMS AND CONDITIONS. IF YOU DO NOT ACCEPT THESE TERMS AND CONDITIONS, YOU MUST NOT ACCESS THE LICENSED CONTENT AND RETURN THE LICENSED PRODUCT TO THOMSON DELMAR LEARNING (WITHIN 30 CALENDAR DAYS OF THE END USER'S PURCHASE) WITH PROOF OF PAYMENT ACCEPTABLE TO DELMAR LEARNING, FOR A CREDIT OR A REFUND. Should the End User have any questions/comments regarding this Agreement, please contact Thomson Delmar Learning at delmar-help@thomson.com.